PHYSICS OF THE SOLAR SYSTEM

Dynamics and Evolution, Space Physics, and Spacetime Structure

by

BRUNO BERTOTTI

Department of Nuclear and Theoretical Physics,
University of Pavia, Italy

PAOLO FARINELLA

Department of Astronomy,
University of Trieste, Italy

and

DAVID VOKROUHLICKÝ

Institute of Astronomy,
Charles University, Prague, Czech Republic

KLUWER ACADEMIC PUBLISHERS
DORDRECHT / BOSTON / LONDON

A C.I.P. Catalogue record for this book is available from the Library of Congress.

ISBN 1-4020-1428-7 (HB)
ISBN 1-4020-1509-7 (PB)

Published by Kluwer Academic Publishers,
P.O. Box 17, 3300 AA Dordrecht, The Netherlands.

Sold and distributed in North, Central and South America
by Kluwer Academic Publishers,
101 Philip Drive, Norwell, MA 02061, U.S.A.

In all other countries, sold and distributed
by Kluwer Academic Publishers,
P.O. Box 322, 3300 AH Dordrecht, The Netherlands.

Printed on acid-free paper

All Rights Reserved
© 2003 Kluwer Academic Publishers
No part of this work may be reproduced, stored in a retrieval system, or transmitted
in any form or by any means, electronic, mechanical, photocopying, microfilming, recording
or otherwise, without written permission from the Publisher, with the exception
of any material supplied specifically for the purpose of being entered
and executed on a computer system, for exclusive use by the purchaser of the work.

Printed in the Netherlands.

Contents

Introduction		x
Useful physical quantities		xvii

1.	DYNAMICAL PRINCIPLES	1
	1.1 Gravitational equilibrium	1
	1.2 The equation of state	6
	1.3 Dynamics of fluids	9
	1.4 Dynamics of solid bodies	12
	1.5 Transport	16
	1.6 Magnetohydrodynamics	21
	1.7 Conservation of magnetism and vorticity	25
	1.8 Kinetic theory	28
2.	THE GRAVITATIONAL FIELD OF AN ISOLATED BODY	35
	2.1 Spherical harmonics	35
	2.2 Harmonic representation of the gravity field	42
	2.3 The gravity field of the Earth	46
	2.4 Gravity fields of planetary bodies	54
3.	PLANETARY ROTATION	63
	3.1 Measurements of time and distance	63
	3.2 Frames of reference	67
	3.3 Slowly rotating bodies	74
	3.4* Equilibrium shapes of bodies in fast rotation	78
	3.5 Rigid-body dynamics: free precession	82

	3.6*	The rotation of the Earth and its interior structure	87
4.	**GRAVITATIONAL TORQUES AND TIDES**		95
	4.1	Lunisolar precession	95
	4.2	Tidal potential	99
	4.3	Tides in a non-rigid Earth	102
	4.4	Tidal harmonics	104
	4.5*	Equilibrium shape of satellites	108
5.	**THE INTERIOR OF THE EARTH**		115
	5.1	Seismic propagation	115
	5.2	Internal structure of the Earth	120
	5.3	Heat generation and flow	127
	5.4	Tectonic motions	135
6.	**PLANETARY MAGNETISM**		145
	6.1	The main dipole field	146
	6.2	Magnetic harmonics and anomalies	147
	6.3	Secular changes and reversals	152
	6.4*	The generation of planetary magnetic fields	155
	6.5	Planetary magnetic fields	161
7.	**ATMOSPHERES**		169
	7.1	Structure of the atmosphere and climate variations	169
	7.2	Planetary atmospheres	182
	7.3*	Radiative transfer	186
	7.4	Dynamics of atmospheres	191
	7.5	Turbulence	198
8.	**UPPER ATMOSPHERES**		211
	8.1	General properties	211
	8.2	Ionosphere	216
	8.3	Waves in an electron plasma	220
	8.4	Atmospheric refraction	225
	8.5	Evolution of atmospheres	229
9.	**THE SUN AND THE SOLAR WIND**		239
	9.1	The Sun	239
	9.2	The solar wind	247

9.3	The heliosphere	252
10. MAGNETOSPHERES		**261**
10.1	The solar wind and the magnetosphere	261
10.2	Motion of charged particles in a strong magnetic field	268
10.3	Trapped particles	274
10.4	Magnetohydrodynamic waves	278
10.5	The bow shock	281
10.6 *	Cosmic rays and the magnetosphere	284
10.7	Planetary magnetospheres	286
11. THE TWO-BODY PROBLEM		**297**
11.1	Reduction to a central force problem	297
11.2	Keplerian orbits	300
11.3	Bound orbits	302
11.4	Unbound orbits	307
12. PERTURBATION THEORY		**313**
12.1	Gauss perturbation equations	313
12.2	Lagrange perturbation equations	319
12.3	Atmospheric drag	325
12.4	Oblateness of the primary	331
12.5	Approximation methods	333
13. THE THREE-BODY PROBLEM		**345**
13.1	Equations of motion and the Jacobi constant	346
13.2	Lagrangian points and zero-velocity curves	349
13.3	Stability analysis and motion near the Lagrangian points	355
13.4	Hill's problem	364
13.5 *	Tisserand's criterion	371
14. THE PLANETARY SYSTEM		**377**
14.1	Planets	377
14.2	Satellites	382
14.3	Asteroids	391
14.4	Transneptunian objects and Centaurs	405
14.5	Comets	412
14.6	Interplanetary material and meteorites	422

14.7	Planetary rings and circumplanetary dust	433
14.8 *	From planets to boulders	449

15. DYNAMICAL EVOLUTION OF THE SOLAR SYSTEM — 459

15.1	Secular perturbations and the stability of the solar system	459
15.2 *	Resonances and chaotic behaviour	473
15.3	Tidal orbital evolution	487
15.4	Dynamics of small bodies	493

16. ORIGIN OF THE SOLAR SYSTEM — 509

16.1	Mass and structure of the solar nebula	510
16.2	Growth of solid grains and formation of planetesimals	515
16.3	Formation of planetary embryos and accretion of terrestrial planets	521
16.4	Formation of giant planets and satellites	526
16.5	Extrasolar planetary systems	537

17. RELATIVISTIC EFFECTS IN THE SOLAR SYSTEM — 557

17.1	The equivalence principle	558
17.2	Curvature of spacetime	563
17.3	The nature of gravitation	565
17.4	Weak fields and slow motion	569
17.5	Doppler effect	575
17.6	Relativistic dynamical effects	578
17.7 *	Gravitomagnetism	581

18. ARTIFICIAL SATELLITES — 587

18.1	Launch	587
18.2	Spacecraft and their environment	591
18.3	Forces acting upon an artificial satellite	603
18.4	Space navigation	613

19. TELECOMMUNICATIONS — 621

19.1	The power budget	621
19.2	Spectra	625
19.3	Noise	631
19.4	Phase and frequency measurements	634
19.5	Propagation in a random medium	637

20. PRECISE MEASUREMENTS IN SPACE		641
20.1	Planetary imaging	641
20.2	Space astrometry	648
20.3	Laser tracking	657
20.4	Testing relativity in space	665
20.5	Signals and data analysis	674

Index 687

Introduction

This book is a direct sequel to: B. Bertotti and P. Farinella, "Physics of the Earth and the Solar System, Dynamics and Evolution, Space Navigation, Space-Time Structure" (Kluwer Academic Publishers, 1990). Nearly 15 years after its publication it became evident that the volume was in need of a new edition to keep up with the outstanding progress and the changing perspectives in this field. David Vokrouhlický agreed to collaborate on the project and be the third author. On March 25, 2000, after a long illness and a heart transplant, Paolo Farinella passed away. We then decided that, rather than aiming at a second edition, it made more sense to rewrite the book anew. While its basic content and the structure of the chapters are the same, important new topics have been added, including the extrasolar planetary systems, transneptunian objects, accurate determination of reference frames and new space projects. Greater relevance has been given to semiquantitative discussions before introducing formal developments; many figures have been added and updated and several errors corrected. More emphasis has given to the solar system, whereas geophysical topics have been left at a less advanced level. To mark this change the slightly different title "Physics of the Solar System" was chosen.

We wish to dedicate this book to the memory of Paolo Farinella, an outstanding scientist, an invaluable collaborator and a dear friend.

Bruno Bertotti and David Vokrouhlický.   Pavia and Prague, January 2003

* * *

In the solar system, gravitation, the main force for the binding of large bodies and their dynamics, has the role of protagonist. Classical (i.e., non-relativistic) celestial mechanics had, already at the end of the 19th century, reached an extraordinary degree of perfection, thanks especially to the French and English schools; new mathematical tools had been discovered, including perturbation theory and the three-body problem, which were subsequently developed in great detail and are now the standard basis for current work. Celestial mechanics has provided an excellent, and basically correct, understanding of the dynamics of the solar system, allowing accurate predictions of the positions of planetary bodies in the sky. The comparison and the agreement of these theories with painstaking telescope observations are a great success of modern science.

Before the last World War, four different fields in the parallel area of geophysics were developed mainly through intelligent and extensive collection and interpretation of data: gravity anomalies, magnetic and ionospheric disturbances, meteorological and climatic observations, and seismic oscillations. Progress occurred through slow accumulation and change within the framework of long-established paradigms, according to the well known patterns of a "normal" science, but often without an underlying unifying theory. This process favoured extreme specialization and the growth of closed scientific communities whose main purpose was the development and the application of a particular experimental technique and the recording of its results.

Following the Second World War, owing in part to the great progress made in military technology (in particular in radar and rockets), new and much more expensive experimental techniques were developed and put to scientific use. They include scientific spacecraft, radio telescopes, Synthetic Aperture Radar, radio and optical interferometry, and solid state radiation detectors. They have not only ushered the exploration of bodies in the solar system other than the Earth, but have also greatly increased the accuracy of relevant physical quantities. Laser ranging to the Moon and to Earth satellites now provides their distance to cm accuracy, or even less; coherent and very stable microwave beams can now measure relative velocities in the solar system to accuracies of μm/s. Before the last World War, it was known that the Newtonian picture of (Euclidean) space and (uniform) time was not adequate and needed to be supplemented with Relativity Theory (both Special and General); this theory, however, had essentially unobservable consequences in the solar system. At present this scenario is utterly inadequate and celestial mechanics must be set up in spacetime, taking into account the relevant relativistic corrections. Our model of the world has changed, even for practical applications like the Global Positioning System.

The above-mentioned instrumental developments together with several paradigmatic changes in theoretical understanding have also led to a merging between geophysical and solar system sciences. Indeed, the Earth is best understood as just one of the planets; this is surely the case with major geophysical developments, i.e., plate structure, polar motion, the origin of the dipolar magnetic field (now understood with the dynamo theory) and global climatology, all of which are relevant for the whole solar system. Magnetic fields in other planetary bodies, in particular in Jupiter's satellites, have been discovered. A recent, major example of this widening of horizons is the discovery of extrasolar planetary systems which seem to have properties different from our own and may call for major revisions of our understanding of its formation and evolution. The origin of life is still not understood: although the present biochemical and biophysical processes leading to replication and evolution are well established, the early origin and history of elementary organisms and the need for particular atmospheres and places (i.e., near the hot vents on the bottom of the oceans) is unknown. This topic is not addressed in the book.

The new instruments are generally characterized by their much greater size, complexity and economic value; in this, as in other fields, we have seen a qualitative transition from a "little" to a "big" science, where experimental programmes require long and often inflexible planning and large investments of human resources and money. The outer space near the Earth is getting crowded and the danger of collisions of spacecraft with artificial debris produced in launches and other space activities, both civilian and military, is relentlessly increasing. Furthermore, as far as we know, present military satellites do not carry weapons: by tacit restraint space activities of all kinds are accepted and respected by all nations; but, lacking international agreements, this may not last for ever and the safety of space activity may be endangered.

One must also recognize the great intellectual impact of space exploration which has shifted the boundary of unknown lands to natural satellites, planets and interplanetary space. Now our machines can actually go out there and report back; the magic spell of exploration, so well described in the story of Ulysses in Dante's *Divina Commedia* (Inferno, Canto 26), is still with us. This fascination produces a convergence of wills and establishes and feeds well-organized scientific communities, commanding a large economic power.

* * *

We now highlight some recent advances. By means of several experimental techniques, the geopotential of the Earth is now known up to degree and order ≈ 360, and further improvements and validations are possible with space gradiometers measuring the difference in gravitational acceleration. Accurate descriptions of the gravity field of other solar system bodies, in particular of the Moon, Venus and Mars, are available. With multiple seismic arrays and better instrumentation, our knowledge of the interior of the Earth has much improved, in particular with the discovery of deviations from spherical symmetry, especially at the boundary between core and mantle, leading to a better understanding of convective motions and their consequences for long-wave surface topography.

The space astrometric mission HIPPARCOS attained a level of accuracy of a milliarcsecond; other astrometric satellites will improve this accuracy by nearly three orders of magnitude, with major consequences for the search for extrasolar planetary systems and astrophysics. Routine monitoring of plate motions and Earth rotation is accomplished with several techniques, in partic-

Introduction xiii

ular laser ranging to Earth spacecraft, Very Long Baseline Interferometry and the Global Positioning System.

In 1994, the D/Shoemaker-Levy 9 comet in a close encounter with Jupiter broke up into 21 pieces; subsequently the fragments had a spectacular fall into the planet. At that time, the Galileo space probe was nearby, serendipitously positioned to observe the event. During its trip to Jupiter, Galileo had provided the first close-up images of two asteroids, Gaspra and Ida, and discovered a small moon of the latter. Tens of binary asteroidal systems (including several in the transneptunian region) were discovered with ground-based observations. The most detailed investigation was obtained for Eros, where a probe landed in February 2001. Asteroids, both near the Earth and in the main belt, have also been extensively observed with ground-based techniques, in particular radar, providing shape, physical properties, rotation and orbits of more than 100 objects. Evaluating and monitoring the risk of (potentially dangerous) impacts with the Earth is the object of several projects, including Spaceguard. A new population of objects, hundreds of km in size, has been discovered beyond the orbit of Neptune, in the zone called the Edgeworth-Kuiper belt, providing a plausible source for the short-period comets; Pluto is now regarded as the largest member of a vast population, rather than an individual small planet. Objects on transitory orbits, most likely escaping from the transneptunian belt and wandering in the zone of the large planets, have also been discovered.

In the early 1990's, the Magellan mission obtained high-resolution radar maps of the surface of Venus that showed that a planet similar in size to the Earth has had a very different geological evolution. Several lunar missions have unravelled long-standing puzzles of this most extensively known cosmic body; in particular, the Lunar Prospector spacecraft detected a thin magnetosphere. The Mars Pathfinder and the Mars Global Surveyor missions brought about significant advances in our understanding of Mars, in particular concerning the likelihood of vast water reservoirs in the geological past. The in situ exploration of Mars, in particular with the Pathfinder mission which landed a small vehicle on the planet in 1997, is in full swing today.

In 1996, the Galileo mission dropped a probe into Jupiter to explore its ionosphere and atmosphere. The Hubble Space Telescope provided images of outstanding resolution of small and/or faraway objects, like Pluto and other transneptunians, the asteroids Ceres and Vesta and the Jovian satellite Io. In 1979, on the basis of dynamical considerations, volcanic activity was predicted for Io and subsequently observed with the Voyager spacecraft. Since the beginning of the 1990's, the Ulysses spacecraft has been providing outstanding measurements of the heliosphere, for the first time unravelling its complete, three-dimensional structure and its astonishing dynamics. In October 1997, a big spacecraft – Cassini – was launched to explore the Saturnian system from 2004 to 2008 and to release the Huygens probe into the opaque atmosphere of

Titan. With ground instruments, it was discovered that Uranus and Neptune have rings. Similarly, ground-based observations have recently allowed discovery of many distant and irregular satellites of giant planets (34 of Jupiter, 12 of Saturn and 5 of Uranus). Mercury and Pluto, the least explored planets, are the future targets of several new missions.

* * *

The subject matter is developed starting from concrete objects; the relevant physical principles are in part discussed in Ch. 1 and then revisited or developed as the need arises. Mathematical rigour, morphological details and descriptions of instrumentation are not a priority. We assume that the reader is acquainted with undergraduate physics, in particular with dynamics, electromagnetism, Special Relativity, thermodynamics and statistical physics; starting from this basis the reader will hopefully be led up to the threshold of current research. In choosing the sequence of topics we have tried, when possible, to satisfy the criterion of increasing difficulty; in many cases a substantial effort was made to simplify and clarify the material. The earlier book has been the basis of some courses of lectures, where its didactic structure, aimed at the qualitative and mathematical understanding of very rich and new physical processes, has been tested. Ch. 17 provides an introduction to General Relativity with as little formalism as possible and to the extent required for the study of the solar system. Particular attention has been given to precise measurements in space and the structure and evolution of the solar system; in the last Chapter we discuss several interesting measurement techniques, with a focus on their principles and the accuracy they can attain. The final Section in this Chapter is devoted to noise and data analysis, a subtle topic that is so important in space experiments. A large amount of data can now be found on the Web, implying a new kind of organization of scientific work, from which this book has largely profited.

Throughout the book we use the cgs system of units; Maxwell's equations are written in esu, non-rationalized units (see eqs. (1.70) and (1.71)). Bold symbols denote Cartesian vectors. Sometimes, in a diadic notation, a sans serif and capital letter denotes a tensor of second rank (e.g., eq. (1.40)); $i = \sqrt{-1}$ is the imaginary unit, but I is the orbital inclination. To help the reader, we have indicated with a star the more difficult Sections, which can be left out in a first reading. The problems at the end of each chapter are in no particular order; they are also starred to denote difficulty. At the end of each chapter some books and review articles are listed for further reading. Here we quote a few relevant general textbooks (in alphabetic order of authors):

- A.N. Cox (ed.), Allen's Astrophysical Quantities (4th edition), Springer-Verlag (2000);

- W.K. Hartmann, Moons and Planets, Wadsworth (1983);

- M. Hoskin (ed.), The General History of Astronomy, Vol. 2: *Planetary Astronomy from the Renaissance to the Rise of Astrophysics* (R. Taton and C. Wilson, eds.) Cambridge University Press (1995);

- W.M. Kaula, An Introduction to Planetary Physics, Wiley (1968);

- A. Morbidelli, Modern Celestial Mechanics: Aspects of Solar System Dynamics, Taylor and Francis (2002);

- C.D. Murray and S.F. Dermott, Solar System Dynamics, Cambridge University Press (2000);

- I. de Pater and J.J. Lissauer, Planetary Sciences, Cambridge University Press (2001);

- A.E. Roy, Orbital Motion, Hilger (1978);

- F.D. Stacey, Physics of the Earth, Brookfield Press (1992);

- P.R. Weissman, L.A. McFadden and T.V. Johnson (eds.), The Encyclopedia of the Solar System, Academic Press (1999).

Acknowledgements

The revision for the new version was mostly done by B.B. and D.V.; chapter 14 on the planetary system is mainly the work of D.V. We are particularly grateful to M. Brož for his competent help with the editing, in particular for all of the figures, which have been prepared with a special software. Z. Knežević helped us with the LaTeX version of the book and suggested many improvements. A number of colleagues advised us regarding issues in particular chapters; we would like to thank them (in alphabetic order): B. Bavassano (Ch. 10), J. Bičák (Ch. 17), W.F. Bottke (Ch.14), S. Breiter (Ch. 12), I. Ciufolini (Ch. 17), I. Hubený (Sec. 7.3), G. Laske (Sec. 5.2), F. Mignard (Sec. 20.2) and P. Tanga (Ch. 16).

Poetic note: Uriel's aria in Haydn's 'The Creation'

Nun schwanden vor dem heiligen Strahle
Des schwarzen Dunkels gräuliche Schatten:
Der erste Tag enstand.
Verrwirrung weich, und Ordnung keimt empor.

Now vanish before the holy beams
The gloomy shades of ancient night;
The first of days appears.
Now chaos ends, and order fair appears.

These verses, and F.J. Haydn's (1732-1809) music, offer a beautiful expression of the emergence of the order and structure of the planetary system from formless interstellar gas after the sudden formation of the Sun. The libretto of this Oratorio (in German) is the work of van Swieten; he translated for this purpose an English text by Mr Lindley, based on Milton's Paradise Lost, which has been used here.

Useful physical quantities

(a) Fundamental and defining constants (see also http://physics.nist.gov/)

arcsec	=	$1'' = 4.848 \times 10^{-6}$ rad
mas	=	$0.001''$ corresponds to 3.1 cm on the surface of the Earth
c	=	2.998×10^{10} cm/s $= 2.998 \times 10^{8}$ m/s velocity of light
e	=	4.803×10^{-10} esu $= 1.602 \times 10^{-19}$ Coulomb magnitude of electron charge
eV	=	1.602×10^{-12} erg $= 1.602 \times 10^{-19}$ Joule $= 11{,}600$ K energy and temperature associated with 1 electron volt
G	=	6.674×10^{-8} dyne cm^2 g^{-2} $= 6.674 \times 10^{-11}$ N m^2 kg^{-2} gravitational constant
h	=	6.626×10^{-27} erg s $= 6.626 \times 10^{-34}$ J s Planck constant
k	=	1.381×10^{-16} erg/K $= 1.381 \times 10^{-23}$ J/K Boltzmann constant
ly	=	9.46×10^{17} cm $= 9.46 \times 10^{15}$ m light year
m_e	=	9.109×10^{-28} g $= 9.109 \times 10^{-31}$ kg electron mass
m_p	=	1.673×10^{-24} g $= 1.673 \times 10^{-27}$ kg proton mass
pc	=	3.086×10^{18} cm $= 3.086 \times 10^{16}$ m $= 3.26$ ly parsec
λ_D	=	$\sqrt{kT/(4\pi e^2 n_e)}$
	=	$743 \sqrt{(kT/\text{eV})/(n_e \text{ cm}^3)}$ cm $= 7.43 \times 10^{-3} \sqrt{(kT/\text{eV})/(n_e \text{ m}^3)}$ m Debye length
σ	=	$2\pi^5 k^4/15c^2 h^3 = 5.67 \times 10^{-5}$ erg/(cm^2 s K^4)
	=	5.67×10^{-8} W/(m^2 K^4) Stefan–Boltzmann constant

ω_p = $\sqrt{4\pi n e^2/m_e}$ = $5.64 \times 10^4 \sqrt{n\,\text{cm}^3}$ rad/s = $56.4 \sqrt{n\,\text{m}^3}$ rad/s
plasma frequency

Ω_e = $eB/m_e c$ = 1.76×10^2 (B/nT) rad/s = 1.76×10^{11} (B m^2/weber) rad/s
electron cyclotron frequency

(b) Quantities related to specific objects

AU = 1.496×10^{13} cm = 1.496×10^{11} m
Astronomical Unit (mean heliocentric distance of the Earth)

H = 55 km/s/Mpc (with an uncertainty of $\approx$ 15%)
Hubble constant

J_2 = $(C - A)/M_\oplus R_\oplus^2$ = 0.0010826
oblateness coefficient of the Earth

$L_\odot$ = 3.845×10^{33} erg/s = 3.845×10^{26} J/s
solar luminosity

$M_\odot$ = 1.988×10^{33} g = 1.988×10^{30} kg
solar mass

$m_\odot$ = 1.48 km = 2.2×10^{-6} $R_\odot$
solar gravitational radius

$M_\oplus$ = 5.972×10^{27} g = 5.972×10^{24} kg
Earth mass

$m_\oplus$ = 0.44 cm = 6.97×10^{-10} $R_\oplus$
Earth gravitational radius

M_J = 317.9 $M_\oplus$ = $M_\odot$/1047.35
Jupiter mass

$n_\oplus$ = 1.991×10^{-7} rad/s
mean motion of the Earth

$R_\oplus$ = 6.371×10^8 cm = 6.371×10^6 m
mean Earth radius

$R_\odot$ = 6.958×10^{10} cm = 6.958×10^8 m
mean solar radius

v_{esc} = $\sqrt{2GM_\oplus/R_\oplus}$ = 11.2 km/s
escape velocity from the Earth

y = 3.15×10^7 s
a year

Useful physical quantities xix

ϵ = 23° 26′ 21″
obliquity of the ecliptic for J2000.0

F_0 = 1.366×10^6 erg cm^{-2} s^{-1} = 1.366×10^3 J m^{-2} s^{-1}
solar constant (solar radiation flux at 1 AU)

$\omega_\oplus$ = $(7.2921 \times 10^{-5} - 1.4 \times 10^{-14}\, T)$ rad/s
sidereal angular velocity of the Earth and its secular decrease (T in y);

Ω_{pr} = 50.291″ /y = 7.726×10^{-12} rad/s
lunisolar precession constant

(c) Units of power and flux

Luminosities, powers and power fluxes are given in cgs units (sometimes in Watts (10^7 erg/s) and W/m^2). Telecommunication engineers commonly measure ratios of powers in a logarithmic scale, with the quantity

$$10 \times \log_{10} \frac{P_2}{P_1} ,$$

whose unit is the *decibel (dB)* (Sec. 19.1).

For luminosity, or brightness, astronomers use the concept of magnitude, also based on a logarithmic scale; here is a brief summary. The *apparent magnitude m* of a star (with total flux F) relative to a reference star (with total flux $F_\star$ and magnitude $m_\star$) is

$$m - m_\star = -2.5 \log_{10} \frac{F}{F_\star} .$$

To roughly describe the colour of a star and to account for the instrumental bandwidth, magnitudes obtained with standardized filters and a restricted passband are also used; the most popular reference in the optical regime is the Johnson-Morgan-Cousins system (UBVRI; see http://obswww.unige.ch/gcpd/system.html).

Name of the filter	Passband (nm)	Effective wavelength (nm)
U ultraviolet	300 - 420	365
B blue	360 - 560	445
V visual	460 - 700	550
R red	530 - 950	660
I infrared	700 - 1150	805

The *absolute magnitude M* of a star is its apparent magnitude at the distance of 10 pc. Since the flux decreases with the distance r as $1/r^2$,

$$m - M = 5 \log_{10} \frac{r}{10\,\text{pc}} .$$

The absolute magnitude of a star of luminosity L and the absolute magnitude of the Sun are related by

$$M - M_\odot = -2.5 \log_{10} \frac{L}{L_\odot} .$$

The origin of the magnitude system is chosen so that a star with luminosity $L = 3.055 \times 10^{35}$ erg/s has the zero absolute (bolometric) magnitude.

Due to the obvious reasons, for planets and minor objects in the Solar system the absolute magnitude (henceforth denoted H) is referred to a distance of 1 AU from the Earth and the Sun and a zero phase angle (i.e., at opposition).

Chapter 1

DYNAMICAL PRINCIPLES

In this chapter we review the dynamical equations that govern the motion and determine the structure of a planetary body and its environment. The basic model is that of a continuum, whose local properties are described by the matter density ϱ, the flow velocity **v** and the temperature T. If the fluid is electrically conductive, we also need the magnetic field **B**. These quantities obey a set of partial differential equations which will be discussed in the relevant approximations, with examples and illustrations taken from the physics of the solar system. Gravitation plays an essential role here: as it is a long range force, it becomes more and more important with increasing size. It determines the shape, size and structure of the planets, the stars, the galaxies and the Universe itself. It is balanced primarily by thermal pressure and, secondarily, by centrifugal and other forces: this balance is responsible for the nearly spherical shape of big bodies. When a body is sufficiently small and cold, however, it solidifies; its structure is then determined by interatomic forces. In cosmic bodies, different transport processes play important roles: in particular, the transport of momentum determines viscosity and the transport of energy is the basis for heat conduction. In an electrically conductive medium, we have a large variety of electric and magnetic processes, such as the generation of large scale magnetic fields in rotating bodies, particle acceleration, and the propagation of different kinds of electromagnetic waves. When local thermal equilibrium does not hold, a description in phase space is required, giving rise to a great variety of phenomena, as in ionized gas.

1.1 Gravitational equilibrium

The structure of a big cosmic body is usually determined by the balance between the gravitational pull and the internal pressure; this pressure, in turn, depends upon the state of matter, the composition, the heat flow, the magnetic field, and so on. These factors are influenced by the way the body was formed and its history. Their determination requires the laws of microscopic physics: the evaluation of the state of stress, determined by the pressure tensor P, for example, provides the main link between the microscopic state of each part and the overall structure and shape. If, as in a fluid, this microscopic state has no privileged direction, in particular no stratification, the pressure tensor is isotropic and is determined by a single scalar, the pressure. Then, if there are no specific reasons for asymmetry, like rotation or an external gravitational field, a configuration of spherical equilibrium is achieved: the pressure is a function of the radial distance only and thus is able to compensate the radial gravitational pull. A deviation from the spherical equilibrium will also occur if the body, or part of it, is a solid, capable of supporting a force tangential to its surface: a situation described by a strain and a stress which are not isotropic.

Spherical configurations. Let us first derive the condition of equilibrium of a spherical mass of density $\varrho(r)$ and pressure $P(r)$. The total mass within the distance r from the centre is

$$M(r) = 4\pi \int_0^r dr' \, r'^2 \varrho(r') \,. \qquad (1.1)$$

The symbol $M(= M(R))$, with no arguments, denotes the total mass and R the radius. The gravitational acceleration $g(r)$ at a given distance r from the centre of the body has a radial inward direction and is equal to the one produced by a point mass $M(r)$ placed at the centre

$$g(r) = GM(r)/r^2 \,. \qquad (1.2)$$

This result follows from the analogue of Gauss' theorem in electrostatics. The gravitational acceleration **g** produced by a point mass m is formally obtained from the electrostatic field produced by a charge q by substituting q with $-Gm$; the mathematical properties of the two quantities are the same[1]. Therefore we have a gravitational analogue of Gauss' theorem

$$\int_S dS \, \mathbf{n} \cdot \mathbf{g} = -4\pi G M(r) \,, \qquad (1.3)$$

where S is a spherical surface of radius r with external normal **n**.

Let us apply eq. (1.3) to a sphere of radius r inside our body, with mass $M(r)$. The acceleration $\mathbf{g} = -g\mathbf{r}/r = -g\hat{\mathbf{r}}$ has, by symmetry, a radial inward

---

[1] Note that there is also a correspondence between the electrostatic force $qq'\mathbf{r}/r^3$ between two charges and the gravitational force $-Gmm'\mathbf{r}/r^3$ between two masses: one obtains the latter by replacing the charge q with $m\sqrt{-G}$. This correspondence embodies much physics of both fields of force. One could say, gravitation is electrostatics with an imaginary charge; for this reason, for example, electrostatic plasma oscillations correspond to the unstable gravitational modes that may result in fragmentation of an initially homogeneous medium.

1. Dynamical principles

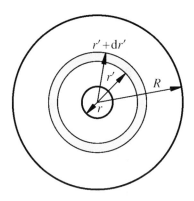

Figure 1.1 The calculation of the gravitational potential energy inside a spherical body. The contribution from the outer layers is calculated by summing the terms due to each spherical shell $dM = 4\pi r'^2 \varrho(r')dr'$; for the inner layers we use Gauss' theorem.

direction; hence the left-hand side is $-4\pi r^2 g$ and the theorem (1.2) follows at once. The elegance of this argument can be compared with the complicated three-dimensional integration which one must perform with the naïve, direct method of summing the vector contributions to **g** from all constituent elements of the body; but the argument works only when the body has spherical symmetry. For the case of a density distribution with ellipsoidal symmetry another direct method is very efficient (*Newton's canals*; see Sec. 3.4).

The gravitational potential energy per unit mass U, related to the acceleration by $\mathbf{g} = -\nabla U$, is $-GM/r$ outside the body; but inside it is *not* equal to $-GM(r)/r$ and a theorem similar to the one proved for the acceleration does not hold. To evaluate the potential energy, note that inside a thin spherical shell of mass dM and radius r' the gravitational pull vanishes; there the potential energy is constant and equal to its value just outside: $-GdM/r' = -4\pi G\varrho(r')r'dr'$. Therefore the total value of $U(r)$ is the sum of the contributions from the shells outside r and those inside, amounting to $-GM(r)/r$. Hence

$$U(r) = -G\left[4\pi \int_r^R dr'\, r'\varrho(r') + \frac{M(r)}{r}\right]. \tag{1.4}$$

For a sphere of uniform density

$$U(r) = -\frac{GM}{2R}\left[3 - \frac{r^2}{R^2}\right], \tag{1.5}$$

which correctly reduces to $-GM/R$ on the surface and fulfils $g(r) = -(\partial U/\partial r)$ inside.

An infinitesimal volume element between r and $r + dr$, with a base dS, is subject to three forces (Fig. 1.2):

- the pressure from below, $dS\, P(r)$;
- the pressure from above, $-dS\, P(r+dr) = -dS\, P(r) - dS\, dr\, dP/dr$;
- the gravitational pull, $df = -dS\, dr\, \varrho(r)\, g(r) = -dS\, dr\, \varrho(r)\, GM(r)/r^2$.

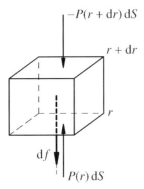

Figure 1.2 The equilibrium of a spherical body under gravity.

Because of the spherical symmetry the lateral forces do not contribute. In a static situation their sum must vanish:

$$\frac{dP}{dr} + \frac{GM(r)}{r^2}\varrho(r) = 0 \ . \tag{1.6}$$

This is the condition for *static equilibrium* of a body with spherical symmetry. The determination of the complete equilibrium requires an understanding of the pressure forces, to be discussed in the following Section. Here we follow with order-of-magnitude estimates for a homogeneous body.

Since at the surface $r = R$, $P(R) = 0$, we get the pressure at the centre

$$P(0) = \int_0^R dr\,\varrho(r)\,\frac{M(r)}{r^2} \ . \tag{1.7}$$

For a uniform density

$$P(0) = \frac{2\pi}{3}G\varrho^2 R^2 = \frac{3}{8\pi}\frac{GM^2}{R^4} \ , \tag{1.8}$$

equal to $\simeq 1.7 \times 10^{12}$ dyne/cm^2 for the Earth. When a dense core is present, as in all planets, the effective radius is smaller and a larger value results; the estimated value terrestrial value is $P_\oplus(0) \simeq 3.6 \times 10^{13}$ dyne/cm^2, much larger than the pressures attainable in laboratory experiments. As a consequence, the thermodynamical properties of matter, in particular the phase transitions, in the deep Earth interior are not well known; only the density profile can be measured directly with seismic propagation (Sec. 5.1).

The gravitational binding energy is

$$E_B = -\frac{1}{2}G\int\int\frac{dm(\mathbf{r})\,dm(\mathbf{r}')}{|\mathbf{r}-\mathbf{r}'|} = 2\pi\int_0^R dr\,r^2\varrho(r)U(r) \ ; \tag{1.9}$$

for a uniform density, using (1.5), this is

$$E_B = -\frac{3}{5}\frac{GM^2}{R} \ , \tag{1.10}$$

1. Dynamical principles

equal to -2.25×10^{39} erg for the Earth. Note that this value differs by only 10 % from the actual value -2.46×10^{39} erg.

The density profile determines also the *moment of inertia*

$$I = \frac{8\pi}{3} \int_0^R dr\, r^4 \varrho(r) = \kappa MR^2 . \quad (1.11)$$

For a homogeneous body the dimensionless *concentration coefficient* κ is $2/5 = 0.4$; a core decreases its value, which for the Earth is 0.331, 21% less (see Problem 1.2).

Sometimes it is useful to measure the mass of a body in terms of its *gravitational radius*

$$m_g = \frac{GM}{c^2}, \quad (1.12)$$

equal to 0.44 cm for the Earth and 1.5 km for the Sun. This is the distance from a point of mass M at which the Newtonian energy of a particle falling in from infinity equals its rest energy; relativistic corrections are then essential. In ordinary bodies m_g is always much smaller than the radius and relativistic effects, of order m_g/R, are very small (Ch. 17).

Deviations from spherical symmetry. The main reasons for violation of spherical symmetry for a planetary body are rotation and an external gravitational field. The rotational deformation, symmetric around the rotation axis, is driven by the ratio

$$\mu_c = \frac{R^3 \omega^2}{GM} \quad (1.13)$$

between the rotational energy and the gravitational binding energy of the body, sometimes called *Helmert's parameter*. On the surface the deformation is measured by the *oblateness* (assumed small)

$$\epsilon = \frac{R_e - R_p}{R} \approx \mu_c , \quad (1.14)$$

in terms of the equatorial (R_e), the polar (R_p) and the mean (R) radii (Sec. 3.3).

This deformation is of great importance also because it is enhanced by collapse or compression. If a body is not coupled to the environment or to other bodies, its angular momentum $\mathbf{L}$, of the order of magnitude $MR^2\omega$, remains constant, and so is the mass; hence ω is proportional to $1/R^2$ and the oblateness to $1/R$.

In relativistic physics the angular momentum per unit mass L/M also determines a length

$$a_g = \frac{L}{cM} = \frac{GL}{mc^3} , \quad (1.15)$$

equal to 0.28 km for the Sun and to 3.2 m for the Earth. We show in Ch. 17 that it characterizes a peculiar correction to the gravitational force, called *gravitomagnetism*, corresponding to an acceleration of order $c^2 m a_g/r^3$.

1.2 The equation of state

A gravitationally bound body is supported by its internal pressure P, which depends on the thermal kinetic energy of its particles and tends to expand it. Only if P is independently known or related to ϱ, the equilibrium equation (1.6) can be solved for the density profile. In general, this requires an understanding of the thermodynamical behaviour and the microscopic interactions between the particles, often based on quantum physics.

The paradigmatic case is a *perfect gas*, in which the pressure

$$P = nkT \tag{1.16}$$

is determined by the absolute temperature T and the total number density n. It should be noted that the mass of each particle is irrelevant here; for a mixture of two gases of molecular masses m_1 and m_2

$$n = \frac{\varrho}{(1-\alpha)m_1 + \alpha m_2} = \frac{\varrho}{m_{\text{mol}}}. \tag{1.17}$$

Here $\alpha = n_2/(n_1+n_2)$ is the concentration (in number) of the second constituent and $m_{\text{mol}} = Am_p$ is the mean molecular mass in terms of the mean atomic weight A. A gas behaves in this way if its particles spend most of the time in free motion, between rare collisions; in practice this hold for pressures below, say, 50 dyne/cm^2. At higher pressures complex interparticle forces (such as the van der Waals molecular force) play a role and the determination of the equation of state may require laboratory measurements.

The state near the centre of a self-gravitating perfect gas can be obtained from the equilibrium condition (1.6). The pressure gradient dP/dr can be approximated by $-P/R$, where P is a typical pressure and R is the size; similarly, $M(r)/r^2 \approx M/R^2$. Therefore

$$\frac{P}{R} \approx \frac{GM}{R^2}\varrho \approx \frac{GM}{R^2} Am_p n = \frac{GM}{R^2} m_{\text{mol}}. \tag{1.18}$$

With the perfect gas law (1.16)

$$kT \approx \frac{GM}{R} Am_p. \tag{1.19}$$

This is a consequence of the *virial theorem* and says that the thermal and the gravitational energies of a particle are of the same order of magnitude. This relation holds in the Sun; in the Earth, with an average value $A \approx 25$, it gives $T = 0.65 A$ eV $\approx 3\times 10^5$ K. However, the Earth's interior is dominated by intermolecular forces and is very far from a perfect gas. Roughly speaking, we can say that the gravitational binding energy is shared not only by the (three) translational degrees of freedom which determine the pressure, but also by many other ones. Moreover, the temperature profile of a planet strongly depends also on the thermodynamical state; for instance, large amounts of latent heat are exchanged in melting or freezing; the past history, complex energy transport processes and the uncertainties of matter at high pressure all contribute to make the estimate of the interior temperature difficult. For the Earth core the estimate is $T \approx 4,000$ K.

In general the equation of state expresses the fluid pressure $P(\varrho, s_e)$ as a function of the matter density and the *entropy per unit mass* s_e;

$$\frac{\partial P(\varrho, s_e)}{\partial \varrho} = c_s^2 \qquad (1.20)$$

is the square of the *sound speed* c_s, as briefly discussed in the following Section. If a volume of matter undergoes a change in which no heat is exchanged with the environment, its entropy remains constant and we have an *adiabatic transformation*; generally, this is the case if it happens so quickly that conductivity and other processes of heat exchange do not have the time to affect the state. In this case we have a *barotropic regime* in which the pressure $P(\varrho)$ is a function of the density ϱ alone.

In Sec. 5.3 a general thermodynamical discussion of barotropic transformations is given; here, at an easier level, we deal with a perfect gas, in which case they are called *polytropic transformations*. If its molecules have f degrees of freedom (e.g., three translations and other internal freedoms, like rotations and oscillations), the *polytropic index*

$$\gamma = \frac{f+2}{f} = \frac{c_p}{c_v} \qquad (1.21)$$

is defined in terms of the specific heats c_p and c_v at constant pressure and constant volume. For point molecules $\gamma = 5/3$; the addition of more internal degrees of freedom gives a smaller value which tends to 1 when $f \gg 1$. Since to every degree of freedom the same energy $kT/2$ is attributed, the internal energy per unit mass is

$$c_v T = \frac{fkT}{2\,m_{\mathrm{mol}}} \, ; \qquad (1.22)$$

this gives the specific heat at constant volume $c_v = fk/(2\,m_{\mathrm{mol}})$.

It turns out that an adiabatic transformation follow the differential law

$$d(\ln P) = \gamma \, d(\ln \varrho) \, , \qquad (1.23)$$

which becomes

$$P \varrho^{-\gamma} = \mathrm{const} \qquad (1.24)$$

when γ is constant. In that case the sound speed is

$$c_s = \sqrt{\frac{\gamma P}{\varrho}} = \sqrt{\frac{\gamma kT}{m_{\mathrm{mol}}}} \, . \qquad (1.25)$$

Similar relations hold for the other pairs of thermodynamical variables:

$$d(\ln P) + \frac{\gamma}{1-\gamma} d(\ln T) = 0, \qquad (1.26a)$$

$$d(\ln T) - (\gamma - 1) d(\ln \varrho) = 0. \qquad (1.26b)$$

They show that when $\gamma \to 1$ the transformation is isothermal, an obvious consequence of the fact that a very large number of degrees of freedom share the total heat input; from the gas law (1.16) the pressure is a linear function of the density. Such a linear relation holds also for a gas of photons or, in general, of massless particles, characterized by a relativistic energy-momentum tensor with a vanishing trace; this means

$$P = \frac{\varrho c^2}{3} \qquad (1.27)$$

and the sound speed is $c/\sqrt{3}$ (c is the light velocity).

Spherical models. The simple polytropic regime, even when the adiabatic assumption is not warranted, often provides very useful models for the interior of planetary bodies. The equilibrium is completely determined by eq. (1.6), a non-linear, second order differential equation for $\varrho(r)$; Problem 1.13 deals with a case with a self-similar solution which can rigorously describe the extraordinary behaviour when $\gamma \to 4/3$.

In general, a more complicated situation occurs. In order to completely describe the structure of a static, spherical planetary body we need the pressure $P(r)$, the matter density $\varrho(r)$, the temperature $T(r)$ and the composition. Their solution requires a clear understanding and appropriate modelling of the processes inside a body. Leaving aside variations of composition, to determine these three quantities we must have: (i) the equilibrium condition (1.6); (ii) an equation of state $P(\varrho, T)$; and (iii) a heat transport equation for the energy balance. In planetary bodies three processes are responsible for energy transport: conventional heat conduction (eq. (1.59)), radiative transport (particularly important for atmospheres; Sec. 7.2) and convection. Further complication arises if there are hidden reservoirs of energy in phase transitions. Finally, the problem must be supplemented with the appropriate boundary conditions (like finite values at the centre, vanishing mass density and pressure at the surface). In real cases one should also take account the time evolution of the body, which results from primordial accretion, possible slow contraction, decay of internal radiogenic heat sources, impact of other bodies, etc.

Stability analysis.– Having succeeded to construct a static configuration, the question of its stability arises. An instability, in fact, collapsed an interstellar cloud into the Sun and gave rise to the solar system. If, for example, a random disturbance slightly decreases the radius, unless the temperature is appropriately increased as well, further compression may follow and the initial unbalance between gravity and pressure may increase also. The critical factor, hence, is the degree to which the initial compression heats the interior. A simple, but powerful way to understand this is with a formal polytropic equation $P \propto \varrho^\gamma$ and an order of magnitude form (1.18) of the equilibrium condition (1.6). If density and pressure are allowed to vary, keeping M constant, the locus of equilibria in the (ϱ, P) plane is a curve $P \propto \varrho^{4/3}$. If a change in density momentarily displaces the body from equilibrium in an adiabatic way and $\gamma > 4/3$, the body is brought to a pressure above the corresponding equilibrium pressure; then an unbalance will arise and bring back the body towards its initial density and beyond, with stable oscillations; if, on the contrary $\gamma < 4/3$, the final pressure will be smaller than the corresponding equilibrium pressure and collapse will inexorably ensue until another, stronger form of pressure is met. For a perfect gas of point particles, $\gamma = 5/3$, we have a stable configuration; but if internal degrees of freedom appear, lowering γ (eq. (1.21)), the energy released by gravitation will be shared with them as well, leaving a smaller part for the translational random motion, which determines the pressure; 3 additional degrees of freedom bring the system to the threshold $4/3$. Another possible way to collapse occurs when the body is not opaque and radiates away whatever additional thermal energy it acquires from the compression; formally, this can be described with $\gamma = 1$, in the unstable range. A large opacity favours the stability of a gravitationally bound body; the initial gas cloud which gave rise to the Sun likely was transparent, hence unstable. Other processes by which γ can decrease below $4/3$ take place at the end of thermonuclear burning of a main sequence star, leading to collapse to a white dwarf, a neutron star or a black hole.

1.3 Dynamics of fluids

When there is motion, we have a *dynamic equilibrium* and eq. (1.6) must be supplemented with the inertial force (the equilibrium is referred to the co-moving frame with the fluid). In spherical symmetry a fluid element of mass $\varrho \, dr \, dS$, which at the time t has a radial velocity $v(r, t)$, at time $t+dt$ has a velocity $v(r+dr, t+dt)$, where $dr = v(r, t) \, dt$ is the radial displacement undergone by the fluid element. Its acceleration results from the change of the velocity field at a given place and the change due to the fact that the fluid element moves

$$\frac{v(r + v\,dt, t + dt) - v(r, t)}{dt} \to \frac{dv}{dt} = \frac{\partial v}{\partial t} + v \frac{\partial v}{\partial r} \,. \tag{1.28}$$

The first term on the right-hand side is the ordinary, *Eulerian derivative* and the whole expression defines the *Lagrangian derivative*. More generally, the Lagrangian derivative, which defines the rate of change suffered along with the moving fluid, is given by

$$\frac{d}{dt} = \frac{\partial}{\partial t} + \mathbf{v} \cdot \nabla \,. \tag{1.29}$$

Neglecting viscosity we have, in place of eq. (1.6),

$$\varrho\left(\frac{\partial v}{\partial t} + v\frac{\partial v}{\partial r}\right) + \frac{\partial P}{\partial r} + \frac{GM(r)}{r^2}\varrho = 0, \qquad (1.30)$$

with the first term being the inertial acceleration $\varrho\, dv/dt$ experienced in the frame comoving with the fluid. In a *stationary flow* v is a function of r alone, and this reads

$$v\frac{dv}{dr} + \frac{dP}{\varrho\, dr} + \frac{GM(r)}{r^2} = 0. \qquad (1.31)$$

In the polytropic regime, when the pressure P is a given function of the density ϱ, this integrates to *Bernoulli's equation*

$$\frac{v^2}{2} + F(\varrho) - \frac{GM}{r} = \text{const}; \qquad (1.32)$$

here the function $F(\varrho)$ is defined by

$$\varrho\, dF = dP = c_s^2\, d\varrho,$$

in terms of the speed of sound c_s (eq. (1.20)). It shows how the flow velocity is affected by gravity and by variations in density. For a perfect adiabatic gas, in terms of the polytropic index γ,

$$F = \frac{\gamma}{\gamma - 1}\frac{P}{\varrho}$$

and Bernoulli's equation reads, in terms of the initial conditions,

$$\frac{v^2}{2} + \frac{c_s^2}{\gamma - 1} - \frac{GM}{r} = \frac{v_0^2}{2} + \frac{c_{s0}^2}{\gamma - 1} - \frac{GM}{r_0}. \qquad (1.33)$$

We may notice that as $\gamma \to 1$ the sound speed tends to a constant ($c_s^2 = dP/d\varrho$), corresponding to the isothermal regime. Radial flows are important for cosmic bodies; in particular, a radial wind of charged particles is emitted by the Sun and produces striking phenomena in interplanetary space (Ch. 9). Bernoulli's law is the gist of their dynamics.

For moving fluids, *mass conservation* must be also taken into account. A shell between r and $r + dr$ contains the mass $4\pi r^2 \varrho\, dr$. In an infinitesimal time dt a particle at r moves by $v(r, t)\, dt$, so that

$$r^2 \to r^2 + 2\, r v\, dt.$$

The shell thickness dr and the mass density ϱ change according to

$$dr \to dr + dr\frac{\partial v}{\partial r} dt, \qquad \varrho \to \varrho + \left(\frac{\partial \varrho}{\partial t} + v\frac{\partial \varrho}{\partial r}\right) dt.$$

1. Dynamical principles   11

The total change in the quantity $r^2\varrho\,dr$ must vanish; hence

$$2rv\varrho + r^2\left(\frac{\partial\varrho}{\partial t} + v\frac{\partial\varrho}{\partial r}\right) + r^2\varrho\frac{\partial v}{\partial r} = \frac{d\varrho}{dt} + \varrho\left(\frac{\partial v}{\partial r} + \frac{2v}{r}\right) = 0 \,. \tag{1.34}$$

The last bracketed quantity is just the divergence of the radial flow $\mathbf{v} = v\mathbf{r}/r$. In the form

$$\frac{d\varrho}{dt} + \varrho\nabla\cdot\mathbf{v} = \frac{\partial\varrho}{\partial t} + \nabla\cdot(\varrho\mathbf{v}) = 0 \,, \tag{1.34'}$$

the mass conservation equation is also valid in the general case, when no spherical symmetry is assumed. It can be easily interpreted by recalling that, if in a time dt a fluid element is displaced from $\mathbf{r}$ to $\mathbf{r} + \mathbf{v}dt$, an infinitesimal volume dV associated with it changes into $dV(1 + dt\,\nabla\cdot\mathbf{v})$ (see Problem 1.1). Since the total mass of the volume remains constant, the change $d\varrho$ in mass density fulfils

$$\varrho\,dV = (\varrho + d\varrho)\,dV\,(1 + \nabla\cdot\mathbf{v}\,dt) \,,$$

equivalent to eq. (1.34'). In an incompressible flow (ϱ is constant) the velocity is free of divergence ($\nabla\cdot\mathbf{v} = 0$).

To describe a generic fluid motion, we need the appropriate generalization of eq. (1.30) to account for the conservation of momentum[2]

$$\varrho\frac{d\mathbf{v}}{dt} + \nabla P + \varrho\nabla U = \varrho\left(\frac{\partial\mathbf{v}}{\partial t} + \mathbf{v}\cdot\nabla\mathbf{v}\right) + \nabla P + \varrho\nabla U = 0 \,. \tag{1.35}$$

Here U is the gravitational potential energy per unit mass, solution of the *Poisson's equation*

$$\nabla^2 U = 4\pi G\varrho \,, \tag{1.36}$$

which vanishes at infinity. It is given by *Poisson's integral*

$$U(\mathbf{r}) = -G\int d^3r'\,\frac{\varrho(\mathbf{r}')}{|\mathbf{r}-\mathbf{r}'|} \,. \tag{1.37}$$

For typical applications in planetary science, the integration in (1.37) is extended over a finite number of compact supports occupied by cosmic bodies (i.e. where the density does not vanish). For a point mass M at the origin this gives the monopole $U = -GM/r$. The linearity of Poisson's equation ensures the *superposition principle*: the potential of N bodies is the sum of the individual potentials. Hence a field of N monopoles is just $U = -G\sum M_i/r_i$.

---

[2] Without the gravitational term $\varrho\nabla U$, this equation has been first derived by L. EULER in 1755 and is usually called *Euler equation*.

The geoid.– The gravitational potential of a body in uniform and constant rotation, at the angular velocity ω, is best described in a rotating (rest) frame. The equilibrium of a rotating fluid body of arbitrary shape is determined by

$$\nabla P + \varrho \nabla W = 0 , \qquad (1.38)$$

where

$$W = U - \frac{1}{2}\omega^2 r^2 \sin^2 \theta$$

includes the potential of the centrifugal force; θ is the angle from the rotation axis. $P = 0$ defines an equipotential surface for W and is called the *geoid*; on the Earth this is physically approximated by the mean surface of the oceans. Changes in this surface due to the variable atmospheric pressure and large scale currents in the oceans must be eliminated (Sec. 2.3). The normal to the geoid, along ∇W, is the *vertical*, whose direction differs from the radial direction by $O(\mu_c)$ and slightly changes from place to place due to the inhomogeneities of the interior. At a low altitude z above the geoid we can approximately write $W(z) = W(0) + gz$; g is the *gravity acceleration*, which varies a little from place to place.

Wave propagation.– A fluid supports a fundamental mode of propagation, *acoustic waves*. In the simplest version, gravity is neglected and a barotropic relation $P(\varrho)$ determines the pressure. A uniform background state at rest has a small perturbation in the density $\varrho = \varrho_0 + \varrho'$ and the velocity $\mathbf{v}'$; the linearized eqs. (1.34′) and (1.35) then read

$$\frac{\partial \varrho'}{\partial t} + \varrho_0 \nabla \cdot \mathbf{v}' = 0 , \quad \varrho_0 \frac{\partial \mathbf{v}'}{\partial t} + \left(\frac{dP}{d\varrho}\right)_0 \nabla \varrho' = 0 .$$

This gives the wave equation

$$\frac{\partial^2 \varrho'}{\partial t^2} - \left(\frac{dP}{d\varrho}\right)_0 \nabla^2 \varrho' = 0 \qquad (1.39)$$

with the propagation velocity c_s (eq. (1.20)).

1.4 Dynamics of solid bodies

In Sec. 1.1 we have shown that the radius of a gaseous cosmic body, like the Sun, is determined by the inner state and temperature, therefore by the previous history, by the active energy sources (e.g., radioactive materials) and the energy transport processes (see Sec. 1.5). The pressure P is a scalar; in a perfect gas (eq. (1.16)) it is a simple function of density and temperature.

At a temperature below the freezing point (i.e., when the thermal energy is smaller than the binding energy between neighbouring molecules), the body becomes solid and a different analysis is called for. In general, the force exerted

1. Dynamical principles

by the medium upon an infinitesimal volume dV is determined by the *stress tensor* P_{ij}. If we remove the material on one side of a surface element dS with outward normal **n**, to keep the body in the same state of motion (or of rest) we must exert from the empty side (where the normal lies) a force, say, $P_{ij}n_j dS$. The stress tensor has the dimension of a pressure, or an energy density. In a fluid this force is orthogonal to the surface element, corresponding to an isotropic

$$P_{ij} = -P\delta_{ij} \quad (\text{or} \quad \mathsf{P} = -P\mathsf{I}); \tag{1.40}$$

the scalar P is the *pressure* (note the minus sign, corresponding to the fact that pressure tends to expand the volume element dV). In general, however, there is also a force parallel to the surface element (*shear stress*). From the conservation of angular momentum it can be shown that P_{ij} is symmetric.

The total stress force exerted upon a volume element V bounded by a surface S by the surrounding medium is given by

$$\int_S dS\, n_i\, P_{ij},$$

or, by Gauss' theorem,

$$\int_V dV\, \partial_j P_{ij}.$$

Therefore the stress force per unit volume is

$$f_i = \frac{\partial P_{ij}}{\partial r_j} \tag{1.41}$$

(r_j are Cartesian coordinates).

Solidity means that each element, in absence of external forces, has a reference position **r**. The state of motion of a solid body is described by a small displacement field $\mathbf{s}(\mathbf{r}, t)$ of an element from the reference position. The dynamical effects of the displacement, unaffected by a translation, must depend upon the derivative tensor, sum of a symmetrical and an antisymmetric part

$$\frac{\partial s_j}{\partial r_i} = \frac{1}{2}\left(\frac{\partial s_j}{\partial r_i} + \frac{\partial s_i}{\partial r_j}\right) + \frac{1}{2}\left(\frac{\partial s_i}{\partial r_j} - \frac{\partial s_j}{\partial r_i}\right). \tag{1.42}$$

The antisymmetric part, second term on the right-hand side above, expresses an infinitesimal rotation and does not generate any stress. For small deformations the squares of the quantities in eq. (1.42) can be neglected; then the (symmetric) stress tensor is expected to be a linear function of its symmetric part

$$e_{ij} = \frac{1}{2}\left(\frac{\partial s_j}{\partial r_i} + \frac{\partial s_i}{\partial r_j}\right), \tag{1.43}$$

the *strain tensor*. It can be split into a diagonal part with trace

$$\theta = e_{ii} = \frac{\partial s_i}{\partial r_i} = \nabla \cdot \mathbf{s} \tag{1.44}$$

and in a left over, trace-free part

$$e_{ij}^{\text{TF}} = e_{ij} - \frac{1}{3}\delta_{ij}\theta = \frac{1}{2}\left(\frac{\partial s_j}{\partial r_i} + \frac{\partial s_i}{\partial r_j}\right) - \frac{1}{3}\delta_{ij}\theta . \tag{1.45}$$

θ measures the *dilatation* of an infinitesimal volume element produced by the deformation: a volume element dV becomes $dV(1+\theta)$. The trace-free part e_{ij}^{TF} conserves the volume.

Hooke's law. A solid under a small deformation behaves *elastically*, to wit, the elastic stress is a *linear* function of the strain (*Hooke's law*; when deformations are large, a non-linear relation between stress and strain must be introduced). This is expressed by a fourth rank (Hooke's) tensor C_{ijkh}

$$P_{ij} = C_{ijkh} e_{kh} . \tag{1.46}$$

It can be shown that the energy density w of the deformation is the quadratic form

$$w = \frac{1}{2} C_{ijkh} e_{ij} e_{kh} , \tag{1.47}$$

which determines the stress

$$P_{ij} = \frac{\partial w}{\partial e_{ij}} . \tag{1.48}$$

In a generic material Hooke's tensor – symmetric under the exchange of i, j and k, h, and the two pairs – has only 21 independent components. Symmetries related to the internal structure of the body, such as the appropriate crystalline structure, decrease the number of its independent components. The simplest case is that of an isotropic material, for which the energy (1.47) must be entirely determined by scalars. It turns out that there are only two quadratic invariant quantities which can be constructed with a symmetric tensor like e_{ij}, that is, θ^2 and

$$e_{ij}e_{ij} = (e_{11})^2 + (e_{22})^2 + (e_{33})^2 + 2(e_{12})^2 + 2(e_{23})^2 + 2(e_{31})^2 .$$

In this case the elastic energy must be of the form

$$w = \frac{1}{2}\lambda\theta^2 + \mu\, e_{ij}e_{ij} ; \tag{1.47'}$$

the scalars λ and μ are *Lamé's constants*, with the dimension of a pressure. From eq. (1.48) we get Hooke's law for an isotropic medium

$$P_{ij} = \lambda\theta\delta_{ij} + 2\mu\, e_{ij} . \tag{1.49}$$

1. Dynamical principles

Since molecular forces in the body tend to restore its initial state, the parameters λ and μ are positive; contrary to the gas pressure (eq. (1.40)), for positive components of the stress tensor, $P_{ij} n_j dS$ is directed outward from the volume element dV. Two typical examples are useful to illustrate the physical meaning of Lamé's constants. A one-dimensional compression, with only $e_{11} = \partial_x s_x$ different from zero, produces a stress

$$P_{11} = (\lambda + 2\mu) \partial_x s_x . \qquad (1.50)$$

For shortness, sometimes a suffix to ∂ denotes the independent variable. A shear deformation, with only the component $e_{12} = \partial_y s_x / 2$, produces a shear stress

$$P_{12} = \mu \partial_y s_x . \qquad (1.51)$$

The *rigidity* μ expresses the resistance to shear. Under a three-dimensional expansion, with $e_{ij} = \delta_{ij} \theta / 3$, a scalar stress arises

$$P_{ij} = \delta_{ij} (\lambda + 2\mu/3) \theta = \delta_{ij} K \theta = -\delta_{ij} P_{\text{el}} , \qquad (1.52)$$

where P_{el} is the effective pressure generated by elasticity. The expansion changes the density by $d\varrho = -\varrho \theta$; hence

$$\frac{dP}{d\varrho} = \frac{K}{\varrho} \qquad (1.53)$$

is the *compressibility* and $K = \lambda + 2\mu/3$ the *bulk modulus*, equal to about $5\mu/3$ in the Earth's interior. In terms of the trace-free part (1.45) of the strain energy, (1.47') reads

$$w = \frac{1}{2} K \theta^2 + \mu e_{ij}^{\text{TF}} e_{ij}^{\text{TF}} , \qquad (1.47'')$$

which clearly shows the role of K.

Dynamics of a solid body. The inertial acceleration is the second Lagrangian derivative of the displacement **s**: the displacement, and not the fluid velocity, is the appropriate field variable. Equating the inertial force to the sum of the stress force and the gravitational force we get

$$\varrho \frac{d^2 \mathbf{s}}{dt^2} - \nabla \cdot \mathbf{P} + \varrho \nabla U = 0 . \qquad (1.54)$$

In synthetic vector notation we denote by **P** the stress tensor. If the equilibrium state of the body is referred to a non-inertial frame, the deformation is also affected by the apparent forces. An outstanding example is the Coriolis acceleration $-2\varrho \, \boldsymbol{\omega} \times (d\mathbf{s}/dt)$, which must be taken into account to deal with a non-rigid Earth rotating with angular velocity $\boldsymbol{\omega}$; this term produces the important effect of removing the degeneracy of the oscillation spectrum of the Earth (Sec. 5.2). The centrifugal force must also be taken into account.

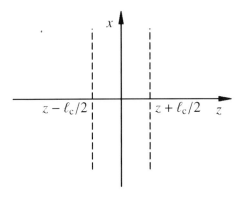

Figure 1.3 A physical quantity Q is transported across a surface z = const. by the molecules which travel freely from $z - \ell_c/2$ to $z + \ell_c/2$, and vice versa.

If λ and μ are constant, the momentum equation (1.54) reads

$$\varrho \frac{d^2 \mathbf{s}}{dt^2} - \mu \nabla^2 \mathbf{s} - (\lambda + \mu)\nabla\theta + \varrho\nabla U = 0 . \tag{1.55}$$

This medium is capable of propagating both transversal and longitudinal waves with velocities determined by Lamé's constants (Ch. 5). The dynamics of a solid body, as described by (1.55), is generally coupled with its thermal state, since part of the change in the volume θ may be related to the temperature gradients, e.g., $\nabla\theta = \alpha\nabla T$, with α the coefficient of thermal expansion. The mechanical and thermal state of a body (e.g. planet) must be thus solved together.

The deformation of solid planetary bodies is in fact governed by still more complicated (*viscoelastic*) laws. Under a suddenly imposed external stress the final elastic deformation is reached with a delay; when the stress ceases the recovery is not immediate and generally does not bring it back to the original shape. Moreover, the timescale of external forces is crucial: it is possible that a material, elastic at high frequencies, behaves plastically under a prolonged stress and shows flow phenomena called *creep*. This is indeed the case of the interior of the Earth and the terrestrial planets.

1.5 Transport

With simple arguments, and skipping the complicated machinery of kinetic theory, we now come to the microscopic heart of transport theory. Let a quantity $Q(\mathbf{r}, t)$ be associated to an average particle of the medium with mass m_{mol} and density $\varrho = n m_{\text{mol}}$, at the position $\mathbf{r}$ and the time t. Collisions transfer large values of Q to regions where it is smaller, thereby implementing the general tendency to a uniform state required by the second law of thermodynamics and introducing a privileged direction of time. Q can be $c_p T$, corresponding to the mean thermal energy and energy transport; or, for the momentum, $m_{\text{mol}} \mathbf{v}$, leading to viscosity; or the concentration of a solute, for diffusion processes. If Q depends only on z, consider the surfaces $z - \ell_c/2$ and

1. Dynamical principles

$z + \ell_c/2$, separated by a mean free path ℓ_c (Fig. 1.3); one can say, a molecule which had a collision at $z - \ell_c/2$ suffers the next collision at $z + \ell_c/2$ and transfers to this place the value of Q belonging to $z - \ell_c/2$, with a net transfer of $Q(z + \ell_c/2) - Q(z - \ell_c/2) \simeq \ell_c \partial Q/\partial z$, and *vice versa* in the opposite sense. Roughly $nv_T/2$ particles cross the surface z in each sense and each group carries the same flux (v_T is the thermal velocity); the total flux is then

$$F_Q = nv_T \left[Q(z - \ell_c/2) - Q(z + \ell_c/2)\right] \approx -nv_T \ell_c \frac{\partial Q}{\partial z} \; .$$

In general

$$\mathbf{F}_Q = -nv_T \ell_c \nabla Q \; . \tag{1.56}$$

The mean free path ℓ_c is used here only in order of magnitude; the precise numerical coefficient, taking into account the three-dimensional velocity distribution, requires kinetic theory. When, in terms of the specific heat at constant pressure, $Q = c_p T$ this says that heat flows to the cooler parts; then the quantity $\chi = v_T \ell_c$ is the *heat transport coefficient*, or *heat diffusivity*. When $Q = m_{\mathrm{mol}} v_x(z)$, the sheared velocity field $v_x(z)$ is smoothed out and

$$\eta = \varrho v_T \ell_c \tag{1.57}$$

is the *viscosity coefficient*.

Transport of a scalar quantity. Let $\int dV \, Qn$ be the amount of Q contained in a volume V, whose boundary S has the outer normal $\mathbf{n}$. If there are no sources, its change, using Green's theorem, is

$$\frac{d}{dt}\int_V dV \, Qn = -\int_S dS \, \mathbf{n} \cdot \mathbf{F}_Q = -\int_V dV \, \nabla \cdot \mathbf{F}_Q \; ,$$

leading to the conservation law

$$\varrho \frac{dQ}{dt} = \nabla \cdot (\varrho v_T \ell_c \nabla Q) \; . \tag{1.58}$$

We have used the fact that mass is conserved, so that the Lagrangian derivative $d/dt = \partial/\partial t + \mathbf{v} \cdot \nabla$ of $\varrho \, dV$ vanishes. This derivative takes also into account the velocity $\mathbf{v}$ of the medium.

Heat transport.– For heat transport $Q = c_p T$ is the thermal energy per unit mass and we get the *heat transport equation*

$$\varrho \frac{d(c_p T)}{dt} = \nabla \cdot (\varrho \chi \nabla c_p T) + S \; . \tag{1.59}$$

The contribution of a source (or a sink) is taken into account by adding to the right-hand side the production rate of Q per unit volume, called S. Heat sources are important in the early phases of planetary formation, for instance, due to radioactive decays. When ϱ, χ and c_p are constant, the previous equation reduces to

$$\varrho c_p \frac{dT}{dt} = \kappa \nabla^2 T + S ,$$

where $\kappa = \varrho \chi c_p$ is the *thermal conductivity*.

The heat transport equation must be supplemented with the boundary conditions at the surface (e.g., radiative loss) and with the continuity condition across a sharp transition (e.g., between the core and the mantle; Fig. 5.4). With an energy source F (with the dimension of a flux) at the transition, we expect a jump in the flux $\varrho \chi \mathbf{n} \cdot \nabla(c_p T)$ along the normal $\mathbf{n}$ to the surface. The jump condition is obtained by integrating (1.59) over a thin and small layer with boundaries just inside and outside the surface. Indicating with $[s]$ the jump in a quantity s, we get

$$[\varrho \chi \mathbf{n} \cdot \nabla (c_p T)] = F . \tag{1.60}$$

The physical nature of heat transport is best illustrated with the simplest example in which $T(z,t)$ is the only variable quantity and the fluid is at rest:

$$\frac{\partial T}{\partial t} = \chi \frac{\partial^2 T}{\partial z^2} . \tag{1.61}$$

This is a *parabolic* partial differential equation. An initial temperature spike spreads after a time t over a distance of order $\sqrt{\chi t}$ and there is no wave propagation; the corresponding solution, which conserves the total thermal energy $\propto \int dz\, T = K$,

$$T(z,t) = \frac{K}{2\sqrt{\pi \chi t}} \exp\left(-\frac{z^2}{4\chi t}\right) .$$

When $t \to 0$ it becomes Dirac's function $K\delta(z)$. This characteristic behaviour betrays the deep relation between diffusion processes and random walk.

Viscosity. A general expression of the viscous force can be derived from first principles, similarly to the stress in a solid. A fluid element of volume V is acted upon by the surrounding elements through the velocity gradients which can transport momentum through the bounding surface S. The *viscous stress tensor* q_{ij} defines the force $q_{ij} n_j dS$ exerted on the element dS with an outer normal $\mathbf{n}$; the corresponding force per unit volume is $\partial_j q_{ij}$.

Like the stress tensor, q_{ij} is symmetric; moreover, for small velocity gradients, it must be linear in the derivative tensor $\partial_i v_j$. The antisymmetric part of this tensor corresponds to a rotational velocity of the fluid element and can be shown to have no dynamical effects; what counts is its symmetric part. Just like for eq. (1.49), when the microscopic properties of the fluid have no privileged

1. Dynamical principles

directions, the only admissible form of the viscous stress tensor is

$$q_{ij} = \eta\left(\frac{\partial v_j}{\partial r_i} + \frac{\partial v_i}{\partial r_j}\right) - \frac{2}{3}(\eta - \zeta)\delta_{ij}\frac{\partial v_k}{\partial r_k}. \tag{1.62}$$

If the scalar coefficients η and ζ are constant, the viscous force per unit volume is

$$\frac{\partial q_{ij}}{\partial r_j} = \eta \nabla^2 v_i + \frac{1}{3}(\eta + 2\zeta)\frac{\partial}{\partial r_i}(\nabla \cdot \mathbf{v}). \tag{1.63}$$

Kinetic gas theory shows that the second coefficient ζ vanishes or is very small. We are therefore able to complete eq. (1.35) as follows:

$$\varrho\left(\frac{\partial \mathbf{v}}{\partial t} + \mathbf{v}\cdot\nabla\mathbf{v}\right) + \nabla P + \varrho\nabla U = \eta\left[\nabla^2\mathbf{v} + \frac{1}{3}\nabla(\nabla\cdot\mathbf{v})\right]. \tag{1.64}$$

For an incompressible flow – $\nabla \cdot \mathbf{v} = 0$ – this reduces to

$$\varrho\left(\frac{\partial \mathbf{v}}{\partial t} + \mathbf{v}\cdot\nabla\mathbf{v}\right) + \nabla P + \varrho\nabla U = \eta\nabla^2\mathbf{v}, \tag{1.65}$$

the *Navier-Stokes equation*; the quantity η/ϱ is the *kinematic viscosity*. Eq. (1.35) is time-reversible: if $\mathbf{v}(\mathbf{r},t)$ is a solution, so is $-\mathbf{v}(\mathbf{r},-t)$; in agreement with the irreversible character of transport processes, the addition of viscosity destroys this property.

Viscosity results in dissipation; the kinetic energy of the fluid motion is thus transformed into heat and, in general, should be accounted for in the heat transfer equation (1.59) (this happens when convection is active). We shall now compute the appropriate source term S, the density of energy dissipation due to the motion of a viscous fluid. For sake of simplicity we neglect here the effects of the gravitational potential U. The total kinetic energy of the fluid is then

$$E_{\text{kin}} = \frac{1}{2}\int dV \varrho v^2, \tag{1.66}$$

where integration is performed over the volume occupied by the fluid. A straightforward, but lengthy, computation using the Navier-Stokes and the continuity equations yields

$$\frac{dE_{\text{kin}}}{dt} = -\int dS_i \left[\varrho v_i\left(\frac{v^2}{2} + \frac{P}{\varrho}\right) - v_j q_{ij}\right] - \int dV\, q_{ij}\frac{\partial v_i}{\partial r_j} \tag{1.67}$$

for an incompressible fluid. When the fluid velocity vanishes at infinity or at the boundary of the finite volume where the flow is considered, the first integral vanishes and we have

$$\frac{dE_{\text{kin}}}{dt} = -\frac{1}{2}\int dV\, q_{ij}\left(\frac{\partial v_i}{\partial r_j} + \frac{\partial v_j}{\partial r_i}\right), \tag{1.68}$$

which, for an incompressible motion, turns out to be

$$\frac{dE_{\text{kin}}}{dt} = -\frac{\eta}{2} \int dV \left(\frac{\partial v_i}{\partial r_j} + \frac{\partial v_j}{\partial r_i} \right)^2 . \tag{1.68'}$$

Since dissipation decreases the energy, η must be positive. The integrand in eq. (1.68′) provides an estimate of the density by which viscous motion transforms energy to heat, and thus the source term S in eq. (1.59) for an appropriate physical situation.

Viscosity also increases the order of the system of partial differential equations; therefore the determination of the solution requires more boundary conditions than the inviscid flow. It can be expected, therefore, that the limit $\eta \to 0$ is catastrophic, in the sense that it collapses the set of solutions into one of fewer dimensions. The full boundary conditions cannot be satisfied by the inviscid limiting solution and in general discontinuities arise. This is the origin of *boundary layers*: the viscosity has negligible effects almost everywhere, except in very thin regions (i.e., near solid bodies surrounded by the flow), where the velocity gradient is very large, in order to accommodate for the extra boundary conditions. An important boundary layer occurs in the atmosphere near the ground, where, besides the condition of vanishing velocity at the surface, complex heat transfer and radiative processes take place. Mathematically, boundary layers must be treated with the methods of *singular perturbation theory*.

The relevance of viscosity is generally described by the dimensionless *Reynolds number*

$$Re = \frac{\varrho L v}{\eta} ; \tag{1.69}$$

L is a characteristic macroscopic length and v a typical velocity of the flow. When Re increases beyond a critical value (much larger than unity) a drastic change takes place: the flow from smooth and laminar becomes turbulent with a hierarchy of eddies spanning a large range of wave numbers up to Re/L, at which dissipation prevails.

Planetary bodies are often solid, with a shape determined when their temperature became lower than the freezing point. If they are far from each other they are spherical; in practice they show a semi-permanent deformation due to rotation and tidal forces. If these forces change in time, however (e.g. due to polar motion or the variability of the tides in an eccentric orbit), their response is mediated by the elastic constants of the interior. Important examples are discussed in Chs. 3 and 4. When this happens the shape of the body changes continuously, and viscous forces play a role and generate heat. A dynamical model which includes elasticity and viscosity is necessary. The formal similarity between the two stress tensors (1.49) and (1.62) can then be used. In this

1. Dynamical principles 21

case the velocity is just the time derivative of the displacement and, for a given frequency ω,

$$\mathbf{v} = i\omega\mathbf{s}.$$

One can then combine the two stress tensors in one by taking a *complex* set of elastic constants; for example, for an incompressible flow we have $(\mu + i\omega\eta)$ in place of μ. This provides an appropriate framework to describe dissipation.

1.6 Magnetohydrodynamics

At a sufficiently high temperature and low density and, possibly, under a strong ultraviolet radiation from the Sun, a gas may become partially or totally ionized (Ch. 8). When the number of particles in a Debye length λ_D (eq. (1.102)) is very large and for scales larger than λ_D, approximate charge neutrality holds. In this case one can still describe it as a single, neutral fluid; however, the relative motion of electrons and ions produce electric currents and magnetic fields.

Maxwell's equations. We first briefly recall the governing equations for the electromagnetic field (in Gaussian units). A distinction must be made between the microscopic level, at which charges and currents are very small and isolated, and the macroscopic level, at which a medium affects the electric and magnetic fields with its response. In the former the fields $\mathbf{E}$ and $\mathbf{B}$ are determined by a charge density σ_c and a current density $\mathbf{J}$ by means of the equations

$$\nabla \cdot \mathbf{E} = 4\pi \sigma_c, \quad \nabla \times \mathbf{B} - \frac{1}{c}\frac{\partial \mathbf{E}}{\partial t} = \frac{4\pi}{c}\mathbf{J}. \tag{1.70}$$

It is important to recognize that the magnetic field is generated not only by the current density, but also by the *displacement current* $\partial \mathbf{E}/(c\partial t)$. In addition, we have *Faraday's law* and the absence of free magnetic poles:

$$\nabla \times \mathbf{E} + \frac{1}{c}\frac{\partial \mathbf{B}}{\partial t} = 0, \quad \nabla \cdot \mathbf{B} = 0. \tag{1.71}$$

Faraday's law ensures that the divergence of $\mathbf{B}$ vanishes for ever, if it does so at the beginning; the same is true for the first pair of the Maxwell's equations (1.70), namely the second equation – Ampère's law – ensures validity of the divergence equation provided it is satisfied by the initial data. As a result, with any charge and current distribution which fulfils the conservation law

$$\frac{\partial \sigma_c}{\partial t} + \nabla \cdot \mathbf{J} = 0, \tag{1.72}$$

an arbitrary initial electromagnetic field can be propagated in time.

At a *macroscopic level*, in presence of matter, the macroscopic electric field generated by a charge distribution is affected also by the electric polarization it induces and becomes **D**, the electric induction. Similarly, magnetic fields – now denoted by **H** – are not only generated by the currents directly, but also through the polarization they induce in the medium; the first set of Maxwell's equations (1.70) then reads

$$\nabla \cdot \mathbf{D} = 4\pi \sigma_c, \quad \nabla \times \mathbf{H} - \frac{1}{c}\frac{\partial \mathbf{D}}{\partial t} = \frac{4\pi}{c}\mathbf{J} ; \qquad (1.73)$$

B in the second set of equations (1.71) is usually called magnetic induction. At this level electromagnetism requires an appropriate description of the response of matter, in form of "constitutive equations", which connect **B** to **H** and **E** to **D**; this, however, is beyond our scope. We can only say that when the fields are weak and matter is isotropic, they are given in terms of the *dielectric constant* ϵ and the *magnetic permeability* μ:

$$\mathbf{B} = \mu \mathbf{H}, \quad \mathbf{D} = \epsilon \mathbf{E} . \qquad (1.74)$$

In normal modes of oscillations the electric and magnetic response depends on the mode; the constitutive equations are expressed in terms of the Fourier components of the fields. In this book, for simplicity, we use Maxwell's equations in empty space (1.70) and (1.71), and we neglect the dielectric constant and the permeability, a reasonable assumptions in rarefied ionized gases; but, for example, the derivation of the dispersion relation for electron plasma waves in Sec. 8.3 can easily be rephrased in terms of a dielectric tensor, expressed in terms of the index of refraction.

Ohm's law. The current density **J** is determined by *Ohm's law*: currents in a medium at rest flow at a rate proportional to the electric field:

$$\mathbf{J} = \sigma \mathbf{E} . \qquad (1.75)$$

σ, the *electrical conductivity* (with the dimension of a frequency), can be expressed in terms of microscopic quantities as follows. A current arises when the electron gas moves with respect to the ions with a velocity **u**; this generates a friction, which in steady state balances the electric force. The frictional force upon a single electron has the form $-m_e \nu_c \mathbf{u}$, where $\nu_c \approx v_T/\ell_c$ is the collision frequency. Then

$$\mathbf{u} = -\frac{e}{m_e \nu_c}\mathbf{E} ;$$

or, in terms of the current density,

$$\mathbf{J} = -ne\mathbf{u} = \frac{ne^2}{m_e \nu_c}\mathbf{E} = \sigma \mathbf{E}, \qquad (1.75')$$

providing the conductivity σ.

1. Dynamical principles

In a moving conductor Ohm's law must be modified. If a wire segment crosses a magnetic field, an electric field $\mathbf{v} \times \mathbf{B}/c$ arises – Lenz's law – and eq. (1.75) must be replaced with

$$\mathbf{J} = \sigma \left(\mathbf{E} + \frac{1}{c} \mathbf{v} \times \mathbf{B} \right). \tag{1.75''}$$

When there is a strong external magnetic field, or in presence of other factors which destroy the isotropy of the fluid, the conductivity is, in general, a tensor quantity and the previous equation must be appropriately generalized. Often the conductivity is large enough to allow replacement of (1.75'') with

$$\mathbf{E} + \frac{1}{c} \mathbf{v} \times \mathbf{B} = 0. \tag{1.76}$$

MHD equations. In an electrically conductive fluid the magnetic force must be added to the momentum equation (1.35), leading to the self-consistent model of magnetohydrodynamics (MHD). A charge q moving with velocity $\mathbf{u}$ in a magnetic field $\mathbf{B}$ suffers the Lorentz force $q\mathbf{u}\times\mathbf{B}/c$. Summing these elementary forces for all the charged particles in a volume element dV we get

$$\sum q \mathbf{u} \times \mathbf{B} = \mathbf{J} \times \mathbf{B} \, dV,$$

where $\mathbf{J} = \sum q\mathbf{u}/dV$ is the current density. Therefore, the force per unit volume $\mathbf{J} \times \mathbf{B}/c$ must be added to momentum conservation law; in the case of non-viscous fluid we have

$$\varrho \frac{d\mathbf{v}}{dt} + \boldsymbol{\nabla} P + \varrho \boldsymbol{\nabla} U = \frac{1}{c} \mathbf{J} \times \mathbf{B}. \tag{1.77}$$

Of course, the pressure P must be expressed in terms of the matter density ϱ, either directly with a polytropic equation, or indirectly, through the equation of state and the energy equation. It should also be noted that in this fluid model electric and magnetic fields are produced by matter moving at velocities v much smaller than c; hence their time and space variations are related by $\partial/\partial t \approx v\nabla \ll c\nabla$ and in the second Maxwell's equation (1.70), Ampère's law, the displacement current can be neglected, resulting in a simpler form

$$c\boldsymbol{\nabla} \times \mathbf{B} = 4\pi \mathbf{J}; \tag{1.78}$$

but this means that the current density is solenoidal and net charges are neglected. This is in agreement with the general property of a plasma, in which there no charge fluctuations in volumes much larger than the Debye length λ_D. The small charge density can be recovered by taking the divergence of Ohm's law (1.75''), but it does not play any role in the MHD system of equations.

Note also that this approximation cannot deal with electromagnetic waves, for which E and B are of the same order of magnitude.

This completes the MHD equations, which carry forward in time the velocity with (1.77), and the magnetic field with (1.71), where for the current density Ohm's law (1.75″) is used.

Analysis of the magnetic force. Using the vector identity

$$(\nabla \times \mathbf{B}) \times \mathbf{B} = (\mathbf{B} \cdot \nabla)\mathbf{B} - \frac{1}{2}\nabla B^2$$

and eq. (1.78), we can write the magnetic force per unit volume as

$$\mathbf{f} = \frac{1}{c}\mathbf{J} \times \mathbf{B} = \nabla \cdot \left(-\frac{B^2}{8\pi}\mathbf{I} + \frac{\mathbf{BB}}{4\pi}\right) = \nabla \cdot \mathsf{P}^{\text{mag}} . \quad (1.79)$$

The significance of the *magnetic stress tensor* P^{mag} is best illustrated with two paradigmatic examples. When the magnetic field is directed along z, but has an arbitrary intensity $B(x,y)$, P^{mag} is a scalar and a flow in the (x,y) plane is governed by the total pressure $P + B^2/(8\pi)$. $B^2/(8\pi)$, the *magnetic pressure*, has the effect of pushing the flow away from the high intensity regions; this happens, for example, in the interaction between the supersonic solar wind and the Earth's dipole field, which acts like an obstacle with a pressure $B^2/8\pi$, giving rise to a shock front. The magnetic pressure prevails over the fluid pressure P when the ratio

$$\beta = \frac{8\pi P}{B^2} ; \quad (1.80)$$

is small. Secondly, consider the case in which B is uniform, but its direction $\mathbf{n}$ is not. The scalar term in P^{mag} has no effect whatsoever; only the *positive* component $B^2/(4\pi)$ along $\mathbf{nn}$ is relevant. The same occurs in a medium made up of elastic strings directed along $\mathbf{n}$, with a tension per unit (orthogonal) surface element $B^2/4\pi$; they are stretched and can oscillate, propagating transversal waves with *Alfvén speed* (see Sec. 10.4)

$$V_A = \sqrt{\frac{B^2}{4\pi\varrho}} . \quad (1.81)$$

The induction equation. By means of Ohm's law (1.75″) we can eliminate the electric field from the Ampère's law in eq. (1.78). When the conductivity σ is constant we get

$$\frac{\partial \mathbf{B}}{\partial t} = \nabla \times (\mathbf{v} \times \mathbf{B}) + \frac{c^2}{4\pi\sigma}\nabla^2 \mathbf{B} , \quad (1.82)$$

sometimes called the *induction equation*. The *magnetic diffusion coefficient*

$$\lambda = \frac{c^2}{4\pi\sigma} \qquad (1.83)$$

is the magnetic analogue of viscosity (eq. (1.65)). In a medium at rest ($v = 0$) the induction equation is formally equivalent to the heat equation (eq. (1.59)), with a conductivity λ; an initial magnetic field spike after a time t spreads over a distance of order $c\sqrt{\lambda t}$. This shows that any dipole field generated inside a planet requires an active regenerating mechanism (see Ch. 6).

Infinite conductivity limit. Finally, note that in a perfectly conductive fluid ($\sigma \to \infty$) the relation (1.76) must hold. Contrary to Newtonian dynamics, in this case the electromagnetic field determines the component of the velocity orthogonal to the lines of force

$$\mathbf{v}_\perp = c\,\frac{\mathbf{E}\times\mathbf{B}}{B^2}, \qquad (1.84)$$

not its time derivative; the velocity along **B**, however, is governed by the momentum law (1.77). Eq. (1.76) shows also that the electric and magnetic fields are orthogonal; any initial electric field along the lines of force is quickly neutralized by the free motion of charges.

Eq. (1.76) is of great importance in the motion of a conductive body through the plasma in the magnetosphere or interplanetary plasma. For example, Io, a satellite of Jupiter, moving through the dipole magnetic field of the planet, develops a potential difference ≈ 400 kV and carries a current of ≈ 5 million amperes across its body and its ionosphere. This has great consequences for the whole magnetosphere of Jupiter (see Ch. 10). Artificial, Earth-bound satellites also show a similar effect.

For convenience, we write the full MHD system of equation in this limit, neglecting gravity and viscosity:

$$\varrho\left[\frac{\partial \mathbf{v}}{\partial t} + (\mathbf{v}\cdot\boldsymbol{\nabla})\mathbf{v}\right] + \boldsymbol{\nabla}P = \frac{1}{4\pi}(\boldsymbol{\nabla}\times\mathbf{B})\times\mathbf{B}, \qquad (1.85a)$$

$$\frac{\partial \varrho}{\partial t} + \boldsymbol{\nabla}\cdot(\varrho\mathbf{v}) = 0, \qquad (1.85b)$$

$$\frac{\partial \mathbf{B}}{\partial t} = \boldsymbol{\nabla}\times(\mathbf{v}\times\mathbf{B}). \qquad (1.85c)$$

1.7 Conservation of magnetism and vorticity

In this Section we deal with the conservation of fluxes in fluid motion. First, we need to determine the rate of change of a surface element dS dragged along by a velocity field **v**. A line element $d\mathbf{r}$ following the flow changes at the rate

$$\frac{d(d\mathbf{r})}{dt} = (d\mathbf{r}\cdot\boldsymbol{\nabla})\mathbf{v}. \qquad (1.86)$$

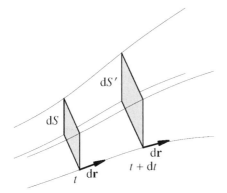

Figure 1.4 The divergence in the flow lines of a given velocity field **v** induce changes in the surface element dS (see eq. (1.87)).

A volume element $dV = d\mathbf{S} \cdot d\mathbf{r}$ can be expressed as the scalar product of a surface element $d\mathbf{S} = \mathbf{n}\,dS$ and a vector $d\mathbf{r}$. Taking into account the rate of change of $dV = dm/\varrho$ (eq. (1.34′)) and noting that $d\mathbf{r}$ is arbitrary, we find

$$\frac{d(dS_i)}{dt} = dS_i\,\boldsymbol{\nabla}\cdot\mathbf{v} - dS_j\,\frac{\partial v^j}{\partial r^i}\,. \tag{1.87}$$

Conditions for conservation of magnetic flux and circulation. The Lagrangian change of the magnetic field is easily obtained from eq. (1.82):

$$\frac{d\mathbf{B}}{dt} = \frac{\partial\mathbf{B}}{\partial t} + (\mathbf{v}\cdot\boldsymbol{\nabla})\mathbf{B} = (\mathbf{B}\cdot\boldsymbol{\nabla})\mathbf{v} - \mathbf{B}(\boldsymbol{\nabla}\cdot\mathbf{v}) + \frac{c^2}{4\pi\sigma}\nabla^2\mathbf{B}\,. \tag{1.88}$$

Combining the last two equations, we finally get the Lagrangian change in the magnetic flux through $d\mathbf{S}$

$$\frac{d(\mathbf{B}\cdot d\mathbf{S})}{dt} = \frac{c^2}{4\pi\sigma}\,d\mathbf{S}\cdot\nabla^2\mathbf{B}\,. \tag{1.89}$$

When the conductivity is infinite, the flux through a surface attached to the fluid is constant and dragged along with the fluid (this is sometimes called the *Alfvén's frozen flux theorem*). This is important, for instance, when the fluid motion compresses the material across the lines of force, thereby increasing the field intensity B. In the collapse of a cosmic body of a size R, B increases as $1/R^2$, a process which can produce a large amplification of the magnetic field. A similar property holds when a charged particle moves in a slowly varying magnetic field: the flux embraced in a Larmor gyration remains almost constant (see Sec. 10.2).

It is also interesting to calculate the time change along the flow of the vector product $d\mathbf{r}\times\mathbf{B}$. Combining eqs. (1.88) and (1.86) we obtain

$$\frac{d(d\mathbf{r}\times\mathbf{B})}{dt} = -\mathbf{B}\times(d\mathbf{r}\cdot\boldsymbol{\nabla})\mathbf{v} + d\mathbf{r}\times[(\mathbf{B}\cdot\boldsymbol{\nabla})\mathbf{v} - \mathbf{B}(\boldsymbol{\nabla}\cdot\mathbf{v})] + \frac{c^2}{4\pi\sigma}\,d\mathbf{r}\times\nabla^2\mathbf{B}\,. \tag{1.90}$$

1. Dynamical principles

Let us now consider two points $\mathbf{r}$ and $\mathbf{r} + d\mathbf{r}$ initially lying on the same line of force, so that $d\mathbf{r} \times \mathbf{B} = 0$, $d\mathbf{r}$ and $\mathbf{B}$ being parallel. In this case, because of the antisymmetry of the vector product, the first two terms on the right-hand side cancel each other; hence in an infinitely conductive plasma the condition $d\mathbf{r} \times \mathbf{B} = 0$ holds for ever: a line of force is tied to the fluid elements lying on it. This is the *freezing* theorem, of great importance in cosmic physics, in particular for the solar wind.

Conservation of vorticity. A result similar to eq. (1.89) can be deduced for the *vorticity*

$$\mathbf{w} = \nabla \times \mathbf{v} \tag{1.91}$$

of a flow governed by the Navier-Stokes equation; it represents the rotational velocity of a fluid element. The simplest example of vorticity is a straight and very thin vortex line: $\mathbf{w} = C\delta(x)\delta(y)\mathbf{e}_z$. The velocity field it generates is formally the same as the magnetic field generated by a line current along z: the flow-lines are circles $x^2 + y^2 = \text{const.}$, with $v \propto 1/\sqrt{x^2 + y^2}$ ("Biot-Savart law"). Since $\nabla \cdot \mathbf{w} = 0$ the strength C of the vortex must be constant. Stokes' theorem

$$C = \int_S d\mathbf{S} \cdot \mathbf{w} = \int_s d\mathbf{s} \cdot \mathbf{v}$$

then relates the flux of $\mathbf{w}$ through a surface S to the circulation C of $\mathbf{v}$ over its boundary s. To obtain the rate of change of the elementary flux $\mathbf{w} \cdot d\mathbf{S}$ we take the curl of eq. (1.65) divided by the density ϱ, and note that

$$v^j \partial_j v_i = v^j(\partial_j v_i - \partial_i v_j) + \frac{1}{2}\partial_i v^2 .$$

The first term on the right-hand side is $-(\mathbf{v} \times \mathbf{w})$; the last term gives no contribution to the final result. In eq. (1.65) divided by ϱ, $\nabla P/\varrho$ is also cancelled by the curl operator when the flow is incompressible or barotropic (i.e., when P is a function of ϱ only), with a constant sound speed. We therefore end up with the equation

$$\frac{\partial \mathbf{w}}{\partial t} = \nabla \times (\mathbf{v} \times \mathbf{w}) + \nabla \times \left(\frac{\eta}{\varrho}\nabla^2 \mathbf{v}\right) , \tag{1.92}$$

with the same structure as (1.82). Hence, when the viscosity is negligible, the vorticity, as the magnetic field, fulfils (*Kelvin's vorticity theorem*)

$$\frac{d}{dt}\int_S d\mathbf{S} \cdot \mathbf{w} = 0 , \tag{1.93}$$

i.e., its flux through a surface dragged along by the flow is constant; moreover, two neighbouring fluid elements lying on a vorticity line will stay on it. In a barotropic, inviscid fluid the vorticity lines are "anchored" to the matter. If a compression takes place orthogonally to a vortex line, dS decreases and its strength increases.

1.8 Kinetic theory

In a rarefied medium we must question the main assumption used so far, that the flow can be completely described with the mean quantities ϱ, **v** and T (in addition, if needed, to the electromagnetic field). This assumption is based upon the hypothesis that the material is in *local thermodynamical equilibrium*; that is to say, that collisions have had the time and the space to produce a local Boltzmann distribution of velocities.

A set of N point particles is described by their $6N$ coordinates in phase space, which evolve in time according to the interactions and the external forces. If N is very large, and the system is isolated and occupies a finite volume V (e.g., a box), random collisions will eventually destroy any initial order and bring the gas in the state of maximum probability consistent with the constants of motion of the system, in particular the energy. This is the state of *thermodynamical equilibrium*, for which we have a Maxwellian distribution f in velocity space, with coordinates **u**,

$$f = n\left(\frac{m}{2\pi kT}\right)^{3/2} \exp\left[-\frac{mu^2}{2kT}\right], \tag{1.94}$$

uniform throughout the box ($f\, d^3r\, d^3u$ is the number of particles in the phase space volume element $d^3r\, d^3u$). Here $n = N/V$ is the number density and m is the particle mass.

Radiation.– In passing we note that the kinetic approach may be also applied to radiation ("photonic gas"). Obviously, the mass of particles and the velocity distribution is now replaced with the frequency distribution, hence the spectrum of the radiation. In the case of thermodynamical equilibrium, the frequency distribution of the photon density is again uniquely determined by the temperature in terms of *Planck's distribution function*

$$n_\nu\, d\nu\, dV = \frac{8\pi\nu^2}{c^3}\frac{d\nu\, dV}{\exp(h\nu/kT) - 1}. \tag{1.95}$$

The corresponding spectral energy density

$$u_\nu = h\nu\, n_\nu, \tag{1.96}$$

for frequencies much less than $\approx kT/h$ (*Rayleigh limit*), is proportional to ν^2, since

$$u_\nu = \frac{8\pi kT}{c^3}\nu^2. \tag{1.97}$$

This is the *Rayleigh-Jeans law*, often used in radioastronomy (Fig. 1.5).

The spectral energy density (1.96) has a maximum at the wavelength λ_m given by

$$\lambda_m = \frac{b}{T}, \tag{1.98}$$

1. Dynamical principles

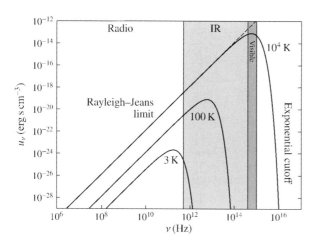

Figure 1.5 Spectral density u_ν of radiation in thermodynamical equilibrium (Planck's state) vs frequency ν for three different temperatures of astronomical interest: (i) $T = 10^4$ K (roughly matching the solar spectrum), (ii) $T = 100$ K (approximately matching planetary thermal radiation), and (iii) $T = 3$ K (for the cosmological microwave background).

where $b \simeq 0.290$ cm K (*Wien displacement law*). The exponential in the denominator of the Planck law (1.95) makes the total energy density

$$u = \int_0^\infty d\nu\, u_\nu = aT^4 \qquad (1.99)$$

finite; $a = 4\sigma/c \simeq 7.56 \times 10^{-15}$ erg cm^{-3} K^{-4}. Note that quantum effects place an effective cutoff to this integral at a frequency of order kT/h, since the Rayleigh-Jeans function would give a divergence in u (so called "ultraviolet catastrophe"). The elimination of this divergence had a crucial historical role in the introduction of Planck's constant h and the birth of quantum physics in general. More details about description of radiation fields are in Sec. 7.3.

Local thermodynamical equilibrium. In general, we associate to a fluid a distribution function in phase space $f(\mathbf{r}, \mathbf{u}, t)\, d^3r\, d^3u$, which gives the number of particles in the infinitesimal volume $d^3r\, d^3u$ of phase space around the point $(\mathbf{r}, \mathbf{u})$. The five fluid quantities n, $\mathbf{v}$ (mean velocity) and T (effective temperature) are obtained from f by means of appropriate averaging:

$$n(\mathbf{r}, t) = \int d^3u\, f(\mathbf{r}, \mathbf{u}, t)\,, \qquad (1.100\text{a})$$

$$n(\mathbf{r}, t)\, \mathbf{v}(\mathbf{r}, t) = \int d^3u\, \mathbf{u} f(\mathbf{r}, \mathbf{u}, t)\,, \qquad (1.100\text{b})$$

$$3n(\mathbf{r}, t)\, k\, T(\mathbf{r}, t) = m \int d^3u\, u^2 f(\mathbf{r}, \mathbf{u}, t) = mn v_T^2\,. \qquad (1.100\text{c})$$

$v_T = \sqrt{kT/3m}$ is the *thermal speed*.

Kinetic theory shows that the final thermodynamical equilibrium is attained in two stages. First, after a few collision times the fluid is described by a locally Maxwellian distribution function which depends on space and time only through the mean quantities $n, \mathbf{v}$ and T:

$$f = n\left(\frac{m}{2\pi kT}\right)^{3/2} \exp\left[-\frac{m|\mathbf{u} - \mathbf{v}|^2}{2kT}\right]. \quad (1.94')$$

It is easy to check that the relationships (1.100) are fulfilled. Collisions occur at the rate (*collision frequency*)

$$\nu_c = n \langle u\sigma_c \rangle, \quad (1.101)$$

obtained by averaging the relevant cross section σ_c with the velocity distribution. The gas is brought into a *local thermodynamical equilibrium* and is described by a fluid model. Its conservation laws for mass, momentum and energy provide the time evolution for $n, \mathbf{v}$ and T. After this, if the system is closed and isolated, on a space scale L and a timescale $L^2/(v_T \ell_c)$, it will be slowly brought into the uniform state (1.94) by diffusion and dissipative processes. For radiation, we also have a state of local thermal equilibrium, to be discussed in Ch. 7. In a diffusion process particles undergo random encounters which change their momentum with a probability determined by the cross section; therefore they undergo a *random walk*, in which the mean displacement after a time t is proportional to $\sqrt{t}$. This explains why diffusion equations like (1.65) are closely connected with stochastic Brownian processes. In order for local thermodynamical equilibrium to hold, the typical scales for macroscopic variations in time and in space (produced, for example, by external agents) must be, respectively, much larger than the collision time $1/\nu_c$ and the mean free path $\ell_c = v_T/\nu_c$.

Collisionless limit. In the opposite approximation, when collisions are unimportant, we have the *free-molecular flow* (*Knudsen regime*), in which particles move independently of each other. When none of these two extreme approximations hold, and we wish study general deviations from the Maxwell's distribution (1.94'), a dynamical description in phase space must be adopted, using a differential equation for the particle distribution function $f(\mathbf{r}, \mathbf{u}, t)$. This is *Boltzmann's equation*.

Plasma. In the solar system, charged particles are generally created by ionization processes and the resulting ionized gas is, in the bulk, neutral; if neutrals are present as well, we have a *partially ionized gas*. The very strong attraction between opposite charges makes any local deviation from local charge neutrality difficult. To see this in detail, consider a fully ionized hydrogen gas, with of charge e, and electrons of charge $-e$, with the same mean density $n_p = n_e$; it is a *plasma* if the number of electrons in a cube of radius

$$\lambda_D = \left(\frac{kT}{4\pi e^2 n_e}\right)^{1/2} = 743 \left(\frac{T}{n_e \, \text{eV cm}^3}\right)^{1/2} \text{cm}, \quad (1.102)$$

the *Debye length*, is very large (i.e., if λ_D is much greater than the mean interparticle distance $n_e^{-1/3}$). Note that a cube of size s containing a large number N_e of electrons will be, in the average, in its lowest energy state, with the same number $\langle N_e \rangle = s^3 n_e$ of electrons and protons; but statistical fluctuations will temporarily produce deviations from charge neutrality. A deviation $\delta N_e = N_e - \langle N_e \rangle$ in the total number of electrons needs an electrostatic energy per particle of order $e^2 \delta N_e / s$; its average order of magnitude can be estimated by equating this energy to the thermal energy $\approx kT$. Hence, at a size of the order of the Debye length λ_D we can have $\delta N_e \approx \langle N_e \rangle$, with a large charge depletion or excess. Statistical fluctuations can produce substantial charge voids only within regions whose size is smaller than λ_D; but at scales larger than λ_D the condition of charge neutrality $n_e = n_p$ generally holds. The Debye length describes also the screening property of the plasma: the potential of a charge is screened by opposite charges beyond a distance $\approx \lambda_D$ (Problem 1.8).

When – as usual in the solar system – there is more than one nuclear species, their relative concentration must be taken into account. For example, in the solar wind helium is present, with a relative mass concentration $Y = 4n_{He}/(n_p + 4n_{He}) \approx 0.16$, lower than in the photosphere (Table 9.1) for reasons not yet known; charge neutrality requires

$$n_H = n_e \frac{2(1-Y)}{2-Y}, \qquad n_{He} = n_e \frac{Y}{2(2-Y)}. \tag{1.103}$$

The dynamics of a plasma is determined by long-range, collective interactions between currents and charge fluctuations, and is governed by Maxwell's equations; they affect also the particle velocity distribution and must be described in the phase space $(\mathbf{r}, \mathbf{v})$, using the correct electromagnetic forces due to the other particles, even those at large distances. This is characteristic of all long range interactions, like those of a large number of gravitating point masses (the stars in a galaxy or particles in planetary rings). To describe the plasma dynamics, a non-linear, first order differential equation for each species is needed (*Vlasov equation*); complex effects, like wave-wave coupling, high velocity streams, temperature differences between electron and nuclei, collisionless dissipation, cyclotron resonant interactions, etc., can be described in this way. In many cases the Vlasov equation can be approximated with propagation equations for the densities, the mean velocities and the energy densities of the present species, leading to the magnetohydrodynamical approximation. At this level plasma dynamics in phase space is time-reversible. However, particle encounters at impact parameters smaller than the Debye length produce collisional disordering processes, which tend to establish a Maxwellian velocity distribution (1.94) and local thermal equilibrium. If they are relevant, they require an additional, irreversible term in Vlasov equation.

Table 1.1. Typical plasma parameters in the equatorial solar wind at 1 AU.

Electron density n_e (cm^{-3})	Temperature T (K)	Electron mean free path ℓ_c (cm)	Debye length λ_D (cm)	Plasma frequency ω_p (sec)	$n_e \lambda_D^3$
5	2×10^5	10^{13}	1,400	1.2×10^5	1.4×10^{10}

Some relevant plasma processes will be discussed in Chs. 8 (wave propagation), 9 and 10 (fluid approximation); here we only mention the three main characteristic frequencies of a plasma. The *plasma frequency*

$$\omega_p = \sqrt{\frac{4\pi n_e e^2}{m_e}} = 56,400 \sqrt{n_e \text{ cm}^3} \text{ rad/s} . \quad (1.104)$$

characterizes the effect of the plasma on electromagnetic propagation, which is unaffected for frequencies $\gg \omega_p$. Coulomb attraction also acts as the restoring force of collective longitudinal plasma oscillations; their basic frequency is ω_p, but they are affected by thermal motion and magnetic field. The motion of a *single* electron in a magnetic field **B** is governed by the *electron cyclotron frequency*

$$\Omega_e = \frac{eB}{m_e c} = 1.76 \times 10^7 \frac{B}{10^5 \text{ nT}} \text{ rad/s} ; \quad (1.105)$$

for ions we have the *ion cyclotron frequency*

$$\Omega_i = \frac{ZeB}{m_i c} = 9,580 \frac{Z}{A} \frac{B}{10^5 \text{ nT}} \text{ rad/s} . \quad (1.106)$$

$m_i = A m_p$ is the ion mass, expressed in terms of the proton mass m_p. They are the "gyration" frequencies with which a particle spirals around a magnetic line of force (Sec. 10.2). Inhomogeneity of the magnetic field results in additional oscillations or drifts. Any particle so accelerated emits the resonant *cyclotron radiation* at this frequency and, if relativistic, also all its harmonics (*synchrotron radiation*).

– PROBLEMS –

1.1. Prove the relation
$$\frac{d \ln(dV)}{dt} = \nabla \cdot \mathbf{v}$$
used in the derivation of the mass conservation law (eq. (1.34′)).

1.2.* Consider a planetary body (of fixed mass M and radius R) with a core. Evaluate, as function of the dimensionless parameters, the ratio of the following quantities to their values without the core: central pressure, binding energy and moment of inertia.

1. Dynamical principles

(i) In terms of a real positive number n, consider the density profile

$$\varrho(r) = \varrho(0)\left[1 - \left(\frac{r}{R}\right)^n\right] ;$$

it reaches its median value $\varrho(0)/2$ at $R/2^{1/n} = R_c$, which may be called the radius of the core. If $n \ll 1$ the core is very small; if $n \gg 1$ the density is practically constant.

(ii) The core, with radius R_c, has a uniform density ϱ_c; the mantle has a uniform density ϱ_m determined in terms of the mean density $\bar{\varrho}$ by

$$\varrho_m R^3 + (\varrho_c - \varrho_m) R_c^3 = \bar{\varrho} R^3 .$$

This is a more realistic, two-parameter model determined by R_c/R and $\varrho_c/\bar{\varrho}$.

1.3. Calculate the ratio a/m for the main planets and the Sun (eqs. (1.12) and (1.15)). This ratio is important in determining the final black hole state of a collapsing body (see Ch. 17).

1.4. Calculate Helmert's parameter for the planets, the Sun and the Moon.

1.5. Determine the distribution of stress and the lengthening of (i) a freely hanging rod subject to its own weight, and (ii) a rod rotating around a perpendicular axis. Compute the stress field in a spherical body due to the tidal gravitational field of a close by mass. (Hint: Consider the tidal gravitational potential (4.17).)

1.6.* Find the velocity profile of an incompressible and viscous flow along a flat surface.

1.7. Estimate the electric potential difference between the ends of a long conductive wire orbiting vertically around the Earth.

1.8.* Find the potential around a spherical conductor in an isothermal electron plasma. Such a plasma can be simply described by the equation of hydrostatic equilibrium for a isothermal electron fluid (with pressure $P_e = n_e kT$)

$$-\nabla P_e + n_e e \nabla V = 0 ;$$

thus the electron density n_e is proportional to $\exp(eV/kT)$. When the exponent is small, Poisson's equation can easily be solved for the electrostatic potential V. For a simpler version of this problem, consider an infinite, plane conductor. (Hint: Use appropriate dimensionless variables and the Debye length.)

1.9. Using Gauss' theorem, find the gravitational field of an infinite, circular cylinder and an infinite plane slab.

1.10.* Assuming that the third order moments of the distribution function $f(\mathbf{r}, \mathbf{u}, t)$ vanish, derive the differential equations for the number density n and the mean velocity $\mathbf{v}$ (eqs. (1.100a) and (1.100b)). The second order moments provide a (generally anisotropic) pressure tensor.

1.11. Estimate the viscosity and Reynolds number of the atmosphere near the ground.

1.12. Show that the velocity of an incompressible plane fluid fulfils $v_x = -\partial_y \psi$, $v_y = \partial_x \psi$; the Laplacian of ψ is the vorticity $w = w_z$. Find the flow corresponding to two point-like vortices.

1.13.* The equilibrium condition (1.6) for a polytropic (1.24) gravitating body acquire a simple form when $n = 1/(\gamma - 1)$ is a positive integer. Using the dimensionless variables $\theta = [\varrho/\varrho(0)]^n$ and $\xi = r/a$, derive the universal *Lane-Emden equation*

$$\frac{1}{\xi^2} \frac{d}{d\xi} \left(\xi^2 \frac{d\theta}{d\xi} \right) + \theta^n = 0 ,$$

to be supplemented by the boundary conditions $\theta(0) = 1$ and $\theta'(0) = 0$. The body has a radius $a\xi_0$ determined by the first zero of $\theta(\xi)$; find out its dependence from the central pressure and density for different values of γ, in the range $4/3$ to $5/3$. Show also that the Lane-Emden equation possesses analytical solutions for $n = 0, 1, 5$.

1.14.* Prove that the Navier-Stokes equation (1.64) can be expressed as a continuity equation for the fluid momentum density $\varrho \mathbf{v}$:

$$\frac{\partial}{\partial t}(\varrho v_i) + \frac{\partial}{\partial x_j} \Pi_{ij} = 0 ,$$

where

$$\Pi_{ij} = \varrho v_i v_j + P\delta_{ij} - q_{ij} + g_{ij}$$

is the momentum flux tensor, q_{ij} is the viscous stress tensor (1.62) and

$$g_{ij} = -\frac{1}{4\pi G} \left(\frac{\partial U}{\partial x_i} \frac{\partial U}{\partial x_j} - \frac{1}{2} \delta_{ij} \frac{\partial U}{\partial x_k} \frac{\partial U}{\partial x_k} \right)$$

the "gravitational stress tensor".

Chapter 2

THE GRAVITATIONAL FIELD OF AN ISOLATED BODY

Deviations of the shape of a cosmic body from spherical symmetry manifest themselves in its gravitational field; conversely, corrections to the *monopole* term GM/r provide important, though incomplete, information about the interior. The main, axially symmetric deviation is due to rotation and corresponds to a correction in the gravitational potential proportional to $1/r^3$ (the *quadrupole* term). The potential induced by the tides raised by a nearby body, also $\propto 1/r^3$, is symmetric around the axis joining the centres. Smaller disturbances in the mass distribution produce smaller corrections; they decrease faster with the distance from the centre. The gravitational field of a generic, non-spherical body is appropriately described by a powerful mathematical tool, the multipole series of the *spherical harmonic* functions. This chapter introduces this method and applies it to the Earth. Gravity anomalies due to small scale inhomogeneities are also discussed. The last section briefly deals with the gravity fields of planetary bodies.

2.1 Spherical harmonics

The theorem (1.2) shows that the external gravitational field of a spherically symmetric body is determined only by its mass and does not show any trace of its density distribution. Outside the body the internal structure manifests itself only when the spherical symmetry is violated; in this case the full Poisson's integral (1.37) is needed. The appropriate mathematical tool to compute this integral and to describe the gravitational field of an isolated body of arbitrary shape is the expansion of the potential energy U in *spherical harmonics*; we now briefly recall its main properties, of fundamental importance in many fields of mathematical physics.

If $p_\ell(\mathbf{r})$ is a homogeneous polynomial of degree ℓ, solution of Laplace's equation, the function $U_\ell(\mathbf{r}) = p_\ell(\mathbf{r})/r^{2\ell+1}$ is also harmonic (i.e. a solution of Laplace's equation). This result can easily be proved by direct calculation, recalling that a homogeneous polynomial fulfils[1]

$$\mathbf{r} \cdot \nabla p_\ell = \ell \, p_\ell \, .$$

The functions

$$Y_\ell = \frac{p_\ell(\mathbf{r})}{r^\ell}$$

depend only on the direction $\mathbf{r}/r$ and are termed *spherical harmonics of degree* ℓ. It can be shown that there are just $2\ell + 1$ independent, harmonic and homogeneous polynomials of degree ℓ, hence just $2\ell + 1$ independent spherical harmonics $Y_{\ell m}(\theta, \phi)$, functions of the colatitude θ and the longitude ϕ in a polar coordinate system. The index m, ranging over $2\ell + 1$ values, denotes the *order*. p_0 is a constant, and so is Y_{00}; p_1 is an arbitrary linear combination of x, y and z, and the three functions

$$\frac{x}{r} = \sin\theta \cos\phi, \quad \frac{y}{r} = \sin\theta \sin\phi, \quad \frac{z}{r} = \cos\theta$$

are a basis for Y_1. For Y_2 we can choose, for example, the five-dimensional set

$$\frac{xy}{r^2} = \sin^2\theta \cos\phi \sin\phi, \quad \frac{yz}{r^2} = \sin\theta \cos\theta \sin\phi \, ,$$

$$\frac{zx}{r^2} = \sin\theta \cos\theta \cos\phi, \quad \frac{x^2 - y^2}{r^2} = \sin^2\theta (\cos^2\phi - \sin^2\phi) \, ,$$

$$\frac{2z^2 - x^2 - y^2}{r^2} = 3\cos^2\theta - 1 \, .$$

The last function is independent of the longitude.

One of the possible realizations of spherical harmonics, often used in mathematical physics, uses the *Legendre polynomials* $P_\ell(u)$. Their importance in gravitation physics stems from their definition: $P_\ell(u)$ is the coefficient of α^ℓ in the power expansion of the function ($|u| \leq 1$ and $\alpha < 1$)

$$\left(1 - 2\alpha u + \alpha^2\right)^{-1/2} = \sum_{\ell=0}^{\infty} P_\ell(u) \alpha^\ell \, . \qquad (2.1)$$

This expansion shows up in the gravitational potential energy produced at a point $\mathbf{r}$ by a mass point placed at $\mathbf{r}'$

$$\frac{1}{|\mathbf{r} - \mathbf{r}'|} = \frac{1}{r}\left(1 - 2\frac{r'}{r}\cos\psi + \frac{r'^2}{r^2}\right)^{-1/2} = \frac{1}{r}\sum_{\ell=0}^{\infty}\left(\frac{r'}{r}\right)^\ell P_\ell(\cos\psi) \, . \qquad (2.2)$$

---

[1] Obtained by differentiating with respect λ the defining property $p_\ell(\lambda \mathbf{r}) = \lambda^\ell p_\ell(\mathbf{r})$, and taking $\lambda = 1$.

2. The gravitational field of an isolated body

Here ψ is the angle between $\mathbf{r}$ and $\mathbf{r}'$. The first four polynomials read

$$P_0 = 1, \quad P_1 = u, \quad P_2 = \frac{1}{2}\left(3u^2 - 1\right), \quad P_3 = \frac{1}{2}u\left(5u^2 - 3\right). \tag{2.3}$$

For a polynomial of a generic degree ℓ, we have *Rodrigues' formula*

$$P_\ell = \frac{1}{2^\ell \ell!} \frac{d^\ell}{du^\ell}\left(u^2 - 1\right)^\ell. \tag{2.4}$$

We also define the *associated Legendre functions* by

$$P_{\ell m}(u) = (1 - u^2)^{m/2} \frac{d^m P_\ell(u)}{du^m}, \tag{2.5a}$$

$$P_{\ell,-m}(u) = (-1)^m \frac{(\ell - m)!}{(\ell + m)!} P_{\ell m}(u) \tag{2.5b}$$

with $0 \leq m \leq \ell$, so that

$$P_{11} = \sqrt{1 - u^2},$$
$$P_{21} = 3u\sqrt{1 - u^2}, \quad P_{22} = 3(1 - u^2), \tag{2.5'}$$
$$P_{31} = \frac{3}{2}\sqrt{1 - u^2}(5u^2 - 1), \quad P_{32} = 15u(1 - u^2), \quad P_{33} = 15(1 - u^2)^{3/2}.$$

They reduce to P_ℓ when $m = 0$. It can be shown that $P_{\ell m}(u)$ (including $m = 0$) has $\ell - m$ zeros in the relevant interval $(-1, 1)$. For other mathematical details we refer the reader to appropriate textbooks; we just note that occasionally one can find the associated Legendre functions defined with an additional factor $(-1)^m$ in (2.5a), unnecessary for our purposes.

The spherical harmonics generalize to the unit sphere the exponential functions $\exp(im\phi)$ on the unit circle; indeed, the polynomials of degree m-th $r^m \exp(im\phi)$ are harmonic in the plane. The expansion in spherical harmonics generalizes the Fourier expansion on the unit circle. In our case, too, it is often convenient to exploit the formal advantages of complex numbers to define a complex basis

$$Y_{\ell m}(\theta, \phi) = N_{\ell m} P_{\ell m}(\cos \theta) \exp(im\phi). \tag{2.6}$$

The *order* m ranges from $-\ell$ to ℓ; the *degree* ℓ from 0 to ∞. $N_{\ell m} = N_{\ell,-m}$ are dimensionless normalization coefficients. The functions (2.6) fulfil

$$Y_{\ell m}(\theta, \phi) = Y^\star_{\ell,-m}(\theta, \phi) \tag{2.6'}$$

(the star $\star$ means complex conjugate); moreover, they are orthogonal with respect to the scalar product

$$\langle Y_1 | Y_2 \rangle = \frac{1}{4\pi} \int d\Omega \, Y_1^\star \, Y_2; \tag{2.7}$$

Table 2.1. Normalized spherical harmonics $Y_{\ell m}$ up to $\ell = 3$ in the complex representation (degree ℓ in columns, order m in rows).

	$\ell =$	0	1	2	3
$m =$	0	1	$\sqrt{3}\cos\theta$	$\frac{\sqrt{5}}{2}(3\cos^2\theta - 1)$	$\frac{\sqrt{7}}{2}\cos\theta(5\cos^2\theta - 3)$
	1		$\sqrt{\frac{3}{2}}\sin\theta\, e^{i\phi}$	$\sqrt{\frac{15}{2}}\cos\theta\sin\theta\, e^{i\phi}$	$\frac{3}{4}\sqrt{\frac{7}{3}}(5\cos^2\theta - 1)\, e^{i\phi}$
	2			$\frac{1}{2}\sqrt{\frac{15}{2}}\sin^2\theta\, e^{2i\phi}$	$\frac{15}{2}\sqrt{\frac{7}{30}}\sin^2\theta\cos\theta\, e^{2i\phi}$
	3				$\frac{5}{4}\sqrt{\frac{7}{5}}\sin^3\theta\, e^{3i\phi}$

the integral is performed on the unit sphere[2] ($d\Omega = \sin\theta\, d\theta d\phi$). The choice of the normalization coefficients completes the orthonormality conditions

$$\langle Y_{\ell' m'} | Y_{\ell m} \rangle = \frac{1}{4\pi} \int d\Omega\, Y^{\star}_{\ell' m'}(\theta, \phi) Y_{\ell m}(\theta, \phi) = \delta_{\ell \ell'}\, \delta_{m m'}\,. \quad (2.8)$$

To obtain these coefficients, note that the square modulus is independent of the longitude and its average over the sphere is half the integral $\int du$ from -1 to 1, so that

$$N^2_{\ell m} \int_{-1}^{1} du\, P^2_{\ell m}(u) = 2\,.$$

It can then be shown, using (2.4) and (2.5), that

$$N_{\ell m} = \sqrt{2\ell + 1} \left[\frac{(\ell - m)!}{(\ell + m)!} \right]^{1/2}. \quad (2.9)$$

In the language of complex vector spaces, we can say that the set (2.6) (with $-\ell \leq m \leq \ell$) is a *unitary basis* in a $(2\ell + 1)$-dimensional space with respect to the scalar product defined by the average on the unit sphere. It can also be shown that the set (2.6) is complete: any regular function f on the unit sphere can be expressed as an expansion in spherical harmonics

$$f(\theta, \phi) = \sum_{\ell} \sum_{m} f_{\ell m} Y_{\ell m}(\theta, \phi)\,. \quad (2.10)$$

---

[2] In theoretical physics, in quantum mechanics in particular, the 4π factor in the normalization (2.7) is moved to the definition of the spherical harmonics, so that $N_{\ell m}$ changes to $N_{\ell m}/\sqrt{4\pi}$. Similarly, the factor $(-1)^m$ is often introduced in (2.6) and (2.6') (the so-called Condon-Shortley phase). Here we stick to the definitions common in geodesy.

2. The gravitational field of an isolated body

Table 2.2. Normalized spherical harmonics $Y^c_{\ell m}, Y^s_{\ell m}$ up to $\ell = 3$ in the real representation (degree ℓ in columns, order m in rows).

	$\ell =$	0	1	2	3
$m =$	0 c	1	$\sqrt{3}\cos\theta$	$\frac{\sqrt{5}}{2}(3\cos^2\theta - 1)$	$\frac{\sqrt{7}}{2}\cos\theta(5\cos^2\theta - 3)$
	1 c		$\sqrt{3}\sin\theta\cos\phi$	$\sqrt{15}\cos\theta\sin\theta\cos\phi$	$\frac{3}{4}\sqrt{\frac{14}{3}}(5\cos^2\theta - 1)\cos\phi$
	1 s		$\sqrt{3}\cos\theta\sin\phi$	$\sqrt{15}\cos\theta\sin\theta\sin\phi$	$\frac{3}{4}\sqrt{\frac{14}{3}}(5\cos^2\theta - 1)\sin\phi$
	2 c			$\frac{1}{2}\sqrt{15}\sin^2\theta\cos 2\phi$	$\frac{15}{2}\sqrt{\frac{7}{15}}\sin^2\theta\cos\theta\cos 2\phi$
	2 s			$\frac{1}{2}\sqrt{15}\sin^2\theta\sin 2\phi$	$\frac{15}{2}\sqrt{\frac{7}{15}}\sin^2\theta\cos\theta\sin 2\phi$
	3 c				$\frac{5}{4}\sqrt{\frac{14}{5}}\sin^3\theta\cos 3\phi$
	3 s				$\frac{5}{4}\sqrt{\frac{14}{5}}\sin^3\theta\sin 3\phi$

Hereinafter the summation limits shall be understood (if not stated explicitly). The expansion coefficients are obtained by means of eq. (2.8)

$$f_{\ell m} = \langle Y_{\ell m} | f \rangle = \frac{1}{4\pi} \int d\Omega \, Y^\star_{\ell m}(\theta, \phi) f(\theta, \phi) \,. \tag{2.11}$$

If f is real

$$f_{\ell,-m} = f^\star_{\ell m} \,. \tag{2.12}$$

Real representation of function on the sphere.– In place of the complex basis (2.6) the real and orthonormal set

$$Y^c_{\ell 0}(\theta, \phi) = N_{\ell 0} P_\ell(\cos\theta) = \bar{P}_\ell(\cos\theta) \,, \tag{2.13a}$$

$$Y^c_{\ell m}(\theta, \phi) = \sqrt{2} N_{\ell m} P_{\ell m}(\cos\theta) \cos(m\phi) = \bar{P}_{\ell m}(\cos\theta) \cos(m\phi) \,, \tag{2.13b}$$

$$Y^s_{\ell m}(\theta, \phi) = \sqrt{2} N_{\ell m} P_{\ell m}(\cos\theta) \sin(m\phi) = \bar{P}_{\ell m}(\cos\theta) \sin(m\phi) \tag{2.13c}$$

is more commonly used ($m = 1, 2, \ldots, \ell$ in the last two equations; obviously $Y^s_{\ell 0} = 0$). The symbols $\bar{P}_{\ell m}(\cos\theta)$ (including $m = 0$, for which $\bar{P}_{\ell 0} = \bar{P}_\ell$) are called *fully normalized* Legendre functions; $\bar{P}_{\ell m}(\cos\theta)$ have much more suitable numerical properties than the (unnormalized) functions $P_{\ell m}(\cos\theta)$ and are thus used for high-degree studies of planetary fields. Note that in the real representation (2.13) the index m ranges from 0 to ℓ only. Since the complex basis

$$Y_{\ell 0} = Y^c_{\ell 0}, \quad Y_{\ell m} = \frac{1}{\sqrt{2}}(Y^c_{\ell m} + iY^s_{\ell m}) \quad (m > 0)$$

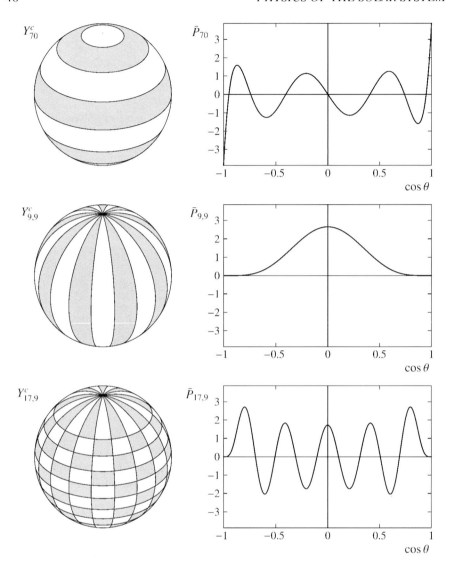

Figure 2.1. The qualitative behaviour of the spherical harmonics is indicated by the meridians and the parallels on which they vanish (white/black zones correspond to positive/negative values); here they are drawn for $Y^c_{70} = \bar{P}_{70}$, $Y^c_{9,9} = \bar{P}_{9,9}\cos(9\phi)$ and $Y^c_{17,9} = \bar{P}_{17,9}\cos(9\phi)$. *Zonal* harmonics ($m = 0$) divide the sphere along parallels into $\ell + 1$ zones; *sectorials* ($\ell = m \neq 0$) divide the sphere in $2m$ sectors along meridians; *tesserals* ($\ell \neq m \neq 0$) change sign $\ell - m$ times from North to South and $2m$ times along parallels.

is normalized to unity, so are its real and imaginary parts multiplied by $\sqrt{2}$ (except for $\ell = 0$)

$$\langle Y^c_{\ell m} | Y^c_{\ell m} \rangle = \langle Y^s_{\ell m} | Y^s_{\ell m} \rangle = 1 \ .$$

2. The gravitational field of an isolated body

The corresponding expansion

$$f(\theta, \phi) = \sum_{\ell} \sum_{m \geq 0} \left(f_{\ell m}^c Y_{\ell m}^c + f_{\ell m}^s Y_{\ell m}^s \right), \tag{2.14}$$

when compared with (2.10), gives the new coefficients

$$f_{\ell 0}^c = f_{\ell 0}, \quad f_{\ell m}^c = \frac{f_{\ell m}^\star + f_{\ell m}}{\sqrt{2}}, \quad f_{\ell m}^s = \frac{f_{\ell m}^\star - f_{\ell m}}{\sqrt{2}i}. \tag{2.15a}$$

and the old ones

$$f_{\ell 0} = f_{\ell 0}^c, \quad f_{\ell m} = \frac{f_{\ell m}^c - i f_{\ell m}^s}{\sqrt{2}}. \tag{2.15b}$$

The tables 2.1 and 2.2 list the normalized spherical harmonics up to the third degree in the complex and real representations.

Theoretical complement.– An arbitrary rotation T determines a map $P(\theta, \phi) \to P'(\theta', \phi')$ of the unit sphere onto itself, transforming $Y_{\ell m}(P)$ into $Y'_{\ell m}(P'(P)) = Y_{\ell m}(P)$. The defining properties of the harmonic functions (the degree of the polynomial $r^\ell Y_{\ell m}$ and the harmonic property) are invariant under rotations and therefore are not affected by this mapping; hence $Y'_{\ell m}$ is also a spherical harmonic of degree ℓ and must be expressible as a linear combination of the spherical harmonics of the same degree

$$Y'_{\ell m}(P') = \sum_{m'} D^{(\ell)}_{mm'}(T) Y_{\ell m'}(P). \tag{2.16}$$

For each ℓ, the set of square matrices $D^{(\ell)}(T)$, each of dimension $2\ell + 1$, obey the multiplication table of the rotation group and therefore constitute a representation of the group itself; for $\ell = 1$ it reduces to the set of ordinary rotations. It can be shown that this representation $D^{(\ell)}$ is *unitary*, to wit, it leaves the scalar product (2.7) invariant. This group-theoretical argument provides a justification for the choice (2.6) for the basis, against the "naïve", and in geophysics more usual, real set (2.13). It can also be shown that these representations are *irreducible*, that is to say, the $2\ell + 1$-dimensional space S_ℓ over which they operate does not have any invariant subspace. Moreover, every single-valued matrix representation of the rotation group is a linear combination of the set $D^{(\ell)}(T)$; in other words, the spherical harmonics as a whole provide the basis vectors whose transformations generate every single-value matrix representation of the group itself. There are also infinite, two-valued spinor representations in which two different matrices correspond to each rotation T; they are labelled by half integer values of the degree ℓ and describe elementary particles with half integer spin. These important group-theoretical properties provide powerful tools to deal with the spherical harmonics expansions.

2.2 Harmonic representation of the gravity field

The spherical harmonics provide a representation of the gravitational potential energy per unit mass produced outside by an isolated body

$$U(\mathbf{r}) = -\frac{GM}{r} + \delta U = -\frac{GM}{r}\left[1 + \sum_{\ell \geq 1}\left(\frac{R}{r}\right)^{\ell}\sum_{m}U_{\ell m}Y_{\ell m}(\theta, \phi)\right] ; \quad (2.17)$$

here

$$U^{\star}_{\ell m} = U_{\ell,-m}$$

(note that $U(\mathbf{r})$ thus satisfies Laplace equation). We denote by U_ℓ the part of U of degree ℓ. It can be shown that an arbitrary harmonic function which vanishes at infinity has the form (2.17). Each ℓ corresponds to a different kind of deviation from spherical symmetry; larger values correspond to more complex and, at large distances, smaller components of the gravitational field: the potential of degree ℓ is $O(1/r^{\ell+1})$. In a real, dimensionless representation the potential is usually described by the coefficients $C_{\ell m}$ and $S_{\ell m}$

$$U(\mathbf{r}) = -\frac{GM}{r}\left[1 + \sum_{\ell \geq 1}\left(\frac{R}{r}\right)^{\ell}\sum_{m \geq 0}\left(C_{\ell m}Y^c_{\ell m}(\theta, \phi) + S_{\ell m}Y^s_{\ell m}(\theta, \phi)\right)\right] , \quad (2.17')$$

which is standard in geophysics. An easy computation shows the relation between the coefficients of the real and the complex representation of the geopotential, namely $C_{\ell 0} = U_{\ell 0}$ and $C_{\ell m} - iS_{\ell m} = \sqrt{2}\, U_{\ell m}$ ($m > 0$). If the spherical harmonics in (2.17') are not normalized, namely are $P_{\ell m}(\cos\theta)\,[\cos,\sin](m\phi)$, the coefficients $C_{\ell m}$ and $S_{\ell m}$ become

$$C'_{\ell m} = \sqrt{2}\, N_{\ell m} C_{\ell m} , \quad (2.18a)$$

$$S'_{\ell m} = \sqrt{2}\, N_{\ell m} S_{\ell m} , \quad (2.18b)$$

for $m \neq 0$ and $C'_{\ell 0} = N_{\ell 0} C_{\ell 0} = \sqrt{2\ell + 1}\, C_{\ell 0}$ (note that $S_{\ell 0} = 0$).

It is often convenient to describe each term in the potential with an amplitude $J_{\ell m}$ and a phase $\phi_{\ell m}$

$$C'_{\ell m} = J_{\ell m}\cos m\phi_{\ell m} , \qquad S'_{\ell m} = J_{\ell m}\sin m\phi_{\ell m} , \quad (2.19)$$

so that

$$U(\mathbf{r}) = -\frac{GM}{r}\left[1 + \sum_{\ell \geq 1}\left(\frac{R}{r}\right)^{\ell}\sum_{m} J_{\ell m} P_{\ell m}(\cos\theta)\cos m(\phi - \phi_{\ell m})\right] . \quad (2.17'')$$

We therefore have the complex representation (2.17), the sine and cosine representation (2.17') and the amplitude and phase representation (2.17'').

2. The gravitational field of an isolated body

Note that we can construct another class of harmonic functions which diverge at infinity

$$V(\mathbf{r}) = \sum_{\ell \geq 1} r^\ell \sum_m V_{\ell m} Y_{\ell m}(\theta, \phi) . \qquad (2.20)$$

They are useful to study the gravitational field inside a body and, together with (2.17), cover all the harmonic functions. The function $V(\mathbf{r})$ in (2.20) can also express a tidal (external) gravitational field, for which we have $\ell \geq 2$ (see Ch. 4).

Low-degree gravity field. The $\ell = 0$ term in the expansion (2.17) is the main, monopole term GM/r. Note that the origin of the coordinates is still undetermined and under a translation $\mathbf{r} = \mathbf{r}' + \mathbf{a}$ we have

$$\frac{1}{r} = \frac{1}{|\mathbf{r}' + \mathbf{a}|} = \frac{1}{r'}\left[1 - \frac{\mathbf{r}' \cdot \mathbf{a}}{r'^2} + O\left(\frac{a}{r'}\right)^2\right] ;$$

hence a displacement of the origin produces in a harmonic function U an additional dipole term, and other terms of higher degree as well; recall that $r Y_{10}^c = \sqrt{3}\, z$ and $r(Y_{11}^c + iY_{11}^s) = \sqrt{3}\,(x + iy)$. Therefore, the three coefficients U_{1m} can, and will, always be taken equal to nil with a suitable choice of the origin. The first relevant deviation from spherical symmetry is the *quadrupole* term, with $\ell = 2$, described by five arbitrary coefficients; if the body is axially symmetric around the z-axis, only $C'_{20} = \sqrt{5}\, C_{20} = -J_2$ survives (note the minus sign, which makes J_2 positive for oblate bodies):

$$U(\mathbf{r}) = -\frac{GM}{r}\left[1 - J_2\left(\frac{R}{r}\right)^2 \frac{3\cos^2\theta - 1}{2}\right] . \qquad (2.21)$$

In the general case the integral form of eq. (1.37) (Poisson's integral) reads

$$U(\mathbf{r}) = -G \int dV' \frac{\varrho(\mathbf{r}')}{\sqrt{r^2 + r'^2 - 2\,\mathbf{r}\cdot\mathbf{r}'}} . \qquad (2.22)$$

Expanding the denominator with respect to r'/r we get

$$U(\mathbf{r}) = -\frac{G}{r} \int dV' \varrho(\mathbf{r}')\left[1 + \frac{\mathbf{r}\cdot\mathbf{r}'}{r^2} + \frac{3(\mathbf{r}\cdot\mathbf{r}')^2 - r^2 r'^2}{2\, r^4} + O\left(\frac{r'}{r}\right)^3\right] . \qquad (2.22')$$

The first term is the monopole contribution $-GM/r$; the second term is the dipole contribution and, as stated earlier, vanishes in a suitable coordinate system. From (2.22') we see that

$$U_1 \propto \int dV' \, \mathbf{r}' \varrho(\mathbf{r}') = 0 \qquad (2.23)$$

when the origin of the coordinates is placed at the centre of mass. The third term reads

$$U_2(\mathbf{r}) = -\frac{G}{r^5}\mathbf{r} \cdot \left[\int dV' \varrho(\mathbf{r}') \frac{3\mathbf{r}'\mathbf{r}' - \mathbf{1}r'^2}{2} \right] \cdot \mathbf{r} = -\frac{3}{2} G \frac{\mathbf{r} \cdot \mathsf{Q} \cdot \mathbf{r}}{r^5} \quad (2.24)$$

and, being $O(1/r)^3$, must be the quadrupole term. A sans serif symbol denotes a second rank tensor;

$$\mathsf{Q} = \int dV' \varrho(\mathbf{r}') \left(\mathbf{r}'\mathbf{r}' - \frac{1}{3}\mathbf{1}r'^2 \right) \quad (2.25)$$

is the symmetric and trace-free *quadrupole tensor* (**1** is the unit matrix). Hence U_2 can be written as S_2/r^5, where $S_2 = \mathbf{r} \cdot \mathsf{Q} \cdot \mathbf{r}$ is a quadratic harmonic form. Thus U_2 depends on five arbitrary coefficients and must be of the form

$$U_2(\mathbf{r}) = -\frac{GM}{R}\left(\frac{R}{r}\right)^3 \sum_m \left[C_{2m} Y^c_{2m}(\theta, \phi) + S_{2m} Y^s_{2m}(\theta, \phi) \right] \quad (2.24')$$

in agreement with (2.17'). From definition of the *inertia tensor* I

$$\mathsf{I} = \int dV' \varrho(\mathbf{r}') \left(\mathbf{1}r'^2 - \mathbf{r}'\mathbf{r}' \right) ; \quad (2.26)$$

comparing this with eq. (2.25) we easily find

$$\mathsf{Q} = \frac{1}{3}\operatorname{Tr}\mathsf{I} - \mathsf{I}, \quad (2.27)$$

which results in *MacCullagh's formula*

$$U_2(\mathbf{r}) = \frac{G}{2r^5}\left[3\mathbf{r} \cdot \mathsf{I} \cdot \mathbf{r} - r^2 \operatorname{Tr}\mathsf{I} \right]. \quad (2.28)$$

With respect to the principal axes of inertia it reads:

$$U_2(\mathbf{r}) = \frac{G}{2r^5}\left[x^2(2A - B - C) + y^2(2B - C - A) + z^2(2C - A - B) \right]. \quad (2.29)$$

It is easy then to express the quadrupole harmonic coefficients in terms of the moments of inertia

$$C_{20} = \frac{1}{2\sqrt{5}} \frac{A + B - 2C}{MR^2}, \quad C_{22} = \frac{1}{2}\sqrt{\frac{3}{5}} \frac{B - A}{MR^2}. \quad (2.30)$$

The other coefficients vanish. In axial symmetry we recover (2.21), with

$$J_2 = \frac{C - A}{MR^2} = -C'_{20}. \quad (2.31)$$

Note that J_2 does not give any information about the moment of inertia itself, but is a measure of the degree of concentration in the core (Problem 1.2).

Higher multipoles can also be defined, as a generalization of eq. (2.25); a general multipole is a symmetric and trace-free tensorial quantity with ℓ indices, corresponding to the spherical harmonics of degree ℓ.

Harmonic coefficients of arbitrary degree. We now generalize eq. (2.30) to $\ell > 2$ and express $U_{\ell m}$ in terms of the spherical harmonic representation of the mass density

$$\varrho(\mathbf{r}) = \sum_\ell \sum_m \varrho_{\ell m}(r) Y_{\ell m}(P) . \qquad (2.32)$$

This equation describes how the asymmetry in the matter distribution of type ℓ, m changes with depth; P is a point on the unit sphere. We use in eq. (2.22) the expansion (2.2) and the addition theorem for spherical harmonics

$$P_\ell(\cos\psi) = \frac{1}{2\ell+1} \sum_m Y_{\ell m}(P) Y^\star_{\ell m}(P') , \qquad (2.33)$$

where ψ is the angle between the unit vectors OP and OP'. Recalling the orthonormality property (2.8), we find for the coefficients $U_{\ell m}$ in eq. (2.17)

$$U_{\ell m} = \frac{4\pi}{(2\ell+1)MR^\ell} \int_0^{R_m} dr' \, r'^{\ell+2} \varrho_{\ell m}(r') . \qquad (2.34)$$

The upper limit R_m in the radial integral is the largest distance from the centre at which there is matter. For low degrees, $\ell \leq 2$, we obviously recover the previous results.

Potential of a rotating body. For a planet rotating with a constant angular velocity ω_0 around the z-axis the total potential energy W includes also the centrifugal potential energy

$$W = U - \frac{1}{2}\omega_0^2 r^2 \sin^2\theta \qquad (2.35)$$

and, up to $\ell = 2$, reads

$$W = -\frac{GM}{r}\left[1 - J_2\left(\frac{R}{r}\right)^2 \frac{3\cos^2\theta - 1}{2}\right] - \frac{1}{2}\omega_0^2 r^2 \sin^2\theta . \qquad (2.35')$$

Of course, in this frame the force on a moving particle includes the Coriolis force, as needed, for example, for fluid dynamics (Ch. 7).

If the oblateness, measured by J_2, is due to rotation (see Sec. 3.3), the two terms in this equation responsible for the deviation from spherical symmetry must be, on the ground, of the same order of magnitude; this justifies the estimate (1.14). The higher harmonic coefficients describe more complicated deviations from axial symmetry. In particular, the third degree zonal term ($\ell = 3, m = 0$) in eq. (2.17') is proportional to $\cos\theta\,(5\cos^2\theta - 3)$ and describes a North-South asymmetry.

Another important effect of the deviations of the gravity field from spherical symmetry is the change δg in the value of the acceleration of gravity (*gravity anomaly*). The direction of the vector ∇W defines the local vertical, while its modulus is $g + \delta g$ ($g = GM/R^2$ is the reference value). To lowest order we may neglect the difference between the vertical and the radial direction and set $r = R$ when computing the correction δg. From eq. (2.17') we find

$$g + \delta g = \frac{\partial W}{\partial r} = \frac{GM}{R^2} + \frac{GM}{R^2}\sum_{\ell \geq 2}(\ell+1)\sum_m (C_{\ell m} Y^c_{\ell m} + S_{\ell m} Y^s_{\ell m}) - \omega_0^2 R \sin^2\theta \quad (2.36)$$

and thus

$$\frac{\delta g}{g} = \sum_{\ell \geq 2}(\ell+1)\sum_m (C_{\ell m} Y^c_{\ell m} + S_{\ell m} Y^s_{\ell m}) - \frac{\omega_0^2 R}{g}\sin^2\theta. \quad (2.37)$$

The parameter $\omega_0^2 R/g$ has been introduced as the Helmert's parameter μ_c in eq. (1.13). However, in geodetic practice it is convenient to refer gravity anomalies not to a sphere, as above, but to a reference ellipsoid represented by the solution for the surface of a slowly rotating liquid planet (Sec. 3.3), a good approximation to the geoid (see below). In this way the effects of rotational flattening, the main cause of variation, are absorbed and the anomalies, both for gravity and geoid height, are small. In particular, gravity anomalies are represented by a difference between the observed acceleration on the geoid and the computed acceleration on the ellipsoid of reference propagated to the geoid height.

A gravity anomaly means that the weight of objects is different and, hence, makes a mountain sink less or more into the mantle. This phenomenon called *isostasy*, produces a correlation between gravity anomalies and topography. Deviations from this correlation provide a diagnostic about the local mechanical properties and the state of stress of the crust material. Precise gravity data allow such comparison not only for the Earth, but for all terrestrial planets (except for Mercury so far).

2.3 The gravity field of the Earth

An accurate knowledge of the gravity field of the Earth, even for large degrees, has a great importance for geodesy, space physics and practical applications. The overall shape and orography of the Earth is, of course, intimately related to its potential. With the space age and the possibility of very accurate determination of the orbits of Earth satellites, an accurate and periodically

2. The gravitational field of an isolated body

updated harmonic model has become essential. Analogously, for the investigation of irregularities of the crust, in particular for mineral concentrations, precise measurements of local gravity anomalies are needed.

Traditionally, the geopotential was measured on the ground with *gravimeters*, essentially an inverted pendulum whose period is proportional to the square root of the gravity acceleration. Their accuracy is limited, about $10 - 100 \ \mu$Gal (1 Gal = 1 cm/s^2); besides, it is difficult to use it on a ship, and a truly global coverage is not possible. Moreover, they can provide only changes of g from place to place, yielding thus relative measurements; absolute gravimeters, based on laser measurements of the free-fall time of a mass, reach an accuracy of about 1 μGal, or even 1 nGal for superconducting experimental instruments, but are seldom available.

Satellite tracking (both from the ground and the other satellites) have now reached great sophistication and accuracy. Electromagnetic measurements of the distance or the relative velocity of a satellite are currently used to determine its orbit. Analysis of such data is, of course, unavoidably connected with polar motion and the motion of the ground stations due to tides, plate tectonics and other effects, and represents thus a very complicated problem.

In the body-fixed frame of reference the geopotential of an isolated rigid body is constant; however, as discussed in Sec. 3.2, the rigid Earth (and its inertia tensor) is an idealization and special care is required to define it and to implement it. The figure axis $\mathbf{e}_z$ can be empirically determined from the polar motion, as explained in Sec. 3.2; in the Cartesian system based on the principal axes of inertia the coefficients C_{21}, S_{21} and S_{22} vanish, leaving only C_{20} and C_{22}, determined by the inertia tensor through eq. (2.30). In practice, however, the principal axes $\mathbf{e}_x$ and $\mathbf{e}_y$ are difficult to determine directly and the longitude φ is defined geographically with the origin at Greenwich. In other words, the Greenwich direction defines the vector $\mathbf{e}'_x$ of the system in which the harmonic coefficients of the geopotential are expressed. Then S_{22} does not vanish and determines the longitude of $\mathbf{e}_x$.

Moreover, it is convenient to define the third axis $\mathbf{e}'_z$ along a direction whose determination is accurate and stable. As discussed in Sec. 3.2, an inertial reference system tied to extragalactic sources can be set up; its third axis $\mathbf{e}'_z$, chosen for simplicity near the figure axis, is the Conventional Reference Pole. It is located at 1.44×10^{-6} rads, or 9 m on the Earth surface, from the figure axis. This is why the harmonic coefficients C_{21} and S_{21} have a small, non-vanishing values.

The geopotential coefficients (Table 2.3) have been determined up to degree 360 (EGM96 model) using a large set of observations, including satellite and lunar tracking, ground and astronomical measurements. This determination requires delicate and complex numerical analysis to deal with different instruments and a large number of unknown parameters ($360^2 \simeq 130,000$ for the

Table 2.3 EGM96 values of the geopotential coefficients and their formal uncertainty up to the fourth degree for the Earth; the real and normalized coefficients are listed in units of 10^{-6}, the uncertainties in units of 10^{-10}. The Cartesian frame used is the one defined by the International Earth Rotation Service and differs from the principal axes frame, as explained in the text.

ℓ	m	$C_{\ell m}$	$S_{\ell m}$	$\delta C_{\ell m}$	$\delta S_{\ell m}$
2	0	−484.1654	−	0.356	−
2	1	−0.0002	0.0012	0	0
2	2	2.4391	−1.4002	0.537	0.543
3	0	0.9573	−	0.181	−
3	1	2.0300	0.2485	1.397	1.365
3	2	0.9046	−0.6190	1.096	1.118
3	3	0.7211	1.4144	0.952	0.933
4	0	0.5399	−	1.042	−
4	1	−0.5363	−0.4734	0.857	0.824
4	2	0.3507	0.6627	1.600	1.639
4	3	0.9908	−0.2009	0.847	0.827
4	4	−0.1886	0.3089	0.873	0.879

harmonic coefficients only). Satellite orbital data, through laser tracking, constrain the geopotential only up to degree $\ell \approx 70$ and are insensitive to higher degrees (with the current tracking precision of ≈ 1 cm). This is why gravity models (like JGM3) based only on satellite measurements do not go beyond this degree. Adding large sets of surface gravity and altimetry data, high degree models for special geodetic purposes have been constructed. The Table 2.3 lists only the constant part of the geopotential coefficients; small time-dependent effects occur due to various causes, e.g. tides (Ch. 4) and global variations of the atmospheric mass. The indicated uncertainty in the geopotential coefficients should be considered as formal, since many correlations among different groups of parameters occur and the real error can be larger.

The main harmonic coefficient is, of course, $C_{20} \simeq -484 \times 10^{-6}$, which corresponds to the gravitational oblateness; the other coefficients (except for C_{21} and S_{21}) are about 10^3 times smaller or more, so that the approximation of axial symmetry is a good one. It has been shown that $J_2 = -\sqrt{5}C_{20}$, as well as the other low-degree zonal coefficients, depend on time (see Sec. 20.3).

The quadrupole potential of the Earth has a small deviation from axial symmetry, measured by C_{22} and S_{22}, about 10^{-6} (Table 2.3). It is interesting to see their effect in the equatorial plane ($\theta = \pi/2$)

$$\begin{aligned} \delta U_2 &= -\frac{\sqrt{15}}{2}\frac{GM}{r}\left(\frac{R}{r}\right)^2 (C_{22}\cos 2\phi + S_{22}\sin 2\phi) \\ &= -3\frac{GM}{r}\left(\frac{R}{r}\right)^2 J_{22}\cos 2(\phi - \phi_{22}) \ . \end{aligned} \quad (2.38)$$

From Table 2.3 we obtain $J_{22} \simeq 1.81 \times 10^{-6}$ and $\phi_{22} \simeq -15°$. A satellite on an equatorial circular orbit is subject to a potential with two rotating minima at the opposite longitudes $-15°$ and $165°$. A geostationary satellite keeps a constant

2. The gravitational field of an isolated body

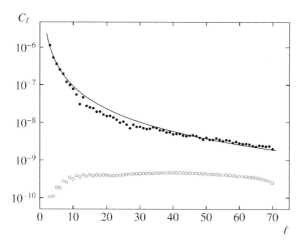

Figure 2.2 Full circles: the mean values C_ℓ (eq. (2.39)) of the geopotential coefficients as a function of their degree ℓ; open circles: the formal errors of the geopotential coefficients (both from the EGM96 geopotential model). The curve gives the empirical approximation (2.40; Kaula's rule).

longitude in the monopole field, but it oscillates back and forth around these two minima if the quadrupole part of the field is considered (Sec. 18.2).

Kaula's rule and the fine structure of the gravity field. Generally the coefficients of the same degree ℓ have the same order of magnitude, which decreases with ℓ. This trend is best described by the mean

$$C_\ell^2 = \frac{1}{2\ell + 1} \sum_m (C_{\ell m}^2 + S_{\ell m}^2), \quad (2.39)$$

analogous to the energy spectrum for a two-dimensional scalar field. It has been noticed that the empirical law (*Kaula's rule*)

$$C_\ell^2 = A_K \frac{10^{-10}}{\ell^4} \quad (\ell \gg 1) \quad (2.40)$$

is followed fairly well for the Earth (with $A_K = 0.7$; see Fig. 2.2) and for terrestrial planets. Its justification is not well known.

It is interesting to evaluate the geopotential at the angular resolution $\Delta \ll 1$ obtained by averaging the potential U over (not too elongated) angular domains of size Δ. Since the spherical harmonics of degree ℓ vary over latitude (and possibly over longitude) with the angular scale $1/\ell$, this average cuts off the summation (2.17′) at the degree $1/\Delta$, and gives:

$$\delta_\Delta U = -\frac{GM}{R} \sum_{\ell \leq 1/\Delta} \sum_m (C_{\ell m} Y_{\ell m}^c + S_{\ell m} Y_{\ell m}^s). \quad (2.41)$$

Averaging the square of this quantity over the sphere, using the orthonormality relations (2.8), and applying Kaula's rule we obtain

$$\langle (\delta_\Delta U)^2 \rangle = \frac{1}{4\pi} \int d\Omega \, [\delta_\Delta U(\mathbf{r})]^2 = \left(\frac{GM}{R}\right)^2 \sum_{\ell \leq 1/\Delta} (2\ell + 1) C_\ell^2 \quad (2.42)$$

$$\simeq 2\left(\frac{GM}{R}\right)^2 A_K \, 10^{-10} \sum_{\ell \leq 1/\Delta} \frac{1}{\ell^3} = \left(\frac{GM}{R}\right)^2 A_K \, 10^{-10} \Delta^2 \; .$$

The gravity anomaly over a given angular scale is obtained in order of magnitude dividing U_ℓ by its smallest spatial scale R/ℓ, so that

$$\delta_\Delta g \approx -\frac{GM}{R^2} \sum_{\ell \leq 1/\Delta} \ell \sum_m (C_{\ell m} Y^c_{\ell m} + S_{\ell m} Y^s_{\ell m}) \; . \tag{2.43}$$

Its mean value is given by

$$\langle (\delta_\Delta g)^2 \rangle \simeq \left(\frac{GM}{R^2}\right)^2 \sum_{\ell \leq 1/\Delta} \ell^3 C_\ell^2 \approx 2\left(\frac{GM}{R^2}\right)^2 A_K \, 10^{-10} \ln \Delta \; . \tag{2.44}$$

The fact that it diverges at vanishing resolution ($\Delta \to 0$) suggests that Kaula's rule should have a cutoff.

In order to obtain gravity information about crustal structures with size less than 30 km, it is necessary to extend the gravity model (derived through a single and global procedure) up to degree $\ell \approx R/(30 \text{ km}) \approx 200$ or more. It is easy to see that any harmonic function of a *large degree* ℓ decreases exponentially with height; indeed, when $h = r - R \ll R$, we get

$$\left(\frac{R}{r}\right)^{\ell+1} \simeq \left(1 - \frac{h}{R}\right)^\ell = \exp\left[\ell \ln\left(1 - \frac{h}{R}\right)\right]$$

$$= \exp\left[-\ell \frac{h}{R} + O\left(\ell \left(\frac{h}{R}\right)^2\right)\right] \; . \tag{2.45}$$

Similarly, we can evaluate the contribution of density inhomogeneities to the potential at a large degree ℓ. In the eq. (2.34) the power in the integrand can be written in terms of the depth $d = R - r \ll R$ as

$$\left(\frac{r}{R}\right)^{\ell+2} \simeq \exp\left(-\ell \frac{d}{R}\right) \; ; \tag{2.46}$$

therefore, the density inhomogeneity at depth much greater than d does not contribute.

The flat-Earth approximation.– At large ℓ and low altitude the curvature of the Earth may be neglected and a simpler representation of the gravity field is useful. Consider a flat Earth with an inhomogeneous mass distribution in the half space $z < 0$. For a given z (< 0) the gravitational potential and the matter distribution can be described by their Fourier transforms $U_\mathbf{k}(z)$ and $\varrho_\mathbf{k}(z)$ with respect to x and y; $\mathbf{k} = (k_x, k_y)$ is a two-dimensional wave vector. Poisson's equation (1.36) reads

$$\frac{d^2 U_\mathbf{k}(z)}{dz^2} - k^2 U_\mathbf{k}(z) = 4\pi G \varrho_\mathbf{k}(z) \quad (k^2 = k_x^2 + k_y^2) \; . \tag{2.47}$$

2. The gravitational field of an isolated body

The solution which vanishes when z is very large reads

$$U_{\mathbf{k}}(z) = -\frac{2\pi G}{k} \int_{-\infty}^{0} dz' \varrho_{\mathbf{k}}(z') \exp[k(z'-z)]. \tag{2.48}$$

This shows that, for a given wave number $\mathbf{k}$, the potential is determined by the density inhomogeneities within a layer of depth $\approx 1/k$; and that the potential decreases exponentially with height with the same characteristic scale. Therefore $k \approx \ell/R$.

An important way to get a global mapping of the gravity field of the Earth at a very high resolution is by tracking a low flying spacecraft. Consider, for simplicity, a spacecraft flying over a "flat Earth" at a constant altitude $z = h$ along the x direction, with an unperturbed velocity v_0. Conservation of energy requires

$$v_0 \, \delta v_x(t) = \sum_{\mathbf{k}} U_{\mathbf{k}}(h) \exp(ik_x v_0 t) \, ; \tag{2.49}$$

its Fourier transform has only the frequency $\omega = k_x v_0$, with amplitude

$$v_0 \, \delta \tilde{v}_x(\omega) = \sum_{k_y} U_{\omega/v_0, k_y}(h) \, .$$

Eqs. (2.48) and (2.49) clearly show the main difficulty we have to face in a measurement of this kind: the potential perturbation decreases exponentially with height with a characteristic scale $1/k = R/\ell$. The component of degree 200, for example, at 180 km is attenuated by the factor $\exp(-180 \times 200/6{,}378) \simeq 0.0036$. With this attenuation the ground value $gR\sqrt{0.7}\,10^{-5}/\ell^2 = 130$ cm^2/s^2 of the potential perturbs the spacecraft velocity $v_0 \simeq 5$ km/s by $\approx 10^{-6}$ cm/s.

The GRACE (Gravity Recovery and Climate Experiment of NASA and DLR, the German Space Agency) mission, was launched in March 2002; it consists in two equal satellites on the same, slightly eccentric, low-altitude polar orbit, with a separation D between 150 and 300 km; they can transmit and receive coherent and stable microwave carriers, thus measuring their relative velocity Δv_x, with a corresponding precision in the change in their distance of $\simeq 5\,\mu$m. Since they fly over the same point with the delay D/v_0, according to eq. (2.49), the contribution to Δv_x from a given wave vector $\mathbf{k}$ has an amplitude proportional to $\sin(k_x D/2)$; therefore they are sensitive only to the Fourier components of the geopotential with $k_x D$ not much smaller than one. This differential measurement will provide the fine structure of the Earth gravity field with an unprecedented accuracy; it is especially suitable for monitoring the time variations of the geopotential, since components of degree $\ell \simeq 150$ or larger, depending on the altitude, will be available on a monthly basis.

The geoid. We finally return to the complex issue of the relation between the gravity field and the modelling of the topography and marine surface, a topic of geodesy. The main concept here is that of the *geoid* (and the gravity anomalies; Sec. 2.2). The geoid is defined as an equipotential surface $W = W_0$ (eq. (2.35)) of an isolated planetary body; it describes the surface of an ideal ocean, with no waves, currents or atmospheric influence, hence at constant pressure (eq. (1.38)); as commented below, with the current precision its determination must also take into account the tidal potential (Ch. 4). Any deviation of the surface of the oceans from the geoid contains important information about their dynamics; several space projects are currently devoted to their measurement. The geoid also defines the zero of the topographic altitude h. We now outline this concept and later discuss its current precise modelling.

Since the gravitational potential on the ground differs from GM/r by terms of order J_2 or smaller, we expect the geoid to deviate from a sphere by the same order, so that in computing these corrections $\delta r = r - R$ we can take $r = R$. Therefore from eq. (2.17'), supplemented by the centrifugal potential in the rotating frame (2.35), we get

$$\frac{\delta r(\theta, \phi)}{R} = \text{const} + \sum_{\ell \geq 2} \sum_m (C_{\ell m} Y^c_{\ell m}(\theta, \phi) + S_{\ell m} Y^s_{\ell m}(\theta, \phi)) + \frac{\mu_c}{2} \sin^2 \theta \ . \quad (2.50)$$

The value of the arbitrary constant can be chosen so that the average of $\delta r(\theta, \phi)$ over the Earth surface is nil.

Though the above approach is a correct and possible description of the geoid, its height is traditionally related not to a sphere, as above, but to an axisymmetric ellipsoid. Since the geoid, and the Earth topography, is fitted much better with an ellipsoid, its deviations from this reference surface are conveniently much smaller. The best current global geodetic model, the refined World Geodetic System 1984 (WGS84; Fig. 2.3), defines the reference ellipsoid with four data: (i) the semimajor axis (i.e. equatorial radius) $R_e = 6378.137$ km; (ii) the inverse of the oblateness $1/\epsilon \simeq 298.2572$ (eq. (1.14)); (iii) the Earth mass (including the atmosphere), in terms of $GM \simeq 3.986 \times 10^{14}$ m^3/s^2; (iv) the nominal rotation rate of the Earth ($\omega_0 = 7.292115 \times 10^{-5}$ rad/s), corresponding to the sidereal rotation period 86164.0989 s. The ellipsoid is defined in the Conventional Terrestrial Reference Frame (CTRF; see Sec. 3.2). This fact is important, since with the WGS84 high quality astronomical observations are enabled from any place on the Earth surface (the agreement between the WGS84 and CTRF realizations is better than ≈ 10 cm). From these data a number of parameters are derived, including the geoid potential W_0, the even degree zonal harmonic coefficients C_{20}, C_{40}, etc. The normal acceleration on, and close to, the reference ellipsoid may be also computed, and its comparison with the observed values gives the gravity anomalies. The Earth gravity model associated with the WGS84 is EGM96; as mentioned above, it is formally complete up to $\ell = 360$. High degree terms have been determined from ground and altimetry data, and from orbit determination of a number of

2. The gravitational field of an isolated body

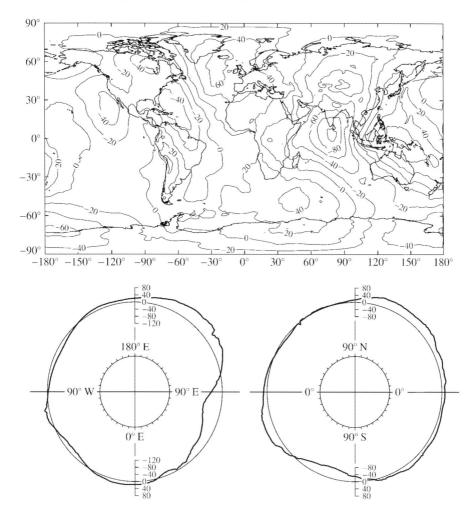

Figure 2.3. Level lines of the geoid, with its height over the mean ellipsoid in meters. The WGS84 parameters and the EGM96 geopotential coefficients are assumed. The tides due to the Moon and the Sun are not taken into account here; the features shown in the map are due to large-scale and deep-lying density inhomogeneities, corresponding to a tide-free planet. The bottom plots show the sections of the geoid through (i) the equatorial plane (left), and (ii) the 0° and the 180° meridians (right). Data from http://www.nima.mil/GandG/wgs-84/.

artificial satellites. Interestingly, since the parameters of the WGS84 ellipsoid have been taken from previous solutions, in effect the mean value of the geoid height is not zero, but $\simeq -0.57$ cm. Its maximum height is -108 m, at 8.25 S latitude and 147.25 E longitude; its estimated average precision is $\approx 0.5 - 1$ m.

As mentioned in Sec. 3.2 for the case of CTRF, the realization of the WGS84 is complementary to, but equally fundamental as, its definition. Significant improvements have been achieved with the Global Positioning System (GPS; Sec. 18.2); the WGS84 is realized by the coordinates assigned to the selected set of GPS tracking stations (with precision better than 5 cm). Broadcasting the geodetic coordinates to any point on the Earth surface is then also enabled with the GPS system.

2.4 Gravity fields of planetary bodies

Planetary bodies of size larger than a few hundred km are bound by gravity and, if isolated, are almost spherical in shape (Sec. 14.8); smaller bodies are bound by solid state forces and, as a consequence of their formation, have a substantial quadrupole moment, with dimensionless coefficients of order unity. The quadrupole moment of bigger bodies is determined by rotation and tides. In the first case it is given, in order of magnitude, by the ratio $\mu_c = \omega^2 R^3/(GM)$ of rotational to gravitational energy (eq. (1.13); see Secs. 3.3 and 3.4); in the second the characteristic order of magnitude is (Ch. 4)

$$\mu_T = \frac{M_p}{M}\left(\frac{R}{D}\right)^3, \qquad (2.51)$$

where M_p is the mass of the acting body (the "planet"), at a distance D. In the case of the Earth both the Moon and the Sun have a comparable tidal effect, namely

$$\frac{\mu_T(\text{Sun})}{\mu_T(\text{Moon})} \simeq 0.46. \qquad (2.52)$$

In both cases the response depends on the elastic properties of the body. The case of *synchronous* satellites, in which the rotational velocity ω equals the orbital velocity n around the primary, is important because $\mu_c \simeq \mu_T$ and the two deformations have the same order of magnitude. This is the case of the Moon and of most satellites of the giant external planets, in particular Jupiter and Saturn.

The perturbations induced by the quadrupole and higher moments of the gravity fields of bodies in the solar system on the orbits of spacecraft have provided important information on their interiors (see Secs. 12.4 and 20.3); however, note that measurements of the quadrupole field give the differences in the moments of inertia, while the values of the moments of inertia themselves must be obtained with additional information. For instance measurements of the lunar librations or precession of the Mars spin axis can yield such additional (and algebraically independent) constraints on the moments of inertia; their combination with J_2 can eventually yield the exact values of the inertia moments.

No direct measurement is available yet for the solar oblateness $J_{2\odot}$, but, in the assumption of uniform rotation, with the known density profile one obtains $J_{2\odot} \approx 10^{-7}$. An improved solar interior model, combined with ground-based observations of the solar oblateness and the SOHO/GONG measurements of the proper modes of solar oscillations and their splitting, results in $J_{2\odot} \simeq 2\times10^{-7}$ (the latter technique is similar to that for Earth, briefly discussed in Sec. 5.2). An accurate knowledge of the solar quadrupole is important for testing general relativity using the perihelion drift of Mercury (Sec. 20.4).

Terrestrial bodies. Density and gravity anomalies of the terrestrial planets, in which a lighter crust is supported by a denser mantle, are in general correlated with the topography by the phenomenon of *isostasy*. If an elevation h of the free level is observed, its additional weight must be supported by a modification of the density distribution underneath. The *Airy compensation mechanism* is what happens to a floating iceberg: the crust, of density ϱ_c smaller than the density ϱ_m of the mantle, has a root in the mantle of thickness b, say, and equilibrium is maintained if

$$\varrho_c h = (\varrho_m - \varrho_c) b \,. \tag{2.53}$$

For a given h, with a thicker root the gravity anomaly of the feature is smaller (compensation). In the compensation mechanism the additional weight of the elevation is supported by a lower density of the crust, resulting in a decrease of the gravity anomaly. In general, in the terrestrial planets we expect a correlation between topography and density/gravity anomalies; their study is important for understanding the structure of the crust and the isostatic mechanism.

The best way to measure the gravitational field of planetary bodies is by accurately tracking the orbits of spacecraft in their vicinity; if the spacecraft is bound to the body, secular perturbations increase their effect in time and much better results are expected. The Magellan (and, before, the Pioneer Venus) mission has provided a very good model of Venus' gravity field up to $\ell = 180$ (this is a formal solution, since the errors are smaller than the solved-for parameters only for $\ell < 70$). The Moon, of course, has been extensively investigated, and the best gravity model has been obtained in 1998 with the Lunar Prospector mission. Its gravity field has been resolved up to $\ell \approx 75$ (a formal solution, combining all previous orbital data, up to $\ell = 165$ has also been obtained). Similarly to the lunar case, Mars has also received a wide attention. The analysis of Mars Global Surveyor data resulted in a degree $\ell = 80$ model of the Martian gravity field. Such a high-degree solutions allow analysis of fine (short-wave) structure of the gravity field. This is especially interesting when a high-resolution topography model is also available. For instance, due to isostasy one expects, and usually finds, areas of larger gravity at low-lying terrains, such as the maria in the lunar case (these features are called *mascons*). Lunar mascons are generally well-understood in terms of infiltration of basaltic lava with a density higher than that of crustal rocks. This situation may occur after a large impact, when the crust fractures down to the crust-mantle interface. Martian gravity data are somewhat peculiar, since they suggest a large difference between the history of the northern and southern hemispheres. The analysis of gravity anomalies shows that the northern hemisphere likely has a thin and strong lithosphere, while the southern hemisphere is covered with a thick and weak one. Thus the southern crust, having more time to reach isostasy, may be older. Not all gravity anomalies correlate with topographic features, especially in the polar regions, that may hide large deposits of frozen water with seasonal (or longer) periodicities.

However, even the low-degree fields of the terrestrial bodies contrast with the Earth data. The synchronous rotation of the Moon causes an anomalous equatorial ellipticity of the lunar gravitational field; this is shown by the fact that J_{22} ($\approx 22.36 \times 10^{-6}$) is only about one tenth of J_2 ($\approx 0.203 \times 10^{-3}$). Such anomaly is observed also for Mars and Mercury (see below), and especially Venus, where J_{22} ($\approx 0.557 \times 10^{-6}$) is even larger than one tenth of J_2 ($\approx 4.404 \times 10^{-6}$) (for Venus this is due to the very slow rotation). In the Martian case, the anomaly consists in a pronounced equatorial ellipticity, corresponding to an anomalously large value of J_{22} ($\approx 63.17 \times 10^{-6}$) with respect to J_2 ($\approx 1.959 \times 10^{-3}$) (compare with the Earth coefficients in Table 2.3). This is caused by the Tharsis Rise, a huge region near the equator resulting from a massive accumulation of volcanic material; since Mars has no tectonic motion, large volcanoes like the Olympus Mountain could have been depositing material in this zone for billions of years.

For Mercury, a poorly explored planet, only the mass and the quadrupole coefficients are known; however, the latter are affected by large errors. Even with this poor resolution, the data again confirm a high equatorial ellipticity (as in the case of other terrestrial bodies, except the Earth) due to Mercury's slow rotation. The analysis of ranging and Doppler data in the forthcoming ESA and NASA Mercury missions should provide its gravity field up to $\ell \approx 25$.

A rule similar to Kaula's for the C_ℓ coefficients (eq. (2.40)) seems to hold for all terrestrial planets and the Moon, for which we have detailed enough measurements (Fig. 2.4). Its universality is somewhat astonishing, especially with regard to the discussion in Sec. 2.3. One may also notice that the coefficient A_K in (2.40) seems to be inversely proportional to the surface gravity (and thus the size).

Outer planets and their satellites. The outer planets are – possibly with the exception of a small core – gaseous and do not support topographic features. Their gravity field is determined by the density profile $\varrho(r)$ and the condition of hydrostatic equilibrium of each layer under the gravitational and centrifugal forces. Since rotating configuration has axial symmetry and reflection symmetry with respect to the equatorial plane, only the harmonics of zero order ($m = 0$) and even degree ℓ are present, with coefficients

$$J_\ell = -\sqrt{2\ell + 1}\, C_{\ell 0} \quad (\ell = 2n \text{ even}). \tag{2.54}$$

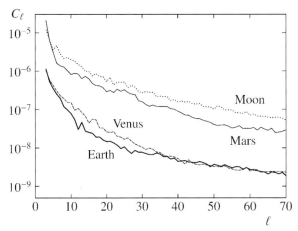

Figure 2.4 The mean values C_ℓ (eq. (2.39)) of the gravity field coefficients as a function of their degree ℓ for the terrestrial planets (except Mercury) and the Moon.

They all vanish, of course, when there is no rotation. In the next chapter we will see that, to the linear approximation in the forcing parameter μ_c (eq. (1.13)) only J_2 does not vanish. In the outer planets μ_c is not very small (0.1 for Jupiter and 0.2 for Saturn), and higher order terms are well within measurement accuracy. We have the expansion:

$$J_2 = \Lambda_{20}\mu_c + \Lambda_{21}\mu_c^2 + \ldots, \quad (2.55a)$$
$$J_4 = \Lambda_{40}\mu_c^2 + \Lambda_{41}\mu_c^3 + \ldots, \quad (2.55b)$$
$$J_6 = \Lambda_{60}\mu_c^3 + \Lambda_{61}\mu_c^4 + \ldots, \quad (2.55c)$$

and so on. The Λ coefficients (smaller than unity) are determined by the density profile and the barotropic relation $P(\varrho)$. In Sec. 3.3, Λ_{20} will be determined for the case of uniform density. In fact, the measurement of J_{2n}, together with the total mass and radius, places important constraints on models of the interior structure and on the evolution of giant planets.

Planetary encounters of the Pioneer and Voyager spacecraft allowed for the first time to determine the J-coefficients of the outer planets (except for Pluto). A combined analysis of their Doppler tracking provided coefficients up to J_6 for Jupiter and Saturn (Table 2.4) and J_4 for Uranus and Neptune. These data also allowed to place a tight constraint on J_{22} for Jupiter and Saturn, confirming their axial symmetry. Note the high value of J_2 due to fast rotation of the giant planets. The Galileo spacecraft, while orbiting in the Jovian system, significantly improved our knowledge of the satellites gravity fields; in particular, it provided measurements of the quadrupole fields of Europa, Ganymede and Io (putting also constraints on the octupole parameter J_3). Combining all these data, the total mass of the Jovian system is known with the astonishing precision of $\approx 8 \times 10^{-9}$ (fractional). From 2004 to 2008 the Cassini spacecraft will orbit more than 60 times around Saturn, measuring the gravity fields of the planet and its satellites, in particular Titan, with very high accuracy.

	Jupiter	Saturn
$M/M_\oplus$	317.83	95.147
R_e ($\times 10^{-9}$ m)	7.1492(4)	6.0268(4)
J_2 ($\times 10^{-2}$)	1.4697(1)	1.6332(10)
J_4 ($\times 10^{-4}$)	−5.84(5)	−9.19(40)
J_6 ($\times 10^{-4}$)	0.31(20)	1.04(50)

Table 2.4 Gravitational fields of Jupiter and Saturn determined with satellite flybys; $M/M_\oplus$ is the mass scaled with the Earth mass and R_e the reference (equatorial) radius of the planet. The values in brackets indicate uncertainty.

Minor bodies. Doppler tracking of the NEAR/Shoemaker probe, put in orbit around the near-Earth asteroid Eros in February 2000, allowed to resolve the harmonic coefficients of its gravitational field to the degree $\ell = 4$. Given the very elongated shape of this 30-km size asteroid, their hierarchy is peculiar and significantly differs from the planetary fields. The non-vanishing quadrupole coefficients (with the rotation axis along the z-direction) are $C_{20} = -0.052$, $C_{22} = 0.083$ and $S_{22} = -0.028$. It should be noticed that the precise knowledge of Eros' gravity field is still exceptional among small bodies in the solar system. If a model of the surface topography is available (for example, with a flyby), assuming a constant density the coefficients of the gravity potential can be obtained; this technique has been applied to Phobos, Deimos and several asteroids.

It might be also mentioned that the dynamics of objects very close to an irregular body (like an asteroid), for instance low-velocity ejecta or a landing trajectory, requires advanced methods to describe its gravitational field; the multipole expansion may be unsuitable because of lack of convergence.

Appendix: *External and internal problems of gravitational motion

The fluid model, possibly with the inclusion of viscous and elastic effects (Sec. 1.5), is appropriate to describe the dynamics of single bodies in the solar system (the *internal problem*). Their – mainly gravitational – interaction is usually dealt with the point particles model (the *external problem*). In fact, however, the matter distribution in each body is affected, mainly through tidal forces, by other bodies; on turn, the external gravitational force is modified by the shape of each acting body. Thus the internal and the external problems are coupled. At a fundamental level, this coupling can be obtained from the rigorous dynamical equations of a fluid, distributed in separate lumps with a size much smaller than their distances. It turns out that the main coupling is described by the quadrupole moment (2.25); it is appropriate, therefore, to briefly discuss it here. For each body, of volume V,

$$m = \int_V dV \varrho, \quad m\mathbf{z} = \int_V dV\, \mathbf{r}\varrho \tag{a}$$

give the mass and the centre of mass $\mathbf{z}$. If $\mathcal{F}$ is the force density, since the mass conservation (Sec. 1.3) implies $d(dV\varrho)/dt = 0$, we obtain for the centre of mass of the chosen body

$$m\frac{d^2\mathbf{z}}{dt^2} = \int_V dV\, \mathcal{F}. \tag{b}$$

We consider, for simplicity, the ideal fluid approximation (eq. (1.35))

$$\mathcal{F} = -\nabla P - \varrho \nabla U. \tag{c}$$

2. The gravitational field of an isolated body

$\mathcal{F}$ splits in an internal contribution due the mass density and the pressure fields in the body (the self force) and an external contribution due to the gravitational potential of other bodies (the external force). Since at the boundary of the body the pressure vanishes, the pressure contribution to the self force vanishes. It can also be shown, on the basis of Newton's third law, that the gravitational contribution to the self force $-\int_V dV \varrho \nabla U^{(\text{int})}$ vanishes, too. These results can be generalized to more realistic viscous fluids.

Thus, the acceleration of the centre of mass depends only on the external gravitational potential $U^{(\text{ext})}$ and reads

$$m\frac{d^2\mathbf{z}}{dt^2} = -\int_V dV' \varrho(\mathbf{r}') \nabla U^{(\text{ext})}(\mathbf{z}+\mathbf{r}'),\qquad(d)$$

where $\mathbf{r}' = \mathbf{r} - \mathbf{z}$ ($r' \ll r$) are the coordinates of the integration point relative to the centre of mass and $\mathbf{r}$ has the order of magnitude of the distances between the bodies. A Taylor expansion gives then

$$m\frac{d^2\mathbf{z}}{dt^2} = -m\nabla U^{(\text{ext})} - \frac{1}{2}\mathbf{Q} : \nabla\nabla\nabla U^{(\text{ext})}\left(1+O(r'/r)\right).\qquad(e)$$

Here, gradients ($\nabla = \partial/\partial\mathbf{z}$) of $U^{(\text{ext})}$ are computed at the centre of mass. We have used the definitions (a), (2.25) of the centre of mass and the quadrupole tensor, and the harmonic property of the external potential $U^{(\text{ext})}$. The quadrupole correction is $\approx Q/(mr'^2)$ smaller than the main monopole term.

For space probes close to planets, the external potential $U^{(\text{ext})}$ must be retained to high multipole order; the internal structure of the planet strongly affects the orbital motion of the satellite, though the coupling is one-way only. To explicitely bring out the quadrupole coupling in the general case with several interacting bodies, labelled with p, we must generalize the definition (2.25) to an arbitrary number of bodies; outside them we have

$$U^{(\text{ext})}(\mathbf{r}) = -G\sum_p \left\{ \frac{m_p}{|\mathbf{r}-\mathbf{z}_p|} + \frac{1}{2}\mathbf{Q}_p : \nabla\nabla\frac{1}{|\mathbf{r}-\mathbf{z}_p|} + \ldots \right\}\qquad(f)$$

with the gradients $\nabla = \partial/\partial\mathbf{r}$. Of course, the sources change shape and move around; this potential is also a function of time. Inserting the last equation into eq. (e) we finally obtain

$$m_a\frac{d^2\mathbf{z}_a}{dt^2} = G\sum_{b\neq a}\left\{ m_a m_b \frac{\partial}{\partial\mathbf{z}_a}\frac{1}{|\mathbf{z}_a-\mathbf{z}_b|} + \frac{1}{2}(m_a\mathbf{Q}_b + m_b\mathbf{Q}_a) : \frac{\partial^3}{\partial\mathbf{z}_a\partial\mathbf{z}_a\partial\mathbf{z}_a}\frac{1}{|\mathbf{z}_a-\mathbf{z}_b|} + \ldots \right\}.\qquad(g)$$

The quadrupole moments are produced by rotation, tidal forces due to the external potential and other causes forcing a deviation from spherical symmetry; they are determined, as a function of time, by means of the internal equations of motion. For a rigid rotation, the quadrupole can be obtained from Euler's equation (Sec. 3.5) and the torques by external bodies (Sec. 4.1). The elastic and plastic behaviour of cosmic bodies, however, must be taken into account and the internal dynamical equations may require taking into account higher order moments. The monopole-quadrupole interaction in the second term is the main level at which the external and the internal dynamics are coupled. If it can be neglected, decoupling ensues and one has the usual "point-mass model". Conversely, if very precise measurements are available, higher order terms may be necessary; for instance, to accurately describe the motion of the Moon, one must take into account multipoles of the Earth and the Moon up to degree 5. In (g) we have neglected the quadrupole correction to the potential, leading to the quadrupole-quadrupole interaction. It is usually negligible and it should be included only in special cases, such as the dynamics of binary asteroids.

– PROBLEMS –

2.1. The second degree harmonics (2.13) generate a 5 by 5 representation of the rotation group. Construct the matrices corresponding to the rotations around the coordinate axes.

2.2.* A symmetric tensor of the second rank with vanishing trace has five independent components. Show that it transforms under rotations according to the representation $D^{(2)}$ defined by eq. (2.16).

2.3. Calculate the potential perturbation and the gravity anomaly on a flat Earth due to (i) a uniform sphere and (ii) a uniform infinite cylinder buried under the surface and with a different density. (Hint: Use Gauss' theorem.)

2.4.* Following the previous problem, evaluate the change in the velocity of a spacecraft flying horizontally along x at a low altitude $z = h$ when it passes over a underground cylinder of radius $r < d$ with the axis at $x = 0$, $y = -d$ and with a mass density ϱ' different from the local density ϱ. Using reasonable values for a mineral deposit and keeping h variable, discuss the possibility of measuring the velocity change with an accurate frequency standard (see Sec. 3.1).

2.5. Find the conditions under which an inhomogeneous, axially symmetric density distribution $\varrho(r, \theta)$ in a sphere gives exactly the external gravitational potential $-GM/r$. Determine simple models that fulfil these conditions.

2.6. Using the harmonic character of the harmonic functions, find the ordinary differential equation fulfilled by the associate Legendre functions (the *Legendre equation*).

2.7.* The angular momentum of a body, in terms of the positions $\mathbf{r}'$ and velocities $\mathbf{v}'$ of its parts, is $\mathbf{s} = \int_V dV' \varrho(\mathbf{r}') (\mathbf{r}' \times \mathbf{v}')$. Compute its time derivative and study at which level external bodies affect it; in particular, show that the contribution to $d\mathbf{s}/dt$ from the internal dynamics (the self-moment of the force) is zero, in analogy with the self-force (see the Appendix).

2.8. Evaluate the 5 quadrupole coefficients U_{2m} ($m = -2, \ldots, 2$) of the gravitational field of a body oblate along the z axis, with $C - (A + B)/2 = \delta_1 C$ and prolate along the x direction, with $(B+C)/2 - A = \delta_2 A$. With $\delta_1 \approx \delta_2$, this configuration corresponds to a synchronous natural satellite, deformed by rotation and tidal forces.

2.9. At frequencies much less than its orbital period, the gravitational effect of the Moon can be ascribed to a ring of matter. Calculate the corresponding J_2 of the effective lunar potential.

2.10. Consider the gravitational potential of two fixed centres (with masses m) at $x = 0$, $y = 0$, $z = \pm c$ and determine its harmonic coefficients, which correspond to prolate equipotential surfaces. Oblate equipotentials are obtained by formally taking $z = c(\sigma \pm i)$. When the masses get an imaginary part, $m \to m(1 \pm i\sigma)$, we can produce a potential with an arbitrary J_2 and J_3. Determine $J_2(c, \sigma)$ and $J_3(c, \sigma)$.

2.11.* Compute the gravity anomaly across a continental margin in a complete isostatic compensation: a semi-infinite continental-crust slab is in contact with a semi-infinite ocean layer above a thinner crust. Is the anomaly measurable with ordinary gravimeters (Sec. 2.3)?

– FURTHER READINGS –

A classical and still good textbook on geodesy and theoretical gravimetry is W.A. Heiskanen and H. Moritz, *Physical Geodesy*, Freeman (1967); more recent are W. Torge, *Geodesy*, Walter de Gruyter and Co (1980), A. Anderson and A. Cazenave, eds., *Space Geodesy and Geodynamics*, Academic Press (1986), K. Lambeck, *Geophysical Geodesy*, Clarendon Press (1988), and P. Vanicek and N. Christou, eds., *Geoid and its Geophysical Interpretations*, CRC Press (1994). We also quote W.M. Kaula, *Theory of Satellite Geodesy*, Blaisdell, Waltham, Massachusetts (1966). Topics of Chs. 2 and 5 are described in more depth by D.L. Turcotte and G. Schubert, *Geodynamics*, Cambridge University Press (2002). On the transformation properties of the spherical functions under rotations and their relation to irreducible representations of the corresponding groups, mentioned in Sec. 2.1, see A. Edmonts, *Angular Momentum in Quantum Mechanics*, Princeton University Press (1957) or E.P. Wigner, *Group Theory and its Applications to Quantum Mechanics*, Academic Press (1959). The EGM96 gravity model is described in detail, both conceptually and practically, by F. Lemoine et al., The development of the joint NASA GSFC and the National Imagery and Mapping Agency (NIMA) geopotential model EGM96, *NASA/TP-1998-206861*, July 1998. High-degree gravity models of the terrestrial planets (including the Moon) from satellite tracking can be obtained from the Web pages of Planetary Data System maintained at the Washington University, St. Louis (http://pds-geophys.wustl.edu/pds/). Special issues of the Icarus journal are devoted to results of recent space missions to planetary bodies: *Icarus* **139**, Nr. 1, (1999) for Venus (Magellan radar tracking), *Icarus* **155**, Nr. 1, (2002) for Eros (the NEAR/Shoemaker mission), *Icarus* **144**, Nr. 2, (2000) for Mars (the Mars Global Surveyor mission), *Icarus* **150**, Nr. 1, (2001) for the Moon (the Lunar Prospector mission).

Chapter 3

PLANETARY ROTATION

The present and the following chapter are devoted to a complex subject which brings together three different fields: astrometry, geophysics and celestial mechanics. The Earth provides the basic frame of reference in which most of our celestial observations are expressed: the determination of its (variable) rotation with respect to very distant matter (e.g., quasars) is a fundamental task of astrometry, the discipline that determines the position, the distance and the motion of celestial objects. In this way an inertial reference system is provided. This chapter is also an appropriate place to introduce the problem of time and distance measurement in space physics. The Earth's rotation produces an oblateness and, at the same time, is affected by its internal motion, an important topic in geophysics. The presence of other bodies in the solar system also deeply influences the structure and rotation of a planet through tidal phenomena.

In this chapter we shall also develop two idealized models which are of great help in understanding the rotational dynamics of an isolated body. We can rigorously compute the effect of a uniform and slow rotation upon a liquid mass; at the opposite extreme, the rotation of a rigid body shows precession and, possibly, nutation. The effects of external gravitating bodies, in particular tides, are considered in the following chapter.

3.1 Measurements of time and distance

The definition of the second of time uses very accurate frequency standards provided by radio resonances in atomic and nuclear systems: they provide a clock which counts the number $N = T\nu$ of periods elapsed in a time interval T. The fractional accuracy in T is equal to the fractional accuracy in the frequency ν. The main frequency standard currently available and operational

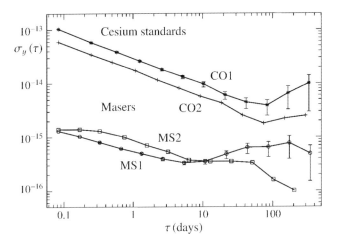

Figure 3.1. Measured Allan deviation $\sigma_y(\tau)$ as function of the integration time τ of frequency standards based upon hydrogen masers (MS1 and MS2) and cesium oscillators (CO1 and CO2), relative to the main frequency standard kept at the National Bureau of Standards (Boulder, USA). The Allan deviation (eq. (19.36)) is a measure of the fractional frequency fluctuations $y(t) = \Delta\nu(t)/\nu$. If $y_\tau(t)$ is the (running) average of $y(t)$ in the interval $(t - \tau/2, t + \tau/2)$, $\sigma_y(\tau) = \langle (y_\tau(t + \tau) - y_\tau(t))^2/2 \rangle$. Long-term drifts have been fitted out for the hydrogen maser signals. These results are obtained in the laboratory comparing several standards; operational H-masers have a worse performance at long integration times, with a minimum in Allan deviation at about $10,000$ s. Data kindly provided by J. Lipa and T. Parker.

is the *hydrogen maser*, a microwave cavity tuned to the hyperfine frequency $\nu = 1,420,405,751.68$ Hz of the ground level of atomic hydrogen due to the spin interaction between the electron and the nucleus. For averaging times of the order of one day its frequency can be stabilized to better than a part in 10^{15} (Fig. 3.1).

It has been found that different atomic frequency standards agree within their present accuracy and are stable over long timescales; in agreement with the equivalence principle, this justifies the choice of the atomic time as the *fundamental time variable*. This variable is identified with the *proper time* of general relativity and provides the geometrical structure of spacetime. Another important time variable – mainly concerning astronomical observations – is the phase $\varphi(t)$ of the Earth with respect to distant stars ("Earth rotation angle"). Conventionally its origin is the intersection of the equatorial plane with the plane of the ecliptic at a fixed epoch ("γ point"; see Fig. 3.3). Instead of the angle φ one often uses the Universal Time

$$\mathrm{UT1}(t) = \varphi(t)/\omega_0, \qquad (3.1)$$

obtained with a nominal and constant Earth rotation frequency ω_0.

3. Planetary rotation

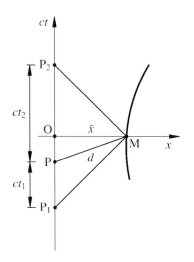

Figure 3.2 The chronometric measure of distance. In the spacetime frame (ct, x), we have $M = (0, \bar{x}), P = (ct_1 - \bar{x}, 0)$. In special relativity the proper length of the hypotenuse MP is the square root of the *difference* of the squares of the time-like and the space-like sides: $d^2 = (\bar{x} - ct_1)^2 - \bar{x}^2 = c^2 t_1 t_2$.

Traditionally the unit of length – the centimetre – was defined, through interferometric measurements, as a given multiple of the wavelength λ_0 of a stable optical line. By transferring the frequency standard from the microwave to the optical band it is possible to measure the frequency c/λ_0 and obtain the velocity of light c (with the dimension cm/s). This transfer over a frequency range of about 7 orders of magnitude, however, is subject to relevant errors; moreover, the stability of the lasers used in interferometric techniques is much worse than that of atomic frequency standards. As a consequence, the *standard of length has been foregone* and the velocity of light c is a conventionally fixed quantity. If $c = 1$, lengths are measured in light seconds; in this book this convention is used only in Ch. 17.

This point of view is quite appropriate to space physics, in which distances are usually measured with the transit times of light or radio pulses, timed with atomic clocks, or by means of the Doppler effect. The latter method uses a radio frequency standard and, by comparing received and transmitted frequencies, determines the relative velocity of the source and the receiver; integration over time then gives the change in distance. In both cases, only the time standard is used (interferometric measurements in space have been planned, but not performed yet).

The theory of relativity provides another argument in support of dropping the standard of length. Space and time are unified in a single geometrical entity, spacetime, a Riemannian manifold with indefinite signature, where an invariant number – the proper time interval ds – is assigned to any infinitesimal displacement from an event x to an event $x + dx$. If dx is time-like, ds is measured with a good clock; for a point moving along a time-like curve the integral $\int ds$ measures the elapsed proper time. $\sqrt{-ds^2}$ is the proper length if the displacement is space-like (see Ch. 17); however, rigid bodies cannot be defined in a generic state of motion with relativistic velocities and there is no fundamental way, independent of frequency standards, to measure spatial distances. In this framework, to use two different units for time intervals and space distances is similar to have in ordinary physics two independent units of distances, one for horizontal distances and one for vertical heights.

Fig. 3.2 shows how to formulate in a relativistic context the measurement of distance with atomic clocks. A short electromagnetic pulse is emitted at the event P_1 on a world line ℓ and received at the event P_2 after reflection by a mirror M; the proper distance between M and an event P on ℓ separated from P_1 and P_2 by the time intervals t_1, t_2 is

$$d = c \sqrt{t_1 t_2} \,. \tag{3.2}$$

Eq. (3.2) accounts for the relativistic corrections due to the lack of simultaneity; d is the ordinary distance in the frame in which P and M are simultaneous.

Time and frequency can be transferred from one point to another by electromagnetic signals. The observed frequency ν_2 depends on the velocity of the source $\mathbf{v}_1$, the emitted frequency ν_1 and the velocity of the observer $\mathbf{v}_2$; in flat spacetime the frequency shift is given by (Sec. 17.5)

$$\frac{\nu_2}{\nu_1} = \left[\frac{1-(v_1/c)^2}{1-(v_2/c)^2}\right]^{1/2} \frac{1 - \mathbf{v}_2 \cdot \hat{\mathbf{k}}/c}{1 - \mathbf{v}_1 \cdot \hat{\mathbf{k}}/c} \,; \tag{3.3}$$

$\hat{\mathbf{k}}$ is the unit wave vector. The linear terms give the ordinary Doppler effect; the square root corresponds to the relativistic transversal Doppler effect, which does not vanish even when the motion is orthogonal to the ray. For a typical planetary velocity of 30 km/s, the relativistic corrections are about a part in 10^8 and easily observable (Problem 3.6). We also have a gravitational shift, which can be interpreted as the change in energy, hence of frequency, that a photon suffers when its potential energy changes. For example, if the source is at an altitude $h \ll R_\oplus$ above the ground, the received photon is blue-shifted by gh/c^2. Note that the distinction between the gravitational shift and the transversal Doppler effect is not invariant: in a freely falling frame there is no gravity acceleration, but the velocities of the two points are different. In interplanetary communications, where the potential and the kinetic energy are generally of the same order of magnitude, the two effects are comparable.

With microwave links it is currently possible to measure the absolute distance of a spacecraft in the solar system with an accuracy of a few meters and its relative velocity with an accuracy of $\sigma_v = c\sigma_y = 3 \times 10^{10} \times 10^{-14}$ cm/s =3 μm/s, or better. This extraordinary precision can be compared to the one when only optical observations and, therefore, only angles between celestial objects were available; their distance D could be deduced from the known value of the Earth radius $R_\oplus$, if the daily parallax $R_\oplus/D$ was measured. The distance of the Moon, which has a parallax of about 1°, was known in antiquity; but for the Sun, with a parallax of about 8.8″, and the planets, things were much more difficult[1]. To achieve the required accuracy it was necessary to know well the law of atmospheric refraction, to use telescopes of good optical and mechanical qualities and to be able to locate precisely the image in the focal plane with a micrometric screw.

3.2 Frames of reference

Traditionally the position of a celestial object on the sky was defined by its celestial coordinates, the declination and the right ascension. The equator and the ecliptic plane trace in the sky two great circles and intersects in two points, the vernal equinox (γ point) and its antipode. The angle between the circles is the *obliquity* ϵ. As the Sun moves relative to the stars, at the equinoxes it lies in the direction orthogonal to the rotation axis of the Earth, so that days and nights have the same duration (Fig. 3.3). The *right ascension* of point P in the sky is the angle between its projection P_0 on the equator and the vernal equinox; the *declination* is the angle P_0P.

The pole of the Earth is not fixed. Its main motion is a rotation around the pole of the ecliptic, at a constant obliquity, with a period of $\simeq 25,770$ y (Sec. 4.1); as a consequence, the equinox moves westward and the celestial coordinates of a star change. The motion is due to the torque exerted by the Moon and the Sun on the oblate Earth (Sec. 4.1). The angular velocity

$$\Omega_{\mathrm{pr}} \simeq 50.29\,''/\mathrm{y} \tag{3.4}$$

is the *precession constant*. There is also a secular decrease $d\epsilon/dt \simeq -0.468\,''/\mathrm{y}$ in the obliquity ϵ due to planetary perturbations, and smaller periodic corrections in the motion of the pole. Due to these and other effects, in traditional astronomy an accurate definition of the celestial coordinates were achieved by referring to the equator and the ecliptic at a conventional time, usually noon, January 1st, 2000; this requires a careful integration of planetary orbits. Large and precise stellar catalogues, in particular FK5, gave the coordinates of each object in this frame, called J2000.0.

New concepts. With the great increase in the accuracy of measurements of angles and distances the previous concepts and definitions have become inad-

---

[1] The measurement of the parallax of Mars and, indirectly, of the Sun, was carried out for the first time in 1672 by J. FLAMSTEED in England and by J. RICHTER – an assistant of G.D. CASSINI in Paris – in Cayenne. The object of the measurement was the displacement of the image of Mars in a few hours due to the rotation of the Earth (see Problem 3.1). By Kepler's third law the value of $GM_\odot$ was then also obtained. Another famous determination of the solar parallax was achieved in 1761 and 1769 by observing transits of Venus across the solar disc. Accurate measurements of the ingress and egress times from different places over the Earth allowed a determination of the solar parallax with an uncertainty of about 5%.

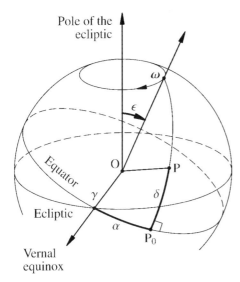

Figure 3.3 The equatorial frame of reference on the celestial sphere: the right ascension α and the declination δ of a direction OP are defined with respect to the instantaneous equator at a given, conventional epoch. The precession moves the axis of rotation on a circular cone, whose axis is the pole of the ecliptic and whose aperture is the obliquity ϵ.

equate and a deeper analysis is required. In I. NEWTON's view absolute space and absolute time are at the foundations of dynamics and provide an absolute standard of rest, in which distant stars do not move. A class of frames of reference connected to each other by a Galilei transformation – the *inertial frames* – is identified, in which the equations of motion have a "privileged" form, with no apparent forces, in particular no Coriolis and centrifugal forces. However, in his seminal work *Die Mechanik in ihrer Entwicklung Historisch-kritisch Dargestellt* (1893; English translation published by Open Court in 1960 as *The Science of Mechanics: A Critical and Historical Account of its Development*), E. MACH pointed out that Newton's formulation is epistemologically inconsistent. He forcefully stressed that there are no absolute kinematical quantities: the motion of a body can be described only in relation to other bodies and the reckoning of time is nothing but the comparison between two time keeping devices. The splitting of a force in a "real" and an "apparent" component is arbitrary. The definition of inertial frames as those with respect to which the acceleration and the rotation of distant matter in the Universe vanish is a vague statement and implies that local forces depend on how such matter moves; for example, rotating distant matter should create locally a Coriolis force, which is cancelled in a suitable rotating frame. According to Mach's view, local dynamics and the motion of the Universe should be deeply related. This criticism of the foundations of dynamics led Einstein to search for a theory valid in any frame, with arbitrary acceleration, and to the construction of General Relativity (Ch. 17); this theory is currently accepted for the motion of macroscopic bodies. In General Relativity both near and distant matter affect the local metric structure, hence the motion of bodies, through the field equations; it is widely regarded as a satisfactory implementation of Mach's view.

Dynamical and kinematical inertial frames. – In both Mach's view and in General Relativity the absolute rotation of a local Cartesian frame can be operationally defined; in the former, by the absence of Coriolis and centrifugal forces in the equations of motion, in the latter by the geometric construction of parallel propagation (Sec. 17.3). This is the *dynamical* method, used in traditional astronomy. In practice, observations of the gravitational motion of bodies near the Earth or in the solar system are compared with the predictions computed on the basis of both known physical forces and apparent forces (mainly Coriolis'), in which the angular velocity ω of the frame is not well known. This unknown vectorial parameter is fitted to the data, together with all other relevant parameters (in particular, the initial conditions), and its value is determined. But, in a Machian way, one can also say that a local frame does not rotate if very distant matter is at rest with respect to it; this is the *kinematical* method. Stars of the Galaxy, the traditional choice, are not appropriate, since they rotate around a normal to the galactic plane with a velocity in principle unknown. With a galactic rotation period of $\approx 10^8$ y, the angular velocity ≈ 30 mas/y is far larger than the current accuracy ≈ 0.5 mas/y in ω. Extragalactic sources must be used. It is now possible to use quasars, point-like objects at cosmological distances, which emit both in the radio and the optical band. Very accurate radio positions can be obtained with VLBI measurements. Quasar coordinates for about 608 objects are included in 1994 catalogue of the International Earth Rotation Service (IERS); 212 of them are called "defining", since they provide the officially adopted celestial frame – called *Celestial Reference System* (CRS) – that has superseded the traditional J2000.0 frame. The remaining objects are "candidates", to be used if they prove to be stable over long time. The kinematical method has the advantage of being independent from the complex planetary dynamics and is currently maintained with an accuracy of 0.1 mas (and even 0.02 mas for relative orientation); for a more extensive discussion see Sec. 20.2. Formally, as the J2000.0 frame, it consists in the assignment, with respect to known sources, of a *conventional* equator and a *conventional* ecliptic; for convenience they are taken near the equator and the ecliptic of the J2000.0 frame, although they have no dynamical significance, nor are they related to the motion of the Earth. A Cartesian set of coordinates is thereby defined, with the origin at the centre of gravity of the solar system. The time variable t is one provided by current atomic standards, or equivalent[2] and is maintained by the Bureau Internationale de l'Heure (BIH).

---

[2]Relativistic effects make exact and global synchronization impossible; however (see end of Sec. 17.5), proper clocks at rest on the rotating geoid are synchronized to $O(v^2/c^2)$. International Atomic Time (TAI) is obtained as a weighted mean of atomic clocks around the world, reduced to the geoid; it is used as the independent variable for geocentric coordinates. For the dynamics of bodies in the solar system it is more appropriate to use the proper time at its barycentre, called TDB; it differs from TAI due to relativistic effects.

In this way one can define and measure the angular velocity ω_U of the Universe relative to the local, dynamically defined inertial frame. The fact that ω_U is small is a fundamental feature of our physical world and it would be interesting to test it. In the accepted cosmological framework of General Relativity ω_U can be defined and is constrained to be much smaller than the Hubble constant $H \simeq 10^{-10}/\text{y} \simeq 2 \times 10^{-2}$ mas/y; the present accuracy in the angular velocity of the CRS (≈ 0.5 mas/y), even taking into account the expected future improvements, is not sufficient to measure it directly. But on a more fundamental level, at which *standard cosmology is not assumed*, it would be interesting to measure ω_U with an accuracy better than H; at present this is not possible, but future astrometric missions, in particular Gaia, should realize this test.

Though the HIPPARCOS satellite (Sec. 20.2) could not observe the IERS "defining" extragalactic objects because of their optical faintness, a correspondence between its catalogue and the IERS celestial reference system has been achieved with a small number of specifically selected objects, namely radio stars that emit both in optical and radio bands. They can thus be observed in both systems and provide their link. The precision of this method is ≈ 0.6 mas at the mean epoch of the HIPPARCOS catalogue and ≈ 0.25 mas/y for the time evolution. Such a calibration of the star positions is very important and allowed determination of local galactic dynamics with an unprecedented precision. An analogous link of the extragalactic frame with the solar system ephemerides of the Jet Propulsion Laboratory has been realized with lunar laser ranging; a precision of ≈ 3 mas is expected for this tie.

The Celestial Reference System defined by quasars must be related to the Earth-bound reference frame; indeed the majority of astronomical and space observations are carried out with ground-based instruments. The Earth, however, is not rigid; in particular, its crust is broken up in tectonic plates which move with respect each other with velocities of the order of cm/y (Sec. 5.4); since a cm corresponds to 0.3 mas on the Earth surface, observations over several years cannot forgo this effect. Fortunately their motion has a simple geometrical characterization: each plate can be regarded as rigid and moves horizontally with respect to another plate with a constant angular velocity around a known pole. One can therefore define a conventional frame, as close as possible to the real Earth, with the origin at its centre of mass, called *Conventional Terrestrial Reference System* (CTRS). For a satisfactory definition, an accurate modelling is required for all the motions on the surface of the Earth in the inertial CRS. They are reviewed below; most of them are discussed elsewhere in this book. In principle, the CTRS orientation in CRS could be described by three independent parameters (e.g., the Euler angles); however, traditionally the Earth rotation is separately described as seen from space and from the Earth itself, using five parameters – generally called *Earth orientation parameters* – grouped three classes:

3. Planetary rotation

- The *celestial coordinates* of the Earth rotation pole change in time because of precession and nutation, and so does the colatitude of a given place. The currently best CRS realization of the rotation pole is the *Celestial Intermediate Pole* (CIP); conventionally, it must take into account all celestial motions of the Earth pole with periods longer than 2 days (shorter variations of the pole position in the inertial space are described as the polar motion in the terrestrial reference system; see also Sec. 3.5). The spatial position of the CIP is well modelled with a $\simeq 0.5$ mas accuracy with the lunisolar (and to a less extent planetary) torques on the Earth equatorial bulge (small additional impacts of atmospheric and oceanic effects are also included). However, the current precession-nutation model still cannot take into account the variable components due to atmospheric, oceanic and Earth internal processes. Differences of the polar CRS position from the model are monitored using accurate tools, in particular VLBI and the precise tracking of Earth satellites in the orbits (like LAGEOS) and of the Moon, and they are reported by the IERS as the *celestial pole offsets*.

- For an accurate Earth-bound reference frame the variation with time of the angular velocity of the Earth $\omega(t) = 2\pi/\mathrm{LOD}$ must be taken into account (LOD is the *length of the day*). The rotation of the Earth suffers a mean deceleration due to the tidal torque exerted by the Moon, with a minor non-tidal component probably due to atmospheric effects; records of solar and lunar eclipses since 700 BC, from the ancient and medieval civilizations of Babylon, China, Europe and the Arab world, yield

$$\frac{d\omega}{dt} \simeq -4.5 \times 10^{-22} \text{ rad/s}^2 \; ; \qquad (3.5)$$

the tidal component, accurately known from the laser ranging to satellites and the Moon, amounts to -6.1×10^{-22} rad/s^2. This mean effect corresponds to a quadratic term in the universal time (3.1). Thus the LOD increases by about 1.7 ms/cy (and 2.3 ms/cy due to tides), a substantial amount when its accumulation over millennia is considered (Problem 3.5). Besides the secular acceleration, the LOD exhibits a complex behaviour with several spectral lines, in particular at the frequencies corresponding to the lunar month, half a lunar month, half a year, a year and ≈ 28 y; part of the power in the first four is due to tides and atmospheric effects. Figure 3.4 illustrates these variations; on this timescale, periodic and irregular terms have larger amplitude than the secular term (3.5). Recent re-evaluation of astronomical observations back to the mid 17th century also revealed large long-term changes of LOD; though irregular, a possible periodic term of ≈ 80 y was suggested. The explicit value of the Universal Time, taking into account these effects, is provided by IERS.

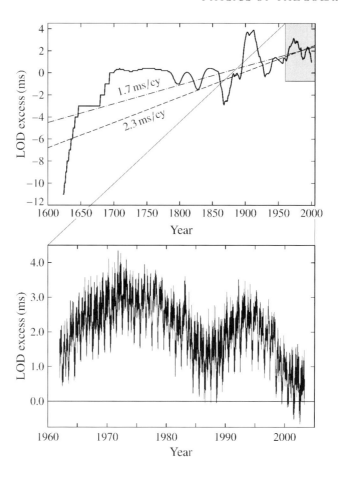

Figure 3.4. The variation of the length of the day (LOD), traditionally expressed as the excess over a mean period, vs time. Top: long-term variations, mainly from observations of stellar occultations by the Moon and solar eclipses; the 1.7 ms/cy and 2.3 ms/cy linear trends are discussed in the text (dashed lines). Bottom: modern, and much more precise, determination of LOD since the late 50s, when atomic standards provided a fundamental scale of time and space techniques become available. Short-period (monthly, semiannual, annual, etc.) and decadal irregularities can be noticed. The secular deceleration of the Earth rotation, dashed, is masked here by these periodic and irregular terms. Data from IERS (http://hpiers.obspm.fr/).

- On the ideal rigid Earth it is convenient to adopt as base the principal axes of inertia; the moments of inertia fulfil $(C - A) \gg (B - A)$ and are constant. In this frame – CTRS – the coordinates of all relevant geodetic stations are available; e_z, the principal axis related to C, is the *International Conventional Pole* (ICP). The position of CIP in the *terrestrial frame* is described by *polar motion*, roughly circular wandering of the rotation axis with main periods of $\simeq 433$ d (Chandler's period; see Sec. 3.6) and one year; as men-

3. Planetary rotation

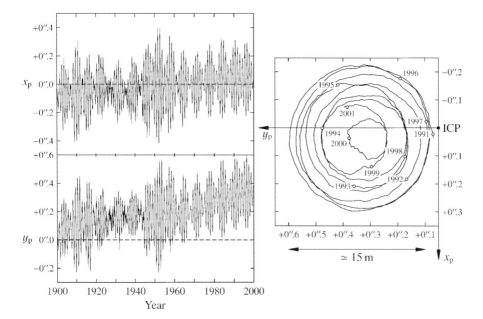

Figure 3.5. The motion of the instantaneous rotation axis in the ideal, body-fixed coordinates (x_p, y_p) near the pole, with the origin at the International Conventional Pole (ICP). The rotation pole moves counterclockwise on a roughly circular path with a (Chandler) period of about 14 months, somewhat longer than the free precession period $\simeq$ 305 d; in addition, there is also evidence of a drift, approximately in the y_p direction. The x_p axis is along the Greenwich reference meridian and the y_p axis is in the direction 90° East. Data since 1962 from the IERS Web page (http://hpiers.obspm.fr/); earlier from J. Vondrák, *Surveys in Geophysics* **20**, 169 (1999).

tioned above, the CIP motion in this frame should conventionally contain all variations with frequencies longer than a day. It depends upon the dynamics of the Earth interior, its atmosphere and oceans; since they cannot be predicted, it is continuously monitored and provided every day (even every 6 h during the last years) by IERS. On the average, the position of the CIP coincides with the figure axis $\mathbf{e}_z$, making just a small angle of about 0.3″; however, a linear drift of the pole of rotation was also found in the data, probably due to horizontal readjustment of material in the present post-glacial rebound (see Fig. 3.5).

Finally, the determination of the actual position of a station must take into account the following effects:

- The velocity of the plate where the station is located relative to the CTRS (Ch. 5).

- Solid Earth tides (Ch. 4).

- Ocean and atmospheric loading, deformations of the crust due to the changing weight of oceans and atmosphere.

- Post-glacial rebound (Sec. 20.3), the residual upward lift of the high-latitude regions due to the decrease of the ice weight at the end of last glaciation.

As far as the first item is concerned, note that, as discussed in Sec. 5.4, the CTRS is realized by solving for the positions and velocities ($\mathbf{r}_0, \mathbf{v}_0$) at a given time t_0 (accelerations are not used so far) of a set of selected geodetic sites on the Earth. This is analogous to the defining quasars of the CRS. The positions and velocities at any other time t depend on the velocity of the plate where the site is and several other contributions from tidal effects, ocean and atmospheric loading, etc; their measured values and careful modelling are used for the fit.

3.3 Slowly rotating bodies

During the remainder of this chapter we quantitatively investigate the rotation of a gravitating body in empty space, neglecting the effect of external forces (to be discussed in Ch. 4). In this section, we determine the shape and the gravitational potential of a slowly rotating body, leading to a small deformation; the following section deals with bodies in fast rotation; in the last two sections we discuss the rotation of a body in the opposite approximation of rigidity. Both these approximations are useful to understand the behaviour of the Earth, an intermediate case.

When the rotation is uniform, the problem is best set in the rotating frame, where it is reduced to the effect of the centrifugal potential energy (per unit mass)

$$U_c = -\frac{1}{2}\omega^2 r^2 \sin^2\theta = \frac{\mu_c GM}{3R}\left(\frac{r}{R}\right)^2 (1 - Y) \ . \tag{3.6}$$

It is expressed in terms of *Helmert's parameter* μ_c (eq. (1.13)) and the spherical harmonic

$$Y = Y_{20}/\sqrt{5} = \frac{1}{2}\left(3\cos^2\theta - 1\right) \tag{3.7}$$

of degree $\ell = 2$ and order $m = 0$. In a spherical harmonics expansion U_c is sum of a vector in the space S_0 corresponding to $\ell = 0$, and a vector in S_2 corresponding to $\ell = 2$. We confine ourselves to slow rotation ($\mu_c \ll 1$), in which the induced deformation is small.

We need the perturbation δU in the gravitational potential due to the rotational deformation produced by the forcing potential U_c. Outside the body it has the general expression (2.17'); it is determined, for each degree ℓ', by the corresponding $2\ell' + 1$ components in the space $S_{\ell'}$. In the linear case we have a linear mapping from the spaces S_0 and S_2 onto $S_{\ell'}$, solely determined by the

3. Planetary rotation

intrinsic properties of the body in the unperturbed state; they do not contain any privileged direction, but can depend only upon scalar quantities (like the elastic coefficients). Such mapping operators, and the matrices which determine them, must therefore be invariant: their numerical values are independent of the choice of the base. Now, it can be shown that an invariant linear mapping from S_ℓ onto $S_{\ell'}$ exists only if $\ell = \ell'$; and it is a simple proportionality relation. This is a particular case of the *Clebsch-Gordan theorem* and hinges upon the crucial fact that the spaces S_ℓ are irreducible under the rotation group. We therefore conclude that δU can have components only in S_0 and S_2, and they are proportional to the corresponding components of U_c.

Outside the body U is harmonic and axially symmetric. Therefore it must be of the form (see eq. (2.21))

$$U = -\frac{GM}{r}\left[1 - J_2\left(\frac{R}{r}\right)^2 Y\right], \qquad (3.8)$$

with the constant J_2 proportional to μ_c. The quadrupole correction is sufficient to describe the effect, and no higher terms are present. The total external potential energy W per unit mass is

$$W = U + U_c = -\frac{GM}{r}\left[1 - J_2\left(\frac{R}{r}\right)^2 Y\right] - \frac{\mu_c GM}{3R}\left(\frac{r}{R}\right)^2 (1-Y). \qquad (3.9)$$

The surface of the body, deformed from R to $R + \delta r$, is an equipotential for W (eq. (1.38)); neglecting non-linear terms in $\delta r/R$ we have

$$\frac{\delta r}{R} + J_2 Y - \mu_c \frac{1-Y}{3} = \text{const}. \qquad (3.10)$$

The constant on the right-hand side is undetermined because the unperturbed radius R has not been defined. We require R to be the mean radius; in other words, the average of $\delta r(\theta)$ over the sphere is to vanish. Since $Y(\theta)$ also has a vanishing average, the constant is $-\mu_c/3$, and the condition becomes

$$\frac{\delta r}{R} + \left(J_2 + \frac{\mu_c}{3}\right)Y = 0. \qquad (3.10')$$

The difference between the equator and poles gives the estimate

$$\epsilon_r = \frac{R_e - R_p}{R} = \frac{1}{2}(3J_2 + \mu_c). \qquad (3.11)$$

The gravity acceleration is almost radial and reads (in modulus)

$$g = \frac{\partial W}{\partial r} = \frac{GM}{r^2} - 3J_2 \frac{GM}{r^2}\left(\frac{R}{r}\right)^2 Y - \frac{2\mu_c}{3}\frac{GM}{R^2}\frac{r}{R}(1-Y)$$

(it is easy to show that the latitudinal component of the acceleration contributes to g with a term $O(\mu_c^2)$). On the surface of the body there is a small change $\delta g(\theta)$ in g

$$\frac{\delta g}{g_0} = -2\frac{\delta r}{R} - 3J_2 Y - \frac{2\mu_c}{3}(1-Y) + O(\mu_c^2), \qquad (3.12)$$

with the reference value $g_0 = GM/R^2$.

Combining eqs. (3.12) and (3.10′) we can eliminate J_2 and obtain the condition

$$\frac{\delta g}{g_0} - \frac{\delta r}{R} = \mu_c \frac{5Y-2}{3}, \qquad (3.13)$$

which refers only to quantities measurable on the ground. Taking the difference between the poles and the equator, we get the oblateness (1.14) as a function of the gravity anomaly

$$\frac{g_p - g_e}{g_0} + \frac{R_e - R_p}{R} = \frac{5\mu_c}{2}. \qquad (3.14)$$

This relation is due to A.C. CLAIRAUT.

Uniform density model. To proceed further and to determine J_2, the internal density profile must be specified. With a *uniform density* (as for a liquid) δU must be harmonic also inside the body, since the unperturbed potential (1.5) takes care of the internal mass; but then the known angular dependence determines the radial dependence. We cannot have a function of the type (2.17), which diverge at the origin; we must use instead the class (2.20) of harmonic functions. Therefore, for $r < R$,

$$U = -\frac{GM}{R}\left[\frac{3R^2 - r^2}{2R^2} - a\left(\frac{r}{R}\right)^2 Y + b\right]. \qquad (3.15)$$

The requirement of continuity with the exterior solution (3.8) at $r = R + \delta r(\theta)$ implies $a = J_2$ and $b = 0$:

$$U = -\frac{GM}{R}\left[\frac{3R^2 - r^2}{2R^2} - J_2\left(\frac{r}{R}\right)^2 Y\right]. \qquad (3.15')$$

Let us note an important consequence of this relation, discovered by I. NEWTON. The last equation shows that the gravitational potential energy inside the body is a quadratic function of the Cartesian coordinates (plus a constant); hence the gravity acceleration is linear. Along the coordinate vectors to the poles and the equator, respectively, we have, therefore, in terms of the surface values g_p and g_e,

$$g = g_p \frac{r}{R_p}, \quad g = g_e \frac{r}{R_e}; \qquad (3.16)$$

3. Planetary rotation

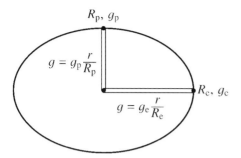

Figure 3.6 The "canals" method due to Newton: one computes the work needed to extract from the centre of the planet a test body along two canals drilled from the pole and from the equator. The work done is the same in the two cases; hence the gravity accelerations are inversely proportional to the radii (eq. (3.16′)).

the work to perform to bring a unit mass from the centre to the surface in the two cases is, respectively, $g_p R_p/2$ and $g_e R_e/2$. On the other hand, the surface being an equipotential, they must be equal (see Fig. 3.6):

$$g_e R_e = g_p R_p . \quad (3.16')$$

Therefore the surface oblateness (1.14) is determined by the gravity anomaly

$$\epsilon = \frac{R_e - R_p}{R} = \frac{g_p - g_e}{g_0} = \frac{5\mu_c}{4} ; \quad (3.17)$$

we have made use of Clairaut's relation (3.14). Using eq. (3.11) we finally get

$$J_2 = \frac{\mu_c}{2} . \quad (3.18)$$

The centrifugal potential energy U_c (3.6) (or, rather, its part with vanishing mean) generates a deformation and a change in the potential δU (the second term in eq. (3.9)); their ratio on the surface is a constant (eq. (2.31))

$$k_s = \frac{3J_2}{\mu_c} = 3\frac{G(C-A)}{\omega^2 R^5} , \quad (3.19)$$

called *secular Love number*. We have just shown that for a liquid planet with uniform density this number is 3/2. For a body of finite rigidity μ (see eq. (1.51)) and uniform density ϱ, Love's number is generalized to

$$k_s = \frac{3}{2}\frac{1}{1 + 19\mu R/2GM\varrho} , \quad (3.20)$$

which becomes $k_s \approx 3GM\varrho/(19\mu R)$ in the high rigidity case, when the deformation is small. The value of the secular Love number for the Earth, Jupiter and Saturn may be computed from the Table 3.1 and eq. (3.19); for the Earth $k_s \simeq 0.94$. It is interesting to note the difference between k_s and the Love number $k_2 \simeq 0.3$ for quadrupole tidal deformations (eq. (4.23)); clearly the Earth responds differently to different stresses (see also Sec. 3.6). Several hypotheses have been made, but a detailed discussion is beyond the scope of this book.

Table 3.1. Parameters of interior models for the Earth, Jupiter and Saturn. The first three columns give the measured values of J_2, the zonal quadrupole coefficient, μ_c, the Helmert parameter, and ϵ, the surface oblateness. In the fourth column, the fact that $\Lambda_{20} = J_2/\mu_c$ is less than 0.5 signals the presence of a core. Finally, in the last column, $\epsilon_t = (3J_2 + \mu_c)/2$ is the theoretical value of the oblateness for a generic model in terms of the measured values of J_2 and μ_c.

Planet	J_2 (in 10^{-3})	μ_c (in 10^{-3})	ϵ (in 10^{-3})	Λ_{20}	ϵ_T (in 10^{-3})
Earth	1.08	3.45	3.35	0.314	3.35
Jupiter	14.7	89.0	64.9	0.165	66.6
Saturn	16.3	155.0	98.0	0.105	101.9

Comparison with observations. The separate measurements of the three independent quantities ϵ, J_2 and $(1 - g_e/g_p)$ and the comparison with their ideal values as functions of μ_c yield valuable information about the internal structure of a cosmic body. A core, always denser than the mantle, is expected to give, for a given μ_c, a smaller J_2 (Table 3.1). The change in gravity near the equator combines three different effects (eq. (3.12)): a larger distance from the centre (negative), more mass near the equator (positive) and the centrifugal force (negative). A denser core mainly attenuates the second and a smaller gravity anomaly is expected. Finally, the surface oblateness is a superficial phenomenon and should not be greatly affected by the core.

The experimental and the theoretical values of these parameters are shown in Table 3.1 for the Earth, Jupiter and Saturn. We note that $\Lambda_{20} = J_2/\mu_c$ significantly deviates from the value 0.5 predicted by the uniform density model (eq. (3.18)), a clear signature of a dense core. On the other hand, the general constraint (3.11) between the surface oblateness ϵ, Helmert's parameter μ_c and J_2 – for an arbitrary density distribution – is fairly well satisfied. A remarkable agreement between theory and experiment holds also for the Clairaut's relation (3.14): for Earth this formula yields $1 - (g_e/g_p) = 0.00528$, while the observed value is 0.00529 (the simplified formula (3.17) for the uniform density model would give 0.00431). These results indirectly suggest that the differential rotation in the interior of these planets, especially in the Earth, is not significant.

3.4 * Equilibrium shapes of bodies in fast rotation

When the parameter μ_c (and, as a consequence, also ϵ and J_2) is not much smaller than unity, the deformation is large. We restrict our analysis to *ellipsoidal* shapes of equilibrium of fluid bodies with uniform density (namely, self-gravitating incompressible "liquids"), but the qualitative results are more general. In this case, an obvious solution of Laplace's equation for the gravitational potential energy (per unit mass) U which generalizes (3.15') is a quadratic function of the coordinates:

$$U = \pi G\varrho (A_1 x^2 + A_2 y^2 + A_3 z^2) + \text{const}. \qquad (3.21)$$

(x, y, z) are Cartesian coordinates along the principal axes of inertia, with the origin at the centre of the ellipsoid; A_1, A_2, A_3 are dimensionless constants. Following a procedure similar to the one used in Sec. 3.3, it can be shown that this solution fulfils the continuity requirement for the potential energy at the surface provided that

$$A_1 = abc \int_0^\infty du\, (a^2+u)^{-3/2}(b^2+u)^{-1/2}(c^2+u)^{-1/2}, \qquad (3.22a)$$

$$A_2 = abc \int_0^\infty du\, (a^2+u)^{-1/2}(b^2+u)^{-3/2}(c^2+u)^{-1/2}, \qquad (3.22b)$$

$$A_3 = abc \int_0^\infty du\, (a^2+u)^{-1/2}(b^2+u)^{-1/2}(c^2+u)^{-3/2}; \qquad (3.22c)$$

a, b and c are the semiaxes of the ellipsoid. While in general these integrals can be expressed in terms of elliptic integrals, in particular cases they can be evaluated analytically. For instance, if $a = b = c$, $A_1 = A_2 = A_3 = 2/3$, corresponding to a sphere (eq. (1.5)); if $a = b > c$ we have

$$A_1 = A_2 = \frac{\sqrt{1-e^2}}{e^3}\left(\arcsin e - e\sqrt{1-e^2}\right), \qquad (3.23a)$$

$$A_3 = \frac{2}{e^3}\left(e - \sqrt{1-e^2}\arcsin e\right), \qquad (3.23b)$$

where $e = (1 - c^2/a^2)^{1/2}$ is the eccentricity of the oblate, axisymmetric ellipsoid. From eq. (3.21) the gravity components are still linear functions of the coordinates;

$$g_x = -\frac{\partial U}{\partial x} = -2\pi G\varrho A_1 x, \quad \text{etc.} \qquad (3.24)$$

Maclaurin spheroids. To obtain the equilibrium shape we can apply again the method of Newton's canals (Fig. 3.6). The work done against gravity to bring the unit mass from the centre to the surface along the three axes is $\pi G\varrho A_1 a^2$, $\pi G\varrho A_2 b^2$ and $\pi G\varrho A_3 c^2$, respectively. If the ellipsoid spins about the shortest axis c, for the first and second axis the (negative) contribution of the centrifugal force must be added; since the surface is equipotential, the work is the same in the three cases:

$$\pi G\varrho A_1 a^2 - \omega^2 a^2/2 = \pi G\varrho A_2 b^2 - \omega^2 b^2/2 = \pi G\varrho A_3 c^2. \qquad (3.25)$$

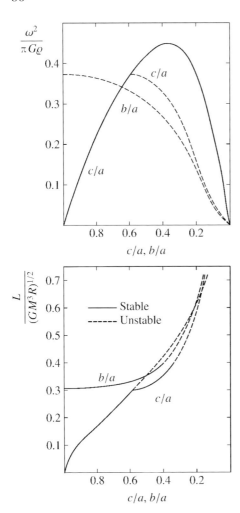

Figure 3.7 The figure shows the variation of the axial ratios c/a and b/a as a function of the dimensionless parameter $\omega^2/\pi G\varrho$ for Maclaurin spheroids (axisymmetric, $a = b > c$; solid line) and for Jacobi ellipsoids (triaxial, $a > b > c$; dashed lines); c is along the spin axis. These ellipsoidal shapes are equilibrium solutions for a self-gravitating, spinning and "liquid" body.

Figure 3.8 A plot of the angular momentum of rotation L versus the axial ratios c/a and b/a for Maclaurin spheroids (for which b/a is always 1) and for Jacobi ellipsoids (whose function $L(c/a)$ bifurcates at $c/a \simeq 0.583$ from the Maclaurin curve). Solid and dashed lines correspond here to stable and unstable equilibrium figures of both types.

For $a = b$, using eqs. (3.23) and $c = a(1 - e^2)^{1/2}$, we obtain a relationship between ω^2 and e (or, equivalently, between ω^2 and c/a):

$$\frac{\omega^2}{\pi G\varrho} = \frac{2\sqrt{1-e^2}}{e^3}\left[\left(3 - 2e^2\right)\arcsin e - 3e\sqrt{1-e^2}\right]. \tag{3.26}$$

The dependence of ω^2 on c/a is illustrated in Fig. 3.7 (solid line), while Fig. 3.8 shows the dependence on c/a and b/a of the rotation angular momentum L. Notice that for slow rotation ($\mu_c \ll 1$) $\omega^2/\pi G\varrho = 4\mu_c/3$, $\epsilon = e^2/2$ and we recover eq. (3.17).

The angular momentum L is given by

$$L = \sqrt{GM^3R}\,\frac{\sqrt{3}}{5}\left(\frac{a}{R}\right)^2\omega, \tag{3.27}$$

where $R = (abc)^{1/3}$ is the radius of the sphere with the same volume as the spheroid. We notice that, although L increases monotonically with the flattening (owing to the rapidly growing moment of inertia), $\omega^2/\pi G\varrho$ reaches the maximum value of 0.449 for $c/a \simeq 0.367$ (namely $e \simeq 0.930$), solution of the equation $d(\omega^2)/de = 0$, i.e.

$$9 - 2e^2 - \frac{9 - 8e^2}{e\sqrt{1-e^2}} \arcsin e = 0 . \tag{3.28}$$

The fact that in the limit $\omega \to 0$ we obtain not only a quasi-spherical equilibrium solution, but also a very flattened one, can intuitively be understood referring to Newton's method. In the flattened spheroidal solution, gravity is much stronger along the polar axis, and this compensates both for the fact that the equatorial axis is longer and for the centrifugal term. Linear stability analysis shows that triaxial deformations of one of these axisymmetric solutions (called *Maclaurin spheroids*) become unstable at $e \simeq 0.813$ (i.e., $c/a \simeq 0.583$), corresponding to $\omega^2/\pi G\varrho \simeq 0.374$ and $L/(GM^3R)^{1/2} \simeq 0.304$. Since this instability lies on the ascending branch of the ω^2 versus e relationship, for each value of ω we have only one stable Maclaurin spheroid.

Jacobi ellipsoids. The fact that eq. (3.25) is symmetrical with respect to the exchange $a \leftrightarrow b$ does not mean that $a = b$ is a necessary condition for equilibrium. In fact, as pointed out by K.G.J. JACOBI in 1834, if we assume $a \neq b \neq c$ we obtain from eqs. (3.25) the ω versus shape relationship

$$\frac{\omega^2}{\pi G\varrho} = 2\frac{A_1 a^2 - A_2 b^2}{a^2 - b^2} , \tag{3.29}$$

together with a geometric condition, which yields b/a if c/a is known, or vice versa:

$$\frac{a^2 b^2}{a^2 - b^2}(A_2 - A_1) = c^2 A_3 . \tag{3.30}$$

This is a sequence of solutions with $a \geq b > c$ (the so-called *Jacobi ellipsoids*), which "bifurcates" from the Maclaurin sequence ($a = b$) at the point where the latter become unstable. Figs. 3.7 and 3.8 show also how the axial ratios c/a and b/a vary as a function of ω^2 and L along the Jacobi sequence. The existence of triaxial equilibrium shapes can be understood in the same way as that of slowly rotating, very flattened ones: for a cigar-shaped body, gravity is much stronger along the b and c axes, compensating for the smaller centrifugal contribution. Along the Jacobi sequence, as the axial ratios decrease and the angular momentum grows, the spin rate becomes slower and slower, owing to the rapid increase of the moment of inertia.

The linear stability analysis can also be applied to Jacobi ellipsoids, for example, to a small, "pear-shaped" deformation. As shown by H. POINCARÉ in 1885, such deformations are unstable when $b/a < 0.432$, $c/a < 0.345$,

$\omega^2/\pi G\varrho < 0.284$, $L/(GM^3R)^{1/2} > 0.390$. Thus no stable ellipsoidal equilibrium shape is possible for a homogeneous, self-gravitating body when its angular momentum exceeds the threshold $0.390\,(GM^3R)^{1/2}$; for higher values only binary or multiple configurations, with a fraction of the angular momentum accounted for the orbital motion, may be consistent with equilibrium. Therefore, when a nearly homogeneous self-gravitating body receives an angular momentum input (e.g., by an off-centre impact) such that the threshold is exceeded, fission into two or more components is the expected outcome. We shall discuss in Ch. 14 some applications of these results to the origin of some solar system bodies. It is important to note that it is possible to waive the assumption of uniform density and to analyze more general cases, in particular when the isodensity surfaces are still ellipsoids, but with variable eccentricity, and when hydrostatic equilibrium with a polytropic law (eq. (1.24)) holds. The same qualitative features (transition from axisymmetric to triaxial solutions, fission instability, etc.) have been found. The arising of triaxial shapes in a rotating collapsing body is well known in astrophysics, e.g., in the formation of galaxies, where barred and other triaxial structures are often seen.

3.5 Rigid-body dynamics: free precession

Were the Earth rigid, its inertia tensor would be constant and define with its eigenvectors a body-fixed basis. This is a simple, but very useful model and a prerequisite for further generalizations; it will also be used in relation to the rotation of small solar system bodies (Ch. 14). As discussed in Sec. 3.2, in fact the real Earth is not a rigid body, due to the complex structure of its interior, the oceans and the atmosphere. A more refined analysis, outside our scope, is needed, relying not only on theoretical models, but also on observationally adjusted parameters.

In the body-fixed frame Euler's equations for the angular velocity ω of a rigid body read

$$A\frac{d\omega_x}{dt} + (C - B)\omega_y\omega_z = \Gamma_x, \qquad (3.31a)$$

$$B\frac{d\omega_y}{dt} + (A - C)\omega_z\omega_x = \Gamma_y, \qquad (3.31b)$$

$$C\frac{d\omega_z}{dt} + (B - A)\omega_x\omega_y = \Gamma_z \qquad (3.31c)$$

(A, B and C are the eigenvalues of the inertia tensor). The quadratic terms in the angular velocity are due to the centrifugal force. When the external torque Γ vanishes there are three fundamental solutions in which only one of the three components of the angular velocity differs from zero and is constant in time. This shows that a body is free to rotate around a principal axis of inertia; however, the rotation around the axis corresponding to the middle moment

3. Planetary rotation

of inertia is unstable (Problem 3.10). The general solution can be expressed analytically in terms of Jacobi elliptic functions; but its complexity does not bring qualitatively new effects, compared to the simple case discussed below; moreover, the effect of triaxiality, in which $A \neq B$, is small for planets and their satellites.

Free precession: description in the body-fixed frame. When the body is axially symmetric ($A = B$), it is termed *oblate* if $C > A$ and *prolate* if $C < A$; ω_z is then constant and the other two components fulfil

$$A \frac{d\omega_x}{dt} + (C - A)\omega_y \omega_z = 0, \quad (3.32a)$$

$$A \frac{d\omega_y}{dt} + (A - C)\omega_x \omega_z = 0. \quad (3.32b)$$

One can easily see that both ω_x and ω_y satisfy the equation of a harmonic oscillator; it is convenient at this point to introduce the dimensionless complex variable

$$m = (\omega_x + i\omega_y)/\omega_z. \quad (3.33)$$

The solution

$$m = m_0 \exp(i\sigma t) \quad (3.34)$$

of eq. (3.32) describes a uniform rotation in the (ω_x, ω_y) plane with the *free precession frequency* $\sigma = \omega_z (C - A)/A$ and arbitrary complex amplitude m_0; it is easily seen that the real amplitude is given in terms of the conserved kinetic energy T and the angular momentum L by

$$|m_0| = \sqrt{\frac{TC - L^2}{A(C - A)}}.$$

The radius $(\omega_x^2 + \omega_y^2)^{1/2}$ is constant, and, therefore, so is ω. The axis of rotation describes a circular cone about the axis of symmetry z; the motion is prograde in the usual case of an oblate body. In the case of the Earth, the aperture of the cone is about $0.2''$ and $|m_0| \approx 10^{-6}$, corresponding to about 10 m in distance at the poles (see Sec. 3.2 and Fig. 3.5). From the expression (2.31) of J_2 and the value

$$A = 0.329 \, MR^2 \quad (3.35)$$

we get

$$\frac{C - A}{A} = \frac{J_2}{0.329} \approx \frac{1}{305},$$

corresponding to a period $2\pi/\sigma_E \approx 305$ days. For the Earth the frequency σ_E is usually termed *Euler frequency*, after L. EULER for his pioneering work in this field in 1765.

Free precession: description in the inertial-fixed frame. The motion of an axially symmetric body in an inertial frame can be interpreted with a geometrical construction due to L. POINSOT (see Fig. 3.9). Let $\mathbf{e}_x, \mathbf{e}_y$ (determined to within a rotation) and $\mathbf{e}_z$ (the symmetry axis) be the unit vectors along the principal axes of the body. The (constant) angular momentum

$$\mathbf{L} = A\,\boldsymbol{\omega} + (C - A)\,\omega_z \mathbf{e}_z \qquad (3.36)$$

is the sum of two vectors of constant size and separated by a constant angle. Hence the vectors $\boldsymbol{\omega}$ and $\mathbf{e}_z$ make a constant angle with $\mathbf{L}$ and describe two cones (called a and b) having $\mathbf{L}$ as axis, and semi-apertures α and β determined by (see eqs. (3.33) and (3.34))

$$\cos(\alpha + \beta) = \omega_z/\omega = (1 + |m_0|^2)^{-1/2}, \qquad (3.37a)$$

$$\cos\beta = L_z/L = C\omega_z\,[A^2\omega^2 + (C^2 - A^2)\,\omega_z^2]^{-1/2}. \qquad (3.37b)$$

If $C > A$, the vectors $\boldsymbol{\omega}$ and $\mathbf{e}_z$ lie on opposite sides of $\mathbf{L}$.

In its motion the body – on which the z-axis is fixed – rotates around the instantaneous axis of rotation $\boldsymbol{\omega}$. The cone c, which has $\mathbf{e}_z$ as axis and contains $\boldsymbol{\omega}$, rolls without skidding around a: in fact, this motion maintains the required congruence of the triangle and consists of successive infinitesimal rotations around the instantaneous axis $\boldsymbol{\omega}$. This axis, a generatrix common to a and c, rotates around $\mathbf{L}$ in the same sense as the main rotation. Seen from the body, instead, the cone c is fixed and the cone a rolls on it, without skidding, with the angular velocity σ, dragging around the angular momentum and the rotation axis. In the usual case $C > A$ this motion occurs in the same sense as the main rotation; when $C < A$ (prolate body) the sense is opposite.

Writing

$$A = C\,(1 - \delta), \qquad (3.38)$$

for a small flattening ($|\delta| \ll 1$) eqs. (3.37) reduce to

$$\cos\beta = \frac{\omega_z}{\omega}\left[1 + \delta\left(1 - \frac{\omega_z^2}{\omega^2}\right) + O(\delta^2)\right], \qquad (3.39a)$$

$$\alpha = \delta \sin\beta \cos\beta\,[1 + O(\delta)]\,; \qquad (3.39b)$$

α is small ($\propto \delta$), while β is arbitrary. This shows that the rotations involved in the problem fall into two classes: the fast class, corresponding to the rolling of the big cone c on the small cone a (approximately the daily rotation), and the slow class, corresponding to the slow drift of the axis of the cone a due to its (fast) rolling on c. In the inertial frame the symmetry axis z rotates with the velocity ω around the axis $\boldsymbol{\omega}$, almost aligned along the fixed $\mathbf{L}$; such axis precesses around $\mathbf{L}$ with the velocity σ. In the body frame the plane $(\mathbf{L}, \boldsymbol{\omega})$ rotates around z with angular velocity σ.

3. Planetary rotation

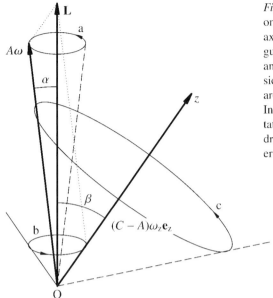

Figure 3.9 Poinsot's cones: the geometrical disposition of the symmetry axis, the angular momentum and the angular velocity for an oblate body. In an inertial frame the parallelogram with sides $A\omega$ and $(C-A)\omega_z \mathbf{e}_z$ rotates rigidly around its diagonal $\mathbf{L}$, fixed in space. In a body-fixed frame the cone c rotates without skidding along the cone a, dragging around the principal axis of inertia.

In the case of the Earth, in addition, the amplitude of this motion

$$|m_0| = \beta + O(\delta) \simeq 0.2'' \simeq 10^{-6} \tag{3.40}$$

is small as well, and $\alpha \simeq \beta\delta \simeq 3 \times 10^{-9} \ll \beta$ is even smaller. The axis of instantaneous rotation, as seen by an inertial observer, describes in about a day a small circle near the pole with radius $3 \times 10^{-9} R \simeq 2$ cm (R is the Earth radius).

Poinsot's geometrical construction also allows recovering the relevant angular velocities. Consider the motion of the interceptions Z and R of the symmetry axis and the rotation axis with the unit sphere, taking $\mathbf{L}$ as a pole. When $|\delta| \ll 1$ the motion is planar. In the inertial frame, since Z, R and $\mathbf{L}$ are coplanar, their displacements around $\mathbf{L}$ read $dZ = \beta\omega' dt$, $dR = \alpha\omega' dt$, say; also, $dZ = (\alpha + \beta)\omega dt$. In the body fixed frame the velocity of the free precession is given by $dR = (\alpha + \beta)\sigma dt$. Hence, taking ratios,

$$\sigma = \omega\alpha/\beta = \omega\delta \tag{3.41}$$

as derived above. The angular velocity ω' in the inertial frame is

$$\omega' = \omega(\alpha + \beta)/\beta = \omega(1 + \delta). \tag{3.41'}$$

Excitations. Perturbing torques (sometimes called "excitations") and/or lack of rigidity modify this simple model in a complex way; to describe them, one still has the choice between inertial and body-fixed frame. The first is suitable with external torques (e.g., to deal with precession and nutation); the second is useful for internal forces and displacements (e.g., to deal with polar motion). Euler's equations (3.31) in one of the two frames are used to calculate the change $\delta\omega$ in the angular velocity; we show below an example using the body-fixed frame (eqs. (3.47)). To illustrate the appropriate choice we start with simple kinematical considerations.

The body-fixed basis rotates with the instantaneous velocity $\omega(t)$, so that

$$\frac{d\mathbf{e}_x}{dt} = \omega \times \mathbf{e}_x, \quad \text{etc.} \tag{3.42}$$

Consider a perturbation $\delta\omega(t)$ of the equilibrium rotational state in which $\omega_0 = \omega_0 \mathbf{e}_z$ is constant (we have seen that this is a very good approximation for the Earth); the corrections of the body-fixed components

$$\delta\omega_x = \delta\omega \cdot \mathbf{e}_x + \omega_0 \mathbf{e}_z \cdot \delta\mathbf{e}_x, \quad \text{etc.}$$

now depend also on the changes of the body-fixed basis

$$\frac{d\,\delta\mathbf{e}_x}{dt} - \omega_0 \mathbf{e}_z \times \delta\mathbf{e}_x = \delta\omega \times \mathbf{e}_x, \quad \text{etc.} \tag{3.43}$$

This equation can be formally solved introducing the periodic matrix R which in a time t rotates vectors by an angle $\omega_0 t$ around $\mathbf{e}_z$ (this obviously describes the unperturbed motion of the body axes); it fulfils

$$\frac{d\mathsf{R}}{dt} - \omega_0 \mathbf{e}_z \times \mathsf{R} = 0.$$

The operation $\delta\mathbf{e}_x \to \mathsf{R}^{-1} \cdot \delta\mathbf{e}_x = \delta\mathbf{a}_x$ "unwinds" the rotation; using $\delta\mathbf{e}_x = \mathsf{R} \cdot \delta\mathbf{a}_x$ we have from (3.43)

$$\frac{d\,\delta\mathbf{a}_x}{dt} = \mathsf{R}^{-1} \cdot \delta\omega \times \mathbf{e}_x, \quad \delta\omega_x = \delta\omega \cdot \mathbf{e}_x + \omega_0 \mathbf{e}_z \cdot \mathsf{R} \cdot \delta\mathbf{a}_x, \quad \text{etc.}$$

If the perturbation $\delta\omega(t)$ has a frequency $\Omega \ll \omega_0$ in the inertial frame (as in the case of lunisolar precession), in the quantity $\delta\mathbf{a}_x$ only the frequencies Ω and $2\omega_0 \pm \Omega$ appear. Hence, $\delta\omega_x$ has the frequency components $\omega_0 \pm \Omega$, $2\omega_0 \pm \Omega$ and $3\omega_0 \pm \Omega$. The lunisolar precession appears in the body-fixed frame as a side line of the main rotational frequency and its two next harmonics. Vice versa, a low-frequency component $\sigma \ll \omega_0$ (such as the free oscillation discussed above) in the body-fixed frame is transferred in the inertial frame to side-bands of ω_0, $2\omega_0$ and $3\omega_0$. Thus the free pole motion appears in the inertial frame as a spectral line at the annual frequency bandwidth around ω_0. This result corresponds to eq. (3.41'), where it was obtained geometrically.

Rotation of minor bodies. When the δ parameter is not small, or the body is triaxial, the difference between the proper rotation frequency and the free

3. Planetary rotation

motion of the pole (in the body-fixed frame) may not be as large as for the Earth. Then the $\Omega = \sigma$ sidebands around the proper frequency ω_0, by which the rotation manifests itself in the inertial frame, may produce quite different frequencies. Such situation was observed for comets (e.g., P/Halley) and some small asteroids (e.g., Toutatis) of irregular shape, for which the rotational motion may not be in the state of lowest energy (i.e. around the principal axis of the inertia tensor; see also Ch. 14).

3.6 * The rotation of the Earth and its interior structure

In spite of its simplicity, the previous model of polar motion gives a remarkably good qualitative picture of what happens; but, due the imperfect rigidity of the Earth, this motion, called *Chandler's wobble*, is not quite circular, nor periodic. Its approximate period, about 433 days (Chandler's period), is *greater* than 305 days. Refinements due to oceanic effects and core-mantle interaction are also needed to explain its exact value; there are also other Fourier components, in particular one with a period of a year, produced by seasonal effects in the atmosphere and in the oceans. We now present a simple model to describe the effect of lack of rigidity on free polar motion. This is an illuminating example of how rotational dynamics in the planetary system can provide deep insights into their interior structure.

Chandler's wobble in a non-rigid Earth. To deal with the rotational dynamics of a non-rigid body, Euler's equations (3.31) must be generalized. In this case there is nothing to anchor a body-fixed frame to, and one must deal with a degree of arbitrariness. With an orthonormal base e_i at the centre of mass, a frame S is defined, and so are the coordinates $r^i = \mathbf{r} \cdot \mathbf{e}_i$ of a point of the body. In this frame, let $I_{ij} = \int dm\,(r^2 \delta_{ij} - r_i r_j)$ be the inertia tensor (eq. (2.26)), of course, in general time dependent. If ω is the (inertial) angular velocity of the base, under a torque $\boldsymbol{\Gamma}$ Euler's equations read

$$\frac{d\mathbf{L}}{dt} + \omega \times \mathbf{L} = \boldsymbol{\Gamma} . \tag{3.44}$$

The angular momentum $\mathbf{L}$ of the body

$$\mathbf{L} = \mathbf{I} \cdot \omega + \mathbf{h} \qquad \left(\mathbf{h} = \int dm\,\mathbf{r} \times \mathbf{v}\right) \tag{3.45}$$

is the sum of two terms: the first corresponds to the case in which all its parts are at rest in the frame S; $\mathbf{h}$, the second, takes into account the motion of its elements, of mass dm and relative velocity $\mathbf{v}$, with respect to S. This formulation was originally given by J. LIOUVILLE in 1858 and (3.44) is usually called *Liouville equation*. To endow this abstract formalism with a physical content, a criterion for the choice of the frame is needed. Following F.F. TISSERAND, one may require $\mathbf{h} = 0$, so that in the course of time S continuously adjusts itself

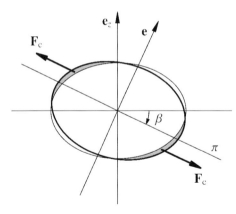

Figure 3.10 In an elastic body, due to the (small) angle β between the rotation axis **e** and the symmetry axis $\mathbf{e}_z$ axes, the equatorial bulge is displaced towards the plane π orthogonal to **e**, as shown by the shaded areas. Consider the torque Γ that the centrifugal force $\mathbf{F}_c$, acting on the displaced matter, exerts on the body. If this matter is symmetric around π, $\Gamma = 0$; but, before the displacement, when it lies in the (x, y) plane, there is a torque which tends to increase β. Problem 3.14, a simple application of gyroscope theory, shows that the displacement brings about an effective reduction in the oblateness.

to the internal motions to reproduce the rigid body expression of the angular momentum; the corresponding $\mathbf{e}_i$ are called *Tisserand mean axes*. But for the Earth we do have a well defined frame, based on geodetic measurements on the crust, in which **h** can account also for the kinematics of the mantle; this is our choice.

Consider now the unperturbed state of the Earth in which the inertia tensor $\mathsf{I}_0 = \mathrm{diag}(A, A, C)$ is diagonal and the rotation is uniform, with angular velocity ω_0 around $\mathbf{e}_3$; one would like to include small perturbations due to internal motions and an external torque Γ. The inertia tensor becomes

$$\mathsf{I} = \mathsf{I}_0 + \begin{pmatrix} c_{11} & c_{12} & c_{13} \\ c_{21} & c_{22} & c_{23} \\ c_{31} & c_{32} & c_{33} \end{pmatrix} \quad \left(|c_{ij}| \ll A, C\right), \quad (3.46)$$

while the angular velocity can be written as $\omega = \omega_0 (\mathbf{e}_3 + \mathbf{m})$, with $|\mathbf{m}| = O(|c_{ij}|) \ll 1$. m_1 and m_2 represent polar motion, while m_3 gives the change in the LOD. Similarly to what was done with eq. (3.33), polar motion is best described with the complex variables $m_+ = m_1 + i m_2$, $h_+ = h_1 + i h_2$, $\Gamma_+ = \Gamma_1 + i \Gamma_2$ and $c_+ = c_{13} + i c_{23}$. Liouville's equation (3.44) give, to first order:

$$\omega_0 A \frac{dm_+}{dt} - i \omega_0^2 (C - A) m_+ = \Gamma_+ - \left(\frac{d}{dt} + i\omega_0\right)(h_+ + \omega_0 c_+), \quad (3.47a)$$

$$\omega_0 C \frac{dm_3}{dt} = \Gamma_3 - \frac{d}{dt}(h_3 + \omega_0 c_{33}). \quad (3.47b)$$

In the linear approximation polar motion and LOD are decoupled. These formal global equations can be used to study the response of a non-rigid body to a small external torque and to a rotation; but, of course, they must be supplemented with a description of its inner dynamics and mass redistribution, in particular, the change in the centrifugal force due to a change in ω.

Rotational deformation.– Our main interest is the effect of lack of rigidity on Chandler's period, in particular, how the **m**-dependent part in c_+ affects polar

3. Planetary rotation

motion (it turns out that the LOD is not affected). As shown in Fig. 3.10, when there is an angular separation β between the rotation axis $\mathbf{e}$ and the symmetry axis $\mathbf{e}_z$, in an elastic body the centrifugal bulge is slightly displaced; we concentrate on the correction c_{ij} so produced in the inertia tensor I, neglecting the relative angular momentum $\mathbf{h}$ (and, of course, the external torque $\boldsymbol{\Gamma}$). Rather than a detailed analysis of the elastic response of the interior, we just note that I, through MacCullagh's formula (2.28), determines the total external quadrupole gravitational potential $U_{T,2}$; but, on turn, this is related, by Love's number k_2 of degree 2, to the forcing centrifugal potential $U_c = -\omega^2 r^2 \sin^2 \theta/2$ (eq. (4.23)):

$$U_{T,2}(\mathbf{r}) = k_2 \left(\frac{R}{r}\right)^5 U_{c,2}(\mathbf{r}) . \qquad (3.48)$$

where

$$U_{c,2}(\mathbf{r}) = \frac{1}{6}\left[3(\boldsymbol{\omega}\cdot\mathbf{r})^2 - \omega^2 r^2\right] \qquad (3.49)$$

is the quadrupole part of U_c, the only one relevant in this context. To deal with the correction to eq. (3.48) due to the change in the direction of the axis, in $\delta\boldsymbol{\omega} = \omega_0 \mathbf{m}$ we set $m_3 = 0$ and note that

$$\delta U_{c,2} = \omega_0^2 z (xm_1 + ym_2) . \qquad (3.50)$$

From MacCullagh's formula we get

$$c_+ = -k_2 \frac{R^5 \omega_0^2}{3G} m_+ , \qquad (3.51)$$

the needed correction to the inertia tensor. With $\Gamma_+ = 0$ and $h_+ = 0$, Eq. (3.47a) acquires the simple form

$$\frac{dm_+}{dt} - i\sigma_c m_+ = 0 \quad \left(\sigma_C = \sigma_E \frac{1 - \frac{k_2}{k_S}}{1 + \frac{k_2}{k_S}\frac{\sigma_E}{\omega_0}}\right). \qquad (3.52)$$

k_S is the secular Love number from (3.19) and σ_E is the rigid-Earth Euler frequency of the free wobble (eq. (3.34)). The obvious solution $m_+ = m_0 \exp(i\sigma_c t)$ shows that σ_C is the free wobble frequency corrected for lack of rigidity; it reduces to σ_E when $k_2 = 0$. As suggested by Fig. 3.10 and Problem 3.14, elasticity decreases the frequency. For the Earth, with a reasonable value of the ratio k_2/k_S, $2\pi/\sigma_C \simeq 441$ days, much closer to the observed 433 d than the Eulerian value of 305 days: a striking confirmation of this simple theoretical model. Further refinements include the oceans and the atmosphere, which contribute a term h_+ depending on $\mathbf{m}$ and the interaction between core and mantle. In general, polar motion is affected by differential rotation in the interior of the planet.

Free-wobble dissipation. The energy of a precessing body is higher than that of a pure rotation around the principal axis $\mathbf{e}_z$ of the inertia tensor. Since the Earth has a non-vanishing viscosity, it can be expected that, in absence of energy sources, the wobble will eventually dissipate away (as in Sec. 4.4, one may formally obtain this result by assuming an imaginary part of the Love number k_2). It is therefore important to assess the power needed to maintain the observed amplitude $\beta \simeq 0.2''\,(\gg \alpha)$ of Chandler's wobble (here we refer to the rigid model from Sec. 3.5). To estimate its energy we consider, besides the real configuration with angular momentum (eq. (3.36))

$$L = C\omega_0 (1 - \beta^2 \delta), \qquad (3.53)$$

an axially symmetric configuration with the same angular momentum (along z) and therefore with angular velocity

$$\omega_r = \omega_0 (1 - \beta^2 \delta). \qquad (3.54)$$

Internal dynamics conserves $\mathbf{L}$ and therefore will tend to make the energy difference E_w between the two configurations as small as possible. This difference

$$E_w = \frac{1}{2} C \omega_0^2 \beta^2 \delta \qquad (3.55)$$

is the wobble energy, for Earth $\approx 10^{-12}/300 \simeq 3 \times 10^{-15}$ times the rotational energy $C\omega_0^2/2 \simeq 2 \times 10^{36}$ erg. In the simple model of a uniform, viscous Earth the dissipation of the wobble is due to the periodic strain; the relevant frequency is Chandler's σ_C (a more precise calculation indicates that higher multiples, notably $2\sigma_C$, are also present; Problem 3.12). A dissipative material, in absence of external forces, will damp the mode in a time $T_d = Q/\sigma_C$, where Q is its quality factor (see Sec. 4.4). Its value is uncertain, but, typically, for solid bodies it ranges from ≈ 10 to 200 (as we shall see in Sec. 15.3, for the Earth's response to lunisolar tides $Q \simeq 30$). The power $P = E_w \sigma_C / Q$ required to sustain Chandler's wobble is about 10^{13} erg/s (eq. (3.55)). This is much smaller than the power dissipated by tides ($\simeq 3 \times 10^{19}$ erg/s), predominantly at the ocean bottoms; but it should be noted that this energy must be supplied globally and just at Chandler's frequency. There are indications that earthquakes may provide the appropriate energy source; recent analysis of the 1985 to 1996 data of the atmospheric and oceanic circulation suggested that, during this period, Chandler's wobble was dominantly excited by them. However, a detailed understanding of the processes that sustain on a long-term the free modes of the Earth rotation is still unsolved geophysical problem; for instance, in the late 1920s Chandler's term nearly disappeared for about a decade and reappeared only in the 1940s, with a slow increase (Fig. 3.11).

Dissipation in minor bodies.– In the case of objects which do not have enough internal energy sources to sustain the free wobble, viscosity leads to its dissipation. Approximating $\sigma \simeq J_2 \omega \simeq k_s \mu_c \omega$, and assuming a rigidity large

3. Planetary rotation

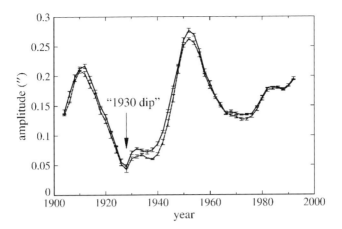

Figure 3.11. Semimajor and semiminor axes of Chandler's wobble as a function of time (their small difference is due to the small triaxiality of the Earth). Note the apparent irregularity, with a significant decrease around 1930. Results obtained from latitude variations of many observing stations during the last century. Data kindly provided by J. Vondrák; see also J. Vondrák, *Surveys in Geophysics* **20**, 169 (1999).

enough so that Love's secular number reads $k_s \simeq 3GM\varrho/(19\mu R)$ (eq. (3.20)), we obtain a dissipation timescale

$$T_d = \kappa \frac{\mu Q}{\varrho R^2 \omega^3} \,. \tag{3.56}$$

The numerical factor $\kappa \simeq 1$ depends on the shape of the body. This result has interesting applications for the rotation of minor bodies in the solar system, in particular asteroids and comets (Secs. 14.3 and 14.5). Mutual collisions between asteroids, and the recoil torque due to the non-isotropic outgassing processes for comets, can result in a rotation axis not aligned along a principal axis; in this case the solution of Euler's equations (3.31) is not sinusoidal. Eq. (3.56) shows that the lowest frequencies decay faster, so that eventually a harmonic solution will result around a principal axis. This happens more quickly (in $\approx$ My or less) for $R \geq 5$ km and is in accordance with the fact that in the rotational curves of most large asteroids one spectral line prevails. The torques on small asteroids ($R \leq 5$ km) and comets may be sufficient to change their rotational state on a timescale shorter than (3.56) and prevent the settling of rotation in the lowest energy state.

– PROBLEMS –

3.1. Discuss the parallactic motion of an object in the solar system due to the rotation of the Earth. For instance, consider an observation of 6 hours centred around midnight, when the source is at the zenith of the observer.

3.2. Evaluate the angle between the vertical and the radial direction.

3.3.* Show that eq. (3.21) for the gravitational potential energy inside a homogeneous ellipsoid (with A_1, A_2, A_3 given by eqs. (3.22) or (3.23)) reduces to eq. (3.15′), with $J_2 = 0$, for a sphere.

3.4.* The prevailing right-hand circulation of cars and lorries creates a net angular momentum, whose change slightly alters the angular velocity of the Earth. Construct a rough model to evaluate this effect.

3.5. Estimate the cumulative effect of the mean Earth rotation braking (3.5) on the phase of day for timing a solar eclipse observed by Babylonian astronomers around 700 BC.

3.6.* A radar tracking of the asteroid Golevka in June 1995 provided the following values for the round-trip light-time Δt and the Doppler frequency shift in X-band ($\nu = 8.51$ GHz):

$$\Delta t = 4.4 \times 10^7 \text{ μs} \quad (\sigma_{\Delta t} = 0.2 \text{ μs}),$$
$$\Delta \nu = -3.5 \times 10^5 \text{ Hz} \quad (\sigma_{\Delta \nu} = 0.35 \text{ Hz}).$$

Find the distance and the relative radial velocity of the telescope and the asteroid, with their errors? Was the measurement sensitive to the transversal Doppler effect (3.3)?

3.7. A model of the 1964 big earthquake in Alaska consists in the displacement by 22 m of a block with mass 10^{23} g. Find the change in LOD it produced if the displacement was along the meridian.

3.8.* Study the polar motion of a rigid Earth with small deviations from axial symmetry: $(C - A) \gg (A - B) > 0$. Prove, in particular, that the polar motion (3.34) becomes slightly elliptic and find the difference between the major and minor axis.

3.9. Show that eq. (3.26) reduces to (3.17) when $e \ll 1$.

3.10. Show that in a triaxial rigid body with $A < B < C$ the rotation around the middle axis of inertia B is unstable.

3.11.* The rotation of the Earth is affected by atmospheric phenomena. When a strong wind from East impinges upon the Rocky Mountains range in the United States, it slows down the Earth rotation. Estimate this effect by assuming reasonable values for the length of the range (from North to South), its slope, the velocity of the wind and its duration.

3.12.* Compute the stress tensor and the elastic energy of an elastic body with uniform density which rotates according the idealized Chandler's wobble (3.52).

3.13.* Prove that a point-like body with mass M at the origin causes on a planet, whose centre of mass is at $\mathbf{r} = r\mathbf{u}$, the gravitational torque:

$$\mathbf{\Gamma} = -\mathbf{u} \times (\partial U/\partial \mathbf{u}) \,;$$

3. Planetary rotation 93

here U is their gravitational potential energy. What is the torque when the planet is small relative to r? (Hint: Use $U = M U_2$, with U_2 given by MacCullagh formula (2.28).)

3.14. The rotation axis **e** of a rigid and axially symmetric rotating body makes an angle β with the symmetry axis $\mathbf{e}_z$ (see Fig. 3.10); describe the effect of a small torque along the direction orthogonal to both vectors when it tends to decrease β.

– FURTHER READINGS –

The classical textbook on rigid rotation is E.J. Routh, *Advanced Dynamics of a System of Rigid Bodies*, Dover Editions (1955). On the Earth's rotation an old, but still good introductory book is W.H. Munk and G.J.F. Macdonald, *The Rotation of the Earth. A Geophysical Discussion*, Cambridge University Press (1960). More recent and complete are the following: K. Lambeck, *The Earth's Variable Rotation: Geophysical Causes and Consequences*, Cambridge University Press (1980), H. Moritz, *Theories of Nutation and Polar Motion*, Ohio State University Press (1980), J.M. Wahr, The Earth Rotation, *Annu. Rev. Earth Planet. Sci.* **16**, 231 (1988) and V.N. Zharkov et al., *The Earth and its Rotation: Low Frequency Geodynamics*, H. Wichmann Verlag (1996); see also *Reference Frames in Astronomy and Geophysics*, J. Kovalevsky, I.I. Mueller and B. Kolaczek, eds., Kluwer (1989). The International Astronomical Union periodically sponsors colloquia relevant to this topic: J. Kovalevsky, ed., *Reference Systems*, Reidel (1988); A.K. Babckock and G.A. Wilkins, eds., *The Earth's Rotation and Reference Frames for Geodesy and Geodynamics* (IAU Symposium 128), Kluwer (1988); J.A. Hughes, C.A. Smith and G.H. Kaplan, eds., *Reference Systems* (IAU Colloquium 127), US Naval Observatory Publ. (1991); S. Dick and D. McCarthy, eds., *Polar Motion: Historical and Scientific Problems* (IAU Colloquium 178), ASP Conf. Ser. **208** (2001). The latest nutation model, officially adopted by IAU, is P.M. Mathews et al., Modeling nutation-precession: New nutation series for nonrigid Earth, *J. Geophys. Res.* **107**, 165 (2002). IERS produces and maintains several documents related to the subjects of Chs. 2 and 3; they can be found at http://hpiers.obspm.fr/ or http://www.iers.org. On Sec. 3.4 (figures of rapidly rotating, self-gravitating bodies), see S. Chandrasekhar, *Ellipsoidal Figures of Equilibrium*, Yale University Press (1969). More astrophysically oriented is J.-L. Tassoul, *Stellar Rotation*, Cambridge University Press (2000), with a fine discussion of (magneto)hydrodynamics of viscous fluids.

Chapter 4

GRAVITATIONAL TORQUES AND TIDES

In the previous chapter the effects of rotation on an isolated planetary body (fluid or rigid) have been discussed; here we study the changes in rotation and shape due to an external gravitating mass. When its distance is much larger than the size of the body, what counts is the difference in the acceleration of different points; this is the *tidal acceleration*. Tidal processes are ubiquitous in the solar system and, as discussed in Ch. 15, are responsible for long-term changes in the rotational and orbital parameters of planetary bodies; they produce time-dependent deformations and energy dissipation, owing to friction in sheared flows. The gradients of the accelerations due the Sun and the Moon produce a torque on the rotational bulge of the Earth and a long-term change in its axis **e**; as a result, **e** moves with a period of 25.7 ky on a cone, whose axis lies approximately along the direction normal to the ecliptic. This major effect, the *lunisolar precession*, was already known to ancient Greek astronomers.

4.1 Lunisolar precession

As we have already anticipated in Sec. 3.2, the rotation vector ω of the Earth is variable at different timescales due to disparate reasons. Here we discuss *lunisolar precession*, which changes the direction of ω in an inertial frame very significantly, but with a very long timescale: the *vernal equinox*, or γ-point, intersection between the celestial equator and the ecliptic, moves along the latter at a rate of about 50 ″/y. Such an effect is easily measured with the techniques of positional astronomy and was already known to Greek astronomers: HIPPARcos (190-120 BC) discovered the precession of the equinoxes by comparing his own observations with a stellar catalogue 150 y earlier.

The bulge of the Earth consists in an equatorial belt of mass $\approx \mu_c M_\oplus$ (see eq. (1.14)); an external gravitating body of mass M at a distance R exerts on the bulge an acceleration different from that on the centre by $\approx GMR_\oplus/R^3$ and, if

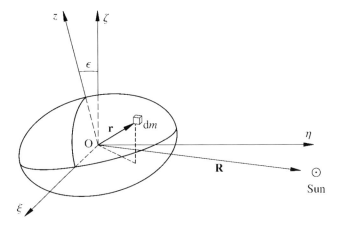

Figure 4.1. Geometry of the solar torque on the flattened Earth. The Sun moves in the ecliptic plane (ξ, η), while the rotation axis z of the oblate Earth is displaced by the angle ϵ (the obliquity) from the normal to the ecliptic ζ. We assume that the rotation and symmetry axes coincide.

it does not lie in the equatorial plane, a torque $\Gamma \approx GM\mu_c M_\oplus R_\oplus^2/R^3 \approx GM(C-A)/R^3$. This is also the rate of change of the angular momentum $\mathbf{L} = C\omega \mathbf{e}$, so that the direction of the axis $\mathbf{e}$ changes at a rate $\approx GM(C-A)/(CR^3)$. Note that the ratio GM/R^3 for the Sun is about half the one for the Moon, so that in the case of the Earth the two contributions are comparable.

For an accurate theory, let $\mathbf{R}(t)$ be the inertial position of the external body related to the centre of the Earth; the elementary force on a mass element $dM_\oplus$ at the geocentric position $\mathbf{r}$ is

$$\begin{aligned} d\mathbf{F} &= GM\, dM_\oplus \frac{\mathbf{R}-\mathbf{r}}{|\mathbf{R}-\mathbf{r}|^3} = GM\, dM_\oplus \frac{\mathbf{R}-\mathbf{r}}{R^3}\left(1 + \frac{r^2}{R^2} - 2\frac{\mathbf{R}\cdot\mathbf{r}}{R^2}\right)^{-3/2} \\ &= GM\, dM_\oplus \frac{\mathbf{R}-\mathbf{r}}{R^3}\left[1 + 3\frac{\mathbf{R}\cdot\mathbf{r}}{R^2} + O\!\left(\frac{r^2}{R^2}\right)\right] ; \end{aligned} \quad (4.1)$$

since $R \gg r$, we have used the multipole expansion of the force up to $O(r/R)$ (see eqs. (2.2) and (2.3)). Integrating the corresponding torque $d\Gamma = \mathbf{r} \times d\mathbf{F}$ over the Earth's volume V we obtain

$$\Gamma = \int_V \mathbf{r} \times d\mathbf{F} = \frac{GM}{R^3}\int_V \mathbf{r}\times\mathbf{R}\, dM_\oplus + \frac{3GM}{R^5}\int_V (\mathbf{r}\cdot\mathbf{R})(\mathbf{r}\times\mathbf{R})\, dM_\oplus . \quad (4.2)$$

Since the origin is at the centre of mass, the last but one integral is zero.

For long-term effects we need the torque averaged over an orbital period. We assume that $\mathbf{R}(t)$ is a circular orbit with unit normal $\mathbf{N}$, along the ζ axis of Fig. 4.1. The matrix $\overline{\mathbf{RR}}$ can be evaluated with a powerful symmetry argument.

4. Gravitational torques and tides

It must be a linear combination of **NN** and **1**, the only symmetric matrices that appear in the problem; hence $\overline{\mathbf{RR}} = R^2(\alpha \mathbf{1} + \beta \mathbf{NN})$, where α and β are two dimensionless numbers. Since $\overline{\mathbf{RR}}$ is orthogonal to **N**, $\alpha + \beta = 0$; taking the trace, we see that $3\alpha + \beta = 1$, so that $\alpha = -\beta = 1/2$ and

$$\overline{\mathbf{RR}} = \frac{1}{2} R^2 (\mathbf{1} - \mathbf{NN}) . \tag{4.3}$$

The first term, proportional to the unit matrix, does not contribute to the torque (4.2); we are left with

$$\overline{\Gamma} = -\frac{3}{2} \frac{GM_\odot}{R^3} \mathbf{N} \cdot \mathbf{Q} \times \mathbf{N} = \frac{3}{2} \frac{GM_\odot}{R^3} \mathbf{N} \cdot \mathbf{I} \times \mathbf{N} , \tag{4.4}$$

where **Q** is the quadrupole tensor (eq. (2.25)) and **I** is the inertia tensor, defined in (2.26); they differ by the sign and an irrelevant term proportional to the unit matrix. For a body axially symmetric around the unit vector **e** (which makes with **N** the angle ϵ, the *obliquity*), $\mathbf{I} = A\mathbf{1} + (C - A)\mathbf{ee}$, so that

$$\overline{\Gamma} = -\frac{3}{2} \frac{GM_\odot}{R^3} (C - A) \cos \epsilon \, \mathbf{N} \times \mathbf{e} . \tag{4.5}$$

A straightforward computation shows that when the perturbing body has an elliptic orbit with semimajor axis a and eccentricity e, R^3 should be replaced with $(a\eta)^3$, where $\eta = \sqrt{1 - e^2}$ (Problem 4.9); similarly, when the planet is triaxial, the $(C-A)$ factor should be replaced by $[C-0.5(A+B)]$ (Problem 4.10). The angular momentum of the Earth $\mathbf{L} = C\boldsymbol{\omega} = C\omega \mathbf{e}$ and the mean torque $\overline{\Gamma}$ are orthogonal, so that no work is done and ω is constant; the axis of rotation changes according to

$$\frac{d\mathbf{e}}{dt} = -\Omega_{\text{pr}} \mathbf{N} \times \mathbf{e} , \tag{4.6}$$

where

$$\Omega_{\text{pr}} = \frac{3}{2} \frac{GM_\odot}{R^3} \frac{(C - A)}{C} \frac{\cos \epsilon}{\omega} . \tag{4.7}$$

Since $\cos \epsilon = \mathbf{N} \cdot \mathbf{e}$ is also constant, **e** moves at the constant rate Ω_{pr} on the cone with axis **N** and aperture ϵ. The motion is retrograde, i.e., is seen as clockwise by an observer standing along **N**; it causes stellar longitudes to increase with time. For the Sun, this rate is 15.95 "/y, with a period of 81.3 ky. Since

$$\frac{M_\odot}{M_{\text{Moon}}} \left(\frac{R_{\text{Moon}}}{R_\odot} \right)^3 \simeq 0.46 ,$$

the Moon and the Sun give comparable contributions. The combined precession vector is

$$\Omega_{\text{pr}} \mathbf{N} = \Omega_{\text{pr}\odot} \mathbf{N}_\odot + \Omega_{\text{pr,Moon}} \mathbf{N}_{\text{Moon}} , \tag{4.8}$$

where $\mathbf{N}_\odot$ is the normal to the ecliptic and $\mathbf{N}_{\text{Moon}}$ the normal to the orbital plane of the Moon around the Earth; $\mathbf{N}$ is closer to $\mathbf{N}_{\text{Moon}}$ than to $\mathbf{N}_\odot$. Neglecting the small inclination $5.15°$ of the lunar orbit and, therefore, the difference between $\mathbf{N}_\odot$ and $\mathbf{N}_{\text{Moon}}$, the total precession rate is

$$\Omega_{\text{pr}} = \Omega_{\text{pr}\odot} + \Omega_{\text{pr, Moon}} \simeq (1.46/0.46)\,\Omega_{\text{pr}\odot} \simeq 50.4\,''/y\,, \qquad (4.8')$$

corresponding to the smaller period of 25.7 ky (see eq. (3.4)).

Planetary perturbations (principally by Jupiter and Venus), higher multipoles of the Earth gravitational field and general relativistic terms contribute to the mean precession of the terrestrial equatorial plane by a much smaller total amount $\approx 0.02\,''/y$. Since the γ-point is referred to the ecliptic of date, there is also a contribution from long-term changes of the ecliptic plane (due to the planetary perturbations) with typical periods of the order 10^5 y (Sec. 15.1). Moreover, the axes of the celestial reference system are non-rotating, so that there is also a small relativistic contribution called geodetic precession (some $\approx 0.02\,''/y$; Sec. 17.4). The resulting mean value of the precession, henceforth called *general precession*, is $50.29\,''/y$.

Nutations. Since the shape and the orientation of the lunar and the solar orbit with respect to the Earth change, the main torques have small, but faster variations which affect the precessional motion. Moreover, the perturbation due to the Sun causes the Moon's nodal line to make a revolution in 18.6 y, much smaller than the precessional period; this changes the direction $\mathbf{N}_{\text{Moon}}$ and produces a small variation in the obliquity (called *nodding*) with the same period and amplitude $9.2''$. This is the largest of several similar effects, collectively called *nutations*, with amplitudes ranging from about $0.01''$ to about $1''$, and periods from fractions of a lunar orbital period (27.3 d) to several years. Precession and nutation change the right ascension and the declination of celestial objects (Fig. 3.3) and can easily be observed.

Planetary rotation. The determination of precession and nutation of the spin axes of other bodies in the solar system is more difficult; note that even the rotation periods – much more easily observed – have not been measured for Mercury and Venus till about 1965, when they were obtained by radar ranging from the ground (see Sec. 20.1). It is much better to have a radio transponder or a transmitter on the surface of the planet and to use range or Doppler measurements. In the case of Mars, the Viking and the Pathfinder landers have provided long and accurate runs of such observations and allowed a determination of the planet's precession rate $- 7.58\,''/y$ – with an accuracy better than 1%. This value is lower than the Earth's due to its greater distance from the Sun and to the fact that the small Martian moons (Phobos and Deimos) contribute very little. Since its obliquity is well known, from eq. (4.7) the dynamical ellipticity parameter $(C - A)/C$ was determined; with a good measurement of the Martian $J_2 = (C - A)/(MR^2)$ (eq. (2.31)) from the perturbations of satellite orbits, one obtains A and C, placing important constraints on the models of its interior.

4.2 Tidal potential

The relative acceleration of two test bodies whose distance is much less than the distance to the sources of the gravitational potential U is best described in the *tidal approximation*. This is a situation which often occurs in celestial mechanics and in the solar system, e.g., in the theory of tides, in the motion of a planet-satellite pair in the field of the Sun, and in the dynamics of accretion. The difference $\mathbf{r}_2 - \mathbf{r}_1 = \mathbf{r}$ in the coordinates of two bodies fulfils the linear differential equation

$$\ddot{\mathbf{r}} = -\nabla U(\mathbf{r}_2) + \nabla U(\mathbf{r}_1) = -\mathbf{r} \cdot \nabla\nabla U(\mathbf{r}_1) + O(r^2) . \tag{4.9}$$

Their relative motion is solely determined by the initial values of the relative distance and relative velocity and in no way depends on their mass and nature.

Tides are the deformations of a planetary body with insufficient rigidity, subject to the gravitational pull of an external, far away body; they are determined by the difference in the force per unit mass (4.9) between a given point and the centre of mass. This difference, of course, attains its maximum at two points on the surface, along the line $\mathbf{p}$ joining the centre of mass and the outer body. The boundary of a spherical body is raised to a configuration which is axially symmetric around $\mathbf{p}$. On the Earth the main tides are determined by the Sun and the Moon; these two points slowly move following their motion. Seen from the (fast) rotating Earth, tides have a periodicity of 12 hours; however, this periodicity is not exact, because what counts is the difference between the angular velocity of the Earth and the mean motion of the Sun and the Moon; in addition, many other and longer periods are present.

The order of magnitude h_T of this rise can easily be established if rigidity is neglected, which is the case for the oceans. There are two forces at play: the external agent, the difference between the force per unit mass exerted by the mass m on a point on the surface and at the centre, and the attraction of the Earth $GM_\oplus/R_\oplus^2$. We assume that the work done on a fluid element by these two forces is of the same order. The external force displaces the fluid by h_T; the resisting force moves fluid masses to the plane perpendicular to $\mathbf{p}$, with a displacement of order $R_\oplus$. Hence

$$\frac{h_T}{R_\oplus} \approx \frac{m}{M_\oplus} \left(\frac{R_\oplus}{R}\right)^3 = \mu_T . \tag{4.10}$$

This is the characteristic *tidal parameter*; this estimate will be substantiated in Sec. 4.3.

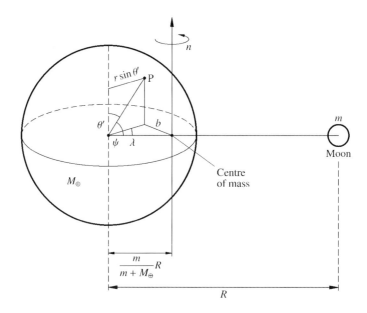

Figure 4.2. Geometry of the tidal effect of the Moon at a point P on the Earth's surface (sizes and distances not to scale).

Tidal potential in a bound system. Let us consider the external potential U at a point P on the Earth's surface caused by the Moon's gravitational field. The Moon has an angular velocity (mean motion) n about an axis perpendicular to the lunar orbit through the centre of mass of the system. Then

$$U = -\frac{Gm}{|\mathbf{R} - \mathbf{r}|} - \frac{1}{2} n^2 b^2 , \qquad (4.11)$$

where $\mathbf{r}$ and $\mathbf{R}$ are the same as in Sec. 4.1, m is the mass of the Moon, b is the distance of P from the axis of orbital rotation (see Fig. 4.2). By Kepler's third law we get

$$n^2 R^3 = G(M_\oplus + m) . \qquad (4.12)$$

Since, as in Sec. 4.1, $r \ll R$, we can expand $1/|\mathbf{R} - \mathbf{r}|$ (eq. (2.2)):

$$\frac{1}{|\mathbf{R} - \mathbf{r}|} = \frac{1}{R} \sum_{\ell \geq 0} \left(\frac{r}{R}\right)^\ell P_\ell(\cos\psi) , \qquad (4.13)$$

where we call ψ the angle between $\mathbf{r}$ and $\mathbf{R}$ ($\mathbf{r} \cdot \mathbf{R} = rR\cos\psi$). By trigonometry (see again Fig. 4.2) we easily get

$$\cos\psi = \sin\theta' \cos\lambda , \qquad (4.14)$$

4. Gravitational torques and tides

where θ' is the colatitude of P with respect to the lunar orbit plane and λ is its instantaneous longitude distance from the sublunar point. Moreover,

$$b^2 = \left(\frac{mR}{m+M_\oplus}\right)^2 + (r\sin\theta')^2 - 2\left(\frac{mR}{m+M_\oplus}\right)r\sin\theta'\cos\lambda =$$
$$= \left(\frac{mR}{m+M_\oplus}\right)^2 + r^2\sin^2\theta' - 2\left(\frac{mR}{m+M_\oplus}\right)r\cos\psi . \quad (4.15)$$

Substituting eqs. (4.12) to (4.15) in eq. (4.11) we obtain

$$U = -\frac{Gm}{R}\left(1 + \frac{1}{2}\frac{m}{m+M_\oplus}\right) - \frac{1}{2}n^2r^2\sin^2\theta' - \frac{Gm}{R}\sum_{\ell\geq 2}\left(\frac{r}{R}\right)^\ell P_\ell(\cos\psi). \quad (4.16)$$

The first term in eq. (4.16) is just a constant, while the second term can be interpreted as a rotational potential about the axis through the Earth's centre normal to the orbital plane. It has an effect similar (although much smaller, since $n^2 \ll \omega^2$) to that of the diurnal rotation of the Earth: it gives rise to a permanent "equatorial bulge" and contributes to the second degree geopotential. The last term in eq. (4.16) is the *tidal potential*. Since we assume $r \ll R$, the quadrupole ($\ell = 2$) term

$$U_2 = -\frac{Gm}{R}\left(\frac{r}{R}\right)^2 P_2(\cos\psi) = -\frac{Gm}{2R}\left(\frac{r}{R}\right)^2(3\cos^2\psi - 1) \quad (4.17)$$

dominates and we focus on it. It deforms the equipotential surfaces in a prolate, axisymmetric ellipsoid aligned with the position of the Moon in the sky. This means that tidal effects have a dominant component of nearly semidiurnal periodicity (U_2 depends on λ through $\cos 2\lambda$), exceeding 12^h by about 25^m because of the Moon's orbital motion around the Earth. An asymmetry between the two diurnal tides (*tidal inequality*) arises when the Moon is displaced from the Earth's equatorial plane, and at high latitudes this may even cause one of the two diurnal waves to disappear (see Fig. 4.3). As we shall see, the ellipticity of the lunar orbit and the superposition of solar tides (smaller, but only by a factor ≈ 2; see Sec. 4.1) results in a complex wave pattern with many periodicities (Sec. 4.4).

Let us first assume that the Earth is rigid, so that its mass distribution is not changed by the tidal potential. Then the fractional tidal variation in the gravity acceleration is

$$\frac{\delta g_R}{g} = \left(\frac{R_\oplus^2}{GM_\oplus}\right)\left(-\frac{\partial U_2}{\partial r}\right)_{r=R_\oplus} = -\frac{m}{M_\oplus}\left(\frac{R_\oplus}{R}\right)^3(3\cos^2\psi - 1), \quad (4.18)$$

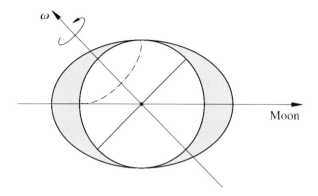

Figure 4.3 The asymmetry in the diurnal tides when the Moon lies off the equatorial plane can be seen by considering two points about 180° apart along a parallel.

of order 10^{-7} and detectable only with very sensitive gravimeters. The tidal deflection of the vertical $\delta \alpha_R$ is given by

$$\tan \delta \alpha_R = \left(\frac{1}{g}\right)\left(-\frac{1}{R_\oplus} \frac{\partial U_2}{\partial \psi}\right) = \frac{3}{2} \frac{m}{M_\oplus} \left(\frac{R_\oplus}{R}\right)^3 \sin 2\psi \,, \quad (4.19)$$

$\approx 0.01''$. The variation in height of the equipotential surface is

$$\delta h_R = \frac{U_2}{g} = \frac{1}{2} \frac{m}{M_\oplus} \left(\frac{R_\oplus}{R}\right)^3 R_\oplus \left(3\cos^2\psi - 1\right) \,, \quad (4.20)$$

of the order of one meter. The higher degree terms ($\ell \geq 3$) in the development of the last term in (4.16) contribute less, by approximately $R_\oplus/R \simeq 1/60$ in the lunar case.

4.3 Tides in a non-rigid Earth

We now abandon the assumption of rigidity and take into account the Earth's deformation. When $R \gg R_\oplus$ we can assume that the response of the solid Earth is linear in the external potential (see Sec. 1.4); moreover, since cause and effect must share the same group transformation properties (as in Sec. 3.3, in relation to the rotational deformation), on the Earth the additional potential $U_T(R_\oplus)$ generated by the deformation must be proportional to $U_2(R_\oplus)$ and we can set $U_T(R_\oplus) = k_2 U_2(R_\oplus)$. Similarly, h_2 is the ratio of the total tidal deformation δh_T of the surface to the displacement of the geoid δh_R (4.20). The index $_2$ indicates a response to a quadrupole deformation. The dimensionless quantities h_2 and k_2 are two *Love's numbers*, which appear also in the theory of a rotating Earth, see eq. (3.20). Hence

$$\delta h_T = h_2 U_2/g \,, \quad (4.21)$$

$$\delta g_T = -\frac{\partial (U_2 + U_T)}{\partial r} + \frac{\partial g}{\partial r} \delta h_T \,. \quad (4.22)$$

4. Gravitational torques and tides

Since $U_T = k_2 U_2$ is a second degree harmonic[1] arising from the masses within $R_\oplus$, for $r > R_\oplus$ it must be $\propto 1/r^3$, and

$$U_T(r) = \left(\frac{R_\oplus}{r}\right)^3 U_T(R_\oplus) = \left(\frac{R_\oplus}{r}\right)^3 k_2 U_2(R_\oplus) = -k_2 \frac{Gm}{R_\oplus}\left(\frac{R_\oplus}{R}\right)^3 \left(\frac{R_\oplus}{r}\right)^3 P_2(\cos\psi) \,. \tag{4.23}$$

Substituting eqs. (4.18), (4.21) and (4.23) into eq. (4.22) we obtain

$$\delta g_T = -\left(1 - \frac{3}{2}k_2 + h_2\right)\left(\frac{\partial U_2}{\partial r}\right) = \left(1 - \frac{3}{2}k_2 + h_2\right)\delta g_R \,. \tag{4.24}$$

Another measurable quantity is the variation in height δz of an equipotential surface with respect to the Earth's surface (e.g., the surface of a pond with respect to its bottom). The tide would rise the equipotential by $(1 + k_2)U_2/g$, but $h_2 U_2/g$ must be subtracted to account for the rise of the solid bottom, to yield

$$\delta z = (1 + k_2 - h_2) U_2/g \,. \tag{4.25}$$

Finally, the tilt $\delta\alpha_T$ of the vertical is different from $\delta\alpha_R$ because of the enhancement by the factor $(1 + k_2)$ of the potential, and of a horizontal displacement of the Earth's crust which is taken into account by a third coefficient l_2 (called also *Shida's coefficient*). In analogy with eq. (4.19), we have

$$\delta\alpha_T = (1 + k_2 - l_2)\left(-\frac{1}{gR_\oplus}\frac{\partial U_2}{\partial \psi}\right) \,. \tag{4.26}$$

Measurements of these tidal effects on the Earth's surface, together with the observed perturbations caused by U_T on the orbits of artificial satellites, have yielded for the Earth's Love numbers the approximate values $h_2 \simeq 0.60$, $k_2 \simeq 0.30$ and $l_2 \simeq 0.08$. The solid-tide vertical movements are thus typically tens of centimetres. This static equilibrium model, which assumes perfect elasticity, is approximate for two reasons. First, the Earth includes fluid parts – the core and the oceans – in which the tidal response may be rate-dependent and cannot be represented by a single set of constants for all kinds of deformations. For instance, marine tides must be treated with a dynamical theory. They produce

---

[1] For not too distant satellites, terms with a degree $\ell > 2$ may be needed. They are still axially symmetric around the line joining the two centres, but depend on the direction as $P_\ell(\cos\psi)$ (eq. (4.16)); with the appropriate dependence on the distance and a different Love's number k_ℓ for each degree, the previous formula can be generalized:

$$U_T(r) = -\frac{Gm}{R_\oplus}\sum_\ell k_\ell \left(\frac{R_\oplus}{R}\right)^{\ell+1}\left(\frac{R_\oplus}{r}\right)^{\ell+1} P_\ell(\cos\psi) \,. \tag{4.23'}$$

shallow water waves of wavelength much greater than the water depth H, with a propagation velocity of order $(gH)^{1/2}$. Were this velocity larger than the speed $(\omega_\oplus - n)R_\oplus$ of the driving tidal force, one would have, like in forced oscillators, a quasi-equilibrium tidal wave in phase with the tidal potential; but since the opposite holds, we rather have out-of-phase *inverted tides*, namely, low tides where the equilibrium theory predicts high ones and vice versa. In reality the situation is much more complicated because of the complexity of the sea floor and coastal lines, resulting in marine tides which depend very much on the location and, in particular geographical scenarios, can undergo strong enhancements. The second reason why the equilibrium model is inadequate is friction, which always occurs in real materials and is often particularly large at the interfaces between solid and fluid regions. Like in forced oscillations with friction, the net result is a *phase lag* (again, because $\omega_\oplus \gg n$), implying that the maximum tidal bulge at some fixed position occurs after that point has passed the sublunar longitude. For the Earth, this lag is about $3°$ (about 12 minutes of time), most of the energy being dissipated in tidal currents in shallow seas. Although small, this dissipative lag causes important, long-term changes in the dynamical state of the Earth-Moon and other planet-satellite systems (Sec. 15.3).

4.4 Tidal harmonics

The external gravitational potential U_2 given by eq. (4.17) and the related effects (both static and dynamical) contain many different periodicities, since: (i) the angle ψ depends on time through the geographical coordinates of the point P at **r**, which rotates with the Earth, and the celestial coordinates of the Moon and the Sun; (ii) the $1/R^3$ factor also changes with time due to their orbital eccentricities and perturbations. Using the addition formula (2.33) for the spherical harmonics in the complex representation, the forcing external potential can be written as

$$U_2(\mathbf{r}) = -\frac{Gm}{5R}\left(\frac{r}{R}\right)^2 \sum_m Y^\star_{2m}(\mathbf{R}/R)\, Y_{2m}(\mathbf{r}/r)\ . \quad (4.27)$$

For an almost circular orbit of the external body, the main frequency dependence of this potential, which is written in the body-fixed frame, comes from the argument $\mathbf{R}/R$ of $Y^\star_{2m}$ and is mainly given by the fast rotation of the Earth appearing in the exponential $\exp(im\phi)$ of that harmonic function. Therefore, the main variation will be at the diurnal frequency (for $m = 1$) and at the semi-diurnal frequency (for $m = 2$). In addition, there are also low-frequency lines (for $m = 0$). The slower changes of the inertial coordinates of the perturbing bodies due, in particular, to their eccentricity and inclination will then produce a cluster of lines around those frequencies.

4. Gravitational torques and tides

As a very simple illustration, consider the case in which the external object is at rest at a declination δ and right ascension α; then one can show that the coefficients of the external potential (4.27) are

$$Y_{20}^\star = \frac{\sqrt{5}}{2}\left(3\sin^2\delta - 1\right), \tag{4.28a}$$

$$Y_{21}^\star = \sqrt{\frac{15}{2}}\sin\delta\cos\delta\,\exp[-i(\lambda-\alpha)], \tag{4.28b}$$

$$Y_{22}^\star = \frac{1}{2}\sqrt{\frac{15}{2}}\cos^2\delta\,\exp[-2i(\lambda-\alpha)]; \tag{4.28c}$$

$\lambda = \omega_0 t$ is the Greenwich mean time (up to a phase). We see that there are three lines at frequencies 0, ω_0 and $2\omega_0$; the second disappears when the body is on the equator (see Fig. 4.3 and Problem 4.7).

The situation becomes more complex when the external object moves, hence α and δ change in time. Traditionally, the overall time dependence of U_2 appears through a linear combination with integer coefficients of the *Doodson variables*[2]:

τ : The mean hour angle of the Moon, namely the time elapsed from its passage at the meridian; $2\pi/\dot\tau$ is the mean lunar day, equal to 1.03505 mean solar days (about $24^\text{h}\,50^\text{m}$).

s : The mean longitude of the Moon in a geocentric, inertial reference system; $2\pi/\dot s$ is the orbital period of the Moon, or *tropic month*, equal to 27.3216 days.

q : The mean longitude of the Sun in a geocentric, inertial reference system; $2\pi/\dot q$ is the orbital period of the Earth, or *tropic year*, equal to 365.242 days.

p : The mean longitude of the lunar perigee; $2\pi/\dot p = 8.847$ y is the *apsidal period* of the Moon.

---

[2]One can easily relate the Doodson variables to other astronomically important quantities:

$(\tau + s) = \omega_0 t'$, where t' is the sidereal time; and

$(\tau + s - q) = \omega_0 t' - q = \omega_0 t$, where t is the mean solar time. Moreover:

$2\pi/[d(s-N)/dt] = 27.2122$ days is the *draconic month*, i.e., the period between two passages of the Moon at a node;

$2\pi/[d(s-p)/dt] = 27.5546$ days is the *anomalistic month*, the period between two passages of the Moon at perigee;

$2\pi/[d(s-q)/dt] = 29.5306$ days is the *synodic month*, the period of lunar phases;

$2\pi/[d(s-2q+p)/dt] = 31.812$ days is the so-called *evection* period; and

$2\pi/[2d(s-q)/dt] = 14.765$ days is the so-called *variation* period.

The two last periods are those of the main solar perturbations on the Moon's eccentricity. It is worth noting that solar and lunar eclipses are periodic, with a period of ≈ 18 years and 11 days ≈ 223 synodic months ≈ 239 anomalistic months. This is *Saros period*, already known (and used to predict eclipses) in antiquity.

Table 4.1. The main Fourier components of the tidal potential (the terminology is due to G.H. Darwin).

Origin	Symbol	Argument	Periodicity
Lunar *principal* wave	M_2	2τ	Semidiurnal
Solar *principal* wave	S_2	$2t = 2\tau + 2s - 2q$	Semidiurnal
Lunar *eccentricity* wave	N_2	$2\tau - s + p$	Semidiurnal
Lunar *declination* wave	K_{2m}	$2t' = 2(\tau + s)$	Semidiurnal
Lunar *declination* wave	K_{1m}	$t' = \tau + s$	Quasi-diurnal
Lunar *principal* wave	O_1	$\tau - s$	Quasi-diurnal
Solar *principal* wave	P_1	$\tau - q$	Quasi-diurnal
Solar *declination* wave	K_{1s}	$t' = t + q$	Quasi-diurnal
Lunar *eccentricity* wave	Q_1	$(\tau - s) - (s - p)$	Quasi-diurnal
Lunar *declination* wave	M_f	$2s$	Fortnightly
Lunar *eccentricity* wave	M_m	$s - p$	Monthly
Solar *declination* wave	S_{sa}	$2q$	Semiannual
Lunar *constant flattening*	M_0	–	Permanent deformation
Solar *constant flattening*	S_0	–	Permanent deformation

N : The mean longitude of the ascending lunar node; $\dot{N}$ is negative, $2\pi/|\dot{N}| = 18.613$ y is the Moon's nodal period.

p_s : The mean longitude of the Earth's perihelion. $2\pi/\dot{p}_s = 20,940$ y is the Earth's apsidal period around the Sun.

Apart from τ (which is measured from the meridian), all these angles are measured in an equatorial reference frame with respect to a conventional origin (the instantaneous vernal equinox). At least over a time not much longer then a century, the Doodson variables are linear functions of time, so that they appear in the Fourier expansion for U_2. The most important harmonics are listed in Table 4.1.

The response of the Earth then determines the tidal potential U_T, already introduced at a preliminary level with (4.23). In reality this response is frequency-dependent and a separate Love number k has to be defined for each frequency. In addition, the response of an oblate Earth is slightly different for different values of the degree ℓ and the order m; moreover, in order to take into account deviations from a dissipationless elastic behaviour, a small imaginary part has to be assumed in each Love number. Therefore one can write for *each frequency f* the component of the tidal potential as

$$U_T^{(f)}(\mathbf{r}) = -\frac{Gm}{5r}\left(\frac{R_\oplus}{r}\right)^2 \sum_m k_{2m}^{(f)} \left\{\left(\frac{R_\oplus}{R}\right)^3 Y_{2m}^\star(\mathbf{R}/R)\right\}^{(f)} Y_{2m}(\mathbf{r}/r) ; \quad (4.29)$$

the total quadrupole tidal potential is then the sum of terms like (4.29) for all possible frequencies f (the upper index (f) indicates frequency dependence).

Love's numbers k of degree 2 are about 0.3 and they differ by a few tens of a percent for different orders. This shows how staggering is the complexity of solid tides in the real Earth; as qualitatively described above, oceanic tides are even more complicated because they arise in a fluid in an irregular environment.

Comparing the formula (4.29) with the internal geopotential field (2.17) and (2.17'), we realize that the Earth tidal potential can be interpreted through the change of the second degree coefficients

$$\Delta C_{20}^{(f)} = \frac{1}{5}\frac{m}{M} k_{20}^{(f)} \left\{ \left(\frac{R_\oplus}{R}\right)^3 Y_{20}^\star(\mathbf{R}/R) \right\}^{(f)}, \qquad (4.30a)$$

$$\Delta C_{2m}^{(f)} - i\Delta S_{2m}^{(f)} = \frac{\sqrt{2}}{5}\frac{m}{M} k_{2m}^{(f)} \left\{ \left(\frac{R_\oplus}{R}\right)^3 Y_{2m}^\star(\mathbf{R}/R) \right\}^{(f)} \quad (m > 0). \quad (4.30b)$$

Though we have worked out these formulæ for $\ell = 2$, the same approach applies also to higher degree terms.

Energy loss due to anelastic tides. As with all response coefficients in physical processes, one can introduce an imaginary part to describe energy dissipation in a phenomenological way. We now discuss the effect of a complex Love's number.

A component of the external potential U_2 at the frequency f has the real representation

$$\mathrm{Re}\, U_2 = AV_2 \cos(2\pi ft + \alpha), \qquad (4.31)$$

where V_2 is a function of $\mathbf{r}$ and $A\exp(i\alpha)$ is a complex amplitude. With the complex Love's number $k_2 = |k_2|\exp(i\phi)$ (with $|\phi| \ll 1$), the velocity field $\delta \mathbf{v}$ caused by U_2

$$\mathrm{Re}\, \delta\mathbf{v} = -\mathrm{Re}\left(\frac{k_2 \nabla U_2}{2\pi i f}\right) = -\frac{|k_2|\nabla V_2 A}{2\pi f} \sin(2\pi ft + \alpha + \phi) \qquad (4.32)$$

is in quadrature with the force unless $\phi \neq 0$. The energy change ΔE in a period is then

$$\begin{aligned}
\Delta E &= -\int_0^{2\pi/f} dt \int_V dm\, \mathrm{Re}\,\delta\mathbf{v} \cdot \nabla \mathrm{Re}\, U_2 \\
&= \frac{|k_2|A^2}{2\pi f} \int_0^{2\pi/f} dt\, \cos(2\pi ft + \alpha) \sin(2\pi ft + \alpha + \phi) \int_V dm\, \nabla V_2 \cdot \nabla V_2 \\
&= \frac{|k_2|A^2 \phi}{2f^2} \int_V dm\, \nabla V_2 \cdot \nabla V_2 \qquad (4.33)
\end{aligned}$$

and vanishes if $\phi = 0$. Dissipation corresponds to a *negative* ϕ. This is usually described in a dimensionless way with a *quality factor*

$$Q = \frac{2\pi \bar{E}}{|\Delta E|}, \qquad (4.34)$$

where $\bar{E}$ is the average over a period of the energy perturbation

$$E = -\int_V dm \, \text{Re} \, \delta\mathbf{r} \cdot \nabla \text{Re} \, U_2 = \frac{|k_2| A^2}{(2\pi f)^2} \sin^2(2\pi f t + \alpha) \int_V dm \, \nabla V_2 \cdot \nabla V_2, \qquad (4.35)$$

i.e.,

$$\bar{E} = \frac{|k_2| A^2}{2(2\pi f)^2} \int_V dm \, \nabla V_2 \cdot \nabla V_2. \qquad (4.36)$$

Therefore

$$Q = \frac{1}{2\pi |\phi|}. \qquad (4.37)$$

Dissipation mechanisms in solid bodies arise because of a variety of processes, including unpinning of dislocations or sliding at grain boundaries. The quality factor of geophysical phenomena decreases slowly with frequency, going from $Q \simeq 30$ at tidal frequencies to $Q \simeq 100 - 500$ for the free-oscillations of the Earth (Sec. 5.2); a power law $Q \propto f^{-0.15}$ is usually assumed. Note that $Q \simeq 30$ implies an effective phase lag ϕ of about 0.25 degrees, showing that the solid Earth tide contributes much less to the tidal braking of the lunar motion. This latter is mainly due to energy dissipation in the oceans, for which the phase lag is larger (Sec. 4.3).

It is worth mentioning that the energy produced in the Earth by the tidal-deformation cycle is negligible with respect to the internal radiogenic sources (Sec. 5.3). However, the situation is quite different for some satellites in the solar system. In the case of the Galilean satellite Io of Jupiter, the tidal energy production currently sustains volcanism. Similarly, Ganymede – another Galilean satellite – might have been resurfaced due to tidal melting during the high-eccentricity episodes of its complex orbital evolution (Sec. 14.2).

4.5 * Equilibrium shape of satellites

In Sec. 3.4 we have derived the equations for the ellipsoidal equilibrium shapes for isolated, spinning objects of uniform density and no internal strength. For satellites orbiting near their planet, provided their spin rate is synchronized with the orbital angular velocity (as we shall see in Ch. 15, this is in most cases the natural end-product of tidal evolution), the same kind of theory can be applied to derive the ellipsoidal equilibrium shapes corresponding to different values of the density, the orbital distance (i.e., by Kepler's third

4. Gravitational torques and tides

law, of the orbital and rotational angular velocity) and the satellite-to-planet mass ratio (usually much smaller than unity). We can qualitatively expect that satellites with a slow rotation, and consequently a large orbital distance, are almost spherical, because tidal and centrifugal forces are negligible. On the other hand, satellites close to the planet will be distorted more if they are less dense. As we shall see in Sec. 14.7, it is also easy to show that below some minimum orbital distance (the so-called *Roche limit*), no equilibrium shape is possible and a satellite with negligible tensile strength would be disrupted by tidal forces. Of course, all these results, including the concept of Roche limit, cannot be applied in a straightforward way to bodies with a finite material strength. We shall now derive quantitatively the equilibrium shapes of homogeneous "fluid" satellites (which are called *Roche ellipsoids*, because of a relevant pioneering work by M. ROCHE carried out in 1847) by applying the same type of argument used in Sec. 3.4.

Let us consider a homogeneous ellipsoidal satellite of density ϱ, semiaxes a, b and c (a directed towards the planet's centre and c orthogonal to the orbital plane), orbital radius R and mass ratio $p = m/M$. Its angular velocity ω is directed along c and, from the spin-orbit synchronization condition, its magnitude is given by

$$\omega^2 = G(M+m)/R^3 = (1+p)n^2, \qquad (4.38)$$

where $n^2 = GM/R^3$. In a reference system with the origin at the satellite's centre and the axes (x, y, z) directed along (a, b, c), respectively, the centrifugal potential energy (per unit mass) (3.6) due to the orbital motion is

$$U_c = -\frac{1}{2}\omega^2\left[\left(x - \frac{MR}{M+m}\right)^2 + y^2\right], \qquad (4.39)$$

where we have corrected for the distance between the planet's centre and the centre of mass. As for the planet's gravitational potential energy U, we assume that the planet has a spherical shape (i.e., is not rapidly spinning and is not distorted by the satellite's gravitational field – this latter assumption being consistent only for small values of p and/or large orbital separations), and expand U about the satellite's centre:

$$U = -\frac{GM}{R}\left(1 + \frac{x}{R} + \frac{2x^2 - y^2 - z^2}{2R^2} + \ldots\right). \qquad (4.40)$$

Thus, if $W = U_c + U$, the satellite's gravity must include the tidal terms

$$-\frac{\partial W}{\partial x} = \frac{GM}{R^2} + \frac{2GM}{R^3}x + \omega^2\left(x - \frac{MR}{M+m}\right) = (3+p)n^2 x, \qquad (4.41a)$$

$$-\frac{\partial W}{\partial y} = -\frac{GM}{R^3}y + \omega^2 y = pn^2 y, \qquad (4.41b)$$

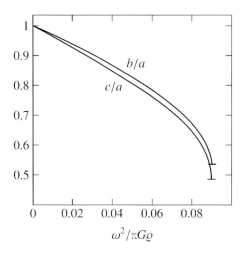

Figure 4.4 The axial ratios c/a and b/a vs. $\omega^2/\pi G\varrho$ for the sequence of the $p = 0$ Roche ellipsoids, which ends at the bars representing the so called Roche limit.

$$-\frac{\partial W}{\partial z} = -\frac{GM}{R^3} z = -n^2 z \,. \tag{4.41c}$$

Note that for the x-component, the gravitational pull of the planet and the centrifugal force balance each other at the satellite's centre (where we have used eq. (4.38)). Since these additional gravity terms are still linearly dependent upon the coordinates, we can consider, like in Sec. 3.4, three Newton's canals along the semiaxes a, b and c, and impose the condition that the work to bring a unit mass to the equipotential surface is the same in the three cases. We have

$$-2\pi G\varrho A_1 a^2 + (3+p) n^2 a^2 = -2\pi G\varrho A_2 b^2 + p n^2 b^2 \tag{4.42a}$$
$$= -2\pi G\varrho A_3 c^2 - n^2 c^2 \,, \tag{4.42b}$$

where A_1, A_2 and A_3 are the same integrals defined in Sec. 3.4, and depend on a, b and c. From the previous equations we can get the two relationships

$$[(3+p) a^2 + c^2] n^2 = 2\pi G\varrho (A_1 a^2 - A_3 c^2) \tag{4.43a}$$
$$(p b^2 + c^2) n^2 = 2\pi G\varrho (A_2 b^2 - A_3 c^2) \,. \tag{4.43b}$$

Given the mass ratio p, by inverting these equations numerically, it is possible to obtain the two axial ratios c/a and b/a as a function of n (and, via eq. (4.38), of ω or R; see Fig. 4.4).

Figure 4.5 shows the behaviour of the sequences of the Roche ellipsoids for $p = 0$, 1 and $p \to \infty$ (the last case, provided $n \to 0$ and $n^2 p \to \omega^2$, is that of a spinning, isolated body, and we find again the Maclaurin spheroids – for $c/a > 0.5826$ – and the Jacobi ellipsoids – for $c/a < 0.5826$). In the case $p = 1$ the Roche figures are good approximations to the real equilibrium shapes of the two binary components only for $\omega^2/\pi G\varrho \ll 1$, otherwise the mutual distortion effects cannot be neglected. From the figure it is apparent

4. Gravitational torques and tides

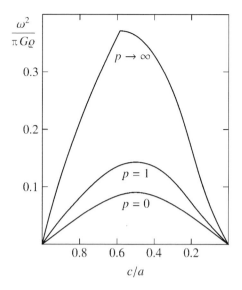

Figure 4.5 The variation of $\omega^2/\pi G\varrho$ as a function of c/a along the sequences of the Roche ellipsoids corresponding to $p = 0$, $p = 1$ and $p \to \infty$. The maxima of the curves define here the Roche limit.

Table 4.2 The properties of the Roche ellipsoids.

p	$\omega^2_{max}/\pi G\varrho$	$n^2_{max}/\pi G\varrho$	$(c/a)_{cr}$	$(b/a)_{cr}$
0	0.09009	0.09009	0.4826	0.5113
1	0.14132	0.07066	0.4903	0.5408

that the Roche ellipsoids qualitatively behave as anticipated; in particular, the fact that $\omega^2/\pi G\varrho$ attains a maximum along the Roche sequence (depending on p) implies that for distances R less than that corresponding to this maximum no equilibrium solution is possible. This is called the Roche limit, and represents, in other words, the distance of closest approach between the two components consistent with equilibrium. Table 4.2 shows the properties of the "critical" ellipsoids located at the Roche limit, for $p = 0$ and $p = 1$. We shall discuss some important applications of this concept to planetary rings in Sec. 14.7.

The stability of the Roche sequence of equilibria – necessary to apply them to astronomical situations – has been widely studied; they were found stable up to a critical point always preceding the maximum of $\omega^2/\pi G\varrho$; thus the unexpected branch in which ω decreases for increasing deformation is unstable. When a satellite orbit decays (e.g., by tides) towards its primary, the dynamical instability sets in before the last possible configuration (Roche limit) is attained; however, the two cases, especially when viscosity is taken into account, are very close, so that this limit is a good approximation to the last possible state of the satellite. Obviously, close to this point a more accurate analysis is needed than our simplified approach; in particular, the cubic and higher-degree terms in (4.40) cannot be omitted and the satellite ceases to have an ellipsoidal shape (it resembles a water drop). Hydrodynamical simulations have been performed to study these last stability stages.

– PROBLEMS –

4.1. Evaluate the Coriolis force on a low orbit satellite produced by an error in the lunisolar precession of 0.002 ″/y and its effect on its position for a mission lasting 3 y.

4.2. How high is the tide raised by the Earth on the Sun? And that raised on Saturn by its largest satellite, Titan? (Hint: Use the data tabulated in Ch. 14.)

4.3. Compute in order of magnitude the total energy residing in the Earth's tidal bulges. Assuming that $\approx 10^{-3}$ of this energy is dissipated during one day, what is the corresponding power?

4.4. What is the critical water depth for inverted marine tides, out of phase with the tidal potential?

4.5. Using the satellite data listed in Sec. 14.2 and considering Figs. 4.4 and 4.5, find the natural satellites for which the expected tidal deformation exceeds 1% of the size.

4.6.* Study the dynamics – equilibria and small oscillations – of two mass points connected by a short, rigid wire, whose centre of mass orbits on a circular path around the Earth.

4.7.* Compute the external gravitational potential U_2 due to a body which moves on a circular orbit at an arbitrary inclination (eqs. (4.27) and (4.28)). Show, in particular, that in the lunar case there are forthnightly sidebands in the zero and second degree terms in (4.27).

4.8.* Study the solutions of eqs. (4.42) formally adopting $p = -1$ (this corresponds to the case of a satellite linearly accelerated toward the planet). Prove that in this case $c = b$ is a solution that corresponds to a prolate spheroid (with the longest axis directed to the planet). Find the Roche limit for this sequence of (so called Jeans) spheroids. (Hint: Beware that A_1 and $A_2 = A_3$ should be evaluated for prolate spheroids.)

4.9. Compute the following average over an orbital period for an elliptical motion:

$$\overline{\frac{\mathbf{RR}}{R^5}} = \frac{1}{2(a\eta)^3} \left(\mathbf{e}_P \mathbf{e}_P + \mathbf{e}_Q \mathbf{e}_Q\right) = \frac{1}{2(a\eta)^3} (\mathbf{1} - \mathbf{NN}) .$$

(Hint: Use the two-body problem, in particular eqs. (11.6), (11.13), (11.36a).)

4.10. Prove that for a triaxial planet, with principal moments of inertia $A < B < C$, the $(C - A)$ factor in the precession rate (4.7) is replaced with $[C - 0.5(A + B)]$.

– FURTHER READINGS –

P. Melchior, *The Tides of the Planet Earth*, Pergamon Press (1983) provides a comprehensive and systematic treatment of Earth tides. On marine tides see H.W. Schwiderski, Ocean tides, *Marine Geodesy* **3**, 161 and 219 (1980); C. Le Provost et al., Spectroscopy of the world ocean tides from a finite-element hydrodynamics model, *J. Geophys. Res.* **99**, 24,777 (1994); astronomical consequences of oceanic tides, with emphasis to the dissipative phenomena, are discussed in K. Lambeck, Tidal dissipation in the oceans: astronomical, geophysical and oceanographic consequences, *Phil. Trans. R. Soc. London* **A287**, 545 (1977). The theory of the Roche ellipsoids is discussed in S. Chandrasekhar, *Ellipsoidal Figures of Equilibrium*, Yale University Press (1969).

Chapter 5

THE INTERIOR OF THE EARTH

The interior of the Earth (and, a fortiori, of the other planets) is much less known than the interior of the Sun, and also of stars light years away. The physics and chemistry of solid and liquid phases at high pressures, essential for the understanding of terrestrial planets and the core of the outer planets, are much more complex than those of nearly perfect gases (which form stars) and very little known. For the Earth, the detailed study of seismic propagation has provided, especially for the mantle, excellent measurements of its density distribution and is discussed in this chapter. The model of its interior so obtained is quite relevant for the study of its geological history, as well as for other planets. Planetary interiors are characterized by two major features: a fluid, conductive core, which generally produces a global magnetic field by the dynamo process (Ch. 6); and the interaction between the crust and the slow motion of convective eddies in the mantle. Due to this interaction, the crust undergoes continuous deformations, responsible for continental drifts, earthquakes and the uplifting of mountain chains. They are called plate tectonics and are discussed in some detail here, also in view of the fact that they can be measured directly from space.

5.1 Seismic propagation

As briefly discussed in Sec. 1.4, solid bodies respond *elastically* and linearly to deforming forces when they are small enough and of short enough duration. In other words, the *strain* (measure of deformation) is proportional to the *stress* (force per unit area). Within the Earth, transient forces arise mainly due to earthquakes: sudden fractures in its crust can release as much as 10^{26} ergs and the corresponding disturbances can travel to large distances throughout the planet. The propagation of seismic waves is as a powerful tool to investigate the physical properties of the Earth's interior; with this tool, the Earth is the most extensively studied planet. Using the arrays deployed by the Apollo programme, the lunar interior has also been constrained with this technique, although with far worse data. While on the Earth the primary source of seismic waves are deep earthquakes, on the Moon they are mainly produced by tides raised by the Earth, and by surface impacts. Future missions to Mars (e.g., the NetLander mission scheduled for 2007) plan to deploy an array of seismometers on its surface. Apart from rare impacts, the expected quakes might be induced by thermoelastic cooling of the lithosphere; normal modes may have a permanent excitation due to tides.

Neglecting the gravitational term in eq. (1.55) and taking the divergence, we obtain (when μ and λ are constant):

$$\varrho \, \partial^2 \theta / \partial t^2 = (\lambda + 2\mu) \nabla^2 \theta \,, \tag{5.1}$$

where $\theta = \nabla \cdot \mathbf{s}$. Since we are working in the linear approximation, the difference between Eulerian and Lagrangian derivatives is negligible. Alternatively, the curl of eq. (1.55) gives

$$\varrho \, \partial^2 (\nabla \times \mathbf{s}) / \partial t^2 = \mu \nabla^2 (\nabla \times \mathbf{s}) \,. \tag{5.2}$$

The divergence and the curl of $\mathbf{s}$ thus propagate as waves, with the velocities $v_P = [(\lambda + 2\mu)/\varrho]^{1/2}$ and $v_s = (\mu/\varrho)^{1/2}$, respectively, which depend upon the Lamé's constants of the material. The first (eq. (5.1)) are longitudinal *P-waves*, a succession of dilatations and compressions along the direction of propagation. The second (eq. (5.2)) are transversal *S-waves*, consisting, in general, in a rotational and a shear deformation; they are described by the vector $\nabla \times \mathbf{s}$ perpendicular to the propagation vector and have two components, corresponding to two polarization modes. In the Earth $\lambda \simeq \mu$, so that the P waves are faster by about a factor $\sqrt{3}$.

Refraction of seismic waves. Let us consider now a wave front travelling through a material where ϱ, λ and μ *change*, as in the interior of a planet. If – as it normally happens – λ and μ increase with depth faster than ϱ, the seismic velocities increase with depth also; therefore refraction makes the ray paths concave toward the planetary surface (recall Fermat's least time principle; Sec. 8.4). The difference in travel times of the waves from a given source to different points on the surface, which can be directly measured, depends on the variation of velocity with depth, and hence on the physical properties of the planetary interior. In a spherically symmetric body it can easily be seen that the quantity (with the dimension of time)

$$p = r \sin i / v(r) \tag{5.3}$$

5. The interior of the Earth

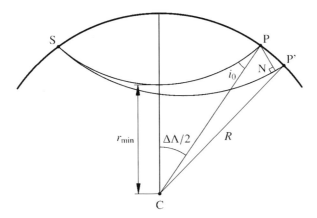

Figure 5.1. Consider two neighbouring rays starting at the source S and emerging at P and P'. If at the surface $r = R$, $i = i_0$ and $v = v_0$, we have $\sin i_0 = NP'/PP' = [v_0\, d(\Delta t)] / [R\, d(\Delta\Lambda)] = (v_0/R)\, d(\Delta t)/d(\Delta\Lambda)$. Since $p = r \sin i / v$ is constant along each ray, we obtain: $p = d(\Delta t)/d(\Delta\Lambda)$.

is constant along any given ray[1] (here r is the distance from the centre, v is the propagation velocity – either v_P or v_S, since both waves can be detected independently – and i is the angle of the ray with the radial direction; see Problem 5.1). If Λ is the angle at the centre measured along the ray and ds is the arc length element, we have $\sin i = r\, d\Lambda/ds$ and therefore

$$p = \frac{r^2}{v(r)} \frac{d\Lambda}{ds}. \tag{5.4}$$

Using the relationship $ds^2 = dr^2 + r^2 d\Lambda^2$, and defining R and $r_{\min}$ as the radius of the planet and the smallest radius reached by a given ray, from eq. (5.4) we obtain the total angular distance $\Delta\Lambda$ from the source and the travel time Δt

$$\Delta\Lambda = 2p \int_{r_{\min}}^{R} \frac{dr}{r\,[(r/v)^2 - p^2]^{1/2}}, \tag{5.5}$$

$$\Delta t = 2 \int_{r_{\min}}^{R} \frac{ds}{v} = 2 \int_{r_{\min}}^{R} \frac{(r/v)^2\, dr}{r\,[(r/v)^2 - p^2]^{1/2}}$$

$$= p\,\Delta\Lambda + 2 \int_{r_{\min}}^{R} \frac{[(r/v)^2 - p^2]^{1/2}}{r}\, dr. \tag{5.6}$$

From Fig. 5.1, comparing two neighbouring rays, we also see that $p = d(\Delta t)/d(\Delta\Lambda)$; this quantity, therefore, can be measured directly (if the location

---

[1] This is an elementary application of refraction theory; for a more complete discussion, see Sec. 8.4.

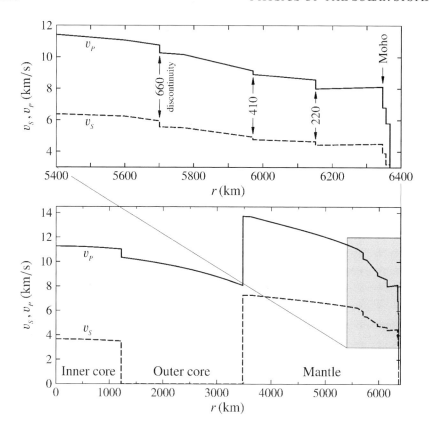

Figure 5.2. The compressional (v_p) and the shear (v_s) velocities as functions of the distance from the centre for the spherically symmetric reference Earth model PREM. On the bottom, in the outer core $v_s = 0$ is due to its lack of rigidity ($\mu = 0$). On the top, in the upper mantle, prominent discontinuities are shown, including Moho's at the bottom of crust. Seismic data elaborated by A.M. Dziewonski and D.L. Anderson, *Phys. Earth Planet. Int.* **25**, 297 (1981).

of the source of seismic waves is known) by comparing a set of values of $\Delta\Lambda$ and Δt provided by a suitable array of seismometers. The location of an earthquake is obviously not known a priori, but can be solved for if a large enough amount of data is available. Putting together data obtained from many earthquakes and using eqs. (5.5) and (5.6), the function $v(r)$ can be reconstructed (see Fig. 5.2) and the internal structure investigated (Sec. 5.2).

Boundary and surface effects. The mathematical description of seismic waves uses the decomposition of a generic vector field in an irrotational (P-wave) and a solenoidal (S-wave) component:

$$\mathbf{s} = \nabla u + \nabla \times \mathbf{a} \qquad (\nabla \cdot \mathbf{a} = 0). \qquad (5.7)$$

5. The interior of the Earth

In agreement with the transversal character of the S-waves, the vector potential **a** can, without loss of generality, be chosen solenoidal; **a** is perpendicular to the direction of propagation **k** and there are two states of polarization. From eqs. (5.1) and (5.2) we get:

$$\nabla^2 \left(\frac{\partial^2 u}{\partial t^2} - v_P^2 \nabla^2 u \right) = 0 , \quad \nabla^2 \left(\frac{\partial^2 \mathbf{a}}{\partial t^2} - v_S^2 \nabla^2 \mathbf{a} \right) = 0 . \qquad (5.8)$$

These equations must be supplemented with the appropriate boundary conditions at the surface and any discontinuity. If one looks for solutions of eqs. (5.8) regular in the whole space, one can use the fact that a harmonic function is a constant and can be taken to be zero. Hence u and **a** satisfy d'Alembert equation for the propagation velocities v_P and v_S, respectively. In the more general case of a solution in a compact region, u and **a** satisfy inhomogeneous wave equations with regular harmonic functions on the right-hand side.

As we shall see (Sec. 5.2 and Fig. 5.2), in the Earth interior there are several discontinuities in the wave velocities, the main one between the outer core and the mantle; they deeply affect seismic wave propagation, which, on turn, provides important evidence about their structure and location. Similarly to what happens with electromagnetic waves, discontinuities in the material produce reflection, refraction, exchange of energy between different modes, and evanescent modes (which diminish exponentially on one side); in this context, the two modes of propagation of an S-wave are called SH (horizontal) if **a** is parallel to the discontinuity, and SV (vertical) if it is perpendicular. This topic, extensively developed by seismologists, is not discussed here in detail. In an elastic medium, however, there is an important novelty: *surface waves* can propagate below the surface in a thin exponential layer, whose thickness of the order of a wave length. Since they spread energy over two, rather than three, dimensions, they are less attenuated than P- and S-waves, and provide important evidence about the sources. In the Earth they arise from a coupling between P and SH surface waves, and are called *Rayleigh waves*. It can be shown that pure SH waves – *Love waves* – can propagate and can always be confined to the neighbourhood of a discontinuity in the interior. We only discuss Rayleigh waves.

In an infinite space eqs. (5.8) are solved by exponential functions $\exp[i(\omega t - \mathbf{k} \cdot \mathbf{r})]$, with $\omega = \pm k v_P$ (P-waves) and $\omega = \pm k v_S$ (S-waves); the phase velocity is constant and there is no dispersion. But near the boundary $z = 0$ of a "flat Earth" $z > 0$, solutions with a complex wave vector are possible. In contrast with the usual stability analysis, in which one looks for the frequency ω as function of **k**, in a propagation problem the frequency is fixed by the Fourier decomposition of the source, and the wave vector is sought. For propagation along the x-direction, the functions

$$u = u_1 e^{-\sigma_P z} \exp[i(\omega t - k_x x)] , \quad \mathbf{a} = a \mathbf{e}_y = a_1 \mathbf{e}_y e^{-\sigma_S z} \exp[i(\omega t - k_x x)] \quad (5.9)$$

are solutions of eq. (5.8) if

$$\sigma_p = k_x\sqrt{1-\xi q^2}\,,\quad \sigma_s = k_x\sqrt{1-\xi}\quad \left[\xi = (\omega/v_s k_x)^2\right]. \tag{5.10}$$

We have introduced the ratio $q = v_s/v_p = \sqrt{\mu/(\lambda+2\mu)}$ (< 1) of the wave speeds and the dimensionless measure $1/\sqrt{\xi}$ of the wave vector k_x. The reciprocals $1/\sigma$ are the penetration depths (different for the two waves).

The solution is acceptable if the boundary conditions at the discontinuity $z = 0$, with normal $\mathbf{n} = (0, 0, -1)$, are fulfilled. Since there is no matter below, the stress vector

$$P_{ij}n_j = P_{i3} = \lambda\theta\delta_{i3} + 2\mu e_{i3} \tag{5.11}$$

must vanish[2]. In general this translates into the boundary conditions at $z = 0$

$$2\frac{\partial^2 u}{\partial x \partial z} + \frac{\partial^2 a}{\partial x^2} - \frac{\partial^2 a}{\partial z^2} = 0,\quad 2\mu\left(\frac{\partial^2 u}{\partial z^2} + \frac{\partial^2 a}{\partial x \partial z}\right) + \lambda\left(\frac{\partial^2 u}{\partial x^2} + \frac{\partial^2 u}{\partial z^2}\right) = 0. \tag{5.12}$$

With the exponential solution this yields the homogeneous linear system

$$2ik_x\sigma_p u_1 - (k_x^2 + \sigma_s^2)a_1 = 0, \tag{5.13a}$$

$$\left[(2\mu+\lambda)\sigma_p^2 - \lambda k_x^2\right]u_1 + 2i\mu k_x \sigma_s a_1 = 0. \tag{5.13b}$$

Its determinant is real and its vanishing guarantees a solution. Some algebra leads to the cubic equation:

$$f(\xi) = \xi^3 - 8\xi^2 + 16(3/2 - q^2)\xi - 16(1 - q^2) = 0. \tag{5.14}$$

Since $f(0) = -16(1-q^2) < 0$ and $f(1) = 1 > 0$, there is always a real root ξ_0 in the interval $(0, 1)$, corresponding to a real value of k_x, uniquely determined by the frequency. The penetration depths (5.10) are of the order of $1/k_x$. It can be shown that, for realistic values of μ and λ, the velocity of propagation of Rayleigh waves always lies between $0.874\,v_s$ and $0.955\,v_s$. A typical seismogram thus looks like in Fig. 5.3. In this model, in which the dependence of the propagation speeds on the depth is neglected, there is no dispersion and no deformation of the pulse, a simplification which in reality does not hold. The study of the distortion of Rayleigh waves provides important information about the structure of the crust and the upper layers of the mantle.

5.2 Internal structure of the Earth

Our knowledge of the interior of the Earth is almost entirely derived by the analysis of seismic wave propagation and its oscillations. Qualitatively, the

---

[2] Instead, across on a discontinuity inside the Earth $P_{ij}n_j$ must be continuous, corresponding to the fact that the two sides do not slip relative to each other; see Problem 5.3.

5. The interior of the Earth   121

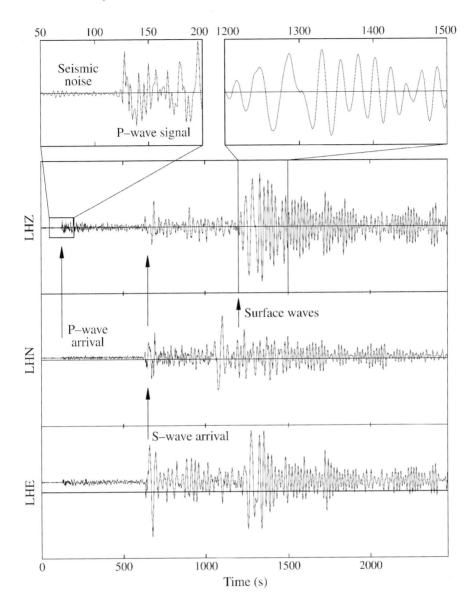

Figure 5.3. Seismogram of a 6.8 magnitude earthquake, with the epicentre in Venezuela, recorded on July 9, 1997 at Dublin (Ireland) ≈ 7,000 km far away. The amplitude of the signal (in arbitrary units) vs time is shown for three displacement components: (i) vertical (LHZ), (ii) North-South (LHN), and (iii) East-West (LHE); L stands for long period. The arrival of P-, S- and surface-waves is indicated by arrows; the fastest P-waves arrive the first, and the slowest surface waves last. Note the high amplitude of surface (Rayleigh) waves, corresponding to their small damping; they provide the magnitude of the earthquake. Data acquired from the University of Michigan at Ann Harbor (http://www.geo.lsa.umich.edu/~MichSeis/).

main result is that, whereas P-waves propagate throughout the entire volume of the planet, S-waves cannot enter a spherical shell, whose inner and outer surface can be located at $r \simeq 0.20 R$ and $r \simeq 0.54 R$, respectively. This is consistent with an overall structure where a solid mantle encloses a liquid outer core (liquids are, of course, unable to withstand shear deformations), which itself encloses a solid inner core.

The fundamental Earth model with spherical symmetry is called PREM (Preliminary Reference Earth Model); it has been adjusted to a large number of seismic data and free oscillation modes in the early 1980s and assumes a mean Earth radius of 6,371 km. Below we briefly comment on the equations involved in the model construction (except for the free oscillation data). A number of discontinuities, that reflect and transmit seismic waves in a particular way, must be also included. Here are the most prominent ones (see also Figs. 5.2 and 5.4):

- *Mohorovičić discontinuity* (or Moho), which separates the Earth crust from the mantle. There are strong variations in its depth, but the most important difference is between continents and oceans. Beneath a typical oceanic crust the Moho lies below a $\approx$ 6.5 km thick crystalline crust ($\approx$ 10.5 km below the surface); the crustal thickness of continents is significantly larger, from $\approx$ 25 km to over $\approx$ 75 km (in areas of large mountain building, like the Himalaya).

- *The core-mantle boundary* (CMB), at a depth of 2,890 km. The CMB is suspected to have some variation in depth (not larger than $\approx$ 5 km) from place to place, an appealing source for the fluctuations in the length of the day over the timescale of a decade.

- *Inner core/outer core boundary* (ICB) at a depth of 5,150 km. The uncertainties in the depth of CMB and ICB may be $\approx$ 10 km.

There are more, still well documented, but less prominent, global discontinuities at 410 km and 660 km depth. Both are very sharp (possibly as sharp as $\approx$ 4 km) and associated with phase transformations of the olivine-rich upper mantle material. At the depth of $\approx$ 220 km the *Lehmann discontinuity* has been observed at numerous places, especially beneath the continents; perhaps it marks the lithosphere-asthenosphere transition. However, this feature is not global and it might be abandoned in the next generation of spherically symmetric Earth models.

PREM cannot fit all most recent data and, for this reason, there is currently an effort to replace it with REM, a full, three-dimensional model. To this end, spherically symmetric models, such as PREM, serve as a reference for local structures (typical local fluctuations of the seismic velocities v_p and v_s amount up to 2% of the PREM value). A considerable number of these structures has

5. The interior of the Earth

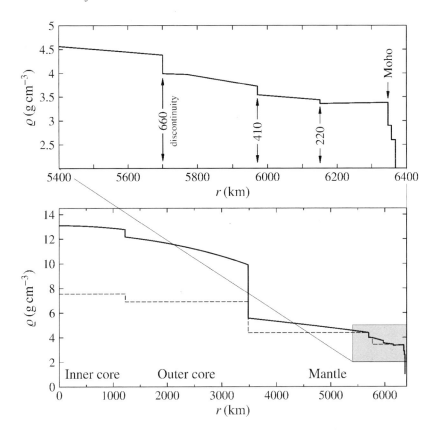

Figure 5.4. Bottom: density profile for the Earth's interior in the spherically symmetric PREM model. The upper mantle region is zoomed at the top. Prominent discontinuities and an approximate boundary between the asthenosphere and the lithosphere are marked. The dashed line gives the corresponding "uncompressed" densities, extrapolated to zero pressure and room temperature. Data from A.M. Dziewonski and D.L. Anderson, *Phys. Earth Planet. Int.* **25**, 297 (1981).

been reported over the last decade, with a general consensus between different models as far as the largest ones are concerned. The upper mantle variations are largely correlated with tectonic features seen at the surface. There are also two major variations on a quasi-global scale, related to the subsiding of ancient tectonic plates in the western Pacific ocean and along the west coast of South America. Another feature on a planetary scale is a huge, hot plume beneath the African plate, that originates from the outer surface of the core and expands upward through the mantle. The detected raising of the African plate is most likely related to this major feature. Very extensive advances have been made towards a full three-dimensional model, but the picture is still far from complete.

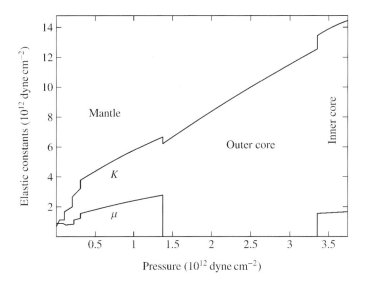

Figure 5.5. PREM variations of the elastic constants with pressure in the Earth's interior; K is the bulk modulus and μ the rigidity.

Construction of the radial model of the Earth. Here we give an elementary insight on how a simple Earth model can be constructed. Using the known propagation velocities $v_s(r)$ and $v_p(r)$, we get:

$$\frac{dP}{d\varrho} = \frac{K}{\varrho} = v_P^2 - \frac{4}{3}v_S^2 \, , \quad \frac{\mu}{\varrho} = v_S^2 \, . \qquad (5.15)$$

Since the bulk modulus $K = \lambda + 2\mu/3$ (eq. (1.53)) and μ are not known a priori, in order to get $\varrho(r)$, we can use the condition of hydrostatic equilibrium (eqs. (1.1) and (1.6))

$$\frac{d\varrho}{dr} = \frac{d\varrho}{dP}\frac{dP}{dr} = -\frac{\varrho^2}{K}\frac{GM(r)}{r^2} \, , \qquad (5.16)$$

sometimes referred to as the *Adams-Williams equation*. Differentiating, this gives a second-order differential equation for the unknown $\varrho(r)$; inserting the inferred values of K/ϱ as a function of r, it can be integrated numerically, using as initial conditions the values at the surface $\varrho(R) \simeq 3.3$ g/cm^3 and $d\varrho(R)/dr$, provided by eq. (5.16). Thus the density profile is obtained and, from eq. (5.15), one gets $K(r)$, $\mu(r)$ and $P(r)$ (Figs. 5.4 and 5.5).

Planetary oscillations. The free oscillations of planetary body provide another, independent method to obtain information about its interior (especially the deeper structure). On the Earth they are excited by earthquakes, as a bell rings when struck with a hammer; they are the low frequency and global counterpart of the high-frequency seismic waves already discussed, which propagate according to geometrical optics along well-defined rays. Planetary modes are eigensolutions of the linearized dynamical equations and have well-defined eigenfrequencies; their classification is based upon expansions in spherical harmonics (Sec. 2.1). Following up the decomposition (5.7), the solenoidal (incompressible) part $\nabla \times \mathbf{a}$ of the displacement $\mathbf{s}$ can be uniquely expressed as the sum of a *toroidal* and a *poloidal* component; this corresponds to the following decomposition of a solenoidal vector:

$$\mathbf{a} = \mathbf{a}_T + \mathbf{a}_P = w(\mathbf{r})\mathbf{r} + \nabla \times (v(\mathbf{r})\mathbf{r}) . \tag{5.17}$$

In the following chapter this theorem will be used for the dynamo theory of the planetary magnetic field. For the displacement $\mathbf{s}$ we now have three components, spheroidal (compressible), toroidal and poloidal:

$$\mathbf{s} = \mathbf{S} + \mathbf{T} + \mathbf{P} = \nabla u + \nabla \times (w(\mathbf{r})\mathbf{r}) + \nabla \times [\nabla \times (v(\mathbf{r})\mathbf{r})] . \tag{5.18}$$

Consider first the case in which the planet does not rotate and has a spherical symmetry; the density and the elastic constants are given as functions of r. Each of the three scalar functions u, w, v can be expanded in spherical harmonics, dependent on the degree ℓ and the order m; the coefficients $u_{\ell m}(r)$, $w_{\ell m}(r)$, $v_{\ell m}(r)$ are function of the distance only and fulfil coupled ordinary differential equations. In the $m = 0$ case, for example, the displacement field lines are along the parallels for the toroidal component, and in the meridian plane for the toroidal part; it can be shown that the spheroidal modes are coupled with the poloidal modes, but not with the toroidal ones. The oscillation frequency $\omega_{\ell m}$ of each mode results from this eigenvalue problem; but with spherical symmetry there is degeneracy, and the eigenfrequencies do not depend on m (Problem 5.11). This decomposition also provides the basis for the study of the normal modes of the Sun.

For the Earth, in the last 30 years many different modes have been measured with increasing accuracy. $\ell = 0$ corresponds to the fundamental, radial oscillation mode, with period $P_0 = 20.46$ minutes, in a good agreement with theoretical calculations based on Earth's models. For the low frequency modes the main restoring force is gravity; hence, on the basis of dimensional analysis, P_0 must be given, in order of magnitude, by

$$2\pi/P_0 \approx (4\pi G \bar{\varrho})^{1/2} ,$$

where $\bar{\varrho}$ is the mean density. This is also Jeans' frequency (eq. (a) in the Appendix of Ch. 16) and, approximately, the mean motion of a low-orbiting spacecraft. As for the geopotential, the next mode $\ell = 1$ is not present if the origin of the coordinates is at the centre of mass; $\ell = 2$ corresponds to successive prolate and oblate deformations and has the longest period, ≈ 53.89 minutes. As ℓ increases, only the upper layers are involved.

Especially for the low frequency modes, whose period is not much shorter than a day, an accurate treatment should take into account the Coriolis force due to the rotation. Waves travelling westward and eastward have different speeds, and the azimuthal degeneracy is removed. Due to the formal correspondence between the Coriolis force and the Lorentz force on an electron, such multiplets are analogous to the Zeeman effect of spectral lines of atoms in an external magnetic field. This frequency splitting has been actually observed for a few, very strong earthquakes. A similar effect is produced by the ellipticity. In these cases spheroidal and toroidal modes are coupled, with great mathematical complication. Earthquake data allowed the analysis of the Earth proper frequencies up to degree $\ell \approx 40$. For some of these lines the attenuation has been measured and may be expressed in terms of the corresponding quality factor Q: $2\pi/Q$ is the fraction of the energy dissipated in a cycle (eq. (4.34)). For the $\ell = 1 - 6$ modes (periods of $10 - 50$ minutes) we have $Q \approx 250 - 400$, while for $\ell = 6 - 40$ (periods of $3 - 10$ minutes) $Q \simeq 100 - 250$. Interestingly, the purely radial oscillations ($\ell = 0$) are attenuated less ($Q \approx 800 - 6,800$); this confirms that the energy damping mainly depends on the shear deformation.

Normal modes of oscillations are important for all planetary bodies, for the Sun and the stars. The seismometers deployed by the Apollo spacecraft on the lunar surface recorded P- and S-waves, propagating inside the Moon, and also a limited amount of low-frequency free oscillations. The fundamental quadrupole and octupole ($\ell = 2, 3$) spheroidal modes were detected after impacts of large meteoroids on the surface.

The study of free oscillations has reached a high degree of sophistication in the case of the Sun, with the birth of a new scientific discipline – *helioseismology* – based upon the detailed spectroscopical decomposition of the velocity field on its surface. With this method a large number of eigenmodes have been identified. In this way space missions (in particular, the SOlar and Heliospheric Observatory (SOHO)) and ground-based projects (in particular, the Global Oscillation Network Group (GONG)) have recently contributed a good deal to our knowledge of the solar interior. For the Earth and the Sun, for which the data are most extensive, fitting of the normal mode frequencies has allowed discriminating between various theoretical models.

Composition issues. If we aim at inferring something about the composition from the density profile, we have to extrapolate densities to conditions observable in the laboratory, i.e. to a smaller pressure. For this purpose we cannot apply conventional elasticity theory, because, as shown by Fig. 5.5, the elastic constants, under pressures comparable to their value, change appreciably. A simple model can be set up by assuming a linear relationship, the so-called *Murnaghan's equation*

$$K = K_0 + K'_0 P + O(P^2) \,, \tag{5.19}$$

where subscript $_0$ refers to values at zero pressure; we can see from Fig. 5.5 that, at least for the mantle, this is a good approximation. Since $K = \varrho\, dP/d\varrho$, we have then:

$$\frac{\varrho}{\varrho_0} = \left(1 + \frac{K'_0}{K_0} P\right)^{1/K'_0} = \left(\frac{K}{K_0}\right)^{1/K'_0} ; \qquad (5.20)$$

taking for the lower mantle $K_0 \simeq 2.25 \times 10^{12}$ dyne/cm^2 and $K'_0 \simeq 3.35$, we obtain an "uncompressed" density ϱ_0 of about 4 g/cm^3 (the extrapolation to room temperature produces a much smaller correction). This is consistent with a substantially homogeneous chemical composition of the whole mantle, formed by a mixture of high pressure phases of iron, magnesium and silicon oxides.

For the outer core the same method is less accurate, but the result ($\varrho_0 \simeq$ 6.3 g/cm^3) appears to imply that the main constituent is iron, which at the melting point (a reasonable estimate for the temperature there) has a density of about 7 g/cm^3. The $\approx 10\%$ difference between these densities implies that some lighter elements are also present as minor constituents: good candidates are sulphur, which alloys with iron even at low pressures and temperatures, and oxygen, which can be incorporated into iron only at high pressures. The cosmic abundance of sulphur is $\approx 30\%$ that of iron by mass, but it was a volatile in the preplanetary nebula. The outer core is a very good candidate for the liquid, electrically conducting region where the Earth's magnetic field is generated (see Ch. 6). It might be noted that additional light elements in the core are suspected to favour the molten outer core state, an effect that is necessary to explain magnetic fields of some small bodies (e.g., Mercury).

The interpretation of inner core data is more difficult. Solidity is suggested by the finite rigidity, but μ is abnormally low with respect to K for a close-packed crystalline solid. The material is very probably a solid nickel-iron alloy, but its phase is still an open question. It is interesting to notice that the boundary between the solid mantle and the liquid outer core coincides with a gross difference of composition; this is probably not the case for the boundary between inner and outer core, which more likely results just from the dependence of the melting point of ferrous materials on the pressure. Finally, it is worth noting that the Earth's core contains about 32% of the total mass of the planet, a fraction several times larger than that expected on the basis of the cosmic abundance of iron with respect to all other elements, except hydrogen and helium (see Table 9.1). This is a significant clue for the processes through which the material, that subsequently formed the Earth, condensed from the primordial solar nebula.

5.3 Heat generation and flow

Although the overall equilibrium of a planet does not explicitly involve its thermal state, the conditions prevailing in a planetary interior are strongly con-

Table 5.1. Radioactive elements in the Earth and parameters related to their decay (ppm in abundances means "part per million").

Isotope	Half-life (Gy)	Heat production (erg/(s g))	Abundance in natural rocks (ppm) Crust	Upper mantle
^{235}U	0.70	5.69	$10^{-3} - 10^{-2}$	10^{-4}
^{238}U	4.47	0.95	$10^{-1} - 1$	10^{-2}
^{232}Th	13.90	0.26	$1 - 10$	$10^{-2} - 10^{-1}$
^{40}K	1.25	0.29	$10^{-1} - 10$	$10^{-2} - 10^{-1}$

nected with the way heat is generated and transferred, with the resulting temperature distribution and with its evolution in time. If the planet is initially hotter than the surroundings, heat will be released at the surface, causing the temperature to increase with depth (the energy input received from the Sun, though very important in determining the state of the surface and of the atmosphere, makes a negligible contribution to the total heat content of a planet). The detailed internal distribution at any time depends on three factors:

- *The initial state*, where for "initial" we intend at the time the planetary accumulation process was completed, about 4.5 Gy ago. As we shall see, for a planet as big as the Earth, the memory of the initial state is probably not yet lost. Note that if the heat content is proportional to the planet mass, since the dispersion of energy is proportional to the surface and the density of solid bodies does not vary strongly in the solar system, the time needed to cool the interior (in absence of energy generation) is roughly proportional to the body size.

- *The energy generation mechanism*. The temperatures in the interior of large planetary bodies are of the order of $10^3 - 10^4$ K, too low for the onset of thermonuclear reactions; however, the planetary material contains small fractions of long-lived radioactive elements whose decay causes a continuous energy input.

- *The energy transfer mechanism*. Thermal energy can be transported by radiation, conduction and convection. At least the two latter mechanisms are important in the interior of planets.

All of these factors are briefly discussed below.

Heat sources. As a "boundary condition", we can use the direct measurements of the heat flow through the Earth's crust, carried out on a global scale during the last few decades. The heat flux is typically 60 erg/(s cm^2), with a variability by a factor 2 or 3, depending on the type of crust and the intensity of geologic activity. However, this is not sufficient to determine uniquely the thermal state of the interior, since a wide range of combinations of "primordial" and radioactive heating, with different isotopes, may lead to the same surface conditions. For instance, the fact that the heat flux through the continental crust and the ocean floors has the same order of magnitude is somewhat surprising; in the former case the measured concentrations of radioactive elements in the rocks imply that a large fraction of the heat is generated in the crust itself, while this is not possible for the much thinner oceanic crust, where virtually all the flux must come from the mantle.

The decay of radioactive nuclei in a given mass of material occurs according to the equation

$$dN = -\alpha N dt ,\qquad(5.21)$$

where dN is the number of nuclei decayed in the time dt, N is their total number and α is a constant characteristic of every radioactive species. Integration yields

$$N(t) = N(0) \exp(-\alpha t) = N(0) \exp(-t \ln 2/T_{1/2}) ,\qquad(5.22)$$

where $T_{1/2}$ is the half-life of the species. The decay can involve the emission of massive particles, gamma rays and neutrinos. Apart from the latter, for which the Earth is practically transparent, the energy of the emitted particles is rapidly absorbed by the surrounding material, which is thereby heated. Four radioactive nuclei are of particular importance in planetary interiors: ^{235}U, ^{238}U, ^{232}Th and ^{40}K. Their properties are listed in Table 5.1, including approximate abundances in natural rocks.

Taking the upper mantle abundances as representative for the whole Earth, we obtain an energy production rate of the order of 10^{-7} erg/(s g), which is just the order of magnitude needed to give the observed average surface flux quoted earlier. As mentioned above, the abundances of radioactive elements in crustal rocks are in general much higher, so that in the continents (where the crust is thicker) most radioactive heat probably comes from the crustal layers. On the other hand, the abundance values found in the upper mantle are of the same order as those of meteorites, and thus are probably typical of the primordial solid material out of which the planets accreted. The reason why the surface layers of the Earth (and of the Moon at the sites sampled in the Apollo missions) are enriched in radioactive material is not yet understood.

Apart from the long-lived radionuclides, which are now the principal heat sources, the early planets, their embryos and even smaller planetesimals, might have possessed at least two more heat sources: (i) energy released by short-lived radionuclides (now extinct), and (ii) the gravitational energy of accreting planetesimals. The most important short-lived radionuclides are listed in Table 5.2. Obviously, their presence in the protoplanetary nebula requires their early implantation. This would be consistent with the assumption that its collapse has been triggered by a shock wave produced by a nearby supernova

Table 5.2 The short-lived radionuclides present in the early solar system; their daughter nuclides and half-lives.

Parent nuclide	Daughter nuclide	Half-live (My)
^{41}Ca	^{41}K	0.1
^{26}Al	^{26}Mg	0.7
^{60}Fe	^{60}Ni	1.5
^{10}Be	^{10}B	1.5
^{53}Mn	^{53}Cr	3.7
^{107}Pd	^{107}Ag	6.5
^{182}Hf	^{182}W	9.0
^{129}I	^{129}Xe	15.7

or Wolf-Rayet stars (their creation in the solar system by the energetic cosmic particles after collapse began is less likely); the gas that produces shock waves from these sources carries freshly synthesized short-lived radionuclides. The extinct radionuclides provide also very important constraints on the early chronology in the solar system. Assuming a uniform distribution, the daughter abundances in meteorites and planetary samples set the relative dating of their origin. This concerns mainly objects that were formed very early, such as meteorites and their specific parts (e.g., calcium-aluminium-rich inclusions or chondrules; Sec. 14.6), but interesting constraints were also obtained for the lunar formation, some 30 – 50 My after the origin of the solar system (from ^{182}W abundance in lunar samples), and for the origin of the primitive Earth atmosphere, some 4.46 Gy ago (from ^{129}Xe atmospheric abundance).

The heat deposited by the gravitational energy of accreting planetesimals during the early phase of planetary formation appears to be an important mechanism for the giant planets, in particular to explain the excess of their infrared emission (Sec. 6.5); but it is likely minor for the terrestrial planets. It is also assumed to be a significant factor for the large transneptunian bodies with size $\geq$ 200 km, possibly increasing their central temperature above the phase transition of amorphous to crystalline ice.

Heat transfer processes. Contrasting with stars, in planetary interiors, where the temperature is lower and the opacity very large, radiative transfer is not important.

Conduction.– We discuss first the role of conduction, governed by the heat equation (1.59), in terms of the heat transport coefficient χ and the heat source S due to radioactivity. The (outward) energy flux on the surface is

$$F_\varrho = -c_p \varrho \chi \frac{dT}{dr} = -\kappa \frac{dT}{dr}, \qquad (5.23)$$

where c_p is the specific heat at constant pressure and $\kappa = c_p \varrho \chi$ the *thermal conductivity*. In the lithosphere of the Earth the heat flux at the surface is estimated by measuring the temperature gradient and κ; on the aver-

5. The interior of the Earth

age $-dT/dr \approx 15$ K/km and $\kappa \approx 4 \times 10^5$ erg/(s cm K), with a heat flow of ≈ 60 erg/(s cm^2). To gain an idea about the timescale for this energy loss, note that the time for the heat to diffuse over a distance L is $\tau \approx L^2/\chi$. Since for terrestrial-type materials $\chi \approx 0.01$ cm^2/s, τ is of the order of the age of the solar system for $L \approx 400$ km; in the inner planets the initial accretion heat did not have time to escape by conduction. It is also interesting to note that the cooling time of the oceanic lithosphere, with thickness $h \approx 70$ km, is about h^2/χ, of the order of a few hundreds My; this is also the order of its surface residence time in the continental spreading (see Sec. 5.4) between the formation at a ridge and sinking at a subduction zone. Apparently, the oceanic lithosphere remains at the surface just about the time needed to give up its residual heat and sink into the mantle again, in agreement with the hypothesis that plate tectonics is driven by convection currents in the mantle.

Convection is the other relevant process of heat transfer in planetary interiors. We need first to recall some thermodynamical principles. Consider a unit mass of material at the temperature T and pressure P and let a quantity dQ of heat enter its volume $V = 1/\varrho$. From the first principle of thermodynamics

$$dQ = dU + PdV = dH - VdP = c_p dT + \left[\left(\frac{\partial H}{\partial P}\right)_T - V\right] dP, \quad (5.24)$$

where U is the internal energy, $c_p = (\partial H/\partial T)_P$ is the specific heat at constant pressure, and $H = U + PV$ the enthalpy. Introduce the entropy S, with $dQ = TdS$, and the free energy $F = U - TS + PV$, with $dF = -SdT + VdP$; then

$$\left(\frac{\partial H}{\partial P}\right)_T = T\left(\frac{\partial S}{\partial P}\right)_T + V = V - T\left[\frac{\partial}{\partial T}\left(\frac{\partial F}{\partial P}\right)_T\right]_P = V - T\left(\frac{\partial V}{\partial T}\right)_P. \quad (5.25)$$

Substituting in (5.24) we obtain

$$dQ = c_p dT - T\left(\frac{\partial V}{\partial T}\right)_P dP. \quad (5.26)$$

In an *adiabatic transformation* $dQ = 0$, and eq. (5.26) becomes

$$\left(\frac{dT}{dP}\right)_S = \frac{T}{c_p}\left(\frac{\partial V}{\partial T}\right)_P = \frac{T}{c_p}\frac{\beta_V}{\varrho}, \quad \left[\beta_V = \frac{1}{V}\left(\frac{\partial V}{\partial T}\right)_P\right]. \quad (5.27)$$

β_V is the *volume expansivity* at constant pressure. For a perfect gas $\beta_V = 1/T$ and eq. (5.27) can be derived more simply using the equation of state ((1.16); see also Problem 5.7).

In hydrostatic equilibrium $dP + \varrho g\, dr = 0$ and

$$-\left(\frac{dT}{dr}\right)_S = -\left(\frac{dT}{dP}\right)_S \frac{dP}{dr} = \left(\frac{T}{c_p}\frac{\beta_V}{\varrho}\right)\varrho g = \frac{g\beta_V}{c_p} T. \quad (5.28)$$

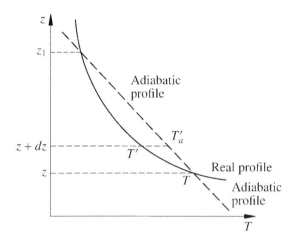

Figure 5.6 The convective instability produced by a temperature gradient steeper than the adiabatic profile: if a fluid element quickly rises from z to $z + dz$ its temperature T'_a is larger than the temperature T' of the surrounding and its motion continues. At the other intersection z_1 the temperature gradient is less steep than the adiabatic profile and we have a stable situation.

This is the *adiabatic temperature gradient*, the largest gradient that a fluid at rest under gravity can support in a stable situation. Indeed, if the temperature increases with depth faster than specified by eq. (5.28), an accidental upward displacement of a fluid element, if quick enough to be adiabatic, brings it to a region where it is hotter than its new surroundings (see Fig. 5.6); its density is lower and the buoyancy force will rise it further. The motion then continues in an unstable way until, eventually, the heat exchanged with the environment brings it back to the temperature equilibrium. The motion then stops and, as a result, a net amount of heat is carried upward[3]. A similar process occurs if the fluid element initially moves downward. Convection is of great importance in cosmic bodies, where gravity is the main macroscopic force and there are often internal energy sources which heat a fluid from below. The evaluation of heat transport in a convective – and turbulent – regime is a difficult and still little understood problem. Convection is a very efficient way to transport heat and, with a self-regulating process, tends to keep the temperature gradient just above the adiabatic gradient; in a convective medium the assumption of an adiabatic temperature profile is often a good approximation.

In a perfect gas the adiabatic gradient (5.28) is $-g/c_p$; but the interiors of terrestrial planets behave as solids and dilatation, as described by the volume expansivity β_V, has a different nature; we have a more complex problem. Since the volume expansivity in (5.27) is defined at constant pressure $P(V,T)$, it can also be written as

$$\beta_V = -\frac{1}{V}\frac{\left(\frac{\partial P}{\partial T}\right)_V}{\left(\frac{\partial P}{\partial V}\right)_T} = \frac{\left(\frac{\partial P}{\partial T}\right)_V}{K}, \qquad (5.29)$$

---

[3] If the temperature gradient is less than the adiabatic gradient any initial motion will be reversed by the buoyancy force and an oscillatory mode will set in at the Brunt-Väisälä frequency (see Sec. 7.4).

where $K = -V(\partial P/\partial V)_T = \varrho(\partial P/\partial \varrho)_T$ is the bulk modulus (1.53). We need the rate $(\partial P/\partial T)_V$ at which the pressure increases with temperature. A solid is essentially a set of oscillators, whose restoring forces are due to the interactions between the lattice centres. For a lattice spacing d, it is described by a potential $U(\delta d)$, function of its change δd from the reference position. If the oscillators were strictly harmonic, heating would just increase their mean energy, but leave the interparticle distance d and the pressure unaffected; but if the elementary interaction potential $U(\delta d)$ is not even, a temperature change also displaces the equilibrium positions and generates a pressure. Therefore, in a solid, thermal expansion is a consequence of the anharmonic character of the interaction, in particular of the term $b(\delta d)^3$ in the potential (Problem 5.12). This state of affairs is best described by the dimensionless *Grüneisein's parameter*

$$\gamma_G = \frac{\beta_V K}{\varrho c_P}, \qquad (5.30)$$

mainly determined by the interaction term $-b(\delta d)^3$. It turns out that γ_G is close to unity for virtually all geologically interesting materials, and almost constant for the range of temperatures and pressures found in the Earth's interior (in contrast with the variable factors β_V, K and ϱ). From eq. (5.28) we have then

$$\left(\frac{dT}{dr}\right)_S = -\frac{g\gamma_G \varrho T}{K}; \qquad (5.31)$$

or, in thermodynamic space,

$$K\,dT - \gamma_G T\,dP = \varrho\,dT - \gamma_G T\,d\varrho = 0, \qquad (5.32)$$

formally the same equation (1.26b) for a polytropic gas with $\gamma = \gamma_G + 1$. This shows how the adiabatic temperature profile is constructed with the density $\varrho(r)$.

What happens in the real Earth? Of course, in the (solid) lithosphere no convection is possible and heat is transferred only by conduction. In the mantle, on the contrary, convection plays a dominant role. At first glance this may seem surprising, since the propagation of shear elastic waves in it (see Fig. 5.2) guarantees that the material is not liquid; on the other hand, with the lithospheric temperature gradient, at the bottom of the lithosphere ($\approx$ 70 km in depth) the temperature becomes $\approx 1,500$ K, close to the melting point of most minerals. In fact, the mantle is in a state intermediate between the fluid and the solid. At temperatures somewhat below the melting point, crystalline materials respond to shear stresses in a way depending on the timescale: they behave elastically for rapid deformations (such as seismic waves), while for very long times they are "soft enough" to *creep*, i.e. undergo a slow, steady deformation. In other words, rocky materials show a finite viscosity when their temperature approaches the melting point. In the Earth's mantle the average melting temperature of the various mineral components increases with depth (because of the increasing pressure) at a rate of ≈ 0.5 K/km, only about a factor 2 higher than the adiabatic temperature gradient (5.28) ($T \approx 2,000$ K, $c_p \approx 10^7$ erg/(g K), $\beta_V \approx 10^{-5}$ K^{-1}) or (5.31) (with $\gamma_G \approx 1$ and $K \approx 5 \times 10^{12}$ dyne/cm^2 – see Fig. 5.5). On the other hand, since the conductivity does not depend much on the temperature, the adiabatic gradient is too small to allow radioactive heat to be transferred upwards by conduction only. As a consequence, very slow convective currents arise, compatible with creep deformations, and the temperature increases with depth at a rate which is close both to that of the melting temperature (allowing a low enough viscosity) and to that of the adiabatic gradient (making convection possible)[4].

There is indirect evidence of convection phenomena in the mantle of other terrestrial bodies. With a rotation in 27 d, one would expect the Moon to have a centrifugal oblateness $(C - A)/B$ of order $\mu_c \approx 3 \times 10^{-6}$ (eq. (1.13)); but in fact $(C - A)/B \simeq 6.28 \times 10^{-4}$ is much larger. This anomalous equatorial bulge may have been formed by an ancient two-cell convection in the mantle in presence of a solid core. As the Moon cooled, the bulge was preserved. This scenario is also supported by the evidence of an ancient, strong lunar magnetic field (Sec. 6.5).

For the liquid outer core, the situation is still different. Here, the material is fluid and the only condition required for convection is a sufficiently strong energy generation. This condition is more demanding because: (i) the abundance of radioactive elements in the iron core is uncertain, but probably lower than in the mantle for chemical reasons; (ii) the thermal conductivity is higher, being dominated (like for the electrical conductivity of metals) by electrons in the conduction band. However, an alternative heat source is provided by the latent heat and gravitational energy released by the growth of the inner solid core. Gravitational energy becomes available because the liquid outer core contains $\approx 10\%$ of light, alloying constituents (like sulphur or oxygen – see Sec. 5.2), which are at least partially excluded from the solid phase and displaced upwards as the inner core grows. The formation and growth of the inner core in planetary centres essentially occur because for iron and iron-rich alloys the rate of increase of the freezing temperature with pressure is larger than the adiabatic gradient $(\partial T/\partial P)_S$, which approximately holds there. Moreover, in general, the addition of impurities depresses the freezing point of liquids (if the corresponding solid is purer); if the core starts as a liquid and then gradually cools, when some solid iron forms, the coexisting fluid is richer in lighter

---

[4]The real situation is still more complicated, since – as mentioned in Sec. 5.2 – there are significant lateral inhomogeneities in the Earth mantle. Some of them are clearly caused by local intrusions of core magma, much hotter than the surrounding material in the mantle.

constituents, and therefore less dense, than the overlying fluid (which has not frozen yet) and therefore tends to rise and to start a compositionally driven convection. In any case, the onset of convection in the liquid core – with characteristic velocity much faster than the slow creep in the mantle – appears to be an essential prerequisite for the generation of a significant magnetic field on a planetary scale (see Sec. 6.4).

How was this situation established? Even assuming that the Earth started out cold (an unlikely case, as we shall see in Sec. 16.3), the radioactive heat release was probably sufficient to bring it close to melting in $\approx 10-100$ My (at present this release is $\approx 10^{-7}$ erg/(s g), but it was higher in the past by about an order of magnitude – see Table 5.1; the specific heat is $\approx 10^7$ erg/g); moreover, it is possible that short-lived radionuclides released more heat than the long-lived ones. Convection then set up and the situation became self-stabilized: a slower convection would heat up the Earth, but then the material would become "softer" and convection would be accelerated again. Convection stops only when the heat flux will become so low that conductive transfer along a quasi-adiabatic temperature gradient will match the flux itself.

5.4 Tectonic motions

A scientific revolution has affected geology during the 1960s and the 1970s, transforming this discipline from a mostly descriptive one, providing explanations only for small scale features and phenomena, to a mature science, based on a paradigmatic theory (the so-called *plate tectonics*) which explains on a global scale the most prominent characteristics and the evolution of the Earth's outermost layers. The evidence for extensive horizontal displacements of large sections of the Earth's crust relative to each other comes from disparate types of data and techniques:

- Impressive similarities exist between continental margins now separated by wide oceans, both in shape (an outstanding example are the Atlantic coast lines of Africa and South America) and in geological features, paleoclimatic histories, and fossil records of ancient flora and fauna. In the first decades of the last century this evidence was systematically collected and used by A. WEGENER to support his theory of continental drift and of the origin of the present continents from a primordial, common land mass. This theory, though much debated, was rejected for a long time by the majority of geophysicists, mainly because no plausible physical mechanism was deemed capable of causing the proposed drift.

- The morphology of oceanic floors, first globally mapped in the 1950s, shows many prominent linear features associated with intense volcanic and seismic activity, which appears to divide the Earth's crust into a number of huge "plates", including the continents. These linear features are either long ridges, where the sea floor appears to be generated from below and then to diverge, or deep trenches, where the surface layer appears to plunge back into the mantle. They are frequently broken into many shorter sections somewhat displaced horizontally, and connected by *transform faults*, i.e., shear lines parallel to the direction of relative motion between the boundaries. Thus the continents do not "float" like rafts on the mantle, but move about as a consequence of the continuous generation and destruction of the oceanic crust.

The new data on oceanic floors also showed clearly that the bipartition of the Earth's crust into oceans and continents is not just the result of the presence of large amounts of liquid water: the oceanic crust is thinner, younger, different in mineralogical composition (less acidic) and lies on the average some 4.6 km lower than the continental shelfs, with a comparatively steep slope in between. This dichotomy appears to be unique among the terrestrial-type planets.

- The ideas about continental drift and sea-floor spreading received a detailed and quantitative support from paleomagnetic data, collected in the 1950s and the 1960s. As it will be discussed in Ch. 6, most crustal rocks retain a record of the local geomagnetic field at the time they cooled and solidified, or were deposited as sediments. In particular, the orientation that the geomagnetic pole should have had to produce the local field can be obtained. For rocks younger than ≈ 20 My, these locally determined polar directions are consistent and, to a good approximation, parallel or antiparallel to the rotational pole (Sec. 6.3); but for earlier epochs the locally determined magnetic poles are not consistent among each other, even when one assumes that the rotational pole moves around in the rigid Earth (Sec. 3.5). They can be reconciled only if large relative drifts have occurred on a timescale of ≈ 100 My, in a way fully consistent with Wegener's picture, derived from entirely different data. Moreover, on the oceanic crust, sea-floor spreading from ridges is confirmed by the observation of symmetric, linear stripes of alternately positive and negative magnetic anomalies on the two sides of the ridges, as expected from the known occurrence of reversals of the field polarity, which are recorded by cooling volcanic rocks (Fig. 6.3). These data are consistent with spreading rates of $1 - 10$ cm/y, and allow the determination of the relative velocity of neighbouring plates.

- Finally, an increase of accuracy in laser ranging to artificial satellites (like LAGEOS; see Secs. 18.2 and 20.3), the Global Positioning System satellites (Sec. 18.2) and extensive observations of extragalactic radio sources with VLBI (Sec. 20.3) resulted over the past 20 years in direct evidence for the motion of the tectonic plates. With a precision in the determination of the position of geodetic sites, now improved to a few centimetres, the direct measurement of the plate motion is achievable with accuracy of a few cm/y or better. This is comparable to that attained by above mentioned analysis of geological records.

Other large scale crustal features can be interpreted in this framework, provided the mechanism for horizontal motions of the crust is recognized in their coupling with vertical motions in the underlying mantle, associated with convective currents (see Sec. 5.3). The basaltic ocean crust is then seen as a veneer on the *lithosphere*, the rigid layer ≈ 70 km thick and broken in different plates, moving and evolving due to convection in the deeper, plastic *asthenosphere*[5]. At the ocean trenches (*subduction zones*), the lithosphere bends down forming a rigid, inclined slab which plunges into the mantle, producing earthquakes with foci at intermediate and great depths (contrasting with earthquakes associated with transform faults at ocean ridges, which are generated closer to the surface), until, at ≈ 700 km of depth, it is re-assimilated in the mantle. But during this process the lighter material differentiates and emerges again as *andesitic magma*, resulting into extensive volcanism and forming island chains (e.g., Japan) or building up continental edges (e.g., the Andes). On the other hand, when a collision between two continental blocks occurs, the continental crust is too light and thick for sinking into the mantle, and thus folds and forms extensive mountain systems (e.g., the Himalaya and the Alps).

The only type of volcanic and orogenetic activity which is unrelated to plate motions and to the upper mantle convection pattern is associated with the so-called *hot spots*. Some chains of volcanic islands (e.g., the Hawaii), lying far from the plate boundaries, show a progressive increase of age, which suggests plate motion across a steady volcanic source, possibly a narrow convective plume originating in the lower mantle. Some 120 hot spots, active in the last 10 My, have been recognized, both inside the continents (e.g., the volcanoes Nyamlagira and Nyiragongo in central Africa) and in the oceans, either superposed on ridges (e.g., Iceland) or well removed from them (e.g., Hawaii). A map of the main plates and hot spots is shown in Fig. 5.7. Figure 5.8 shows epicentral locations of the major earthquakes during a single year (nearly 14,000 events of magnitude > 3); comparison with Fig. 5.8 confirms that the earthquakes are largely restricted to the plate boundaries (about 95% of all earthquakes), following mainly ocean trenches or ridges. The obvious source of the earthquake energy is a sudden, and largely unpredictable, release of the strain between colliding plates.

---

[5]The Earth is the only known terrestrial body that has developed a global-scale plate tectonics; both Mars and Venus lack it. A plausible – and interesting – reason may be related to a sufficient amount of water. Subducted water, through sinking ocean plates, may have softened the upper asthenosphere, thus allowing horizontal motion.

Kinematic description of plate motion. As mentioned in Sec. 3.2, an accurate body-fixed reference system requires the definition and the realization of an ideal rigid body, with respect to which the motion of all stations on the continental plates can be referred. This is the Conventional Terrestrial Reference System (CTRS; Sec. 3.2); it is realized by the assignment and the accurate maintenance of the positions and the velocities in this frame of a set of special sites, usually tracking stations or related ground marks. The position **r**(*t*) of any site results from (i) its motion Δ**r**(*t*) relative to its own plate, which includes all the relevant time-dependent effects (solid Earth tides, oceanic and atmospheric loading, post-glacial rebound, etc.), obtained with suitable modelling and (ii) the motion of the plate itself in the CTRS, assumed rigid around the centre of mass of the Earth. This is a purely rotational motion, entirely determined by its angular velocity $\mathbf{\Omega}_{pl}$; if the acceleration is neglected, the plate component of the velocity of a point **r** is

$$\mathbf{v}_{pl} = \mathbf{\Omega}_{pl} \times \mathbf{r} . \qquad (5.33)$$

For small displacements during the geologically short time in which observations are made,

$$\mathbf{r}(t) = \mathbf{r}_0 + \mathbf{\Omega}_{pl} \times \mathbf{r}_0 \, (t - t_0) + \Delta\mathbf{r}(t) , \qquad (5.34)$$

where ($\mathbf{r}_0, \mathbf{v}_0$) are the initial position and velocity at the time t_0. In turn, the angular velocities $\mathbf{\Omega}_{pl}$ are obtained by fitting them to the measured positions and taking into account all the relevant geophysical information, in particular magnetic measurements (Table 5.3). By definition, to ensure that there is no average rotation of the frame relative to the crust, the appropriate weighted average of $\mathbf{\Omega}_{pl}$ must vanish.

Two independent techniques are used to obtain plate motion: (i) geological records of the local magnetization of crustal rocks associated with the oceanic ridges, and (ii) direct measurements with satellite tracking and VLBI. The first gives averages over geological time spans of at least a few My; the second provides present-day values. Any acceleration could show up in a discrepancy between them, with important geophysical consequences; for example, the accumulation of stress energy produced by the interaction between neighbouring plates may suddenly be released in an earthquake and modify the resistance to the motion. Recent comparison between the two techniques shows an overall agreement, but in some particular cases there are discrepancies. It should also be noted that in regions where the structure (and the dynamics) of the crust is complex on a local scale and many micro-plates interact (e.g., in the Mediterranean area), coarse global models are inadequate. The acquisition of longer series of satellite and VLBI observations will certainly be of great help for a better understanding of the causes of earthquakes and predicting their occurrence.

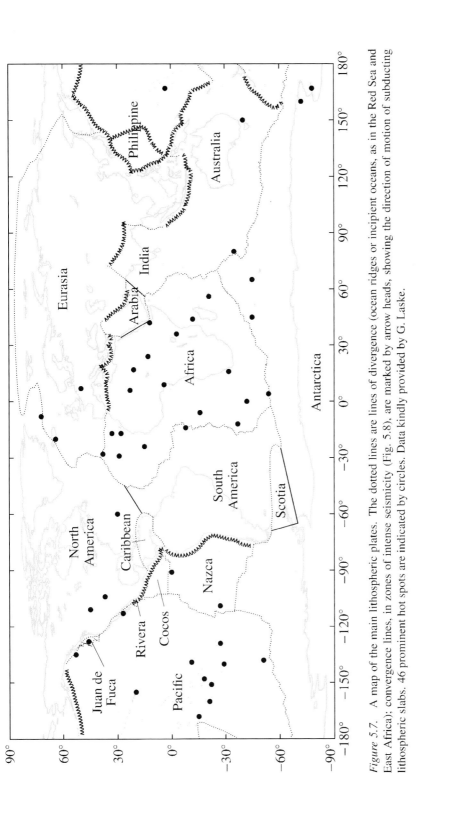

Figure 5.7. A map of the main lithospheric plates. The dotted lines are lines of divergence (ocean ridges or incipient oceans, as in the Red Sea and East Africa); convergence lines, in zones of intense seismicity (Fig. 5.8), are marked by arrow heads, showing the direction of motion of subducting lithospheric slabs. 46 prominent hot spots are indicated by circles. Data kindly provided by G. Laske.

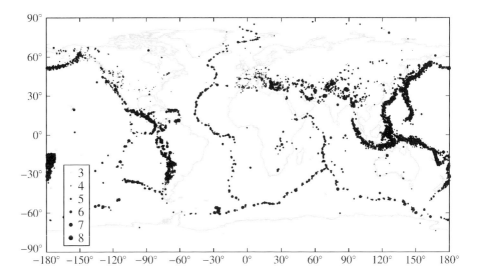

Figure 5.8. Epicentres of earthquakes in 1997 with magnitude > 3. Note their correlation with the boundaries of the tectonic plates (Fig. 5.7). Data acquired from the University of Berkeley (http://quake.geo.berkeley.edu/cnss/).

Plate	Ω_x (mrad/My)	Ω_y (mrad/My)	Ω_z (mrad/My)
Pacific	-1.510	4.840	-9.970
Cocos	-10.425	-21.605	10.925
Nazca	-1.532	-8.577	9.609
Caribbean	-0.178	-3.385	1.581
South America	-1.038	-1.515	-0.870
Antarctica	-0.821	-1.701	3.706
India	6.670	0.040	6.790
Australia	7.839	5.124	6.282
Africa	0.891	-3.099	3.922
Arabia	6.685	-0.521	6.760
Eurasia	-0.981	-2.395	3.153
North America	0.258	-3.599	-0.153
Juan de Fuca	5.200	8.610	-5.820
Philippine	10.090	-7.160	-9.670
Rivera	-9.390	-30.960	12.050
Scotia	-0.410	-2.660	-1.270

Table 5.3 The global kinematical model NNR-NUVEL1A with 16 lithospheric plates in the CTRS according to DeMets et al., *Geophys. Res. Lett.* **21**, 2, 191 (1994). The components of the rotation vector are given for each plate.

The CTRS ideal rigid body is tied to the lithosphere and is not directly related to the interior of the planet; in particular, it can not provide a determination of the principal axes of inertia. Any information about motions in the upper mantle with respect to the CTRS would be a very valuable contribution to our understanding of the *dynamics* of the mantle-crust system. "Markers"

5. The interior of the Earth

of this kind are provided by hot spots; as said earlier, they probably are plumes of hot material ascending from the deep mantle, unrelated to the lithospheric motions; in fact, they have almost fixed positions in the CTRS. The abundance and prominence of hot spot volcanism in regions like Africa and Eastern Asia (as compared, for instance, with America) are consistent with the very small (≤ 1 cm/y) velocities of the corresponding plates (see Fig. 5.7). In general, the CTRS plate velocities are not correlated with their size, but, rather, with the length of the subduction margins; they also seem to anticorrelate with the extension of their continental parts.

The energy needed to feed the observed seismic, volcanic and orogenetic activities in the lithosphere is $\approx 10^{18} - 10^{19}$ erg/s; the thermal energy upward flux from the Earth's interior, about two orders of magnitude greater, is an amply adequate source. However, the detailed interrelation between tectonic motions and the system of convective currents in the upper mantle is still largely unknown, despite significant improvements in three-dimensional modelling of the crust and the mantle.

– PROBLEMS –

5.1. Prove that within a spherically symmetric planet, where the propagation velocity v of the wave is a function only of the radius r, along any given ray the quantity $p = r \sin i / v(r)$ is a constant (here i is the angle between the ray and the radial direction). (Hint: Consider three homogeneous spherical layers and apply Snell's law at the two boundaries.)

5.2. Deduce the constancy of p (previous problem) from the ray equation (8.39).

5.3.* Compute as a function of the incidence angle, the energy conversion factor from S- to P-waves and from P- to S-waves at a free boundary.

5.4. What is the heat flux needed to balance radioactive heat production in the Earth?

5.5. Estimate the global kinetic energy of plate motion.

5.6. On the basis of the model of DeMets et al. for plate motions (Table 5.3), compute the relative velocity between some important cities, i.e., New Delhi with respect to Moscow.

5.7. Obtain eq. (5.27) for a perfect gas with a constant polytropic index γ.

5.8.* In a spherical planet of constant initial temperature T_0 a radiogenic heat source with half-life $t_{1/2}$ is located at the centre. Assume homogeneous planetary model with thermal diffusivity χ and conductivity κ. Neglecting convection, compute the surface temperature of the planet as a function of time. Discuss also the case in which source is uniformly distributed throughout the planet.

5.9. Verify that the $\simeq 25$ km mean difference of the crust-mantle boundary (MOHO) below continents and oceans is in a rough agreement with the Airy compensation mechanism (Sec. 2.4). (Hint: Assume a mean ocean depth of $\simeq 4$ km and mean crust and mantle densities of $\simeq 2.8$ g/cm^3 and $\simeq 3.3$ g/cm^3, respectively; Fig. 5.4.)

5.10.* A spherical and homogeneous planet with radius R and negligible albedo rotates with an angular velocity ω. Solve for the temperature distribution assuming a thermal emission from the surface according the Stefan-Boltzmann law (1.99). (Hint: In terms of the thermal diffusivity χ, the lack of temperature uniformity is determined by the dimensionless parameter $\chi R^2/\omega$; expand in spherical harmonics.)

5.11.* In a simplified model of a homogeneous, non-rotating and spherical planet spheroidal and toroidal modes of deformation are represented, respectively, by $\mathbf{s} = \boldsymbol{\nabla} u + \boldsymbol{\nabla} \times (\boldsymbol{\nabla} \times v\mathbf{r})$ and $\mathbf{s} = \boldsymbol{\nabla} \times (w\mathbf{r})$ (eq. (5.18)). Developing the scalar functions (u, v, w) in spherical harmonics and neglecting self-gravity, prove that spheroidal and toroidal deformations decouple. Compute the spectrum of proper spheroidal and toroidal oscillations of the planet at the lowest degrees ($\ell \leq 2$).

5.12. Evaluate the mean value $\langle \delta d \rangle$ of the change δd of the interatomic distance d in a lattice for the interaction potential $U(\delta d) = a(\delta d)^2 - b(\delta d)^3$ (where a and b are positive) and the linear expansion coefficient $\langle \delta d \rangle/T$.

– FURTHER READINGS –

Three useful, though somewhat older, books exploring in more details the subject of this chapter, are F.D. Stacey, *Physics of the Earth*, Brookfield Press (1992); W.M. Kaula, *An Introduction to Planetary Physics: the Terrestrial Planets*, J. Wiley and Sons (1968) and Y.N. Zarkov, *Internal Structure of the Earth and the Planets*, Nauka-Mir (1986). The second part of the last book also contains discussions of the interiors of the Moon and planets. More recent are C.M.R. Fowler, *The Solid Earth: An Introduction to Global Geophysics*, Cambridge University Press (1990), I. Jackson, *The Earth mantle: Composition, structure, and evolution*, Cambridge University Press (2000), and A. Udias, *Principles of seismology*, Cambridge University Press (2000). L. Brekhovskikh and V. Goncharov, *Mechanics of Continua and Wave Dynamics*, Springer-Verlag (1985) contains a clear discussion of wave propagation in a solid, while A. Ben-Menahem and S. Jit Singh, *Seismic waves and sources*, Dover Publications (2000) gives a very complete discussion of this topic. The issue of convection in the planetary mantle is discussed by G. Schubert, D.L. Turcotte and P. Olson, *Mantle convection in the Earth and planets*, Cambridge University Press (2001). On the normal modes, there is E.R. Lapwood and T. Usami, *Free Oscillations of the Earth*, Cambridge University Press (1981). The efforts to construct a new Reference Earth Model (REM) are described at http://mahi.ucsd.edu/Gabi/rem.html. A good, though older, textbook on planetary interiors, not discussed here, and their evolution is W.B. Hubbard, *Planetary Interiors*, Van Nostrand Reinhold Co. (1984).

Chapter 6

PLANETARY MAGNETISM

The fact that a small needle of lodestone suspended at its centre of mass maintains an almost constant orientation with respect to the Earth's body has been known for thousands of years and widely used in compasses for navigation purposes. Indeed, the magnetic compass has been employed in China at least since the 4th century AD and used by Amalfi's seamen since the 11th century. In 1600 W. GILBERT associated this effect with the magnetic properties of the Earth itself. In his *De Magnete*, one of the earliest scientific treatises ever written, Gilbert stated that the Earth behaves like a great permanent magnet, similar to a uniformly magnetized sphere. In spite of these early practical and theoretical achievements, the study of the Earth's magnetic field has progressed slowly. Among its difficult aspects are: (i) its complex structure and rapid time variability; (ii) the fact that it is produced by at least three different sources (currents in the conductive, rotating and convective core; magnetization of crustal rocks; plasma currents in the ionosphere and further out); and (iii) the problem of its origin and permanence in a highly dissipative medium. The latter problem was described by A. EINSTEIN as one of the three most difficult, unsolved issues of physics. Today, thanks mainly to space missions, we have a much more detailed knowledge of the Earth's field and also of the magnetic properties of other planetary bodies. There is a wide consensus that in a rotating planet with a fluid and conductive core, a dipole magnetic field parallel or antiparallel to the rotation axis arises; however, the details of this complex process, and its main model, the *dynamo theory*, are still not completely understood.

Magnetic fields provide important evidence about the structure of planetary interiors and, for the Earth, about the motions of its crust. In addition, the interaction of a planetary magnetic dipole with the solar wind determines the complex and peculiar properties of the *magnetosphere* (Ch. 10).

6.1 The main dipole field

The magnetic field $\mathbf{B}_0$ on the Earth's surface can be expressed in terms of its radial component A (positive downward), its horizontal component H and the deflection angle D between the latter and the geographical North (positive eastward); D is the *magnetic declination*. Also traditionally used is the angle $I = \arctan(H/A)$ between the line of force and the horizontal plane (the *magnetic inclination*). In Cartesian coordinates x (northward), y (eastward) and z (downward),

$$\begin{pmatrix} B_{0x} \\ B_{0y} \\ B_{0z} \end{pmatrix} = \begin{pmatrix} H \cos D \\ H \sin D \\ A \end{pmatrix} = B_0 \begin{pmatrix} \cos I \cos D \\ \cos I \sin D \\ \sin I \end{pmatrix}. \tag{6.1}$$

Since no free magnetic monopole is known to exist ($\nabla \cdot \mathbf{B} = 0$), if local currents and time varying electric fields are neglected, so that $\nabla \times \mathbf{B} = 0$, the magnetic field can be expressed as gradient of a harmonic scalar potential V_M: $\mathbf{B} = -\nabla V_M$. The theory is then formally identical with the theory of the gravitational field (Ch. 2), with the difference that the main term in the spherical harmonics expansion is the dipole, not the monopole. This component is determined by the dipole vector $\mathbf{d}$, which we now assume at the centre of the Earth:

$$\mathbf{B} = -\nabla \left(\frac{\mathbf{d} \cdot \mathbf{r}}{r^3} \right), \quad V_M = \frac{\mathbf{d} \cdot \mathbf{r}}{r^3}. \tag{6.2}$$

In magnetic spherical coordinates

$$B_\theta = -\frac{1}{r} \frac{\partial V_M}{\partial \theta_M} = \frac{d}{r^3} \sin \theta_M, \quad B_r = -\frac{\partial V_M}{\partial r} = \frac{2d}{r^3} \cos \theta_M. \tag{6.3}$$

(θ_M, the *magnetic colatitude*, is the angle between $\mathbf{d}$ and $\mathbf{r}$ and $d = |\mathbf{d}|$). On the surface $r = R$

$$B_{0\theta} = -H = B_e \sin \theta_M, \quad B_{0r} = -A = 2 B_e \cos \theta_M, \tag{6.4}$$

where $B_e = d/R^3 \simeq 31,000$ nT (corresponding to $d \simeq 7.8 \times 10^{30}$ nT cm^3)[1]. The axis of the magnetic dipole $\mathbf{d}$ makes an angle $\delta_M = 169°$ with the Earth's polar axis. Although, as recognized by Gilbert, a dipole field can be generated by a uniformly magnetized sphere, such large value of d cannot be justified. The proper source of the magnetic field, residing deep in the Earth interior, will be discussed in Sec. 6.4.

---

[1] Our unit for the magnetic field is the *nanoTesla* (nT) which is the commonly used SI unit; the older literature uses mostly the cgs unit Gauss (G), with the transformation: 1 G = 10^5 nT.

6. Planetary magnetism

From the previous equation we get

$$B_0 = (B_{0\theta}^2 + B_{0r}^2)^{1/2} = B_e (1 + 3\cos^2\theta_M)^{1/2} ; \qquad (6.5)$$

the field at the magnetic poles is two times larger than at the equator and

$$\tan I = B_{0r}/B_{0\theta} = 2 \cot\theta_M . \qquad (6.6)$$

Since $B_r/B_\theta = dr/rd\theta_M$, the lines of force $r(\theta_M)$ are solutions of the differential equation

$$\frac{1}{r}\frac{dr}{d\theta_M} = 2\cot\theta_M , \qquad (6.7)$$

readily integrated to

$$r = L\sin^2\theta_M . \qquad (6.8)$$

Here L is the radial distance at which the line of force crosses the (magnetic) equatorial plane. The distribution of charged particles trapped in the Earth magnetic field (Sec. 10.3) is conveniently described in *McIlwain's coordinates* $(\beta = B/B_1, L)$, a one-to-one map of the magnetic polar coordinates (r, θ_M) on one hemisphere (B_1 is the equatorial value of the magnetic field strength on a given field line; $\beta \geq 1$). Notably, from (6.3) and (6.8) we have

$$\beta(\theta_M) = \frac{(1 + 3\cos^2\theta_M)^{1/2}}{\sin^6\theta_M} , \qquad (6.9a)$$

$$L(r, \theta_M) = \frac{r}{\sin^2\theta_M} . \qquad (6.9b)$$

Note that the L = const. coordinate lines are just the field lines of the dipole magnetic field, to first approximation followed by trapped charged particles (Sec. 10.3). The inverse transformation $(\beta, L) \to (r, \theta_M)$ cannot be accomplished analytically, but relies on the numerical solution of

$$\beta\left(rL^{-1}\right)^3 - \left(4 - 3rL^{-1}\right)^{1/2} = 0 . \qquad (6.10)$$

Obviously, r/L in (6.10) is just $\sin^2\theta_M$.

6.2 Magnetic harmonics and anomalies

If **B** is curl-free the spherical harmonic expansion of the potential V_M in general includes terms proportional to $1/r^{\ell+1}$ (due to internal sources; see eq. (2.17)) and to r^ℓ (for an external source like the ionosphere; see eq. (2.20)). We thus have

$$V_M = \frac{1}{R}\sum_{\ell \geq 1}\sum_{m=0}^{\ell} \bar{P}_{\ell m}(\cos\theta) \times$$

$$\left\{ \left[c_{\ell m}\left(\frac{r}{R}\right)^{\ell} + (1-c_{\ell m})\left(\frac{R}{r}\right)^{\ell+1} \right] C_{\ell m} \cos m\varphi + \right.$$
$$\left. \left[s_{\ell m}\left(\frac{r}{R}\right)^{\ell} + (1-s_{\ell m})\left(\frac{R}{r}\right)^{\ell+1} \right] S_{\ell m} \sin m\varphi \right\}. \quad (6.11)$$

Here θ and φ are the geographical colatitude and longitude and R the mean Earth radius; the dimensionless coefficients $c_{\ell m}$ and $s_{\ell m}$ (lying between 0 and 1) describe the fractional contributions of the external sources. Unlike for the geopotential (Ch. 2), in geomagnetic work associate Legendre functions $\bar{P}_{\ell m}(\cos\theta)$ are conventionally normalized with the definition

$$\bar{P}_{\ell m}(\cos\theta) = N'_{\ell m} P_{\ell m}(\cos\theta) = \sqrt{(2-\delta_{m0})\frac{(\ell-m)!}{(\ell+m)!}}\, P_{\ell m}(\cos\theta)\,,$$

usually called quasi-normalization. Comparing this with eq. (2.13), we see that the factor $(2\ell+1)$ is missing. Given V_M from (6.11), the magnetic field reads

$$B_\theta = -\frac{1}{r}\frac{\partial V_M}{\partial \theta}\,, \quad B_\varphi = -\frac{1}{r\sin\theta}\frac{\partial V_M}{\partial \varphi}\,, \quad B_r = -\frac{\partial V_M}{\partial r}\,. \quad (6.12)$$

With the traditional method, using magnetometers on the surface of the Earth $r = R$, the four sets of coefficients c, s, C, S can, in principle, be measured (in fact, B_θ and B_φ do not contain c and s, but B_r does). C.F. GAUSS first applied this method in 1839 and with the data available at that time showed that $c_{\ell m} = s_{\ell m} = 0$, i.e. the geomagnetic field is entirely internal. Today we know that this is only a good approximation, since there are corrections up to $B_e/100$ (with typical timescales ranging from milliseconds to 11 y, the solar cycle period) due to currents in the ionosphere and further out in the magnetosphere. Such external fields induce electric currents in the upper mantle and temporal variations of the internal field which depend upon the conductivity and can be used to investigate its structure. All these electric currents and time-dependent electric fields produce a curl of the magnetic field and require, besides the scalar potential, a vector potential as well. Thus the representation (6.12), with (6.11), has only a limited value.

The internally generated field (from eq. (6.11)) can be written

$$V_M = R \sum_{\ell m} \left(\frac{R}{r}\right)^{\ell+1} \bar{P}_{\ell m}(\cos\theta)(g_{\ell m}\cos m\varphi + h_{\ell m}\sin m\varphi)\,,$$

where the *Gauss' coefficients* $g_{\ell m} = (1-c_{\ell m})C_{\ell m}/R^2$, $h_{\ell m} = (1-c_{\ell m})S_{\ell m}/R^2$ have the dimension of a magnetic field. The $\ell = 1$ terms give the dipole field:

6. Planetary magnetism

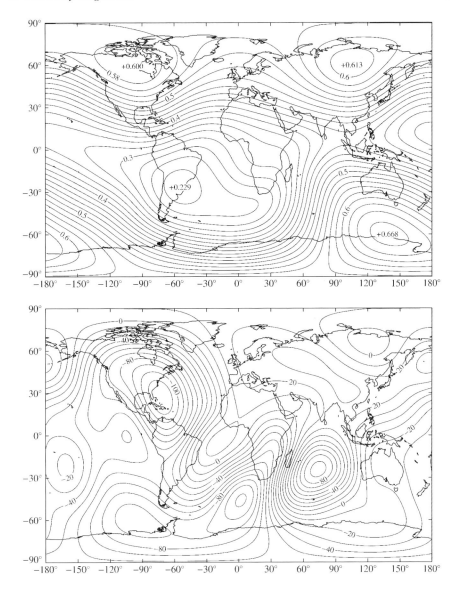

Figure 6.1. The total intensity of the geomagnetic field at the surface (in 10^5 nT or Gauss) in the upper plot and its secular change (in nT/y) at the bottom, both for epoch 2000.0; data taken from the International Geomagnetic Reference Field (IGRF2000) available at http://www.ngdc.noaa.gov/.

$g_{10} = d_z/R^3$, $g_{11} = d_x/R^3$, $h_{11} = d_y/R^3$ (with d_x, d_y, d_z being geographic components of the magnetic dipole); the tilt angle is

$$\delta_M = \arctan[(g_{11}^2 + h_{11}^2)^{1/2}/g_{10}] .$$

ℓ	m	$g_{\ell m}$	$h_{\ell m}$	$dg_{\ell m}/dt$	$dh_{\ell m}/dt$
1	0	−29,615	0	14.6	0
1	1	−1,728	5,186	10.7	−22.5
2	0	−2,267	0	−12.4	0
2	1	3,072	−2,478	1.1	−20.6
2	2	1,672	−458	−1.1	−9.6
3	0	1,341	0	0.7	0
3	1	−2,290	−227	−5.4	6.0
3	2	1,253	296	0.9	−0.1
3	3	715	−492	−7.7	−14.2

Table 6.1 IGRF2000 Gauss' coefficients (and their secular changes) for the dipole, quadrupole and octupole components (mean Earth radius $R =$ 6371.2 km assumed). Units: nT and nT/y.

The International Association of Geomagnetism and Aeronomy (IAGA) has established a working group to elaborate various geomagnetic field representations, from which the Earth's main field and its secular variation can be computed. IAGA also updates about every 5 years Gauss' coefficients, which provide the International Geomagnetic Reference Field (IGRF); because of the mentioned limitations of the potential representation, they are usually truncated at $\ell = 10$. In Table 6.1 we list them up to $\ell = 3$ for the IGRF2000 model, based on the measurements of low flying Danish satellite Oersted (a follow-up of the successful MAGSAT mission launched in 1979). Oersted, with its coverage and data quality, provided a model up to $\ell = 20$; however, the degrees $\ell > 13$ have only a statistical relevance, while for $\ell \leq 13$ the individual coefficients are reliable (Fig. 6.2). In (6.11), however, it is necessary to take into account the external contribution; the Oersted data showed that the main coefficient $c_{10} C_{10}/R^2$ is about 22.4 nT, well above the measurement uncertainty (≈ 1 nT). With the accurate data from the forthcoming space missions (Oersted 2 and Champ), relevant improvements are expected in modelling and quantitative determination.

Figure 6.1 shows the intensity of the geomagnetic field at the surface and its secular change. Although the dipole coefficients are larger than the quadrupole ones (Table 6.1), the total field intensity at the surface deviates from the pure dipole pattern. In particular, if we operationally define the "magnetic pole" as the point where the magnetic field vector is normal to the Earth surface (i.e., $I = 90°$), the north pole is at 79.0° N and 105.1° W (geographic coordinates). The north pole computed from the dipole field coefficients only ($\ell = 1$) is at 79.3° N and 71.5° W. A more spectacular peculiarity of the Earth magnetic field is the South Atlantic Anomaly, a significant decrease of the field intensity to the east of South America (see Fig. 6.1). This feature allows penetration of energetic cosmic ray particles deep into the ionosphere in this region, which may affect electric circuits of spacecraft.

In the gravitational case a shift of the origin changes simultaneously the monopole term and all higher harmonics; in the magnetic case a shift $\mathbf{r} \to \mathbf{r}+\mathbf{a}$ in the origin produces an additional quadrupole term. Starting with the pure

6. Planetary magnetism

dipole field we obtain

$$V_M = \frac{\mathbf{d}\cdot\mathbf{r}}{r^3} + \mathbf{d}\cdot\frac{r^2\mathbf{1}-3\mathbf{rr}}{r^5}\cdot\mathbf{a} + \ldots = \frac{\mathbf{d}\cdot\mathbf{r}}{r^3} + \mathsf{q}_d : \frac{r^2\mathbf{1}-3\mathbf{rr}}{r^5} + \ldots \quad (6.13)$$

The (traceless) dipole-originated quadrupole is then

$$\mathsf{q}_d = \mathbf{da} - \frac{1}{3}\mathbf{1}\,(\mathbf{d}\cdot\mathbf{a})\,. \quad (6.14)$$

From the known quadrupole field we can determine the actual displacement **a** of the dipole from the centre of the Earth as follows. A general quadrupole term is of the form (see (2.24))

$$\mathsf{q} : \frac{r^2\mathbf{1}-3\mathbf{rr}}{r^5},$$

where **q** is a symmetric, trace-free tensor. We can decompose it with respect to the orthonormal basis $\mathbf{e}_1, \mathbf{e}_2, \mathbf{e}_3 = \mathbf{e}_M$

$$\mathsf{q} = q_{11}\mathbf{e}_1\mathbf{e}_1 + q_{22}\mathbf{e}_2\mathbf{e}_2 - (q_{11}+q_{22})\mathbf{e}_M\mathbf{e}_M + q_{12}(\mathbf{e}_1\mathbf{e}_2+\mathbf{e}_2\mathbf{e}_1)$$
$$+q_{23}(\mathbf{e}_2\mathbf{e}_M+\mathbf{e}_M\mathbf{e}_2) + q_{31}(\mathbf{e}_M\mathbf{e}_1+\mathbf{e}_1\mathbf{e}_M)\,; \quad (6.15)$$

q_d is then given by

$$3\,q_{d11} = 3\,q_{d22} = -a_M d, \quad q_{d12} = 0, \quad q_{d23} = a_2 d, \quad q_{d31} = a_1 d\,. \quad (6.16)$$

A simple, but interesting example of a conclusion about the core can be thus obtained from the surface magnetic field. Since the quadrupole terms are about 10% of the main dipole (Table 6.1), one expects that $a \approx 0.1\,R_\oplus$; direct evaluation shows that $a \simeq 0.07\,R_\oplus$.

Fine structure of the geomagnetic field. Similarly to the gravity field, the internally produced magnetic field is dominated by the lowest degree multipole component (dipole in this case) far away from the planet; higher degree corrections are more and more important as the source is approached. It is interesting to consider the average of the square of the magnetic field components of degree ℓ on the ground, which can be called its "spectrum" or "mean magnetic energy density", just as (2.39) is for the gravity field; it reads

$$R_\ell = (\ell+1)(R/r)^{2\ell+4}\sum_m (g^2_{\ell m} + h^2_{\ell m})\,. \quad (6.17)$$

Components of degree ℓ have a scale of variation $\approx R/\ell$; R_ℓ has a very steep dependence on the degree

$$R_\ell \simeq 1.5\times 10^9\,(3.7)^{-\ell}\,\mathrm{nT}^2\,, \quad (6.17')$$

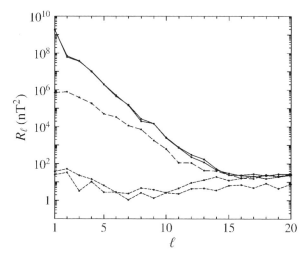

Figure 6.2 The spectra R_ℓ of the Oersted and MAGSAT magnetic models (solid), their errors (short-dashed) and their difference (long-dashed). The decrease in R_ℓ in 20 y, well above the noise (short-dashed lines), may be an indication of an incipient reversal. Data kindly provided by G. Hulot; see also G. Hulot et al., *Nature* **416**, 620 (2002).

to be contrasted with the power law $C_\ell \propto \ell^{-2}$ of Kaula's rule for gravity. This behaviour is observed up to degree $\ell \approx 14$, after which the spectrum levels off (Fig. 6.2); however, the validity of the potential representation for large degrees is uncertain, in particular because of external and time-dependent components. As r/R decreases, each spectral component increases proportionally to $(R/r)^{2\ell}$:

$$\log R_\ell = \ell\,[2\log(R/r) - \log 3.7] + \text{const} ;$$

at $r = R\,(3.7)^{-1/2} \simeq 3{,}300$ km the spectrum is approximately independent of the scale R/ℓ. Note that for a power law $\log R_\ell = -\alpha \log \ell +$ const – a common case for disordered systems – and the logarithm of the spectrum is practically independent of the degree; it is then natural to ascribe the steep dependence (6.17′) not to its intrinsic spectral "colour", but to the depth of the source. This argument strongly suggests that the source lies below 3,300 km, in the conductive and fluid Earth's core.

Interesting results can be obtained when the power spectrum R_ℓ is compared for two models constructed at different epochs; this has been recently done for geopotential models derived from the Oersted data in 2000 and those from the MAGSAT mission, flown in 1979-1980. Both share the same characteristic behaviour (6.17′), but a small and systematic decrease of the low degree components was found (Fig. 6.2). In this way a model of the change in the flow structure in the Earth core was obtained, including the emergence of an asymmetric patch under South Africa (see also Sec. 5.2). It is possible that these changes are premonitory signs of a polarity reverse.

6.3 Secular changes and reversals

Already in the 17th century it was known that the geomagnetic field is not steady, but varies in strength and orientation in a coherent way over large areas

6. Planetary magnetism

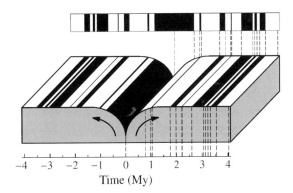

Figure 6.3 The alternate magnetization of the sea floor on both sides of an ocean ridge reveals the magnetic polarity at the time at which rocks emerged from the mantle; black and white intervals correspond to opposite polarities. Adapted from R.T. Merrill et al., *The Magnetic Field of the Earth*, Academic Press, San Diego (1998).

of the Earth (see Fig. 6.1). These variations are much greater than transient changes[2] due to phenomena in the ionosphere and the magnetosphere (or reflecting sudden changes in the Earth core) and probably occur on all timescales from ≈ 10 to $\approx 10^9$ y. Direct observations are available for the last few centuries, showing timescales up to $\approx 10^4$ y. The main observed secular effects are the following:

- The dipole field decreases in strength by about 0.05% per year (≈ 15 nT/y) and drifts westward in an almost precessional way at a rate of about 0.06 °/y, corresponding to a period of $\approx 6,000$ y.

- The non-dipole part of the field changes more rapidly, by some 50 nT/y, but with a large scatter. These variations give rise to an average westward drift of its features of about 0.2 °/y (corresponding to a rotation period of $\approx 1,800$ y with respect to the solid Earth), but with a strong variability between different latitudes and different harmonic components. In fact, these features deform, shift and disappear; they behave more like eddies in a fluid stream than permanent or quasi-permanent structures.

As said in Sec. 5.4, the alternating magnetic polarity seen in the magnetization in the sea floor spreading around a mid ocean ridges can reveal important information about the ancient configuration of the terrestrial magnetic field: as rocks cool below the Curie temperature, they permanently acquire the magnetization of that time. Corresponding layers of alternate magnetization are seen at the same distance from the ridge and provide a measure of the velocity of the continental drift (Sec. 5.4). Figure 6.3 illustrates this important concept.

Another important source of information about the ancient geomagnetic field is the analysis of deep-sea sedimentary sequences (obtained by deep dri-

---

[2]The most spectacular of these abrupt events are called *geomagnetic jerks*; they appear as sudden changes in the second derivatives of the geomagnetic potential, separating periods of constant secular variation. Prominent jerks occurred in 1969, 1978, 1991 and 1999. Their origin is uncertain, but since they imply a reorganization of the secular variation in the dipole field, they are probably due to processes in the core.

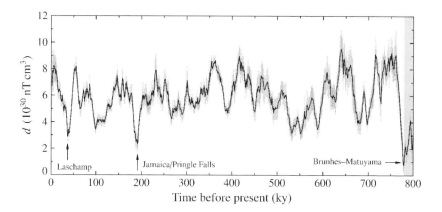

Figure 6.4. Continuous sequence of the Earth dipole intensity d over the last magnetic chron (Brunhes); time before present (in ky) in abscissa and d (in 10^{30} nT cm^3) in ordinate. The last reversal occurs at $\simeq 0.78$ My, when the signal reaches near zero. The peak value is $\simeq 9 \times 10^{30}$ nT cm^3, while the time-averaged value over the Brunhes chron is $\simeq 6 \times 10^{30}$ nT cm^3. Note the significant variations related to excursions of the magnetic axis orientation. Model Sint-800 based on a large number of magnetostratigraphic measurements. Data kindly provided by J.P. Valet; see also Y. Guyodo and J.P. Valet, *Nature* **399**, 249 (1999).

lling), a method called *magnetostratigraphy*. An absolute dating of the extracted specimens can be obtained with radiometric means (K-Ar isotope system is rather precise up to $\simeq 5$ My) or by correlating microfossils in geologic layers on different places on the Earth.

Data collected with these methods form the basis of *palaeomagnetism*, which investigates the past orientation of the main dipole and, to a lower accuracy, its strength, on long timescales. Were the field purely dipolar, the remnant magnetic vector at a given place would determine a fictitious dipole in the centre of the Earth. Since there are higher degree components, however, the dipole moments so determined differ from place to place. To eliminate them one assumes that they have fast variations; averaging the fictitious dipoles over several thousands years produce a reasonable determination of a well defined, global dipole. The Earth has been found to possess a dipole magnetic field since more than 3 Gy and, except during polarity reversals, its intensity has not varied by a factor larger than 3 (Fig. 6.4). Measurements have shown that dipole directions strongly cluster within $\approx 20°$ around the rotation axis. This leads to the *axial dipole hypothesis*: for most of the time the main magnetic field is represented by a geocentric dipole, aligned along the rotation axis in a parallel or antiparallel state. These polarity states – each called *chron* – are separated by quick reversals (Fig. 6.5). The sequence of chrons has been reconstructed for the last ≈ 160 My; older palaeomagnetic data exist – back to ≈ 1 Gy – but the picture is less continuous. The state of vanishing dipole

6. Planetary magnetism

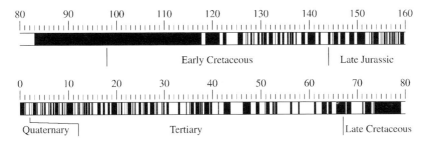

Figure 6.5. The sequence of normal (black) and reversed (white) polarity states in the last 160 million years (scale in My). Compilation of data from different sources published by R.T. Merrill et al., *The Magnetic Field of the Earth*, Academic Press, San Diego (1998).

is metastable; the present tilt angle of 169° may be an indication of such a transient state.

Therefore the Earth's rotation must be an essential ingredient of the mechanism generating the field. In reality the reconstruction of the past dipole must take into account also the different position of the continental plates; this does not affect much the accuracy of the reconstruction, but, given the history of the magnetic reversals, it provides information the past positions of the continents (as it has been discussed in Sec. 5.4).

The possibility that the intensity of the geomagnetic field undergoes variations modulated by the Earth's orbital and rotational parameters (see Sec. 7.1) has been repeatedly suggested; the periodicities of 41 ky and 100 ky, known from the eccentricity and obliquity variations, were announced. Nevertheless, the last studies do not seem to unambiguously confirm this hypothesis; it is thus not known whether the Earth orbital parameters have a direct influence on the geodynamo. Similarly, magnetic reversals display no clear periodicity or regularity (see Fig. 6.5). The current reversal rate is a few times per My (the last one occurred $\simeq 0.78$ My ago), but this rate has varied a great deal over geological times: two very long periods with no reversal, lasting 36 and 70 My, began about 118 and 320 My ago respectively (the former is usually called *Cretaceous superchron*). Reversals take place in $\approx 10^3$ y; on this timescale there are also many fluctuations, incomplete or temporary reversals, referred to as *palaeomagnetic excursions*. The timescale of 100 My in the reversal *rate* may, on the other hand, reflect changes in the pattern of convection currents in the fluid outer core.

6.4 * The generation of planetary magnetic fields

Under some conditions, which have been reviewed in Ch. 5, the heat content and the internal pressure of a planet are so high that its core melts and behaves as a very viscous fluid. When it contains metals, it is also an electric conductor and it is in principle possible that the kinetic energy of its internal motion is transformed, through magnetohydrodynamical (MHD) effects, into magnetic energy; this is usually called a *dynamo problem*. Characteristic of the dynamo problem is the requirement that there are no sources of the magnetic field outside the body (a special variant of a magneto-convective dynamo will be mentioned in Sec. 6.5). The criterion for a successful dynamo is to actively maintain the magnetic field as long as the energy sources of internal motion exist.

Details of a working dynamo, resulting in the main dipole field, approximately aligned with the angular momentum, are still to some degree a poorly understood problem in planetary physics, although its mathematical basis appears – at a first glance – straightforward and deceptively simple. A given velocity field **v** (determined by external forces and convection) determines the magnetic field through the magnetic diffusion equation (1.82):

$$\frac{\partial \mathbf{B}}{\partial t} = \lambda \nabla^2 \mathbf{B} + \nabla \times (\mathbf{v} \times \mathbf{B}) \;. \tag{6.18}$$

Here $\lambda = c^2/4\pi\sigma$ is the magnetic diffusion coefficient and σ is the electrical conductivity. Mathematically, we wish to characterize the velocity fields **v** which can amplify and maintain a large scale, aligned dipole field and thus provide the dynamo. This is the *kinematic dynamo problem*: "given **v**, find **B**" neglecting the effect of the magnetic field on the velocity (as such the problem is linear in the unknown **B** and thus apparently simple). This kinematic approach was found unsatisfactory and it was recognized that any fruitful framework entails a characterization of the energy sources for the fluid motion. Then, in principle, the solution for both **v** and **B** are coupled and we have the full *MHD dynamo problem*. Such a problem is in general distinctly non-linear and much more difficult, and must rely on numerical solutions. Still, an acceptable approach is that of partly decoupling the large-scale ("macroscopic") and small-scale ("microscopic") field patterns, and approximate the influence of microscopic effects on the large-scale field (a domain of mean-field electrodynamics; see below).

The induction equation and the failure of the laminar dynamo. The first term in the right-hand side of (6.18) describes the irreversible trend to uniformity, as in the heat conduction. The effect of the last term in eq. (6.18) has been discussed in Sec. 1.7: it tends to anchor the field lines to the fluid elements, dragging them along with the flow (*frozen-in field*). In the interior of the Earth $\lambda \simeq 2 \times 10^4$ cm^2/s; for a dynamo of size R_c ($\approx$ the Earth core radius 3.5×10^8 cm), therefore, the magnetic decay time is

$$\tau_m = \frac{R_c^2}{\lambda} \approx 2 \times 10^5 \text{ y} \;, \tag{6.19}$$

much less than the age of the Earth. This is the key problem: the field cannot be a residual primordial magnetization, but must be continuously regenerated. It is important to note that in terrestrial materials the heat transport coefficient is $\approx 10^5$ times smaller, thus allowing the permanence over geological times of internal heat sources (Sec. 5.3). In the enormous difference between these two diffusion processes lies the peculiarity of dynamo processes. Magnetic diffusion processes over a scale ℓ are characterized by the *magnetic Reynolds number*

$$Re_m = \frac{\ell v}{\lambda} = \frac{4\pi \ell v \sigma}{c^2}, \qquad (6.20)$$

which gives the order of magnitude of the ratio between the two terms in the right-hand side of eq. (6.18). We can estimate the velocity v in the Earth's molten core from the known westward drift of 0.2 °/y of the non-dipole field. At the radius R_c this corresponds to $v \approx 0.01$–0.1 cm/s, giving a magnetic Reynolds number of ≈ 300. Note that this velocity is several orders of magnitude higher than the typical convection velocity in the mantle ($\approx 10^{-7}$ cm/s) and that the kinematical timescale $\tau_R = R_c/v \approx 10^3$ y is much shorter than τ_m.

Before coming to physics, we discuss some general mathematical properties of eq. (6.18). *Axially symmetric* vector fields

$$\mathbf{B} = [B_r(r, \theta), B_\theta(r, \theta), B_\varphi(r, \theta)]$$

can be *poloidal* ($B_\varphi = 0$) or *toroidal* ($B_r = B_\theta = 0$) (Sec. 5.2). It is easy to see that the curl operator transforms a poloidal field into a toroidal field, and vice versa. In fact, consider the fluxes Φ_m and Φ_p through a surface on a meridian plane and through a surface on a plane of parallels, respectively. For poloidal field, of course, $\Phi_m = 0$; for a toroidal field $\Phi_p = 0$. Similarly, consider the circulations c_m and c_p along a closed circuit on a meridian plane and a closed circuit on a plane of parallels, respectively; for the poloidal and toroidal fields $c_p = 0$ and $c_m = 0$, respectively. Since the circulation of the curl of a vector equals its flux, the theorem is proved. Note also that the vector product obeys the simple multiplication rules

$$\mathbf{P}_1 \times \mathbf{P}_2 = \mathbf{T}, \quad \mathbf{P}_1 \times \mathbf{T} = \mathbf{P}_2, \quad \mathbf{T}_1 \times \mathbf{T}_2 = 0,$$

with $\mathbf{P}_1$ and $\mathbf{P}_2$ arbitrary poloidal vectors and $\mathbf{T}$, $\mathbf{T}_1$ and $\mathbf{T}_2$ arbitrary toroidal vectors. In contrast to the curl operator, the diffusion operator in eq. (6.18) does not change the type of the magnetic field.

The possibility that slowly varying, axially symmetric flows can maintain or amplify an overall magnetic field has been investigated, with no success; this is the *laminar dynamo problem*. As a simplest "no go" theorem of this kind, we show that a toroidal velocity field $\mathbf{v}$ (including differential rotation), as required by the axial dipole hypothesis, is not able to compensate diffusion. The induction operator $\nabla \times (\mathbf{v} \times \mathbf{B})$ gives 0 or a toroidal field and can never offset the diffusion operator for a poloidal magnetic field; its generation by a dynamo process cannot be driven by an axially symmetric toroidal velocity flow. Laminar dynamos are inadequate also factually: convection induces in the core of the planet turbulent motions, whose elementary eddies are not axially symmetric and change in time. The permanent magnetic field over a planetary scale can arise from the interaction and interference of elementary magnetic eddies driven by convection. In fact, a thorough mathematical analysis culminates in *Cowling's theorem*, which states that an axisymmetric **B** cannot be sustained by a kinematic dynamo action as an admissible solution of the induction equation (6.18) for *any given* velocity flow. Some degree of self-interaction between the magnetic and velocity fields is necessary, and simple examples are shown below.

Turbulent dynamo. Some intuitive insight on this interaction can be gained by considering the effect of differential rotation and convection on a line of force (Fig. 6.6). In general one expects that the core of rotating celestial bodies, which have suffered a greater compression, has a larger angular velocity. In the fluid upper core of the Earth this differential rotation drags around to the East a poloidal line of force (a) with respect to the mantle and generates a toroidal field with a North-South antisymmetry (b, c); this process is usually denoted as ωPT. The heat generated below (e.g., as a result of latent heat released by freezing at the boundary of the solid inner core; Sec. 6.5) is transported upward by hotter and less dense convective cells (15 – 20 in number, say), which rise because of buoyancy. A pressure gradient may then arise in the horizontal plane, which pushes in fluid from outside the rising column to compensate for the vertical displacement. When planetary rotation is added, one expects *cyclonic* and *anticyclonic* flows, a very general feature in rotating atmospheres (Sec. 7.4). The pressure gradient is in part balanced by a Coriolis force due to a circulation in the horizontal plane, with a velocity given by eq. (7.50); for a low pressure trough – our case – this circulation is counter-clockwise in the northern hemisphere and vice versa. A rising and rotating convective cell distorts the toroidal lines of force into small scale Ω-shaped kinks (d, e); note that, due to the high value of the magnetic Reynolds number Re_m (eq. (6.20)), the magnetic field can be dragged along by the velocity flow (see Alfvén's frozen-flux theorem in Sec. 1.7). Their right sides are displaced to deeper levels by rotation and generate a poloidal, northward component; the left sides of the kinks are displaced to lower depths and generate an opposite poloidal component by reconnection (this is, in turn, allowed by a finite value of Re_m and the fact that diffusion of the magnetic field lines is largest where strong field gradients occur). The first, however, prevails because of the smaller volume available, so that the initial poloidal field is amplified. This local transformation of a toroidal in a poloidal field is called αTP; we can also

6. Planetary magnetism

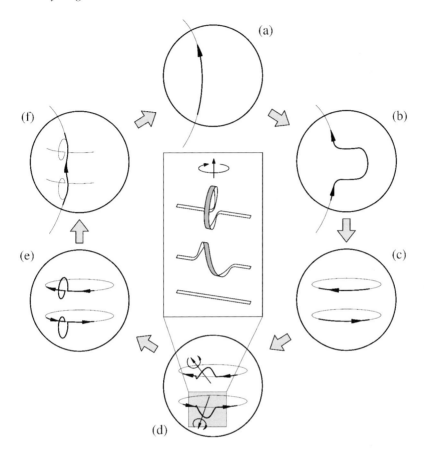

Figure 6.6. A sequence of configurations allowing to sustain a large-scale poloidal field through the $\omega\alpha$ process; details in the text.

have αPT. Note that, contrary to the ordered, large scale ωPT process, this is a local random and turbulent mechanism. Local poloidal loops then coalesce (f) into a large-scale poloidal field that enhances the original poloidal "seed". It has been found that the same elementary processes can be combined in other ways to regenerate the initial poloidal field; for instance, in the α^2 dynamo, the poloidal field is reconstructed with αPT and αTP processes operating in the core turbulent cells (see below).

After explaining the principle in physical terms, we now briefly turn to the mathematical side of *mean-field electrodynamics*. As often assumed in the theory of turbulence (Sec. 7.5), we split the fields in a part $\mathbf{B}_0$, $\mathbf{v}_0$, which changes over the planetary scale R, and a fast varying part $\mathbf{B}'$, $\mathbf{v}'$, with a scale $\ell \ll R$. The primed quantities are stochastic variables with zero mean and describe an ensemble of planetary configurations with the same large scale fields. By averaging over the ensemble (indicated with angular brackets), the properties of the mean planetary field $\langle \mathbf{B} \rangle = \mathbf{B}_0$ are obtained. The average of eq. (6.18) is

$$\frac{\partial \mathbf{B}_0}{\partial t} = \lambda \nabla^2 \mathbf{B}_0 + \nabla \times (\mathbf{v}_0 \times \mathbf{B}_0 + c \mathbf{E}_0) \qquad (6.21)$$

and, by difference, we get

$$\frac{\partial \mathbf{B}'}{\partial t} = \lambda \nabla^2 \mathbf{B}' + \nabla \times (\mathbf{v}_0 \times \mathbf{B}' + \mathbf{v}' \times \mathbf{B}_0 + c \mathbf{E}') \qquad (6.22)$$

for the stochastic field. Here we introduced an effective electric field (also called electromotive field)

$$\mathbf{v}' \times \mathbf{B}' = c \left(\mathbf{E}_0 + \mathbf{E}' \right), \qquad (6.23)$$

with an average $c \mathbf{E}_0$ ($= \langle \mathbf{v}' \times \mathbf{B}' \rangle$) and a stochastic part $c \mathbf{E}'$. In (6.22) the term $\nabla \times (\mathbf{v}' \times \mathbf{B}_0)$, which drives the turbulence, must be of the same order of magnitude as the diffusion term $\lambda \nabla^2 \mathbf{B}'$; that is to say, the magnetic Reynolds number of the turbulence $\ell v'/\lambda$ must not be smaller than B'/B_0. Eqs. (6.21) and (6.22) are coupled through the term $c \nabla \times \mathbf{E}_0$. The effective electric field $\mathbf{E}_0$ describes a macroscopic effect of the interplay between the velocity and magnetic perturbations; similarly to the theory of ordinary isotropic turbulence, one can postulate its expression in terms of $\mathbf{B}_0$ on the basis of symmetry arguments. Since $\mathbf{B}_0$ changes little over the turbulence scale ℓ, one can expand

$$c E_{0i} = \alpha_{ij} B_{0j} + \beta_{ijk} \partial_j B_{0k} + \ldots, \qquad (6.24)$$

in terms of the tensors α_{ij} and β_{ijk}, which depend on the average properties of the turbulence. In the simplest case in which the medium is on average homogeneous, isotropic and at rest, $\alpha_{ij} = \alpha \delta_{ij}$ and in the second term the magnetic gradient $\partial_j B_{0k}$ can only appear with the vector $\nabla \times \mathbf{B}_0$; any other choice would require an anisotropic background. Hence the electromotive, large scale field can be written as

$$c \mathbf{E}_0 = \alpha \mathbf{B}_0 - \lambda^\text{t} \nabla \times \mathbf{B}_0, \qquad (6.24')$$

where λ^t is the *turbulent magnetic diffusivity*. The large-scale field then fulfils

$$\frac{\partial \mathbf{B}_0}{\partial t} = \lambda \nabla^2 \mathbf{B}_0 + \nabla \times (\mathbf{v}_0 \times \mathbf{B}_0 + \alpha \mathbf{B}_0) - \nabla \times \left(\lambda^\text{t} \nabla \times \mathbf{B}_0 \right). \qquad (6.21')$$

This is a generalized induction equation to which Cowling's theorem does not apply, and – in principle – one can have non-trivial axisymmetric solutions when $\alpha, \lambda^\text{t} \neq 0$.

To see under what conditions α does not vanish, consider the transformation properties under a reflection of the axes, which leaves the electric vector

6. Planetary magnetism

unchanged, but changes the sign of the magnetic pseudo-vector; hence α is a *pseudo-scalar* and goes into $-\alpha$. This shows that the turbulent velocity field $\mathbf{v}'$ must allow the construction of pseudo-scalar quantities. A typical quantity of this kind is the *kinematical helicity* $\mathbf{v}' \cdot \nabla \times \mathbf{v}'$, which occurs when the motion has a helical structure and it is not possible to find surfaces orthogonal to the velocity field (the helicity vanishes when the velocity is proportional to the gradient of a scalar). A configuration of this kind arises, for instance, when a convective cell rises in a rotating planet: when the rotation vector is directed towards the north pole the helicity is positive in the northern hemisphere and negative in the southern one (see Problem 6.5).

As regard the α-term in eq. (6.21'), whose existence and importance was first noticed in the 1950s by E.N. PARKER, we now have the source

$$\nabla \times (\alpha \mathbf{B}_0) = \nabla \alpha \times \mathbf{B}_0 + \alpha \nabla \times \mathbf{B}_0 .$$

With a toroidal large-scale field $\mathbf{B}_0$, the second term is again toroidal and the planetary magnetic field may be sustained in a steady state (i.e. satisfy $\partial \mathbf{B}_0 / \partial t = 0$). This is the second step in the previously discussed $\omega \alpha$ process. Analogously, with a poloidal magnetic field $\mathbf{B}_0$ the second term is of no use, but if $\nabla \alpha$ is toroidal, magnetic amplification will result. This corresponds to the α^2 process, in which the large scale velocity plays no role to sustain $\mathbf{B}_0$.

The advancements of the computational facilities and physical modelling of the core has recently allowed to construct self-consistent MHD dynamo models directly from the boundary conditions for the energy input. This formulation requires to solve consistently for both the velocity $\mathbf{v}$ and the magnetic $\mathbf{B}$ field. Conceptually, the principal role of the non-null helicity for sustaining the large-scale magnetic field, as described above, has thus been confirmed. It appears however, that small-scale turbulence is not enough to drive the field, but convective motions of every scale with non-vanishing helicity are formed in a rotating fluid, and create a sufficient electromotive field. The geodynamo is maintained principally by large-scale flows.

6.5 Planetary magnetic fields

As mentioned earlier, in planetary bodies the magnetic decay time is much shorter that the cooling time; the latter, often comparable with the age of the solar system, allows conservation of primordial heat from internal sources (radiogenic for terrestrial planetary bodies and gravitational for the giant planets). In planetary cores, therefore, high temperatures and fluid conditions may enable dynamo processes to transform rotational and thermal energy in magnetic energy. Table 6.2 shows that in the Earth and the giant planets (Jupiter, Saturn, Uranus and Neptune, all rotating fast) they are likely to operate.

Table 6.2. Parameters of the main dipole field for planets and some satellites (indicated with S); B_e is the surface equatorial field and data in the last column compare the dipole moment to that of the Earth.

	R (km)	B_e (nT)	$d = B_e R^3$ (nT cm^3)	Tilt to polar axis (°)	$d/d_\oplus$
Mercury	2,440	330	? 4.8×10^{27}	169	6.2×10^{-4}
Venus	6,052	< 2	< 4.3×10^{26}	–	–
Earth	6,378	31,000	7.8×10^{30}	169	1
Moon (S)	1,760	< 0.2	< 10^{24}	–	–
Mars	3,390	5	< 2×10^{26}	–	–
Jupiter	71,398	428,000	1.6×10^{35}	10	2×10^5
Ganymede (S)	2,634	750	1.4×10^{28}	10	1.6×10^{-3}
Io (S)	1,821	1,300	8×10^{27}	small	10^{-8}
Saturn	60,330	21,000	5×10^{33}	≤ 1	641
Uranus	25,600	23,000	3.8×10^{32}	59	50
Neptune	24,765	14,000	2.2×10^{32}	47	28

General considerations and outer planets. To get a deeper insight in the planetary data in Table 6.2, we recall again that a dynamo can operate only when three (independent) conditions are satisfied:

- there is an electrically conductive
- and fluid region,
- and energy is available to drive an irregular flow.

Conductivity is fairly easy to obtain, either by the presence of metallic iron – a common element in the terrestrial planets – or by pressure. The largest energy of free electrons in a solid is their Fermi energy, which increases with the density; if the pressure in the deep interior raises it above the band gap, the material becomes conductive. Typically, a pressure of at least 2.5 Mbar is required, which is attained in planet with the mass of the Earth or larger.

As far as fluidity is concerned, one has to compare the melting temperatures T_m of the major constituents with the interior temperature T_c (Sec. 1.1). In the giant planets, mainly composed of hydrogen and helium, $T_m < T_c$ and, possibly with the exception of a small rocky core, the interior is fluid, indeed they are a "bottomless atmosphere". Jupiter, Saturn and, to a lesser degree, Neptune, are known to lose energy in the infrared band: $\simeq 5,440$, $\simeq 2,010$ and $\simeq 430$ erg/cm^2/s, respectively (the case of Uranus is less clear, but an intrinsic flux of $\simeq 60$ erg/cm^2/s is consistent both with measurements and theoretical models). Such large energy losses, probably due to primordial heat stored during the formation of the planet (Sec. 16.4), are larger than the flux that the adia-

6. Planetary magnetism

	Jupiter	Saturn	Uranus	Neptune
g_{10}	4.242	0.215	0.119	0.097
g_{11}	-0.659		0.116	0.032
h_{11}	0.241		-0.157	-0.099
g_{20}	-0.022	0.016	-0.060	0.075
g_{21}	-0.711		-0.126	0.007
g_{22}	0.487		0.002	0.045
h_{21}	-0.403		0.061	0.112
h_{22}	0.072		0.048	-0.001
R	71,372	60,000	25,600	24,765
ω	0.1759	0.1638	0.1012	0.1083

Table 6.3 Low-degree Gauss' coefficients of the giant planet inner magnetic field (units in $G = 10^5$ nT); the reference values of the planetary radius (R, in km) and rotation rate (ω, in mrad/s) are given. From J.E.P. Connerney, *J. Geophys. Res.* **98**, 18,659 (1993).

batic temperature gradient can support and, as discussed in Sec. 5.3, convective instabilities must ensue. Another possibility to power convection is a release of the gravitational energy of denser material, such as drops of liquid helium, settling to the centre of the planet. It is not surprising, therefore, that a dynamo is present in all the giant planets. For Jupiter the ratio of the quadrupole and octupole to the dipole field (0.25 and 0.20, respectively) is greater than for the Earth (0.14 and 0.10), evidence of a larger source region. While Jupiter's magnetic tilt is similar to that of the Earth, for Saturn it is less then $\approx 1°$. On the other hand, for the icy giants (Uranus and Neptune; Sec. 14.1) the tilt angles are large (some 59° and 47°, respectively), in contradiction with the axial field hypothesis. These planets are *oblique rotators*; the reason why they have magnetic fields so much more asymmetric than the gas giants (Jupiter and Saturn) is unknown. Another peculiarity is a possible large spatial offset of the magnetic dipole (see eq. (6.14)): e.g. for Neptune, the bulk of the quadrupole field may be explained with a spatial offset of the dipole as large as $a \simeq 0.55\,R$ (where R is the planetary radius; in case of Uranus $a \simeq 0.3\,R$). Such an extreme value indicates a source near the surface, probably in the ionized, ice-rich mantle. Planetary rotation, dragging around the dipole field in oblique rotators, creates in the magnetosphere a time-dependent magnetic field and, hence, an electric field (sometimes called the "corotational electric field"; a similar effect happens in the pulsars). The latter generally has a component along the lines of force and can accelerate particles (Problem 6.7). It also has a radial component that can efficiently eject small charged particles from the planetary magnetosphere if their Larmor radius is large enough (Sec. 10.7).

For the interiors of the terrestrial planets and the Moon, mainly made of iron and silicate compounds, the melting temperature of pure material is generally raised by internal pressure to values $> T_c$; but one must take into account the basic thermodynamical fact that the addition of another compound X (e.g., an alloy of iron and sulphur) *decreases* the melting temperature, until a critical concentration is reached at the *eutectic point*. Consider now what may happen when a (fluid) iron alloy cools after the initial formation phase with a large gravitational heating and reaches the freezing point T_m corresponding to its composition: solid iron separates out and, being generally heavier than the other constituents, deposits at greater depths and forms a solid core; above it the concentration of X increases, a lower melting temperature and a fluid state is maintained. This is indeed the case of the Earth, in which the density of the fluid outer core is about 10% less than the density of pure iron at those conditions. Of course, the formation of a solid inner core is critically dependent upon the size (which allows a hotter interior) and the initial composition; it is thought that it is present only in the Earth, Mercury and the Moon.

This very separation of light and heavy constituents, which occurs at the outer boundary of a solid iron core and make it grow continuously, brings about a decrease in gravitational binding energy which must somehow be allowed to flow out and may increase the total energy flux above the adiabatic value corresponding to the local temperature gradient; then convection ensues and provides an energy source for a dynamo. To quote from Stevenson's paper mentioned in Further Readings, *present-day convection in the Earth's outer core is probably a consequence of the growth of the inner core.* This can explain the absence of a dynamo in Mars and Venus; Mercury's outer core is probably thin and it is not clear if a dynamo is operating there capable of explaining the observed magnetic field.

Terrestrial bodies. The Hermean magnetic field was discovered and (poorly) measured in 1974-75 with two flybys of the Mariner 10 spacecraft; its slow rotation and probably frozen iron core make a dynamo an unlikely, but still possible candidate. The analysis of the few Mariner data is complicated by a possible large contribution of the quadrupole (and higher degree) components, due to proximity of Mercury's residual dynamo region to its surface. The Messenger mission of NASA and the BepiColombo mission of the ESA will visit again the planet in 2009 and 2011, respectively, and certainly will clarify this important issue.

Venus is assumed to possess an active dynamo since early after its accretion, perhaps of the same strength as the Earth one. However, after the energy due to the heat left over from the accretion phase diminished, there was apparently no source to replace it. Venus either lacked the necessary internal ingredients (chemical or physical) for solid core formation, or a complete core solidification occurred; contrary to a popular belief, its very slow rotation is a much less important factor for the absence of a magnetic field. It is also notable that Venus would not have maintained any remanent crustal magnetic fields from its proposed early period of dynamo activity (in contrast to the Moon and Mars) because the temperatures in the crust are expected to be above the Curie point.

On the Moon no global dipole was detected by the Apollo orbiters, but small scale magnetic features and even some large scale anomalies are present; their flux tubes reflect electrons which can be detected from space. Returned lunar rock samples show significant remnant magnetization, corresponding to magnetic fields larger in the past, and even reaching values of the order of a Gauss near formation some 4 Gy ago. It is not clear whether these phenomena are related to "local" events such as big impacts, or whether in the past in the core of the Moon a large dynamo was operating, which magnetized the crust. This latter hypothesis requires a reorientation of the lunar crust by a large polar wander, perhaps as a result of large impacts. Support for the dynamo hypothesis comes from direct detection of a very tiny – residual – internal magnetic field by the Lunar Prospector mission. Its strength is compatible with a small (perhaps still molten) lunar core of radius $\approx 250-400$ km (this is in agreement with its rotation state as determined from the analysis of the lunar laser ranging data; Sec. 20.3).

Interestingly, a similar situation holds for Mars. The tightest constraint on the weak internal field of the planet has been obtained with the magnetometer measurements of Mars Global Surveyor. So far this is the only spacecraft that probed the Martian magnetic field at ≈ 150 km altitude, well beneath the ionosphere, which may produce a confounding signal. An upper limit of about 5 nT has been thus obtained. Models of Mars' composition and thermal history suggest several possibilities for a past dynamo, in agreement with the evidence for an iron-rich core of $\approx 1,500$ km radius. The magnetization of Mars' meteorites seems to favour this hypothesis. Mars Global Surveyor also discovered significant local magnetic anomalies (with peak magnitudes of ≈ 500 nT), which appear to be concentrated around the most heavily cratered and oldest terrains. A possible origin of these anomalies thus involves an iron-rich magma (intrusive or due to impact), cooling in the presence of a strong primordial magnetic field. The acquired permanent component may thus record the orientation of the global field at that epoch. All these facts seems to indicate that the once strong field was later inhibited; the exact time at which the Martian dynamo turned on, and how long it lasted, is still a debated problem. The situation is further complicated by the analysis of Mars Global Surveyor orbit indicating that Mars may still possess a molten core.

Jupiter satellites. The magnetometer on board of the Galileo spacecraft revealed very interesting, and sometimes unexpected, facts about the internal magnetic fields of the four Galilean satellites of Jupiter. Ganymede, the largest satellite in the solar system ($R \approx 2,634$ km), is a differentiated body with a nearly certain ongoing internal dynamo. The estimated core radius is about 660 km and the internal dipole moment is 1.4×10^{28} nT cm^3.

Similarly, Io – the Galilean satellite closest to Jupiter – likely has an active dynamo, though in this case the fact that it is embedded in the rather strong ambient magnetic field of Jupiter and the surrounding plasma seems to modify the classical dynamo process (called then *magneto-convective dynamo*). The ambient magnetic field facilitates the α-effect (conversion of the toroidal field to local poloidal loops) because the corresponding Lorentz force, acting on the raising elements of the core, supports appropriate turbulent motion. Note that without this field the turbulence in the core requires a strong enough buoyancy force, which poses a constraint on the minimum temperature gradient in the core. Its radius is estimated to be about 950 km, a fair fraction of the whole satellite, with radius of 1,821 km, and the internally generated dipole moment is 8×10^{27} nT cm^3. In both cases of Ganymede and Io the magnetic dipole is roughly aligned with the external magnetic field of Jupiter at the satellite position.

Another interesting case is Europa. Galileo's flyby indicated rather large perturbations of the background Jovian magnetic field, but interpretation is complicated. Most likely, this satellite does not have an internal dynamo, since its core is solidified; the observed effect may be due to Europa's interaction with surrounding plasma. Another possibility is an induced magnetic field by its salty subsurface ocean, about 100 km deep. Such ocean is supported also by observations of Europa's icy surface, which shows many faults and cracks, resembling moving ice blocks on a liquid layer.

Finally, the diversity of the Galilean satellites, as far as their physical characteristics are concerned, is stressed by Callisto, which does not show any trace of internal magnetic field; the very weak signal of Galileo's magnetometer during its flyby is most likely related to changes in the Jovian plasma around this satellite. Gravity data also indicate a low degree of differentiation, despite of its large size (its 2,403 km radius is comparable to Ganymede's). The large difference between Ganymede and Callisto has been for some time one of the major puzzling facts of the Galilean satellites. It is possible that a different orbital history is responsible for their different physical properties. The orbit of Ganymede, involved in the 2/1 orbital resonance with Europa, and in the Laplacian resonance with Io and Europa, might have undergone periods of significantly increased eccentricity, later damped by tidal effects (see Sec. 14.2). In the high-eccentricity phases Ganymede could suffer a tidally-driven melting that triggered differentiation. On the contrary, Callisto's orbit is the only one, among the Galilean satellites, which is not resonant. Its past orbital eccentricity could have been low enough to prevent such melting. It is, however, also possible that conditions in the protonebula around Jupiter favoured Ganymede's melting, while left Callisto undifferentiated (Sec. 16.4). In both cases, the solidified core of Callisto thus does not allow dynamo processes.

– PROBLEMS –

6.1. Show that a uniformly magnetized sphere produces the same field of a suitable dipole situated at its centre. (Hint: Consider two spheres made of monopoles of opposing polarities, with a small shift.)

6.2. Find the vector potential of which the dipolar field is the curl.

6.3. Find the total energy of the dipole field of the Earth and compare it with its rotational energy.

6.4. Suppose that the dipole field of the Earth is produced by a uniformly magnetized sphere of radius r. If the magnetic energy is equal to the kinetic energy of convection, estimate its rms speed.

6.5. Calculate the helicity of the velocity field composed of a radial, uniform flow and an axial rotation with uniform angular speed.

6.6. Find, as a function of latitude and longitude, the declination D and the deflection from the horizontal plane I of a dipole field of given axis.

6.7. Estimate the energy gain of a proton due to the corotational electric field in the magnetospheres of giant planets.

6.8. Draw a grid of the McIlwain coordinates $L = $ const. and $\beta = B/B_0 = $ const. in the region $R \geq 1$ and $\theta_M \in (0, \pi/2)$ (see eqs. (6.9)).

6.9.* A simple model for the source of a planetary magnetic field is an azimuthal current density $J_\phi = I F(r) G(\cos\theta)$, with total current I; for example, for an equatorial ring at a distance a from the centre $F(r) = \delta(r - a)$ and $G(x) = \delta(x)$. Determine Gauss' coefficients $g_{\ell m}$ and $h_{\ell m}$. What happens if the ring has a small displacement orthogonal to the equatorial plane.

– FURTHER READINGS –

The standard textbooks on geomagnetism are R.T. Merrill, M.W. McElhinny and P.L. McFadden, *The Magnetic Field of the Earth*, Academic Press (1998) and W. Campbell, *Introduction to Geomagnetic Fields*, Cambridge University Press (1997). Older, but still valuable, is E.N. Parker, *Cosmical Magnetic Fields*, Clarendon Press (1979). Palaeomagnetism is discussed in J.A. Jacobs, *Reversals of the Earth's Magnetic Field*, Cambridge University Press (1994). On dynamo processes and planetary magnetism in general, see H.K. Moffat, *Magnetic Field Generation in Electrically Conducting Fluids*, Cambridge University Press (1978), and F. Krause and K.H. Rädler, *Mean-field Magnetohydrodynamics and Dynamo Theory*, Pergamon Press (1980). The most up-to-date reviews of the Earth dynamo are P.H. Roberts and G.A. Glatzmaier, Geodynamo theory and simulations, *Rev. Mod. Phys.* **72**, 1081 (2001) and G.A. Glatzmaier, Geodynamo simulations – How realistic are they?, *Ann. Rev. Earth Planet. Sci.* **30**, 237 (2002); planetary magnetic fields are discussed in D.J. Stevenson, Planetary magnetic fields, *Rep. Prog. Phys.* **46**, 555 (1983), with an update for Mars in D.J. Stevenson, Mars' core and magnetism, *Nature* **412**, 214 (2001). Magneto-convection and related issues are discussed in *Lectures on Solar and Planetary Dynamos*, M.R.E. Proctor and A.D. Gilbert, eds., Cambridge University Press (1994). D.P. Stern, A millenium of geomagnetism, *Rev. Geophys.* **40**, 1 (2002) is a good historical review. Many useful data related to the Earth magnetic field and the magnetosphere are available from http://www.ngdc.noaa.gov/seg/potfld/geomag.shtml.

Chapter 7

ATMOSPHERES

In gaseous bodies like the outer planets the distinction between the interior and the atmosphere is not sharp; in the inner, solid planets the atmosphere terminates at the ground and is affected by complex interactions with it. In this and the following chapters we mainly deal with terrestrial atmospheres, taking the Earth as our focus of attention. As the density decreases with height, collision mean-free paths get larger than spatial scales and the assumption of local thermodynamical equilibrium ceases to hold; in this chapter we deal with the lower atmosphere, where the fluid description is appropriate. In the upper part, atmospheres gradually escape to interplanetary space, which makes them fragile structures; this issue, together with the geological history of atmospheres, is discussed in Ch. 8. The atmospheric energy balance is governed by how solar radiation is absorbed and re-emitted in the infrared spectral band, as well as by the ensuing greenhouse effect; radiative transfer models are discussed, to provide a tool for the understanding of these complex phenomena. Planetary atmospheres are exceedingly varied and complex according to their chemical composition, energy sources, phase changes, and so on. The rotation of a planet drastically affects their dynamics through Coriolis' force and leads to an important flow regime (the *geostrophic flow*) and the generation of vorticity and turbulence on a global scale. The atmospheric dynamics of the Earth is, of course, known in great detail, but lies beyond our scope.

7.1 Structure of the atmosphere and climate variations

The atmospheric density profile is determined by eq. (1.38); but when the thickness is much smaller than the radius R of the planet, the gravitational acceleration $g = GM/R^2$ can be considered constant and the "flat Earth" (or plane-parallel) approximation

$$dP + g\varrho\, dz = 0 \qquad (7.1)$$

is adequate (with the atmosphere confined to the half space $z > 0$). Integrating from the ground to infinity we get

$$P(0) = g \int_0^\infty dz \, \varrho(z) ,$$

which gives an estimate of the total mass of the atmosphere

$$M_{\text{atm}} = \frac{4\pi R^2 P(0)}{g} \qquad (7.2)$$

as function of the average ground pressure $P(0)$. A perfect gas atmosphere (see eq. (1.16)) has the pressure profile

$$P(z) = P(0) \exp\left(-\int_0^z \frac{dz'}{H(z')}\right) , \qquad (7.3)$$

with the *pressure scale height* given by

$$H = \frac{kT(z)}{m_{\text{mol}}(z) \, g} ; \qquad (7.4)$$

m_{mol} is the mean molecular weight. For the Earth, on the ground, $H(0) \approx 8$ km and

$$P(0) \simeq 1.013 \times 10^6 \text{ dyne/cm}^2 ,$$
$$\varrho(0) \simeq 0.00129 \text{ g/cm}^3 ,$$
$$m_{\text{mol}}(0) \simeq 29.2 \text{ amu} = 29.2 \times 1.66 \times 10^{-24} \text{ g}$$

on average, with the pressure undergoing the largest variations. The density ϱ decreases exponentially as in (7.3), but the *density scale height* H' is generally different from the pressure scale height H:

$$\frac{1}{H'} = \frac{1}{T}\frac{dT}{dz} + \frac{1}{H}$$

(here we have neglected small variations of the mean molecular weight). The second term is, however, usually unimportant in the troposphere and $H' \simeq H$ (see also Fig. 8.1). It is also interesting to note that, despite larger differences in other parameters, the pressure scale height ranges from 8 to about 20 km for all planets.

The temperature profile $T(z)$ is determined by heat sources and transport. On the Earth, where the solar radiation is little absorbed by the atmosphere and heats the ground directly, we have a fluid heated from below and, if the temperature gradient, in absolute terms, is larger than the adiabatic gradient

7. Atmospheres

(Sec. 5.3), the convective instability arises; in this case the former settles a little above the adiabatic level (eq. (5.28) and Problem 5.7)

$$-\frac{dT}{dz} = \frac{g}{c_p} = -\left(\frac{dT}{dz}\right)_{ad}. \tag{7.5}$$

For the Earth the adiabatic gradient[1] c_p/g is about 10 K/km and gives a good model for the actual temperature profile below about 100 km. Above this adiabatic layer, heat, transport and absorption processes are much more complex and efficient and, as a result, the temperature is higher and the density decrease slower (see Figs. 8.1 and 8.2). The composition also changes because of chemical reactions, mainly driven by the solar radiation and phase changes.

Solar radiation, the main external energy source, is the determining factor of the structure of planetary atmospheres. The solar spectrum, peaked at the wave length $\lambda \approx 0.5$ μm (see eq. (1.98) and Fig. 7.3), has nearly a black body profile with the temperature $T_\odot \simeq 5,770$ K, corresponding, in terms of the Stefan-Boltzmann constant σ, to the total luminosity

$$L_\odot = 4\pi R_\odot^2 \sigma T_\odot^4 \simeq 3.9 \times 10^{33} \text{ erg/s}. \tag{7.6}$$

At a distance D the solar radiation carries an energy flux

$$F = \frac{L_\odot}{4\pi D^2} = \left(\frac{R_\odot}{D}\right)^2 \sigma T_\odot^4 \simeq 1.366 \times 10^6 \text{ erg cm}^{-2}\text{s}^{-1}\left(\frac{AU}{D}\right)^2, \tag{7.7}$$

which, at 1 AU, is called the *solar constant* F_0 (Fig. 7.1).

For a planetary body *Bond's albedo* is the fraction of the incoming solar radiation which is not absorbed, so that the fraction $(1 - A)$ contributes to the thermal energy of the body. For the local radiative balance, appropriate to the individual surface elements, one uses the *hemispheric albedo* A_h, equal to the fraction of the flux F which is diffused or reflected; it greatly varies with the incidence angle ι, latitude, longitude, meteorological conditions, hour of the day, seasons and other factors (e.g., is increased by snow and ice). The insolation is the flux $(1 - A_h)F \cos\iota$ absorbed per unit area; it vanishes on the night side of the *terminator*, lying in the plane orthogonal to the Sun. Another local albedo coefficient, the *clear sky albedo* A_{cs}, is obtained from A_h by eliminating the effect of the variable cloud cover, an important climatic factor. Bond's albedo is a weighted average of the hemispheric albedo over the incidence angle, position and time.

In equilibrium the absorbed power $(1 - A)\pi R^2 F$ must be re-emitted into space; for an isothermal surface

$$4\sigma T_{eq}^4 = (1 - A) F, \tag{7.8}$$

---

[1] Meteorologists call *lapse rate* the temperature drop $-dT/dz$ in the troposphere.

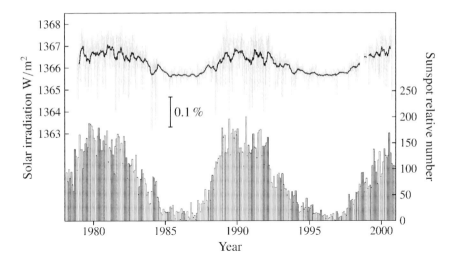

Figure 7.1. In the upper plot, temporal variations of the solar constant during the last two solar cycles, at high (grey curve) and low time resolution (black curve). This is a compilation from different space measurements (mainly with the spacecraft NIMBUS-7, SMM, ERBE and SOHO, since 1996), with the appropriate corrections to account for calibration and degradation of the instruments, operational noise, etc. The lower plot shows the monthly-averaged sunspot numbers R_W (Sec. 9.1). Data obtained from http://www.pmodwrc.ch/-solar_const/solar_const.html.

and we get the equipartition temperature

$$T_{eq} = \left(\frac{1-A}{4}\right)^{1/4} F^{1/4} = \left(\frac{1-A}{4}\right)^{1/4} \left(\frac{R_\odot}{D}\right)^{1/2} T_\odot . \quad (7.9)$$

This gives an estimate of the temperature of planets and satellites. Of course, actual values can be very different, in particular, due to the effect of rotation: the cooler night side requires the day side to be warmer than T_{eq}. This daily variation is a dominant effect for terrestrial bodies with thin atmosphere and insulating surface (such as the regolith layer): Mercury, Moon and Mars (and the asteroids). The mean temperatures of Jupiter and Saturn are higher (see Table 7.1), an indication of an active internal heat source, either "fossil" thermal energy from the primordial accretion phase, or a slow gravitational shrinking. The temperature is also increased by the greenhouse effect.

At $4 R_\odot \simeq 0.02$ AU the equipartition temperature for $A \simeq 0.3$ would be $\simeq 1,800$ K, which would make it impossible to place a spacecraft in a bound orbit at that distance from the Sun. To explore in situ this very interesting region, where the solar wind arises, the concept of the Solar Probe has been devised, with an almost parabolic trajectory with perihelion at $4 R_\odot$. In addition to sophisticated thermal shielding, its thermal safety is ensured by the fact that

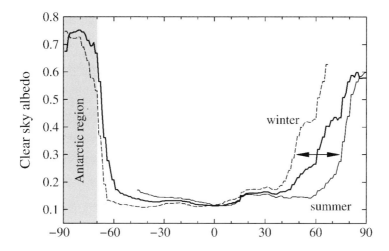

Figure 7.2. The local clear sky albedo of the Earth A_{cs} averaged over longitude vs latitude (in degrees). The solid line shows the yearly average, the thinner curves indicate summer and winter values. Note the significant increase in the albedo near the South Pole and seasonal variations at high latitudes (indicated by the arrow), due to the decrease of the arctic ice cover in winter. ISLSCP data acquired from http://daac.gsfc.nasa.gov/CAMPAIGN_DOCS/ISLSCP/islscp_i1.html.

during the fast transit near the Sun the thermal inertia prevents reaching the equipartition temperature (Problem 11.2).

The detailed radiative properties of the Earth atmosphere and their variations are best studied from space. The International Satellite Land Surface Climatology Project (ISLSCP), with about a dozen dedicated satellites, has provided climatological data with a remarkable level of detail, accuracy and completeness. For instance, a net energy loss from the polar regions (due the cloud cover and higher albedo) and an energy gain by the equatorial belt have been measured (Fig. 7.2); this unbalance is compensated by atmospheric motions and ocean currents. The solar constant has been monitored from space and found to change by $\approx 0.1\%$ over many timescales, from minutes to years (Fig. 7.1). A correlation with the 11 y solar cycle has been observed; the variability over $15 - 30$ d is probably related to the solar rotation. It is unknown whether these minute variations have any climatic effects; but it is remarkable that the so-called *Maunder minimum* in the number of sunspots observed at the end of the 17th century corresponds to the "little ice age" (see Sec. 9.1).

Climatic effects of planetary dynamics. A striking evidence of astronomical effects on the climate was obtained with the samples extracted from deep (a few km) drillings in Antarctica (at the Vostok station) and Greenland. Since each annual snow deposition is made up of seasonal layers, chronological tagging is possible. The isotopic composition of these samples has provided a record of the mean temperature of the Earth during the last 420 ky; its Fourier transform was found to have peaks corresponding to the periods of 19 ky, 23 ky and 41 ky; all of them, as we discuss below, are also present in the orbital and rotational perturbations of the Earth. Similarly, the abundance of ^{18}O in oceanic sediments also provides climatological information to nearly 1 My back in time; it is dominated by a period of 100 ky, found also in the CO_2 record of the Vostok ice core.

First, for a planet with orbital eccentricity e ($\simeq$ 0.017 for the Earth; see Fig. 15.1) the equilibrium temperature trivially changes by $\simeq e\,T_{eq}$ from aphelion to perihelion (see eqs. (11.18b) and (11.18c)); but, more interesting, any long-term change of the eccentricity produces a corresponding modulation of the temperature, as for the Earth, which has small changes with the periods of 95 and 125 ky due to planetary perturbations. This might be the source of the 100 ky palaeoclimatic signal, but the Earth eccentricity undergoes even larger variations with a period of 412 ky, entirely missing in the climatic data. Moreover, the $\simeq$ 100 ky variations are too small to be responsible for sizeable climatic changes and their origin remains an unsolved problem in climatology. Different hypotheses and models assume a non-linear, or otherwise complex, response to the insolation variations or a periodic change of the influx in the atmosphere of particles in the dust bands (Sec. 14.6), since the 100 ky is also found in variations of the Earth inclination.

Milanković hypothesis.– As systematically advocated by M. MILANKOVIĆ in the 1920s (following an earlier work of J. CROLL in the 1870s), much more important than the eccentricity variations are the variations in the obliquity ϵ, and the precession rate of the spin axis. For the Earth ϵ changes from 22°30′ to 24°30′ around the mean 23°26′ with a period of 41 ky (see the Appendix), while the precession of the Earth rotation axis induces Fourier terms at 23 ky and 41 ky, also found in the ice and sediments records (remarkably, the relative amplitude of the 23 ky and 41 ky signal is the same in the obliquity-precession series and in the palaeoclimatic data). This raises a general question (important also for Mars) fundamental for the *astronomical theory of the palaeoclimate*: how obliquity and precession affect the climate, and how its change arises. The details of the relevant processes are still not known, especially after striking, and unexplained, climatic variations have been recently discovered also on the much shorter timescale of $\simeq$ 1 – 3 ky in ice records in Greenland. Here we only comment on the gross role of the obliquity.

The obliquity determines the way how the solar radiation is distributed at different latitudes during the year. Let Φ be the ecliptic longitude of the Sun relative to the γ-point; $\Phi = 0$ and π correspond to the equinoxes, $\Phi = \pi/2$ and $\Phi = 3\pi/2$, respectively, to the summer and the winter solstices. If ϕ is the rotational phase and F_s the solar flux (7.7), the average fraction of the radiative flux impinging on a point at a given latitude is given by

$$F(\theta, \Phi; \epsilon) = \frac{F_s}{2\pi} \int d\phi \cos\iota , \qquad (7.10)$$

where the integral covers only the lit part of the parallel ($\cos\iota > 0$); in palaeoclimatology $F(\theta, \Phi; \epsilon)$ is sometimes termed *daily insolation*. When $\epsilon = 0$, $\cos\iota = \sin\theta\cos(\phi - \Phi)$, so that $F(\theta, \Phi; 0) = \sin\theta/\pi$ is independent of the season and the poles are not illuminated at all. A number of insolation indices have been used in the climatological literature, like the daily averaged insolation at the solstices and equinoxes at a given latitude (e.g., 65° N); another choice is the annual average $\bar{F}(\theta; \epsilon)$. It turns out (see Problem 7.1) that, as ϵ increases, the seasonal modulation increases, but the difference between the average insolation at the equator and the poles at first diminishes, reaches 0 at the critical value $\epsilon_\star \simeq 54°$ and becomes negative beyond (this is the case of Uranus and its satellites, and Pluto). When $\epsilon = \pi/2$, $\bar{F}(0; \pi/2) = (\pi/2)\bar{F}(\pi/2; \pi/2)$; the poles are warmer than the equator by the factor $(\pi/2)^{1/4} \simeq 1.12$. At a large obliquity there are no ice polar caps, but an ice equatorial belt.

The fact that changes in obliquity may have a profound effect on the climate is supported by the analysis of the well established sequence of recent glaciations; the last occurred $\approx$ 21 ky ago, as documented by geological records of major advances of glaciers in Europe and America. Interestingly, during these periods the Earth's obliquity was near its minimum and, due to the secular motion of the apsidal line, aphelion was near the winter solstice. Not only the poles received less light, but winter was colder in the North because the Sun was further away. The possibility that the last glaciation was triggered by this coincidence is forceful; more generally, there is no surprise that the 41 ky periodicity is present in the ice records.

Large (in particular, > 54°; Problem 7.1) changes in the obliquity would drastically affect the climate of a planet. For the Earth, several complex factors, including the different land extent in the North and the South, should be taken into account; but the seasonal temperature excursion could be so large (e.g., from 270 K to 350 K at the latitude 85°) to make the persistence of life, as we know it, impossible.

Secular effects of spin-orbit interaction.– In the derivation of the lunisolar precession (eq. (4.6)) for the axis **e** of an oblate planet, the normal to the orbital plane was assumed fixed; but when there is a secular change of the node and/or the inclination (Sec. 15.1), in the right-hand of (4.6) a resonance occurs when the nodal frequency is equal the precession frequency and a large change in the obliquity can be expected. We must now distinguish between $\mathbf{N}_p$, the (moving) normal to the orbital plane with an inclination I, and **N**, the fixed normal to the invariable plane of the solar system. To single out the time dependent terms, it is convenient to write (4.6) in the form

$$\frac{d\mathbf{e}}{dt} = -f_{\rm pr}\left(\mathbf{N}_{\rm p} \cdot \mathbf{e}\right)\left(\mathbf{N}_{\rm p} \times \mathbf{e}\right) \quad (\mathbf{N}_{\rm p} \cdot \mathbf{e} = \cos\epsilon), \quad (7.11)$$

where

$$f_{\rm pr} = \frac{3}{2} \frac{C - (A+B)/2}{C} \frac{n^2}{\omega \eta^3}$$

is a constant positive frequency; A, B, C are the principal moments of inertia, ω the mean rotation frequency and $\eta = \sqrt{1-e^2}$ (assumed constant). The rigorous and general discussion of this interaction can be done with the powerful Hamiltonian formalism, as discussed in the Appendix; here we confine ourselves to the simple case of a nodal precession with the single frequency $\dot{\Omega}$:

$$\frac{d\mathbf{N}_{\rm p}}{dt} = \dot{\Omega} \mathbf{N} \times \mathbf{N}_{\rm p} \quad (\mathbf{N} \cdot \mathbf{N}_{\rm p} = \cos I). \quad (7.12)$$

The rotation axis $\mathbf{e}$ is determined by its two components: $\cos\epsilon$ along $\mathbf{N}_{\rm p}$ and $\sin\epsilon \cos\phi$ along the ascending node $(\mathbf{N} \times \mathbf{N}_{\rm p})/\sin I$. Traditionally the precession angle is defined as $\psi = \pi/2 - \phi - \Omega$, consistently with the definition (11.34) of the true longitude, so useful when the inclination is small and ψ is reckoned from a fixed direction; note also that for a retrograde motion of the axis ψ increases with time. Using (7.11) and (7.12) we easily find

$$\frac{d\epsilon}{dt} = -\dot{\Omega} \sin I \cos\phi; \quad (7.13)$$

when $I \ll 1$ the obliquity does not change much, unless there is a secularity. The rate of change of the component $\mathbf{e} \cdot (\mathbf{N} \times \mathbf{N}_{\rm p})/\sin I$ provides, after a lengthy, but straightforward algebra, the other equation

$$\frac{d\phi}{dt} = -f_{\rm pr} \cos\epsilon - \left(\cos I - \frac{\cos\epsilon}{\sin\epsilon}\sin I \sin\phi\right)\dot{\Omega}; \quad (7.14)$$

when the inclination is small and the secular change in ϵ is neglected, it gives the expected precession in presence of a nodal drift

$$\frac{d\phi}{dt} = -f_{\rm pr} \cos\epsilon - \dot{\Omega} + O(I) = -\Omega_{\rm pr} - \dot{\Omega} + O(I). \quad (7.15)$$

The minus sign signals that the precession is prograde. Integrating (7.13) we note that a resonance occurs when $\dot{\Omega} = -\Omega_{\rm pr} = -f_{\rm pr}\cos\epsilon$, that is, (if, as usual, $\cos\epsilon > 0$), when the nodal motion is retrograde. In the exact resonance ϵ and ϕ are stationary; $\phi = \pi/2$ and ϵ satisfies a transcendental equation. In its neighbourhood, ϕ librates and the obliquity has large variations ($\propto \sqrt{\sin I}$; Problem 7.13). Outside the resonance, ϕ circulates and variations of the obliquity are smaller ($\propto \sin I$). Description of the resonance regime profits from the Hamiltonian formulation outlined in the Appendix.

The precession frequency of the Earth, $\Omega_{pr} = 50.4''/\text{y} = 34.8''/\text{y} + 15.6''/\text{y}$, is due to the combined, and comparable, influence of the Moon (first term) and the Sun (second term). With a small inclination $I_0 \simeq 1.58°$ to the invariable plane of the planetary system, at present its main nodal frequency $\dot{\Omega} \simeq s_3 = -18.9''/\text{y}$ is (in absolute value) smaller than the total precession, but larger than the solar contribution. The main effect of the spin-orbit coupling is a small amplitude ($\simeq 0.9°$) modulation in obliquity with the period $\simeq 2\pi/(\Omega_{pr} + s_3) \simeq 41$ ky, just that found in the ice records. But this will not last for ever. Since the rotation frequency ω is greater than the Moon's orbital frequency, angular momentum is transferred by tidal friction to the latter; the length of the day increases at the rate (3.5), and so does its distance D. The lunar contribution diminishes as ω/D^3; it can be estimated that in ≈ 2 Gy the resonant state (in which the precession will mainly be of solar origin) will be reached with large obliquity variations (between $0°$ and up to $80°$) and drastic consequences for the climate. It is also interesting to note that without the Moon the precession frequency would be $15.6''/\text{y}$, close to the s_3 resonance, producing large variations in obliquity and, perhaps, making life impossible.

Mars is farther away from the Sun and its small moons produce a small torque, so that the precession is slower ($7.58''/\text{y}$), but near the nodal frequency $s_2 = -6.57''/\text{y}$; large obliquity changes (from $15°$ to $45°$) are taking place at present. Saturn's large ($\simeq 27°$) obliquity may result from capture in the s_8 secular resonance.

Long-term changes of the solar luminosity.– Another interesting aspect of palaeoclimatology is the increase of the luminosity of the Sun. Its internal temperature is determined by its mass and radius, but is proportional to the number A of nucleons in a nucleus (eq. (1.19)). As hydrogen is converted into helium, A increases, the core becomes hotter and more energy is produced in thermonuclear reactions. This trend can be approximately described with

$$L_\odot(t) \simeq L_\odot(t_0) \left[1 + 0.3\left(1 - \frac{t}{t_0}\right)\right]^{-1}, \qquad (7.16)$$

where $t_0 = 4.5$ Gy is the approximate age of the Sun. From eqs. (7.7) and (7.9) a fractional change in the luminosity produces a fractional change in the surface temperature four times smaller; as a consequence, in absence of a greenhouse effect, the mean temperature of the Earth would have been below the freezing point, in contrast with the absence of glaciations in the first 2.7 Gy; this is called the *faint young Sun problem*. Similarly, this lower insolation, and a temperature well below 273 K, is in apparent conflict with the geologic evidence of liquid water on the surface of Mars. For both planets one must infer that the atmospheric composition at that time was different, in particular with a higher concentration of CO_2. In fact, there is evidence that on the Earth oxygen was significantly scarser up to ≈ 2 Gy ago. The geological balance of

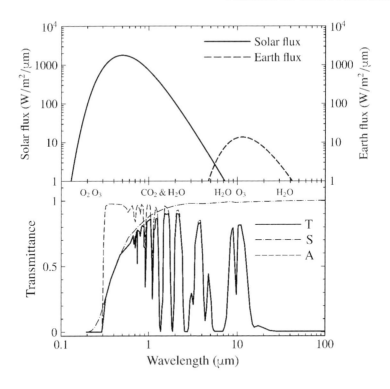

Figure 7.3. The upper panel shows the black body spectral energy flux for the temperatures 5,770 K and 288 K, corresponding to the solar radiation impinging on the Earth and to its thermal emission; the lower panel shows the atmospheric transmittance (T), the absorption (A) and the scattered fraction (S), with the indication of the main molecule responsible for the absorption feature.

atmospheric CO_2 results from depletion due, e.g., to weathering processes in which carbon replaces silicon in rocks, and from production by the hot material emerging from the mantle along plate boundaries (in the Earth case). The situation is further complicated by the possible large concentration of ozone, which preferentially absorbs the UV solar radiation and may be responsible for an *anti-greenhouse effect*, which lowers the temperature. Extensive glaciations, by reducing the planetary albedo, could also lower the temperature and set up a runaway effect.

The greenhouse effect. The mean surface temperature T_s of the terrestrial planets is generally larger than T_{eq} (see Table 7.1). This is due to the *greenhouse effect*: the solar radiation re-emitted by the ground in the infrared may be partially absorbed by the atmosphere, in particular by water vapour and CO_2, thereby affecting the radiative balance (7.8). For the Earth the main causes of atmospheric infrared absorption are O_2, O_3, H_2O and CO_2. The absorption is

7. Atmospheres

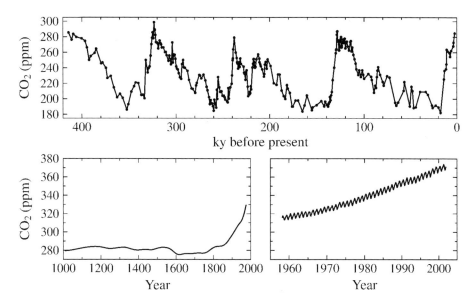

Figure 7.4. Upper panel: long-term evolution of the CO_2 concentration in the Earth atmosphere. This record has been obtained from the analysis of the gas trapped in air bubbles in the 3,623 m deep drilling at the Vostok station in Antarctica. At the bottom: the CO_2 concentration on a short and very short timescale (on the left, data from the Antarctic hole at Law Dome; on the right, atmospheric data from the Mauna Loa station). Data acquired from http://cdiac.ornl.gov/trends/co2/contents.htm.

very strong below 0.3 μm and beyond 15 μm, with complex systems of lines and bands in between (see Fig. 7.3). The CO_2 lines are particularly important because are affected by combustion processes, especially petrol and coal burning. As a consequence of the current increase in the CO_2 content (some % per year; Fig. 7.4), it is widely believed that in the next 20 – 40 years the mean temperature of the Earth will increase and the global climate will be deeply affected, with negative consequences for life conditions. However, as mentioned above, a weak greenhouse effect, making T_s larger than T_{eq}, has been essential for the emergence and evolution of life as we know it. The prediction of a temperature increase due to human activity was made by S. ARRHENIUS in 1896. The complex radiation budget of our planet summarized in Fig. 7.5 is the starting point for models of the greenhouse effect; they must take into account many factors, including the large heat capacity of the oceans and various feedbacks, such as the additional opacity due to a greater evaporation, leading to an increased cloud cover.

The simplest model of the greenhouse effect consists in an atmosphere transparent to the solar radiation (concentrated below 4 μm) and partially opaque in the infrared band (peaked at ≈ 11 μm); then less energy is lost to space and a

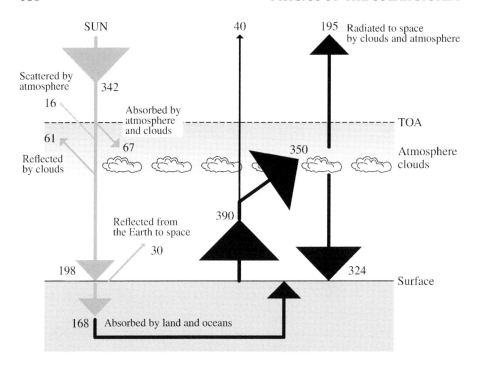

Figure 7.5. The radiation budget of the Earth. The solar flux F, when distributed uniformly over the Earth, corresponds to a flux $F/4 = 342$ W/m^2 = 100% (we use here the unit 1,000 erg/cm^2/s = 1 W/m^2), of which 107 = 31% are reflected back by the atmosphere (16 = 5%), by clouds (61 = 18%) and the surface (30 = 9%). The remaining 235 = 69% is absorbed by the atmosphere (67 = 20%) and the ground (168 = 49%). The ground infrared emission, 390, corresponds to the mean temperature T_s = 288 K; of this only 40 escape to outer space, while the rest, 350, is trapped in the atmosphere, producing the greenhouse effect. The atmosphere, being cooler than the ground, sends down a flux of 324, smaller than the upward flux from the ground of 350, and radiates to outer space 195, with a net loss of $324 + 195 - 67 - 350 = 102$. This loss is compensated by turbulent heat transfer and evaporation heat. Obviously, the numerical values are rough and subject to variation. Gray and black distinguish optical and infrared bands; TOA is the top of the atmosphere.

higher ground temperature T_s results. To quantify this model, consider a thin stratified atmosphere, in which every quantity is a function only of the height z. The radiative flux, generally dependent on the direction, is described by the *specific intensity* I (power per unit area and unit solid angle); absorption means that this quantity, on passing through a layer of thickness dz, changes by

$$dI = -\alpha I \, ds = -I\kappa\varrho \, ds . \qquad (7.17)$$

The quantity κ (the *opacity*) is introduced to stress the fact that usually the *absorption coefficient* $\alpha = \kappa\varrho$ is proportional to the material density ϱ (note that in all other chapters the symbol κ denotes the thermal conductivity). In

7. Atmospheres

a medium with n absorbers per unit volume and a photon mean free path λ_c, $\alpha = n\lambda_c$. If α is constant, the intensity decreases exponentially.

Consider now a thin stratified atmosphere, in which every quantity is a function only of the height z. For a vertical ray, introduce the *optical depth*

$$\tau(s) = \int_s^\infty ds'\, \kappa(s')\varrho(s') \quad (d\tau = -ds\,\kappa\,\varrho). \tag{7.18}$$

It vanishes on the top of the atmosphere $z = \infty$ and is determined by the material above z. In a crude approximation, we also neglect scattering and assume solar rays parallel; in this case the radiation flux F through the area perpendicular to them is just equal to the intensity I. A transparent and opaque atmosphere correspond, respectively, to $\tau(0) \ll 1$ and $\tau(0) \gg 1$. In this model the outgoing infrared flux reads $F(z) = F\exp[-\tau(z)]$; the flux escaping from the surface is thus approximately $F\exp[-\tau(0)]$. This introduces a corresponding attenuation factor in the estimate of the surface temperature, namely $T_s \approx T_{eq}\exp[\tau(0)/4]$, larger than T_{eq} (we neglect the surface albedo). A substantiated model of the greenhouse effect is given in Sec. 7.3.

The cloud cover (in the average, little more than half the Earth's surface) has profound effects on the global climate. First, due to the large latent heats of[2] H_2O, it affects the thermodynamics of the atmosphere. In an ascensional current of damp air the mixing ratio (ratio of the number of water vapour molecules to the number of other molecules) remains constant, but, as the pressure and the temperature decrease, the ratio of the water vapour pressure to the saturation water vapour pressure (the *relative humidity*) increases; when it reaches unity condensation occurs and the water droplets or ice grains precipitate to the ground. Secondly, scattering, absorption and emission by clouds affect the energy budget. Generally speaking, a cloud cover can work both ways: it may cool the planet by reducing the amount of sunlight that reaches the ground, but it may also increase the surface temperature by increasing the infrared opacity. Although the cloud structure is very complex, one can generally distinguish between low (below 7 km, say) clouds, with a large albedo (0.6, say) and nearly black body behaviour in the infrared; and high clouds, at lower temperature and made of thinly distributed ice particles, with a much lower albedo and an emissivity smaller than unity (0.75, say). Calculations show that generally the surface temperature is increased high clouds and reduced by low clouds.

---

[2] The specific heat of water at 273 K is 4.218×10^7 erg/g/K, while the latent heats of fusion and evaporation are, respectively, 3.3×10^9 erg/g and 2.5×10^{10} erg/g.

Table 7.1. The main properties of the atmospheres of six planets. A is Bond's albedo; T_{eq} (eq. (7.9)) is the equilibrium temperature; T_{rad} is the radiation temperature, given by the total radiative emission; T_s the mean ground temperature, all in K; the equatorial surface gravity g and the mean ground pressure $P(0)$ are given in cm/s^2 and bar, respectively; the mass of the atmosphere M_{atm} (eq. (7.2)) is in grams; $2\pi/\omega$ is the rotation period in days. The adiabatic temperature gradient $-(dT/dz)_{ad}$ is in K/km. The last two lines give the two main components, with their abundances in molecular densities. Jupiter and Saturn have no ground and the "surface" temperature refers to the 1 bar level. For the inner planets the difference $T_s - T_{eq}$ is due to the greenhouse effect. For Jupiter and Saturn an intrinsic radiation flux produces a positive difference $T_{rad} - T_{eq}$ greater than the error (see http://atmos.nmsu.edu/).

	Mercury	Venus	Earth	Mars	Jupiter	Saturn
A	0.11	0.75	0.31	0.25	0.34	0.34
T_{eq}	440	232	255	216	113	83
T_{rad}	440	230	250	220	124	95
T_s	440	731	288	215	165	135
g	370	887	978	369	2,312	896
$P(0)$	10^{-15}	92	1	0.006	–	–
M_{atm}	2×10^6	4.8×10^{23}	5.3×10^{21}	2.4×10^{19}	–	–
$2\pi/\omega$	59	243	1	1.03	0.41	0.43
$-\left(\frac{dT}{dz}\right)_{ad}$		9	9.8	4.5	1.9	0.8
	O (42%)	CO_2 (96%)	N_2 (78%)	CO_2 (95%)	H_2 (86%)	H_2 (96%)
	Na (29%)	N_2 (3%)	O_2 (21%)	N_2 (3%)	He (14%)	He (3.3%)

7.2 Planetary atmospheres

The Table 7.1 summarizes the main properties of planetary atmospheres; their geological history is discussed in Sec. 8.5. In general, the temperature and density profiles are similar to the Earth's: a negative, adiabatic temperature gradient in the lower layers and a gradual and irregular temperature increase in the upper layers.

Mercury.– Mercury, with its low gravity and large impinging solar flux, has practically no atmosphere; traces of He, O, K, Na and other species have been detected by Mariner 10 in 1974 and 1975, and later observed from the ground. The lifetime of a single atom in such a thin and hot gas is very short, and it is likely that bombardment of the Mercury's surface by the solar wind provides the necessary resupplying mechanism.

Venus.– The main constituent of Venus' atmosphere is CO_2, a strong infrared absorber. A deep cloud layer, consisting of a featureless and ubiquitous aerosol haze between 45 and 65 km, with its large opacity prevents any optical observation from outside. The main direct investigation of the lower Venusian atmosphere is difficult, but much has been achieved with the two Pioneer probes,

7. Atmospheres

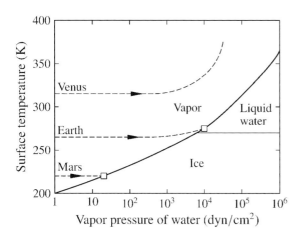

Figure 7.6 The runaway greenhouse effect on Venus. In the phase plane (P, T) the addition of more water vapour enhances the ground temperature so much that it escapes the transition to a mixed phase with liquid water; presumable evolutionary tracks of Venus, Earth and Mars are shown by the dashed lines (see text).

which encountered the planet in December 1978, and with the Magellan mission in the early 1990s. The large albedo, mainly due to the cloud cover, more than offsets the smaller distance from the Sun ($\simeq 0.72$ AU), so that T_{eq} is smaller than the Earth's value; but, because of a runaway form of the greenhouse effect, the actual surface temperature is 730 K, higher than the terrestrial value. In the (P, T) phase diagram of water vapour (Fig. 7.6) a monotonically increasing curve separates the pure vapour phase from the phases in which the vapour coexists with liquid water. As more water vapour was released in the past from the ground by degassing, its pressure increased and so did, through the greenhouse effect, the infrared absorption and the surface temperature. In the case of the Earth and Mars this process stops when the transition curve is reached and water precipitates in oceans or freezes in polar regions; in fact most water on the Earth is liquid or solid. But on Venus the equilibrium temperature above the cloud cover is larger and its increase is sufficient to avoid the transition line; no liquid phase can form and the addition of more water vapour can only increases the surface temperature. Observations have shown that the solar radiation able to penetrate the cloud cover is responsible for heating the surface (the internally generated heat is negligible); moreover, the lower temperature profile – up to some 45 km, the base of clouds – is characterized by a mean lapse rate of $\simeq 8$ K/km, smaller than the adiabatic value of $\simeq 10.5$ K/km (eq. (7.5)), corresponding to a stable or a neutral situation (Fig. 7.7).

Large scale convective heat transport seems unlikely. The long Venus day (117 Earth days) would suggest a large diurnal modulation in pressure and temperature; but this is not what the observations show, with only ≈ 10 K variations in latitude and longitude. In fact, heat is transferred horizontally by strong, zonal winds, in particular by a *superrotation* of the atmosphere, with a period of $\approx 4 - 5$ d, against the (sidereal) rotation period of 243 d. On the equator the wind velocity is ≈ 100 m/s. The reason for this striking fact is unknown. It should also be borne in mind that, as for the Earth (Sec. 8.1),

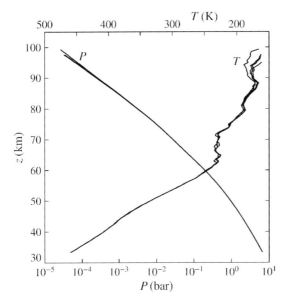

Figure 7.7 Temperature T and pressure P profiles for the atmosphere of Venus. At lower altitudes the temperature gradient is very near the adiabatic value, while in the stratosphere, above 60 km, it drops below. Data from http://atmos.nmsu.edu/ were derived from radio occultation measurements using the Magellan orbiter.

the atmospheric heat capacity is large and these winds mainly affect the upper atmosphere (thermosphere).

Mars.– Mars suffers greater changes in eccentricity and obliquity than the Earth and, similarly to the Milanković cycles, may have undergone greater changes in climate. There is extensive evidence of ancient flows of liquid water and long-period changes in the polar caps; a much denser atmosphere, with a high concentration of CO_2, may have been present in the past. This has led to important exploration programmes, also aimed at the search for evidence of earlier life systems (in particular with the two Viking spacecraft, which landed on Mars in 1976). At present, the thin atmosphere has a small heat capacity and, with the low thermal conductivity of the ground, there are significant latitudinal, diurnal and seasonal variations of the surface temperature. On the polar caps this causes the seasonal appearance of frozen CO_2. Near the surface the thermal lapse varies largely, but it is typically smaller than the adiabatic value of $\simeq 4.5$ K/km; thus convection is usually ineffective. As a response to the large day-to-night changes in the ground temperature, there are strong thermal tide winds which, with occasional superadiabatic thermal gradients, may cause global dust storms throughout the whole atmosphere.

Giant planets.– Their atmospheres are primordial and differ substantially from the fragile gas covers of the terrestrial planets. They show a strikingly large excess of heavy elements relative to the solar composition, an evidence that they formed also under the impact of planetesimals; this can only be understood in relation to their geological history.

Their present state mainly results from the equilibrium between different sources and sinks of radiative energy (Fig. 7.8). In the lower layers (around

7. Atmospheres

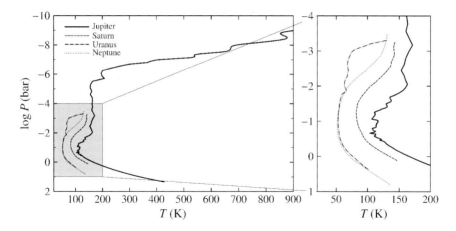

Figure 7.8. Temperature profiles for the giant planets, showing two features similar to the Earth and Venus: in the lower part a linear dependence of $\log P$ from T, typical of an adiabatic layer; higher up the temperature gradient levels off, possibly with a temperature inversion. The Galileo probe data for Jupiter were kindly provided by R. Young; for Saturn, Uranus and Neptune we used data from G.F. Lindal, *Astron. J.* **103**, 967 (1992).

Table 7.2. The main properties of Titan's atmosphere. The same symbols and units as in Table 7.1. Titan's rotation period is 16 d, with negligible effects.

Albedo	T_{eq}	T_s	g	$P(0)$	Composition	$-(dT/dz)_{ad}$
0.20	86	94	135	1.5	N_2 (main), Ar, CH_4	1.30

1 bar), with opacity near one, the absorbed solar radiation is in part re-emitted in the far infrared and in part produces a nearly adiabatic layer; heat is transported upward by convection. The lapse rate of Jupiter (2 K/km) is larger than the other giant planets because of its stronger gravity. In the stratosphere the incoming solar spectrum is absorbed mainly by methane and aerosols, producing (except on Uranus) a temperature inversion and an irregular behaviour, depending on the minor constituents. The chemical equilibrium is ensured by the solar radiation and, possibly, also by lightnings. The far-infrared radiation emitted by the lower layers, heated by the solar radiation and also by the internal energy source, is mainly absorbed by H_2.

Titan.– The atmosphere of Titan, the main satellite in the Saturnian system and the only satellite in the solar system possessing an extensive atmosphere, is opaque in the visible; it was investigated from outside in the flyby of the Voyager 1 mission in November 1980. Spectroscopic observations revealed a peculiar composition, with aerosols and a complex carbon chemistry, includ-

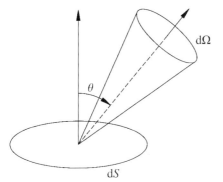

Figure 7.9 The geometry of radiative transfer: ad definition of the specific intensity.

ing molecules (like HCN) which are known to polymerize and produce amino acids, the building blocks of organic molecules. While the very low temperature does not encourage thinking about life, this environment may have prebiotic features, where elementary duplication processes might take place. Recently microwave observations, ground telescopes with adaptive optics (Sec. 19.5) and the infrared instruments of the Hubble Space Telescope were able to penetrate Titan's atmosphere from the Earth with improved angular resolution (Titan's angular size is $\simeq 0.8''$); a few optically thin methane lines in the infrared were used. A global prograde circulation and a north-south asymmetry in brightness (attributed to the seasonal change of the haze density) were detected. Titan will be extensively studied with the Cassini mission from 2004 to 2008. In July 2004, after insertion on a Saturnian orbit, the spacecraft will release the Huygens probe, which will slowly descend into the atmosphere of Titan and fully investigate its properties. During the subsequent tour of the Saturnian system, Titan will be encountered again 43 times and more data about its atmosphere will be collected.

7.3 * Radiative transfer

For a complete description of the atmospheric radiative budget, tools and models of the theory of radiative transfer, a complex and well developed area of mathematical physics, are necessary. The basic quantity is the *specific intensity* I_ν, defined as follows:

$$dE_\nu = I_\nu \cos\theta \, d\nu \, dt \, dS \, d\Omega \tag{7.19}$$

is the energy in the frequency band $(\nu, \nu + d\nu)$ which flows in the time interval $(t, t + dt)$ within the solid angle $d\Omega$ and through the projection of a surface element dS in the plane orthogonal to the ray (Fig. 7.9). In general, I_ν is a function of frequency, position, time and direction.

In the geometrical optics approximation, which we adopt, we have rays, defined in terms of the refractive index n_r. In a stationary situation the position **r** and the wave vector $\mathbf{k} = \hat{\mathbf{k}} k$ are given in terms of the arc s along the ray;

the frequency is $\omega = 2\pi\nu = ck/n_r$. The specific intensity describes the photon distribution in the space $(\mathbf{r}, \mathbf{k})$; its change is mainly due to the interaction with matter and consists of four different processes. Photons in the beam can be eliminated by absorption or created by emission; they can also be scattered away to a different ray and transferred to a different frequency; finally, scattering and frequency changes can bring photons on the ray and in the frequency interval. Emission and absorption often occur in resonant lines, when the photon energy $h\nu$ is equal to the energy difference between quantum levels. Their width and their displacement from the laboratory value, affected by thermal and fluid motion, is an important part of radiative transfer, the theory of line formation. While absorption and emission are described by a loss and a gain function, with the other two processes we have an integro-differential equation for the specific intensity, where an integral over all wave numbers contributes to the rate of change dI_ν/ds along the ray; all rays are coupled, with a considerable complexity. It should also be noted that in general the thermal equilibrium between the radiation and the matter may not be guaranteed: in the non-equilibrium case, the occupation of the atomic energy levels – affected by both interaction with radiation and mutual collisions – should be solved consistently together with the radiative transfer problem. In what follows most of these problems – in particular, refractivity, scattering, energy exchange and line formation – are neglected; thus only a gross description of the greenhouse effect is possible, with no detailed spectral information.

Emission is described by the *emission coefficient* j_ν: the energy

$$dE_\nu = j_\nu \, dV \, dt \, d\Omega \, d\nu \tag{7.20}$$

is emitted from the volume dV in the time interval dt within the solid angle $d\Omega$ along the ray and in the frequency band $d\nu$; for example, for an isotropic emitter, with a radiated spectral power per unit volume P_ν, $j_\nu = P_\nu/(4\pi)$. In advancing a distance ds through a section $dS\cos\theta$, a beam fills a volume $dV = dS \, ds \cos\theta$, so that the contribution to the change in specific intensity is $j_\nu \, ds$. If α_ν is the absorption coefficient (generally frequency dependent), the steady state transport equation reads

$$\frac{dI_\nu}{ds} = -\alpha_\nu I_\nu + j_\nu = -\alpha_\nu (I_\nu - S_\nu) \,, \tag{7.21}$$

where $S_\nu = j_\nu/\alpha_\nu$ is the *source function*. We see that I_ν increases (decreases) along the ray if $I_\nu < S_\nu$ ($I_\nu < S_\nu$); radiative transfer can be seen as a relaxation process, in which I_ν "tries to approach" S_ν. Just as for an ordinary gas (Sec. 1.8), we assume that matter is in *local thermodynamical equilibrium*, with a temperature T generally function of space and time, and the bulk of the photons have a Planck distribution (1.96) at the same temperature. The emission coefficient of a black body is an isotropic and universal function of ν and

T, namely

$$B_\nu(T) = \frac{c u_\nu}{4\pi} = \frac{2h\nu^3}{c^2} \frac{1}{\exp(h\nu/kT) - 1} , \qquad (7.22a)$$

$$B(T) = \int_0^\infty d\nu\, B_\nu(T) = \sigma T^4 . \qquad (7.22b)$$

The neglect of all scattering processes may seem a serious deficiency, but the assumption of local thermodynamical equilibrium takes into account, in a gross way, their overall disordering effect. Now the source function S_ν of a slab of a black body of thickness ds along the beam must be just B_ν; in fact, were this not so, when the radiation is in full thermal equilibrium with matter, the specific intensity would change along the beam, in contradiction with the assumption. Thermal equilibrium is the state that radiation tries to attain. This is *Kirchhoff's law*: a portion of a black body with an absorption coefficient α_ν has an emission coefficient $j_\nu = \alpha_\nu B_\nu$. The transport equation reads then

$$\frac{dI_\nu}{ds} = -\alpha_\nu (I_\nu - B_\nu) . \qquad (7.23)$$

To solve a radiative transfer problem, starting from an initial surface, we propagate I_ν with eq. (7.21) or (7.23) along all the rays to the whole volume of interest.

We now apply the general theory to a thin and stratified atmosphere, variable along z, and label the angular distance from the vertical with $\mu = \cos\theta$ ($-1 \leq \mu \leq 1$); upward and downward fluxes correspond, respectively, to positive and negative values. In the following the limits in integrals over μ are understood. The surface element on the unit sphere is $d\Omega = 2\pi\, d\mu$. The arc length along a ray is $ds = dz/\mu$ and the transport equation (7.23) reads

$$\mu \frac{dI_\nu}{dz} = -\alpha_\nu (I_\nu - B_\nu) . \qquad (7.24)$$

The vertical spectral flux is

$$F_\nu(z) = \int_{4\pi} d\Omega\, \mu I_\nu(z) = 2\pi \int_{-1}^{1} d\mu\, \mu I_\nu(z) . \qquad (7.25)$$

Stars and outer planets are similar in that their atmospheres continue downwards to very large optical depth, where one can expect nearly thermal equilibrium and an almost isotropic specific intensity. In these deep layers we can use $I_\nu = B_\nu$ as a starting solution, and solve (7.24) with the expansion

$$I_\nu = B_\nu - \frac{\mu}{\alpha_\nu} \frac{dB_\nu}{dz} + \dots . \qquad (7.26)$$

7. Atmospheres

The first term carries no flux; and

$$F_\nu(z) = -\frac{2\pi}{\alpha_\nu} \int_{-1}^{1} d\mu\, \mu^2 \frac{dB_\nu}{dz} = -\frac{4\pi}{3\alpha_\nu} \frac{dB_\nu}{dT} \frac{dT}{dz}. \tag{7.27}$$

This is a heat transport equation for each frequency; for the total flow

$$F(z) = \int_0^\infty d\nu\, F_\nu(z) = -\frac{4\pi}{3\alpha_R} \frac{dB}{dT} \frac{dT}{dz} = -K_R \frac{dT}{dz}, \tag{7.28}$$

where

$$\frac{1}{\alpha_R} \frac{d}{dT} \int_0^\infty d\nu\, B_\nu = \int_0^\infty \frac{d\nu}{\alpha_\nu} \frac{dB_\nu(T)}{dT} \tag{7.29}$$

gives a mean opacity, called *Rosseland opacity*. K_R is the radiative thermal conductivity coefficient (Sec. 1.5); if α_R is temperature independent, $K_R \propto T^3$.

Integrating (7.26) over frequency we get

$$I = \int_0^\infty d\nu\, I_\nu = B - \frac{\mu}{\alpha_R} \frac{dB}{dT} \frac{dT}{dz} + \ldots. \tag{7.30}$$

This shows an important feature of the radiation emitted by planets and stars: if, as usual, $dT/dz < 0$, the centre of the disc, at $\mu = 0$, is brighter than the limb, at $\mu = 0$. This is the *limb darkening* effect, due to the fact that near the centre the radiation comes from deeper and hotter layers.

Grey atmosphere. In a terrestrial planet the atmosphere terminates at the ground at $z = 0$ and the assumption of a *grey atmosphere*, in which the opacity is frequency independent, is useful; in this case the optical depth τ can be used as independent variable in place of z, with the surface at τ_s; τ_s is the total atmospheric optical depth, while the top is at $\tau = 0$. We now assume that the atmosphere is transparent to the direct (mostly visible) solar radiation and concentrate on the *infrared radiation*, with the transfer equation

$$\mu \frac{dI_\nu(\tau)}{d\tau} = I_\nu - B_\nu(\tau). \tag{7.31}$$

Note that $I_\nu(\tau)$ is a also function of μ, but $B_\nu(\tau)$ is not. Formal integration from the surface gives

$$I_\nu(0, \mu) = I_\nu(\tau_s) \exp\left(-\frac{\tau_s}{\mu}\right) + \int_0^{\tau_s} B_\nu(\tau) \exp\left(-\frac{\tau}{\mu}\right) \frac{d\tau}{\mu} \tag{7.32}$$

and provides the outgoing specific intensity at the top of the atmosphere, resulting from the attenuated flux from the ground and the thermal emission. The total flux is obtained by integrating μI_ν over angles and frequencies

$$F = 2\pi \int_{-1}^{1} d\mu\, \mu I \quad \left(I = \int_0^\infty d\nu\, I_\nu\right). \tag{7.33}$$

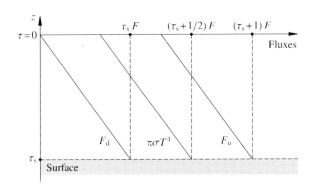

Figure 7.10 A simple model of the greenhouse effect (see the text). The upward and downward fluxes in the atmosphere as functions of the optical depth τ, and their mean, equal to $\pi \sigma T^4$. The ground temperature is greater than the temperature T_s at the bottom of the atmosphere.

In steady state F is independent of τ, so that, using (7.31), we get

$$\frac{1}{2}\int_{-1}^{1} d\mu \, I_\nu = B_\nu \qquad (7.34)$$

and (7.31) becomes an integro-differential equation.

Integrating (7.31) over the frequency, we finally obtain

$$\mu \frac{dI}{d\tau} = I - B = I - \frac{1}{2}\int_{-1}^{1} d\mu \, I = I - J. \qquad (7.35)$$

The arguments τ and μ are understood. On the surface, in thermal contact with the ground, the upward radiation is isotropic and I is constant for $\mu > 0$ (*Lambert's law*); at the top there is no incoming infrared flux and $I(0) = 0$ for $\mu < 0$. As the total optical depth τ_s increases, more radiation flows downward, increasing the surface temperature T_s.

A simple model of the greenhouse effect. Rather than solving the integro-differential equation (7.35), we use an intuitive model based on the upward and downward fluxes (both positive)

$$F_u = 2\pi \int_0^1 d\mu \, \mu I, \quad F_d = -2\pi \int_{-1}^0 d\mu \, \mu I = F_u - F; \qquad (7.36)$$

from eq. (7.35) they fulfil

$$\frac{1}{\pi}\frac{dF_u}{d\tau} = \frac{1}{\pi}\frac{dF_d}{d\tau} = \int_0^1 d\mu \, I - \int_{-1}^0 d\mu \, I;$$

this implies F is constant. To close the system an assumption must be made about the function $I(\mu)$. In a simple model, one can assume $I(\mu)$ constant and equal to I_u (I_d) for $\mu > 0$ ($\mu < 0$), an approach similar to the *two-stream approximation* developed in the radiation transfer theory by A. SCHUSTER and K. SCHWARZSCHILD around 1900 (Problem 7.17). Then, from eq. (7.36), $F_u =$

7. Atmospheres

πI_u, $F_d = \pi I_d$ and, from the previous equation, we get

$$\frac{dF_u}{d\tau} = \frac{dF_d}{d\tau} = F_u - F_d = F . \tag{7.37}$$

Both fluxes are linear functions of τ; since the at the top of the atmosphere ($\tau = 0$) we assume boundary conditions $F_u(0) = F$ and $F_d(0) = 0$ (for simplicity, we assume that all optical radiation absorbed by the surface is re-radiated in infrared), the solution is

$$F_u = (\tau + 1)F , \qquad F_d = \tau F \tag{7.38}$$

(Fig. 7.10). The temperature is given by (eq. (7.35))

$$2\pi B(\tau) = 2\pi\sigma T^4(\tau) = F_u + F_d = (1 + 2\tau)F ; \tag{7.39}$$

at the bottom of the atmosphere

$$2\pi\sigma T_s^4 = (1 + 2\tau_s)F . \tag{7.40}$$

T_s increases with the optical depth as $(2\tau_s + 1)^{1/4}$, a quantitative measure of the greenhouse effect.

This model has an unphysical feature: the ground temperature T_g is larger than the temperature T_s of the bottom of the atmosphere. In fact, the ground, which can emit only upward, gets the flux $F_d + F = F_u = \pi\sigma T_g^4$, so that the temperature ratio

$$\frac{T_g}{T_s} = \left(\frac{2\tau_s + 2}{2\tau_s + 1}\right)^{1/4} \tag{7.41}$$

tends to unity only for large optical depths and is $2^{1/4} \simeq 1.19$ when $\tau_s = 0$. In effect, the large temperature gradient that this discontinuity suggests produces convection, which establishes a nearly adiabatic gradient. In the boundary layer near the ground, where heat and moisture are exchanged with complex processes, static radiative transport is not adequate.

7.4 Dynamics of atmospheres

On the Earth, an extensive and complex global circulation is present, depending also on the distribution of lands and mountain ranges; the larger albedo at high latitudes (Fig. 7.2) requires an efficient process to transfer heat from low latitudes to the polar regions. In the stratosphere and above, global dynamics and heat exchange are strongly affected by complex chemical reactions and diffuse aerosols. Monitoring, understanding and predicting this global flow is the major task of large-scale meteorology and is crucial for predicting changes in climate. This topic is not discussed here, but the main underlying concept of *geostrophic flow*, of much wider relevance, is introduced. Also, we do not deal with vertical heat exchange; here a major role is played by thermodynamics, in particular phase transformations of water and the cloud cover. The effects of human activity are also beyond our scope. We concentrate instead on the basic physical processes, generally relevant for the solar system, in particular on stratified flows and the role of planetary rotation (geostrophic regime), and, in the following Section, on turbulence.

Gravity waves. Among the large variety of waves which can propagate in the atmosphere there is not much to say about acoustic waves, which propagate with the speed of sound (1.20); more interesting are the waves, at much lower frequencies, whose restoring force is gravity. They are already implicitly mentioned in Sec. 5.3 in relation to the convective instability: in the opposite situation in which the density gradient is less than the adiabatic value, hence stable, an initially displaced fluid element will just oscillate around its equilibrium position.

Consider a parallely stratified atmosphere in equilibrium, with a generic equation of state $P(\varrho, S)$ in terms of the entropy per unit mass S. The equilibrium condition (7.1) reads, in terms of the speed of sound (1.20),

$$-\varrho g = \frac{dP}{dz} = c_s^2 \frac{d\varrho}{dz} + \frac{\partial P(\varrho, S)}{\partial S} \frac{dS}{dz} . \qquad (7.42)$$

A small fluid element of volume V and mass ϱV at z is acted upon by two forces: its weight $-g\varrho(z)V$, and Archimedes' force, equal and opposite to the weight of the fluid it displaces. If the element is displaced to $z + \delta z$, an unbalance arises due to the change in Archimedes' force:

$$\delta F = -g\varrho(z)V + g\varrho(z + \delta z)(V + \delta V) .$$

As with the convective instability, we assume that the volume changes adiabatically ($\delta S = 0$), namely

$$\frac{\delta V}{V} = -\frac{\delta \varrho}{\varrho} = -\frac{\delta P}{\varrho c_s^2} = -\frac{\delta z}{\varrho c_s^2} \frac{dP}{dz} = \delta z \frac{g}{c_s^2} .$$

The restoring force

$$\delta F = \varrho V g \left(\frac{g}{c_s^2} + \frac{1}{\varrho} \frac{d\varrho}{dz} \right) \delta z \qquad (7.43)$$

corresponds to the *Brunt-Väisälä frequency* ω_v, given by

$$\omega_v^2 = g \left(\frac{g}{c_s^2} + \frac{1}{\varrho} \frac{d\varrho}{dz} \right) = \frac{gb}{\varrho c_s^2} \frac{dS}{dz} . \qquad (7.44)$$

In the troposphere of the Earth $\omega_v \approx 0.05$ Hz. When S decreases with height this is negative, leading to convective instability (the negative sign corresponds

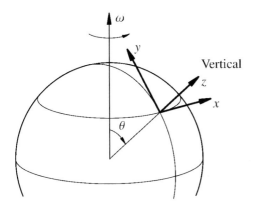

Figure 7.11 The local frame of reference for the atmosphere of a rotating planet. The z axis is along the vertical.

to a density gradient stronger than adiabatic, as explained in Sec. 5.3); otherwise we have stable oscillations. In this simple model we have neglected propagation, which can be obtained by adding to the restoring force (7.43) the inertial force (see Problem 7.8).

Horizontal approximation of the momentum equations. Atmospheric dynamics in rotating planets (with angular velocity ω) is deeply affected by Coriolis force, which changes the momentum equation (1.64) to

$$\varrho\left(\frac{\partial \mathbf{v}}{\partial t} + \mathbf{v} \cdot \nabla \mathbf{v}\right) + \nabla P + \varrho \nabla W + 2\varrho\,\omega \times \mathbf{v} = \eta\left[\nabla^2 \mathbf{v} + \frac{1}{3}\nabla(\nabla \cdot \mathbf{v})\right]. \quad (7.45)$$

$W = U - |\omega \times \mathbf{r}|^2/2$ (eq. (1.38)) includes the centrifugal potential. Following the same method used for the conservation of the vorticity $\mathbf{w} = \nabla \times \mathbf{v}$ (eq. (1.93)) and using the same approximations, it is easily seen that what is conserved is the *absolute vorticity* (Bjerknes theorem)

$$\mathbf{w}_a = \mathbf{w} + 2\,\omega = \nabla \times \mathbf{v} + 2\,\omega . \quad (7.46)$$

Angular momentum is exchanged between planetary rotation and kinematical vorticity. As before, this law is violated when there is viscosity. The absolute vorticity of an inviscid flow fulfils the equation

$$\frac{\partial \mathbf{w}_a}{\partial t} = \nabla \times (\mathbf{v} \times \mathbf{w}_a) + \frac{1}{\varrho^2}\nabla\varrho \times \nabla P : \quad (7.47)$$

vorticity is generated by the *baroclinic vector* $\nabla\varrho \times \nabla P/\varrho^2$, which vanishes in a barotropic flow, when $P = P(\varrho)$.

Planetary atmospheres are typically thin and it is appropriate to use a local Cartesian frame (Fig. 7.11), neglecting the curvature. Large scale, prevailing horizontal velocities

$$\mathbf{u} = \mathbf{v} - v_z\,\mathbf{e}_z$$

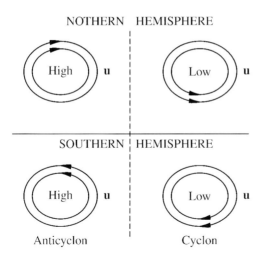

Figure 7.12 Cyclonic and anticyclonic flows.

are common. In planetary global circulation the vertical motion is hindered by the strong pressure gradient and gravity; horizontal motion prevails. On the Earth, typically the horizontal scale is $L \approx 1,000$ km, while the vertical scale $H \approx 10$ km; $u \approx 10^3$ cm/s is much larger than $v_z \approx 1$ cm/s. We also assume $v_z L \ll uH$. The centrifugal force amounts to a slight change in the gravity acceleration $\mathbf{g} = -\nabla U$ and its contribution to the definition of the vertical can be neglected. This leads to the (two-dimensional) *geostrophic equation* (neglecting viscosity)

$$\frac{\partial \mathbf{u}}{\partial t} + \mathbf{u} \cdot \nabla \mathbf{u} = -\frac{1}{\varrho} \nabla_\perp P - f \mathbf{e}_z \times \mathbf{u} . \quad (7.48a)$$

$$f = 2\omega \cos\theta = 2\omega_z . \quad (7.48b)$$

is the *Coriolis parameter*. The vorticity field $\mathbf{w}$ is now straight and vertical, an essential difference from the three-dimensional case; the vorticity conservation law now says that, if dS is a horizontal surface element tied to the flow, $w_{az} dS = (w_z + f) dS$ is conserved. This shows how *cyclonic* and *anticyclonic* features arise in a rotating planet at mid latitudes. A fluid initially at rest is drawn into a pressure low and the associated horizontal area diminishes from dS_0 to dS, say; hence a vertical vorticity

$$w_z = f \left(\frac{dS_0}{dS} - 1 \right)$$

arises. In the northern hemisphere, where $f > 0$, $w_z > 0$ and the flow is counterclockwise; in the southern hemisphere $f < 0$ and $w_z < 0$; this is the cyclonic flow. Around a pressure high we have an anticyclon: clockwise in the North and counterclockwise in the South (Fig. 7.12).

7. Atmospheres

In eq. (7.48a) the horizontal pressure gradient can be balanced either by inertia or by the Coriolis acceleration; their ratio is of order

$$Ro = \frac{u}{|f|L}, \tag{7.49}$$

the *Rossby number* (on the Earth, not too near the equator, ≈ 0.1). When $Ro \ll 1$ the pressure gradient is balanced by a flow along isobaric lines; in stationary conditions we have, to the lowest approximation and contrary to intuition,

$$\mathbf{u} = -\frac{1}{f\varrho}\boldsymbol{\nabla}_\perp P \times \mathbf{e}_z, \tag{7.50}$$

sometimes called geostrophic wind. This gives a quantitative description of cyclonic and anticyclonic flows. If there is also a frictional acceleration $-\nu_c \mathbf{u}$ (in particular, with the ground), eq. (7.48a) gives

$$\mathbf{u} = -\frac{1}{f\varrho}\boldsymbol{\nabla}_\perp P \times \mathbf{e}_z - \frac{\nu_c}{f^2\varrho}\boldsymbol{\nabla}_\perp P, \tag{7.51}$$

with the terms $\propto \nu_c^2$ neglected. There is a flow also along the pressure gradient, towards the low pressure side; air flows into cyclonic troughs and out from anticyclonic peaks, producing a vertical flow to balance the mass.

To see what happens in general, consider a streamline with arc s and unit tangent vector $\hat{\mathbf{s}} = \mathbf{u}/u$. If R_c is its radius of curvature,

$$\hat{\mathbf{n}} = \frac{d\hat{\mathbf{s}}}{ds}R_c$$

is the unit vector orthogonal to $\hat{\mathbf{s}}$ in the direction of the centre of curvature. The decomposition of the momentum equation (7.48a) along the local directions $\hat{\mathbf{s}}$ and $\hat{\mathbf{n}}$ (with coordinate n) gives

$$\frac{du}{dt} + \frac{1}{\varrho}\frac{\partial P}{\partial s} = 0, \qquad \frac{u^2}{R_c} + \frac{1}{\varrho}\frac{\partial P}{\partial n} + fu = 0. \tag{7.52}$$

The pressure gradient along a streamline only changes the speed; the second equation recovers the geostrophic solution when $Ro \ll 1$. Near the equator, where the Coriolis parameter (7.48b) is small, or when u is large, the Coriolis force may be neglected and

$$\frac{u^2}{R_c} + \frac{1}{\varrho}\frac{\partial P}{\partial n} = 0; \tag{7.53}$$

this is the *cyclostrophic approximation*, for which the pressure is balanced by the centrifugal force. The fluid can still flow along the isobaric lines around a pressure low, but in an arbitrary sense; indeed, small-scale vortices, like dust

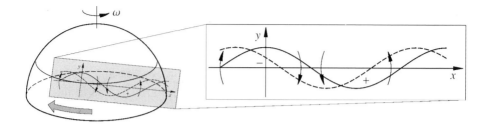

Figure 7.13. Rossby waves: in the local frame of the figure, in the northern hemisphere the Coriolis parameter (7.48b) increases with y. Suppose that a streamline, initially along a parallel $y = 0$, is accidentally displaced into a sinusoidal shape (heavy line). Since the absolute vorticity (7.46) is conserved, the fluid vorticity of a fluid element decreases (increases) if it moves to higher (lower) latitudes; as a result, different parts of the streamline are subject to different rotations and the initial sinusoid moves to the dashed line, with a westward propagation.

devils or water spouts, can be both cyclonic and anticyclonic and do not have a preferred direction of rotation. The situation is different for *tornadoes*, with large u, for which a cyclonic (counterclockwise) sense of rotation is preferential in the northern hemisphere. Apparently, in the initial stages of its development the Coriolis force deflects air parcels accelerating toward the centre of low pressure to the right.

In general, one should consider all terms in the geostrophic inviscid equations (7.52), obtaining thus the *gradient flow approximation*. Denoting with $u_{\text{geo}} = -(\partial P/\partial n)/(\varrho f)$ the geostrophic wind speed (7.50), we obtain (Problem 7.7)

$$\frac{u_{\text{geo}}}{u} = 1 + \frac{u}{fR_c} . \qquad (7.54)$$

Note that the factor (u/fR_c) is just the Rossby number (7.49); since for tropical tornadoes this factor is of order unity, the gradient wind approximation must be used instead of the simpler geostrophic approximation.

Rossby waves. As shown in Fig. 7.13, Rossby waves arise because the Coriolis parameter (7.48b) changes with latitude. For disturbances whose extension in latitude is not negligible, the assumption of constant f is invalid. In a local Cartesian frame (Fig. 7.11) $u_y = -Rd\theta/dt$ and

$$\frac{df}{dt} = \frac{2\omega \sin \theta}{R} u_y .$$

In this formula, for small amplitudes, θ can be considered constant; it amounts to approximating the Coriolis parameter with

$$f = f(\theta) + \beta y , \qquad \beta = -\frac{2\omega \sin \theta}{R} ; \qquad (7.55)$$

… this is usually termed the *β-plane approximation*. In the thin atmosphere approximation and when $Ro = O(1)$, what matters is the vertical component w_{az} of the absolute vorticity, of order u/L. Estimating all the terms in eq. (7.47) in a barotropic flow we get

$$\frac{\partial w_{az}}{\partial t} + \mathbf{u} \cdot \nabla_\perp w_{az} + (\nabla_\perp \cdot \mathbf{u}) w_{az} = 0 \,. \tag{7.56}$$

Before proceeding to the dispersion relation, we introduce a considerable and elegant simplification applicable to an incompressible (two-dimensional) flow; this is a good approximation as long as the fluid velocity is much smaller than the sound speed. A two-dimensional, divergence-free vector field (u_x, u_y) can always be represented in terms of a *stream function* $\psi(x, y)$

$$u_x = -\frac{\partial \psi}{\partial y}, \quad u_y = \frac{\partial \psi}{\partial x} \,. \tag{7.57}$$

The vorticity $\mathbf{w} = w\,\mathbf{e}_z$ lies in the vertical direction and

$$w = w_z = \frac{\partial^2 \psi}{\partial x^2} + \frac{\partial^2 \psi}{\partial y^2} = \nabla_\perp^2 \psi \,. \tag{7.58}$$

$\nabla_\perp^2$ denotes the two-dimensional Laplace operator. An elementary, point-like vortex (eq. (1.91)) at the origin with circulation C is represented by the harmonic stream function

$$\psi = \frac{C}{2\pi} \ln r \,, \tag{7.59}$$

with velocity field

$$u_x = -\frac{C}{2\pi r^2} y \,, \quad u_y = \frac{C}{2\pi r^2} x \,. \tag{7.59'}$$

Just like for the gravitational field, an arbitrary superposition of these elementary solutions provides the general flow in stationary conditions.

Consider now a small perturbation of a uniform flow with velocity u_0 along a parallel, for which

$$\psi = -u_0 y + \psi' \,;$$

the vorticity equation (7.56) reads

$$\left[\frac{\partial}{\partial t} + u_0 \frac{\partial}{\partial x}\right]\left(\frac{\partial^2 \psi'}{\partial x^2} + \frac{\partial^2 \psi'}{\partial y^2}\right) + \beta \frac{\partial \psi'}{\partial x} = 0 \,. \tag{7.60}$$

As discussed in Sec. 8.3, a wave solution is obtained by taking $\psi' \propto \exp i(\sigma t - k_x x - k_y y)$, which gives the dispersion relation for planetary *Rossby waves*

$$\sigma = k_x u_0 - \beta \frac{k_x}{k_x^2 + k_y^2} \,. \tag{7.61}$$

The negative sign shows that they propagate westward with respect to the background flow, with a phase velocity which *increases* with the wavelength. For a wave number of order $1/1,000$ km the period is $\approx 1/3$ of a day and the phase velocity about 20 m/s.

The geostrophic approximation can also be used for accretion discs, planetary rings and the primordial solar nebula. In equilibrium, each particle in the plane at the distance r from the central mass M moves with the Keplerian angular velocity $\omega_K = (GM/r^3)^{1/2}$. In a rotating frame, using the same symbols as before, its neighbourhood can be described by the elevation z over the plane and the displacements x and y in the tangential and the radial direction, respectively. For motion in the plane, Coriolis' force is $f\mathbf{e}_z \times \mathbf{u}$, with $f = 2\omega_K$; eq. (7.55) is replaced by

$$f = f(r) + \beta y \quad \left(\beta = \frac{d\omega_K}{dr} = -\frac{3}{4}\sqrt{GM}r^{-5/2}\right); \tag{7.62}$$

the absolute vorticity is conserved and the same dispersion relation holds. Rossby number (7.49) is often smaller than unity.

7.5 Turbulence

In general terms, a flow becomes fully turbulent if the ratio of inertial to viscous terms in the equations of motion, as described by the Reynolds number (1.69), is sufficiently large; such large velocities arise as an instability develops to a non-linear regime. Fluid turbulence widely occurs in the solar system. It governs heat transfer in atmospheres when the temperature gradient is steeper than the adiabatic value. Although on a much longer timescale, convective turbulence may also be active in planetary interiors, if they are liquid or plastic, and heated from below. Near the ground of the Earth, in a boundary layer a few hundred metres thick, the turbulent heat exchange with the surface is strongly affected by friction, moisture and surface properties. The solar wind is also turbulent, with an important role played by non-linear Alfvén waves (Sec. 9.3). In planetary atmospheres the flow is often mainly horizontal and a peculiar kind of two-dimensional turbulence develops, where Coriolis' force has a crucial role: this is the *geostrophic turbulence*, to be discussed later.

In the troposphere turbulent inhomogeneities in temperature and density affect the refractive index n_r and optical propagation. $n_r - 1$, proportional to the density, fluctuates in space and time; a plane surface of constant phase is warped (see Sec. 19.5). The image of a point source gathered by an objective larger than the size of the warps is diffuse and, moreover, moves around in time ("dancing"). This phenomenon is called *scintillation*, or *seeing*, and is of crucial importance in limiting optical astronomical observations from the ground. In practice, the apparent diameter of a celestial point-like source is larger than

1″; a special technology of adaptive optics has been developed to suppress this disturbance (Sec. 19.5).

In a turbulent flow the velocity

$$\mathbf{v} = \langle\mathbf{v}\rangle + \mathbf{v}' \qquad (\langle\mathbf{v}'\rangle = 0) , \qquad (7.63)$$

is the sum of a mean value $\langle\mathbf{v}\rangle$ and a fluctuating value $\mathbf{v}'$, similarly to what was assumed in the dynamo theory (Sec. 6.4): $\langle\mathbf{v}\rangle$ is determined by the macroscopic conditions, but $\mathbf{v}'$ changes for different realizations of the same macroscopic flow and can be studied only by statistical methods. The angular brackets denote an average over a Gibbs ensemble of realizations of the flow with the same macroscopic conditions; it is a formal operator which commutes with the time and space derivatives.

Turbulent viscosity. The statistical decomposition (7.63) leads to the *turbulent viscosity*, with a simple and interesting application to the turbulent atmospheric dynamics near the ground of a terrestrial planet. While $\mathbf{v}'$ has vanishing average, its product $\mathbf{v}'\mathbf{v}'$ does not; hence the mean change of momentum of a fluid element has a turbulent contribution $\varrho\nabla\cdot\langle\mathbf{v}'\mathbf{v}'\rangle$ which acts as a pressure tensor.

Consider, for example, the case in which the turbulent energy is drawn from the differential motion along the x-direction: $\langle\mathbf{v}\rangle = (v(z), 0, 0)$; this is the case, for example, when the powerful *Kelvin-Helmholtz instability* is present. Since the undisturbed flow is not homogeneous, the usual linear stability analysis in terms of sinusoidal functions is not possible and an eigenvalue problem for ordinary differential equations must be solved, with a great variety of results. To avoid this complication, consider a thin transition layer of thickness δ and vorticity of order $2v/\delta$, separating two homogeneous flows with $v(\infty) = -v(-\infty) = v > 0$. If, under a small sinusoidal deformation, a section of the layer is displaced upwards, it encounters a flow v in the positive direction; its vorticity is dragged along, enhancing the deformation; similarly for a downward displacement. The amplitude of the deformation and the thickness δ of the transition increase; x-momentum is transferred in the z-direction. When the corresponding Reynolds number (eq. (1.69))

$$Re = \frac{\varrho v \delta}{\eta}$$

is large, turbulence sets is and we expect a turbulent stress (the *Reynolds stress*) to develop, proportional to dv/dz:

$$\varrho\langle v'_x v'_z\rangle = -\eta_t \frac{dv}{dz} ; \qquad (7.64)$$

the minus sign indicates that such a transfer tends to destroy the velocity gradient; η_t is the coefficient of turbulent viscosity (a similar argument has been used in Sec. 6.4 for the turbulent dynamo). For example, near the ground, with $v' = 1$ cm/s and $v = 10^3$ cm/s over a scale of 1 km, $\eta_t/\varrho \approx 10^2$ cm²/s, much larger than the microscopic value $\eta/\varrho \approx 0.1$ cm²/s: the momentum transfer in the boundary layer of the Earth is dominated by turbulence. It is likely that turbulent viscosity has played an important role in the formation of the dust layer in the primordial solar nebula (Sec. 16.2).

A simple description of the effect that the turbulent friction is provided by *Ekman's layer solution*. The simple geostrophic equations (7.48a) acquire a dependence on altitude and become

$$\frac{\partial P}{\varrho \partial x} - f u_y - \eta_t \frac{\partial^2 u_x}{\partial z^2} = 0, \qquad \frac{\partial P}{\varrho \partial y} + f u_x - \eta_t \frac{\partial^2 u_y}{\partial z^2} = 0. \quad (7.65)$$

Introducing the geostrophic velocity components (eq. (7.50))

$$u_x^g = -\frac{1}{\varrho f} \frac{\partial P}{\partial y}, \qquad u_y^g = \frac{1}{\varrho f} \frac{\partial P}{\partial x},$$

and the complex variables $Z = u_x + i u_y$ and $Z^g = u_x^g + i u_y^g$, the geostrophic eqs. (7.65) become

$$\eta_t \frac{\partial^2 Z}{\partial z^2} - if(Z - Z^g) = 0. \quad (7.66)$$

At the surface $z = 0$ friction with the ground requires $Z = 0$; at large altitude ($z \to \infty$) we must recover the geostrophic wind $Z = Z^g$. The solution is

$$Z = Z^g \left[1 - \exp\left(-\sqrt{\frac{if}{\eta_t}} z\right) \right], \quad (7.67)$$

known as the *Ekman's spiral*. Above a layer of thickness $\sqrt{f/\eta_t}$, of the order of 1 km on the Earth, the flow is well matched by the inviscid geostrophic wind.

Kolmogorov's spectrum. The simplest, but still relevant model of fluid turbulence, neglects compressibility and assumes at the macroscopic level isotropic and homogeneous conditions. Consider the mean turbulent kinetic energy per unit mass of a turbulent flow in a (very large, fictitious) volume V

$$E = \frac{1}{2V} \int_V d^3r \, \mathbf{v}'^2 = \frac{1}{2} \langle \mathbf{v}'^2 \rangle = \int_0^\infty dk \, \tilde{E}(k). \quad (7.68)$$

The integral over Fourier space is really a sum, consistently with the finite volume. On its boundary the velocity and its derivatives are assumed to vanish. The two expressions are a consequence of an argument similar to the ergodic theorem: averaging over a large volume is equivalent to averaging over all possible realizations. The energy can be decomposed in a Fourier series, $\tilde{E}(k)dk$ being the contribution to E from wave numbers with modulus between k and $k + dk$; in an isotropic regime the *spectral energy density* $\tilde{E}(k)$ of the microscopic flow is a function only of the modulus of $\mathbf{k}$, not of its direction. Each wave number $\mathbf{k}$ corresponds to eddies of size $1/k$, characteristic velocity $v'(k) = [\tilde{E}(k)k]^{1/2}$ and timescale $\tau_k = 1/[kv'(k)]$; with a viscosity coefficient η, the dissipation time

$$\tau_d = \frac{\varrho}{\eta k^2} \qquad (7.69)$$

decreases with size; the Reynolds number (1.69) (ratio of inertial to dissipative forces) for the eddy is

$$Re(k) = \frac{v'(k)\varrho}{k\eta} = \frac{\tau_d}{\tau_k}.$$

An eddy begins to dissipate away when $\tau_d < \tau_k$ and its wave number approaches k_d, say; larger eddies last for a long time. For example, near the ground, for $\eta/\varrho \simeq 0.1$ cm^2/s and $v' \simeq 1$ cm/s, the two times are equal when $1/k \simeq 10$ cm. To see the effect of non-linear processes, note that the Fourier transform of the product of two functions with wave vectors $\mathbf{k}_1$ and $\mathbf{k}_2$ has a component at $(\mathbf{k}_1 + \mathbf{k}_2)$, generally at a larger distance from the origin in k-space[3]. For an incompressible flow the main non-linearity is due to the convective term $\varrho \mathbf{v} \cdot \nabla \mathbf{v}$ in the momentum equation (1.35). Let $\epsilon_{\text{in}} = k_s[v'(k_s)]^3$ be the power per unit mass fed into the system at the (small) wave number k_s; this energy is slowly transferred in a cascade to higher k's, until, at k_d, dissipation prevails. The excitation of smaller and smaller eddies corresponds to the fact that a string of particles, initially smooth, as it is dragged along by the flow progressively becomes longer and more bent and entangled; this dispersion can be easily visualized with an insoluble dye. When the compressibility is small and the volume is conserved, an area dragged along by the flow decreases and, as a consequence of the conservation of vorticity (1.93), in absence of dissipation the vorticity increases.

The interval (k_s, k_d) of wave numbers where dissipation is negligible is called the *inertial range*. When $k_d \gg k_s$ (large inertial range) the spectrum cannot depend on η and must be a universal function of k and ϵ_{in}; hence, purely on dimensional grounds, we have *Kolmogorov's spectrum*

$$\tilde{E}(k) = A \, \epsilon_{\text{in}}^{2/3} k^{-5/3}, \qquad (7.70)$$

---

[3] As an indication, when the angle between the two vectors is random, their sum has average modulus $(k_1^2 + k_2^2)^{1/2}$, always larger than k_1 and k_2.

where A is a dimensionless constant, usually of order unity; for the same reason, the turbulent velocity has the expression

$$v'(k) = A' \left(\frac{\epsilon_{in}}{k}\right)^{1/3} , \qquad (7.71)$$

where A' is another dimensionless constant of order unity. This law was first put forward by A.N. KOLMOGOROV in 1941. In this way, from eq. (7.69) and the condition $\tau_k = \tau_d$, we obtain the "dissipative" wave number

$$k_d \approx \left(\frac{\varrho}{\eta}\right)^{3/4} \epsilon_{in}^{1/4} . \qquad (7.72)$$

It is pleasing that Kolmogorov's power spectrum (7.70) is indeed observed in the turbulent boundary layer on the ground and in other turbulent flows in the solar system (in particular, in the interplanetary solar wind), in spite of the fact that in all these cases the original assumptions of incompressibility and isotropy are not fulfilled. On a general level, it should also be noted that spectral power laws like (7.70) are typical of distributions arising from non-linear processes.

It is interesting to see what happens when the viscosity tends to zero. Since the total energy input is determined by what happens at large scales, the dissipated power must stay constant. In fact, such power has the general form

$$\eta \langle(\nabla \mathbf{v}')^2\rangle \propto \eta \, k^{4/3}$$

and is independent of η at the viscous wave numbers (7.72): as $\eta \to 0$, the inertial range extends further out, but the velocity gradient (and the vorticity) increases.

Two-dimensional turbulence. When, as common in planetary atmospheres, the flow mainly occurs in the horizontal plane, turbulence has quite a different character. In three dimensions and in absence of viscosity, the lengthening and bending of material lines necessarily increases the vorticity; but in two dimensions the vorticity vector $\mathbf{w} = w \, \mathbf{e}_z$ is always straight and vertical and can increase only as a result of the compression of horizontal surface elements; we now show, with a subtle and powerful argument, that the energy of the flow does not cascade to smaller eddies, but, rather, migrates to *smaller* wave numbers, where there is no dissipation and generates macroscopic flow structures; eventually they escape the turbulent disorder and acquire a permanent character. These ideas developed in the late 1950s, in particular due to G.K. BATCHELOR.

In the simple model of an incompressible flow, and in absence of gravity, the (full) Navier-Stokes equations (1.64) read (with the Lagrangian derivative)

$$\frac{d\mathbf{v}}{dt} + \nabla P = \frac{\eta}{\varrho} \nabla^2 \mathbf{v} . \qquad (7.73)$$

7. Atmospheres

Taking the scalar product with **v** and averaging over the volume V we easily get

$$\frac{dE}{dt} = -\frac{\eta}{\varrho}\langle|\mathbf{w}|^2\rangle . \tag{7.74}$$

We have remarked that in the three-dimensional case the right-hand side remains finite even for a very small viscosity. Eq. (1.92) for the vorticity reads

$$\frac{d\mathbf{w}}{dt} = (\mathbf{w}\cdot\nabla)\mathbf{v} + \frac{\eta}{\varrho}\nabla^2\mathbf{w} . \tag{7.75}$$

In the right-hand side, the first term describes the inertial cascade and generally increases the vorticity; the second is the dissipation and works only at large k's. We introduce, besides the energy (7.68), another global quantity, the *enstrophy* (from the Greek root *streb* or *strob*, to rotate)

$$\mathcal{E} = \frac{1}{2V}\int_V d^3r\,|\mathbf{w}|^2 = \frac{1}{2}\langle|\mathbf{w}|^2\rangle , \tag{7.76}$$

which measures the total amount of vorticity present. Clearly, it is the first term in the right-hand side of (7.75) that allows the enstrophy to increase as the viscosity decreases.

But in *two dimensions* things are different. We consider again incompressible and isotropic (of course, only in the plane (x,y)) turbulence. When the viscosity is negligible, the vorticity $w = w_z = \nabla_\perp^2\psi$ is a conserved scalar (see eqs. (7.56) and (7.58)); the inertial coupling term is absent and it is easily seen that the enstrophy can only decrease

$$\frac{d\mathcal{E}}{dt} = -\frac{\eta}{\varrho}\langle|\nabla_\perp w|^2\rangle . \tag{7.77}$$

For large values of the Reynolds number the total kinetic energy is constant in time; therefore the cascade to large k's into the dissipation hole is excluded. But this argument does not apply to the enstrophy. Consider in the plane (x,y) two closed, neighbouring material isovorticity lines corresponding to w_0 and $w_0 + dw_0$; in an incompressible flow the area between them is conserved. But, since, as a consequence of turbulent diffusion, they become in time more and more bent and longer, their separation must decrease; hence the vorticity w acquires a larger gradient and its spectrum moves to larger k's. The cascade that in three dimension occurs for the energy, in this case takes place for the enstrophy: as η decreases, $|\nabla_\perp w|^2$ increases, in such a way as to keep the right-hand side of the previous equation finite; but the rate of change of the energy goes to zero. In Fourier language and isotropic conditions $2\tilde{E}(k) = k^2|\tilde{\psi}(k)|^2$, while $2\tilde{\mathcal{E}}(k) = k^4|\tilde{\psi}(k)|^2 = 2k^2\tilde{E}(k)$. The spectral contribution to the enstrophy mainly comes from large wave numbers, where the dissipation of enstrophy

	dE/dt	$d\mathcal{E}/dt$
Three-dimensional	$\to$ finite	0
Two-dimensional	0	$\to$ finite

Table 7.3 The different behaviour of the energy E and the enstrophy $\mathcal{E}$ in three-dimensional and two-dimensional incompressible turbulence, as $\eta \to 0$. In both cases dissipation takes place at large k's.

occurs; therefore it is possible for the enstrophy to decrease, leaving E constant; but then the decrease of $\tilde{E}(k)$ at large k's must be compensated by its increase at small k's: in an inverse cascade, the energy moves to smaller eddies.

Planetary global circulation. Deviations from the two-dimensional geostrophic model of global circulation are required, for example, to account for meridional heat transfer. On the Earth this is accomplished with circulation cells, in which hot air raises from where more heat is deposited (e.g., at the subsolar latitude) and transports it to where the thermal energy is deficient. The typical example are *Hadley cells*. For example, in summer a Hadley cell brings hot air coming from $\approx 30°$ N latitude back to the ground at $\approx 30°$ S. Not too near the equator, Coriolis' force on this current pushes the flow in the westward direction and produces the easterly jet (also called *trade wind*); on the southern hemisphere the jet flows eastward.

Venus has a large rotation period (243 days); however, its atmosphere has a much faster, easterly zonal rotation (with a period of 4 – 5 days) with respect to the ground (the zonal superrotation); in its own frame Coriolis effects could be important. The Hadley circulation accomplishes the required meridional heat transfer from the equator to the poles. Another example of slowly rotating planetary body with significant superrotation of the atmosphere is Titan: the zonal wind velocity, recently detected from the Doppler shift of ethane atmospheric lines, exceeds the surface velocity by a factor of ≈ 10.

But we shirk from the complexity of the atmospheres of terrestrial planets and, venturing into a still not well understood area, discuss the relevance, on a conceptual level, of the geostrophic, two-dimensional turbulence in relation to the giant planets. Strong zonal winds, with sharp meridional gradients, are present in Jupiter and Saturn (Fig. 7.14). Large and stable cyclonic and anticyclonic structures are a frequent feature of the atmospheres of giant planets. On Jupiter, what is seen in the visible from outside is a thin cloud layer (the "weather layer"), with features moving in bands along the parallels; their velocity – measured using small clouds as tracers – has large meridional gradients, varying from ≈ -50 to ≈ 175 m/s. The *Great Red Spot* is a large (26, 000 by 13, 000 km, the largest dimension along the parallel) anticyclonic vortex with the high pressure centre at the latitude of $-23°$. It has been observed for the first time by R. Hook in 1665 and G.D. Cassini in 1666 and, recently, by the Voyager and Galileo missions. In this region the vertical scale height is ≈ 20 km and it can be assumed that the spot – a big pancake between the al-

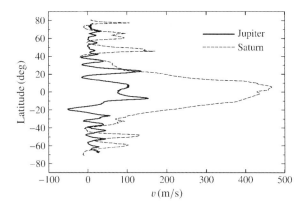

Figure 7.14 The zonal wind speed in the atmospheres of Jupiter and Saturn, in terms of the planetographic latitude. Data from E. Garcia-Melendo and A. Sánchez-Lavega, *Icarus* **152**, 316 (2001) and A. Sánchez-Lavega et al., *Icarus* **147**, 405 (2000).

titudes corresponding to ≈ 0.7 and ≈ 0.3 mbar – has a thickness of that order. The centre of the spot is at rest and the velocity reaches a peak of ≈ 100 m/s, corresponding to a Rossby number ≈ 0.03. The reason for the red colour is uncertain. Other, smaller vortices are present in the outer planets; Neptune has a similar large structure (the *Great Dark Spot*).

On a fast rotating planet the Coriolis force must be taken into account and requires using the *absolute enstrophy* $\mathcal{E}_a$, obtained by replacing in eq. (7.76) **w** with the absolute vorticity **w** + 2ω; the rough model of two-dimensional turbulence discussed above still holds, with the Coriolis term as a "natural" source of vorticity.

The relevant physical quantities for Jupiter and Saturn are comparable: at mid latitudes $f \simeq \omega \simeq 2 \times 10^{-4}$ rad/s, $u_x = u \simeq 10^4$ cm/s (Fig. 7.14), $\beta \simeq \omega/R \simeq 2 \times 10^{-14}$ cm^{-1}s^{-1}. The characteristic scale (along the meridian) is about $10°$, corresponding to $L \simeq 10^9$ cm, so that in both cases $Ro \simeq u/(\omega L) \simeq 0.1 \ll 1$ and a Rossby circulation is expected.

The kinetic energy of turbulent eddies generated by this process and other instabilities moves to smaller k's; what is the threshold k_m at which they become independent of the random turbulence? Let us assume, for simplicity, that the elementary turbulent modes are Rossby waves with non-linear interactions, described by adding in the Lagrangian derivative of eq. (7.60) the inertial term $\mathbf{u} \cdot \nabla_\perp w$, of order $k^2 u^2$. The transition occurs when this term is comparable with the last term, the Rossby restoring force, at $k_m = \sqrt{\beta/u}$. For Jupiter and Saturn its reciprocal is $\approx 10^9$ cm, pleasantly consistent with the meridional scale L. To study the dependence of this transition from the direction of propagation, we note the turbulent frequency scale $1/\tau_k = kv'(k)$ increases with wave number (in Kolmogorov's regime $1/\tau_k \propto k^{2/3}$), while the frequency of Rossby waves (7.61) decreases as $1/k$; let k_m be the value at which the two frequencies are equal. Only for $k < k_m$, when Rossby waves oscillate faster than the turbulent eddies, the inverse energy cascade can be inhibited. Introducing the angle α between the wave vector and the parallel, the inhibition occurs when (eq. (7.61)) $\beta \cos\alpha/k > 1/\tau_k = kv'(k)$, which, taking $v'(k_m) = u$, gives

$$k_m = \sqrt{\frac{\beta}{u}\cos\alpha}, \qquad (7.78)$$

in agreement with the previous estimate. But now we see that k_m becomes small for propagation near the meridional direction, making inhibition very effective; only when α is small the cascade can proceed to small wave numbers and produce coherent structures with great extension along the parallels; they may be the zonal winds actually observed.

Appendix: Effect of secular orbital changes on the rotation axis

In Sec. 7.1 we have shown that when there is a resonance between the lunisolar precession and the nodal motion (i.e., when $\Omega_{pr} + \dot\Omega = 0$), the obliquity ϵ is expected to undergo large changes; for a complete analysis we need the full, non-linear solution. Referring to the main text for the notation, it is convenient to adopt moving coordinates $\mathbf{x}' = \mathbf{A}\mathbf{x}$ tied to the orbital plane, with x'^3 along $\mathbf{N}_p$ and x'^1 along the ascending node; in the general case they are obtained from the inertial coordinates $\mathbf{x}$ with the time-dependent rotation matrix

$$\mathbf{A} = \begin{bmatrix} \cos\Omega & \sin\Omega & 0 \\ -\cos I \sin\Omega & \cos I \cos\Omega & \sin I \\ \sin I \sin\Omega & -\sin I \cos\Omega & \cos I \end{bmatrix}, \qquad (a)$$

Of course, $\mathbf{N}'_p = (0, 0, 1)$; both I and Ω are allowed to be time-dependent as a result of mutual planetary perturbation (Sec. 15.1). In the moving frame the components $\mathbf{e}' = (\sin\epsilon\sin(\psi + \Omega), \sin\epsilon\cos(\psi + \Omega), \cos\epsilon) = \mathbf{A}\mathbf{e}$ of the axis fulfil

$$\frac{d\mathbf{e}'}{dt} = (d\mathbf{A}/dt)\mathbf{A}^{-1}\mathbf{e}' - f_{pr}\left(\mathbf{N}'_p \cdot \mathbf{e}'\right)\left(\mathbf{N}'_p \times \mathbf{e}'\right). \qquad (b)$$

It can be shown that, in terms of $X = \cos\epsilon$ and the precession longitude ψ, the equations (b) follow from the time-dependent Hamiltonian

$$H(X, \psi; t) = \frac{f_{pr}}{2}X^2 - CX + \sqrt{1 - X^2}\left(\mathcal{A}\sin\psi + \mathcal{B}\cos\psi\right); \qquad (c)$$

X is the generalized momentum conjugate to ψ. The functions of time $\mathcal{A} = \cos\Omega(dI/dt) - \sin I \sin\Omega(d\Omega/dt)$, $\mathcal{B} = \sin\Omega(dI/dt) + \sin I \cos\Omega(d\Omega/dt)$ and $C = 2\sin^2(I/2)(d\Omega/dt)$ describe the effect of the orbital perturbations. The solution is by no means trivial, but for a small inclination it simplifies since $\mathcal{A} = (dQ/dt) + O(I^2)$, $\mathcal{B} = (dP/dt) + O(I^2)$ and $C = O(I^2)$, where $P = \sin I \sin\Omega$ and $Q = \sin I \cos\Omega$ are the non-singular orbital elements. In Sec. 15.1 we demonstrate that, for the planet of interest, P and Q are represented by a finite number of Fourier terms: $P = \sum_j S_j \sin(s_j t + \gamma_j)$ and $Q = \sum_j S_j \cos(s_j t + \gamma_j)$, where s_j are the eigenfrequencies of the planetary system; of particular interest for Earth is $s_3 \simeq -18.75''/y$. S_j are small dimensionless constants, typically of order $\simeq 0.01$. Neglecting the second order term $C = O(I^2)$ and those in $\mathcal{A}$ and $\mathcal{B}$, we finally obtain the approximate form of the Hamiltonian (c)

$$H(X, \psi; t) = \frac{f_{pr}}{2}X^2 + \sqrt{1 - X^2}\sum_j S_j s_j \cos\left(\psi + s_j t + \gamma_j\right). \qquad (d)$$

7. Atmospheres

With only one forcing frequency – as when the orbital node precesses with a constant rate – eq. (d) is often referred to as *Colombo's top* (Problem 7.13).

When $S_j = 0$ there is no secular evolution of the orbit; $X = X_0 = \cos \epsilon_0$ is trivially constant and $d\psi/dt = \partial H/\partial X = f_{pr} X_0$. In general the obliquity fulfils

$$\frac{dX}{dt} = \sqrt{1-X^2} \sum_j S_j s_j \sin(\psi + s_j t + \gamma_j), \tag{e}$$

obviously coupled with the conjugated equation for ψ. For small changes $\epsilon' = \epsilon - \epsilon_0$ in the obliquity, we may proceed perturbatively and use $\psi_0(t) \simeq \psi_0 + f_{pr} X_0 t$. With a quadrature, we then get, to lowest order,

$$\epsilon' = \sum_j \frac{S_j s_j}{f_{pr} X_0 + s_j} \cos(\psi_0(t) + s_j t + \gamma_j). \tag{g}$$

When $f_{pr} \gg s_j$ (very slow nodal motion) ϵ is nearly constant and the spin axis has the ability to track the motion of the orbital plane. The interesting case is when $f_{pr} X_0 + s_j = 0$ and a resonance occurs; as the obliquity crosses the resonant value, its variation is amplified and a complete non-linear solution is necessary.

– PROBLEMS –

7.1.* Determine the yearly insolation $\bar{F}(\theta; \epsilon)$ of a spherical planet with fast rotation; consider, in particular, the extreme cases $\epsilon = 0$ and $\pi/2$. Prove that $\bar{F}(0; \epsilon) > \bar{F}(\pi/2; \epsilon)$ for $\epsilon \gtrsim 54°$. (Hint: We need averaging $\cos \iota$ in (7.10) over the rotational phase ϕ and the phase of the Sun Φ in its ecliptic motion. Starting with the latter, we obtain $\overline{\cos \iota} = \sin \alpha/\pi$, where α is the colatitude of a point with respect to the pole of the ecliptic. In the general case, the subsequent daily average requires an elliptic integral.)

7.2.* In a local frame of reference with the axes pointing to the South, East and the Zenith, obtain the motion of the Sun in the sky for each ecliptic longitude Φ. This is the basis for the design of a sundial.

7.3. Find the ratio of the radiative fluxes on the Earth from the full Moon and the Sun.

7.4.* Assuming the $\approx 7\%$ increase of the solar luminosity in a Gy (7.16), estimate approximately when the equilibrium temperature of Mars will increase enough to melt its water supplies (polar caps and underground reservoirs).

7.5. Compute the "solar constant" for all the planets and evaluate their equilibrium temperature; determine the distance from the Sun corresponding to the freezing point of iron, water and hydrogen.

7.6. Calculate, as a function of latitude, the displacement of the vertical from the radial direction due to the centrifugal acceleration.

7.7. Prove eq. (7.54). Show also that for the anticyclonic curvature

$$\frac{\partial P}{\partial n} < \frac{\varrho R_c f^2}{4},$$

and hence for anticyclones the pressure gradient decreases toward the centre (explaining why the pressure gradients are small and winds light near the centre of anticyclone).

7.8.* Using the model of a parallely-stratified, isothermal fluid, derive the dispersion relation for gravity waves.

7.9. Derive Kolmogorov's spectrum (7.70).

7.10.* Compare the value of GM_{atm} for the Earth atmosphere with the current uncertainty $\approx 8 \times 10^5$ m^3/s^2 in GM. Assuming a reasonable value for the seasonal mass redistribution, estimate the corresponding change in the position of the centre of mass of the Earth and the atmospheric loading on the surface for typical values of the crust rigidity (Ch. 5).

7.11.* Compute, in order of magnitude, the angular momentum L_{atm} of the Earth atmosphere and compare it with that of the solid Earth. Which fraction of L_{atm} is variable, due to zonal winds with a typical speed of 10 m/s? And how much can they change the length of the day?

7.12.* Find the dispersion relation of surface water waves, whose restoring force is gravity.

7.13.* Prove that the Hamiltonian $H(X, \psi; t)$ in (d) of the Appendix is integrable when only one Fourier term is present; show that this corresponds to the case when only the node of the planetary orbit drifts with a constant rate $(d\Omega/dt)$ and the inclination stays constant (Sec. 7.1). Find the equilibrium solutions as a function of $f_{pr}/(d\Omega/dt)$ (so called *Cassini states*)?

7.14.* As a model of the seasons without daily temperature excursions, one can take a fast rotating planet, in which the parallels are isothermal, but do not exchange heat between each other. Find the latitudinal temperature profile as a function of the position of the Sun in the sky.

7.15.* Show that the mean radiative intensity $J(\tau) = \frac{1}{2} \int d\mu\, I(\mu, \tau)$ (eq. (7.35)) in a plane-parallel atmosphere of infinite optical thickness has the formal solution (*Milne equation*)

$$J(\tau) = \frac{1}{2} \int_0^\infty J(t)\, E_1(|t - \tau|)\, dt \quad \left(E_1(x) = \int_1^\infty dt\, \exp(-xt)/t \right).$$

7.16.* Show that in a generic radiation field with specific intensity $I_\nu(\mathbf{r}, \hat{\mathbf{k}}, t)$ (eq. (7.19)), $\mathsf{P} = (1/c) \int I(\mathbf{r}, \hat{\mathbf{k}}, t)\, \hat{\mathbf{k}}\hat{\mathbf{k}}\, d\Omega$ is the radiation pressure tensor; and in a plane-parallel atmosphere its vertical component P_{zz} is equal to $K(\tau) = \frac{4\pi}{c} \int d\mu\, I(\mu, \tau)\, \mu^2$.

7.17. Consider another variant of the two-stream approximation for a grey atmosphere, namely $I = I_u\, \delta(\mu - 1) + I_d\, \delta(\mu + 1)$ ($\delta(x)$ is the Dirac function). Compare the result for upward and downward fluxes with those obtained in Sec. 7.3.

7.18.* Consider a two-dimensional flow $\mathbf{v} = \mathbf{e}_z \times \nabla \psi$ with velocity given by a stream function ψ (Sec. 7.4). With elliptic coordinates (ξ, η) defined by $x = \sqrt{a^2 - b^2} \cosh \xi \cos \eta$ and $y = \sqrt{a^2 - b^2} \sinh \xi \sin \eta$ ($a > b$ are arbitrary constants), draw the velocity field associated to

$$\phi_{\text{in}} = -\frac{2}{q}(q+1)\left(\cosh^2 \xi \cos^2 \eta + q^2 \sinh^2 \xi \sin^2 \eta\right)$$

for $\xi \leq \xi_0$ ($\tanh \xi_0 = b/a = 1/q$), and

$$\phi_{\text{out}} = -\frac{q+1}{q-1}(1 + \xi - \xi_0) - \left(q^2 - 1\right)(1 - \cos 2\eta)\sinh^2 \xi \\ - \cos 2\eta \exp\left[2(\xi_0 - \xi)\right]$$

for $\xi \geq \xi_0$; $\psi = (\kappa b^2/4)\phi$ and arbitrary $\kappa > 0$. Compute the corresponding vorticity w.

– FURTHER READINGS –

J.T. Houghton, *The Physics of Atmospheres*, Cambridge University Press (second edition 1986), is an excellent textbook, stressing the physical processes at work. The recent and good textbook G. Visconti, *Fundamentals of Physics and Chemistry of the Atmosphere*, Springer (2001) is relevant for both Chs. 7 and 8. The classical textbook on radiative transfer is S. Chandrasekhar, *Radiative Transfer*, Dover, with several editions; for more recent developments and physical insight, see D. Mihalas, *Stellar Atmospheres*, Freeman (1970) and H.C. van de Hulst, *Multiple Light Scattering*, Academic Press (1980). On fluid instabilities, the main reference is S. Chandrasekhar, *Hydrodynamic and Hydromagnetic Stability*, Clarendon Press (1961). We have found M. Ghil and S. Childress, *Topics in Geophysical Fluids Dynamics: Atmospheric Dynamics, Dynamo Theory and Climate Dynamics*, Springer-Verlag (1987) a clear introduction, suitable also for the non-specialists. A good, tutorial text on two-dimensional vortex dynamics, applied to Jupiter's Great Red Spot dynamics, is P.S. Marcus, Jupiter's Great Red Spot and other Vortices, *Annu. Rev. Astron. Astrophys.* **31**, 523 (1993). For a good introduction to astrophysical fluid dynamics, see M. Lesieur, *Turbulence in Fluids*, Kluwer (1997). About the connection between climate and celestial mechanics, see *Milankovitch and Climate: Understanding the Response to Astronomical Forcing*, A. Berger et al., eds., Reidel (1983) and, for a review of recent work, D. Paillard, Glacial cycles: Toward a new paradigm, *Rev. Geophys.* **39**, 325 (2001). A report on the exceptional results from the ice drillings at Vostok station has appeared in J.R. Petit et al., Climate and atmospheric history of the past 420 ky from the Vostok ice core, *Nature* **399**, 429 (1999); deep ice cores have been extracted also in Greenland (see the issue 12 of *J. Geophys. Res.* **102** (1997)).

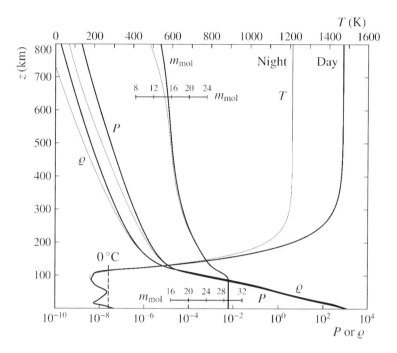

Figure 8.1. Average physical quantities for the upper atmosphere of the Earth as a function of height z (in km) (night, thin lines; day, thick lines): the temperature T (in K); the mean molecular weight m_{mol}; the density ϱ in 10^{-6} g/cm^3; the pressure in 10^3 dyne/cm^2 (from the MSIS-E-90 atmospheric model at http://nssdc.gsfc.nasa.gov/space/model).

example. As explained in the previous chapter, at low altitudes, in the *troposphere*, most of the thermal energy comes from the ground in the infrared; dT/dz is negative and approximately maintained at the adiabatic level. Above ≈ 10 km, in the *stratosphere*, solar heating prevails, the temperature gradient is reversed and the temperature increases from 240 K to 270 K. In the *mesosphere*, between 50 and 85 km, as the gas gets thinner, the absorbed radiation from the Sun becomes inadequate to balance cooling to space and the gradient becomes negative again. Finally, above it in the *thermosphere*, with very low densities, even a small fraction of the solar radiation in the far ultraviolet is sufficient to heat again the gas to high temperatures; this is a direct effect, with large temperature changes between day to night (Fig. 8.2), but not with altitude. An important transition, which separates the *homosphere* and the *heterosphere*, occurs between 100 and 200 km: the composition becomes dependent on the altitude, H and He slowly becoming the major constituents. Fig. 8.1 shows that at all altitudes the density and the pressure are roughly exponential; however, in the heterosphere the scale height (eq. (7.4)) $H = kT/(m_{mol}g) \approx 80$ km is larger, and different for different species (Fig. 8.3).

8. Upper atmospheres

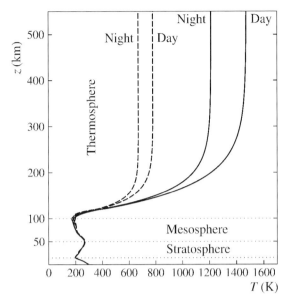

Figure 8.2 The variability of the temperature profile of the atmosphere at the equator: solid lines for June 15, 1989, near the maximum of the solar cycle; dashed lines for February 15, 1996, close to the minimum. In the four regions (as discussed in the text; the troposphere is not indicated) different kinds of radiation balance prevail. Notice the day-night change in the thermospheric temperature and, at lower altitude, the very little sensitivity of the temperature profile to the solar cycle. Data from the MSIS-E-90 model at http://nssdc.gsfc.nasa.gov/space/model.

Due to energetic UV solar photons and scarcity of collisions many unusual and excited molecular and atomic species are present in the upper atmosphere, with a very complex chemistry. For example, even at lower altitudes (15 – 35 km), O_3 (ozone) is produced by collisions and photo-dissociated. Excited molecules, in particular atomic oxygen, produce the *airglow*, a diffuse and structureless radiation present at all latitudes; it is different from the auroræ, which are confined to high latitudes and due to the solar wind.

Around 400 – 600 km, depending on the process, collision mean free paths become comparable with the scale height and local thermal equilibrium does not hold; we have instead the *Knudsen regime* and the fluid approximation is inadequate. Above is the *exosphere*. For each species, with a mean free path ℓ_c, a critical height $z_{\rm eq}$ is defined by

$$\ell_c = H(z_{\rm eq}) \,, \tag{8.1}$$

below which thermal equilibrium can be assumed. Moreover, different components may be thermally isolated from each other and have different temperatures; for example, the distribution of a molecular species in its excited levels is characterized by a *molecular temperature*, often higher than the kinetic temperature. The density profile of each component mainly results from a balance between the outward diffusion and the inward settling down due to gravity; fast molecules may altogether escape to interplanetary space.

The neutral upper atmosphere is affected by many wave phenomena, in particular gravity and tidal waves. In this context the word "tide" is used to describe the general swell and wane of the atmosphere under the influence

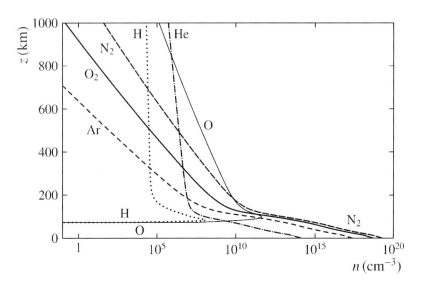

Figure 8.3. Atmospheric abundances of different species for June 15, 1989 (near the solar maximum) at midnight, local time, and above the equator. Molecules (in particular, N_2 and O_2) prevail in the troposphere and the ionosphere, while atomic species (H, He and O) dominate in the thermosphere. Note the flat profile of H and He in the thermosphere; at the solar minimum they begin to dominate at 500 – 600 km (MSIS-E-90 model at http://nssdc.gsfc.nasa.gov/space/model).

(gravitational or thermal) of the Sun. Especially above 100 km, in the thermosphere, tides are mainly due to solar heating. Just like in atmospheric dynamics (Sec. 7.4), the velocity response depends on the balance between the pressure gradient and the Coriolis force. Near the equator the former prevails and the gas rises near the local noon; at mid latitudes we have a large scale, roughly cyclonic flow, rotating counterclockwise (clockwise) in the northern (southern) hemisphere around the point where the Sun is at the Zenith. The velocity reaches its maximum (even 150 m/s in the thermosphere) a few hours earlier and later. As seen from the Earth, the disturbance travels westward with the motion of the Sun in the sky. Besides the obvious component with the period of 24 h, higher harmonics, in particular at 12 h, are present. The flow is much variable with latitude, season and the atmospheric structure.

Besides their effect on neutral molecules, ultraviolet solar radiation and energetic particles ionize molecules and produce a plasma (in the average neutral) made of electrons, positive and negative ions. The terrestrial *ionosphere* is traditionally distinguished in the D, E and F layers, with a peak in the electron density at $\approx$ 300 – 500 km (Fig. 8.4). Since the composition changes with height, different ionization and recombination processes prevail in different layers.

8. Upper atmospheres

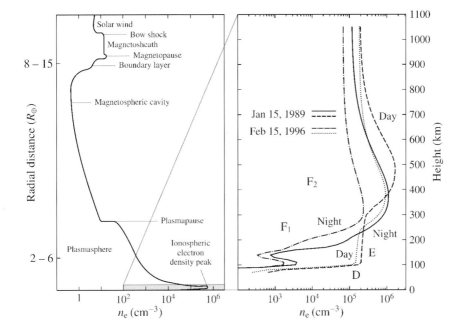

Figure 8.4. Electron concentration in the magnetosphere and ionosphere. The left panel refers to noon and low latitude; the range in the ordinate shows the solar cycle variability. The right panel shows different ionospheric layers, with a maximum in electron density at 400 – 500 km, with the day-night and solar cycle variability; note, in particular, the vertical rise of the peak of the electron density during day. Data from International Reference Ionosphere (IRI-95) at http://nssdc.gsfc.nasa.gov/space/model.

The ionosphere shows great variations with the hour of the day, the latitude, the season and the 11 y solar cycle; during the day and near the maximum of the solar cycle the ionosphere is hotter and bigger (Fig. 8.2). In the polar regions the precipitation of solar wind particles enhances the ionization. Bursts of solar wind produce *magnetic storms*, lasting a few days, with substantial changes in the electron ionospheric density. Because of the friction between ions and neutral atoms, the exospheric dynamics involves complex interactions, especially near the poles, between the charged and the neutral components. There are four relevant energy sources:

- solar ultraviolet and X radiation;
- dissipation of ionospheric electric currents;
- dissipation of tidal and gravity waves;
- energetic particles of the solar wind.

While a minor constituent in terms of particle numbers, the plasma has a profound effect on the propagation of electromagnetic waves. Strong refraction and reflection can turn a radio beam downward and enable communications over the whole globe. Irregularities in the ionosphere, generally following a power law in their wave number, have a complex dynamics and can be investigated with the *scintillation* they produce in radio signals.

8.2 Ionosphere

Three main ionizing agents are active in the upper atmosphere of the Earth: solar radiation, in particular in the ultraviolet band below 0.4 μm and X rays; galactic cosmic rays; and energetic particles from the solar wind. Among the molecular species, NO, with an ionization energy $E_i = 9.25$ eV, is the easiest to ionize; next is O_2, with $E_i = 12.08$ eV. Of course, the *ionization cross section* σ_i abruptly drops to zero below E_i. The rate at which free electrons are produced mainly depends upon the solar ionizing flux F and the number density n' of the relevant molecules. Consider an exponential profile

$$n'(z) = n'(z_0) \exp[-(z - z_0)/H] , \qquad (8.2)$$

with a constant scale height H and z_0 a reference altitude; integrating the photon balance equation of Fig. 8.5 we obtain

$$F(z) = F(\infty) \exp[-\tau(z)] , \qquad (8.3)$$

where

$$\tau(z) = \int_z^\infty dz' \eta \sigma_i n'(z') \sec \chi = \eta \sigma_i n'(z) H \sec \chi \qquad (8.4)$$

is the *optical depth* at the altitude z, $F(\infty)$ is the solar flux outside the atmosphere and η is the *ionization efficiency*, the fraction of the absorbed radiation used for ionization. Below $\tau \simeq 1$ there are much fewer photons; but since the molecular density quickly drops with height, one can expect that most of the ionization occurs around an optical depth of order unity; this is the gist of *Chapman's theory* outlined below. For a typical value $\sigma_i \simeq 10^{-17}$ cm^2 and $H_{N_2} \simeq 40$ km, at $\tau \simeq 1$ the density of nitrogen is $n'_{N_2}(z_i) = 1/(\sigma_i H) \approx 2 \times 10^{10}$ cm^{-3}, corresponding to altitude of about 100 km.

Recombination requires that a free charged particle looses enough energy to become bound to an opposite charge. Of course, an elastic collision between an electron and an ion cannot produce a neutral atom; there must be an energy loss. In the *radiative recombination*

$$e^- + M^+ \rightarrow M + h\nu \qquad (8.5)$$

energy is carried away by *inverse Bremsstrahlung*. A similar, two-body process is *dissociative recombination*,

$$e^- + O_2^+ \rightarrow O^* + O , \qquad (8.6)$$

8. Upper atmospheres

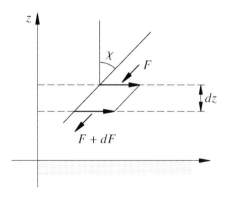

Figure 8.5 A photon flux F impinges at an angle χ from the zenith upon a horizontal surface element dS. The ionization rate in a horizontal layer of thickness dz is $\eta \sigma_i n' F$, so that the flux changes by $dF = \eta \sigma_i n' F dz \sec \chi = -F\, d\tau$.

where one of the resultant atoms is in an excited state. At higher electron densities, in a *collisional recombination*, the extra energy can be removed in a three-body encounter, like

$$e^- + X^+ + Y^+ \rightarrow X + Y^+ . \tag{8.7}$$

It is also possible to have a two-stage process: charge exchange, like $O^+ + N_2 \rightarrow NO^+ + N_2$, followed by dissociative recombination $NO^+ + e^- \rightarrow N + O$; the overall rate is determined by the fastest of these two stages. In these reactions, when there is only one type of ionized molecules, the recombination rate is proportional to $(n_e)^p$, where p is the number of charged particles involved: $p = 1$ in charge exchange, $p = 3$ in a collisional recombination, $p = 2$ in dissociative and radiative recombination.

For sake of illustration, consider the representative $p = 2$ case, with a recombination rate αn_e^2; α is the *recombination coefficient*. In a stationary state, equating the production $\eta n' F \sigma_i \sec \chi$ to the recombination, the electron density

$$n_e = \sqrt{\frac{\eta F \sigma_i n'}{\alpha \cos \chi}} \tag{8.8}$$

is obtained; for simplicity, the ionization cross section σ_i is assumed energy independent and the mass motion is neglected. In this case the electron density profile can be obtained analytically (Problem 8.3); the largest density occurs at $z = \bar{z}$ where Fn' is largest and hence $\tau = 1$. The width δ of the layer can be estimated with a parabolic approximation $n_e(z) = n_e(\bar{z})[1 - (z - \bar{z})^2/\delta^2]$, so that

$$\delta^2 = -4 \left[\frac{Fn'}{d^2(Fn')/dz^2} \right]_{\bar{z}} = 4H^2 . \tag{8.9}$$

The angle χ has a daily and seasonal variation; $d\chi/dt$ is of the order of the angular velocity of the Earth. The altitude $\bar{z}$ of the maximum changes with χ, with the timescale of a few hours, according to

$$\cos \chi \, \exp(\bar{z}/H) = \text{const} , \tag{8.10}$$

while the peak electron density is constant.

To get a feeling of the numbers involved, consider at $z = 200$ km the ionization equilibrium for atomic O, with a density $n_o \simeq 10^{10}$ cm^{-3} (Fig. 8.3). Its ionization energy $E_i = 13.6$ eV corresponds to the wavelength 0.1 μm, at which (Fig. 9.1) we have an average spectral energy flux $\Phi_\lambda = 3 \times 10^{-3}$ W/(m^2 μm) = 3 erg/(cm^2 μm s), corresponding, over the bandwidth 0.05 to 0.1 μm, to the number flux $F = \Phi_\lambda \Delta\lambda/E_i \simeq 10^{10}$ cm^{-2} s^{-1}. With a photo-ionization cross section $\sigma_i \simeq 10^{-17}$ cm^2, $\eta \simeq 1$ and a recombination coefficient $\alpha \simeq 10^{-7}$ cm^3/s

$$n_e = \sqrt{\frac{Fn_o\sigma_i}{\alpha}} \approx 10^5 \text{ cm}^{-3}, \qquad (8.11)$$

with a fractional ionization of $\approx 10^{-5}$.

This simple stationary model is useful only during the day; during the night (or during solar eclipses) slow recombination processes gradually diminish the ionization. Additional ionization may be also caused by more exotic processes, like precipitating electrons, micrometeorite bombardment and mass motion. When a different recombination process with an exponent $p \neq 2$ prevails, the electron density is proportional to $(n'F)^{1/p}$ and we have a similar behaviour (Problem 8.8).

The structure of the ionosphere is greatly affected by the complex chemistry and the variability of ionizing agents and neutral components. At the bottom, in the D and E layers, photo-ionization of molecular oxygen (O_2) and nitrogen (N_2) produces a sudden increase in the electron density with height. In the higher F_1 layer the stronger solar radiation primarily ionizes atomic oxygen O and N_2. The first process prevails also in the upper F_2 region, where the optical thickness is very low.

Planetary ionospheres.– Venus and Mars, as well as the giant planets, possess ionospheres, whose properties, primarily the electron and ion density profiles, have been probed with radio occultation measurements of interplanetary spacecraft. Similarly to the terrestrial E layer, they have a peak electron density at $\simeq 150$ km due to photo-ionization of O_2; an important source is also the far more abundant CO_2 molecule (Table 7.1). Interestingly, the ionosphere Venus, in spite of its very slow rotation, survives in the night side, with only a small drop in electron density. It has been suggested that thermal winds across the terminator may carry sufficient amount of ions from the day to the night side. The Venusian and Martian ionospheres have important implications for their magnetosphere (Sec. 10.7). The atmospheres of the giant planets are dominated by molecular hydrogen, whose photo-ionization may result in either H_2^+, or H+H^+; atomic hydrogen is further ionized to H^+. Radiative recombination of H^+ is very slow and the hydrogen ion is expected to last a long time. The peaks of the electron density are in the range $10^4 - 10^5$ cm^{-3}, but its profile in-

8. Upper atmospheres

dicates many narrow layers with considerable latitudinal, diurnal and temporal variations.

A dynamo effect in the ionosphere. In a weakly ionized gas the velocity **U** of the main neutral component may affect differently the charged components and generate, with a dynamo process, currents and magnetic fields. The best way to describe this effect is an extension of Ohm's law (1.75''), normally used to get the electric current driven by an electric field; here the driver is the Lorentz force. The velocity of each species (electrons and singly charged ions, say), is determined by

$$-m_e \nu_e (\mathbf{v}_e - \mathbf{U}) - \frac{e}{c} \mathbf{v}_e \times \mathbf{B} = 0, \quad (8.12a)$$

$$-m_i \nu_i (\mathbf{v}_i - \mathbf{U}) + \frac{e}{c} \mathbf{v}_i \times \mathbf{B} = 0. \quad (8.12b)$$

We have neglected electric and pressure forces and denoted with ν_e, ν_i the friction frequencies of electrons and ions with the gas. For a uniform magnetic field along **n** we have

$$\mathbf{v}_e + \frac{\Omega_e}{\nu_e} \mathbf{v}_e \times \mathbf{n} = \mathbf{U}, \quad \mathbf{v}_i - \frac{\Omega_i}{\nu_i} \mathbf{v}_i \times \mathbf{n} = \mathbf{U}. \quad (8.13)$$

$$\Omega_e = 5.8 \times 10^6 \frac{B}{B_e} \text{ Hz}, \quad \Omega_i = 2,900 \frac{B}{A\,B_e} \text{ Hz} \quad (8.14)$$

are the *cyclotron frequencies* ((1.105) and (1.106)), referred to the surface magnetic field of the Earth $B_e \simeq 30,000$ nT; A is the mean atomic number of the ions. When $\Omega \ll \nu$ the magnetic force is negligible and the charged fluid is driven along **U**; but in the opposite case $\Omega \gg \nu$ the velocity is orthogonal to both **U** and **n** and changes sign with the charge. As a consequence, there is a current density

$$\mathbf{J} = -e\, n_e (\mathbf{v}_e - \mathbf{v}_i) = -e\, n_e \left(\frac{\nu_e}{\Omega_e} + \frac{\nu_i}{\Omega_i} \right) \mathbf{n} \times \mathbf{U} \quad (8.15)$$

(the velocity components along **n** have been neglected). It turns out that the transition $\nu = \Omega$ occurs at ≈ 75 km for electrons and ≈ 120 km for ions. Below 75 km friction prevails; above 150 km the density drops and there is little effect. This current generates a magnetic field which contributes to the potential expansion (6.11) with the $c_{\ell m}$ and $s_{\ell m}$ terms corresponding to external sources.

As an example, consider what happens at low latitudes, as a consequences of the vertical velocity $\mathbf{U} = U\mathbf{e}_z$ (ascending during the day) due to the thermospheric tide. Neglecting the magnetic declination, we take $\mathbf{n} = \mathbf{e}_x$ along the southern meridian, so that a current

$$J_y = e n_e \left(\frac{v_e}{\Omega_e} + \frac{v_i}{\Omega_i} \right) \tag{8.16}$$

is generated along the parallel. This is the *equatorial electrojet*, a strong current around 100 km of altitude, spread around a few degrees of latitude. It flows eastward in the day and westward in the night, when the thermosphere cools. The corresponding electron velocity is several 100 m/s. The magnetic field so generated is routinely measured with ground magnetometers. The electrojet irregularities in the E layer are detected by radar and rocket soundings.

8.3 Waves in an electron plasma

This is a good place to introduce from first principles the theory of plasma waves, of great importance in space physics. Shirking from an excursion into complex plasma physics, we confine ourselves to a paradigmatic problem, wave propagation in a polytropic electron gas, leading to longitudinal plasma waves and to the profound effect that a plasma has on electromagnetic propagation. We consider a frequency ω so high that ions – singly charged and with density n_i – are unaffected and at rest. Electrons, with charge $-e$, have a density n_e (equal to n_i when there is charge neutrality), a mean velocity $\mathbf{v}$ and a pressure $P_e(\varrho_e)$, function of their mass density $\varrho_e = m_e n_e$ only; $c_s = \sqrt{dP_e/d\varrho_e}$ is constant. Conservation of mass and momentum gives:

$$\frac{\partial n_e}{\partial t} + \nabla \cdot (n_e \mathbf{v}) = 0 , \tag{8.17a}$$

$$\left(\frac{\partial}{\partial t} + v_c + \mathbf{v} \cdot \nabla \right) \mathbf{v} + c_s^2 \frac{\nabla n_e}{n_e} + \frac{e}{n_e m_e} \left(\mathbf{E} + \frac{1}{c} \mathbf{v} \times \mathbf{B} \right) = 0 . \tag{8.17b}$$

For a simple description of momentum dissipation, a frictional acceleration $-v_c \mathbf{v}$ has been added; $v_c = v_{Te} n_s \sigma_c$ is the electron collision frequency, expressed in terms of their thermal speed v_{Te}, the density n_s of scatterers (mostly neutral particles) and the relevant cross section σ_c. Consider a small disturbance of a background, assumed uniform, at rest and with no magnetic field (see Problem 8.2 for a generalization). Denoting with a prime the perturbations, we consider only linear terms in $n'_e = n_e - n_i$, $\mathbf{v}$, $\mathbf{E}$, $\mathbf{B}$, all proportional to $\exp[i(\omega t - \mathbf{k} \cdot \mathbf{r})]$. Each quantity has a formal complex amplitude, but only its real part is meaningful. Then $\partial/\partial t = i\omega$, $\nabla = -i\mathbf{k}$ and the differential operators become algebraic; the dynamical equations (8.17) read

$$\omega n'_e = n_i \mathbf{k} \cdot \mathbf{v} , \qquad n_i(\omega - iv_c) \mathbf{v} - c_s^2 n'_e \mathbf{k} = -\frac{e}{i m_e} \mathbf{E} . \tag{8.18}$$

They give the current density $\mathbf{J} = -e n_i \mathbf{v}$ with a generalization of *Ohm's law*

$$J_i = \sigma_{ij} E_j ; \tag{8.19}$$

8. Upper atmospheres

here

$$\sigma_{ij} = \frac{e^2 n_i}{i(\omega - i\nu_c)m_e}\left[\delta_{ij} + \frac{c_s^2 k_i k_j}{\omega(\omega - i\nu_c) - c_s^2 k^2}\right] \quad (8.20)$$

is the *conductivity tensor*. When $\omega = \mathbf{k} = 0$ this gives the usual, real conductivity

$$\sigma = \frac{e^2 n_i}{m_e \nu_c}. \quad (8.21)$$

The conductivity (8.20) is a scalar also when the phase velocity $v_p = \omega/k$ is much larger than the sound speed c_s:

$$\sigma = \frac{e^2}{i(\omega - i\nu_c)m_e}, \quad (8.22)$$

purely imaginary when $\nu_c = 0$. When the conductivity tensor is complex, the imaginary part increases the phase $\phi = \omega t - \mathbf{k}\cdot\mathbf{r}$ of the electric field by $\pi/2$ and produces a current in quadrature with it. An imaginary part of the conductivity thus does not contribute to the mean of the Joule's dissipation power

$$W = \operatorname{Re}\mathbf{J}\cdot\operatorname{Im}\mathbf{E};$$

in fact, W in this case is a product of a sine and a cosine of the same argument, which in the average vanishes. Only the real part of the conductivity brings about a mean dissipation. In addition, we have Maxwell's equations (1.70) and (1.71), now in algebraic form

$$-i\mathbf{k}\times\mathbf{B} = \frac{1}{c}(i\omega\mathbf{E} + 4\pi\mathbf{J}), \quad \mathbf{k}\times\mathbf{E} = \frac{1}{c}\omega\mathbf{B}; \quad (8.23)$$

they give the current as a function of the electric field ($\hat{\mathbf{k}} = \mathbf{k}/k$)

$$4\pi\mathbf{J} = -i\omega[(1 - n_r^2)\mathbf{E} + n_r^2(\hat{\mathbf{k}}\cdot\mathbf{E})\hat{\mathbf{k}}]. \quad (8.24)$$

The *refractive index*

$$n_r = \frac{ck}{\omega} \quad (8.25)$$

is the ratio of the light speed to the *phase velocity* $v_p = c/n_r$; $v_g = d\omega/dk = v_p[1 - k(d\ln n_r/dk)]$ is the *group velocity*, different from v_p when the medium is dispersive. Note that the two divergence equations for $\mathbf{B}$ and $\mathbf{E}$ follow from (8.23) and the conservation equation (first of eqs. (8.18)). Using Ohm's law (8.19), a system of three linear and homogeneous equations for the electric field

$$D_{ij}E_j = 0, \quad D_{ij} = 4\pi\sigma_{ij} + i\omega\left[(1 - n_r^2)\delta_{ij} + n_r^2\hat{k}_i\hat{k}_j\right], \quad (8.26)$$

Occultations, or near occultations, of a point source, like a star, by a planetary body or by the Sun are an important tool in space physics. If the body has no atmosphere, the ingress and egress times, at which the source is extinguished and reappears again, provide accurate information about the position and the radius of the body. When there is an atmosphere, refraction and absorption, depending on its composition and density profile, the ingress and the egress are not sharp (Problem 8.10). The bending can be usefully described in the *thin screen approximation*, in which one assumes that the refractivity δn_r is confined to the plane (x, y) orthogonal to the ray (along z). Then the unit tangent vector $\mathbf{t} = d\mathbf{r}/ds$ is deflected by

$$\Delta \mathbf{t} = \nabla_\perp \int dz\, \delta n_r = -\frac{1}{2\omega^2} \nabla_\perp \int dz\, \omega_p^2 . \tag{8.41}$$

The explicit expression holds for a plasma (eq. (8.29)), like the solar corona; the deflection is proportional to the gradient of the electron columnar content along the ray and is in the *outward* direction. This effect, and phase fluctuations due to the plasma, make radio propagation through the solar corona difficult, even impossible. In the case of the Sun there is a smaller, *inward* gravitational contribution (Sec. 17.6).

It should also be noted that in geometrical optics *diffraction* is neglected and, for a finite wavelength λ, *Fresnel optics* must be taken into account. If (x, y) is the point P on the screen on the geometrical ray, the phase at the receiver at a distance L is, in effect, the superposition of the phases from all the points on the screen within a circle centred at P with approximate radius $d_F = \sqrt{\lambda L/\pi}$. If the inhomogeneities of the screen do not have a scale larger than d_F, rays are not well defined and diffraction effects must be taken into account.

A *neutral* atmosphere causes an *inward* bending. In the exceptional case in which the occultation is perfect – when the path of the star in the sky crosses the centre of the planet – a dramatic effect, the *central flash*, may take place. In the case of a spherically symmetric atmosphere, when the centre of the planet is on the line joining the star and the observer, all rays are focused on this line, and the star behind the planet suddenly reappears; the limb of its atmosphere acts as a huge convergent lens. This flash is preceded and followed by rays near the ingress and the egress limbs. The perfect focusing is not attained due to the oblateness and atmospheric inhomogeneities. As an example, on July 3, 1989 the star 28 Sgr had an almost central occultation by Saturn (and its rings); the star reappeared for ≈ 15 s with a luminosity $1 - 2\%$ the normal value, and provided important information about the atmosphere. A similar phenomenon occurs due to the gravitational field of a mass between the star and the observer, which acts as a spherically symmetric refractive index, producing a relevant amplification (see Sec. 16.5).

8. Upper atmospheres

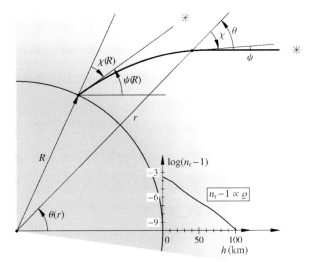

Figure 8.7 Refraction in the atmosphere diminishes the zenithal distance of a star from $-\theta$ to the apparent value $\chi(R)$. This correction ψ (refraction angle) is largest when the star is at the horizon. On the side, a plot of the refraction index n_r with the altitude, according to the (approximate) Gladstone-Dale law $n_r - 1 \propto \varrho$ (the density).

Effect on ground-based astronomical observations. Refraction affects astronomical observations from the ground. In a parallely stratified medium, in which the index of refraction $n_r(z)$ is a function of altitude, the propagation is entirely described by the angle $\chi(z)$ that the ray makes with the vertical direction. By applying the elementary Snell's law to an infinitesimal increment from n_r to $n_r + dn_r$ we get

$$n_r \operatorname{ctg} \chi \, d\chi + dn_r = 0 ,$$

which integrates directly to

$$n_r \sin \chi = \text{const} . \tag{8.42}$$

For an extended atmosphere, or for low elevation, the radius of the Earth must be taken into account. As shown in Ch. 5, eq. (5.3), when $n_r(r)$ has spherical symmetry,

$$p = r n_r \sin \chi \tag{8.43}$$

is constant along the ray, with zenithal distance χ; total reflection occurs when $r n_r(r) = p$. When r does not change much, (8.42) is recovered.

When there is no refraction a star in the x-direction is seen from a point at a distance r from the centre at a zenithal distance $\chi = -\theta$ (positive in the clockwise sense; see Fig. 8.7); to compute the constant p from (8.43), we evaluate $r \sin \chi$ for large r. Then

$$p = R n_r(R) \sin \chi(R) \tag{8.44}$$

gives the zenithal distance $\chi(R)$ on the ground. The *refraction angle* $\psi = \theta + \chi$ measures the difference between the apparent and the true zenithal distance;

we want to see how ψ changes along the ray. From eq. (8.43)

$$\frac{dn_r}{n_r} + \frac{dr}{r} + \operatorname{ctg}\chi \, d\chi = 0 \, ;$$

and from the infinitesimal right triangle with cathetuses $(dr, r\, d\theta)$ and $dr = r \operatorname{ctg}\chi \, d\theta$; we thus obtain

$$d\psi = -\tan\chi \, \frac{dn_r}{n_r} \, .$$

As expected, in a uniform medium the refraction angle does not change; it diminishes when n_r increases. Integrating from $r = R$ to infinity with the index of refraction itself as integration variable, we get

$$\psi(R) = \int_1^{n_r(R)} \frac{dn_r}{n_r} \tan\chi \, , \qquad (8.45)$$

where χ is expressed in terms of r (and n_r) through eqs. (8.43) and (8.44); at infinity $n_r(\infty) = 1$. For a small refractivity this becomes

$$\psi(R) = R\sin\chi(R) \int_0^{\delta n_r(R)} d(\delta n_r)\,[r^2 - R^2 \sin^2\chi(R)]^{-1/2} \, . \qquad (8.46)$$

One needs the radial profile of δn_r, normally proportional to the density $\varrho(r)$ of the atmosphere (Gladstone-Dale law; Fig. 8.1). For higher accuracy, more complex, often empirically adjusted, models are used.

The correction in elevation of an extraterrestrial object is very important for accurate optical observations. The Earth atmosphere has a small effective thickness $H \ll R$; when the elevation $e = \pi/2 - \chi(R) = O(H/R) \ll 1$, the effective length of the path is $\simeq R$, rather than H, and the refraction angle is larger. In Problem 8.9a we propose the calculation of $\psi(R)$ when the refractivity δn_r is constant below H and zero above (single layer model). One finds

$$\psi(R) = \delta n_r(R)\,(e^2 + 2H/R)^{-1/2} \, . \qquad (8.47)$$

When $e = O(1)$ we recover the previous estimate. When $1 \gg e \gg \sqrt{2H/R}$, $\psi = \delta n_r/e$; but when $e \ll \sqrt{2H/R}$ the refraction angle tends to the constant value

$$\psi_0 = \delta n_r \sqrt{\frac{R}{2H}} \, . \qquad (8.48)$$

This result is confirmed with the more accurate model of an exponential atmosphere (Problem 8.9b). For the Earth, with $\delta n_r \simeq 3\times 10^{-4}$ and $H \simeq 8$ km (in the lowest layer in the troposphere), we obtain the estimate $\psi_0 \simeq 6 \times 10^{-3}$, while the actual value is $35' \approx 10^{-2}$ rad (the difference is due to the contribution of higher layers). This corresponds to a delay of about 138 s in the setting time of a star. It should be also noted that the horizontal refraction sensitively depends on physical conditions on the surface (temperature, pressure, amount of water vapour etc.); low elevation observations – $e \leq 10°$, say – are often discarded.

8.5 Evolution of atmospheres

Understanding the structure, the variety and the evolution of the atmospheres of the terrestrial planets is a difficult, and still partly unaccomplished task. This issue is essential to establish the chain of events which led to the emergence and the evolution of biological replicators on our own planet, and to assess the probability of life. It should also be noted that, because of their small mass (for the Earth, $M_{atm} \approx 10^{-6} M_\oplus$, see Table 7.1), these atmospheres do not appreciably affect planetary dynamics and thermal structure. We confine ourselves to a discussion of the main processes at work, stressing physical, rather than chemical aspects; and adopt the commonly accepted view that planets did not form directly from the nebular gas, but out of accretion of small planetesimals settled in the median plane; the resulting planetary embryos then underwent a gradual growth to planets by non-disruptive collisions (Sec. 16.3).

Formation processes. It is useful to preface our main topic with a discussion of compositional differences. At present the terrestrial atmospheres contain no H and very little He, at striking variance with the outer planets; but, since these light gases are quickly lost to the outer regions (see below), no conclusion about their primordial composition can be drawn from the present state. The concentration of noble gases, in particular Ne, Ar, Kr and Xe, in the atmospheres of terrestrial planets and in chondrites is several orders of magnitudes lower than what one would expect from their solar abundances. This is due to the fact that they have no chemical affinity, hence did not participate in the condensation process out of which grains and, later, planetesimals arose; they have been swept away with the gaseous part of the solar nebula. Apparently the present, very scarce inventory of noble gases in the terrestrial atmospheres has been produced later by a variety of processes, including implantation of solar wind molecules, adsorption by grains and gravitational capture before the complete dissipation of the solar nebula.

The identification of a primordial reservoir for an element present in the atmosphere is greatly facilitated whenever its isotopic composition is known. The most important example concerns hydrogen, namely the *D/H isotopic ratio*. Deuterium has been formed in the cosmological nucleosynthesis and subsequently destroyed in stars; the primordial deuterium in the interstellar gas of the Galaxy has been progressively diluted by the deuterium-free hydrogen released in stellar winds and explosions. In the local interstellar medium the D/H ratio has been found to be $\approx 15 \times 10^{-6}$; this is also the ratio in the primordial solar nebula. Deuterium enrichment comes from the isotopic dependence of chemical reactions, in particular those involving water; it is measured by the *deuterium enrichment factor* $f = (D/H)_{H_2O}/(D/H)_{H_2}$, the ratio between the D/H value in water and in hydrogen; $f > 1$ means that deuterium is favoured in the synthesis of water. The $D/H = (20 \pm 4) \times 10^{-6}$ ratio in the solar wind has been measured in lunar soils, exposed to the wind over geological times; a similar value was found for Saturn and Jupiter. The discrepancy with the Earth value $(149 \pm 3) \times 10^{-6}$ is striking. Carbonaceous chondrites (Sec. 14.6) – the oldest objects in the solar system, whose source is at the outer edge of the asteroid belt – contain hydrogen both in clay minerals and organic molecules; their D/H ratio is variable, but higher than the protosolar value and consistent with the Earth's.

While the chemistry of deuterium in the interstellar gas and the solar system is poorly known, two basic facts should be pointed out. First, planetary atmospheres preferentially lose to outer space H and H_2 (see below); in the case of Mars, this is born out by the fact that the D/H ratio is larger in the atmosphere than on the ground. Secondly, the differences in the reaction rates between H and D are important while they are in the atomic or molecular state, but much less in compounds, whose isotopic separation is much more difficult (the ratios of thermal speeds between ordinary and deuterated compounds is $\sqrt{2}$ for H, $\sqrt{3/2}$ for H_2 and $\sqrt{19/18} \simeq 1.027$ for water). In the protonebula, and on Jupiter and Saturn, $f = 1$; but nearer the Sun, with a higher temperature, f is larger, and more deuterium-free hydrogen was produced and swept away by the early solar wind. As a result, as the temperature of the protonebula decreased, a gradient in the D/H ratio developed, decreasing with the distance r. As the water vapour condensed into the planetesimals its D/H ratio became greater than in the early nebula, increasing with r.

In our current understanding, life could not have arisen on the Earth without an adequate amount of water. Certainly volatiles, in particular water, have been delivered to the terrestrial planets by cometary impacts; this is the *late veneer scenario*, widely held in the past. But the ratio D/H of comets is well known and much higher ($\simeq 309 \times 10^{-6}$); this is an important reason to discard this possibility. Carbonaceous chondrites, made of water-rich material (up to $\simeq 10\%$ in mass), do not face this difficulty. Ordinary chondrites, whose source is in the inner part of the asteroid belt, are very dry, with only $\simeq 0.05 - 0.1\%$ water content in mass. These facts suggest that on the Earth H_2O mainly comes from the outer asteroid belt region and that it was delivered in a few big impacts of planetary embryos swept inward by the gravitational influence of Jupiter and their mutual perturbations. Note that the planetesimals and planetary embryos that mainly contributed to the terrestrial planets, formed in the inner part of the solar system (≤ 2 AU, say) and had only a very small water content. As a result, water vapour concentration in early atmospheres evolved through discrete violent events, rather than a gentle increase due to outgassing from the mantle.

In our current understanding terrestrial planets formed late and gradually by violent collisions of planetary embryos (Sec. 16.3). The gaseous composition of planetesimals is uncertain, but a fraction of chemically affine molecules must have condensed and produced trapped volatiles, or compounds with volatile molecules. In the accretion phase of planetary formation, a primitive atmosphere arose from the release of these volatiles and/or the melting of the mantle. Of course, big impacts, ejecting material into space, erode the atmosphere; but the gas itself, by slowing down these impacts (projectiles with size $\leq 0.1 - 1$ km, say), might have prevented erosion and played the role of a regulator. The amount of outgassed volatiles from the early terrestrial planets crucially depends on the available heat sources. Besides the direct solar radiation, which determines the equilibrium temperature $T_{eq} \propto 1/\sqrt{r}$ (eq. (7.9)), the thermal energy generated in the accretion process must be taken into account.

Three processes are at work here: radiative loss, gain in gravitational binding energy and heat exchange with the accreting material. The last one crucially depends on the state and the thermal conductivity of the mantle, very difficult to assess; in a simple model one can assume that the incoming material, at a constant temperature T_0, instantaneously sheds its thermal energy to an upper layer – the "mantle" – of thickness ηR ($\eta < 1$). Its density ϱ and specific heat c_p are also constant. The main part of the planet, which has mass M and radius R, is unaffected by accretion. The (constant) rate $\dot{M}$ of mass inflow is the driving parameter; of course, it does not include material ejected back to space. The energy balance equation reads:

$$\frac{d}{dt}\left(4\pi R^3 \eta c_p \varrho T\right) = \dot{M}\left[\frac{GM}{R} - c_p(T - T_0)\right] - 4\pi R^2 \sigma T^4 ; \qquad (8.49)$$

$dt\,\dot{M}c_p(T-T_0)$ is the heat energy which the accreted mass deposits in a time dt in the layer, assumed with infinite conductivity; $4\pi R^3 \eta c_p \varrho T$ is the total thermal energy of the layer. The first and the last terms on the right-hand side correspond to the gravitational energy acquired through the accretion process and the thermal radiated energy. When there is no mass inflow ($\dot{M} = 0$), the "mantle" slowly cools by radiation. The effectiveness of accretional heating critically depends on the rate $\dot{M}$, rather than on the total accreted mass; and on the three times scales of mass growth, heat exchange and radiative loss. Even the possibility of partial vaporization of the mantle has been entertained. Planetary interiors have also been heated by the differentiation of the core, in which heavier elements sank below; and by the decay of radioactive elements (Sec. 5.3). More exotic processes, such as phase changes in the core, resulting in release of latent heat (Sec. 6.5), have been suggested. The evolution is too complex and uncertain for a clear-cut and reliable picture; moreover, besides a steady, spherically symmetric process, isolated, high energy impacts occurred. Of course, the formation of a terrestrial atmosphere is a differential process, in which different sources produce different volatiles, as in the case of water. After the subsequent cooling phase, only the low melting point compounds remained. It is thus very likely that, due to all these inner and outer heat sources, volatile parts of the mantle have been released and formed a primordial atmosphere.

The origin and the evolution of the atmospheres of the giant planets is closely related to their formation, discussed in Sec. 16.4. In the "baseline" core-instability scenario, the collapsing part of the protonebula, with the component of solid planetesimals present at that time and that place, differentiated in an inner, high density and high Z core, and an outer gaseous mantle, where H and other light gases were prevailing. The present abundances reflect the proportions between the two components of the collapsing gas: H_2 and rare gases from the protonebula; C and N from both the protonebula and the planetesimals; silicates and water from the solid component. The present composition is also affected by the differential destruction of planetesimals and the material vaporized from the core. The accretion radiation emitted by the core (in particular for Jupiter) had also an important role. The C/H ratios, larger than the solar abundance value for all the four giant planets, is evidence of this complex mixing process. The ultimate fate of a planetesimal – whether it is engulfed in the core or remains embedded in the atmospheric envelope – depends on the drag due to the gas and its turbulence, on the evaporation, and other factors dependent on the size and the composition of the body.

In the extensive satellite systems of the outer planets, only Saturn's Titan has a dense atmosphere. This satellite, which will be extensively investigated with the Cassini space probe during its tour of the Saturnian system from 2004 to 2008, has a particularly interesting atmosphere, mainly composed of N_2, Ar, CH_4 and H_2. Since it is optically opaque, it can be investigated only by radio waves or in situ exploration. Under the UV solar spectrum, CH_4 is photo-dissociated to hydrogen and heavier hydrocarbons (like C_2H_2), with a lifetime of about 100 My; the inference of an oceanic source (on the surface or underground) of hydrocarbons is nearly inescapable. One should talk not of an atmosphere, but of a liquid and gaseous system whose parts are in close interaction. In Titan's atmosphere molecules of high molecular weight (like C_4N_2) and compounds subject to polymerization into organic molecules (like HCN) have been detected; although the surface temperature is low ($\approx$ 100 K), it is possible that prebiotic conditions, for example in shallow pools on the surface, are present, where large organic molecules are produced. The organic chemistry of Titan's atmospheric system is a major task of Cassini's mission.

Depletion and evolution processes. While the primordial atmospheres of terrestrial planets have been produced by similar processes and probably were similar, only from the peculiarities of their subsequent evolution the strikingly different present atmospheres could arise. We briefly review here the main processes at work.

8. Upper atmospheres

Thermal (Jeans) depletion.– The main depletion process is a gradual and differential loss to space. Let us first derive the diffusion equilibrium condition for a species with mass m' and number density n' embedded in the main species of molecular mass m with scale height $H = kT/(mg)$. The reasoning of Fig. 1.3 can easily be adapted to get the (upward) diffusion flux $F'_{\text{diff}} = -\ell'_c v'_T \, dn'/dz$; ℓ'_c is the mean free path and $v'_T = \sqrt{kT/m'}$ is the thermal speed of m'; to this we must add the net downward flux due to the fact that a molecule, starting at $z - \ell'_c/2$, crosses the height z with the velocity $v'_T - g\ell'_c/(2v'_T)$, with a (downward) gravitational flux $F'_{\text{grav}} = -n'g\ell'_c/v'_T$. In stationary conditions the total flux vanishes and we get the usual exponential density profile with the appropriate scale height

$$H' = \frac{v'^2_T}{g} = \frac{kT}{m'g} = \frac{m}{m'} H \; : \tag{8.50}$$

heavier species settle at lower altitudes. This analysis shows the basic principle, but the vertical diffusion of atmospheric species may be complicated by several processes. For instance, when the atmosphere is turbulent, e.g., due to a superadiabatic temperature gradient or tidal gravity waves, as it happens for the Earth, molecular vertical diffusion is dominated by turbulence.

The main loss to interplanetary space in the exosphere has been described by J.H. JEANS. For a given species with mean free path ℓ'_c, its *exobase* is defined as the altitude z'_{eq} just one mean free path away from empty space (eq. (8.1)):

$$\ell'_c = H(z'_{\text{eq}}) \; . \tag{8.51}$$

One can say, below this level this species is in local thermodynamical equilibrium, while above collisions are negligible. When a molecule, which had its last collision at the altitude z'_{eq} with a main molecule, travels upwards, it does not find enough molecules to collide with and continues its motion undisturbed. Therefore, above this critical altitude the species has a net upward flow and loses all the time molecules to outer space. In what follows the argument z'_{eq} is understood. To evaluate this loss, assume that all molecules which at this height have a positive upward velocity $u_z = u\cos\theta > 0$ and a positive Keplerian energy, i.e.,

$$u^2 > u_0^2 = \frac{2GM}{R + z'_{\text{eq}}} \; , \tag{8.52}$$

do not collide any more and are lost to space in a hyperbolic orbit. If $f(\mathbf{u})\,d^3u$ is the Maxwellian velocity distribution of the species (eq. (1.94)), the flux of lost particles is easily obtained with an integration over θ (the angle from the vertical) from 0 to $\pi/2$ and over u from u_0 to ∞:

$$F_{\text{esc}} = \frac{2}{\sqrt{\pi}} n' \sqrt{\frac{2kT}{m'}} \left(1 + f_{\text{eq}}\right) \exp(-f_{\text{eq}}) \; ; \tag{8.53a}$$

$$f_{\text{esc}} = \frac{GMm'}{kT(R + z'_{\text{eq}})} = \left(\frac{v_{\text{esc}}}{v_T}\right)^2 \qquad (8.53b)$$

is the *escape parameter*. The factor m' in the exponent makes the flux strongly dependent on the mass (Problem 8.12). Since all the $N' = 4\pi R^2 H' n'$ particles contained in a layer H' deep are subject to this loss, at the rate $4\pi R^2 F_{\text{esc}}$,

$$\tau_{\text{esc}} = \frac{N'}{F_{\text{esc}}} = \frac{H'n'}{F_{\text{esc}}} \qquad (8.54)$$

is the mean loss time. Because of the exponential behaviour in (8.53a), the confinement is more effective for heavier atoms; this is why all the terrestrial planets cannot retain H, He and H_2. This process forcefully points out the fragility of their atmospheres. For the outer planets, where $M/(TR)$ is much larger, even the escape of hydrogen is negligible.

In a more accurate model other processes should be taken into account. At the exobase the outer tail of the velocity distribution is depleted, so that the assumption of thermal equilibrium overestimates the loss. Diffusion transports heavier molecules above their exobase, enhancing the escape. Loss and diffusion do not depend on the atomic number (and chemical properties), but solely on the molecular mass; they are very effective in determining the isotopic composition of the atmosphere. In turn, isotopic abundances (e.g., of deuterium) are an important tool to investigate exchanges between atmospheres and outer space.

Non-thermal escape processes.– There are also non-thermal losses, including:

- *Dissociation* (e.g., photo-dissociation) splits a molecule in two which escape more easily.

- *Sputtering*, the process with which a fast atom or ion (e.g., a cosmic ray particle), colliding with an atom, imparts to it sufficient energy to escape. Very rarefied atmospheres, like the Moon's and Mercury's, may be in part due to this effect. In absence of an internal magnetic field charged particles may also directly interact with the solar wind, a process referred to as *solar wind sweeping* (likely important for the past evolution of the Martian atmosphere).

- In a *charge exchange* a fast ion (e.g., from the solar wind) gains an electron from a neutral atom and, free from any magnetic force, may escape.

Other evolutionary processes.– Even a small concentration of hydrogen makes an atmosphere reducing, favouring reactions like $CO_2 + H_2 \rightarrow CO + H_2O$, $N_2 + 3H_2 \rightarrow 2NH_3$, $CO + 3H_2 \rightarrow CH_4 + H_2O$ and $C + 2H_2 \rightarrow CH_4$;

8. Upper atmospheres

when hydrogen is scarce, the atmosphere is oxidizing and the opposite reactions prevail. A loss of hydrogen, for terrestrial atmospheres inevitable, oxidizes the atmospheric composition with typical end-products such as CO_2, CO and N_2. If the present atmospheres of the terrestrial planets were primordial, hydrogen-depleted remnants, due to the above reactions they would be composed of CO_2 (63%), Ne (22%) and N_2 (10%), very different from what is observed. This points to a significant chemical evolution, under many, still uncertain processes; besides the losses already mentioned, they include:

- The greenhouse effect (Secs. 7.1 and 7.3).

- Exchanges with the mantle, including weathering, in which CO_2 dissolves in water and produces various salts.

- Asteroidal and cometary impacts, which have brought in large amounts of water and hydrocarbons.

- On the Earth, biological photosynthetic activity, which has increased the oxygen content over geological times.

- Effects of human activity on the climate. In the last half a century the production of CO_2 and other greenhouse gases (Fig. 7.4), together with diminished oxygen production due to disappearance of rain forests, may have affected the climate and produced measured rise in the mean atmospheric temperature. Predictions about climate evolution are not easy to make, due to the complex chemical reactions between atmospheric gases and pollutants, the vast and little known heat reservoirs in the oceans and the non-linear, difficult to model, interactions; but a major climate change in the next century, with a global increase of the oceanic water level, is a distinct possibility.

- Impacts of large meteorites up to small-asteroid size. The best known example is an object $10 - 20$ km in size which fell close to the Yucatan peninsula at the transition between the Cretaceous and the Tertiary ≈ 65 My ago. The deposited kinetic energy, $\approx 10^{31}$ erg, corresponding to 3 months of sunshine, spread over the globe a clay layer 1 cm thick, producing a large opacity and evaporating about 1 m of oceanic water. Dinosaurs were extinguished, with a major effect on biological evolution. The Moon has evidence of much bigger, earlier impacts, between 3.8 and 4.1 Gy ago (the late heavy bombardment phase); at that epoch the Earth must have had even bigger impacts. The energy of a body 400 km across, $\approx 10^{35}$ erg, is sufficient to vaporize all terrestrial oceans; it seems difficult that life, if it existed before the bombardment, could have survived such a catastrophe if confined to the surface.

- Deposition of interplanetary dust. This is a much softer process, which, however, may have effects on the atmosphere. The 100 ky climatic variations (Sec. 7.1) may be related – through the changes in the orbital inclination – to the periodic accretion of dust particles from the asteroidal dust bands (Sec. 14.6).

As mentioned above, the main problem of the atmospheres of terrestrial planet is their great diversity. The processes responsible for this evolutionary divergence is a still unsolved problem, though the gross features are likely understood: (i) Venus – very close to the Sun – suffered a runaway greenhouse effect by which liquid water was vaporized and later photo-dissociated in the upper layers; hydrogen quickly escaped and a hot atmosphere was left behind, where the dominant CO_2 molecules trapped the infrared radiation. (ii) Mars, due to its low mass, lost most of its atmospheric compounds via thermal and non-thermal depletion processes (such as solar wind sweeping). Dramatic changes of the Mars climate might also have been enhanced by the chaotic excursions of the planet's obliquity (Appendix of Ch. 7). Many of the above processes have either random features or poorly understood aspects (e.g., as shown by large variations of the atmospheric parameters on a very short time span, Fig. 7.4); reliable models of the atmospheric evolution are still not available.

– PROBLEMS –

8.1. Derive the ray equation (8.39) from Fermat's principle, which says that the integral $\int ds\, n_r$ between two fixed points is a minimum.

8.2.* Generalize the dispersion relation (8.26) and (8.27) for plasma waves to include the effect of an external constant magnetic field; one obtains a biquadratic equation for the refractive index (Booker's equation). Discuss various limits: $\omega \to \infty$; $\omega \to 0$; parallel propagation; near the electron cyclotron frequency when a resonance occurs; etc.

8.3. Derive the altitude dependence of the electron production rate $Q(z)$ with the simplifying assumptions of Sec. 8.2 (an exponential decrease of ionized molecules with constant cross section and ionization efficiency). Prove that

$$Q(z) = Q_m \exp\left[1 - y - \exp(-y)\right],$$

where Q_m is the maximum production rate, at the altitude $\bar{z}$ and $y = (z - \bar{z})/H$; this is *Chapman's function*. Study its mathematical properties for large $|y|$.

8.4. Using appropriate coordinates prove that (8.42) and (8.43) are first integrals of the differential equation (8.39) in plane-parallel and spherical symmetry.

8.5.* In the accretion heating equation (8.49), $T^* = c_{\rm p}TR/(GM)$ is the appropriate dimensionless temperature variable, $\ll 1$ in the late phase. Discuss the solution when the heat exchange term is negligible, in particular the steady state temperature.

8.6.* Estimate the timescale in which Mars lost significant amount of its original atmosphere and the ratio between the escape fluxes of H and D.

8.7. Find the velocity of a charged species with friction with a neutral component for an arbitrary value of the ratio ν/Ω (eqs. (8.13)).

8.8. Discuss Chapman's theory for the recombination rate $\alpha\, n_e^p$. Compare the resulting stationary electron profiles for different values of p.

8.9. a) Evaluate the refraction angle for a uniform atmosphere of thickness $H \ll R$ and a weak refractivity $\delta n_r = n_r - 1 \ll 1$. b) Using eq. (8.46), evaluate the refraction angle for an exponential atmosphere with a scale height H.

8.10.* Study the role of refraction and absorption in a planetary atmosphere on the reduction of the relative intensity I/I_0 of a distant point-like radiation source during an occultation; I is the observed radiation flux and I_0 is the unperturbed radiation flux. Formulate the conditions under which one of the two mechanisms for the flux attenuation prevails. (Hint: Assume an exponential atmosphere with constant scale height H.)

8.11. Faraday rotation: a) Prove eq. (8.35) for the rotation of the plane of polarization. b) Conceptually design a simple measurement of the electron columnar content in the ionosphere using two carriers at 1 and 2 GHz emitted by an Earth orbiting spacecraft. c) What is the order of magnitude of the effect for a source in the Galactic centre?

8.12. From Fig. 8.3 estimate the scale height in the outer atmosphere and the lifetimes of O_2, N_2, H and He.

– FURTHER READINGS –

A good treatise on the ionosphere and the magnetosphere is J.K. Hargreaves, *The Solar-terrestrial Environment*, Cambridge University Press (1992). See also S. Kato, *Dynamics of the Upper Atmosphere*, Reidel (1980) and M.H. Rees, *Physics and Chemistry of the Upper Atmosphere*, Cambridge University Press (1989). A comprehensive text on plasma wave propagation is T.H. Stix, *The Theory of Plasma Waves*, McGraw-Hill (1962). Among the many textbooks of plasma physics we quote the short classic L. Spitzer, Jr., *Physics of Fully Ionized Gases*, Interscience (1962); more recent, F.F. Chen, *Introduction to Plasma Physics and Controlled Fusion. Vol. 1. Plasma Physics*, Plenum Press (1984), and R.A. Treumann and W. Baumjohann, *Advanced Space Plasma*

Physics, Imperial College Press (1997). About wave propagation in the ionosphere, see, e.g., the lectures on *Magnetoionic Theory* by K.G. Budden in *Geophysique Exterieure*, C. De Witt, J. Hieblot and A. Lebeau, eds., Gordon and Breach (1962). For the planets, R.W. Schunk and A.F. Nagy, *Ionospheres: Physics, Plasma Physics and Chemistry*, Cambridge University Press (2000) and the older textbook S.K. Atreya, *Atmospheres and Ionospheres of the Outer Planets and their Satellites*, Springer-Verlag (1986). See also Ch. 7 of *Introduction to Space Physics*, M.G. Kivelson and C.T. Russell, eds., Cambridge University Press (1995).

Chapter 9

THE SUN AND THE SOLAR WIND

Some 4.56 billion years ago, out of the gravitational collapse of a cloud in the interstellar medium with angular momentum **L**, a rotating star – the Sun – was formed, and began to burn hydrogen into helium and heavier nuclei. A gaseous disc in the plane orthogonal to **L** developed; its chemical composition, initially the same as in the solar photosphere, quickly differentiated, giving rise to a variety of planets according to the distance from the Sun. The Sun is an electromagnetically active body, with great variability, including a cycle of $\simeq 11$ y; it deeply affects the entire solar system, not only through its steady radiation, but also through its variable electromagnetic emission and the solar wind, a flow of charged particles heated and emitted at supersonic speed by the corona. This plasma drags the solar magnetic field outward and, coupling with matter in interplanetary space, carries away angular momentum. It interacts with the planets and their dipole magnetic fields in a variety of complex processes and produces local manifestations of solar activity. At still larger distances, beyond the planets, the solar wind merges through a termination shock with the interstellar medium, confining the heliosphere.

9.1 The Sun

The Sun is a gaseous mass $M_\odot \simeq 1.99 \times 10^{33}$ g, 743 times larger than the total planetary mass, with a radius $R_\odot \simeq 6.96 \times 10^{10}$ cm and a mean density $\bar{\varrho} \simeq 1.4$ g/cm^3. Such a high value is consistent with the gaseous state because of the very high pressure in the interior. To a first approximation, and especially in the outer part, the Sun may be considered composed of a perfect gas in hydrostatic equilibrium; this leads to the estimate 3×10^{-9} erg $\simeq 1.8$ keV $\simeq 2 \times 10^7$ K for the central temperature $T(0)$ (eq. (1.19)), compared with the true value $T(0) \simeq 1.5 \times 10^7$ K. The central density $\varrho(0) \approx 150$ g/cm^3 is much larger than the mean value and corresponds to a pressure $P(0) \approx 2.3 \times 10^{17}$ dyne/cm^2. The Sun is mainly composed of ionized hydrogen and helium.

The internal temperature results from a delicate balance between the nuclear fusion reactions in the core, which have a strong temperature dependence, and the large opacity, which makes the loss of photons to the surface a very slow process. The large temperature gradient facilitates convection (Sec. 5.3), which appreciably contributes to the energy transport. This balance regulates nuclear combustion processes in the Sun, which, starting from the centre, slowly turn hydrogen into helium and heavier nuclei. The basic fusion process of four protons into an alpha particle (the *pp cycle*) releases about 0.3% of the initial rest energy; the energy made available in this way is sufficient to ensure the current luminosity $L_\odot \simeq 4 \times 10^{33}$ erg/s for the past life of the Sun and for a comparable time in the future. After this the Sun will move to the red giant stellar phase, significantly extending its radius (perhaps up to the Earth orbit).

The escape velocity

$$v_{\text{esc}} = \sqrt{\frac{2GM_\odot}{R_\odot}} \simeq 620 \text{ km/s} \qquad (9.1)$$

corresponds to the proton thermal energy at the centre $T(0)$, much larger than the surface temperature $T_s \simeq 5,770$ K (Fig. 9.2). The escape velocity is much smaller than the light velocity, corresponding to the fact that the radius $R_\odot$ is much larger than the gravitational radius $GM_\odot/c^2 = 1.43$ km (eq. (1.12)). Relativistic effects in the solar system, at a distance r from the Sun, are of order $GM_\odot/c^2 r$ (see Ch. 17), on the surface 2×10^{-6}.

G. GALILEI published in 1613 a detailed report on his observations of sunspots and a determination of the solar rotation; however, it took other 250 years before R.C. CARRINGTON established, again by tracking the apparent motion of sunspots, that the surface solar rotation rate $\omega_\odot(\theta)$ increases with the heliospheric colatitude θ: the outer layers are in a state of *differential rotation*. Traditionally $\omega_\odot(\theta)$ may be represented by a function symmetric about the equator like

$$\omega_\odot(\theta) \simeq 2.76 \times 10^{-6} (1 - 0.12 \cos^2\theta - 0.14 \cos^4\theta + 0.05 \sin\theta) \text{ rad/s}, \qquad (9.2)$$

obtained from interior solutions using the rotational splitting of acoustic eigenmodes of the Sun (similar values are obtained with radio maps of the solar surface). This corresponds to rotation periods of about 25 d near the equator and 35 d near the poles. As discussed in Ch. 16, solar rotation has an important bearing on the evolution of the angular momentum of the planetary nebula. The observed surface rotation rates of young solar-type stars are up to 50 times that of the Sun, and it is likely that some of its angular momentum was transferred by the magnetized solar wind to the outer parts of the nebula. The rotation axis has an obliquity $\epsilon_\odot \simeq 7° 15'$. Neglecting differential rotation, the rotational energy is $\approx 1.3 \times 10^{43}$ erg and the angular momentum is $\approx 9 \times 10^{48}$ g cm²/s. The deformation parameter (1.13) is

$$\mu_c = \frac{\omega_\odot^2 R_\odot^3}{GM_\odot} = \left(\frac{\omega_\odot R_\odot}{c}\right)\left(\frac{R_\odot}{m_\odot}\right) \simeq 2.3 \times 10^{-5} . \tag{9.3}$$

The quadrupole moment J_2 is much less, due to the large mass concentration near the centre; in the standard, rigidly rotating model of the Sun, $J_2 \simeq 1.8 \times 10^{-7}$; its direct measurement is an important and still unsolved problem (see Sec. 20.4).

The surface undergoes oscillations which have been extensively studied spectroscopically through the Doppler effect. The most extensive observations were obtained by the Global Oscillation Network Group (GONG project), a fifteen-station network of extremely sensitive and stable velocity imaging systems, spread at six different longitudes to obtain a continuous monitoring. High resolution maps of the velocity field as function of heliographic latitude and longitude are obtained; their Fourier spectrum consists of discrete frequencies, each mode corresponding to a spherical harmonic (ℓ, m). This method allowed the identification of thousands of proper modes; one can recognize both deep, gravity-driven modes, and shallow, faster modes, whose restoring force is the pressure; their periods range from minutes to hours. In a way similar to the study of the Earth with its proper modes, solar oscillations have opened up a new discipline, *helioseismology*, for the investigation of the interior of the Sun. The deep interior shows up mainly in the low-frequency modes, which, however, are difficult to observe. Detailed models of the interior and its differential rotation are available and can be confronted with observations; contrary to what one would expect if the mantle has slowed down, there is no substantial increase of the angular velocity with depth, although available data are able to constrain well the solar structure down to a radius of $\simeq 0.3 R_\odot$. Near the base of the convective zone there appears to be a rather abrupt transition from a latitudinally dependent, surface-like rotation, to a nearly latitudinally independent rotation in the radiative interior; this is the *solar tachocline*, at about $0.7 R_\odot$ from the centre. Above, and perhaps near, this surface we have a global meridional circulation of material; beneath the tachocline, the rotation rate is roughly 2.71×10^{-6} rad/s and no significant meridional circulation is detected (although recent models aimed to correctly describe the latitudinal distribution of sunspots during the solar cycle seem to require a deeper meridional flow in the Sun).

Solar radiation. The – prevailingly optical – solar radiation (see Fig. 9.1) has a nearly black body spectrum, corresponding to a surface temperature $T_s \simeq 5,770$ K. This radiation has profound effects on the solar system, as separately discussed in other chapters. It heats bodies to a temperature inversely proportional to the distance from the Sun (eq. (7.9)) and determines the differential condensation and solidification of particles in the solar nebula according to their chemical composition (Ch. 16); it is the main energy source for the atmospheres and the surfaces of planetary bodies, and the essential factor in the appearance and the maintaining of life, and determining its quality (i.e., the evolution of eyes; Ch. 7); it generates thermal tides in the upper atmospheres of planets (Ch. 8) and, with its high energy photons, it produces ionospheres; its momentum exerts an important repulsive force on dust grains (Sec. 15.4) and artificial satellites (Ch. 18); the recoil of an anisotropically re-emitted radiation may affect the long-term dynamics of small asteroids and meteoroids (Ch. 14).

In a *solar eclipse* the Sun is occulted by a planetary body. On the Earth it is a major astronomical and social phenomenon; since the angular diameter of the Sun ($\simeq 30'$) is almost equal to the one of the Moon, if the centres of the three bodies are aligned (*total eclipse*), there is no direct illumination. Before and after this time only a fraction of the solar disc is in view and we have the *penumbra* (Problem 9.2). During a total eclipse, ground astronomical observations near the Sun, otherwise prevented by the scattered radiation, are possible and allow observations of the corona and measurements of the gravitational deflection of light by the Sun (Sec. 17.6). The passage of an artificial satellite in the shadow of a planetary body has a complex effect on its thermal behaviour (depending on its thermal conductivity and thermal capacity) and, through the radiation recoil, on its orbit.

The optical radiation is emitted from the uppermost layer of the solar atmosphere – the *photosphere*, about 500 km thick – between an optical depth of order unity and the limit of transparency. The main microscopic processes responsible for the opacity are photo-excitations and bound-free transitions. There is also a marked limb-darkening effect (see Sec. 7.3). The solar optical spectrum shows many ($\approx 25,000$) absorption lines, the *Fraunhofer spectrum* (Fig. 9.1); they are due to resonant photon absorption, corresponding to bound-bound and bound-free processes in the photosphere, at a smaller depth than the continuum. The Sun is also a powerful ultraviolet and X-ray source, with an effective temperature of 10^5 to 10^6 K. These emissions vary very much according to solar activity and the 11 y cycle (Fig. 9.1). In the radio band there are strong bursts and a steady emission as well.

The properties and the evolution of the solar system crucially depends on the chemical composition of the primordial solar nebula. One way to determine it uses the ratio of the intensities of absorption lines; the photosphere, in fact, is not affected by nucleosynthesis and did not change in composition since the formation of the Sun. Chemical analysis of meteorites – whose composition also reflects the primordial nebula – and the spectroscopy of the solar corona gives very similar results (Table 9.1); in the Sun there is only a deficit

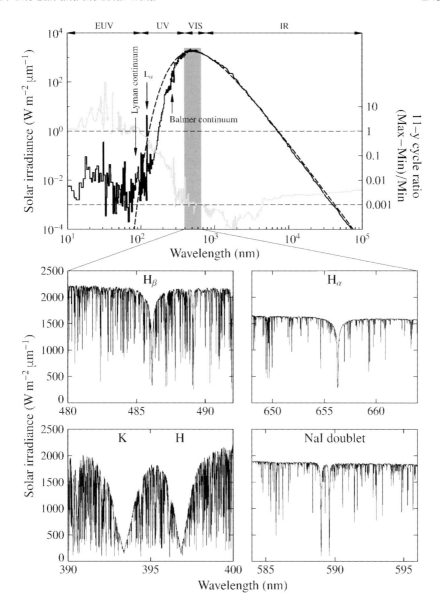

Figure 9.1. Upper part: The solar electromagnetic spectrum (thick curve) and the black body spectrum for a temperature 5,770 K (dashed line); Fraunhofer absorption lines are not shown. The Lyman-α emission in the UV is indicated by the arrow. The extreme UV and X spectra are highly variable, also due to short-term solar activity (e.g., flares). The solar cycle variability is described by the grey curve. Lower part: Examples of solar absorption lines of the quiet Sun obtained with the National Solar Observatory Spectrometer on Kitt Peak, Arizona. We show narrow bands around the following lines: Balmer H_α and H_β, H and K level of ionized calcium and neutral sodium doublet. Data adapted from http://science.nasa.gov/headlines/images/sunbathing/sunspectrum.htm (upper part) and ftp://argo.tuc.noao.edu/pub/atlas/ (lower part).

of lithium, beryllium and boron due to their combustion in thermonuclear reactions. The presence of elements other than hydrogen and helium, which are produced only in stellar nucleosynthesis, is due to the fact that the Sun and most stars in its neighbourhood are not first generation stars and have been formed partly out of material already processed in earlier, more massive stars in the Galaxy and ejected by stellar winds and explosions. The chemical composition of the interstellar medium, especially in our neighbourhood, has also been widely investigated with the spectrometry of the gas near main sequence stars and the composition of cosmic rays. They provide abundance ratios very similar to solar system data, with the discrepancy that neon is more abundant (by about 10%) in interstellar medium. This is evidence that in the last 4.5 Gy the Galaxy has kept in our neighbourhood nearly the same composition, and provides information about its chemical evolution and stellar nucleosynthesis. Most of the helium (accounting for about 0.28 of the total mass[1]), however, has been produced in primordial cosmology. The presence, although in smaller concentration, of heavier nuclei, in particular carbon, in the nebula out of which planets formed, has crucially conditioned their evolution, the appearance of atmospheres and the origin of life: in the early phase of the evolution of galaxies no chemistry, nor life was possible.

Solar activity and variability. Due to its large opacity, in the layer below the surface, the heat generated in the interior is transported upwards by convection. The convection cells are visible (from space) on the surface and form the *solar granulation*. A granule, of size $\approx$ 700 km, consists in an interior with hotter, rising gas (at $\simeq$ 2 km/s) and a roughly hexagonal boundary (a *Bénard cell*, also observed in the laboratory), with descending fluid. The granules appear and disappear with a timescale of $\approx$ 10 minutes. Larger and deeper convection cells, *supergranulation*, are also present; their characteristic scale is about the Earth size; the velocities reach $\simeq$ 500 m/s and their lifetime is about a day. At the edges of supergranule cells we find that the magnetic field is concentrated around $\simeq$ 1 T; such intense flux tubes occur especially at the junction between three cells.

Sunspots, regularly observed on the solar disc since the 17th century, consist of a dark central region (the *umbra*) surrounded by a brighter area (the *penumbra*), composed of light and dark filaments. They are about of the Earth size. Sunspots are dragged around by solar rotation and last one or more periods; their number changes with a period of $\approx$ 11 y (the *solar cycle*, see Fig. 7.1). At the beginning of the cycle they appear at $\pm 20°$ to $\pm 30°$ of latitude; later they drift towards the equator and disappear altogether at the end of the cycle. 22 y,

---

[1] In astronomical literature the mass fractions of H, He and heavier elements are referred to as X, Y and Z. The mean solar values are $X \simeq 0.707$, $Y \simeq 0.274$ and $Z \simeq 0.019$, with errors of a few percent.

9. The Sun and the solar wind

Table 9.1. The primordial abundances of the elements in the solar system (total over all stable nuclides assumed). If N is the number density and m the atomic mass, the fourth and fifth columns give the abundance by number N/N_H and the abundance by weight $mN/(m_H N_H)$ relative to hydrogen in the solar photosphere (hydrogen values normalized to unity). For comparison, the last column shows the numerical abundance of elements in C1 ordinary chondrites (see Sec. 14.6). Data from E. Anders and N. Grevesse, *Geochim. Cosmochim. Acta* **53**, 197 (1989).

Atomic number	Element	Atomic mass	Numerical abundance	Abundance by mass	Numerical abundance in CC
1	H	1.01	1.0	1.0	1.0
2	He	4.00	0.0977	0.3880	0.0975
6	O	16.00	8.51×10^{-4}	0.0135	8.53×10^{-4}
8	C	12.01	3.63×10^{-4}	0.0043	3.62×10^{-4}
26	Fe	55.85	4.68×10^{-5}	0.0026	3.23×10^{-5}
10	Ne	20.18	1.23×10^{-4}	0.0025	1.23×10^{-4}
7	N	14.01	1.12×10^{-4}	0.0016	1.12×10^{-4}
14	Si	28.09	3.55×10^{-5}	9.89×10^{-4}	3.58×10^{-5}
12	Mg	24.31	3.80×10^{-5}	9.16×10^{-4}	3.85×10^{-5}
16	S	32.07	1.62×10^{-5}	5.15×10^{-4}	1.85×10^{-5}
28	Ni	58.69	1.78×10^{-6}	1.04×10^{-4}	1.76×10^{-6}
20	Ca	40.08	2.29×10^{-6}	9.11×10^{-5}	2.19×10^{-6}
13	Al	26.98	2.95×10^{-6}	7.90×10^{-5}	3.04×10^{-6}
24	Cr	52.00	4.68×10^{-7}	2.41×10^{-5}	4.83×10^{-7}
25	Mn	54.94	2.46×10^{-7}	1.34×10^{-5}	3.42×10^{-7}
17	Cl	35.45	3.16×10^{-7}	1.11×10^{-5}	1.89×10^{-7}

and longer, periodicities are also observed. In the plane (time, latitude) the average density of spots appears like a sequence of butterflies aligned along the equator. In the period between about 1645 and 1715 sunspots almost disappeared (*Maunder minimum*); it is remarkable that at this time the temperature of the Earth was lower (the *little ice age*), a correlation still not well understood (Sec. 7.1). A strong magnetic field between 0.1 and 0.3 T may be present in the umbra; the spots often occur in pairs of opposite magnetic polarity, aligned along a parallel. This field may be produced by local dynamo processes in the convection zone; arising cells may bring up to the surface a magnetic tube in the form of an arch, at whose feet two spots with opposite polarity appear. It can be shown that a magnetic field raises the threshold of the temperature gradient required for the onset of convective instability (Sec. 5.3); within a sunspot, therefore, convection is weaker and the gas is cooler and dimmer.

A *solar flare*, a major event in solar physics, mainly manifests itself as a strong and short (say, half an hour) electromagnetic burst from a region on the surface of a typical size of one thousandth the disc; with an estimated thickness of 20 km, the total volume of the source is $\approx 10^{14}$ km^3. Besides the optical continuum and the characteristic Balmer H$_\alpha$ line (656.2 nm) emissions, there are also UV and X; this is evidence of very large temperatures, larger than 10^7 K ≈ 1 keV. Even more interesting, there is a polarized radio emission produced by high energy particles which emit synchrotron radiation spiralling along lines of force, and MeV photons (γ rays) from the excitation of nuclear energy levels and electron-positron annihilation. With a total energy larger than $\approx 10^{32}$ erg and an energy density of 10 erg/cm^3, flares cannot be powered by thermal energy. A standard model consists in a magnetic loop near the surface on which a hot and dense plasma slowly accumulates, until it suddenly blows up, carrying upwards lines of force; the driving mechanism is *magnetic reconnection* – the process by which parallel lines of force of opposite direction annihilate each other, transforming magnetic energy in kinetic energy (see Sec. 10.1) – in the middle section. With such large and fast-varying magnetic fields there is no wonder that intense electric fields along the lines of force arise, capable to accelerate particles to high energies. Other forms of solar activity are the *faculæ*, bright regions in the H$_\alpha$ line, and the *prominences*, dark filaments pushing out into the corona, often associated with sunspots. They will not be discussed here.

Solar flares, beside the radiation and particle burst, are sometimes associated with *coronal mass ejections* (CMEs). Big lumps of plasma, up to 10^{15-16} g, are emitted with velocities falling in a broad range of 20 up to 2,000 km/s with a median value of $\simeq 450$ km/s; a significant fraction of CMEs thus propagates much faster than the solar wind (Sec. 9.2). During periods of high activity one to three may occur each day. With an equivalent mean mass-loss rate of about 10^{11} g/s, CMEs contribute by about 10% to the total mass loss of the Sun, principally the solar wind. Their main significance, however, consist in a complete reconfiguration of the coronal magnetic field on a large-scale. Most CMEs are characterized by a three-part structure when observed in white light: a bright, high-density front moving ahead of a dark, low-density cavity, within which a bright, relatively high-density core is observed. This, in fact, reflects the structure of the helmet streamer associated with the generating solar flare.

Observations of the Sun in X-band show dark and cool patches, lasting several rotation periods; they are semi-permanent, but variable, especially near the poles. These *coronal holes* are the origin of open magnetic field lines and the source of fast components of the solar wind.

The understanding the global structure of the solar activity was boosted mainly by high latitude observations carried out by the spacecraft Ulysses in the mid 1990s. A large scale, dipole-like and much weaker magnetic field is present, with a typical surface value of 2×10^5 nT and an axis tilted by $10° - 20°$ with respect to the rotation axis. Its obvious origin is a dynamo process in the convective and differentially rotating interior. Evidence for this field is obtained with in situ observations at large distances, where the lines of force are stretched out and distorted by the solar wind. It shows up with open field lines at high latitudes, with opposite polarity in the North and the South; at low latitudes (below 20°, say) we have closed field lines, often mixed up with variable and complex local loops arising from the surface. The overall field is particularly evident around solar minima; around solar maxima a much more disordered configuration appears and there is evidence of an inversion of its polarity, with a complete cycle of 22 y. The nature of the solar cycle characterized by the global magnetic field reversal is unknown.

The solar activity has a profound influence on the magnetospheres, ionospheres and atmospheres of planetary bodies. When solar flare particles or CME bursts impinge upon the Earth, they compress the magnetosphere, change the particle density in the ionosphere and, in the ensuing magnetic storm, globally affect radio communications and even electrical power lines. We can say, we have a *"space weather"*, with the problem of its prediction; SOHO (SOlar and Heliospheric Observatory), ACE (Advanced Composition Explorer) and Wind, three spacecraft near the L_1 Lagrangian point between the Earth and the Sun (Sec. 13.2), are very helpful in monitoring the solar activity for this purpose.

Painstaking observations of solar activity during more than a century are summarized in some empirical indices; here we only quote two. The *Wolf sunspot number* R_W is a combination of the number of sunspots and the number of active areas, averaged over a month or a year; sunspot observations are available since 1848 on a daily base, but the annual sunspot numbers have been estimated since the telescope was invented in 1610. Most importantly, the Wolf number is related to the rate of flares. We have also the total radio flux $F_{10.7}$ from the Sun at $\lambda = 10.7$ cm (2, 800 MHz), again highly correlated with R_W. Similar empirical indices of activity have been introduced for the ionosphere (Sec. 8.1) and the magnetosphere (Sec. 10.1); their correlation opens a way to understand the complex effect of solar activity and its relation to magnetospheric dynamics.

9.2 The solar wind

As shown in Fig. 9.2, the temperature in the neighbourhood of the Sun reaches a minimum just above the photosphere; beyond, it increases again by more than two orders of magnitude. This hot region – the *corona* – is visible, especially in an eclipse, up to a few solar radii, due to the solar radiation scattered by free electrons and emission lines of highly ionized atoms. The process by which such large amounts of energy ($\approx 10^5$ erg cm^{-2} s^{-1}) can be deposited far above the surface of the Sun, where the temperature is higher, is still uncertain and will not be discussed in detail in this book, devoted to the solar system. Turbulence and explosive phenomena like flares are evidence of a large reser-

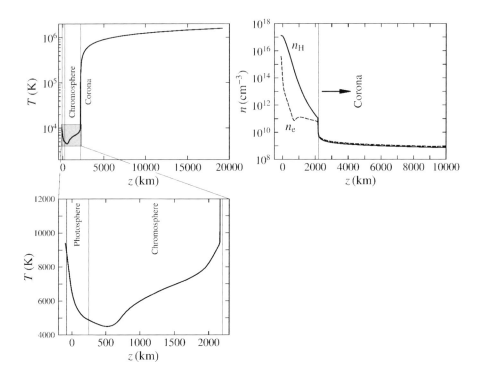

Figure 9.2. The temperature profile in the solar atmosphere and the corona (FAL-C model): the upper plot shows the large increase, which takes place abruptly at $\simeq 2,200$ km, at the bottom of the corona. The altitude z in the atmosphere is conventionally measured from the level at which the optical depth τ at 500 μm wavelength is unity. The bottom panel shows a zoom of the photosphere and chromosphere. On the right, the electron density n_e and the neutral hydrogen density n_H are shown. Data kindly provided by E.H. Avrett and R. Loeser.

voir of energy in the photosphere in form of mass motion, intense magnetic fields and energetic particles far from thermal equilibrium; a number of wave phenomena have been suggested to propagate upward to the chromosphere and the corona, depositing there their energy by some form of dissipation. Obviously, the ultimate reservoir is the system of convective eddies below the solar surface, on turn fed by the large temperature gradient established in the core by thermonuclear burning.

The solar corona is the source of the *solar wind*, a supersonic radial flow of charged particles from the Sun. At 1 AU it has a velocity v between 400 and 800 km/s, always larger than the sound speed c_s and Alfvén velocity V_A. At low latitudes Mach's number $M_s = v/c_s$ is about 8. The temperature $T \approx 10^5$ K is somewhat lower than the coronal value; the directed proton kinetic energy can reach 1 keV, 100 times larger than their thermal energy. The supersonic regime has three important consequences. Since the square of Mach's number gives the ratio between inertial and pressure forces, the solar wind is affected mainly by gravity and, at large distances, has a small acceleration. Shocks arise when it encounters obstacles or a slow fluid. In the moving frame, wave perturbations have a phase velocity v_p less than v; hence a disturbance with a wave number $\mathbf{k}$ in the solar system frame has a frequency $\mathbf{k} \cdot \mathbf{v}$ generally larger than the proper frequency kv_p.

The wind is a fully ionized gas composed of electrons, with density $n_e \approx$ 5 cm^{-3} (and plasma frequency $\omega_p \simeq 1.2 \times 10^5$ rad/s), protons and α-particles at the solar abundance (5% by number); its density is $\varrho \simeq n_e m_p$. These (very variable) values refer to a distance of 1 AU. The solar wind is deflected by planetary bodies and their magnetospheres, with a large variety of phenomena, including bow shocks, wakes, magnetic storms and generation of fast neutral particles by charge exchange (Ch. 10). It sweeps the tails of comets near the Sun in the outward direction. Its fast particles affect the surfaces of orbiting spacecraft and deteriorates their optical properties. Its refractivity, although weak, deeply influences radio propagation in interplanetary space, in particular near the Sun (Sec. 19.5).

Condition for a static atmosphere. In a thin atmosphere, with a temperature profile $T(z)$, the pressure for a perfect gas in *static equilibrium* (eq. (7.1)) is

$$P(z) = P(0) \exp\left[-\int_0^z dz\, \frac{gm_{\mathrm{mol}}}{kT(z)}\right] ;$$

g is the acceleration of gravity and m_{mol} the mean molecular mass. The atmosphere is finite if the pressure tends to zero for $z \to \infty$ and this requires that the integral in the exponent diverges. If at great heights $T(z) \propto z^n$, this holds only for $n < 1$; in particular, an isothermal atmosphere is finite. In spherical symmetry things are different. The equilibrium condition (1.6) for a perfect gas with a radial profile $T(r)$ can be readily integrated

$$P(r) = P(R) \exp\left[-\int_R^r \frac{dr'}{r'^2}\, \frac{GMm_{\mathrm{mol}}}{kT(r')}\right]. \qquad (9.4)$$

If at large distance $T(r) \propto r^s$ the integral in the exponent diverges if $s < -1$; only in this case the atmosphere is finite. A static isothermal atmosphere does not exist. Any process which makes the temperature gradient less steep, such as an efficient energy transport, facilitates the onset of a wind.

Across the sonic point. The previous difficulty can be dissolved only relinquishing the assumption of static configuration and allowing a radial flow; this is the case of the Sun, around which the corona is dynamically heated from below by turbulent processes and the plasma ensures a high thermal conductivity. In the lower corona the sound speed is less than the escape velocity and it can be easily shown that, for typical scales of the order of the distance r, the hydrostatic pressure is still weaker than gravity; notwithstanding the flow is accelerated. One can understand this as follows: when the velocity is independent of distance, mass conservation implies that the density decreases as $1/r^2$; but an acceleration, with $dv/dr > 0$, requires a steeper decrease, and a larger pressure gradient is produced. Since at first the flow is subsonic, the balance between gravity and pressure determines the velocity gradient. This process continues until the sonic point is reached; beyond that the pressure force becomes progressively negligible and the velocity does not change much. The solar wind is a general phenomenon which appears every time the atmosphere of a gravitating body has sufficiently large heat conductivity; it occurs in stars and galaxies as well. A simple, but deep mathematical insight into this striking phenomenon has been developed by E.N. PARKER in 1958.

Let us assume a radial flow, whose conservation requires

$$4\pi \varrho(r) v(r) r^2 = 4\pi \varrho_0 v_0 r_0^2 = -\frac{dM}{dt} \tag{9.5}$$

in terms of the mass loss $-dM/dt$ (see eq. (1.34)). The suffix $_0$ indicates the value at the base $r_0 \simeq 1.01 R_\odot$ of the solar wind, above the heating region (Fig. 9.2). Conservation of momentum requires, in an adiabatic flow with the sound speed $c_s(r)$ (eqs. (1.31) and (1.20)),

$$v \frac{dv}{dr} + c_s^2 \frac{d \ln \varrho}{dr} + \frac{GM}{r^2} = 0 . \tag{9.6}$$

Using eq. (9.5) this reads

$$v \frac{dv}{dr} \left(1 - \frac{c_s^2}{v^2} \right) = -\frac{GM}{r^2} + \frac{2 c_s^2}{r} \equiv H(r) . \tag{9.7}$$

When the temperature does not vary much, the function $H(r)$ near the Sun is dominated by the first term and is negative there, but becomes positive as r increases. Let r_c be the point where H vanishes.

We can guess the general features of the solution by looking at the sign of the derivative dv/dr in the (r, v) plane (Fig. 9.3), which changes at $r = r_c$ and at the sonic value $v = c_s$. Starting at the surface with a low, subsonic value v_0, the velocity cannot penetrate in the double-value region IV (curve e). Region II (curve c), where the velocity decreases, is excluded for the following reason. The density is formally obtained from eq. (9.6) as

$$\varrho(r) = \varrho(r_0) \exp\left[- \int_{r_0}^{r} \frac{dr}{c_s^2} \left(v \frac{dv}{dr} + \frac{GM}{r^2} \right) \right] .$$

For an isolated body the density $\varrho(r)$ must vanish at infinity, so that the integral must tend to $+\infty$. Assuming that the sound speed c_s tends to a constant,

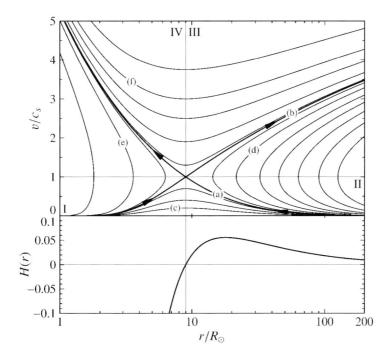

Figure 9.3. Qualitative features of Bernoulli's equation (9.7) for an isothermal flow (c_s = const.). The outgoing solar wind corresponds to the critical curve b); a) describes an inward accretion flow (see Sec. 16.4). The bottom panel shows the function $H(r)$ (eq. (9.7)).

the second term in the integral contributes with a finite constant, so that the required divergence is possible only if $v(r)$ increases at large distances. Therefore the appropriate solution is b), which crosses from I into III, becoming sonic just at the critical distance r_c; beyond it the velocity slowly increases. The time-reversed solution a) corresponds to an inflow of gas and is important for the accretion problem (see Sec. 16.4).

When the sound speed c_s is constant, appropriate for the isothermal model, we obtain

$$r_c = R_\odot \frac{v_{esc}^2}{4\,c_s^2} \simeq 9\,R_\odot \sqrt{\frac{T}{10^6 \mathrm{K}}}, \tag{9.8}$$

in terms of the escape velocity v_{esc} (9.1). In this case eq. (9.7) integrates directly to

$$\frac{v^2 - v_0^2}{2} + c_s^2 \ln\left(\frac{v_0\,r_0^2}{v\,r^2}\right) - \frac{GM}{r} + \frac{GM}{r_0} = 0\,; \tag{9.9}$$

at large distances the velocity grows very slowly; here v_0 is the velocity at the base of the corona r_0.

In this model the flow is radial and the azimuthal velocity v_ϕ is neglected. One would expect that the rotation of the lines of force, which are anchored to the rotating surface, produces such a component of the velocity along a parallel; it has an important role in the balance of angular momentum. However, since the wind speed is much larger than the Alfvén velocity V_A (eq. (10.41)), the radial flow is not much affected.

Corotating interacting regions. The solar wind, far from constant and uniform, is affected by fast streams and sudden explosions; a coronal mass ejection may be emitted from a small area on the surface and persist for several rotations. At a given time its outer boundary separates a fast component from the outer, normal wind. This is a *corotating interacting region* (CIR), often consisting of a leading-edge shock wave with an inner rarefaction wave. To determine its shape, consider an emission from a corotating point at the colatitude θ and the longitude ϕ_0, equal to ωt in the inertial frame; also, neglect differential rotation and, for illustration, the velocity gradient dv/dr. At the inertial longitude ϕ the stream has travelled for the time $(\phi_0 - \phi)/\omega$ and has reached the distance

$$r = r_0 + \frac{v}{\omega}(\phi_0 - \phi) = r_0 + v\left(t - \frac{\phi}{\omega}\right). \tag{9.10}$$

The equatorial projection of this curve is *Archimedes' spiral*. In space it lies in the cone of colatitude θ and makes with the radial direction the angle ψ given by

$$\tan\psi = r\sin\theta\frac{d\phi}{dr} = -\frac{r\omega\sin\theta}{v} \simeq -1.1\left(\frac{r}{1\,\text{AU}}\right)\left(\frac{400\,\text{km/s}}{v}\right)\sin\theta \tag{9.11}$$

(Fig. 9.4). The CIR rotates around the rotation axis of the Sun with its rotation frequency ω. At large distances (≈ 10 AU at low latitudes) ψ tends to 90°, so that different streams merge and coalesce into wider structures.

9.3 The heliosphere

The Ulysses mission of the European Space Agency has for the first time explored the heliosphere in three dimensions, spanning more than a solar cycle, including the last maximum in 2000 (Fig. 7.1). This spacecraft, launched in October 1990 on a solar orbit with an inclination of 79.4° with respect the ecliptic plane and a period of 6.2 y, has passed twice over the South Pole of the Sun (in 1994 and 2000) and twice over the North Pole (in 1995 and 2001), exploring for the first time high latitude regions of the solar system. Its simple, but accurate instruments have measured in situ electromagnetic fields and particle distributions. Using also simultaneous observations carried out with the NASA satellite Wind near the L_1 Lagrangian point of the Earth-Sun system, a rough, but clear picture of the three-dimensional solar wind and its dynamics has emerged (Fig. 9.5).

9. The Sun and the solar wind

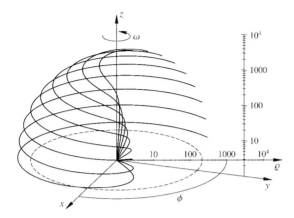

Figure 9.4 The family of three-dimensional spirals which describe the locus of fast streams and of lines of force in the solar wind. The frozen magnetic field is dragged around with the solar angular frequency $\omega_\odot$. Distances are given in a logarithmic scale and the unit is $R_\odot$; the solar surface is folded at the origin. A dashed circle with radius of 1 AU is drawn in the equatorial plane for reference.

Parker's model (Sec. 9.2) neglects the magnetic field in the solar wind and simplistically uses elementary fluid mechanics; before briefly describing the picture that has emerged, we need some theoretical complements.

Theoretical complements. The fluid approximation requires that the collision mean free path be smaller than the characteristic scale ℓ. When the force between particles is inversely proportional to the square of their distance, the relaxation to local thermal equilibrium is due to encounters spanning a large range of the impact parameter b. In a plasma, below $b = e^2/kT$ we have strong collisions; but even when $b > e^2/kT$ many weak collisions produce a cumulative effect of the same order; the largest effective value of the impact parameter is the Debye length λ_D (1.102), beyond which collective interactions prevail. Essentially the same occurs for a "gas" of gravitating particles and is relevant for planetesimal accumulation into planetary embryos (Sec. 16.3). In a plasma the problem is further complicated by the presence of different relaxation processes, including attaining an isotropic velocity distribution, a Maxwellian velocity distribution and temperature equality between different species; each of them has its own relaxation frequency. As in the gravitational case, the collision frequency ν_c is proportional to $n_e/T^{3/2}$; at 1 AU, for protons, $\nu_c \simeq 5 \times 10^{-7}$ Hz and the mean free path $\ell_c = v_T/\nu_c$ is ≈ 1 AU; as a result, at large distances the flow is, by and large, collision-free and dissipation is provided by more complex, small scale wave-wave interactions.

In principle one should use a description in phase space, like Vlasov's equation (Sec. 1.8). In particular, a thermodynamical temperature cannot be defined: not only the mean kinetic energy of electrons and ions are different, but also their velocity distribution is not isotropic around their mean velocity, and the stress tensor is not a scalar. It can be expected, however, that axial symmetry around the direction of the local line of force holds, so that one can define two "temperatures" $T_\parallel$ and $T_\perp$, parallel and orthogonal, respectively, to the magnetic field. It is thus remarkable, that in spite of all these complications required by the lack of local thermal equilibrium, the ordinary fluid approximation, as the one used in the previous Section, can still provide a correct insight into the dynamics of the solar wind.

The interplanetary plasma is a good electric conductor and the question arises, whether the frozen-field approximation (1.76) holds. As discussed in Sec. 6.4, this is determined by the magnetic Reynolds number $Re_m = 4\pi \ell v \sigma/c^2$. $\sigma = n_e e^2/(m_e v_c)$ is the electrical conductivity (eq. (1.75′)), independent of the density and proportional to $T^{3/2}$. In the corona and interplanetary space this number is huge (even 10^{12}) and this approximation can safely be trusted. Near the Sun, especially at low latitudes, where the magnetic field is strong and the wind speed moderate, the latter's flow is thwarted and slowed down if closed lines of force are present. At large distances and above coronal holes, where the lines of force are open, the opposite occurs: the wind flows radially, nearly unaffected by the magnetic field and drags along the lines of force.

We now show that in this case the open lines of force, which are anchored to a point of the rotating Sun, are deformed to the same helical shape previously discussed. In this simple model, if the Sun were not rotating, the lines of force would be stretched to a radial direction; but rotation produces a component along a parallel. With axial symmetry we have only the components $B_r(r, \theta)$ and $B_\phi(r, \theta)$. In a corotating frame the lines of force are at rest and the flow, in addition to the radial velocity v (assumed constant) has the velocity $-\omega r \sin\theta$ along ϕ and makes with the radial direction the angle ψ' defined by $\tan\psi' = -\omega r \sin\theta/v$. Since there is no change in time, the velocity and the field are parallel; a line of force, determined by the function $r(\phi)$, lies on a cone of constant latitude and fulfils

$$\frac{B_\phi}{B_r} = r\sin\theta \frac{d\phi}{dr} = \tan\psi'(r) = -\frac{\omega r \sin\theta}{v}, \qquad (9.12)$$

the same equation (9.11) as before (Fig. 9.4; see also Problem 9.10).

This rotating magnetic pattern extends to large distances and, interacting with conductive bodies, transfers to them part of the angular momentum of the Sun; in this way, in the early protonebula dust might have acquired angular momentum, partly compensating its inward drift by gas drag.

Configuration of the heliosphere. With its high latitude orbit, the Ulysses probe has allowed a thorough investigation of the two modes of the solar wind (Fig. 9.5). In 1995, near the solar minimum, the quick transition, at about $\pm 30°$ of latitude, from the low (≈ 400 km/s) mode to the high velocity (800 or even 900 km/s) mode is clear; these states are also evident in the density (bottom panel), in agreement with the conservation law (9.5). In the polar regions large coronal holes with open field lines are observed, where the wind can be emitted without magnetic impediments. Fast streams are also occasionally ob-

9. *The Sun and the solar wind* 255

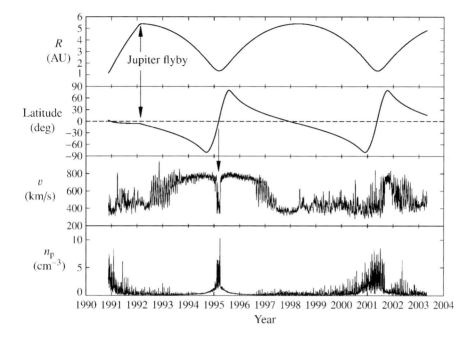

Figure 9.5. Ulysses measurement of the solar wind parameters. The upper panels show the radial heliocentric distance and the latitude of the spacecraft (note the Jupiter flyby in January 1992, used to attain the required large inclination). The bottom panels show the radial speed and the proton density. The two regimes of high speed, low density and, near the equator, low speed, high density, are evident, especially near the solar minimum in 1995. Data from http://helio.estec.esa.nl/ulysses/Archive/.

served at low latitudes, often corresponding to local coronal holes. The large scale, dipole-like lines of force are deformed in a spiral structure, as approximately described by eq. (9.12); since they have opposite directions in the two hemispheres, a *heliospheric current sheet* (HCS) develops near the magnetic equator, where the magnetic field changes sign (Fig. 9.6); there reconnection processes may take place, with sudden transformation of magnetic into kinetic energy. In this schematic model, due to the magnetic tilt, the ecliptic is divided in two *magnetic sectors*, corresponding respectively to the northern and the southern field; this structure rotates with the solar rotation frequency. An observer in the ecliptic plane (whose normal makes with the rotation axis an angle smaller than the magnetic tilt) crosses the HCS twice every solar rotation (Fig. 9.6). In reality these magnetic sectors are more irregular and variable, since the HCS is warped and oscillates around the magnetic equator; they have been extensively investigated with low inclination spacecraft. Ulysses has measured the angle $\delta\psi$ between the magnetic line of force and the direction of the spiral and has confirmed that at low latitudes, for much of the time,

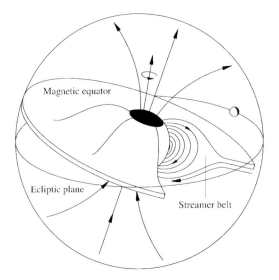

Figure 9.6 A scheme of the magnetic configuration of the solar wind far from the solar maximum. At high latitudes the radially flowing wind stretches away the lines of force from the poles, but near the magnetic equator its flow is slowed down by closed loops. Further out the heliospheric current sheet (HCS) near the magnetic equator separates regions where the radial magnetic field has opposite directions. The (small) tilt of the magnetic axis produces a rotation of the configuration with the solar rotation period.

it is near $0°$ or $180°$; at high latitudes it is always near one of the two states. Ulysses has also confirmed the general three-dimensional structure of corotating interacting regions, which appear up to latitudes of $\approx 30°$.

Fig. 9.5 shows also two other crossings of the $30°$ latitude at times not too far from the solar minimum, at the beginning of 1993 and at the end of 1996; the wind speed shows a clear periodic pattern, evidence of the fact that, as the latitude decreased, the satellite encountered a persistent corotating interaction region each solar rotation period, with a temporary transition between fast and slow modes. The last crossings of the $30°$ latitude, one at the beginning of 1999, two in 2001 and one in 2003, occur near the solar maximum; the transition between a low and a fast wind is still present, but is less evident and with strong fluctuations. Near a solar maximum the ordered large scale structure of the magnetic field is much less evident. Most of the surface is covered with closed loops and, in particular in the polar regions, there are no large scale coronal holes; the fast solar wind is rare and local. CME occur at a faster rate (ten times the solar minimum rate) at all latitudes. The dipole-like magnetic structure slowly disappears, leading to a reversal, in a complex and drawn out process lasting several rotations. The oscillations of the HCS wildly extend to high latitudes.

The solar wind is determined by two source functions near the Sun: the velocity and the magnetic field, which in principle can be estimated from observations. If maps of the coronal holes are available, one can attribute a fast speed to their interiors and a slow speed outside; the magnetic field can also be measured spectroscopically. Just like in Parker's model the velocity near the surface is propagated outwards, in a more realistic approach the MHD equations, with an empirical source term for the energy input in the corona, can be numerically integrated and predict the velocity and the magnetic structure of the solar wind (in particular the shape of the HCS), with the dependence on the phase of rotation and of the solar cycle. It is remarkable that, especially near solar minimum, these numerical models are consistent with observations.

The solar wind plasma is a turbulent medium, with different wave phenomena over a very wide frequency range. Of particular importance are transverse and incompressible Alfvén waves (Sec. 10.4), characterized by the relation (10.48) between the perturbations in the velocity $\mathbf{v}'$ and the magnetic field $\mathbf{B}'$. This relation appears in the data as a correlation between these two vectors and shows which of the two helicities is predominant. The energy spectra of the perturbation energy have been found to approximately follow a power law over several decades of the wave number k, between 5×10^{-8} and 5×10^{-6} km^{-1}; the exponent, especially for a low speed wind, is remarkably close to $-5/3$. This value is the same as in the power law for the energy of isotropic and incompressible turbulence in an ordinary fluid at large Reynolds number (Sec. 7.5, eq. (7.70); Kolmogorov's regime); it is remarkable that this simple law seems to be applicable also in much more complex and inhomogeneous conditions.

Interaction with the interstellar medium. The properties of the interstellar medium are very variable, also due to the energy inputs from nearby stars and supernovæ. The Sun is embedded in a local cloud mainly composed of H and He with the cosmic abundance, a temperature $\approx 7,000$ K, a density ≈ 0.2 atoms/cm^3 and a small ionization fraction of $\approx 5 \times 10^{-2}$; in addition, there is a magnetic field $B_{im} \approx 0.5$ nT. In the rest frame of the Sun, this cloud flows with a supersonic velocity of $V_\odot \simeq 25$ km/s and is separated from the heliosphere by a transition region. The supersonic expansion of the solar wind stops when its kinetic pressure $\varrho v^2/2$ equals the magnetic pressure $P_{im} = B_{im}^2/8\pi \approx 10^{-12}$ dyne/cm^3; referring to the density $n_1 \simeq 5$ cm^{-3} at 1 AU and assuming $v \simeq 400$ km/s constant, this occurs at the distance

$$r_{ts} = \sqrt{\frac{\varrho_1 v^2}{2P_{im}}} \approx 80 \text{ AU} ; \quad (9.13)$$

at this point a (collisionless) *termination shock* abruptly dissipates the kinetic energy of the solar wind in thermal energy, with a compression ratio $\simeq 4$ and a final temperature $\simeq 10^6$ K. On the bow of the heliosphere an *interstellar bow shock* deflects on its sides the flow of the interstellar medium and extends behind in a long, comet-like wake, dragging along also the solar wind; between them, in the *heliopause*, there is a hot plasma in a compressed magnetic field. In gross terms the heliospheric bow shock has a similar structure as in a planetary magnetosphere (Fig. 10.1), but here the inner structure is the solar wind and the pressure of the outer gas is its magnetic pressure, rather than directed kinetic energy. Independent observations locate the termination shock upstream the interstellar flow at 80 – 100 AU, with a variation of ≈ 20 AU over the solar cycle; the distance is smaller after the maximum of solar activity. The spacecraft Voyager 1 and 2 now (in 2002) move in the upstream direction and will cross the termination shock by 2005 (in summer 2002 Voyager 1 was at ≃ 86 AU, moving nearly radially at ≃ 3.6 AU/y). Their simultaneous measurements, together with the observations of Pioneer 10 downstream, will provide important constraints on the interaction between the heliosphere and the solar wind.

Most of the interstellar medium is neutral and penetrates undisturbed inside the heliosphere, providing important evidence about the galactic environment of the solar system. However, some atoms (mainly H and He) get ionized by UV radiation and charge exchange, and are trapped in the interplanetary magnetic field; they show up in velocity space as a nearly isotropic distribution, which bears trace of the temperature and the velocity of the interstellar medium, superimposed to solar wind ions largely displaced in the radial direction from the origin. *Anomalous cosmic rays* are observed in the solar system with a broad spectrum peaked at 100 MeV (for protons), up to more than 1 GeV; since their density does not decrease with the distance, an interstellar origin is likely. One possibility is that they are neutral atoms ionized in the transition region, and subsequently accelerated in the turbulent termination shock.

– PROBLEMS –

9.1.* Evaluate the angular momentum and the rotational energy of the Sun in the following cases: (i) rigid rotation and uniform density; (ii) uniform density and the rotation law (9.2); (iii) uniform rotation and a central core, with the second model of Problem 1.2.

9.2. Evaluate the illumination of a point of the Earth during a total eclipse.

9.3. Evaluate the magnetic Reynolds number in the solar wind.

9.4. Generalize eq. (9.12) for the motion of lines of force to a generic velocity $v(r)$.

9.5. Evaluate the energy flux from the Sun needed to support the coronal loss to the wind.

9.6. Find the Cartesian equation of Archimedes' spiral in three dimensions.

9.7.* Model the ejection of material from the volcanoes of Io. Assuming a free flow under gravity, find the flux of escaped material per unit surface at a given temperature.

9.8. For sake of illustration, in Sec. 9.2 we considered isothermal solar wind flow; derive the corresponding results for a polytropic flow with $P \propto \varrho^\gamma$ (see eq. (1.33)). Show, in particular, that the critical distance

r_c and velocity v_c of the sonic point are given by

$$r_c = \frac{5-3\gamma}{4(\gamma-1)} R_\odot, \quad v_c = \frac{v_{\text{esc}}}{2}\left(\frac{R_\odot}{r_c}\right)^{1/2} = \left[\frac{4(\gamma-1)}{5-3\gamma}\right]^{1/2} \frac{v_{\text{esc}}}{2}.$$

9.9.[*] Study the linear perturbations of the stationary solar wind solution (Sec. 9.2).

9.10.[*] Derive formally eq. (9.12) for the tilt between the interplanetary magnetic field and the radial direction. In the inertial space assume a purely radial and stationary wind and the magnetic field **B** constrained by the Maxwell's equations $\nabla \cdot \mathbf{B} = 0$ and $\nabla \times \mathbf{E} = 0$, with $\mathbf{E} = -(\mathbf{v} \times \mathbf{B})/c$ (eq. (1.76)).

– FURTHER READINGS –

An introductory text on the Sun and its activity is R.J. Tayler, *The Sun as a Star*, Cambridge University Press (1997); P. Lantos, *Le Soleil en Face*, Masson (1997) gives a good overview, including the interrelations between solar activity and magnetospheric variations. Fundamental issues of the activity of the Sun, and the relation with magnetic field, are discussed in E.R. Priest, *Solar Magnetohydrodynamics*, D. Reidel (1984); B.C. Low, Coronal mass ejections, magnetic flux ropes, and solar magnetism, *J. Geophys. Res.* **106**, 25,141 (2001) is a good recent review of the same topic. E.N. Parker, Dynamical Theory of the Solar Wind, *Sp. Sci. Rev.* **4**, 666 (1965) is an excellent early exposition of the theory of the solar wind; more recent and complete is *Cosmic Winds and Heliosphere*, J.R. Jopkii, C.P. Sonett and M.S. Giampapa, eds., The University of Arizona Press (1997). For the corona, besides M. Kuperus and J.A. Ionsons, On the theory of coronal heating mechanisms, *Ann. Rev. Astron. Astrophys.* **19**, 7 (1981), see also L. Golub and J.M. Pasachoff, *The Solar Corona*, Cambridge University Press (1997). For this and the following chapter we suggest *Introduction to Space Physics*, M.G. Kivelson and C.T. Russell, eds., Cambridge University Press (1995). A number of special issues of the journal *Space Science Reviews* have been devoted to the Sun, its variability, large and small scale magnetic field and heliosphere. On the turbulence of the solar wind and non-linear Alfvén waves, see the comprehensive review by C.Y. Tu and E. Marsch, MHD structures, waves and turbulence in the solar wind: Observations and theories, *Sp. Sci. Rev.* **73**, 1 (1995).

Chapter 10

MAGNETOSPHERES

A planetary magnetic dipole acts as an obstacle to the supersonic flow of the solar wind. Upstream, on the bow of this obstacle, a shock wave develops; on the stern, a wake arises, which extends to a very large distance downstream. Solar wind particles can penetrate the magnetic dipole cavity and reach the ionosphere through the polar regions. This complex interaction is highly variable: there is a seasonal effect due to the changing angle between the magnetic axis and the solar wind; the interplanetary magnetic field may invert its direction, triggering reconnection of lines of force, which results in major magnetospheric perturbations; solar activity produces gusts in the wind. The study of planetary magnetospheres is mainly based on spacecraft measurements that use in situ instrumentation, radio propagation and active experiments which affect the state of the plasma. In this chapter, three topics are selected for more detailed discussion, with particular reference to the Earth: the motion of particles in an intense magnetic field in the guiding centre approximation and their trapping in the magnetosphere; the upstream bow shock, which deflects the solar wind on the sides; and Alfvén waves, a generalization of acoustic waves. The magnetospheres of other planetary bodies are reviewed in the last Section.

10.1   The solar wind and the magnetosphere

The *magnetosphere* is the magnetic dipole region as modified by the solar wind (Fig. 10.1). A dipolar magnetic field at a distance r from the centre of the planet of radius R is characterized by a pressure (eq. (1.79))

$$\frac{B^2}{8\pi} \approx \frac{B_e^2}{8\pi}\left(\frac{R}{r}\right)^6 , \qquad (10.1)$$

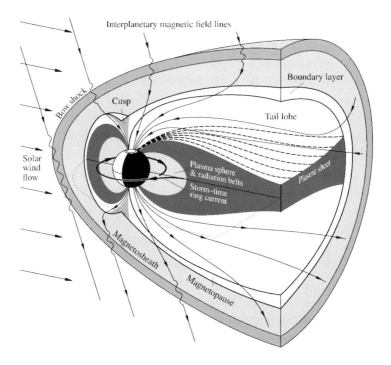

Figure 10.1. A sketch of the Earth magnetosphere, as modified by the solar wind, at magnetic equinox, when the dipole is orthogonal to its direction. Under its powerful kinetic pressure the dipole is compressed upstream and deformed into a long tail downstream. On the bow the flow is deflected sideways and slowed down through a shock; some of its particles are able to penetrate into the ionosphere through the open lines of force in the polar regions.

where $B_e = d/R^3$ is the ground equatorial value for a dipole d. When the wind has penetrated deep enough, so that its kinetic pressure $\varrho_w v^2/2$ is about equal to the magnetic pressure, it behaves as a supersonic flow encountering a soft obstacle; this happens at the *stand-off distance*

$$r_s \approx R \left(\frac{B_e^2}{4\pi \varrho_w v^2} \right)^{1/6} = \frac{d^{1/3}}{(4\pi \varrho_w v^2)^{1/6}} , \qquad (10.2)$$

for the Earth about $10 R_\oplus$.

The perturbation induced upstream by the obstacle generates the *bow shock*, a transition layer behind which the wind is hotter (reaching 5×10^6 K), subsonic and deflected sideways. Thermodynamically, it is characterized by a transformation of directed kinetic energy in random thermal energy, hence by an increase in entropy; but this is not due to collisions, as in ordinary shocks, but to small-scale, non-linear wave-wave interactions which scatter particles at random in velocity space. Downstream, the wind flows around the flanks, along the *magnetopause*, the boundary of the inner magnetic cavity from which the main flow is excluded. The region between the bow shock and the magnetopause – the *magnetosheath*, 2 or 3 Earth's radii thick – it is characterized by strong turbulence and a magnetic field smaller than inside the magnetopause. Inside, in the sunward direction, the dipolar lines of force are generally compressed; however, below about $5\,R_\oplus$ the magnetic field is smaller than the dipole value, by about $\delta B \approx 0.3$ nT or more, due to the *ring current*, to be discussed in the next section. Downstream the lines of force are stretched to very large distances in the *magnetotail*.

The lines of force originating from the magnetic poles, when reaching the magnetopause, "split", so to speak, in opposite directions; at these *neutral points*, lying at the magnetic latitudes $\approx \pm 77°$ and noon when the Sun is at equinox and the local meridian, the magnetic field vanishes. They are the main entry regions for the solar wind; energetic particles, travelling along the lines of force of the *polar cusps*, can reach the polar regions. In this way the ionosphere is directly connected with the solar wind and responds to its variability. In these regions the magnetic structure is complicated by the Earth rotation, which drags around the lines of force. Once penetrated in the magnetic cavity, solar wind particles can eventually reach a region where the dipole field is strong enough; they become bound to a line of force and are trapped in a periodic motion between two points of opposite magnetic latitude. In this way they fill up the *Van Allen belts* and may create safety problems for astronauts in extra-vehicular activity. Some plasma of solar origin is found in a boundary layer below the magnetopause; it transports inside the tail a large amount of momentum.

The magnetotail of the Earth has been observed up to $250\,R_\oplus$, with a diameter of about $5\,R_\oplus$ and a slow expansion. Just like in ordinary wakes in water, vortices with dimensions of several Earth radii develop in the tail, probably due to the Kelvin-Helmholtz instability. Since in the magnetotail the lines of force originate either from the North or the South pole, corresponding to opposite directions, there is a transition region (the *plasma sheet*) where the magnetic field vanishes.

Magnetic reconnection. If opposite field lines coalesce, magnetic energy is lost and made available for heat and particle motion. This is a widespread and often stunning process in astrophysics and space physics, which affects the global topology of field lines. Although its details are complex and still only partially understood, it can be qualitatively described with the simplest model of a planar *current sheet*, in which the magnetic field $\mathbf{B} = B(z)\,\mathbf{e}_x$, directed along x, vanishes at $z = 0$ and tends to $\pm B_\infty$ at large $|z|$ (like $B = B_\infty \tanh(z/\ell)$; Problem 10.14). It is supported by a current layer of thickness ℓ:

$$J_y = \frac{c}{4\pi}\frac{dB}{dz}\,; \tag{10.3}$$

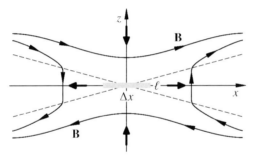

Figure 10.2 Magnetic reconnection in a current sheet in the Sweet-Parker model. As a consequence of the magnetic dissipation in the small shaded box, the plasma flows in, pushing in the lines of force, and escapes to the right and the left (fluid velocities are indicated with heavy arrows). Reconnection occurs by means of shocks (dashed lines in an X-shape) through which the magnetic field changes direction.

the plasma is at rest, balanced by the fluid pressure

$$P(z) = P(0) - \frac{B^2(z)}{8\pi} ,\qquad (10.4)$$

with a maximum at $z = 0$. This configuration is unstable due to magnetic dissipation, which, according to eq. (1.82), is largest at $z = 0$, where the field gradient has its maximum. The disappearance of the field there creates a pressure unbalance, which drives plasma in; with a large conductivity this flow carries along the outer magnetic field, which decreases ℓ and increases the dissipation; above and below the reconnection box the lines of force are pushed in. This occurs in a limited stretch of length Δx along the direction x; outside this stretch the lines of force on the opposite side of the plane $z = 0$ reconnect, eliminating the vanishing field. To conserve mass, the incoming flow from large $|z|$ is squeezed along x, on both sides, and pushes out the reconnected lines of force (Fig. 10.2). This rough model, due to P.A. SWEET and E.N. PARKER, is not very effective when, as usual, the asymptotic value of the magnetic Reynolds number Re_m (6.20) is very large and a very small thickness ℓ is required to have dissipation.

H.E. PETSCHEK has pointed out that a more efficient reconnection can occur outside the small dissipation box (of sizes ℓ and Δx) by means of shock waves. The simple fluid dynamics shock described in Sec. 10.5 can be easily be generalized to the case in which a magnetic field is present; since the flux coming out a small cylinder orthogonal to the shock surface must vanish, the normal component of **B** is constant. The tangential component, however, can change through the transition, supported by a current layer. Petschek considered in the (x, z) plane four shocks, in the shape of an X, issuing from the central dissipation region near the origin and making a small angle δ with the x-axis. Far from the central box, because of the non-vanishing value of δ, the large magnetic field just outside the shock can be turned into the z-direction and reconnects with an opposite line of force on the other side; similarly, the direction of the flow is turned by 90°. Shocks of this kind have in fact been detected in the magnetotail. In fact, the process generally has a fast time variability and often occur in bursts, with the generation of fast particle streams.

Magnetic reconnection is the driving mechanism for the generation of fast particle streams in solar flares. In the magnetosphere it plays an essential role in *magnetic storms* and *substorms*. Storms typically arise when a fast blob of solar plasma (as in a coronal mass ejection; Sec. 9.1) impinges upon the bow shock and compresses the day-side magnetosphere, with a sudden increase of the field at the ground. The magnetotail is also affected and reconnection may ensue, producing an enhanced ring current and a decrease of the ground intensity; the disturbance decays in a few days. This process is crucial for the energy transfer of the solar wind to the ionosphere.

Substorms are produced when the interplanetary magnetic field becomes antiparallel to the magnetospheric dipole field at the bow shock; they last for about an hour and their disturbances are most evident in the polar regions. The initial reconnection on the day side releases field lines, which enhance the magnetotail field and produce new reconnecting "X transitions" at $\approx 15\,R_\oplus$; bursts of energetic particles, with velocities up to $1,000$ km/s, are produced, which impinge upon the night side, producing auroral substorms, with the timescale of tens of minutes. Entering the ionosphere near the poles, they excite neutral molecules and give rise to the *auroral glow*, concentrated in an oval around the pole and turning with the Sun; they also affect the complex electrodynamics in the polar regions. Both processes produce strong fluctuations in the ionospheric plasma density there and underline the strong dependence of the electromagnetic propagation (including radio communications) on solar activity.

The magnetosphere has a large number of oscillation modes. Many of them show up in the ground magnetic field and are called *micropulsations*, with amplitudes below $10^{-4}\,B_e$ and periods from 0.2 s to 10 minutes. The basis for their understanding is the plasma dispersion relation (Sec. 8.3); cyclotron resonances play an important role. At lower frequencies there are magnetic hydrodynamical modes, in particular Alfvén waves (Sec. 10.4).

Quantitative diagnostics of magnetospheric activity. Changes in the magnetic field have been routinely observed and recorded for more than a century with an array of ground stations. The main part of the variation has the obvious period of a day, with an amplitude sensitive to seasonal effects and to long-term solar activity; there is also a smaller component with the lunar day period (nearly 25 d), probably due to gravitational tides. Faster variations, mainly due to storms, are obtained by subtracting out slow components; they change much with latitude and longitude. A great effort has been devoted to empirically correlate ground magnetic measurements with solar activity, notwithstanding our poor understanding of the mechanisms involved. The traditional way summarizes, for each station, the fast variability of the magnetic field with an integer between 0 and 9 (the K index, from the German *Kennziffer*), corresponding to

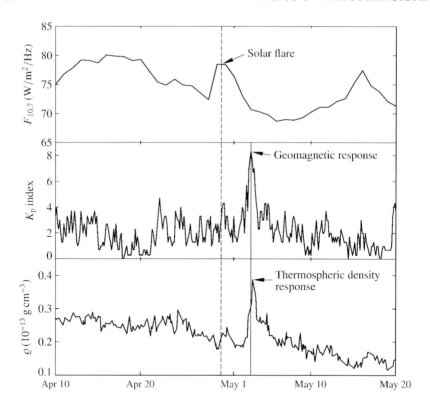

Figure 10.3. Effect of a solar flare on the Earth magnetosphere and thermosphere. Top: Daily values of the solar radio flux at 10.7 cm (the $F_{10.7}$ index, Sec. 9.1). Middle: the geomagnetic index K_p. Bottom: the thermospheric density ϱ measured with a very sensitive accelerometer on board of the French satellite CASTOR in April and May 1976, near the minimum of the solar cycle. The data, obtained near perigee, have been referred to the same altitude of 275 km and the local time 18 h ("evening"). The arrows indicate the major disturbance; note the $\simeq$ 3 days delay between the flare, as seen in the radio data, and the magnetosphere disturbance; this corresponds to the velocity $\approx$ 500 km/s of the equatorial solar wind. The thermospheric response follows with a further delay of about 4.5 h.

a range in the average of the intensity over 3 hours; for each integer this range is approximately double the preceding one. Since the ground response statistically increases with latitude, in order to give a uniform measure of the solar influence, different ranges are adopted for different latitudes; for the Niemegk Observatory, for example, at the latitude 48.8°, the lowest interval $K = 0$ is 0 – 5 nT and the lower bound for the top interval $K = 9$ is 500 nT. At latitudes $> 60°$ the response increases much faster and is less reliable. A number with a planetary significance, the K_p index, also an integer between 0 and 9, is obtained with a suitable averaging over a set of 13 stations. A_p is constructed with the same data, but it is expressed in a linear scale and averaged over a day.

In recent times other better indexes have been introduced, but K_p is still much used. Several correlations with the solar activity have been empirically found; for example, the solar wind speed fulfils the interesting relation

$$v = (8.44 \pm 0.74)\,\bar{K}_p + (330 \pm 17) \text{ km/s}, \qquad (10.5)$$

where $\bar{K}_p$ is the average of K_p over a day.

Space measurements. Our current understanding of the magnetosphere and its interaction with the solar wind is mainly the result of extensive spacecraft measurements, in particular with: electrostatic detectors to determine the density and the velocity distribution of the different kinds of charged particles, sensitive three-dimensional magnetometers, and radio receivers to detect plasma waves. Since the plasma is not in thermal equilibrium, it is important to measure the full three-dimensional distribution in velocity space, with severe requirements on the bit rate. The work done by the International Sun-Earth Explorer (ISEE 1 and 2) satellites is particularly remarkable.

Aside from the general morphological description, an important part of this work requires the identification of the physical nature of the wave phenomena. They appear at the spacecraft as an oscillating quantity with frequency (ω − $\mathbf{k} \cdot \mathbf{v}$), where ω is the frequency in the rest frame of the plasma and $\mathbf{v}$ is the velocity of the spacecraft in this frame. In order to get the frequency ω we need to know the wave vector $\mathbf{k}$; with two spacecraft separated by $\Delta \mathbf{r}$ the relative phase $\mathbf{k} \cdot \Delta \mathbf{r}$ can be measured, leading to the dispersion relation $\omega(\mathbf{k})$ and the identification of the mode. ESA's mission Cluster, launched in summer 2000, consists of four satellites capable of these correlated measurements. During the mission they keep a tetrahedron disposition, allowing the determination of the components of the wave vector along the sides, hence its full value. The orbits of all members of the cluster have an apogee at $20\,R_\oplus$, outside the bow shock or, six months later, in the magnetotail; and a perigee at $4\,R_\oplus$ in the inner magnetosphere. Thus it allows a thorough and significant coverage of magnetospheric wave phenomena.

Active experiments have also been performed, e.g. by releasing in space small quantities of easily ionizable material and observing its diffusion. Flying long (≈ 5 km or more) conducting wires connecting two (*tethered*) satellites has provided an interesting means to produce a large electromotive force (of the order of $v_{\text{orb}} B/c \approx 0.2$ V/m), due to their orbital motion through the dipole field. The current carried by the wire closes through the magnetospheric plasma. The corresponding Lorentz force produces a force along the motion and can lower or raise the system, providing an interesting and natural propulsion system. A similar effect is probably at work in the innermost Jupiter's large satellite, Io (Sec. 10.7).

10.2 Motion of charged particles in a strong magnetic field

The analysis of magnetospheric plasma dynamics is complicated because the appropriate physical description is neither magnetohydrodynamics, where one deals with mean quantities like pressure or bulk velocity, nor a medium where trajectories of individual particles are independent. It is, however, interesting to gain some insight by treating these limiting (and not exactly satisfied) models. We start with the motion of individual particles in given force fields, in particular electromagnetic.

Motion in a constant and uniform electromagnetic field. When the magnetic field is *constant and uniform* a charged particle moves in the plane orthogonal to the line of force with a circular motion around the *guiding centre* and is accelerated by the parallel electric field along the line itself. With these simplifying assumptions, the split in accelerated motion along the magnetic field lines and gyration about the drifting guiding centre is an *exact* solution. When some of the previous assumptions are weakly violated, this solution is still useful as the lowest level description; this is the gist of the *guiding centre approximation*, anchoring the motion to a line of force in a helical trajectory with a drift of the gyration centre.

In the motion of a particle with charge q and mass m,

$$m\frac{d\mathbf{v}}{dt} = q\left(\mathbf{E} + \frac{\mathbf{v}}{c} \times \mathbf{B}\right), \qquad (10.6)$$

only the electric field produces work; the magnetic field only changes the direction of the velocity[1].

With a uniform and constant electromagnetic field, the motion along the lines of force is determined by the electric field $E_\parallel = \mathbf{E} \cdot \mathbf{B}/B$:

$$m\frac{dv_\parallel}{dt} = qE_\parallel . \qquad (10.7)$$

The suffixes $_\parallel$ and $_\perp$ denote respectively the components along, and parallel to, the magnetic field. For the perpendicular motion, consider a frame of reference moving with the *drift velocity*

$$\mathbf{v}_D = c\,\frac{\mathbf{E} \times \mathbf{B}}{B^2}, \qquad (10.8)$$

---

[1] It is interesting to note that when there is no electric field, the relativistic generalization of the equations of motion is simple. In the definition of momentum $\mathbf{p} = m\mathbf{v}$ we must use the relativistic mass

$$m = \frac{m_0}{\sqrt{1 - v^2/c^2}} = \gamma\, m_0 .$$

Since the magnetic force does not perform work the magnitude of the momentum, hence the relativistic factor γ, is unchanged and it all boils down to the new meaning of m, in particular in the cyclotron frequency Ω_c (10.10). This is relevant in the study of the magnetic deflection of cosmic rays.

10. Magnetospheres

which is orthogonal to the magnetic field and does not depend on $E_\parallel$. In this frame the perpendicular velocity $\mathbf{u} = \mathbf{v}_\perp - \mathbf{v}_D$ obeys the simpler equation

$$\frac{d\mathbf{u}}{dt} = \frac{q}{mc} \mathbf{u} \times \mathbf{B} . \tag{10.9}$$

This describes a circular and uniform motion **B** with the angular velocity

$$\Omega_c = \frac{qB}{mc} , \tag{10.10}$$

the *cyclotron frequency* (in order to gain formal simplicity we allow this frequency and, later, the magnetic moment, to have the sign of the charge). The centre of the circle is the *guiding centre* and

$$r_L = \frac{u}{|\Omega_c|} , \tag{10.11}$$

the *Larmor radius*, is its radius. Indeed, from eq. (10.9) the acceleration is normal to the orbit and the speed u is constant. Introducing the tangent unit vector $\hat{\mathbf{u}} = \mathbf{u}/u$ and the arc length s along a line of force, we have

$$u \left| \frac{d\hat{\mathbf{u}}}{ds} \right| = \frac{|q\mathbf{u} \times \mathbf{B}|}{mc} = \frac{|q|uB}{mc} ,$$

which means that the radius of curvature of the orbit is constant and is given by eq. (10.11). When $q > 0$ the circulation is clockwise for an observer who stands along the line of force. The overall motion is helical, with a pitch angle χ (the constant angle between the velocity vector and the magnetic field) given by (Fig. 10.4)

$$\tan \chi = \frac{u}{v_\parallel} = \frac{r_L |\Omega_c|}{v_\parallel} ; \tag{10.12}$$

When there is an additional constant force **F** (like gravity $\mathbf{F} = m\mathbf{g}$), the drift velocity of the gyration centre (10.8) becomes

$$\mathbf{v}_D = c \left(\mathbf{E} + \frac{\mathbf{F}}{q} \right) \times \frac{\mathbf{B}}{B^2} . \tag{10.8'}$$

If the force **F** is not proportional to the charge an electric current is generated; note that drift due to the electric field **E** is the same for both positive and negative charges and thus does not result in a current.

The diamagnetic properties of a plasma become apparent upon noting that a single gyrating charge, with angular momentum $L = mu\, r_L$, generates a magnetic dipole

$$d = \frac{q}{2mc} L = \frac{1}{2B} mu^2 ;$$

Getting a deeper insight.– A more accurate analysis is required to assess the limits of this approximation. It turns out that, in terms of the smallness parameter

$$\epsilon = \frac{1}{\Omega_c^2} \frac{d\Omega_c}{dt}, \qquad (10.21)$$

in a period we have $\delta\mu_m/\mu_m = O(\epsilon^2)$. The proof requires the methods of *singular perturbation theory* to deal with a system of ordinary differential equations, whose solutions are not analytic functions of the small parameter ϵ, as it tends to zero. This is equivalent to the limit $\Omega_c \to \infty$, all other quantities being fixed; the radius of the Larmor circle tends to zero. It is interesting to note that the expression (10.20) of the magnetic moment can be corrected to yield a quantity which is constant to an arbitrary order in ϵ.

The magnetic moment is an *adiabatic invariant*, referring to the slow change of the external magnetic field. This result has a close analogy in the theory of a harmonic oscillator with a slowly varying frequency[2]

$$\ddot{x} + \Omega^2(t) x = 0. \qquad (10.22)$$

If the corresponding ϵ (the fractional change of the frequency Ω in a period, like in (10.21)) is small, the quantity

$$\mu_m = \frac{\text{average kinetic energy}}{\text{frequency}} = \frac{m \overline{\dot{x}^2}}{2\Omega} \qquad (10.23)$$

is indeed an adiabatic invariant (see Problem 10.5).

The appropriate mathematical model for adiabatic invariance is a Hamiltonian quasi-periodic system. With one degree of freedom, consider the *action*

$$S = \oint dx\, p = 4\pi \mu_m, \qquad (10.24)$$

i.e., the area in the phase plane (x, p) enclosed in a period. It can be shown that when the Hamiltonian varies slowly with time, S is adiabatically invariant. An application of this theorem concerns an elliptic Keplerian motion with a slowly varying central mass: the ratio of the total energy to the orbital frequency is an adiabatic constant (see Problem 10.5). Incidentally, the theorem provides also the necessary justification for the Bohr-Sommerfeld quantization rule, which requires the action to be a multiple of Planck's constant. This rule would be

---

[2] The connection between eq. (10.9), with a time-varying magnetic field along the z direction, with the harmonic oscillator, is obtained by introducing $w = u_x + i u_y$; it fulfils

$$\ddot{w} + \Omega_c^2(1 + i\epsilon) w = 0.$$

10. Magnetospheres

inconsistent if small changes of the external conditions affect the action: only discrete jumps in S are allowed by quantum mechanics.

Drift of the gyration centre in a non-uniform magnetic field. The general theory of the motion of the guiding centre is based upon the splitting of the position of the particle

$$\mathbf{r}(t) = \mathbf{r}_0(t) + \mathbf{r}_L(t) \tag{10.25}$$

in the position of the centre $\mathbf{r}_0(t)$ and the Larmor gyration vector $\mathbf{r}_L(t)$, which describes a small circular motion of radius r_L and with frequency Ω_c around $\mathbf{r}_0$. The magnetic field is expanded around $\mathbf{r}_0$ in powers of r_L and the equation of motion is averaged over a gyration period.

Let us now follow the motion of a gyration centre along a line of force, with arc length s. Since the magnetic field does not perform any work, the change in the orthogonal kinetic energy must be compensated by a change in the parallel kinetic energy

$$m v_\parallel \frac{dv_\parallel}{ds} = m \frac{dv_\parallel}{dt} = -m u \frac{du}{ds} = -\mu_m \frac{d\Omega_c}{ds} = -\frac{mu^2}{2B}\frac{dB}{ds} = F_{\text{eff}} \; . \tag{10.26}$$

We have an effective parallel force (independent of the charge) F_{eff} and an effective potential energy

$$U_{\text{eff}}(s) = \mu_m \Omega_c(s) \; . \tag{10.27}$$

The quantity

$$\mathcal{E}_{\text{eff}} = \frac{1}{2} m v_\parallel^2 + \mu_m \Omega_c(s) \tag{10.28}$$

is adiabatically conserved during the motion.

We consider now the case in which there is no electric field and the magnetic field is a generic, slowly varying function of space. As the gyration centre is dragged along a curved line of force it is subject to a centrifugal force along its radius of curvature

$$\mathbf{F}_{c\perp} = -m v_\parallel^2 \frac{d}{ds}\left(\frac{\mathbf{B}}{B}\right) = -m v_\parallel^2 \frac{1}{B^2}\left[(\mathbf{B}\cdot\nabla)\mathbf{B}\right]_\perp \; . \tag{10.29}$$

There is also a correction to the Lorentz force $q\,\mathbf{u}\times(\mathbf{r}_L\cdot\nabla)\mathbf{B}/c$ due to the fact that in its gyration the particle feels a changing magnetic field; this expression, in which $\mathbf{u}$ and $\mathbf{r}_L$ change rapidly and sinusoidally with the cyclotron frequency, must be averaged over a period, with the result

$$\mathbf{F}_{L\perp} = -\frac{q}{2c} u\, r_L \nabla B \; . \tag{10.30}$$

The additional drift velocity reads (eq. (10.8′))

$$\mathbf{v}_D = \frac{mc}{qB^3}\left(\frac{1}{2}u^2 + v_\parallel^2\right)\mathbf{B}\times\nabla B \; . \tag{10.31}$$

This drift changes sign with the charge and generates an electric current. To obtain the current density, note that in thermal equilibrium the averages of $mu^2/2$ and $mv_\parallel^2$ are both kT; summing over all particles in a unit volume we get

$$\mathbf{J}_D = \frac{2cP}{B^4} \mathbf{B} \times (\mathbf{B} \cdot \nabla)\mathbf{B} . \tag{10.32}$$

With an isotropic velocity distribution this current cancels exactly the second term of the diamagnetic current (10.14), so that the total current is

$$\mathbf{J} = \mathbf{J}_D + \mathbf{J}_{\text{dia}} = c \frac{\mathbf{B}}{B} \times \nabla P \tag{10.33}$$

and does not depend on the gradient of $\mathbf{B}$.

On the Earth, this is the *ring current*; for a dipole field with a pressure P growing inward it is directed along a parallel, in the clockwise direction for an observer standing on the North pole. This toroidal current generates a poloidal magnetic field which, with an "antidynamo process", diminishes the dipole field by an amount of order $4\pi P/B \approx \beta B/2$. In effect, this simple description hides the fact that what counts is the particle energy density, which may be mainly determined not by the bulk of the particles, but by their high energy tail. During a *magnetic storm*, when a gust in the solar wind injects more energetic particles into the magnetosphere, the ring current increases and the decrease δB in the magnetic intensity is larger. If the storm is slow enough, the adiabatic invariant (10.20) remains constant and the particle energy in the magnetosphere decreases. Collisions and field inhomogeneities govern the decay of the phenomenon.

10.3 Trapped particles

Due to the constancy of the magnetic moment (10.20), the parallel motion of a charged particle in the dipole field slows down as it moves to higher latitudes, until $U_{\text{eff}} = \mathcal{E}_{\text{eff}}$ and the parallel velocity vanish (see (10.28)): we have there a turning (or mirror) point and the possibility of trapping particles (Fig. 10.5). Since the total speed is constant, it is convenient to describe the effect by means of the pitch angle (10.12). From eq. (10.20) we have

$$\sin^2 \chi = \sin^2 \chi_0 \frac{\Omega_c}{\Omega_{c0}} , \tag{10.34}$$

where the index $_0$ refers to the magnetic equator. The turning point is at the point s on the line of force where

$$\Omega_c(s) \sin^2 \chi_0 = \Omega_{c0} . \tag{10.35}$$

10. Magnetospheres

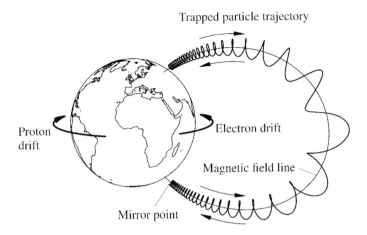

Figure 10.5. As particles approach the ionosphere, where the field is stronger, they have a smaller Larmor radius and a smaller longitudinal velocity; eventually, their motion along the line of force may be turned around. The inhomogeneity of the magnetic field makes them drift around along the parallels in opposite directions.

The largest magnetic field on a given line of force occurs at the top of the ionosphere, where the motion is hindered by collisions and the theory discussed in this section does not apply. If the index $_1$ refers to this maximum, the population of trapped particles is determined by the relationship

$$\sin \chi_0 > \sqrt{\frac{B_0}{B_1}} = \sin \chi_1 . \qquad (10.36)$$

The particles inside χ_1 (the *loss cone*) escape into the ionosphere. This picture, however, does not hold for energetic particles which violate the adiabatic condition (10.15). Notice also that this effect destroys any isotropy in velocity that the population may have; they are far from thermal equilibrium. Those that escape into the polar ionosphere induce, by collisional excitation of the neutral molecules, the auroral glow.

To have a rough idea of the numbers involved, note that – as we shall see later – the kinetic energy of the particles must be appreciably less than that of the trapping field, about 8.4×10^{24} erg (see Problem 10.1); indeed, this energy is about 6×10^{22} erg. Aurorae have a widely varying energy deposition rate, say $\simeq 3 - 3,000$ erg cm^{-2}s^{-1} over $\simeq 0.1$ sterad, corresponding to a power of 10^{18} to 10^{21} erg/s. Therefore the lifetime of trapped particles ranges from $6,000$ s to 6×10^6 s.

The second adiabatic invariant. Another application of the adiabatic invariants concerns the case in which the field is not quite axially symmetric, or changes slowly with time. A particle is turned around in its parallel motion

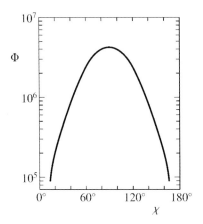

Figure 10.6 A typical profile of the proton flux Φ in particles per cm^2 s sterad as a function of the pitch angle χ for energies of the order of 1 MeV.

at the point given by eq. (10.35); but as the magnetic field experienced by the particle changes because of time variations or the drift of its gyration centre, the effective energy (10.28) cannot be considered as constant any more. In the lowest approximation we have a periodic motion along the line of force with period

$$P = 2 \int_{s_1}^{s_2} \frac{ds}{v_\parallel(s)} , \qquad (10.37)$$

where s_1 and s_2 are the solutions of eq. (10.35). For $\simeq 0.1$ MeV electrons in the inner radiation belt (Fig. 10.7), the corresponding bounce period is about 0.1 s; this is a timescale appreciably longer than the gyration period. If the fractional change of $\Omega_c(s)$ in this time is small, we can apply the general theory of adiabatic invariants, which states that the action (10.24) or, to within a factor,

$$S = \int_{s_1}^{s_2} ds\, v_\parallel(s) , \qquad (10.38)$$

is, to a good approximation, constant in time; S is usually referred to as the *second (longitudinal) adiabatic invariant*. Using the value $B_1 = B_0/\sin^2\chi_0$ of the magnetic intensity at the turning points in terms of the pitch angle χ_0 at the equator, from the constancy of the magnetic moment we get

$$v^2 - v_\parallel^2(s) = v^2 \frac{B(s)}{B_1} , \qquad (10.39)$$

so that

$$S = v \int_{s_1}^{s_2} ds\, \sqrt{1 - \frac{B(s)}{B_1}} . \qquad (10.40)$$

For example, consider particles of the solar wind entering the equatorial region during the day. Their pitch angle χ_0 determines the turning point. The solar wind compresses the equatorial magnetic field, but leaves the field more

10. Magnetospheres

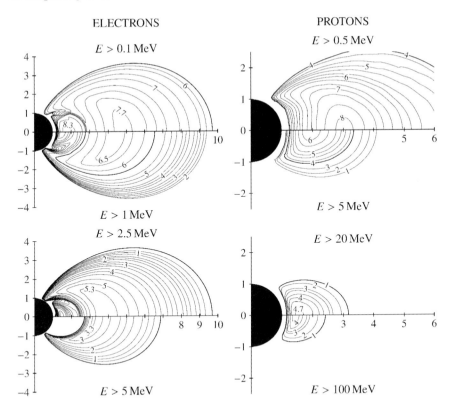

Figure 10.7. Mean isoflux contours in the meridian plane for electrons and protons trapped in the Van Allen belts for different energy thresholds. Distances are measured in $R_\oplus$; the labels on the contours give the logarithm of the flux in particles/cm^2/s. More energetic particles penetrate nearer to the Earth, as expected. Electrons have two maxima, one in the inner and one in the outer belt; protons have a single maximum. The data, from the AP-8 and AE-8 models at solar maximum, are available at http://nssdc.gsfc.nasa.gov/space/model/magnetos/radbelt.html. They take into account large satellite data sets, but do not show the fast response to variations of the Earth magnetic field.

or less unchanged at high latitudes; the velocity v is also unchanged. When night comes the field lines expand and the equatorial pitch angle must change in such a way as to keep the action S constant. One can see from simple examples (Problem 10.3) that the equatorial pitch angle grows with the field, so that at night time particles are able to penetrate deeper into the upper atmosphere at high magnetic latitudes.

Van Allen belts of trapped particles. The belts of high energy particles in the inner Earth magnetosphere were discovered by J.A. VAN ALLEN during the first spacecraft flights in the 1960s and named after him; they have been continuously monitored since then, also for the hazard they pose to astronauts in extra-vehicular activity. These particles, with energies above 100 keV and up to hundreds MeV, are trapped by the relatively undisturbed dipole field in two low latitude regions. There is no clear distinction between these particles and those in the ring current, to which they contribute at low latitudes, where atmospheric losses are small. The Van Allen belts consist of two main regions (Fig. 10.7): in the inner belt, at about $L \simeq 1.5 - 2 R_\oplus$ (in terms of McIlwain's coordinates; see Sec. 6.1), there are electrons up to a few MeV, but protons attain hundreds MeV; in the outer belt, extending up to $L \simeq 4-5 R_\oplus$, protons and electrons reach a few MeV. The ring current, with much larger particle densities, is dominated by oxygen ions below $\simeq 10$ keV. Recent measurements by the SAMPEX satellite on a near-polar orbit showed that the flux of inner belt protons appears to have a solar cycle variation anti-correlated with the sunspot number; the outer radiation belt electrons vary also on semiannual and solar rotation timescales.

At low energies the ionosphere and the solar wind are obvious sources; acceleration processes may be at work in the magnetosphere, in solar flares and in interplanetary shocks associated with high speed solar wind streams emanating from persistent coronal holes near the solar minimum (thus explaining the anti-correlation of the trapped-particles flux and the sunspot numbers). During geomagnetic storms the energies and composition of the belts and the ring current change, a clear signal of their origin. The largest energies, however, require other sources. The remarkable, highest energy protons in the inner belt could be due to neutrons in the secondary cosmic rays produced in the atmosphere by the primary component; subsequently they can generate high energy protons, either by charge exchange, or by spontaneous β-decay in a proton, an electron and an antineutrino. If these protons are produced in the right place with the right velocity, they remain trapped for a long time.

10.4 Magnetohydrodynamic waves

Besides high frequency electron plasma waves, discussed in Ch. 8, at low frequency we have *magnetohydrodynamic waves*, in which the ion dynamics plays an essential role. They are very important in planetary magnetospheres and the solar wind. A background magnetic field B_0 modifies the acoustic waves and introduces two new modes. Shear *Alfvén waves* arise because the stress tensor of the magnetic field has, besides a scalar component which acts like a pressure, a component along the lines of force of opposite sign, which has the effect of stretching them and allowing the propagation of a transverse mode. Its velocity is *Alfvén speed* (1.81)

$$V_A = \sqrt{\frac{B_0^2}{4\pi\varrho_0}} . \tag{10.41}$$

10. Magnetospheres

A simple and exact description of this mode, valid also for large amplitudes, can be gained from the MHD system equations for infinite conductivity (1.85), assuming that all variables depend only on the time t and the space coordinate $z = \mathbf{r} \cdot \mathbf{n}$ along the (constant) unit vector $\mathbf{n}$; moreover, the magnetic field has the form

$$\mathbf{B} = B_0 \mathbf{n} + \mathbf{B}'(t, z) ,$$

where B_0 is constant and $\mathbf{B}'$, not necessarily small, is orthogonal to $\mathbf{n}$; $\mathbf{v}(t, z)$ is also orthogonal to $\mathbf{n}$, so that the flow is incompressible and $\varrho = \varrho_0$ is a constant. The MHD equations then reduce to the simple form

$$\varrho_0 \frac{\partial \mathbf{v}}{\partial t} = \frac{B_0}{4\pi} \frac{\partial \mathbf{B}'}{\partial z} , \qquad \frac{\partial \mathbf{B}'}{\partial t} = B_0 \frac{\partial \mathbf{v}}{\partial z} , \qquad (10.42)$$

so that both $\mathbf{v}$ and $\mathbf{B}'$ fulfil d'Alembert one-dimensional equation with propagation velocity V_A. For propagation in the positive (negative) z direction the only independent variable is $\zeta_- = z - V_A t$ ($\zeta_+ = z + V_A t$); hence the vector $\mathbf{v} + V_A \mathbf{B}'/B_0$ ($\mathbf{v} - V_A \mathbf{B}'/B_0$) is a constant and can be set to zero. The flow velocity is always parallel to the magnetic perturbation. For this reason the dynamics of these waves is usefully investigated using the *Elsasser variables*

$$\mathbf{z}_\pm = \mathbf{v} \pm V_A \mathbf{B}'/B_0 ; \qquad (10.43)$$

one of them vanishes when there is propagation only in one sense. The non-linear interactions between forward and backward waves is described by quadratic terms in these variables. They are a useful tool for the investigation of the turbulence in the solar wind.

Perturbation analysis.– In the general case, the small perturbations of (1.85) ($\varrho' = \varrho - \varrho_0, \mathbf{v}, \mathbf{B}'$) fulfil (with a barotropic relation $dP/d\varrho = c_s^2$)

$$\frac{\partial \varrho'}{\partial t} + \varrho_0 \nabla \cdot \mathbf{v} = 0 , \qquad (10.44a)$$

$$\varrho_0 \frac{\partial \mathbf{v}}{\partial t} + c_s^2 \nabla \varrho' = \frac{B_0}{4\pi} (\nabla \times \mathbf{B}') \times \mathbf{n} , \qquad (10.44b)$$

$$\frac{\partial \mathbf{B}'}{\partial t} = B_0 \nabla \times (\mathbf{v} \times \mathbf{n}) ; \qquad (10.44c)$$

of course, $\nabla \cdot \mathbf{B}' = 0$ from the corresponding Maxwell's equation. With the functional dependence $\propto \exp i(\omega t - \mathbf{k} \cdot \mathbf{r})$ the system becomes algebraic:

$$\omega^2 \varrho_0 \mathbf{v} - c_s^2 \varrho_0 (\mathbf{k} \cdot \mathbf{v}) \mathbf{k} + \omega \frac{B_0}{4\pi} [(\mathbf{k} \cdot \mathbf{n}) \mathbf{B}' - (\mathbf{n} \cdot \mathbf{B}') \mathbf{k}] = 0 , \qquad (10.45a)$$

$$\omega \mathbf{B}' + B_0 [(\mathbf{k} \cdot \mathbf{n}) \mathbf{v} - (\mathbf{k} \cdot \mathbf{v}) \mathbf{n}] = 0 , \qquad (10.45b)$$

and $\mathbf{k} \cdot \mathbf{B}' = 0$. Consider first the case in which both $\mathbf{v}$ and $\mathbf{B}'$ are orthogonal to both $\mathbf{n}$ and $\mathbf{k}$, with components $v_\perp$ and $B'_\perp$. They fulfil

$$4\pi \omega \varrho_0 v_\perp + k B_0 \cos \theta B'_\perp = 0 , \qquad k B_0 \cos \theta v_\perp + \omega B'_\perp = 0 , \qquad (10.46)$$

where $k\cos\theta = (\mathbf{k}\cdot\mathbf{n})$ gives the angle $\theta\,(\neq 0)$ between the propagation direction and the unperturbed magnetic field. There is a solution only if the relation

$$\omega^2 = k^2 V_A^2 \cos^2\theta, \qquad (10.47)$$

is fulfilled; the shear Alfvén wave previously discussed is recovered when $\theta = 0$. For the two modes $\omega = \pm k V_A \cos\theta$ and we also get the relation

$$B_0 v_\perp \pm V_A B'_\perp = 0, \qquad (10.48)$$

leading to the Elsasser variables (10.43) also for $\theta \neq 0$.

These are the specific MHD waves predicted in 1942 by H. ALFVÉN, and later experimentally detected by S. LUNDQUIST in 1949. As an application to magnetospheric physics, we note that under a shear Alfvén wave a dipole line of force oscillates with the period

$$\frac{2\pi}{\omega} = \int \frac{ds}{V_A} = 2\sqrt{4\pi} \int_0^{\pi/2} d\theta_M \, \frac{ds}{d\theta_M} \, \frac{\sqrt{\varrho_0(\theta_M)}}{B_0(\theta_M)}. \qquad (10.49)$$

$B_0(\theta_M)$ is the dipole field expressed in terms of the magnetic colatitude θ_M, while $s(\theta_M)$ is its arc length; their expressions can be obtained from eqs. (6.5) and (6.8), namely $ds = \sqrt{dr^2 + r^2 d\theta_M^2} = r d\theta_M \sqrt{1 + 2\,\mathrm{ctg}^2\theta_M}$. For example, a line of force which crosses the equator at $3 R_\oplus$ has a length L' of about 60,000 km; the magnetic field is about $30,000/27 \simeq 1,000$ nT, with an electron density of, say, 100 cm^{-3}. Then the period is $\simeq 2\sqrt{4\pi\varrho_0}L'/B_0 \simeq 20$ s.

The remaining four equations in (10.45) for the components of $\mathbf{v}$ and $\mathbf{B}'$ along $\mathbf{n}$ and $\mathbf{k}$ lead, after some algebra, to the biquadratic dispersion relation

$$(\omega^2 - k^2 c_s^2)(\omega^2 - k^2 V_A^2) - k^4 c_s^2 V_A^2 \sin^2\theta = 0. \qquad (10.50)$$

It can be shown to have four real roots, given by

$$\omega^2 = \frac{1}{2} k^2 \left[c_s^2 + V_A^2 \pm \sqrt{\left(c_s^2 + V_A^2\right)^2 - 4 c_s^2 V_A^2 \cos^2\theta} \right]. \qquad (10.51)$$

These two modes are compressional and called *magnetoacoustic waves*. The upper sign + corresponds to the *fast mode* (F), whose speed is greater than the largest of c_s and V_A; it reduces to ordinary acoustic waves when $V_A \ll c_s$. The lower sign is a new mode – the *slow mode* (S) – with speed less than the smallest of c_s and V_A; in the same limit it gives $\omega^2 = k^2 V_A^2 \cos^2\theta + O(V_A^4/c_s^2)$. The shear Alfvén wave (10.47) has a phase velocity intermediate between the two, namely

$$\left(\frac{\omega}{k}\right)_F^2 \geq V_A^2 \cos^2\theta \geq \left(\frac{\omega}{k}\right)_S^2. \qquad (10.51')$$

10. Magnetospheres

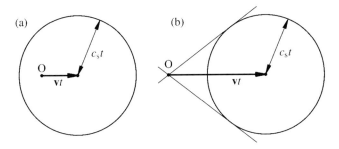

Figure 10.8. The wave front of the disturbance produced by a point obstacle at O in a subsonic (a) and a supersonic (b) flow.

10.5 The bow shock

The disturbance produced in a compressible flow by a point obstacle at O reaches after a time t the surface of a sphere of radius $c_s t$, centred at vt. In the subsonic case eventually the disturbance spreads over all the flow; this is the case, for example, of the Galilean satellites, Io in particular, as they orbit in Jupiter's magnetosphere (Sec. 10.7). If instead $v > c_s$ only the interior of *Mach's cone*, with semi-aperture $\arcsin(c_s/v) = \arcsin(1/M_s)$, is affected (Fig. 10.8). In the solar wind this angle is $\approx 7°$. Mach's cone sharply separates the disturbed from the undisturbed flow and is the locus of a discontinuity. Ahead of a finite obstacle a discontinuity develops upstream, which deflects the flow and slows down to subsonic speed. Mathematically, shock waves arise because the equations for an inviscid fluid cannot fulfil with continuous functions the required boundary conditions. In effect, the neglect of viscosity is allowed only when the scale of the transition is greater (in terms of the kinematic viscosity η) than

$$\delta = \frac{\eta}{\varrho v}, \qquad (10.52)$$

obtained by equating in eq. (1.64) the inertial and the viscous terms. In ordinary fluids this gives, in order of magnitude, the thickness of the shock front.

In the case of the Earth, the obstacle is the compressed magnetic dipole; however, since the mean free path is very large, the thickness and the structure of the shock is not determined by viscosity, but by microscopic turbulence, which has been extensively investigated with in situ detectors. Generally speaking, across the shock the density and the magnetic field increase, while the velocity and Mach's numbers decrease; but one finds a great variety of situations according the values of the upstream Mach's numbers and the angle Ψ_B between the normal to the front and the interplanetary magnetic field. When $\Psi_B \ll 1$ ("parallel" shock), the particles' velocity component normal to the front is little affected by the magnetic field and the structure of the transition is smooth and similar to the hydrodynamic case; but if Ψ_B is large (say, near 90°) ions are reflected back upstream; magnetic and electric turbulence is present there in the *precursor* and the transition is much less sharp.

A collisionless shock is a case of *collisionless dissipation*, in which the large and ordered kinetic energy of the wind is disordered and "thermalized" by small scale, high frequency electric fields: a still little understood process, which requires the collisionless Boltzmann-Vlasov equation (Sec. 1.8) in the non-linear regime. Although a fluid description of the shock transition is clearly poor, it still shows how much one can learn from the laws of conservation, without any assumption about its internal structure, except that it is confined to a (thin and plane) layer. In this model one just requires the conservation of the relevant physical quantities across a layer which separates two stationary states; the transition between them is governed by the *Rankine-Hugoniot relations*. Here we confine ourselves to a paradigmatic and simple case, which is not much affected by the fact that in reality the flow is not stationary. Besides the case of the bow shock, in which the velocity component perpendicular to it does not vanish, we also have *contact discontinuities*, in which the fluid flows parallel to the transition and has a sharp change there; this is the case of the magnetopause.

Hydrodynamic description of a shock. For a plane shock, since there are no forces acting in its plane, the tangential velocity component is constant across the shock and can be cancelled with a Galilei transformation. Mass, momentum and energy conservation then connect density, velocity and temperature on the two sides. While the magnetic field increases across the bow shock, as a result of a tangential current gradient, its dynamical effect is not a drastic one and will be neglected in this illustrative discussion. Denoting upstream and downstream quantities with the suffixes $_1$ and $_2$, respectively, mass and momentum conservation require

$$\varrho_1 v_1 = \varrho_2 v_2 , \qquad (10.53\text{a})$$
$$\varrho_1 v_1^2 + P_1 = \varrho_2 v_2^2 + P_2 \qquad (10.53\text{b})$$

(v is the velocity component normal to the shock plane). In terms of the internal energy per unit mass U the energy flow is $\varrho v (U + v^2/2)$; its change is equal to the power of the work done by the pressure and the magnetic field (through both its pressure and tension) as the fluid moves; therefore

$$\frac{v_1^2}{2} + H_1 = \frac{v_2^2}{2} + H_2 . \qquad (10.53\text{c})$$

Here $H = U + P/\varrho$ is the specific enthalpy, equal to $P\gamma/(\gamma - 1)$ in a perfect, polytropic gas (Sec. 5.3).

While the full solution requires some algebra, in our highly supersonic case, in which P_1 can be neglected in (10.53b) and (10.53c), we easily find the compression ratio

$$\frac{\varrho_2}{\varrho_1} = \frac{v_1}{v_2} = \frac{\gamma + 1}{\gamma - 1} + O(M_1^{-2}), \tag{10.54}$$

equal to 4 for the usual value $\gamma = 5/3$. In terms of the upstream Mach's number $M_1 = v_1/c_{s1} = v_1\sqrt{\varrho_1/(\gamma P_1)}$ we also get the ratios of pressures and temperatures, of order $O(M_1^2)$:

$$\frac{P_2}{P_1} = \frac{2\gamma}{\gamma + 1} M_1^2, \quad \frac{T_2}{T_1} = \frac{2\gamma(\gamma - 1)}{(\gamma + 1)^2} M_1^2. \tag{10.55}$$

In a strong shock ($M_1 \gg 1$) the gas is significantly heated. The Mach number M_2 downstream reads

$$M_2^2 = \frac{\gamma - 1}{2\gamma} + O(M_1^{-2}), \tag{10.56}$$

and the motion becomes subsonic for any value of the polytropic exponent γ.

A simple model of a contact discontinuity. In the previous simplified analysis we neglected the effect of the magnetic field; it must be taken into account when the magnetic stress tensor becomes dominant over the pressure P. The *magnetopause* is the transition surface between the outer, rarefied flow and the stagnant plasma inside, dominated by the magnetic pressure of the compressed dipole. A *contact discontinuity* provides a simple model of this transition; it is similar to the current sheet used to describe reconnection, but here, in addition, the velocity, parallel to the surface, abruptly drops to zero inside. If the discontinuity surface is $z = 0$ (with the half space $z < 0$ inside) and the magnetic field is along x, the model, with thickness ℓ,

$$B_x(z) = B_x(-\infty)\left[1 - \tanh\left(\frac{z}{\ell}\right)\right], \quad P(z) + \frac{B_x^2(z)}{8\pi} = P(\infty),$$

gives a qualitative picture, in which the outer magnetic field is neglected. In the front side of the magnetopause, where a strong velocity gradient is present, instabilities may develop which release the shear kinetic energy available, probably through the *Kelvin-Helmholtz mechanism* (Sec. 7.5). They are probably responsible for the generation of low frequency oscillations, with periods of the order of minutes, resulting also in pulsations of the magnetic field on the ground. The plasma blobs created by this process allow the penetration of solar wind plasma into the magnetosphere at low latitudes. In the magnetopause ions and electrons are deflected tangentially by the inner magnetic field. This deflection cannot occur over a distance smaller than the Larmor radius (10.11), about 100 m for electrons and 20 km for protons; hence a positive charge layer will develop, unless the magnetosphere supplies an electron layer. This charge separation produces a polarization electric field and an electron current in the tangential direction.

Table 10.1. The four main trapping regions of cosmic rays and their parameters.

Trapping region	Size	Magnetic field (nT)	Typical energy (GeV/nucleon)	Flux (cm^{-2} s^{-1})	Energy density (eV/ cm^3)
Magnetosphere	10^9 cm	10,000	< 0.1	10^3	100
Solar neighbourhood	10^3 ly	1	10-100	0.1	1
Galactic halo	10^5 ly	0.1	$10^3 - 10^5$	10^{-6}	10^{-2}
Cluster of galaxies	10^7 ly	10^{-3} (?)	$10^7 - 10^8$	10^{-10}	< 10^{-5}

10.6 * Cosmic rays and the magnetosphere

The main component of cosmic rays are light nuclei: protons and α-particles account for about 97% of all particles; at low energies (up to $\approx 10^{12}$ eV) there are also nuclei of C, N and O. Primary cosmic rays have a composition similar to the interstellar gas (see Sec. 9.1), with fewer α-particles and more heavier nuclei. No antiparticles have been observed. Their differential spectrum (flux per unit energy interval) follows a power law E^{-k} in all the observed range, from 10^{10} to 10^{20} eV. The index k varies, however, from 2.7 at lower energies to the steeper value 3.2 at high energies; the transition at 10^{16} eV is called the knee of the spectrum. At ultrahigh energies (> 10^{19} eV) the spectrum seems to become shallow again. In addition, there is a muonic component (muons and pions) and a soft component (e^+, e^- and γ).

Table 10.1 lists four main trapping regions of cosmic rays and the corresponding parameters. Most cosmic rays are generated in our own Galaxy and confined there in its weak ($\simeq$ nT) and irregular magnetic field, extending also to the halo. In 1949 E. FERMI proposed an efficient acceleration mechanism based on the interaction of charged particles with randomly moving magnetic barriers. If v is their velocity, the change in the particle's energy has a part linear and a part quadratic in v; both of them can accelerate particles, even to relativistic energies. Acceleration processes are at work also in magnetic stars (in particular in pulsars, where very large potential differences are created by the rotation along open lines of force). The primary source and the acceleration mechanisms of ultra-energetic cosmic rays, however, is still an unsolved problem. At lower energies, in the solar system, particles are accelerated in non-thermal processes in interplanetary shocks and reconnection events. Solar flares are also an important source of particles with energies larger than 1 MeV/nucleon, up to hundreds MeV/nucleon; they can be seen as a suprathermal tail of the solar wind.

10. Magnetospheres

In order to reach the ground, primary cosmic rays must penetrate the magnetosphere. Substantial deflection does not occur if the Larmor radius $\approx v/|\Omega_c| = v\gamma m_0 c/(|q|B)$ is larger than the scale r of the magnetosphere at a distance r; this can be expressed in terms of the *rigidity*

$$\mathcal{R} = \frac{m_0 \gamma c v}{|q|} \lesssim Br \ . \tag{10.57}$$

For different particles with the same rigidity the deflection condition at a given place is the same; for relativistic particles, taking $B \simeq 30,000$ nT and $r \simeq R_\oplus$, it gives the approximate lower limit of 60 GeV for undeflected protons or electrons; trapped particles start below, at ≈ 10 GeV. At $10\,R_\oplus$ this limit is two orders of magnitude smaller, 0.1 GeV. Of course, one expects particles to reach the ground more easily at large latitudes, where the lines of force are open. A detailed, but complex analysis of the accessibility of an axially symmetric magnetosphere (in particular, a dipole) is possible with the powerful *effective potential* method for the equations of motion of a relativistic particle. It is similar to the one used for the three-body problem (Problem 10.12) and has been originally developed by C. STØRMER around 1930.

Cosmic ray particles, upon colliding with a nucleus in the atmosphere, lead to three kinds of processes. The production of protons and neutrons proceeds in a cascade with many generations; charged pions produce narrow *muon jets* which may arrive at the ground before they decay, in a time $\approx 2 \times 10^{-6}$ s; neutral pions quickly decay in large numbers of photons and pairs of electron and positrons, which can be detected with large detector arrays on the ground as a *shower*. The ground-based observations of primary cosmic rays is a rapidly evolving experimental field, with several international projects; space experiments, such as satellite observations of fluorescent tracks in the Earth atmosphere, are also planned. The ionizing power of cosmic rays has practical consequences, e.g., for the safety of unprotected astronauts.

The interplanetary cosmic ray flux, about four particles per cm^2 and second, has been painstakingly investigated with ground instruments; in particular, its time variation and its dependence on direction and magnetic latitude were determined. At low energies, besides obvious meteorological changes due to the conditions of the atmosphere, there is a daily variation (about 0.15%, with a yearly modulation), due to the effect of the Earth magnetic field. Much studied are also the *Forbush decreases*, sudden reductions of the flux, often in association to magnetic storms. They occur simultaneously over the world and are due to a sudden increase of the interplanetary magnetic field in the solar wind (see Ch. 9), which denies to energetic particles of galactic origin access to the inner solar system. Solar activity enhances also the direct production of cosmic rays.

Dating with cosmic rays. Radioactivity and nuclear reactions induced by cosmic rays in exposed material in the solar system provide important tools to

Proton density	$n = 5/D^2$ cm^{-3}	
(Ion) temperature	$T = 2 \times 10^4/D^{4/3}$ K	
Wind velocity	$v = 400$ km/s	
Magnetic field	$B = 5/D$ nT	
Alfvénic Mach's number	$M_A = 8$	
Sonic Mach's number	$M_s = 8/D^{2/3}$	

Table 10.2 Typical values for the solar wind in the outer solar system; D is the distance from the Sun in AU.

establish an absolute chronology. An example is the production of a carbon isotope by low energy neutrons: ^{14}N + n → ^{14}C + p; subsequently ^{14}C decays into ^{14}N + e$^-$ with a half lifetime of 5,568 y. It has been shown that the average intensity of cosmic rays has remained constant to within 10% during the last million years and to a factor 2 in the last Gy; we can therefore assume that material exposed to cosmic rays has a fixed activity, resulting from the balance between production and decay. Once ^{14}C is created, it participates in all the geochemical and biological cycles with the same chemical properties of the main isotope ^{12}C; then its activity allows to measure the duration of the exposure and to accurately date the material for timescales of historical and prehistorical relevance. Radioactive dating is very important also for meteorites, where several isotopes act as markers to date their formation and the time they spent on route to the Earth from their parent bodies (Sec. 14.6).

10.7 Planetary magnetospheres

As discussed in Sec. 6.5, six planets and four, possibly five, natural satellites possess an internally generated magnetic field, so that there is enough information for a comparative magnetospheric physics. Most of the present data about planetary magnetospheres, including the Earth's, have been provided by in situ spacecraft measurements. For the giant planets we had the Voyager 1 and 2 missions; for Jupiter, in particular its moons, exceptional observations have been carried out by Galileo and Ulysses.

To estimate the size of a magnetosphere, consider a simple solar wind model in which the radial speed v is constant, so that, in terms of the distance D from the Sun, the density is proportional to $1/D^2$ (Table 10.2). With a polytropic coefficient $\gamma = 5/3$ the temperature decreases as $1/D^{4/3}$ (eq. (1.26b)). In the outer solar system the magnetic field is almost orthogonal to the radial direction and $B \approx B_\phi \propto 1/D$. The flow is always supersonic and super-Alfvénic, and a bow shock is expected. Eq. (10.2) gives the stand-off distance when the kinetic pressure is supported by the planetary magnetic field; remarkably, it is independent of the radius R of the planet and scales with the dipole strength d and the distance D (in AU) as

$$r_s = r_{s\oplus} \left(\frac{d}{d_\oplus}\right)^{1/3} D^{1/3} \quad (r_{s\oplus} \approx 10 R_\oplus). \tag{10.58}$$

Jupiter	Saturn	Uranus	Neptune
60,000	1,000	640	600
84	17	25	24

Table 10.3 Approximate stand-off distances, in 1,000 km and planetary radii, for the giant-planet magnetospheres.

To have a magnetosphere, this quantity must be larger than R, a condition marginally fulfilled for Mercury. For the outer planets, using the Table 6.2, we get the approximate stand-off distances, as given in the Table 10.3. The very large magnetosphere of Jupiter is due to the large dipole strength and the low density of the solar wind. A further factor is the atmosphere of Io, produced by volcanic eruptions; its particles acquire a net charge by impact ionization or photoionization and are driven to corotation with the planet by magnetic forces at its angular velocity ω_J. This mass input and the resulting large plasma density are sufficient to distort the main dipole field in a big magnetic disc (indicated in Fig. 10.9) and a plasma torus around Io's orbit. In Uranus and Neptune and, to a less degree, in Jupiter, the magnetic dipole makes a large angle (59°, 47° and 9.6°, respectively) with the rotation axes, making them oblique rotators, as discussed in Sec. 6.5. For the ice giants, apart from the obvious seasonal variation, there is also an important diurnal variation related to the change of the sunward direction with respect to the magnetic axis, directly affecting magnetospheric dynamics.

A special type of magnetosphere, known as *induced magnetosphere*, develops when the planet does not possess an internal magnetic field, but has a ionosphere (Sec. 8.2); this is the case of Venus and Mars (and, to some extent, comets). There are no belts of trapped radiation, such as the Van Allen belts (Sec. 10.3), and there is no magnetotail with fields of planetary origin. The ionospheric obstacle to the solar wind defines a surface – the *ionopause* – at which the solar wind dynamical pressure outside balances the thermal pressure inside. A bow shock forms upstream, which deflects the solar wind sideways along the magnetosheath (sometimes called the *ionosheath*). Inside the ionopause the plasma is subsonic and of ionospheric character.

For Venus, the interaction between the solar wind and the ionosphere was first investigated with the Pioneer Venus mission. At the solar cycle maximum the ionopause boundary was found upstream at an altitude of ≈ 300 km and ≈ 800 km near the terminator. This boundary moves up and down in response to changes in the external pressure. In the night side the ionosphere is fed by the plasma flow across the terminator and the situation is more complex and fluctuating.

Mars' induced magnetosphere was extensively studied by the Mars Global Surveyor. A close similarity with Venus' magnetosphere was confirmed, with the (well mapped) bow shock, the ionopause, a direct interaction of the solar wind with the ionosphere, etc. In the sunward direction the bow shock stands at $\simeq 0.5\,R$; an asymmetry of the shock about the Mars-Sun line was reported.

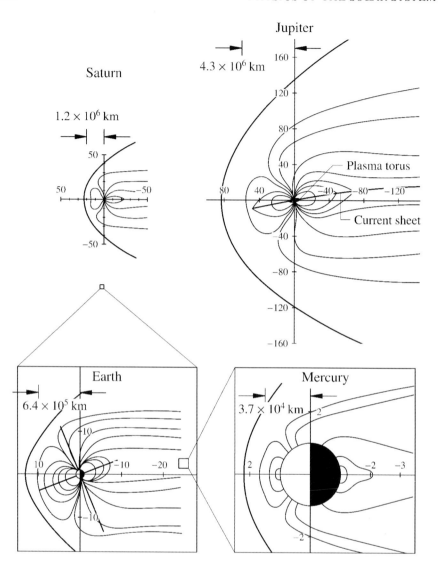

Figure 10.9. Planetary magnetospheres and their relative sizes. The angle between the magnetic dipole and the Sun's direction changes seasonally due to the planetary obliquity, and daily due to the magnetic tilt with respect to the rotation axis. For Jupiter the large magnetic disc produced by Io is also shown. Saturn's dipole is nearly parallel to the pole of the ecliptic. Mercury has a tiny magnetosphere and a very weak field, perhaps not dominated by a dipole. The scales are in planetary radii, but a length in km is also given for reference.

At the terminator, the typical ionopause altitude is between 300 – 500 km. As in the Venus case, near the midnight side streamers of cold plasma detached from the ionosphere were found.

10. Magnetospheres

Individual configurations. Hereafter, we briefly comment on planetary magnetospheres with an internal magnetic field.

Mercury.– Very little is known about the magnetosphere of Mercury, discovered in 1974 by Mariner 10 in two flybys. Although they occurred on the night side, the bow shock and the magnetopause crossing were detected, with storm-like features. Ground observations of the distribution of Na and K in the exosphere bear indirect evidence about the magnetospheric geometry and its processes at high latitudes. No ionosphere has been detected and the mechanism by which electric currents can close is not clear: a thin sheath of photoelectrons close to the sunlit surface may act as a conductor and do the job. Although the radius of the planet is greater than $R_\oplus/3$, the magnetosphere is only 3% that of the Earth; this, and the stronger and more variable solar wind is likely to cause strong fluctuations. This, and the Hermean magnetic field, is a very interesting investigation target and one of the objects of the BepiColombo space mission under development by the European Space Agency.

Jupiter.– Jupiter is particularly interesting because of its huge dipole, its fast rotation and the fact that all its satellites are inside the magnetosphere. The sunward magnetopause, in agreement with the Table 10.3, oscillates between 50 and $100\,R_J$; a very long tail extends downstream to more than $700,000$ km, sweeping lines of force perhaps even beyond Saturn's orbit. Due to the corotation of the magnetospheric plasma, its interaction with satellites is determined by the relative velocity

$$v_{\rm rel} = (\omega_J - n)\,r\,, \tag{10.59}$$

sub-Alfvénic in all cases (see Table 10.4; $n = \sqrt{GM_J/r^3}$ is the corresponding Keplerian frequency). High energy particles are trapped in radiation belts; like for the Earth, they scatter and precipitate into the atmosphere and produce auroral emissions which have been observed by the Galileo spacecraft. The peak electron and proton fluxes at Jupiter are, however, about $1,000$ times greater than at the Earth for energies ≥ 3 MeV, and ≥ 30 MeV respectively.

Contrary to the Earth, much of the matter in the magnetosphere has an internal origin, being supplied by Io's volcanic activity at the rate of 10^6 g/s, mostly in the form of neutral oxygen and sulphur. This fairly dense gas (about 4×10^3 ions per cm^3 and up to 2×10^4 ions per cm^3 in the wake of Io), initially confined near the surface, is subsequently heated and ionized by the fast rotating and energetic magnetospheric plasma and spreads out in a doughnut shape – *Io's torus* – around its orbit. The torus is hot and visible in the UV. Its composition has been determined by the Galileo probe in close Io flybys; once- and twice-ionized oxygen and sulfur ions account for nearly 95%, the remaining being heavier ions. Centrifugal diffusion extends this matter (in to-

Table 10.4. Magnetospheric parameters of the four Galilean satellites (adapted from F.M. Neubauer, J. Geophys. Res. **103**, 19,843 (1998)).

Satellite	Radius (km)	Relative velocity (km/s)	Electron density (cm^{-3})	Magnetic field (nT)	M_A	Mean ion mass	Ion temp. (eV)
Io	1,821	45	3,600	1,800	0.3	20	100
Europa	1,569	85	130	420	0.39	13.7	270
Ganymede	2,631	180	3.7	120	0.48	13	360
Callisto	2,403	175	1.1	35	0.94	15	200

tal, about 10^{12} g) farther out, even beyond Europa's orbit, whose atmosphere is continually maintained by sputtering and recombination. The interaction between Ganymede's dipole field and the corotating magnetosphere is fairly well understood. The satellite magnetic dipole, nearly parallel to Jupiter's dipole, concentrates the outer lines of force within a tube of radius $\approx 2 R_G$; in addition, its upstream side is compressed and the downstream side expanded by the external pressure.

Much more complex is the interaction with Io. Its motion through the Jovian magnetosphere, like a body moving subsonically in a gas, is a source of wave phenomena; it produces Alfvén waves at periods of order $2 R_{Io}/V_A \approx 30$ s and deforms the intersected lines of forces. In our current understanding this deformation and bunching travels to the North (and South) magnetic pole in about 12 minutes and is reflected back to the other pole, completing a few round trips before being dissipated; during this time Io (with an orbital period of 1.769 d) advances by a few degrees. But the key feature of this interaction is a strong and pulsed radio emission in the frequency range 10 to 40 MHz (*decametric radiation*); in 1964 it was found to be highly correlated with Io's orbital phase. This has pointed out a complex and unexpected magnetospheric phenomenon at work, still not completely understood. The radiation is 100% elliptically polarized, as for resonant plasma emission at the cyclotron frequency. With this identification it is possible to locate the source at the feet of the lines of force issuing from Io, where they enter Jupiter's ionosphere. One way to understand this is based on the remark that Io's body, whose core is hot, is electrically conductive. The motion across the lines of force generates a potential difference of order $2 R_{Io} v_{rel} B/c \approx 400$ kV, for an electric field $E = v_{rel} B/c \approx 0.1$ V/m, between the two points of the satellite nearest to, and furthest from, the planet. This produces mildly relativistic electrons which travel along the flux tubes to the magnetic poles, where they emit cyclotron radiation. The radiating electron population, of course, is modulated by Alfvén waves produced by Io's motion;

this shows up in the characteristic arc-like structures of the received pulses in the time-frequency plane. Similarly, Amalthea's motion, closer to the planet, induces Alfvén and other types of waves that affect the inner parts of the Jovian radiation belt, causing relativistic synchrotron radiation with frequencies 0.1 – 15 GHz (*decimetric radiation*).

Saturn.– The magnetosphere of Saturn, probed by the Pioneer and Voyager flybys, will be extensively studied from 2004 to 2008 by the Cassini mission. Its magnetopause extends about 20 R_S, with great variations; inside, several satellites and rings are plasma sources and generate an oxygen-ion rich torus with a density of 1 – 100 particles/cm^3. Satellites and rings are also plasma sinks and produce local minima in the particle density around each orbit; for the same reason Saturn's particle belts terminate at the main rings. A peculiar plasma sheet of hydrogen, helium, oxygen and carbon spreads in the outer magnetosphere as far as Titan's orbit; with its dense atmosphere, Titan supplies ionized particles. Titan's status, at a distance of 20 R_S, is special also because it is embedded in the planetary magnetosphere only part of the time, as its boundary moves in response to the variable solar wind. Kilometric radiation modulated at 10.657 h is emitted, probably in resonance with the rotation period of Saturn.

Uranus and Neptune.– Voyager 2 has explored Uranus in 1985-86 and Neptune in 1990. The large tilts (59° for Uranus and 47° for Neptune) and offsets from the centre (0.3 R_U and 0.55 R_N, respectively) of their dipoles produce complex and unconventional magnetospheres. The large obliquity of Uranus makes the interaction of its major satellites (orbiting in the equatorial plane with a semimajor axis smaller than the stand-off distance) with the magnetosphere very complex, in spite of its low particle density (0.1 – 1 particle/cm^3).

In Neptune, large tilts of both the magnetic and rotation axes cause a very peculiar situation: in a planetary rotation a configuration similar to the Earth alternates with a geometry where the magnetic axis points toward the Sun and the solar wind penetrates directly into the planet's polar cusps. Triton, the largest satellite of Neptune, only modestly contributes to the magnetospheric plasma. Auroral activity has been observed around the magnetic poles of both planets. Several radio emissions in the frequency range from a few to several hundreds kHz are also present.

Charged dust dynamics. Dust grains, widespread throughout the solar system (Sec. 14.6), can acquire a net charge and give rise to several interesting dynamical effects. A solid body initially placed in a plasma in thermal equilibrium at a temperature T at first gets a negative charge, since the electron flux is greater than the ion flux by the square root of the mass ratio. Within about a Debye length λ_D (1.102) a negative electrostatic potential φ develops which depresses the electron density until the collected current vanishes; it can be shown that this happens when $-e\varphi \approx kT$. In a 1 eV plasma with S$^+$ ions, for example, kT/e is 1 V and $\varphi \simeq -3.9$ V. For a spherical body of radius R_s the charge is

10.5.* Show that (10.23) is adiabatically invariant for the harmonic oscillator. Hint: The limit $\Omega \to \infty$ (more accurately, $\Omega^2 \gg |\dot\Omega|$) of eq. (10.22) is a problem of *singular perturbation theory*. To formally describe the limit $\Omega \to \infty$, while keeping finite the timescale $|\Omega/d\Omega/dt|$ of the frequency one can simply write Ω/ϵ in place of Ω and let $\epsilon \to 0$. The resulting equation $\epsilon^2 \ddot{x} + \Omega^2 x = 0$ is then solved with the Ansatz

$$x = \exp\left(\frac{S_0}{\epsilon} + S_1 + \epsilon S_2 + \ldots\right).$$

10.6.* Study the jump conditions for a shock of arbitrary strength; include the magnetic transition.

10.7. Two spacecraft measure the variations in the local plasma density due to a linear wave. What information can one derive about the dispersion relation? (Hint: Study the phase difference in each Fourier component.)

10.8. Estimate the total ring current, assuming a thickness of $2 R_\oplus$ for the region of magnetic depression.

10.9.* A simple model of magnetic reconnection in the limit of infinite plasma conductivity. The induction eq. (1.82) simplifies to $\partial \mathbf{B}/\partial t = \eta \nabla^2 \mathbf{B}$. Assume one-dimensional field $\mathbf{B} = B(z,t)\mathbf{e}_x$ with the initial data $B = B_0$ for $z > 0$ and $B = -B_0$ for $z < 0$ and show that $B = B_0 \,\mathrm{erf}(z/\sqrt{4\pi t})$. Evaluate the related current density $\mathbf{J}$ and the plasma flow.

10.10. Derive the dispersion relation (10.50) of magnetoacoustic waves.

10.11. Write in polar coordinates the dynamical equations (10.61), (10.62) for a dust particle and prove that the angular momentum and the energy are constant. Just like for the two-body problem, the elimination of the longitudinal degree of freedom reduces the problem to quadratures.

10.12.* Study the motion of a relativistic particle with charge q and rest-mass m_0 in dipole magnetic field. Neglecting gravitational effects and planetary rotation, the corresponding Lagrangian reads

$$L = \frac{1}{2}\gamma m_0 v^2 + \frac{q}{c}\mathbf{v}\cdot\mathbf{A},$$

where $\gamma = 1/\sqrt{1 - v^2/c^2}$ is the Lorentz factor (c the light velocity) and $\mathbf{A}$ the vector potential of the magnetic field $A_\phi(r,\theta) = d\sin\theta/r^2$ for the dipole field with moment d. Eliminating the ignorable coordinate ϕ (due to the axisymmetry), determine the forbidden regions of the motion in the meridional plane for particles coming from infinity. (Hint: Derive the energy integral.)

10.13.* The same as in the Problem 10.12, neglecting the relativistic factor γ but considering planetary rotation and taking into account gravity, with the potential energy $-GM/r$. Find the equilibrium solutions (circular orbits); prove that, depending on free parameters, such as the particle's charge and energy, the equilibria may occur off the equatorial plane of the planet (the so-called *halo orbits*).

10.14.* Consider a stationary model of a neutral sheet with the magnetic field **B** and plasma pressure given by

$$\mathbf{B}(z) = B_0 \tanh(z/h) \, \mathbf{e}_x \, , \quad p(z) = p_0 \, \text{sech}^2(z/h) \, .$$

Verify that the total pressure is constant if $p_0 = B_0^2/(8\pi)$ and that there is an equilibrium between the pressure gradient ∇p and $(\mathbf{J} \times \mathbf{B})/c$. Study the motion of test charged particles, in particular near the neutral sheet.

– FURTHER READINGS –

Two good introductory textbooks on plasma physics are P.C. Clemmow and J.P. Dougherty, *Electrodynamics of Particles and Plasmas*, Addison-Wesley (1969) and, more recent, F.F. Chen, *Introduction to Plasma Physics and Controlled Fusion. Vol. 1. Plasma Physics*, Plenum Press (1984); the short, but classic treatise by L. Spitzer, *Physics of Fully Ionized Gases*, Interscience Publishers (1962) is quite useful. Two good, wide-scope books dealing with the specific subject matter of this chapter are J.K. Hargreaves, *The Solar-terrestrial Environment*, Cambridge University Press (1992) and G.K. Parks, *Physics of Space Plasmas*, Addison-Wesley (1991). More advanced are G. Haerendel and G. Paschmann, Interaction of the solar wind with the day-side magnetosphere, in *Magnetospheric Plasma Physics*, A. Nishida, ed., Reidel (1984), L.R. Lyons and D.J. Williams, *Quantitative Aspects of Magnetospheric Physics*, Reidel (1984) and *Introduction to Space Physics*, M.G. Kivelson and C.T. Russell, eds., Cambridge University Press (1995). The recent outstanding results concerning the magnetospheres of outer planets are the subject of the issue E9 of the *J. Geophys. Res.* (1998), and are reviewed by C.T. Russell, The dynamics of planetary magnetospheres, *Planet. Sp. Sci.* **49**, 1005 (2001). For a review on charged dust, see M. Horányi, Charged dust dynamics in the solar system, *Ann. Rev. Astron. Astrophys.* **34**, 383 (1996).

Chapter 11

THE TWO-BODY PROBLEM

For about two thousand years, the efforts of Western astronomers have been largely devoted to the understanding of the motion of the Sun, the Moon and the planets in the sky. More and more complex geometrical and kinematical models, both geocentric and heliocentric, were devised in order to reproduce observational data of increasing accuracy. The final outcome of this effort, and at the same time the starting point of a scientific revolution, can be traced back to J. KEPLER's theory of planetary motions. Giving up the long-standing a priori requirement of circular paths and uniform velocities (or of finite combinations of them, resulting in epicyclic trajectories), Kepler's three laws (elliptic orbits with the Sun at one focus, constant areal velocities, and 3/2-power dependence of orbital periods on semimajor axes) fitted astronomical observations with unprecedented accuracy, and at the same time summarized in a few simple mathematical relationships most of the available observations.

Within less than a hundred years, I. NEWTON discovered the deep dynamical nature of Kepler's laws and put forward a completely new synthesis, in which the motion of celestial bodies, as well as many other physical phenomena, became accessible to understanding, calculation and prediction. Newton's theory of gravitation, together with his formulation of the laws of motion, found its first and simplest challenge in the two-body problem: to find the positions and velocities at any point in time of two massive, point-like particles moving under their gravitational forces, as a function of their initial values. This is the basis of celestial mechanics and the subject of this chapter.

11.1 Reduction to a central force problem

It is a lucky coincidence that the planetary motions are well represented by solutions of the two-body problem; this is due to the fact that the ratios of all the planetary masses to the solar mass are small ($< 10^{-3}$), and also that close encounters among the major bodies of the solar system are very unlikely (see Sec. 15.1 for comments on this issue). The discovery of the universal gravitation law and the development of celestial mechanics would have been much more difficult in a planetary system around a pair of binary stars of comparable masses. The three-body problem (Ch. 13) is hard to solve and to visualize. In addition, most planetary orbits are almost circular and coplanar; for these reasons, historically the motion of planets could be understood and predicted on a geometrical basis, long before the discovery of the law of gravitation. It is interesting to note that the major systems of satellites, in particular Jupiter's and Saturn's, have the same feature: to a good approximation their motions are (coplanar and circular) solutions of the two-body problem in the field of the central planet. Corrections to this approximation, mainly due to the aspherical shape of the central body and interaction with the other bodies, are the subject of *perturbation theory*, to be discussed in Ch. 12.

The gravitational forces $\mathbf{F}_1$ and $\mathbf{F}_2$ acting on the masses m_1 and m_2 are

$$\mathbf{F}_1 = \mathbf{F}_2 = -G \frac{m_1 m_2}{r^2} \frac{\mathbf{r}}{r}, \tag{11.1}$$

Here, $G = 6.674 \times 10^{-8}$ cm^3 g^{-1} s^{-2} is the gravitational constant and $\mathbf{r}$ is the relative vector of the two bodies. If $\mathbf{r}_1, \mathbf{r}_2$ are the position vectors of the two bodies in an inertial reference frame (so that $\mathbf{r} = \mathbf{r}_2 - \mathbf{r}_1$), we have

$$m_1 \ddot{\mathbf{r}}_1 = G \frac{m_1 m_2}{r^2} \frac{\mathbf{r}}{r}, \tag{11.2a}$$

$$m_2 \ddot{\mathbf{r}}_2 = -G \frac{m_1 m_2}{r^2} \frac{\mathbf{r}}{r}. \tag{11.2b}$$

Adding these two equations, we get

$$m \dot{\mathbf{R}} = m_1 \dot{\mathbf{r}}_1 + m_2 \dot{\mathbf{r}}_2 = \text{constant}, \tag{11.3}$$

which expresses the conservation of the total momentum of the system, resulting in the linear and uniform motion of the centre of mass $\mathbf{R}$; $m = m_1 + m_2$ is the total mass. An inertial reference system where $\mathbf{R} = 0$ is referred to as a *centre of mass frame*. Dividing eqs. (11.2a) and (11.2b) by m_1 and m_2, respectively, and subtracting, we obtain

$$\ddot{\mathbf{r}} = -\frac{Gm}{r^3} \mathbf{r}, \tag{11.4}$$

which involves only the relative vector $\mathbf{r}$. The problem is thus completely equivalent to that of a single particle moving under the gravitational pull of a mass m. For any radial force the angular momentum integral per unit mass

$$\mathbf{h} = \mathbf{r} \times \dot{\mathbf{r}} \tag{11.5}$$

11. The two-body problem

is a constant of the motion. Then the motion is planar (it always occurs in the plane perpendicular to **h**) and can be studied using polar coordinates (r, θ). The velocity has a radial component $v_r = \dot{r}$ and a transverse component $v_t = r\dot{\theta}$. The constancy of

$$h = r v_t = r^2 \dot{\theta} \qquad (11.6)$$

is equivalent to *Kepler's second law*, since h is easily shown to be equal to twice the *areal velocity*, the rate of change of the area described by the radius vector **r**. A second integral of the motion is provided by the conservation of energy. Taking the scalar product of $\dot{\mathbf{r}}$ with eq. (11.4) and integrating with respect to time, we obtain the energy of the relative motion per unit mass

$$E = \frac{v^2}{2} - \frac{Gm}{r} = \text{constant}, \qquad (11.7)$$

where $v^2 = \dot{\mathbf{r}} \cdot \dot{\mathbf{r}} = v_r^2 + v_t^2$. Note that the total energy of the system

$$E_{\text{tot}} = \frac{1}{2} m_1 \dot{\mathbf{r}}_1^2 + \frac{1}{2} m_2 \dot{\mathbf{r}}_2^2 - \frac{Gm_1 m_2}{r} = \frac{1}{2} m \dot{\mathbf{R}}^2 + \mu_r E \qquad (11.8)$$

is sum of the kinetic energy of the centre of mass with the mass m and the relative energy with the *reduced mass* $\mu_r = m_1 m_2 / m$.

Since the radial component of the acceleration is $\ddot{r} - r\dot{\theta}^2$, we can write the radial part of eq. (11.4) in the form

$$\frac{d^2 r}{dt^2} = \frac{h^2}{r^3} - \frac{Gm}{r^2} = -\frac{dV(r)}{dr}. \qquad (11.9)$$

Here $\dot{\theta}$ has been eliminated using eq. (11.6). Note that the radial motion is determined by the competition of the gravitational (attractive) force with the apparent *centrifugal* (repulsive) force, the two terms having a different power dependence upon r. Both these forces can be derived from a potential, so that the problem can be qualitatively studied with the usual techniques applicable to one-dimensional conservative systems, i.e., examining the *effective potential energy*

$$V(r) = -\frac{Gm}{r} + \frac{h^2}{2r^2}. \qquad (11.10)$$

The centrifugal and the gravitational contributions prevail, respectively, at small and large distances, so that $V(r)$ has a negative minimum at $r = r_0 = h^2/Gm$. This corresponds to a circular orbit with radius proportional to the $-2/3$ power of the angular velocity (from eq. (11.6)), in agreement with Kepler's third law. Its relative energy (11.7) is $V(r_0)$. In the plane (r, V), at the points where $V(r) = E$, the radial velocity $\dot{r}$ vanishes. When $V(r_0) < E < 0$

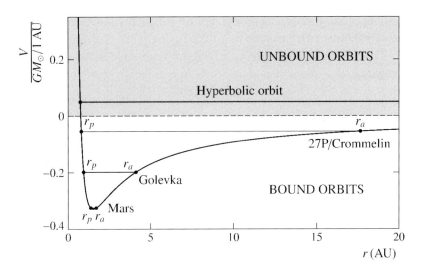

Figure 11.1. Analysis of the Keplerian motion with the effective potential $V(r, h)$ (eq. (11.10)). Provided $h \neq 0$, V has a negative minimum at r_0, which corresponds to a circular orbit. For higher, but negative energy levels E, the motion occurs between the two extrema r_p and r_a. When the energy is positive or zero, the motion is unbound. Three solar system objects are shown with approximately the same angular momentum per unit mass, but different energies (and thus eccentricities): Mars, asteroid Golevka and comet P/Crommelin; a hyperbolic orbit is also indicated.

there are two such points, at which the distance r has a maximum and a minimum; they are called *apocentre* and *pericentre*, and the orbit is bounded. When $E \geq 0$ there is only one such point and the orbit comes from infinity and goes back there (Fig. 11.1).

11.2 Keplerian orbits

The differential equation (11.9) can be directly solved for $u = 1/r$, using (11.6) and the angle θ as independent variable. We obtain

$$\frac{d^2u}{d\theta^2} + u = \frac{Gm}{h^2} \,. \qquad (11.11)$$

This is the equation of the harmonic oscillator with a constant external force; it has the general solution

$$u = \frac{Gm}{h^2} + A\cos(\theta - \omega) \,, \qquad (11.12)$$

where A and ω are two constants of integration. Going back to the variable r, eq. (11.12) becomes

$$r = \frac{h^2/Gm}{1 + e\cos(\theta - \omega)} \,, \qquad (11.13)$$

11. The two-body problem

where we have introduced the new constant (the *orbital eccentricity*)

$$e = A\frac{h^2}{Gm}. \tag{11.14}$$

Eq. (11.13) gives the general form of a conic section in polar coordinates. The value of the eccentricity discriminates among elliptical ($e < 1$), parabolic ($e = 1$) and hyperbolic orbits ($e > 1$). When $e = 0$ the ellipse degenerates into a circle. If $e < 0$, replacing ω with $\omega + \pi$, we go back to the previous cases, so that e can be assumed positive. The minimum value of r (the pericentre distance r_p) is reached when $\theta = \omega$, implying

$$r_p = \frac{h^2}{Gm(1+e)};$$

since at this distance $\dot{r} = 0$, the velocity has a tangential direction and is given by

$$v_p = \frac{h}{r_p} = \frac{Gm(1+e)}{h}.$$

Substituting these values into eq. (11.7) we obtain

$$e^2 = 1 + \frac{2Eh^2}{(Gm)^2}, \tag{11.15}$$

showing that the energy is negative, zero and positive for elliptic, parabolic and hyperbolic orbits, respectively. These results can be obtained in a different way by analyzing the properties of the *Lenz vector* (in some textbooks also called Laplace vector)

$$\mathbf{e} = \frac{\dot{\mathbf{r}} \times \mathbf{h}}{Gm} - \frac{\mathbf{r}}{r}. \tag{11.16}$$

As a consequence of eqs. (11.4,5,7,15), the following relationships can easily be shown to hold

$$\dot{\mathbf{e}} = 0; \qquad \mathbf{e} \cdot \mathbf{h} = 0; \tag{11.17a}$$

$$\mathbf{e} \cdot \mathbf{r} = \frac{h^2}{Gm} - r; \qquad \mathbf{e} \cdot \mathbf{e} = e^2. \tag{11.17b}$$

Therefore **e** is a constant of motion, lies in the orbital plane and its magnitude is the eccentricity. If we define the *true anomaly* f as the angle between **e** and **r**, from the first eq. (11.17b) we find again the polar equation of the conic sections. By comparison with eq. (11.13) we obtain $f = \theta - \omega$; that is to say, the Lenz vector is directed along the *apsidal line*, namely towards the pericentre of the orbit.

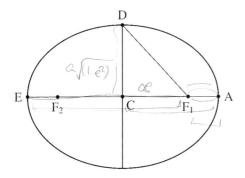

Figure 11.2 Geometry of elliptical orbits. We have the following relationships: $a = CA = F_1D$; $ae = CF_1$; $b = CD = a\sqrt{1-e^2}$; $r_p = F_1A = a(1-e)$; $r_a = F_1E = a(1+e)$.

11.3 Bound orbits

Let us first describe in some more detail the most interesting case, that of elliptic motion. It is easy to show (see Fig. 11.2) that

$$a = \frac{r_p + r_a}{2} = \frac{h^2}{Gm(1-e^2)}, \quad b = a\eta = a\sqrt{1-e^2}, \quad (11.18a)$$
$$r_p = a(1-e), \quad (11.18b)$$
$$r_a = a(1+e). \quad (11.18c)$$

are, respectively, the semimajor and semiminor axes, the pericentre ($f = 0$) and apocentre ($f = \pi$) distances (in astronomical literature, r_p and r_a are often denoted q and Q, respectively). Since the total area of the ellipse πab is swept by the radius vector in an orbital period T equals twice the areal velocity, we obtain (eq. (11.6))

$$T = \frac{2\pi a^2 \eta}{h} = 2\pi\sqrt{\frac{a^3}{Gm}}, \quad (11.19)$$

which shows that the period does not depend on the eccentricity, but only on the power 3/2 of the semimajor axis and the sum of the masses. In the solar system, this sum is always approximately equal (to within $\approx 0.1\%$) to the mass of the Sun, so that *Kepler's third law* (11.19) states that a^3/T^2 is constant for all planets. The same holds within each satellite system. Introducing the *mean motion* $n = 2\pi/T$ (i.e., the mean angular velocity along the orbit), we may write (11.19) in the more usual form

$$n^2 a^3 = Gm. \quad (11.19')$$

Eq. (11.19) points to a widely used way to obtain the product Gm for a celestial body, based on the measurement of the semimajor axis and the orbital period of a satellite (with negligible mass). Note that in astrodynamics and celestial mechanics the mass m never appears separately and is not a relevant quantity. To get the mass, one must use the laboratory value of the gravitational constant G, for instance with Cavendish-type experiments with attracting bodies of known mass. It should also be pointed out that the fractional uncertainty in Gm (for the Earth $\leq 10^{-9}$) is several orders of magnitude smaller than the fractional uncertainty in G, about 10^{-5}.

Conversely, if the semimajor axis is constant or if its variations can be accurately predicted, eq. (11.19) provides a good clock. A nice example is the Moon, which has been used for a long time as an astronomical clock to derive accurate longitude determinations on ships. Another outstanding example is provided by the binary pulsar PSR 1913+16, a binary system in which one member, a pulsar, emits a rotating beacon-like signal modulated by its orbital motion. This effect provides a very precise way to measure the orbital period. It was found that this period ($T = 27,908$ s) decreases at the rate $\dot{T} = -2.1 \times 10^{-12}$; the corresponding energy loss is derived from the decrease in the semimajor axis (see eq. (11.20)) and is most likely due to emission of *gravitational waves*.

The semimajor axis of an elliptical orbit can be related to the energy integral by using eq. (11.15) for the eccentricity:

$$a = \frac{h^2}{Gm(1-e^2)} = -\frac{h^2/Gm}{2Eh^2/(Gm)^2} = -\frac{Gm}{2E}. \tag{11.20}$$

Thus the semimajor axis depends on the energy, but not on the angular momentum. From eqs. (11.7,20) we can obtain the velocity as a function of r in an elliptic orbit

$$v^2 = Gm\left(\frac{2}{r} - \frac{1}{a}\right). \tag{11.21}$$

Note that if the body starts at the distance r, its semimajor axis (and therefore its orbital period) does not depend on the direction, but only on the magnitude of the initial velocity. Of course the eccentricity, and the shape of the orbit, will depend on the initial direction of the motion. The radial and transverse components of the velocity vector are also known as functions of the true anomaly f from eqs. (11.6,13):

$$v_r = \dot{r} = \frac{h^2}{Gm} \frac{e \sin f}{(1 + e \cos f)^2} \frac{df}{dt} = \frac{Gm}{h} e \sin f, \tag{11.22a}$$

$$v_t = r\dot{\theta} = \frac{h}{r} = \frac{Gm}{h}(1 + e \cos f). \tag{11.22b}$$

Besides the geometrical shape of the orbit, we are interested in the time law of the motion, notably the function $f(t)$. To get it, we introduce an additional angular variable u (the *eccentric anomaly*) such that

$$r = a(1 - e \cos u). \tag{11.23}$$

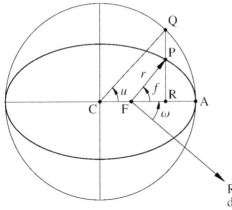

Figure 11.3 Geometric interpretation of the eccentric anomaly u. From the two relationships: $FR = r\cos f = CR - CF = a\cos u - ae$ and $PR = r\sin f = (b/a)QR = \sqrt{1-e^2}\,a\sin u$, eq. (11.23) immediately follows. Note that here we have a parametric representation of the ellipse with the origin at the centre.

Fig. 11.3 gives its geometric interpretation. From the relationships between r and f, and between r and u, it also follows that

$$\tan\frac{f}{2} = \left(\frac{1+e}{1-e}\right)^{1/2}\tan\frac{u}{2}. \tag{11.24}$$

From the conservation of angular momentum (eq. (11.6)), we have now

$$h(t-t_0) = \int_0^f df'\, r^2(f'), \tag{11.25}$$

where t_0 is the time of a passage through pericentre. By eq. (11.24) it can be easily shown that

$$df = \frac{\eta\, du}{1 - e\cos u}, \tag{11.26}$$

and therefore we have

$$h(t-t_0) = a^2\eta \int_0^u du'\,(1 - e\cos u'). \tag{11.27}$$

Since h is twice the areal velocity we have

$$M = n(t-t_0) = u - e\sin u. \tag{11.28}$$

Here M is the *mean anomaly*, by definition a linear function of time, which increases by 2π (as f and u also do) per revolution. If we know n (or T) and the time elapsed since the last pericentre passage, i.e. M/n, by inverting eq. (11.28) (known as *Kepler's equation*) we can obtain $u(t)$. Then we can derive f from eq. (11.24) and, finally, the position along the orbit. Kepler's equation admits analytical solution only for small values of the eccentricity (e.g., the infinite series method of Problem 11.11), while for a generic value of e it must be solved numerically, usually with successive approximations. As long as the eccentricity remains small, their convergence is very fast; care must be paid to situations with a very high eccentricity, for which tailored numerical methods were developed.

Orbits in three dimensions. To entirely specify the orbital motion one needs six quantities. A frequent choice is that of the *Keplerian elements* (discussed below), but note that Hamiltonian perturbation methods may profit from other parameterizations. Three elements are needed to identify the position in the orbital plane. For example, we can give the semimajor axis a, the eccentricity e and the true anomaly $f(t)$ at a specified time t (or, equivalently, the *mean anomaly at epoch* $M(t)$; conventionally, one often chooses the mean anomaly at the origin of time, namely $\sigma = -nt_0$, where t_0 corresponds to an arbitrary passage through the pericentre). In a Cartesian system with origin at the centre of mass, with the x-axis towards the pericentre and the z-axis along[1] $\mathbf{h}$ (Fig. 11.4), the coordinates of the body are $r(\cos f, \sin f, 0)$, with r given by eq. (11.13), or equivalently, $a(\cos u - e, \eta \sin u, 0)$.

Three additional elements are now required to specify the orientation of the orbit with respect to an arbitrary inertial reference frame $(\mathbf{e}_x, \mathbf{e}_y, \mathbf{e}_z)$. Although the choice of the latter is in principle arbitrary, when additional forces are present particular choices are suitable; for example, in the solar system one often takes $\mathbf{e}_z$ toward the pole of ecliptic and $\mathbf{e}_x$ toward the γ point (Fig. 3.3). For satellite dynamics, when planetary oblateness produces the principal perturbation, it is advantageous to identify the $(\mathbf{e}_x, \mathbf{e}_y)$ plane with the equatorial plane of some date.

The rotation between the inertial frame $(\mathbf{e}_x, \mathbf{e}_y, \mathbf{e}_z)$ and the orbit-bound reference frame, introduced above, is characterized by three Euler angles: the *longitude of the ascending node* Ω, the *inclination I* and the *argument of pericentre* ω. Defining a unit vector $\mathbf{e}_a$ along the intersection of the orbital plane with the reference plane $(\mathbf{e}_x, \mathbf{e}_y)$, we thus have

$$\mathbf{e}_a = \mathbf{e}_x \cos \Omega + \mathbf{e}_y \sin \Omega . \tag{11.29}$$

The *inclination I* is then the angle between $\mathbf{e}_z$ and $\mathbf{h}$, namely

$$\cos I = \mathbf{e}_z \cdot \mathbf{h}/h . \tag{11.30}$$

Usually I is assumed to range from 0 to π; prograde and retrograde orbits are those for which $I < \pi/2$ and $I > \pi/2$, respectively. Finally, the argument of pericentre ω is the angle between $\mathbf{e}_a$ and $\mathbf{e}$, from the ascending node to the apsidal line:

$$\cos \omega = \mathbf{e}_a \cdot \mathbf{e}_p , \tag{11.31}$$

---

[1] An observer standing along $\mathbf{h}$ sees the body moving in the counterclockwise sense.

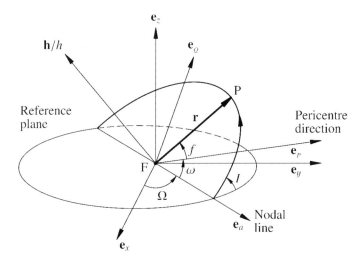

Figure 11.4. The orbit-bound vectors needed to define the angular orbital elements ω, Ω and I with respect to an arbitrary inertial reference system with unit vectors $\mathbf{e}_x$, $\mathbf{e}_y$ and $\mathbf{e}_z$.

with $\mathbf{e}_p = \mathbf{e}/e$ (Fig. 11.4). The five parameters a, e, I, Ω and ω specify the orbit completely; for the same purpose we could also take the two vectors $\mathbf{e}, \mathbf{h}$ (whose six components are not independent because $\mathbf{e} \cdot \mathbf{h} = 0$). They can be obtained from the initial conditions $\mathbf{r}_0$, $\dot{\mathbf{r}}_0$ with the integrals of motion E, $\mathbf{h}$ and $\mathbf{e}$. As a sixth element we can take $\sigma = -nt_0$; the whole set of Keplerian elements is then denoted by

$$c_k = (a, e, I, \Omega, \omega, -nt_0) \quad (k = 1, 2, \ldots, 6) . \qquad (11.32)$$

It should be noticed that any set of six orbital elements is indeterminate for *some* orbits. As an example, the Keplerian elements (11.32) are unsuitable for circular (ω and nt_0 indeterminate), equatorial (Ω and ω indeterminate) and collision orbits (I, Ω and ω indeterminate). In cases where the argument of pericentre ω becomes meaningless (equatorial or circular orbits), it is better to use instead of ω

$$\varpi = \omega + \Omega , \qquad (11.33)$$

the *longitude of pericentre*, reckoned from the origin of longitudes. The angular position along the orbit can be measured with the *true longitude*

$$\Lambda = \Omega + \omega + f = \varpi + f , \qquad (11.34)$$

or, more conveniently, with the *mean longitude*

$$\lambda = \Omega + \omega + M = \varpi + M \qquad (11.35)$$

in terms of the mean anomaly M.

The elements c_k determine the position $\mathbf{r}(t; c_k)$ and the velocity $\mathbf{v} = \dot{\mathbf{r}}(t; c_k)$ on the elliptic orbit at an arbitrary time t. Adopting the true anomaly f as independent variable we have

$$\mathbf{r} = r \left[\cos f\, \mathbf{e}_P + \sin f\, \mathbf{e}_Q\right], \quad (11.36a)$$

$$\mathbf{v} = \frac{na}{\eta}\left[-\sin f\, \mathbf{e}_P + (e + \cos f)\, \mathbf{e}_Q\right], \quad (11.36b)$$

where we have introduced

$$\mathbf{e}_Q = (\mathbf{h} \times \mathbf{e}_P)/h. \quad (11.37)$$

Similarly, in terms of the eccentric anomaly u we have

$$\mathbf{r} = a\left[(\cos u - e)\, \mathbf{e}_P + \eta \sin u\, \mathbf{e}_Q\right], \quad (11.38a)$$

$$\mathbf{v} = na\left(\frac{a}{r}\right)\left[-\sin u\, \mathbf{e}_P + \eta \cos u\, \mathbf{e}_Q\right]. \quad (11.38b)$$

The unit vectors $\mathbf{e}_P$ and $\mathbf{e}_Q$ can be expressed in terms of the angular elements (Ω, I, ω), namely:

$$\mathbf{e}_P = \begin{pmatrix} \cos\Omega\cos\omega - \cos I \sin\Omega \sin\omega \\ \sin\Omega\cos\omega + \cos I \cos\Omega \sin\omega \\ \sin I \sin\omega \end{pmatrix}, \quad (11.39a)$$

$$\mathbf{e}_Q = \begin{pmatrix} -\cos\Omega\sin\omega - \cos I \sin\Omega \cos\omega \\ -\sin\Omega\sin\omega + \cos I \cos\Omega \cos\omega \\ \sin I \cos\omega \end{pmatrix}. \quad (11.39b)$$

11.4 Unbound orbits

Unbound Keplerian orbits, either parabolic ($e = 1$) or hyperbolic ($e > 1$), also play an important role in solar system dynamics (for instance when one deals with the orbital evolution of comets, with a flyby's of a space probe near a planet or a minor body, or with close encounters of two orbiting bodies).

Parabolic orbits. For a parabolic orbit, the equation corresponding to (11.13) is

$$r = \frac{2r_p}{1 + \cos f} = \frac{r_p}{\cos^2(f/2)}, \quad (11.40)$$

where r_p is the distance from the origin (the focus) to the pericentre, namely the vertex of the parabola (see Fig. 11.5) and the true anomaly f is measured from the pericentre; r_p is used here because in this case $e = 1$ and $a \to \infty$.

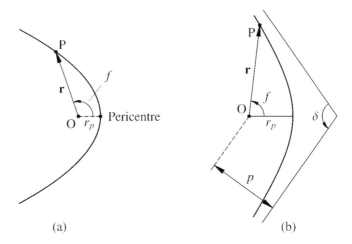

Figure 11.5. Geometry of parabolic and hyperbolic orbits.

The true anomaly f ranges in the interval $(-\pi, \pi)$. At $r = r_p$ the velocity vector is perpendicular to the radius vector, and from eqs. (11.6,7), with $E = 0$, we infer that $h = \sqrt{2Gmr_p}$ and the velocity at infinity vanishes. Then, from eqs. (11.6,40) we get

$$\frac{r_p^2}{\cos^4(f/2)} \frac{df}{dt} = h \qquad (11.41)$$

and, integrating by separating the variables,

$$\tan\left(\frac{f}{2}\right) + \frac{1}{3}\tan^3\left(\frac{f}{2}\right) = \left(\frac{Gm}{2r_p^3}\right)^{1/2} (t - t_0) \,. \qquad (11.42)$$

Here t_0 is the time of the pericentre passage. This cubic equation – called *Barker's equation* – can be solved analytically with a unique real solution.

Hyperbolic orbits. For hyperbolic orbits $E > 0$ and, as a consequence, we obtain $a < 0$ and $e > 1$ (see eqs. (11.15,20))[2]. The eqs. (11.13,18a,18b,20,21) are still valid. Some useful relationships hold for hyperbolic orbits relating the angle δ between the two asymptotes (which defines the range of variation $(\delta/2 - \pi, \pi - \delta/2)$ of the true anomaly), the pericentre distance r_p, the velocity at infinity v_∞ and the impact parameter p (distance of the asymptote from the focus; see Fig. 11.5). We can formally introduce a semimajor axis $a = -Gm/v_\infty^2$, if the energy, in analogy with the elliptical case, is $E = -Gm/2a$. Since for

---

[2] Sometimes a is assumed positive for hyperbolic orbits, too; this requires appropriate changes of sign in the definitions, namely $h^2 = Gma(e^2 - 1)$ instead (11.18a) and $a = Gm/(2E)$ instead of (11.20).

11. The two-body problem

$r \to \infty$ we have $f \to \arccos(-1/e)$, we obtain

$$\delta = 2\arccos(1/e) = 2\arccos\left(\frac{Gm}{Gm + r_p v_\infty^2}\right), \quad (11.43a)$$

$$r_p = \left(p^2 + \frac{G^2 m^2}{v_\infty^4}\right)^{1/2} - \frac{Gm}{v_\infty^2} = a(1-e), \quad (11.43b)$$

$$\cos^2\left(\frac{\delta}{2}\right) = \left(1 + \frac{p^2 v_\infty^4}{G^2 m^2}\right)^{-1}. \quad (11.43c)$$

These relationships can easily be proven using the angular momentum ($h = p v_\infty$) and the energy ($E = v_\infty^2/2$) integrals. Interestingly, these formulæ are valid also for $Gm < 0$, which formally applies to small dust particles ($\approx$ μm size) of interplanetary or interstellar origin, for which the radiation force is larger than the gravitational attraction (Sec. 15.4). Due to gravitational deflection, the effective cross-section σ_{eff} of a planet with radius R moving through a cloud of massless particles with relative velocity v_∞ is larger than the geometric cross-section $\sigma_g = \pi R^2$. From (11.43b) we obtain

$$\sigma_{\text{eff}} = \sigma_g \left(1 + \frac{2Gm}{R v_\infty^2}\right) = \sigma_g \left(1 + \frac{v_{\text{esc}}^2}{v_\infty^2}\right), \quad (11.44)$$

where $v_{\text{esc}} = \sqrt{2Gm/R}$ is the escape velocity from the planet's surface. Particles accreting on the planet with a small velocity at infinity ($v_\infty \ll v_{\text{esc}}$) thus are subject to a much larger effective cross-section; $[1 + (v_{\text{esc}}/v_\infty)^2]$ is usually termed the gravitational enhancement factor. To illustrate this focusing effect, consider dust particles approaching the Earth with $v_\infty \simeq 2$ km/s (corresponding to low-eccentricity, heliocentric and prograde orbits); since $v_{\text{esc}} \simeq 11.2$ km/s, the Earth cross-section is enhanced by a factor $\simeq 30$ (this happens for μm particles in the dust bands discussed in Sec. 14.6). During the runaway phase of the planetary embryo formation the gravitational enhancement factor can be as large as $\simeq 1000$ (Sec. 16.3).

From $dt = r^2 df/h$, using eqs. (11.13,18a), we have also

$$\left[\frac{Gm}{-a^3(1-e^2)^3}\right]^{1/2} dt = \frac{df}{(1 + e\cos f)^2}, \quad (11.45)$$

and integrating from t_0 (pericentre passage) to t with the substitution

$$\tan\left(\frac{f}{2}\right) = \left(\frac{e+1}{e-1}\right)^{1/2} \tanh\left(\frac{F}{2}\right), \quad (11.46)$$

we find that

$$\sqrt{\frac{Gm}{-a^3}}(t - t_0) = e\sinh F - F, \quad (11.47)$$

analogous to Kepler's equation (11.28) for elliptical orbits (recall that $F = f = 0$ at pericentre). From F one can find r via f or, directly, through the equation

$$r = -a(e \cosh F - 1), \qquad (11.48)$$

which is equivalent to (11.23) for elliptic orbits.

– PROBLEMS –

11.1. *The weakness of gravity.* To balance the gravitational force between two bodies one can place on each of them, for every nucleon, a positive charge equal to the fraction ϵ of a proton. What is ϵ?

11.2. *The orbit of the Solar Probe.* An artificial satellite is in a circular orbit at 1 AU. Evaluate the velocity decrement necessary to inject it in an elliptical orbit with perihelion of $4 R_\odot$. Estimate its equilibrium temperature when the distance r from the centre of the Sun is below $8 R_\odot$ and the approximate time spent in this range of r.

11.3. Write a simple program to solve iteratively Kepler's equation with the following procedure: $u_0 = M$, $u_1 = M + e \sin u_0$, etc. Calculate within $5''$ the eccentric anomaly of Jupiter 3 years after its perihelion passage, knowing that $2\pi/n = 11.8622$ y and $e = 0.048$; the same for the asteroid Icarus, with $2\pi/n = 1.1191$ y and $e = 0.823$ (in the second case use the Newton-Raphson method).

11.4. Show that the orbital effect of the solar radiation on a perfectly absorbing body of mass m and cross-section S is equivalent to a change δG in the gravitational constant, and calculate it.

11.5. A *solar sail* is a spacecraft with a ratio $\beta_r = -\delta G/G$ greater than unity; in principle it can be used for interstellar travel. Assuming an instantaneous deployment in a circular solar orbit at 1 AU, evaluate the time needed to reach the nearest star α Centauri, 4.3 light-years away.

11.6.* Radial motion of two point-like bodies (this problem is equivalent to the dynamics of an isotropic and homogeneous universe under its own gravitational force, when its entire mass-energy content is the rest mass). If they are initially at rest at the distance r_0, find $r(t)$ and the time T to the collision; prove the similarity law $T r_0^{-3/2} = $ const for the zero energy case.

11.7. An artificial satellite is placed at the distance r_0 from the Earth's centre with a velocity v_0 orthogonal to the radius. In what conditions it is bound to the Earth? What are its semimajor axis and eccentricity? What is the minimum value of v_0 for which no re-entry occurs?

11.8.* Determine the trajectory of the warhead of an intercontinental ballistic missile released by its booster at a height h above the Earth's surface with a speed v_0 directed at an angle γ from the radial direction. Which is the value of γ which yields the largest range? Show that the same initial conditions correspond, for a given range, to the trajectory of minimum energy.

11.9.* A test body moves around the Sun in an almost circular orbit. Find its motion by linearizing the equations (11.4). What happens if the force depends upon the distance with a power law with an exponent different from -2? (Hint: For the correction in the orbital plane use complex coordinates.)

11.10. Find the rate at which the binary pulsar loses orbital energy and its lifetime (see Sec. 11.3).

11.11.* The eccentricity introduces in a orbit harmonics of the orbital period. Evaluate their amplitude for a small eccentricity e a and large harmonic order p, and show that it decreases as $\exp(-p[\ln(2/e) - 1])$. (Hint: In computing the Fourier transform of, say, the distance r in terms of the eccentric anomaly u, use the property

$$\exp(ipe \sin u) = \sum_{q=-\infty}^{\infty} J_q(ep) \exp(iqu)$$

of the Bessel functions and their asymptotic (Debye's) expansions for a larger order. See H.C. Plummer, *An introductory treatise on dynamical astronomy*, Dover (1960).)

11.12.* The position $\mathbf{r}_0$ and the velocity $\mathbf{v}_0$ at an arbitrary time t_0 define a basis in the orbital plane; the position vector $\mathbf{r}(t)$ on an elliptic orbit can be thus expressed as $\mathbf{r}(t) = f(t, t_0) \mathbf{r}_0 + g(t, t_0) \mathbf{v}_0$. The coefficients are referred to as f- and g-functions; find the coefficients of their polynomial expansions in $(t - t_0)$ in terms of the initial values of $r_0 = |\mathbf{r}_0|$, $\dot{r}_0 = (\mathbf{r}_0 \cdot \mathbf{v}_0)/r_0$ and the semimajor axis.

11.13.* Express the two-body equations of motion (11.4) in terms of the regularizing time τ defined by $dt = r d\tau$ (Sundman's transformation). The variable τ is important for the numerical integration of the N-body problem if collisions are not excluded. Show that the integration of this relationship with respect to τ provides an alternative derivation of Kepler's equation.

11.14.* Evaluate the orbital average of a second order polynomial in $\cos u$ and $\sin u$. At a more general level, prove

$$\frac{1}{\pi} \int_0^\pi \left(\frac{r}{a}\right)^\gamma \cos(ju) \, dM = \left(-\frac{e}{2}\right)^j \frac{\Gamma(\gamma+2)}{j! \, \Gamma(\gamma+2-j)} F\left(\frac{j-\gamma-1}{2}, \frac{j-\gamma}{2}; j+1; e^2\right),$$

where γ and j are arbitrary real and integer numbers, respectively, and e is the eccentricity; Γ and F are the gamma and the hypergeometric functions, respectively.

11.15. Using the adiabatic invariance (Sec. 10.2) show that when the mass m changes slowly with time, the semimajor axis a of an elliptical Keplerian orbit is inversely proportional to m. What is the rate of change of the semimajor axis of the Earth due to the solar wind?

11.16.* The *Lagrange expansion theorem* gives the power expansion

$$y = x + \sum_{j=1}^{\infty} \frac{e^j}{j!} \frac{d^{j-1}}{dx^{j-1}} (\phi(x))^j$$

of a function $y(x, e)$ implicitly defined by $y = x + e\phi(y)$ ($e < 1$ and $\phi(y)$ is an arbitrary function). Use this theorem to obtain the elliptic expansion in powers of the eccentricity of the true anomaly f in terms of the mean anomaly M. (Hint: First express M as a function of f from the angular momentum conservation law (11.6).)

Chapter 12

PERTURBATION THEORY

In the two-body approximation discussed in the previous chapter the systems of planets and their satellites are described by listing their Keplerian elements c_a in relation to their primary; but to investigate the dynamical evolution of the solar system we need to understand the changes in the elements. This chapter is devoted to this crucial problem, with its subtle mathematical aspects. The main concept here is the *osculating orbit*: the motion is represented, using a one-to-one mapping, by a changing elliptical orbit. In order to solve this problem, we need the rates of change of its six orbital elements c_a. For a single body under the influence of a generic perturbing acceleration **F**, they are the famous Gauss equations. Four of them can be derived in a direct and elementary way from the conservation principles (energy and angular momentum) and, as an example, are used to describe the effect of atmospheric drag on a space vehicle and its re-entry. When **F** is conservative and determined by a perturbing function, the techniques of Hamiltonian dynamics are more efficient, leading to the Lagrange equations; as an illustration, we apply them to the effect of the oblateness of the primary. Solving them in terms of the osculating elements requires an understanding of subtle approximations, in particular in relation to long-term secular effects and the problem of resonances. Some of these more complex issues will be discussed in Ch. 15.

12.1 Gauss perturbation equations

When perturbations are present, the variations of the conserved quantities of the two-body problem, in particular the energy and the angular momentum, show up the corresponding variations of the orbital elements upon which they depend. By this technique we can easily obtain four of the six Gauss equations. Below, we use a more formal approach – with a precise definition of the osculating elements – to derive the remaining two Gauss equations.

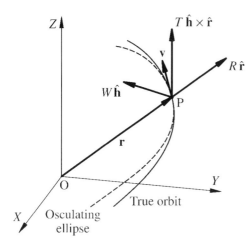

Figure 12.1 The standard decomposition of the perturbing force **F** in three orthogonal components: radial R, transverse T and normal to the orbit W.

Perturbations from conservation laws. We consider a perturbed Kepler motion

$$\frac{d^2\mathbf{r}}{dt^2} = -\frac{Gm}{r^3}\mathbf{r} + \mathbf{F}, \qquad (12.1)$$

where **F** is a generic function of position, velocity and time; the validity of the approximation methods introduced below requires **F** to be in some sense small, but the Gauss equations are correct for any perturbation **F**. Since its effects strongly depend on the relation between its direction and the orbital geometry, it is useful – and practically necessary – to decompose it in three orthogonal components. The most obvious is: R, in the radial direction $\mathbf{r} = r\hat{\mathbf{r}}$; W, along the angular momentum per unit mass $\mathbf{h} = h\hat{\mathbf{h}}$ normal to the orbital plane; and T, transversally in the orbital plane and along $\hat{\mathbf{h}} \times \hat{\mathbf{r}}$:

$$R = \mathbf{F} \cdot \hat{\mathbf{r}}, \quad W = \mathbf{F} \cdot \hat{\mathbf{h}}, \quad T = \mathbf{F} \cdot (\hat{\mathbf{h}} \times \hat{\mathbf{r}}). \qquad (12.2)$$

Other decompositions may be suitable in particular problems; for instance, we may use the components $F_P = \mathbf{F} \cdot \mathbf{e}_P$, $F_Q = \mathbf{F} \cdot \mathbf{e}_Q$ and W, with $\mathbf{e}_P$ toward the pericentre of the osculating orbit and $\mathbf{e}_Q$ from eqs. (11.39). Since the basis $(\mathbf{e}_P, \mathbf{e}_Q, \hat{\mathbf{h}})$ changes slowly, this choice is suitable for perturbations **F** with the same property (e.g., the solar radiative force on a body orbiting a planet).

The unperturbed energy per unit of mass $E = -Gm/(2a) = v^2/2 - Gm/r$ changes at the rate $dE/dt = \mathbf{F} \cdot \mathbf{v}$, the power of the perturbing force **F**. With eqs. (11.22) we obtain

$$\frac{dE}{dt} = R v_r + T v_t = [R (e \sin f) + T (1 + e \cos f)]\frac{Gm}{h}, \qquad (12.3)$$

and for the osculating semimajor axis

$$\frac{da}{dt} = \frac{2a^2}{Gm}\frac{dE}{dt} = \frac{2}{n\eta}[T + e(T\cos f + R\sin f)], \qquad (12.4)$$

where we have used also eqs. (11.20) and (11.18a) for the osculating energy E and the angular momentum h; $n = \sqrt{Gm/a^3}$ is the osculating mean motion and $\eta = \sqrt{1-e^2}$, as in Ch. 11. In the limit $e \to 0$, $da/dt = 2T/n + O(e)$. Thus the transverse component T is the most effective one in changing the size and, by Kepler's third law, the period of the orbit. For instance, in the case of a circular orbit with a small, but constant T, the orbital radius and the orbital period change linearly with time; as a consequence, the change in mean longitude accumulates quadratically and can become quite large. It is intriguing to notice that a drag force, which has a negative T component and causes an overall energy loss, by eq. (12.4) produces a decrease in the semimajor axis; in turn, by Kepler's third law, the orbital velocity and the kinetic energy increase. Thus orbiting bodies, contrary to common sense, *accelerate* when subjected to drag (see Problem 12.1).

The magnitude h of the angular momentum is changed by the torque rT along **h**, namely

$$\frac{dh}{dt} = \mathbf{F} \cdot (\hat{\mathbf{h}} \times \mathbf{r}) = rT . \quad (12.5)$$

Differentiating the first equation in (11.18a), which gives h in terms of the osculating elements, we get

$$rT = \frac{dh}{dt} = \frac{Gm}{2h}\left[\eta^2 \frac{da}{dt} - 2ea\frac{de}{dt}\right]. \quad (12.6)$$

Solving for de/dt, substituting r and da/dt from eqs. (11.23) and (12.4), respectively, we obtain

$$\frac{de}{dt} = \frac{\eta}{na}[R\sin f + T(\cos f + \cos u)] , \quad (12.7)$$

where $\cos u = (e + \cos f)/(1 + e\cos f)$ (see Fig. 11.3). From eqs. (12.4) and (12.7) we see that only forces in the orbital plane can change both the size and the shape of the orbit, namely a and e.

When using the (F_P, F_Q, W) decomposition of the perturbing force, we obtain instead

$$\frac{de}{dt} = \frac{\eta}{na}\left[F_Q\left(2 - \frac{\sin^2 f}{1 + e\cos f}\right) - F_P \sin f \cos u\right]. \quad (12.8)$$

A force with a constant or slowly varying components (F_P, F_Q) causes a secular change in the eccentricity, as, for example, the solar radiation force on a dust particle orbiting around a planet. Since the main change in (F_P, F_Q) is due to the orbital motion of the planet around the Sun, the eccentricity is modulated with the planet's orbital period.

The orientation of the osculating orbital plane, determined by $\hat{\mathbf{h}}$ (or by the inclination I and the nodal longitude Ω), is changed only if $W \neq 0$, producing a torque orthogonal to $\mathbf{h}$:

$$h \frac{d\hat{\mathbf{h}}}{dt} = W \left(\mathbf{r} \times \hat{\mathbf{h}} \right). \tag{12.9}$$

Thus the unit vector $\hat{\mathbf{h}}$ moves with the angular velocity $(W/h)\,\mathbf{r}$. Let us split this torque into the component $r W \sin(\omega + f)$ along the nodal line and the component perpendicular to it in the orbital plane, $r W \cos(\omega + f)$ (recall that the torque is perpendicular to $\mathbf{r}$). The first is perpendicular to $\mathbf{h}$ and $\mathbf{e}_z$ (the reference axis for the inclination) and causes $\mathbf{h}$ to precess about $\mathbf{e}_z$ with angular velocity

$$\frac{d\Omega}{dt} = \frac{r W \sin(\omega + f)}{h \sin I} = \frac{W}{na\eta \sin I} \left(\frac{r}{a} \right) \sin(\omega + f) \, ; \tag{12.10}$$

I is the angle between $\mathbf{e}_z$ and $\mathbf{h}$. The second component turns $\mathbf{h}$ around the nodal line and changes the inclination at the rate

$$\frac{dI}{dt} = \frac{r W \cos(\omega + f)}{h} = \frac{W}{na\eta} \left(\frac{r}{a} \right) \cos(\omega + f) \tag{12.11}$$

(the angle between $\mathbf{h}$ and the nodal line is, by definition, $\pi/2$). The lunisolar precession (4.7) is an example of the nodal drift (12.10); more generally, eqs. (12.10) and (12.11) can be interpreted as the effect of an external torque upon a gyroscope with angular momentum $\mathbf{h}$ aligned along the normal to the orbital plane.

In Sec. 12.3 we apply the previous results to the drag due to neutral molecules experienced by a satellite in an atmosphere.

Osculating elements and the last Gauss' equations. Though the conservations laws might also be used to derive Gauss' equations of the last two elements (Problem 12.19), at this point we take the opportunity to define the osculating elements exactly and derive them with a general mathematical procedure. Consider in the space $(\mathbf{r}, \mathbf{v})$ a generic dynamical system fulfilling the equations

$$\frac{d\mathbf{r}}{dt} = \mathbf{G}_0 + \mathbf{G}, \quad \frac{d\mathbf{v}}{dt} = \mathbf{F}_0 + \mathbf{F}, \tag{12.12}$$

where $(\mathbf{G}(\mathbf{r}, \mathbf{v}, t), \mathbf{F}(\mathbf{r}, \mathbf{v}, t))$ are "perturbations". In the non-relativistic two-body problem $\mathbf{G}_0 = \mathbf{v}, \mathbf{G} = 0$ and $\mathbf{F}_0 = -Gm\,\mathbf{r}/r^3$. Let $(\mathbf{r}(c_a, t), \mathbf{v}(c_a, t); a, b = 1, 2, \ldots, 6)$ be the general solution of the unperturbed problem, where the constants of integration (the *orbital elements*) $c_a = (c_1, \ldots, c_6)$ describe the whole six-dimensional set; they are defined by

$$\frac{\partial \mathbf{r}(c_a, t)}{\partial t} = \mathbf{G}_0, \quad \frac{\partial \mathbf{v}(c_a, t)}{\partial t} = \mathbf{F}_0. \tag{12.13}$$

12. Perturbation theory

We now let c_a be time dependent, so that a Lagrangian (d/dt) and a Eulerian ($\partial/\partial t$) derivative are defined (the summation convention for repeated indexes is assumed)

$$\frac{d}{dt} = \frac{\partial}{\partial t} + \frac{dc_a}{dt}\frac{\partial}{\partial c_a} .$$

If the *osculating conditions*

$$\frac{\partial \mathbf{r}(c_a, t)}{\partial c_a}\frac{dc_a}{dt} = \mathbf{G}, \quad \frac{\partial \mathbf{v}(c_a, t)}{\partial c_a}\frac{dc_a}{dt} = \mathbf{F} \qquad (12.14)$$

are satisfied, the functions $(\mathbf{r}(c_a(t), t), \mathbf{v}(c_a(t), t))$ fulfil (12.12) – whence $c_a(t)$ are the osculating elements. This is similar to the interaction representation of quantum theory. If Ψ is a function of $\mathbf{r}$ and $\mathbf{v}$, it depends on time both directly and, through the orbital elements, indirectly; its Lagrangian and Eulerian derivatives differ only because of the perturbation:

$$\frac{d\Psi}{dt} - \frac{\partial \Psi}{\partial t} = \frac{\partial \Psi}{\partial c_a}\frac{dc_a}{dt} = \left(\frac{\partial \Psi}{\partial \mathbf{r}}\frac{\partial \mathbf{r}}{\partial c_a} + \frac{\partial \Psi}{\partial \mathbf{v}}\frac{\partial \mathbf{v}}{\partial c_a}\right)\frac{dc_a}{dt} = \frac{\partial \Psi}{\partial \mathbf{r}} \cdot \mathbf{G} + \frac{\partial \Psi}{\partial \mathbf{v}} \cdot \mathbf{F} . \quad (12.15)$$

In the Kepler case, if Ψ is a function of $\mathbf{r}$ only,

$$\frac{d\Psi}{dt} - \frac{\partial \Psi}{\partial t} = \frac{\partial \Psi}{\partial c_a}\frac{dc_a}{dt} = 0 . \qquad (12.16)$$

Perturbation of the argument of the pericentre and $\sigma = -nt_0$.– To obtain the motion of the pericentre we take $\Psi = r = h^2/[Gm(1 + e\cos f)]$ in (12.16), and recall $rT\, dt = dh$ (eq. (12.6)). The true anomaly f has a direct dependence on time through the conservation of angular momentum (11.6); but, since it is measured from the pericentre, whose longitude ω, on turn, is reckoned from the node, is affected by both. If $d\theta$ is the absolute change in longitude, from Fig. 11.4 we see that $df = d\theta - d\omega - \cos I\, d\Omega$, and

$$\frac{df}{dt} - \frac{\partial f}{\partial t} = -\frac{d\omega}{dt} - \cos I\, \frac{d\Omega}{dt} . \qquad (12.17)$$

Using the previous equation and eq. (12.16) for $\Psi = r$, we obtain:

$$e \sin f \left(\frac{d\omega}{dt} + \cos I \frac{d\Omega}{dt}\right) = \eta\frac{2T}{na} - \cos f\, \frac{de}{dt} ;$$

using (12.7) for the eccentricity, this boils down to

$$\frac{d\omega}{dt} + \cos I\, \frac{d\Omega}{dt} = \frac{\eta}{nae}\left[-R\cos f + T\left(\sin f + \frac{1}{\eta}\sin u\right)\right] . \qquad (12.18)$$

The pericentre argument ω has no meaning when the eccentricity and/or the inclination vanish; in Sec. 11.3 we pointed out that then the pericentre longitude $\varpi = \Omega + \omega$ provides a better description of its motion. It fulfils

$$\frac{d\varpi}{dt} - 2\sin^2(I/2)\frac{d\Omega}{dt} = \frac{\eta}{nae}\left[-R\cos f + T\left(\sin f + \frac{1}{\eta}\sin u\right)\right]. \quad (12.19)$$

These equations exhibit an apparent singularity when $e \to 0$, a result of the loss of definition of the pericentre itself, hence its argument ω and its longitude ϖ: for a change $\delta\omega$, the actual observable is $e\,\delta\omega$. This anomaly can be remedied by using the non-singular elements h and k given in (12.50); the corresponding Gauss' equations are obtained in an elementary way from eqs. (12.7) and (12.19).

For the last element, the naïve choice is $-nt_0$ (see (11.32)), where t_0 is the time of a given passage at pericentre. However, when the mean anomaly n changes this is not a suitable definition. In the unperturbed case the mean angular advance from the pericentre is the mean anomaly $M = n(t - t_0)$; but if the semimajor axis (and n) is time dependent, such advance is rather $\chi(t)-\chi(t_0)$, where

$$\chi(t) = \int_0^t dt'\, n(t'). \quad (12.20)$$

A satisfactory choice for the last element is then

$$c_6 = \epsilon' = M - \int_0^t dt'\, n(t') = M - \chi(t); \quad (12.21)$$

the origin of time is conventional. Its rate

$$\frac{d\epsilon'}{dt} = \left(\frac{r}{a}\right)\frac{du}{dt} - \frac{de}{dt}\sin u - n. \quad (12.22)$$

is expressed in terms of the total derivative du/dt. Note that in the unperturbed motion we have $(1 - e\cos u)(du/dt) = n$, which directly follows from the Kepler equation $u - e\sin u = n(t - t_0)$. Now we use again (12.16) for $\Psi = r = a(1 - e\cos u)$, this time expressed in terms of the eccentric anomaly u, and we get:

$$\frac{ae}{\eta}\sin f \left[\left(\frac{r}{a}\right)\frac{du}{dt} - n\right] = a\frac{de}{dt}\cos u - \frac{da}{dt}\left(\frac{r}{a}\right).$$

Combining this equation with (12.22), (12.4) and (12.7), we finally obtain the required rate:

$$\frac{d\epsilon'}{dt} = -\frac{2}{na}\left(\frac{r}{a}\right)R - \eta\left(\frac{d\omega}{dt} + \cos I\,\frac{d\Omega}{dt}\right). \quad (12.23)$$

12. Perturbation theory

As mentioned earlier, at small inclinations

$$\epsilon = \varpi + M - \chi(t) = \epsilon' + \varpi \qquad (12.24)$$

is a better choice then ϵ'. Its rate is immediately obtained using (12.10) and (12.18), namely

$$\frac{d\epsilon}{dt} = -\frac{2}{na}\frac{r}{a}R + (1 - \eta)\frac{d\varpi}{dt} + 2\eta \sin^2(I/2)\frac{d\Omega}{dt} . \qquad (12.25)$$

With non-gravitational effects such as the Yarkovsky and Poynting-Robertson drags (Sec. 15.4), or tides (Sec. 15.3), the most important long-term orbital perturbation in the mean anomaly is due to χ (eq. (12.21)), while the effect of the change of ϵ is usually much smaller. Indeed, a secular change Δa in the semimajor axis produces, from Kepler's third law, the fractional change $\Delta n/n = -(3/2)\Delta a/a$ in the mean motion. Integration shows that χ depends quadratically on time, while the change in ϵ is at most linear. When there are only gravitational effects, and no secular drift of the semimajor axis occurs, ϵ and χ may be of the same importance.

12.2 Lagrange perturbation equations

When the perturbing force is conservative, to wit, is given by a scalar $\mathcal{R}$ – the *perturbing function* – as $\mathbf{F} = -\nabla \mathcal{R}$, the use of the R, T and W components is generally not suitable, but the more powerful Hamiltonian approach can be used. We now give a bare compendium, without proofs, of the mathematics we need, to be used also in Ch. 17.

A compendium of analytical mechanics. In an N-dimensional configuration space with coordinates q^i ($i, j = 1, 2, \ldots N$; sometimes, for shortness, the index is understood) we envisage a dynamical system which fulfils a *principle of stationary action*, characterized by a *Lagrange function* $L(q, \dot{q}, t)$ and the corresponding action

$$S[q(t)] = \int_{t_1}^{t_2} dt\, L(q, \dot{q}, t) , \qquad (12.26)$$

which takes values in the space of functions $q(t)$ in the interval (t_1, t_2). The action principle says that $q(t)$ is a possible trajectory of the system if S is stationary on it, i.e., if the change δS vanishes for arbitrary infinitesimal variations $\delta q(t)$ of $q(t)$, null at the boundary: $\delta q(t_1) = \delta q(t_2) = 0$. This is equivalent to the second order *Euler-Lagrange equations*:

$$\frac{\partial L}{\partial q^i} - \frac{d}{dt}\left(\frac{\partial L}{\partial \dot{q}^i}\right) = 0 . \qquad (12.27)$$

Gravitational motion in a potential $U(\mathbf{r})$ (per unit mass) corresponds to the Lagrange function $L = |\dot{\mathbf{r}}|^2/2 - U(\mathbf{r})$. The quantity

$$p_i = \frac{\partial L}{\partial \dot{q}^i} \qquad (12.28)$$

is the *momentum* conjugate to q^i. If L does not depend on q^i, called then an *ignorable coordinate*, the corresponding momentum is a constant of the motion.

To construct the Hamiltonian theory, one assumes that eq. (12.28) can be uniquely solved with respect to $\dot{q}^i$, giving $\dot{q}^i(q,p,t)$; this allows working in a $2N$-dimensional *phase space*[1] (q^i, p_i). We then construct the *Hamiltonian function*

$$H(q, p, t) = \dot{q}^i p_i - L \,. \qquad (12.29)$$

It is easily shown that the orbit $(q(t), p(t))$ in phase space fulfils the set of $2N$ first order *Hamilton equations*

$$\frac{dq^i}{dt} = \frac{\partial H}{\partial p_i}, \qquad \frac{dp_i}{dt} = -\frac{\partial H}{\partial q^i}, \qquad (12.30)$$

which are equivalent to the Lagrange equations (12.27). For any pair F, G of functions in phase space, we define the *Poisson bracket*

$$\{F, G\} = -\{G, F\} = \frac{\partial F}{\partial q^i}\frac{\partial G}{\partial p_i} - \frac{\partial G}{\partial q^i}\frac{\partial F}{\partial p_i} \,; \qquad (12.31)$$

they fulfil the *Jacobi identity*

$$\{E, \{F, G\}\} + \{F, \{G, E\}\} + \{G, \{E, F\}\} = 0 \,. \qquad (12.32)$$

The motion (12.30) generates a flow in phase space; a function $F(q, p, t)$ changes according to

$$\frac{dF}{dt} = \frac{\partial F}{\partial t} + \{F, H\} \,, \qquad (12.33)$$

with an important corollary for the Hamiltonian itself:

$$\frac{dH}{dt} = \frac{\partial H}{\partial t} \,; \qquad (12.34)$$

H is a constant of the motion when it does not depend explicitly on time.

---

[1] We write the index i in the upper and lower position to express the fact that the infinitesimal quantities dq^i and dp_i are, respectively, covariant and contravariant with respect to generic coordinate transformations in phase space; we also understand summation with respect to the same index, if repeated in the two positions.

12. Perturbation theory

The question arises, under which conditions new coordinates $(q'^i, p'_i, i = 1, ..N)$ in phase space still fulfil the Hamiltonian equations? A mapping from (q^i, p_i) to (q'^i, p'_i) with this property is a *contact transformation* and is characterized by a generating function $W(q, q', t)$, such that

$$p'_i = -\frac{\partial W}{\partial q'^i}, \qquad p_i = \frac{\partial W}{\partial q^i} ; \qquad (12.35)$$

the new Hamiltonian is

$$H'(q', p', t) = H(q, p, t) + \frac{\partial W}{\partial t} . \qquad (12.36)$$

The *Hamilton-Jacobi equation* for the generating function W

$$H\left(q, \frac{\partial W}{\partial q}, t\right) + \frac{\partial W}{\partial t} = 0 \qquad (12.37)$$

follows from a requirement $H' = 0$. It turns out that when H is time-independent, $W = -Et + W_0(q, q')$ (in terms of the energy E); then the Hamilton-Jacobi equation reduces to

$$H\left(q, \frac{\partial W_0}{\partial q}\right) = E . \qquad (12.38)$$

The new Hamiltonian H' is constant, and so are the new coordinates q', p'; they provide the solution $q(t), p(t)$ as function of $2N$ arbitrary constants. In the two-body problem this allows the construction of Keplerian orbital elements which are conjugate variables (Problems 12.9 and 12.10). In general, however, the Hamilton-Jacobi equation (12.38) rarely admits a solution and the problem is generally not integrable.

Perturbed systems. We now split the Hamiltonian $H = H_0 + \mathcal{R}$ in an "unperturbed term" H_0 and a "perturbation" $\mathcal{R}$. We just compare two dynamical systems and, so far, make no assumption about smallness. Introduce now $2N$ constants of the motion c_a for the *unperturbed dynamical system*, fulfilling

$$\{H_0, c_a\} = 0 ; \qquad (12.39)$$

then, from eq. (12.33), we obtain the required

$$\frac{dc_a}{dt} = \{c_a, \mathcal{R}\} = \sum_b \{c_a, c_b\} \frac{\partial \mathcal{R}}{\partial c_b} . \qquad (12.40)$$

In addition to the Poisson brackets, we define the *Lagrange bracket* for two functions F and G in phase space as

$$[F, G] = -[G, F] = \frac{\partial q^i}{\partial F}\frac{\partial p_i}{\partial G} - \frac{\partial q^i}{\partial G}\frac{\partial p_i}{\partial F} ; \qquad (12.41)$$

the partial derivatives with respect to F and G imply that the $2N - 2$ other dynamical variables are kept constant. In particular, for the constants of the motion c_a,

$$[c_a, c_b] = \frac{\partial q^i}{\partial c_a} \frac{\partial p_i}{\partial c_b} - \frac{\partial q^i}{\partial c_b} \frac{\partial p_i}{\partial c_b} .$$

Applying the compound differentiation rule, so that

$$\frac{\partial c_a}{\partial c_b} = \delta_a^b = \frac{\partial c_a}{\partial q^i} \frac{\partial q^i}{\partial c_b} + \frac{\partial c_a}{\partial p_i} \frac{\partial p_i}{\partial c_b} ,$$

it is easy to show that the two kinds of brackets generate two reciprocal $2N \times 2N$ matrices

$$\sum_b \{c_a, c_b\} [c_b, c_d] = -\delta_{ad} . \quad (12.42)$$

This is a useful relation, because if the Lagrange brackets are easier to calculate, the Poisson brackets are obtained by inversion. Note that the Lagrange bracket of two constants of motion is also constant.

Perturbation of the Kepler problem. For the perturbation of Kepler's problem $N = 3$, $q^i = r^i$ and $p_i = v^i$ are Cartesian coordinates (the distinction between covariant and contravariant indices is now irrelevant) and

$$H = H_0 + \mathcal{R} = \frac{1}{2} p^2 - \frac{Gm}{r} + \mathcal{R} . \quad (12.43)$$

The first-hand choice of the Keplerian elements is

$$(a, e, I, \Omega, \omega, \sigma = -nt_0) ;$$

the perturbing function $\mathcal{R}$ depends on these elements[2] and on time. Since the Lagrange brackets are constant, one can compute them at the pericentre (e.g., from eqs. (11.36) or (11.38)), considerably simplifying the algebra. It turns out that only six brackets do not vanish, namely:

$$[\Omega, a] = \frac{na\eta}{2} \cos I , \quad [\omega, a] = \frac{na\eta}{2} , \quad [\sigma, a] = \frac{na}{2} , \quad (12.44)$$

$$[e, \Omega] = \frac{na^2 e}{\eta} \cos I , \quad [I, \Omega] = na^2 \eta \sin I , \quad [e, \omega] = \frac{na^2 e}{\eta} .$$

After a brief algebra, one gets from (12.40) the *Lagrange perturbation equations*[3]

$$\frac{1}{a} \frac{da}{dt} = -\frac{2}{na^2} \frac{\partial \mathcal{R}}{\partial \sigma} , \quad (12.45a)$$

---

[2] As already discussed in Sec. 12.1, σ is not very suitable and shall be replaced.
[3] $\mathcal{R}$ in our definition (12.43) has a meaning of the perturbing potential energy per unit mass; many textbooks use a "potential function" with the inverse sign.

12. Perturbation theory

$$\frac{de}{dt} = \frac{\eta}{na^2 e}\left(\frac{\partial \mathcal{R}}{\partial \omega} - \eta \frac{\partial \mathcal{R}}{\partial \sigma}\right), \quad (12.45b)$$

$$\frac{dI}{dt} = \frac{1}{na^2 \eta \sin I}\left(\frac{\partial \mathcal{R}}{\partial \Omega} - \cos I \frac{\partial \mathcal{R}}{\partial \omega}\right), \quad (12.45c)$$

$$\frac{d\Omega}{dt} = -\frac{1}{na^2 \eta \sin I} \frac{\partial \mathcal{R}}{\partial I}, \quad (12.45d)$$

$$\frac{d\omega}{dt} = \frac{\cot I}{na^2 \eta} \frac{\partial \mathcal{R}}{\partial I} - \frac{\eta}{na^2 e} \frac{\partial \mathcal{R}}{\partial e}, \quad (12.45e)$$

$$\frac{d\sigma}{dt} = \frac{2}{na} \frac{\partial \mathcal{R}}{\partial a} + \frac{\eta^2}{na^2 e} \frac{\partial \mathcal{R}}{\partial e}. \quad (12.45f)$$

It is noteworthy that the time derivatives of a, e and I contain only the partial derivatives of with respect to Ω, ω and σ and vice versa.

As for Gauss's equations, the choice as an orbital element of $\sigma = -nt_0$, which appears in the mean anomaly $M = nt + \sigma$, is naïve and not suitable. When the semimajor axis, hence n, changes, M has no physical meaning; the mean phase is really $\chi = \int n\, dt$ (eq. (12.21)). To see the bad consequences of the previous choice in the Hamiltonian formulation, note that $\mathcal{R}$ depends on the semimajor axis both explicitly and, through M and Kepler's law, implicitly

$$\frac{\partial M}{\partial a} = t\frac{dn}{da} = -t\frac{3n}{2a}.$$

Hence the first and the sixth equations read, in fact

$$\frac{1}{a}\frac{da}{dt} = -\frac{2}{na^2} \frac{\partial \mathcal{R}}{\partial M}, \quad (12.46a)$$

$$\frac{d\sigma}{dt} = \frac{2}{na}\left[\left(\frac{\partial \mathcal{R}}{\partial a}\right)_{n=\text{const}} - t\frac{3n}{2a}\frac{\partial \mathcal{R}}{\partial M}\right] + \frac{\eta^2}{na^2 e}\frac{\partial \mathcal{R}}{\partial e}. \quad (12.46b)$$

The coefficient linear in t is unphysical and awkward. The element (equation (12.21))

$$\epsilon' = M - \chi(t) = nt + \sigma - \chi(t)$$

solves the problem. Indeed, since

$$\frac{d\epsilon'}{dt} = \frac{d\sigma}{dt} + t\frac{dn}{dt},$$

the two Lagrange equations read now

$$\frac{1}{a}\frac{da}{dt} = -\frac{2}{na^2} \frac{\partial \mathcal{R}}{\partial \epsilon'}, \quad (12.47a)$$

$$\frac{d\epsilon'}{dt} = \frac{2}{na}\left(\frac{\partial \mathcal{R}}{\partial a}\right)_n + \frac{\eta^2}{na^2 e} \frac{\partial \mathcal{R}}{\partial e}, \quad (12.47b)$$

with the proviso that the mean motion n in $\mathcal{R}$ is an independent variable (the bracket $(\ldots)_n$ reminds that the corresponding derivative takes care of the explicit dependence of $\mathcal{R}$ on a only). It is carried forward in time by

$$\frac{dn}{dt} = \frac{d^2\chi}{dt^2} = -\frac{3n}{2a}\frac{da}{dt} = \frac{3}{a^2}\frac{\partial \mathcal{R}}{\partial \epsilon'}. \tag{12.48}$$

Hereafter we shall stick to this choice. Using eq. (12.48) we replace the time integration in the definition of $\chi(t)$, but this is achieved at the expense of a double integration (since (12.48) is a second-order differential equation). At the first sight, it may thus appear that a superfluous constant of integration was introduced in the system, but this is not true because $d\chi/dt = n$, which is constrained by the third Kepler law.

Especially at low inclination, the element $\epsilon = \varpi + \epsilon'$ (eq. (12.24)) is more appropriate and leads to another form of Lagrange equations for the perturbing function $\mathcal{R}(a, e, I, \Omega, \varpi, \epsilon; t)$. A straightforward calculation leads to:

$$\frac{1}{a}\frac{da}{dt} = -\frac{2}{na^2}\frac{\partial \mathcal{R}}{\partial \epsilon}, \tag{12.49a}$$

$$\frac{de}{dt} = \frac{\eta}{na^2 e}\left[\frac{\partial \mathcal{R}}{\partial \varpi} + (1-\eta)\frac{\partial \mathcal{R}}{\partial \epsilon}\right], \tag{12.49b}$$

$$\frac{dI}{dt} = \frac{1}{na^2 \eta \sin I}\left[\frac{\partial \mathcal{R}}{\partial \Omega} + 2\sin^2\frac{I}{2}\left(\frac{\partial \mathcal{R}}{\partial \varpi} + \frac{\partial \mathcal{R}}{\partial \epsilon}\right)\right], \tag{12.49c}$$

$$\frac{d\Omega}{dt} = -\frac{1}{na^2 \eta \sin I}\frac{\partial \mathcal{R}}{\partial I}, \tag{12.49d}$$

$$\frac{d\varpi}{dt} = -\frac{\eta}{na^2 e}\frac{\partial \mathcal{R}}{\partial e} - \frac{\tan(I/2)}{na^2 \eta}\frac{\partial \mathcal{R}}{\partial I}, \tag{12.49e}$$

$$\frac{d\epsilon}{dt} = \frac{2}{na}\left(\frac{\partial \mathcal{R}}{\partial a}\right)_n - \frac{\eta(1-\eta)}{na^2 e}\frac{\partial \mathcal{R}}{\partial e} - \frac{\tan(I/2)}{na^2 \eta}\frac{\partial \mathcal{R}}{\partial I}. \tag{12.49f}$$

Together with eq. (12.48), which yields $\chi(t)$, the system (12.49) is the standard form of the Lagrange perturbation equations for solar system work.

Some coefficients diverge for $e \to 0$, and for $I \to 0$ or $I \to \pi$. However, this is not due to a physical singularity, but to the geometrical definition of ω and Ω (and ϖ): in this limit, where large changes of these angles affect the orbit very little, they lose their significance. This coordinate effect can be removed by introducing in place of I, Ω, e, ϖ the *non-singular elements*

$$h = e\sin\varpi, \quad k = e\cos\varpi, \tag{12.50a}$$
$$P = \sin I \sin\Omega, \quad Q = \sin I \cos\Omega. \tag{12.50b}$$

If terms quadratic in e and I are neglected, it can be shown that they obey the equations

$$\frac{dh}{dt} = -\frac{1}{na^2}\frac{\partial \mathcal{R}}{\partial k}, \quad \frac{dk}{dt} = \frac{1}{na^2}\frac{\partial \mathcal{R}}{\partial h}, \tag{12.51a}$$

$$\frac{dP}{dt} = -\frac{1}{na^2}\frac{\partial \mathcal{R}}{\partial Q}, \quad \frac{dQ}{dt} = \frac{1}{na^2}\frac{\partial \mathcal{R}}{\partial P}. \tag{12.51b}$$

These equations will be used in Ch. 15 for the analysis of the long-term evolution of the solar system.

Eqs. (12.45) and (12.49) become invalid also for parabolic orbits ($\eta = 0$); their main use is for the elliptic case $e < 1$.

12.3 Atmospheric drag

In this section only the material of Sec. 12.1 is needed. When an object orbits not too far from the Earth's surface, it is subject to the drag due to the atmospheric gas. A knowledge of its orbital effects is important, both to avoid degradation of orbital predictions (see Sec. 18.2) and to obtain information on the upper atmosphere, in particular its density ϱ; low-altitude orbits and their accurate determination are also important to determine high-degree components of the geopotential. If the gas is assumed at rest in an inertial reference frame and the object, of mass m and cross section S (normal to its velocity $\mathbf{v}$), absorbs all the impinging molecules, the drag force – the momentum change per unit time – is $-(Sv\varrho)\mathbf{v}$. In general the drag acceleration is written as

$$\mathbf{F}_D = -\frac{1}{2m}C_D S \varrho v \mathbf{v} = -\frac{1}{2}\gamma \varrho v \mathbf{v}. \tag{12.52}$$

This is *Stokes' drag*; it depends on the body only through the parameter $\gamma = C_D S/m$. The numerical coefficient (of order unity) $C_D/2$ accounts for the real shape of the body and the way it interacts with the molecules. A surface element of the spacecraft can absorb, reflect and diffuse incident molecules in a complex way, depending on its composition and microscopic structure; the ratio between the thermal velocity of the gas and v is also relevant. Note also that the atmosphere is not quite at rest, but rotates with an angular velocity ω_A close to the angular velocity of the Earth; hence $\mathbf{v} - \mathbf{V}_A$, with $\mathbf{V}_A \simeq \omega_A \times \mathbf{r}$, should replace $\mathbf{v}$ in (12.52). As a consequence, in general the drag force depends in a complex way on the geometry and orientation of the orbit and, although it has the order of magnitude (12.52), it may have a different direction. Finally, this model applies to the case in which molecules interact independently with the body, which requires that their mean free path ℓ_c (Sec. 1.8) be larger than the body ("Knudsen regime"). If this does not hold the drag is a fluid-dynamical phenomenon, involving heat transfer, boundary layers and a wake, to be described with the appropriate partial differential equations. In this case the drag is generally a steeper function of the relative velocity. For the moment, we shall neglect these complications, and use eq. (12.52) as such, with γ constant. If the density and the shape of the objects do not change, this quantity in inversely proportional to the size: small objects re-entry first. We also assume that the atmospheric density ϱ is a function only of the altitude $h = r - R_\oplus$ above a spherical Earth.

Perturbations of the semimajor axis and eccentricity. In terms of the force components (12.2), since $\mathbf{F}_D$ is directed along $-\mathbf{v}$, $T = -\gamma\varrho v v_t/2$ and $R = -\gamma\varrho v v_r/2$. Using eqs. (11.22) for the radial and transversal velocities v_r and v_t, we get from eq. (12.4)

$$\frac{da}{dt} = -\frac{2a^2}{GM_\oplus} F_D v = -\gamma \frac{\varrho v^3}{n^2 a} \tag{12.53}$$

for the semimajor axis and from (12.7)

$$\frac{de}{dt} = -(e + \cos f)\varrho \gamma v \tag{12.54}$$

for the eccentricity. We are interested in the case in which the elements change little in an orbital period, so that in the right-hand sides they can be taken constant (as approximated by (12.76)). If the changes of the elements in a period Δc_a differ from zero, over time they steadily accumulate and we have secular effects; this is the crucial point to be heeded. For this purpose, in our case it is simpler to use the eccentric anomaly u. The energy integral, using eq. (11.21), reads

$$v^2 = GM_\oplus \left(\frac{2}{r} - \frac{1}{a}\right) = n^2 a^2 \frac{1 + e \cos u}{1 - e \cos u}. \tag{12.55}$$

Note also that
$$\frac{da}{du} = \frac{da/dt}{du/dt} = \frac{da/dt}{n}(1 - e\cos u),$$

and similarly for e. Substituting into eqs. (12.53) and (12.54), and integrating over one orbital period $2\pi/n$, we obtain

$$\Delta a = -\gamma a^2 \int_0^{2\pi} \left(\frac{1 + e\cos u}{1 - e\cos u}\right)^{1/2} (1 + e\cos u)\varrho(r)\, du, \tag{12.56a}$$

$$\Delta e = -\gamma a(1 - e^2) \int_0^{2\pi} \left(\frac{1 + e\cos u}{1 - e\cos u}\right)^{1/2} \varrho(r) \cos u\, du, \tag{12.56b}$$

with $r = a(1 - e\cos u)$. From these equations we can already draw some qualitative conclusions. First, $\Delta a < 0$: the semimajor axis and the orbital period decrease, so that the speed increases (the "perversity" of gravitation; see Problem 12.1). Secondly, for every density distribution which decreases with altitude – the normal case – $\Delta e < 0$, the eccentricity decreases until the orbit becomes circular. This can be seen from the last integral, since the integrand is positive for $-\pi/2 < u < \pi/2$ and negative for $\pi/2 < u < 3\pi/2$, while both the fraction and ϱ are larger in the orbital segment closer to the pericentre. Since the greatest drag is experienced near the perigee, the satellite does not swing out so far on the opposite side of the orbit and, as a result, the apocentre is lowered, without affecting much the pericentre itself (Problem 12.17). If the geopotential perturbations are included, in particular the odd degree zonal terms, the eccentricity might be forced to acquire a small but non-zero value and an analysis with non-singular elements is necessary.

Re-entry in a locally exponential atmosphere. To further discuss the effects of drag, we introduce a *locally exponential atmosphere*, in which the *scale height* (see Sec. 7.1)

$$H(r) = \left(-\frac{d\ln\varrho}{dr}\right)^{-1} \tag{12.57}$$

changes over a scale $\ell = H/(dH/dr)$ much greater than H itself ($H \approx 40$ km for low-altitude satellites).

For a thin atmosphere, and for a small orbital eccentricity, in eq. (12.53) we can evaluate $(v^3/n^2 a)$ at the planetary radius R and use the simpler form ($h \simeq a - R$)

$$\frac{dh}{dt} = -\varrho(h)\gamma\sqrt{GM_\oplus R}\; ; \tag{12.58}$$

$r \simeq a$ has been replaced by R in the square root term. It turns out that this is a fairly good approximation, given the exponential decrease of the density. We see that the re-entry timescale

$$\tau_r(h) = \left(-\frac{dh}{a\,dt}\right)^{-1} = \frac{1}{\gamma n R \varrho(h)} \tag{12.59}$$

decreases very rapidly with decreasing height h. Of course, when $n\tau_r \approx 1$, the approximation does hold any more and the final, accelerated decay takes place. To obtain $a(t)$, the integral of eq. (12.58) can be approximated by using, in analogy with the optical depth (eq. (7.18)), the dimensionless variable

$$\xi = \int_R^r \frac{dr'}{H(r')}\; ,$$

in terms of which the density is $\varrho = \varrho_0 \exp(-\xi)$ (the index $_0$ refers to the planetary surface). We then get

$$-\gamma t\sqrt{GM_\oplus R} = \int \frac{dh}{\varrho(h)} = \frac{1}{\varrho_0}\int_0^\xi d\xi'\, H(\xi')\exp(\xi')\; ;$$

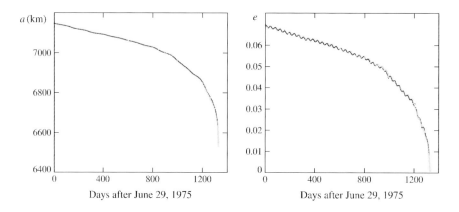

Figure 12.2. Orbital decay due to the atmospheric drag. The panels show the semimajor axis a (in km) and the eccentricity of the French satellite CASTOR vs time (in days after the launch, on June 29, 1975). The short-term wiggles are mainly due to the octupole term J_3 of the Earth gravitational field (Problem 12.4). The data (black curve) are obtained from satellite observations, while the grey curve is a theoretical prediction using the thermospheric model TD88 (for other examples, see http://sirrah.troja.mff.cuni.cz/~ales/density_therm/). No adjustment of the model parameters was performed; see also Problem 12.13. Observations kindly provided by F. Barlier and C. Berger.

the re-entry occurs at $t = 0$. The primitive of $H(\xi)\exp(\xi)$ can be expressed as a series in the derivatives of $H(\xi)$, each term being smaller than the previous one by H/ℓ

$$d[e^\xi(H - H' + H'' - \ldots)] = He^\xi\, d\xi\,.$$

For an exactly exponential atmosphere ($H = H_0 = $ const.), we have a "logarithmic" re-entry

$$\xi = \frac{a - R}{H_0} = \ln\left[1 - \gamma nt\,\frac{\varrho_0 R}{H_0}\right]\,. \qquad (12.60)$$

For a slowly changing scale height H one can solve the equation

$$e^\xi H(\xi) - H_0 = -\gamma nt\,\varrho_0 R^2\,, \qquad (12.61)$$

where terms of order H/ℓ have been neglected. In this approximation the decay rate reaches its maximum at $t = 0$, but in real life, as shown in Fig. 12.2, it is "catastrophic".

Changes of inclination, node and pericentre.– In the simplest model used above, the other orbital elements do not undergo secular perturbations due to drag: for the inclination I and the nodal longitude Ω this corresponds to the fact that a force along the orbital velocity $\mathbf{v}$ has no out-of-plane component ($W = 0$). As for the argument of pericentre ω, substituting in eq. (12.18) $d\Omega/dt = 0$, we obtain

12. Perturbation theory

$$\frac{d\omega}{dt} = \frac{\eta}{nae}\left[-R\cos f + T\sin f\left(\frac{2 + e\cos f}{1 + e\cos f}\right)\right],$$

where T and R are proportional to $\varrho v v_t$ and $\varrho v v_r$, respectively. For v_r and v_t we can use eqs. (11.22) to obtain

$$\frac{d\omega}{dt} = -\frac{\gamma\varrho v}{e}\sin f. \qquad (12.62)$$

Since ϱ and v are both even functions of the true anomaly f, $d\omega/dt$ has a vanishing average.

If a satellite has a complex shape (e.g., due to the antenna and the solar panels), and molecular interactions other than absorption are present, the previous analysis is approximate and typically a lift force with $W \neq 0$ occurs. However, even for a spherical satellite experiencing no aerodynamic lift, the rotation of the atmosphere produces new components of the drag. For example, the out-of-plane component on a circular orbit is, approximately,

$$W = -F_D \frac{\omega_A}{n} \sin I \cos(\omega + f), \qquad (12.63)$$

where we have assumed the ratio $\omega_A/n \leq 0.1$, small. Using eqs. (12.10) and (12.11), we obtain:

$$\frac{d\Omega}{dt} = -\frac{F_D}{na}\frac{\omega_A}{n}\cos(\omega + f)\sin(\omega + f), \qquad (12.64)$$

$$\frac{dI}{dt} = -\frac{F_D}{na}\frac{\omega_A}{n}\sin I \cos^2(\omega + f). \qquad (12.65)$$

Atmospheric rotation has no long-term effect on the node (due to the fact that the torque caused by the spinning atmosphere is directed along the Earth's polar axis); but the change of I in one orbital period

$$\Delta I = -\pi\left(\frac{F_D}{n^2 a}\right)\frac{\omega_A}{n}\sin I \qquad (12.66)$$

does not vanish, and the inclination diminishes. During the last phase, the satellite's inclination typically changes by $\approx 0.1° - 0.4°$, an easily observable effect. Measurements of this decrease have yielded information about the atmospheric rotation rate, which, due to strong eastwardly winds at high altitudes, is significantly greater than that of the Earth (Sec. 8.1).

More realistic models. Our ignorance about the atmospheric density profile is mainly due to its time variability, which may be as large as one order of magnitude; significant daily and seasonal variations are present, in particular below $1,000$ km (Fig. 12.3). Moreover, spherical symmetry is only a crude approximation and there are important latitudinal variations. Over longer times,

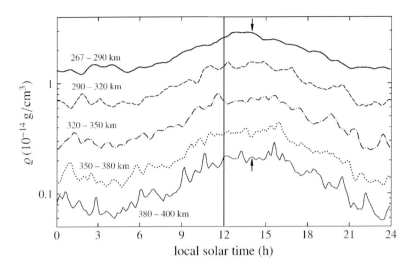

Figure 12.3. Daily variation of the atmospheric density at fixed altitude intervals: the density ϱ (in 10^{-14} g/cm^3) averaged over many satellite observations is given versus the local solar time T in hours ($T = 0$ h at midnight, $T = 12$ h at noon). Note the exponential decrease with altitude and the daily wave: at all altitude bands there is a factor $\simeq 3$ between the minimum and maximum values; the maximum at $\simeq 14$ h (arrows) is delayed due to thermal inertia. Data acquired by the French satellite CASTOR during a minimum of the solar cycle and kindly provided by F. Barlier and C. Berger.

correlations with the solar and geomagnetic activity are observed. In the upper atmosphere density models, these latter phenomena are correlated with the phase of the solar cycle, the solar radio flux $F_{10.7}$ and indices of geomagnetic activity such as K_p (Secs. 9.1 and 10.1). The primary physical processes in the upper atmosphere are sensitive to variations of the solar ultraviolet radiation and the solar wind; none of these are measured directly on ground, but this causal dependence shows up in the correlation with the above parameters. At heights between 100 and 200 km, where the catastrophic decay of a spacecraft starts, reliable and accurate predictions for the atmospheric density are impossible; in general, the exact re-entry time, and its location, for an uncontrolled object cannot be predicted well in advance. Even with a very successful long-term a posteriori model (Fig. 12.2), one gets an error of a few days in the re-entry time; an a priori prediction of the satellite lifetime is usually much worse, since we are not able to accurately predict the solar and geomagnetic activity. The lowest altitude at which operational satellites can be maintained without an excessive fuel consumption is about 180 km; note that there are no means for in situ and semi-permanent exploration of the atmosphere below this altitude and above about 30 km (above which airplanes and balloons cannot fly).

12.4 Oblateness of the primary

In Sec. 4.1 we computed the torque exerted on an oblate planet by an external orbiting body, and found that its average over an orbital period is perpendicular to both the spin axis and the orbital angular momentum. By Newton's third law, an equal and opposite secular torque is exerted by the planet on a satellite. Hence the angular momentum **h** has a constant magnitude and rotates about the planetary spin axis with a rate $d\Omega/dt$ given by

$$\frac{d\Omega}{dt} h \sin I = -\frac{3}{2} \frac{Gm}{r^3} (C - A) \sin I \cos I \qquad (12.67)$$

(compare with eq. (4.7)). Thus over a long time (see (12.73b) below)

$$\frac{d\Omega}{dt} = -\frac{3}{2} J_2 n \left(\frac{R}{a}\right)^2 \frac{\cos I}{\eta^4}, \qquad (12.68)$$

where we have introduced the quadrupole coefficient J_2 by eq. (2.31). The nodal motion is negative (positive) for prograde (retrograde) orbits; for a low spacecraft, unless the inclination is near 90°, it amounts to several degrees per day. The inclination is secularly constant, due to the axial symmetry of the disturbing potential (eq. (2.21)). Indeed, writing the kinetic energy as $[\dot{r}^2 + r^2(\dot{\theta}^2 + \sin^2\theta\,\dot{\phi}^2)]/2$, we note that the longitude ϕ is an ignorable coordinate and

$$\frac{\partial L}{\partial \dot{\phi}} = r^2 \sin^2 \theta \, \dot{\phi}$$

is a constant; but this is just $h_z = h \cos I$, the z-component of the angular momentum.

What about a and e? Since the force can be derived from a time independent potential, the total energy is conserved. The change of the osculating energy, equal to the work done by the perturbing force, vanishes in an orbital period; hence, by eqs. (12.4) the semimajor axis is secularly constant. We have mentioned above that, due to the perturbing interaction between the planet and the satellite, this holds also for the angular momentum; combining this result with the stability of the semimajor axis, we conclude that also the eccentricity has no secular evolution. However, the angular velocity $d\omega/dt$ of the pericentre has a non-vanishing average of the same order as $d\Omega/dt$; as a consequence, the orbit rotates in its own plane and on a long term it is not well approximated by a close ellipse.

We now recover the above results with Lagrange's method and the perturbing function (see Ch. 2)

$$\mathcal{R} = \frac{Gm}{2r} \left(\frac{R}{r}\right)^2 J_2 \left(3\cos^2\theta - 1\right); \qquad (12.69)$$

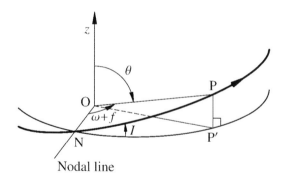

Figure 12.4 The relation $\cos\theta = \sin I \sin(\omega+f)$ is obtained by applying the sine theorem to the spherical triangle NP'P; P is the orbiting body and P' its projection on the equatorial plane.

θ is the colatitude of the test body (see Fig. 12.4). Since $\cos\theta = \sin I \sin(\omega+f)$, we obtain

$$\mathcal{R} = -\frac{Gm}{2a}\left(\frac{R}{a}\right)^2\left(\frac{a}{r}\right)^3 J_2\left[1 - \frac{3}{2}\sin^2 I + \frac{3}{2}\sin^2 I \cos(2\omega + 2f)\right]. \quad (12.70)$$

The functions a^3/r^3 and $a^3\cos(2\omega + 2f)/r^3$ are periodic in the mean anomaly M. As discussed in Sec. 12.5, the secular perturbations are described by the effective potential function $\overline{\mathcal{R}}$, the average of $\mathcal{R}$ over an orbital period (to lowest order, this is equivalent to the simple average over the unperturbed mean anomaly M). We need

$$\frac{1}{2\pi}\int_0^{2\pi} dM \frac{a^3}{r^3} = \frac{1}{\eta^3}, \quad \int_0^{2\pi} dM \frac{a^3}{r^3}\cos(2\omega + 2f) = 0, \quad (12.71)$$

where we have used the relations $r = a\eta^2/(1 + e\cos f)$ and $r^2 df/dt = na^2\eta$ (see Ch. 11). Hence

$$\overline{\mathcal{R}} = -\frac{Gm}{4a}\left(\frac{R}{a}\right)^2 J_2 \frac{2 - 3\sin^2 I}{\eta^3}. \quad (12.72)$$

With the change $\mathcal{R} \to \overline{\mathcal{R}}$, the Lagrange equations (12.45) and (12.47b) now read:

$$\frac{da}{dt} = \frac{de}{dt} = \frac{dI}{dt} = 0, \quad (12.73a)$$

$$\frac{d\Omega}{dt} = -\frac{3}{2}nJ_2\left(\frac{R}{a}\right)^2\frac{\cos I}{\eta^4}, \quad (12.73b)$$

$$\frac{d\omega}{dt} = -\frac{3}{4}nJ_2\left(\frac{R}{a}\right)^2\frac{1 - 5\cos^2 I}{\eta^4}, \quad (12.73c)$$

$$\frac{d\epsilon'}{dt} = -\frac{3}{4}nJ_2\left(\frac{R}{a}\right)^2\frac{1 - 3\cos^2 I}{\eta^3}. \quad (12.73d)$$

We thus confirm that, to lowest order, a, e and I do not have secular changes, while Ω and ω grow linearly with time. For low-altitude satellites, their rates are $\propto n J_2$, so that the corresponding timescale is $\approx 1/J_2 \simeq 10^3$ revolutions. With an orbital period of about two hours, Ω and ω may circulate around the Earth on a timescale as short as ≈ 2 months. The rate of change of M has also a correction due to the increase in the radial pull in the equatorial plane produced by the equatorial bulge.

The equation for $d\omega/dt$ shows that the orbit turns in its own plane with a rate which becomes negative for high inclinations and vanishes at the *critical inclination* $\arccos(1/\sqrt{5}) \simeq 63° 26'$. Near this value the behaviour of the pericentre is determined by higher-order terms, obtained by using in the right-hand sides of the previous equations the first order solution; ω and e become non-trivially coupled and, rather than the steady circulation solution of eq. (12.73c), ω switches to a small libration regime. Orbits at the critical inclination are used by some satellites on very eccentric orbits with orbital period of 12 h (e.g., the Soviet Molniya satellites, Fig. 18.5); every day, at perigee, they transit over the same latitude range.

The previous results can be extended to higher-degree zonal harmonics in a straightforward way. However, major complications arise with tesseral and sectorial harmonics. Axially symmetric perturbations, like the zonal terms in the potential, are time independent; with non-zonal terms the potential becomes time dependent. The resonance phenomena between the planetary rotation and the satellite mean motion then result in computational complexities (qualitatively outlined in Sec. 12.5). The most obvious resonance is 1/1, the geosynchronous orbit (Sec. 18.2). However, the whole spectrum of resonances down to 16/1, at about 240 km, may affect in a non-trivial way the fine details of the satellite dynamics. Moreover, when the resonance orbit has a critical inclination (and a high eccentricity) the orbital problem becomes very complicated.

12.5 Approximation methods

A general discussion of the way how perturbation theory can provide approximate solutions (dealt with in an empirical way in the previous application) is now in order. The osculating elements $c_a(t)$, defined in an abstract way, fulfil equations of the form

$$\frac{dc_a}{dt} = f_a(c_b(t), t) ; \qquad (12.74)$$

if the perturbation is switched off at the time t', they remain constant thereafter and give the Keplerian orbit valid for $t > t'$. Though there is no restriction in their validity, they are useful only if, in some sense, the functions f_a are small.

This can be formalized with the smallness parameter

$$\mu = \frac{\mathcal{R}}{Gm/a} \approx \frac{\mathcal{R}}{n^2 a^2} \approx \frac{\text{perturbing acceleration}}{n^2 a}, \qquad (12.75)$$

ratio of the perturbing to the undisturbed energy in the Lagrangian formalism, or ratio of the perturbing to the Keplerian acceleration in the general case (following from eqs. (12.14)). The rates of change of the elements c_a (for the semimajor axis, the fractional rate of change $da/(a\,dt)$) are of order μn, while for the orbital phase (eq. (12.48)) $d^2\chi/dt^2 = O(\mu n^2)$.

The most straightforward, but naïve, approximation method is Picard's iteration process, in which the argument $c_a(t)$ of the functions f_a is replaced first with its initial value $c_a(0)$, to give the first order approximation

$$c_a(t) = c_a(0) + \int_0^t dt'\, f_a(c_b(0), t') \, . \qquad (12.76)$$

For a finite time span the changes $c_a(t) - c_a(0)$ are formally $O(\mu)$, but this solution will differ from the exact solution by terms $O(\mu^2)$ and the eq. (12.76) becomes invalid for large t. For a higher order solution, we may use $c_a(t)$ from eq. (12.76) in the argument of f_a and integrate again (12.74); this process may be iterated as many times as necessary. Successive iterations, usually performed with a combination of Fourier expansions and power series (also called Poisson series), soon become very cumbersome. Moreover, these "automatically" generated series, although practically used (e.g., for ephemerides predictions), raise serious convergence questions, seldom answered positively. Indeed, since H. POINCARÉ we know that most problems in celestial mechanics are not integrable, and thus series approximations of their solution lead to divergences.

That the assumption of a finite time span may be inadequate is illustrated by the motion around a point mass with a disturbing function $\mathcal{R}$ (or by the R, T and W components) which does not explicitely depend on time. Its variability only depends on the variability of the Keplerian elements and has a timescale determined by the mean motion n. In a finite time the changes in the elements have an amplitude $O(\mu)$ and the same timescale; these are *short-period perturbations*. However, if the average of $\mathcal{R}$ over one orbital period does not vanish, these changes accumulate and eventually, after a time $O(1/(n\mu))$, become of order unity, leading to a significantly different orbit; these are the *secular* or *long-period perturbations*, usually the major feature to investigate. Note that in eq. (12.49a), $\partial/\partial\epsilon = \partial/\partial M$; hence, to lowest order, the semimajor axis has no secular or long-term change. Some examples have been discussed in the previous two sections.

Planetary perturbations. The motion of N interacting planets (or satellites) is the major problem in celestial mechanics. We illustrate, mostly qualitatively,

12. Perturbation theory

the complexities that are encountered in construction of the theory of planetary perturbations. More detailed quantitative discussion of secular effects is found in Sec. 15.1.

Since gravitational forces are conservative, it is natural to use the Hamiltonian approach. It is also convenient to work with the relative coordinate $\mathbf{r} = \mathbf{R} - \mathbf{R}_\odot$ of a planet with respect to the Sun. With only two planets, for the first we have

$$\begin{aligned}\frac{d^2\mathbf{r}_1}{dt^2} &= -G(M_\odot + M_1)\frac{\mathbf{r}_1}{r_1^3} + GM_2\left(\frac{\mathbf{r}_2 - \mathbf{r}_1}{r_{12}^3} - \frac{\mathbf{r}_2}{r_2^3}\right) = \\ &= -\frac{\partial}{\partial \mathbf{r}_1}\left[-\frac{G(M_\odot + M_1)}{r_1} + \mathcal{R}_{12}\right],\end{aligned} \quad (12.77)$$

with $r_{12} = |\mathbf{r}_1 - \mathbf{r}_2|$. For the second planet 1 and 2 are interchanged. The scalar

$$\mathcal{R}_{12} = -GM_2\left(\frac{1}{r_{12}} - \frac{\mathbf{r}_1 \cdot \mathbf{r}_2}{r_2^3}\right), \quad (12.78)$$

a function of $\mathbf{r}_1$ and $\mathbf{r}_2$, is the *perturbing function* for the first planet[4]. Its first term – the *direct perturbation* – is the interaction energy of the two planets; the second term – the *indirect perturbation* – takes into account the motion of the Sun around the centre of mass of the system due to the other planet. If several perturbing bodies are present, $\mathcal{R}_{12}$ is replaced by the sum over all pairs (see Ch. 15). When the distances and the planetary masses are generic (in particular, barring close encounters), the ratio of $\mathcal{R}$ to the main potential energy is of the order of the ratio of the masses $\mu = O(M_1/M_\odot, M_2/M_\odot) \ll 1$.

As usual in the literature, we express $\mathcal{R}_{12}$ in terms of the osculating orbital elements of the two planets. With the choice $(a, e, I, \Omega, \varpi, \epsilon)$ relative to a fixed ecliptic plane (Sec. 12.2), and using the mean longitude $\lambda = \chi + \epsilon$, we note that $\mathcal{R}_{12}$ is periodic in the angular variables $(\Omega_1, \varpi_1, \lambda_1; \Omega_2, \varpi_2, \lambda_2)$; hence it can be expanded in a six-dimensional Fourier series, which can be shown to include only cosine terms:

$$\begin{aligned}\mathcal{R}_{12} &= \sum C_{ijk} \cos(i_1\lambda_1 + i_2\lambda_2 + j_1\varpi_1 + j_2\varpi_2 + k_1\Omega_1 + k_2\Omega_2) = \\ &= \sum C_{ijk} \cos \Phi_{ijk}.\end{aligned} \quad (12.79)$$

The summation extends over all the integer values of the indices $(i, j, k) = (i_1, \ldots, k_2)$, with one important exception: the $\mathcal{R}$ function must not change

---

[4] When dealing with motion of a test-body, whose own gravitational field is negligible, such an asteroids or a comet, the position $\mathbf{r}_2(t)$ of the planet is given and we have to solve only for $\mathbf{r}_1$.

under an arbitrary rotation about the pole of the ecliptic and, since the angles appearing in the argument of (12.79) are reckoned from a common origin, $\sum_{l=1}^{2}(i_l + j_l + k_l) = 0$ (*d'Alembert rule*). The C-coefficients have been traditionally evaluated in terms of very cumbersome series in powers of the eccentricities (e_1, e_2) and the inclinations $(\sin I_1, \sin I_2)$ which, fortunately, in the solar system are generally small (Laplace-type development); when the ratio $\alpha = a_1/a_2$ is small, a power series in α is also used to describe, for instance, the perturbation by an outer planet on an inner one (Kaula-type development). Today efficient algebraic manipulators are used to push such expansions to very high orders and store them in the computer memory. The C-coefficients can also be evaluated numerically for given orbits; the lack of generality is compensated by extremely rapid Fast Fourier Transform codes; much higher orders may thus be attained, which turns out important when eccentricities, inclinations or α become large.

Secular perturbations and averaging. As outlined above, *first order perturbations* are obtained by substituting in the right-hand sides of Lagrange equations (12.49) the constant (initial) values of a, e, I, Ω and ϖ, while $\lambda = nt + \epsilon$ for both planets; the equations are then easily integrated by quadratures. We can see from (12.79) that the terms $i_1 = i_2 = 0$ in the summation give constant phases Φ_{ijk} in (12.79) and therefore produce secular terms, linearly growing with time. This holds for all the elements, but not for the semimajor axes; in fact, except for resonances (see below), $da_1/dt \propto \sum i_1 C_{ijk} \sin \Phi_{ijk}$ vanishes when the phases Φ_{ijk} are constant (and similarly for the second planet). This is an elementary stability theorem: the size of planetary orbits cannot change much in $1/\mu$ orbital periods[5].

When the unperturbed system is periodic, as for an elliptic orbit around an oblate planet, the direct removal of short-period terms is easily accomplished: we just replace $\mathcal{R}$ with its average $\overline{\mathcal{R}}$ over an orbital period, so that the Lagrange equations (12.49) give the change of the orbital elements in the same time. But a two-planet system is not periodic and the problem is more delicate. We deal first with the non-resonant case, in which the ration n_2/n_1 of the mean motions is not near the ratio of two, not large, integers. The removal (to $O(\mu)$) of short-period terms can be done, instead of using the Fourier expansion (12.79), by replacing the perturbing function $\mathcal{R}$ with its average over the torus ($0 < \lambda_1 \leq 2\pi$, $0 < \lambda_2 \leq 2\pi$)

$$\langle \mathcal{R} \rangle_{\text{torus}} = \frac{1}{(2\pi)^2} \int_0^{2\pi} \int_0^{2\pi} d\lambda_1 \, d\lambda_2 \, \mathcal{R}(\lambda_1, \lambda_2; \ldots) \,. \tag{12.80}$$

---

[5] This is a famous finding by J.L. LAGRANGE; later F.F. TISSERAND proved that the planetary semimajor axes are secularly stable to the second order in μ, hence contain no secular term over $1/\mu^2$ orbital periods.

12. Perturbation theory

To understand its physical meaning, note that we are looking for perturbations over a timescale τ much longer than the orbital periods P_1 and P_2, but also shorter than the characteristic time T of the long-term evolution. The equality of the toroidal average and the time average

$$\overline{\mathcal{R}} = \frac{1}{\tau} \int_0^\tau dt\, \mathcal{R} \tag{12.81}$$

is a form of the *ergodic problem*, which studies the conditions under which in a dynamical system the average over phase space equals the average over a time τ when $\tau \to \infty$. In our case, when the two orbital periods are incommensurable, any unperturbed trajectory $\lambda_1 = n_1 t + \epsilon_1$, $\lambda_2 = n_2 t + \epsilon_2$ cannot close on itself and, after sufficiently long time, passes arbitrarily near any point on the torus, so that all its points are expected to *equally* contribute to the time average[6]. The applicability of the toroidal average when τ is finite is a delicate matter and must take into account how near the system is to a resonance. In the case of N planets we have an N-dimensional torus and a similar average applies.

When the mean motions n_1 and n_2 are generically $O(n)$, eq. (12.80) leads to changes of the orbital elements in the period $2\pi/n$ of order $O(\mu)$ and cannot be used for times longer than $2\pi/(n\mu)$. At higher orders the longitudes of the node and the pericentre (and the mean anomaly phases ϵ) acquire slow changes that split the revolution-frequency lines in the spectrum of $\mathcal{R}$ in an infinite number of sidebands. On the other hand, there is a strong motivation to search for higher order secular terms. To second order the change of the elements in a period is $O(\mu^2)$, so that it takes $1/\mu^2$ periods to reach the limit of validity; for $\mu \approx 10^{-3}$, as in the solar system, we need the secular changes to $O(\mu^3)$ for a solution valid for its age $\approx 10^9$ y. From here stems a major difficulty in understanding its long-term evolution. Powerful algebraic methods[7], based on canonical transformations, have been devised to obtain an "equivalent" average disturbing function to represent secular effects to $O(\mu^n)$, $n \geq 2$.

In the general case of the Gauss equations, first order secular perturbations are formally dealt with by averaging over the mean longitude λ. In this case the theory of higher order secular perturbations has been less systematically elaborated; however, the relevant perturbations are often small and the first-order theory is sufficient.

---

[6] A fine example of the use of the average (12.80) is the drift of the node of an outer planet due to an inner one, like Mercury. For circular orbits, $\langle \mathcal{R} \rangle_{\text{torus}}$ describes the interaction between two uniform rings of matter with linear density $M/(2\pi a)$; it can be described with elementary mechanics and, if they are inclined, leads to precession. The inner planet has the same effect as an additional solar oblateness (Problem 12.3).

[7] Given their mathematical complexity, we do not discuss these methods, and suggest to interested readers Morbidelli's book in Further Readings.

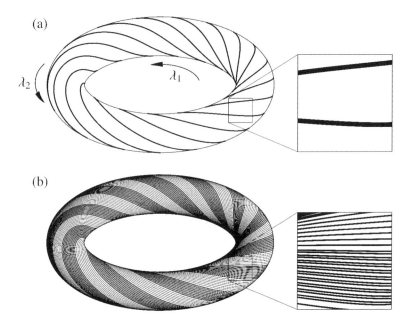

Figure 12.5. Resonances and ergodic behaviour for two bodies in Keplerian orbits around a primary, with mean longitudes λ_1 and λ_2. The perturbing function $\mathcal{R}$ is periodic in both, and can be represented on a torus. When the mean motions are commensurable, the orbit is closed. (a) shows the resonance $n_1/n_2 = 18/7$ (as for the asteroid Pallas and Jupiter); starting at any point on the torus, the orbit closes after 18 periods of λ_1, the mean longitude of Pallas. If they are incommensurable, as in (b), it passes arbitrarily near any point and the time average of $\mathcal{R}$ can be replaced with its average over the torus. The plot of the path has been stopped before the single turns become indistinguishable.

Resonances. Going back to the first order solution of the two-planet problem (12.79), we have an exact resonance when there is a pair of integers (i_1, i_2) such that $(i_1 n_1 + i_2 n_2) = 0$; then the integration produces terms $\propto t$ and, eventually, its validity ceases. The ergodic property on the torus ($0 < \lambda_1 \leq 2\pi, 0 < \lambda_2 \leq 2\pi$) spanned by the mean longitudes provides a geometrical insight on this failure of secular perturbation theory (Fig. 12.5). If the mean motions n_1, n_2 are incommensurable, the unperturbed orbit $\lambda_1 = n_1 t + \epsilon_1, \lambda_2 = n_2 t + \epsilon_2$ is not closed and the ergodic property, used above, holds; but when the mean motions are commensurable, the unperturbed orbit is closed, there is no ergodic behaviour and it is not true that the orbit gets as near as one wishes to an arbitrary point. Toroidal averaging is not applicable and secular perturbation theory fails, even to first order.

However, this is an abstract mathematical argument; since any irrational number can be approximated as well as one likes with a rational number with large enough numerator and denominator, in the infinite Fourier series (12.79)

12. Perturbation theory

no upper bound can be placed to the integral of each term, even when the ratio n_1/n_2 is irrational. The assumption of a power series in μ, valid for an infinite time, requires a deeper discussion. We must deal with *approximate resonances*, when $(i_1 n_1 + i_2 n_2)$ is in some sense small. In general, the integration of a sine or a cosine term gives a cosine or a sine of the same argument; its amplitude is divided by $(i_1 n_1 + i_2 n_2)$ and undergoes a large amplification. In particular, if n is the (common) order of magnitude of n_1 and n_2, and the coefficients C are $O(\mu)$, the changes of the orbital elements in a period $2\pi/n$ are of order μ in the generic case, but near resonance, after a time of order $1/|i_1 n_1 + i_2 n_2|$, become of order $n\mu/|i_1 n_1 + i_2 n_2|$. Hence the approximation linear in μ breaks down if this quantity is $O(1)$, at the same time as the breaking down in the non-resonant case. For the anomaly χ, obtained with a double integration of (12.48), the first order approximation is even worse: with a quadratic divisor $(i_1 n_1 + i_2 n_2)^2$ the solution becomes of order $\mu(n/|i_1 n_1 + i_2 n_2|)^2$ after a time $1/|i_1 n_1 + i_2 n_2|$, and fails before the non-resonant breaking down. The failure occurs after $n/|i_1 n_1 + i_2 n_2| = \sqrt{\mu}$ orbital periods. At higher orders, the perturbations generically contain higher powers of these divisors and the situation is more complex.

For a given (approximately) resonant ratio n_2/n_1, there are two ways in which the terms in the expansion (12.79) may be affected. Given the positive integers (p,q), all pairs $(i_1 = -cp, i_2 = cq)$, with c an arbitrary integer, give small divisors; moreover, the same ratio n_2/n_1 can be (possibly better) approximated by slightly different rational numbers, like $(mp + 1)/(mq + 1)$, where m is a large integer. For example, a $5/2$ resonance amplifies the terms corresponding to the pairs $(-2, 5)$, $(-4, 10)$, $(-6, 15)$, ... and $(-21, 51)$, $(-20, 49)$, etc. In general, there are infinite such terms. The smallness of eccentricities and inclinations in the planetary and the satellite systems "save" the situation. Let $\varepsilon \ll 1$ denote their (common) order of magnitude. In planetary dynamics a large amount of work has been devoted to the expansion of the disturbing function in powers of ε, and the general properties of their coefficients. Each term in the expansion (12.79) is distinguished by the integer $k = |i_1 + i_2|$, called its *order*. For example, the pairs listed above have the orders $3, 6, 9, 30$ and 29. It turns out that the C-coefficients in (12.79) are $O(\varepsilon^k)$; hence the main effect of the resonance is given by the resonant terms of lowest order in eccentricity and inclination. Moreover, the principal failure of secular perturbation theory is important only for resonances with a mean motion ratio p/q whose *order* $k = |q - p|$ is not too large, as $p/q = 1, 1/2, 1/3, 2/3$, etc; then they significantly affect the motion and new stability (or instability) properties arise. In Sec. 15.2 we shall discuss several examples of this phenomenon, which has strongly influenced the dynamical structure of the solar system. However, understanding their long-term evolution (for which even the order μ^3 may be relevant) requires estimating the effect of (i) high order terms and (ii) resonances appearing at higher order of the iteration approximation.

As an example, we mention the role of high order resonances with Jupiter, and multiple resonances with Jupiter and Saturn, that seem to play important role in the long-term ($\approx 0.1 - 1$ Gy) evolution of the asteroid belt (Sec. 14.3).

The resonance condition becomes more complex beyond the lowest order in μ, when secular effects are taken into account. At resonance, a phase Φ in eq. (12.79) is stationary, which gives the condition

$$i_1(n_1 + \dot{\epsilon}_1) + i_2(n_2 + \dot{\epsilon}_2) + j_1\dot{\varpi}_1 + j_2\dot{\varpi}_2 + k_1\dot{\Omega}_1 + k_2\dot{\Omega}_2 = 0 \ . \quad (12.82)$$

Near enough $i_1 n_1 + i_2 n_2 = 0$ the additional terms may be relevant. Since the coefficients $j_1, \ldots, k_2$ are arbitrary (except for d'Alembert constraint), each line is split in a multiplet with infinite members. When the secular terms are due to the mutual planetary perturbations, their separation is very small (in terms of the semimajor axis, typically $\approx 10^{-4}$ AU); multiplet lines may overlap, a typical route toward chaotic motion. A natural satellite presents another interesting situation, when the secular terms in (12.82) are due to the planetary oblateness, much larger than the mass ratio. The lines of a multiplet may be separated much more and occur at significantly different distances from the planet. Jupiter's and Saturn's systems offer several examples in which tidal evolution might have driven a satellite from a multiplet line to another, causing a particular final state in the eccentricity and the inclination (as for Mimas and Tethys, a much studied case).

A new type of resonance occurs for $i_1 = i_2 = 0$ in eq. (12.82); they are called *secular resonances*, since they are caused by secular evolution. Their importance was realized only in the 1970s, when the inner truncation of the main asteroid belt was unambiguously associated with the ν_6 resonance (Sec. 15.1). In Picard's method they appear only at the second iteration, which might suggest their weakness ($\propto \mu^2$ at least); but this conclusion is not true, since the denominator appearing in their integration is the secular rate of the corresponding angle, hence of order μ (from here the difficulty of their description stems). As a result, the principal secular resonances are very important for the accurate understanding of the solar system dynamics. Apart from the motion of minor bodies (asteroids, comets, etc.), their interaction is suspected to trigger an overall chaotic behaviour in the motion of terrestrial planets.

– PROBLEMS –

12.1.* The "perversity" of gravitation shows up in the positive along-track acceleration produced by drag (Sec. 12.3). Find out when an arbitrary radial force $F = -dU/dr$ is "perverse". (Hint: Consider a circular orbit with velocity given by $v^2 = -r\,dU/dr$, $U(r)$ being the potential, and compute its change when the energy E is slowly changed. Show that the sign of $(dv/dt)/(dE/dt)$ is determined by the dimensionless parameter $r(d^2U/dr^2)/(dU/dr)$.)

12.2.* The perturbative force **F** in (12.1) can be expressed in terms of its components W, T' parallel to the velocity and R' perpendicular to the velocity in the orbital plane. This set is the same as (R, T, W) only when $e = 0$. Express R' and T' as a function of R, T, the eccentricity and the true anomaly and write down the new form of Gauss's equations.

12.3. The averaged gravitational effect of Mercury on the Earth motion may be approximated by an effective increase of the solar J_2 value. Evaluate this correction and estimate its effect on the node and the perihelion of the Earth.

12.4. Consider the effect of the geopotential octupole term

$$\mathcal{R} = -\frac{Gm}{2r}\left(\frac{R}{r}\right)^3 J_3 \cos\theta \left(5\cos^2\theta - 3\right)$$

on the motion of an artificial satellite. Evaluate its average $\overline{\mathcal{R}}$ and show that the long-term change in eccentricity is modulated by the period of the pericentre circulation.

12.5.* Calculate the secular precession of the node due to oblateness in an elementary way, using the Cartesian expression (2.28) of the quadrupole potential to obtain the change in the orbital angular momentum.

12.6* A circular orbit is perturbed by a small arbitrary potential $\delta U(r,\theta)$ with axial and mirror symmetry relative to the equatorial plane (i.e., with only even zonal terms). Compute the frequencies of radial and latitudinal oscillations. Using the conservation of angular momentum, derive the constant and the periodic change in the along-track velocity. From the previous analysis, derive the motion of the pericentre and the node.

12.7. Evaluate the secular effects of the perturbing function $\mathcal{R} = (Gm/r) \times (Gm/rc^2)$, where m is the central mass. This perturbation is typical of relativistic corrections (see Ch. 17); in particular, compare the secular effect on pericentre with that of eq. (17.54).

12.8.* Before General Relativity, in order to explain the anomalous advance of the perihelion of Mercury, a gravitational force $\propto 1/r^{2+\delta}$ was assumed. Study the secular perturbations of a Keplerian orbit when $|\delta| \ll 1$. The same for a potential $\propto \exp(-\lambda r)/r$, with $1/\lambda \gg r$.

12.9.* The Hamilton-Jacobi equation (12.38) provides a way to obtain canonically conjugate variables for the two-body problem. One set is *Delaunay variables* $L = \sqrt{Gma}$, $G = L\eta$, $H = G\cos I$, conjugate to $\ell = M$, $g = \omega$, $h = \Omega$, respectively. The traditional symbols are used here.

12.10. Another, derived set (*Poincaré variables*) is: L, $\sqrt{2(L-G)}\cos(g+h)$, $\sqrt{2(G-H)}\cos h$, respectively canonically conjugate to $\ell + g +$

h, $-\sqrt{2(L-G)}\sin(g+h)$, $-\sqrt{2(G-H)}\sin h$. They are non-singular for small eccentricities and inclinations. Show that the transformation between the two sets is a contact transformation.

12.11. With the atmospheric density profile of Fig. 8.1, estimate the height at which the molecular mean free path is 5 m, a typical size of an artificial satellite.

12.12. The resonance paradigm. Discuss the response of a damped harmonic oscillator $d^2x/dt^2 + vdx/dt + \omega_0^2 x = 0$ to an external force $\propto \cos\omega_e t$. (Hint: Use complex variables and Fourier transforms.)

12.13. Compare the observed decay of the CASTOR satellite (Fig. 12.2) with the simple prediction (12.60). The satellite had an aerodynamical coefficient $C_D = 2.1$ and an area-to-mass ratio $S/m = 0.0056$ m²/kg; the initial semimajor axis was $a_0 = 7,147$ km (neglect the small eccentricity $\simeq 0.07$). Use an exponential atmosphere with $H_0 = 40$ km, starting at 150 km with a density $\varrho_0 = 10^{-12}$ g/cm³.

12.14.* Following the Problem 12.12, discuss the response of a harmonic oscillator to a periodic acceleration $a(t) = a(t+P)$; using its Fourier series, analyse the cases in which $P\omega_0/(2\pi)$ is an integer, or nearly an integer. What happens when there is a small dissipation?

12.15. Rewrite Lagrange's eqs. (12.49), and their simplified form (12.51), using the complex variables $\xi = \sin(I/2)\exp(i\Omega)$ and $\zeta = e\exp(i\varpi)$.

12.16.* Using Gauss's equations, evaluate the changes of the Keplerian elements when the perturbing force $\mathbf{F} = \mathbf{F}_\star \delta(t - t_\star)$ occurs instantaneously at time $t_\star$ ($\delta(x)$ is Dirac's delta function). This problem describes the effect of a short thrust on a spacecraft, or a collisional impact.

12.17. The effect of atmospheric drag on an eccentric spacecraft can be described by a decrease δv_T in the transverse (along-track) velocity at the pericentre. Following the previous problem, show that the pericentre is unchanged, while in a revolution the apocentre distance decreases by $4\sqrt{1-e^2}\delta v_T/na$.

12.18.* The flow in phase space defined by the Hamilton eqs. (12.30) changes a function $F(q,p)$ according to $dF/dt = \{F, H\}$. Defining $\mathcal{L}_H^1 F = \{F, H\}$ and $\mathcal{L}_H^i F = \mathcal{L}_H^1 \mathcal{L}_H^{i-1} F$, prove that

$$F(t) = F(0) + \sum_{i=1}^{\infty} \frac{t^i}{i!} \mathcal{L}_H^i F(0),$$

usually called *Lie series* of F under the Hamilton flow.

12.19. The rate of change of argument (or longitude) of pericentre can be straightforwardly derived from variation of the Lenz vector $\mathbf{e}$ (11.16). Find eqs. (12.18) and (12.19) by this method.

– FURTHER READINGS –

The following are useful general textbooks of celestial mechanics, where various perturbation techniques are treated at increasing levels of depth and mathematical rigour: E. Finlay-Freundlich, *Celestial Mechanics*, Pergamon Press (1958); F.R. Moulton, *An Introduction to Celestial Mechanics*, Dover (1970); J. Kovalevsky, *Introduction to Celestial Mechanics*, D. Reidel (1967); A.E. Roy, *Orbital Motion*, A. Hilger (1978); D. Brouwer and G.M. Clemence, *Methods of Celestial Mechanics*, Academic Press (1961). Modern perturbation techniques in celestial mechanics, focusing on both secular and resonant dynamics, are discussed in A. Morbidelli, *Modern Celestial Mechanics: Aspects of Solar System Dynamics*, Taylor and Francis (2002). For stability and chaos, see A.J. Lichtenberg and M.A. Liebermann, *Regular and Stochastic Motion*, Springer-Verlag (1983). For many applications to artificial satellites, see *Le Mouvement du Véhicule Spatial en Orbite*, CNES, Toulouse (1980), and A. Milani, A.M. Nobili and P. Farinella, *Non-gravitational Perturbations and Satellite Geodesy*, A. Hilger (1987). A classical treatise of satellite motion in an atmosphere is D. King-Hele, *Theory of Satellite Orbits in an Atmosphere*, Butterworths (1964). Our derivation of the Gauss equations has been partly based on that by J.A. Burns, Elementary derivation of the perturbation equations of celestial mechanics, *Am. J. Physics* **44**, 944 (1976).

Chapter 13

THE THREE-BODY PROBLEM

The determination of the motion of N point-like masses subject to their mutual gravitational forces is the fundamental problem in celestial mechanics, with important astrophysical applications, including the dynamics of planetary and satellite systems and the evolution of stellar systems (ranging from multiple stars to stellar clusters and galaxies). From the perspective of theoretical physics, this has always been a core problem in gravitation theory; through observations of the motion of celestial bodies it can be tested, quantitatively validated, or even disproved. However, even for N as small as 3, it is impossible to find general solutions in terms of simple analytical functions (as for the two-body problem). A great variety of orbits are possible and it can be shown that only particular initial conditions give rise to periodic behaviour. Even the perturbative techniques worked out in Ch. 12 can be successfully applied only in the particular case of *hierarchical* systems, where the attraction of one body prevails (see Sec. 15.1). As a consequence, the general gravitational N-body problem has been studied by two alternative methods which can yield only partial results, but are in some way complementary. The first is numerical integration, starting from given initial conditions. In this way the motion can be determined in detail, but only for a limited span of time (due to both limitations in computer time and accumulation of numerical errors; see Sec. 15.1); in addition, it is often impossible to determine the region of phase space where a property of the numerically determined orbits holds. The second method is the search for general constraints or criteria regarding qualitative features of the motion, such as its periodic character, its stability and its geometrical and topological properties.

We discuss in detail the simplest N-body problem (for $N > 2$), namely the so-called *restricted three-body problem*, studied for the first time by J.L. LA-

GRANGE in 1772. A test body with a negligible mass moves under the gravitational attraction of two finite masses M and $m < M$ (usually called primaries), which are assumed to follow Keplerian orbits around their centre of mass. When their orbits are circular, we refer to the *circular* restricted three-body problem; otherwise we have the *elliptic* problem. Here we limit ourselves to the former, but will comment on the complexities of the latter.

The three-body problem is thereby highly simplified and reduced to three degrees of freedom, to be compared with $3N$ for the general case. In spite of this the restricted problem has been extensively studied: indeed, many of the important results discussed in the following can be extended, or adapted, to more complex cases (e.g., to the general three-body problem with three finite masses). Understanding the richness and diversity of possible solutions of the restricted three-body problem, including their chaotic behaviour, represents the most important step towards understanding the complexity of N-body dynamics, and strongly contrasts with the "trivial" character of the two-body problem. Moreover, several interesting systems can be approximated with a restricted problem, including: a space probe journeying from the Earth to the Moon, a minor planet under the influence of the Sun and Jupiter, and a planet about a binary stellar system with small eccentricity. In the former two cases, the eccentricity of the relative orbit of the two primaries is ≈ 0.05, hence the circular restricted problem is a rough, but physically illuminating approximation.

13.1 Equations of motion and the Jacobi constant

We recall that the circular orbits of two bodies with masses M and m are coplanar, with their centres at the centre of mass; their radii, in terms of their constant distance a, are $ma/(m+M)$ and $Ma/(m+M)$, respectively. The masses are always aligned with the centre placed between them and, by Kepler's third law (11.19′), rotate with the constant angular velocity $n = \sqrt{G(M+m)/a^3}$.

Let us choose the units of mass, distance and time in such a way that $G(m+M)$, a and the gravitational constant G are equal to unity. Hence the angular velocity n is also equal to unity and the period of revolution is 2π. In these units m is equal to $\mu = m/(m+M)$; time is reckoned with the dimensionless variable $\tau = nt$. The distances of M and m from the centre of mass are, respectively, μ and $(1-\mu)$, and are constant in time. Without loss of generality we also assume that the orbital plane of primaries is the (X, Y) plane of an inertial reference system.

We now write the equations of motion in a frame – usually called *synodic system* – rotating with angular velocity $n = 1$ about the normal to the orbital plane through the centre of mass; m and M are at fixed positions on the x-axis (see Fig. 13.1). If (x, y, z) are the coordinates of the test mass, we have

13. The three-body problem

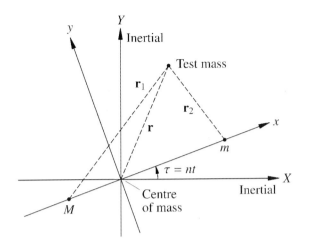

Figure 13.1 The rotating coordinate system for the restricted three-body problem. The x and y-axes rotate with respect to the fixed axes X and Y about the axis Z through the centre of mass with angular velocity n.

$$\ddot{x} - 2\dot{y} = x - \frac{(1-\mu)(x+\mu)}{r_1^3} - \frac{\mu(x-1+\mu)}{r_2^3}, \tag{13.1a}$$

$$\ddot{y} + 2\dot{x} = y - \frac{(1-\mu)y}{r_1^3} - \frac{\mu y}{r_2^3}, \tag{13.1b}$$

$$\ddot{z} = -\frac{(1-\mu)z}{r_1^3} - \frac{\mu z}{r_2^3}, \tag{13.1c}$$

where $r_1 = [(x+\mu)^2 + y^2 + z^2]^{1/2}$ and $r_2 = [(x-1+\mu)^2 + y^2 + z^2]^{1/2}$ are, respectively, the distances of the test mass from M and m. Dots mean derivatives with respect to the (dimensionless) time τ, which is just the phase angle of the synodic x-axis with respect to the inertial X-axis. If $z = 0$ we have a *planar problem*. On the left-hand sides of (13.1a) and (13.1b) we have added the Coriolis acceleration, leaving on the right-hand sides the centrifugal acceleration and the gravitational pull of the two massive bodies. The motion along z, the direction normal to the orbital plane of the primaries, does not contain apparent accelerations. All the right-hand side terms are functions only of the coordinates (x, y, z), and can be derived from the potential function (an effective potential)

$$W(x, y, z) = -\frac{1}{2}(x^2 + y^2) - \frac{(1-\mu)}{r_1} - \frac{\mu}{r_2}; \tag{13.2}$$

eqs. (13.1) can be rewritten as

$$\ddot{x} - 2\dot{y} = -\partial W/\partial x = -\partial \Omega/\partial x, \tag{13.3a}$$

$$\ddot{y} + 2\dot{x} = -\partial W/\partial y = -\partial \Omega/\partial y, \tag{13.3b}$$

$$\ddot{z} = -\partial W/\partial z = -\partial \Omega/\partial z - z. \tag{13.3c}$$

For a more symmetric expression of the potential, the function

$$\Omega(x, y, z) = W - \frac{z^2 + \mu(1 - \mu)}{2} = -(1 - \mu)\left(\frac{r_1^2}{2} + \frac{1}{r_1}\right) - \mu\left(\frac{r_2^2}{2} + \frac{1}{r_2}\right), \quad (13.4)$$

has been introduced. Eqs. (13.3) are similar to the law of motion of a charged particle in a constant and uniform magnetic field and in a conservative electric field.

Jacobi integral. If we multiply eqs. (13.3) by $\dot{x}$, $\dot{y}$ and $\dot{z}$, respectively, and integrate with respect to time, we obtain the *Jacobi integral* of the motion

$$\frac{1}{2}(\dot{x}^2 + \dot{y}^2 + \dot{z}^2) = -W(x, y, z) + C, \quad (13.5)$$

where C is a constant of integration determined by the initial conditions. The derivation of eq. (13.5) has an obvious analogy with the conservation of energy for a point mass in a curl-free force field; however, since we are in a rotating frame, the *Jacobi constant* C is *not* the energy. The Jacobi integral (13.5) is the only first integral of eqs. (13.3). Unfortunately even in the planar problem it is not sufficient to determine the solution. Note also that the Coriolis force does not perform work, nor affects the conservation equation (13.5), but influences the motion in a crucial way.

An important consequence of eq. (13.5) is that, since the "kinetic energy" $v^2/2$ is never negative, the motion can occur only in regions where

$$W(x, y, z) \leq C. \quad (13.6)$$

This constraint depends on the value of the Jacobi constant, hence on the initial conditions. For a given value of C, the equation $W(x, y, z) = C$ defines the boundary of the "allowed" region, i.e. one of *Hill's zero-velocity surfaces* (since the test mass can reach it only when its velocity in the rotating frame vanishes). In the planar problem we have *Hill's zero-velocity curves*.

To understand the topology of the "allowed" regions, note first that $-W$ becomes very large whenever $(x^2 + y^2)$ is large or, alternatively, when r_1 or r_2 are small. Therefore if $-C$ is large the test mass must remain either outside a large cylinder defined by $x^2 + y^2 = -2C$, or inside one of two very small circles surrounding the two massive bodies, defined by $r_1 = (\mu - 1)/C$ and $r_2 = -\mu/C$ (see Fig. 13.3, a). In the former case the test mass "feels" the two gravitating centres approximately as a single one; in the latter only one of the masses is important. Since in this limiting case the three regions where the motion is allowed are disconnected, the test mass will remain for ever in the region where it is initially placed. We shall investigate in some detail the topological character of the zero-velocity curves, especially in the (x, y) plane, for an arbitrary value of C in the next Section.

A note on the elliptic restricted problem. The equations of motion of the elliptic restricted three-body problem may be put in a similar form as (13.3) with the following important difference. The synodic system, in which the primaries are fixed on the x-axis, does not rotate uniformly in inertial space and the scale of its axes changes with time (it is therefore called a *pulsating system*). After a tedious calculations one may show that eqs. (13.3) still hold, provided (x, y, z) are the coordinates in the pulsating synodic system, but the potential function is now $\Omega' = \Omega/(1 + e\cos f)$; f is the true anomaly of the elliptic motion of the primaries, which replaces the scaled time variable τ, and e is the eccentricity of their orbits around their centre of mass. Since Ω' is now an explicit function of f, the Jacobi integral does not exist. Its lack obviously makes the investigation of solutions in the elliptic problem much more complicated.

13.2 Lagrangian points and zero-velocity curves

Starting with a simple matter, we first determine the equilibrium positions of the problem, where the test mass feels a zero net force, and thus can stay at rest in the rotating frame. First, from eq. (13.1c) we conclude that $z = 0$. This is easily understood, since in the synodic frame no inertial acceleration has a z-component and thus the gravitational pull of the primaries toward their orbital plane cannot be compensated.

We thus investigate the equilibrium points in the orbital plane of the primaries. We require that (since $z = 0$)

$$-\nabla W = \mathbf{r} - (1 - \mu)\frac{\mathbf{r}_1}{r_1^3} - \mu\frac{\mathbf{r}_2}{r_2^3} = 0, \qquad (13.7)$$

where $\mathbf{r} = (x, y, 0)$, $\mathbf{r}_1 = (x + \mu, y, 0)$, $\mathbf{r}_2 = (x - 1 + \mu, y, 0)$ (see Fig. 13.1). Since the origin is at the centre of mass, we have

$$\mathbf{r} = (1 - \mu)\mathbf{r}_1 + \mu\mathbf{r}_2 \qquad (13.8)$$

and, substituting into eq. (13.7), we obtain

$$(1 - \mu)\left(\frac{1}{r_1^3} - 1\right)\mathbf{r}_1 + \mu\left(\frac{1}{r_2^3} - 1\right)\mathbf{r}_2 = 0. \qquad (13.9)$$

If the equilibrium points are not on the x-axis, eq. (13.9) is satisfied only for $r_1 = r_2 = 1$, namely at the two *equilateral equilibrium points* which are at the same distance from the two massive bodies and form with them two equilateral triangles. Traditionally, they are called L_4 and L_5, with L_4 preceding the lighter primary with mass m on its orbit around the centre of mass; their position is given by $L_{4,5} = (1/2 - \mu, \pm\sqrt{3}/2, 0)$.

On the x-axis eq. (13.9) is an algebraic, fifth degree equation and its roots do not have a simple expression. The x-axis is divided by the primaries in three intervals; because of the absolute values, its analytical form depends on which of the intervals one considers. We can argue, however, that there are just three real roots, one in each interval, corresponding to the three *collinear equilibrium points* L_1, L_2 and L_3. Consider first the interval $(1-\mu, +\infty)$ on the right of both primaries. The force $-dW/dx$ tends to $-\infty$ as $x \to 1 - \mu$, where the mass m is; for x very large the centrifugal repulsion prevails and $-dW/dx \to +\infty$. Because of the simple functional dependence of both competing forces, it is clear that in this interval there is just one equilibrium point, called[1] L_2. A similar argument can be used for the interval between the primaries, with an equilibrium point called L_1, and for the interval to the left of the primaries, with L_3. When $\mu \ll 1$, satisfied in most cases, the roots of (13.9) can be studied with power expansions; we shall see that, while the L_3 solution is analytic in μ, the other two are analytic in $\mu^{1/3}$.

Starting with the easiest one, for $\mu = 0$ the equilibrium point L_3 is at $x = -1$, under the attraction of the main body and the (equal) centrifugal repulsion; for $\mu > 0$ there is an additional gravitational pull and the point shifts to the left, to use more centrifugal force. In most applications $x = -1$ is all we need; to get the power expansion, we note that in this interval $\mathbf{r}_1 = -r_1 \mathbf{e}_x$ and $\mathbf{r}_2 = -(1+r_1)\mathbf{e}_x$. After some algebra eq. (13.9) now reads

$$\frac{\mu}{1-\mu} = \frac{(1-r_1^3)(1+r_1)^2}{r_1^3(3+3r_1+r_1^2)}. \tag{13.10}$$

We then compute the power series of r_1 in terms of $\mu/(1-\mu)$, by comparing its successive powers on both sides of the eq. (13.10), and get

$$r_1(L_3) = 1 - \frac{7}{12}\frac{\mu}{1-\mu} + \frac{7}{12}\left(\frac{\mu}{1-\mu}\right)^2 - \frac{13,223}{20,736}\left(\frac{\mu}{1-\mu}\right)^3 + O(\mu^4);$$

finally we obtain

$$x(L_3) = -\mu - r_1(L_3) = -1 - \frac{5}{12}\mu + \frac{1,127}{20,736}\mu^3 + O(\mu^4) \tag{13.11}$$

in terms of the small parameter μ.

To deal with L_1 and L_2, note that when $\mu = 0$ there is an equilibrium point at $x = 1$. With a small mass μ, to keep a point at rest at a small distance r_2 from it, we must balance the gravitational pull of the lighter mass μ/r_2^2 with the sum of

---

[1] The notation L_1 and L_2 is not uniform in the astronomical literature; we follow several textbooks on celestial mechanics, including Brouwer and Clemence's. V. SZEBEHELY, in his classical work on the three-body problem, uses the inverted notation.

13. The three-body problem

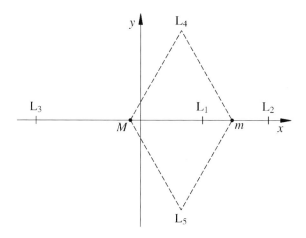

Figure 13.2 The positions of the five Lagrangian equilibrium points in the rotating reference system for $\mu = 0.1$.

the centrifugal force and the attraction by the larger primary $1 - \mu$. This sum, which vanishes at $x = 1 - \mu$, in the neighbourhood of this point is linear and given by kr_2 on the right and $-kr_2$ on the left (k is a positive constant). Hence there are two equilibrium points on each side of the small mass and at the same distance $r_2 = (\mu/k)^{1/3}$. They are L_1 (on the left) and L_2 (on the right). The corresponding power expansion is best obtained by taking r_2 as the unknown. For L_1, $\mathbf{r}_1 = (1 - r_2)\mathbf{e}_x$ and $\mathbf{r}_2 = -r_2\mathbf{e}_x$; substituting in eq. (13.9) we obtain

$$\epsilon^3 \equiv \frac{\mu}{3(1-\mu)} = r_2^3 \frac{1 - r_2 + r_2^2/3}{(1 + r_2 + r_2^2)(1 - r_2^3)}, \qquad (13.12)$$

with the power expansion[2]

$$r_2 = \epsilon - \frac{1}{3}\epsilon^2 - \frac{1}{9}\epsilon^3 + O(\epsilon^4). \qquad (13.13)$$

At L_2, $\mathbf{r}_1 = (1 + r_2)\mathbf{e}_x$ and $\mathbf{r}_2 = r_2\mathbf{e}_x$; the same equation as (13.12) is obtained, with r_2 changed into $-r_2$. To the lowest order in μ, L_1 and L_2 are at the same distance $(\mu/3)^{1/3}$ from the smaller mass; however, this symmetry is violated by the higher order terms. To conclude,

$$x(L_1) = 1 - \mu - \left(\frac{\mu}{3}\right)^{1/3} + O(\mu^{2/3}), \quad x(L_2) = 1 - \mu + \left(\frac{\mu}{3}\right)^{1/3} + O(\mu^{2/3}). \quad (13.14)$$

---

[2] For a systematic approach Lagrange's inversion method of Problem 11.16 can be used. In fact, eq. (13.12) can be written in the same form, as $r_2 = \epsilon + \epsilon\phi(r_2)$, with

$$\phi(r_2) = \left[\left(\frac{1 - r_2 + r_2^2/3}{(1 + r_2 + r_2^2)(1 - r_2^3)}\right)^{-1/3} - 1\right].$$

The same approach can be used to get the power expansion of (13.10).

In Fig. 13.2 we show the positions of the five equilibrium (or *Lagrangian*) points $L_1, \ldots, L_5$ in the rotating frame. The distance between L_1 and the small mass, to the lowest order, is called *Roche's (or Hill's) radius*; in ordinary units (a is the distance of primaries)

$$r_2(L_1) = a \left(\frac{m}{3M}\right)^{1/3} = R_{\text{Roche}} . \tag{13.15}$$

As mentioned above, the equations of motion for the elliptic restricted three-body problem are formally identical to those of the circular problem, provided the coordinates are referred to the pulsating synodic system and the potential function is time-dependent, to wit $\Omega' = \Omega/(1 + e\cos f)$ (in the orbital plane of primaries, $z = 0$, Ω coincides with W up to a constant). Thus it immediately follows that the equilibrium points exist also in the elliptic problem and, moreover, they are located at exactly the same positions as in the circular problem (recall that the coordinates are now referred to the pulsating system). Obviously, their stability may be quite different than in the circular problem (discussed in Sec. 13.3).

Zero-velocity curves. We now turn to the problem of the shape of zero-velocity curves in the orbital plane of the primaries, isolevels of the function $W(x, y, 0)$; actually, it is conventional to work with the function $\Omega = W - \mu(1 - \mu)/2$ and refer the value of the Jacobi constant C to Ω. We first determine the value of Ω at the five stationary points $L_1, \ldots, L_5$ and their topological character. The first derivatives of Ω vanish; we need the second derivatives and the Hessian determinant

$$\mathcal{H} = (\partial_{xx}\Omega)(\partial_{yy}\Omega) - (\partial_{xy}\Omega)^2 \tag{13.16}$$

(for brevity, hereinafter we shall use the notation ∂_{xx} for $\partial^2/\partial x^2$, etc.). This analysis shows that

$$\Omega(L_5) = \Omega(L_4) > \Omega(L_3) > \Omega(L_2) > \Omega(L_1) , \tag{13.17}$$

and also that L_4, L_5 are two maxima, while L_1, L_2 and L_3 are saddle points; moreover, $\Omega(L_5) = \Omega(L_4) = -1.5$ for all values of μ. As C increases from very large and negative values, the topological structure of the zero-velocity curves changes as it crosses the four thresholds (13.17). The entire range $(-\infty, +\infty)$ for C is divided in five intervals (Fig. 13.3):

(a) $C < \Omega(L_1)$: the motion is allowed only within two small regions surrounding the massive bodies and outside a large boundary including both of them (in the former case the test mass can be viewed as a satellite of one of the primaries, in the latter as a planet orbiting around the pair).

(b) $\Omega(L_2) > C > \Omega(L_1)$: a channel opens up at L_1 between the two allowed regions about the massive bodies and the test body may go from one to

13. The three-body problem

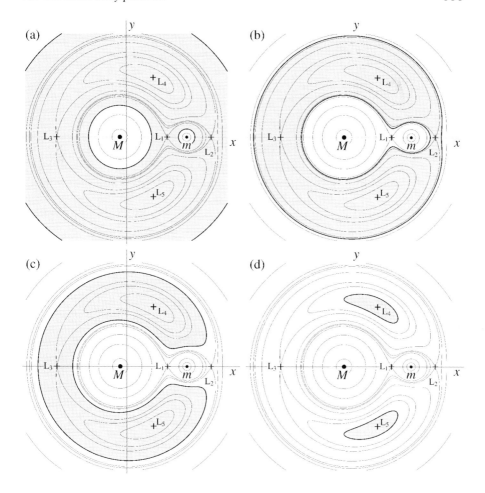

Figure 13.3. The four different topologies of zero-velocity curves in the restricted three-body problem for increasing values of the Jacobi constant C (corresponding to the cases (a)–(e) discussed in the text and $\mu = 0.1$). The positions of the two masses M and m and of the five Lagrangian points are indicated. Case (e), when the motion is allowed in the whole (x, y) plane, is not shown.

another, but is bound to the pair. The orbit of a spacecraft from the Earth to the Moon belongs to this class and has an eight-like shape, with the crossing point near L_1.

(c) $\Omega(L_3) > C > \Omega(L_2)$: the allowed region opens up behind the smaller mass; the test body can pass through this hole either to escape from, or to penetrate into, the neighbourhood of the massive bodies. Several comets (e.g., P/Oterma) belong to this class, with the Sun and Jupiter as primaries; their orbits alternate between the inner and the outer regions, passing through the

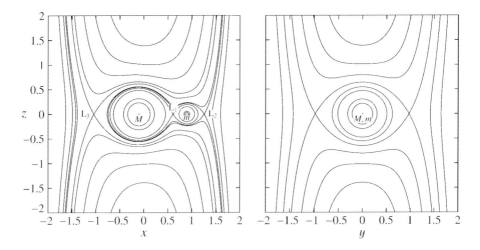

Figure 13.4. Cross sections of the zero-velocity surfaces with the (x, z) and (y, z) planes for the restricted three-body problem for different values of the Jacobi constant C; $\mu = 0.1$ as before.

neck just beyond Jupiter. The forbidden zone is shaped like a horseshoe. The surface $\Omega(x, y)$ has two equal peaks at L_4 and L_5 and a ridge through L_3.

(d) $\Omega(L_5) = \Omega(L_4) = -1.5 > C > \Omega(L_3)$: a channel opens up also behind the bigger primary. Zero-velocity curves are disconnected and closed around L_4 and L_5; the motion is forbidden only in the inner regions. This is a typical situation of the tadpole orbits discussed in Sec. 13.3.

(e) $C > \Omega(L_4) = \Omega(L_5) = -1.5$: the whole (x, y) plane is allowed.

In view of its topological nature, this classification holds for an arbitrary value of μ, of course with the three collinear Lagrangian points at their correct positions on the x-axis, obtained numerically or with a power expansion. In the elliptic problem, when the eccentricity is not too large, this classification is useful to characterize the *hierarchical stability* of the system, namely the possibility that a hierarchical structure like a planet-satellite pair plus a distant disturbing body can take up drastically different configurations (e.g., with the satellite orbiting around the distant body): an obviously interesting question for natural satellites, asteroids and comets. The value of the Jacobi constant can also discriminate whether the ejection of the test mass to infinity is possible or not (e.g., whether an interplanetary probe moving chiefly in the field of a planet and the Sun can or cannot leave the solar system).

Extension to the three-dimensional case. For very large and negative values of the Jacobi constant zero-velocity surfaces (Sec. 13.1 and Fig. 13.4) are nearly spherical around the primaries, with a cylindrical surface at great distance. As the value of C increases toward the critical value $\Omega(L_1)$, the closed surfaces deform to a drop-like shape and can still keep a test mass near one of the primaries, as discussed above (it is interesting to note that this was originally Hill's argument for the long term stability of the lunar orbit around the Earth, and an evidence against the hypothesis of the origin of the Moon by capture). At $C = \Omega(L_1)$ the two deformed surfaces touch at the L_1 point and are called *Roche lobes*. At slightly higher values of C a throat opens between the two regions and a test mass can move from the vicinity of one primary to the vicinity of the other.

This concept has important applications in astrophysics. In ordinary stellar binary systems atmospheric material can spill through the L_1 channel from one component to the other, affecting its chemical, spectroscopical and evolutionary properties. When one component of the binary system is very dense (such as a white dwarf or a neutron star), the accreting material from the other gains much energy in the strong gravity field and, due to the angular momentum conservation, forms an *accretion disc* (a classical model of the cataclysmic variables). The system may become an intense source of X-rays; this is, in fact, a standard model for galactic X-ray sources due to collapsed stars. Even more violent effects may occur when one component is a black hole. Roche's radius (13.15) also gives approximately the largest size of a cloud of particles which can be confined near the position of the secondary; this is also the largest size of a natural satellite, with no cohesion, which is not broken up by the tidal forces of the planet (see Secs. 4.5 and 14.7).

13.3 Stability analysis and motion near the Lagrangian points

Are the five Lagrangian points stable? Does a test mass initially placed in a small neighbourhood near one of them remain there for ever, or does it escape? In view of the importance of *dynamical stability* for the solar system, it is useful to introduce first this problem in a general way. For an N–dimensional dynamical system governed by the equations

$$\frac{dx_i}{dt} = V_i(\mathbf{x}) \quad (i = 1, 2, \ldots N), \tag{13.18}$$

the question of the stability of a solution $\mathbf{x}_0(t)$ consists in the investigation of the long-term evolution of small changes $\delta\mathbf{x}(0)$ in the initial conditions $\mathbf{x}_0(0)$ at $t = 0$. Specifically, we are interested in finding out the behaviour for large t of the magnitude of the displacement $|\delta\mathbf{x}(t)|$ when the initial displacement $|\delta\mathbf{x}(0)|$ tends to zero. It may happen that $|\delta\mathbf{x}(t)|$, after a sufficiently long time, becomes very large, however small we choose $|\delta\mathbf{x}(0)|$: then a small cause produces very large effects. In principle the solution is still uniquely determined by the initial conditions; but, in practice, since the accuracy with which they can be assigned is always finite, determinism is not realizable. This problem is closely connected with *chaotic behaviour*.

In the *linear approximation* the displacement fulfils

$$\frac{d\,\delta\mathbf{x}}{dt} = \mathsf{B}(\mathbf{x}_0(t)) \cdot \delta\mathbf{x} + O\left(|\delta\mathbf{x}|^2\right)\,, \qquad \left(\mathsf{B} = \frac{\partial \mathbf{V}}{\partial \mathbf{x}}\right). \qquad (13.19)$$

The simplest situation occurs when the $N \times N$ matrix B is independent of time; this is the case when the solution $\mathbf{x}_0(t)$ is time-independent, i.e. an equilibrium point. Then the previous equation is solved by a linear combination of N functions $\propto \exp \lambda t$, where λ is one of the N eigenvalues of B. If B is real the eigenvalues are either also real, or appear in complex conjugated pairs. When there is at least one eigenvalue with a positive real part $\kappa = \operatorname{Re}\lambda > 0$, the norm $|\delta\mathbf{x}(t)|$ of the displacement diverges exponentially as $t \to \infty$ and we have (linear) instability. Analytic tools for the linear stability analysis have been also developed for periodic solutions $\mathbf{x}_0(t)$, but the mathematics is much more involved than in the stationary point case.

Lyapunov exponents.– When the corrections $\delta\mathbf{x}$ are large, either because they are so initially, or because the system is unstable, the linear approximation does not hold any more, chaotic behaviour may ensue and more sophisticated tools are needed, which are beyond our scope. However, since this is an important problem in the dynamical evolution of the solar system (Ch. 15), it is appropriate to briefly discuss here a generalization of the concept of stability to the non-linear regime.

We are looking for the conditions under which the size (in Euclidean space) $|\delta\mathbf{x}(t)|$ of a correction, in some sense, diverges exponentially[3]. Note that in the linear approximation, if $\delta\mathbf{x}(t)$ is known, the real part of the exponent λ responsible for the instability can be expressed as

$$\kappa_L = \lim_{t \to \infty,\, |\delta\mathbf{x}(0)| \to 0} \frac{1}{t} \ln \frac{|\delta\mathbf{x}(t)|}{|\delta\mathbf{x}(0)|}\,. \qquad (13.20)$$

For large times the limit kills all oscillatory terms. This quantity gives the *Lyapunov exponents* κ_L relative to the solution $\mathbf{x}_0(t)$; they measure the mean exponential rate of divergence of the trajectories. This definition applies also to the general case in which the basic solution $\mathbf{x}_0(t)$ is time-dependent (and not necessarily periodic) and non-linear terms in the correction are taken into account. It can be shown that there are N (possibly non-distinct) such exponents,

---

[3] As before, the stability of the solution is violated if $|\delta\mathbf{x}(t)|$ becomes "large" for whatever small initial values $|\delta\mathbf{x}(0)|$; however, when $|\delta\mathbf{x}(t)|$ increases exponentially – as in the linear approximation – the instability characterizes the chaotic behaviour of the particular orbit $\mathbf{x}_0(t)$. Since the vector space of $\mathbf{x}$ may be bound (e.g., surface of a sphere), the exponential divergence of close by orbits is always meant locally.

corresponding to initial conditions $\delta\mathbf{x}(0)$ in N different subspaces. In the limit, the component corresponding to the largest among the Lyapunov exponents, the *maximum Lyapunov exponent* κ_{Lm}, prevails and determines the properties of a generic solution. The long-term behaviour of the solution is qualitatively different when $\kappa_{Lm} \leq 0$ or $\kappa_{Lm} > 0$. In practice this is shown by the behaviour of the time-dependent quantity

$$\kappa_L(t) = \frac{1}{t} \ln \frac{|\delta\mathbf{x}(t)|}{|\delta\mathbf{x}(0)|} . \qquad (13.20')$$

As in the linear case, after initial large fluctuations, it decays, roughly as $1/t$, to its asymptotic value, which, when positive, gives, in order of magnitude, the growth rate of the initial correction. The largest of these values provide a measure of the stochastic character of the solution.

The numerical computation of (13.20′) is not obvious, especially for chaotic orbits for which $|\delta\mathbf{x}(t)|$ quickly changes. In practice, numerical integration of the linearized system (13.19) with arbitrary initial conditions is carried out for time T until $|\delta\mathbf{x}(T)|$ reaches a chosen limit; then the difference vector $\delta\mathbf{x}(T)$ is rescaled by a factor s_1, namely $\delta\mathbf{x}(T) \to \delta\mathbf{x}(T)/s_1$, and the integration carried for another time step T; after a second rescaling with the factor s_2 the scheme reiterated. After l steps, the quantity

$$\hat{\kappa}_L(t) = \frac{\sum_{j=1}^{l} \ln s_j}{lT} , \qquad (13.20'')$$

approximates the maximum Lyapunov exponent (13.20′).

Stability of the Lagrangian points. In the case of the restricted three-body problem, were it not for the velocity-dependent Coriolis force, the answer would be easy: none of the Lagrangian points, extrema for the potential W, would be stable, since L_1, L_2 and L_3 are saddle points and L_4, L_5 are maxima. But the Coriolis force modifies this conclusion in an essential way. Let $(x_0, y_0, 0)$ be one of the five equilibrium points and $\delta x, \delta y, \delta z$ a small displacement. Substituting into (13.3) and keeping only first-order terms, we obtain

$$\delta\ddot{x} - 2\,\delta\dot{y} + \partial_{xx}W\,\delta x + \partial_{xy}W\,\delta y = 0 , \qquad (13.21a)$$
$$\delta\ddot{y} + 2\,\delta\dot{x} + \partial_{yy}W\,\delta y + \partial_{xy}W\,\delta x = 0 , \qquad (13.21b)$$
$$\delta\ddot{z} \hspace{3.5cm} + \partial_{zz}W\,\delta z = 0 . \qquad (13.21c)$$

The second derivatives of W are computed at the Lagrangian point. Of course, these equations can be written as a set of six first order equations and the previous discussion holds; but it is convenient to stick to the second order form. Note that the linear motion in the z-direction is decoupled from the motion in the (x, y) plane, a result of the fact that $\partial_{xz}W \propto z$ and $\partial_{yz}W \propto z$. Provided

$\partial_{zz}W > 0$ (as the case is for all Lagrangian points), the last equation is a harmonic oscillator with frequency $\sqrt{\partial_{zz}W}$; for the equilateral points $\partial_{zz}W = 1$ for all values of μ, so that vertical oscillations have a frequency identical to the orbital frequency of the primaries. The fact that vertical motion is always bound is physically clear: the gravity pull of the primaries tends to restore the test mass to their orbital plane.

Thus we consider only motion in the (x, y) plane, as described by the equations (13.21a) and (13.21b). Putting (recall that with the variable τ, the mean motion $n = 1$)

$$\delta x(\tau) = a \exp(\lambda \tau), \quad \delta y(\tau) = b \exp(\lambda \tau), \quad (13.22)$$

we get the linear and homogeneous system

$$a(\lambda^2 + \partial_{xx}W) + b(-2\lambda + \partial_{xy}W) = 0, \quad (13.23a)$$
$$a(2\lambda + \partial_{xy}W) + b(\lambda^2 + \partial_{yy}W) = 0. \quad (13.23b)$$

a and b are arbitrary complex numbers; in obvious analogy with the previous, first-order formulation, we have a solution only if the determinant of the coefficients

$$\Delta = \lambda^4 + (\partial_{xx}W + \partial_{yy}W + 4)\lambda^2 + \mathcal{H}. \quad (13.24)$$

vanishes, with four roots; $\mathcal{H}$ is the Hessian (13.16) computed at the equilibrium point. If a root $\lambda = \kappa + i\sigma$ of this biquadratic equation has a positive real part κ, the corresponding solution diverges and the point is unstable. When the discriminant

$$\Delta' = (\partial_{xx}W + \partial_{yy}W + 4)^2 - 4\mathcal{H} \quad (13.25)$$

of the quadratic equation for $w = \lambda^2$ is positive, its roots w_1 and w_2 are real. This is the case of the collinear points, of the saddle type with $\mathcal{H} < 0$; the roots are also of opposite sign. The positive one gives two real and opposite values for λ, corresponding, respectively, to an exponentially growing ($\lambda_u = \kappa_u > 0$) and an exponentially damped solution; the negative root yields two imaginary conjugate values for λ, corresponding to purely oscillating solutions. Therefore the collinear points are unstable. In the four-dimensional manifold of initial conditions, however, there is a subspace corresponding to stable solutions.

In real situations the inaccuracy in orbit determination and in manœuvres, the time dependence of the eccentricity of the primaries and/or other disturbances prevent this arrangement to ensure confinement. What matters, however, is the e-folding time $1/\kappa_u$ for the equilibrium point; since we have chosen the mean motion n of the primaries as unit of frequency, its order of magnitude is the orbital period divided by 2π; thus it possible to keep a spacecraft near a Lagrangian point with appropriate manœuvres. For the Sun and the Earth this timescale is one or two months, for practical purposes "comfortably" long. A suite of Sun-observing satellites (SOHO, ACE and Wind) was placed near L_1, with only several orbit corrections every year. Similarly, the important and delicate astrometric satellite Gaia is planned to orbit in the vicinity of L_2 for the Sun-Earth system, assuring the environmental (in particular thermal) stability of the spacecraft.

At the two triangular points, on the other hand, eq. (13.24) becomes

$$\lambda^4 + \lambda^2 + 27\mu(1-\mu)/4 = 0 \,, \tag{13.26}$$

with positive coefficients. If the discriminant (eq. (13.25))

$$\Delta' = 1 - 27\mu(1-\mu) \tag{13.27}$$

is negative, we get two complex conjugate solutions for λ^2; the values for λ are two complex conjugate pairs, whose real parts are opposite. They correspond to oscillatory solutions with increasing and decreasing amplitudes and give rise to instability. But if $\Delta' \geq 0$, the solutions for λ^2 are *real and negative*, λ is always imaginary and we obtain stable oscillations. It is easy to show that the condition $\Delta' \geq 0$ is equivalent to

$$\mu \leq \mu_{\rm cr} = (1 - \sqrt{23/27})/2 \simeq 0.03852 \,. \tag{13.28}$$

L_4 and L_5 are linearly stable for a sufficiently small mass ratio. The roots $\lambda = \pm i\sigma$ are imaginary and give the oscillation frequencies

$$\sigma_1 = \pm\sqrt{\left(1 + \sqrt{1 - 27\mu(1-\mu)}\right)/2} \,, \tag{13.29a}$$

$$\sigma_2 = \pm\sqrt{\left(1 - \sqrt{1 - 27\mu(1-\mu)}\right)/2} \,, \tag{13.29b}$$

of which σ_2 is smaller (Fig. 13.5). At $\mu = \mu_{\rm cr}$ the two frequencies coincide, so that the eq. (13.26) has two pairs of degenerate eigenvalues; as in a critically damped oscillator, terms $\propto \tau \exp(i\sigma\tau)$ appear and the amplitude increases.

Motion around the equilateral points in the solar system. As mentioned above, the hierarchical structure of the solar system offers several examples in which a small body moves in the gravitational field of two principal bodies (primaries) and the effect of other objects may be considered as perturbation. For the Sun-Jupiter pair, with $\mu \simeq 1/1,047$, we have some asteroids, short-period comets, etc.; for the Earth-Moon pair, with $\mu \simeq 1/81$, there are space probes; for other satellite systems of the giant planets (e.g., the Saturn-Tethys pair with $\mu \approx 10^{-6}$), other small satellites may be close to the equilateral points. The mass ratio is typically smaller than the critical value $\mu_{\rm cr} \simeq 0.03852$, with the exception of the Pluto-Charon system, for which $\mu \approx 1/10$. The possibility of motion near the equilateral points may be thus confronted with reality; but we should take into account more realistic conditions, in particular, the effect

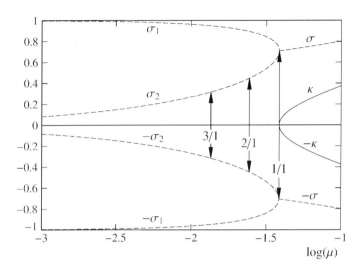

Figure 13.5. Eigenfrequencies for displacements around the equilateral points vs the mass ratio μ. For small enough values of μ ($\mu < \mu_{cr}$), the eigenvalues are purely imaginary and correspond to oscillations with the frequencies σ_1 and σ_2. At $\mu = \mu_{cr}$ they are in 1/1 resonance; they acquire real parts $\pm\kappa$ and linear instability sets on. In the non-linear regime we have instability also for $\mu < \mu_{cr}$, when the eigenfrequencies are resonant and their ratio is 2/1 or 3/1 (their positions are indicated).

of non-linear terms, the eccentricity of the orbit of the primaries and other disturbing forces, both gravitational and non-gravitational. We now comment on some of them.

First, in the elliptic problem, when the eccentricity increases, small oscillations around the equilateral points typically become unstable even for $\mu < \mu_{cr}$; this may be expected, since the distance from the primaries pulsates and eventually may get small enough to produce instability. However, in the solar system the eccentricity of the primaries is generally small, with little effect on the instability region. A subtler effect arises from the secular evolution of the orbit of the less massive primary due to interaction with other bodies. This is, for instance, the case of Jupiter, whose node precesses due Saturn and other planets. The *Trojan asteroids*[4], the most outstanding example of motion near

---

[4]These asteroids oscillate around two points on Jupiter's orbit, leading and trailing the planet by 60°. Sometimes the leading group is called "Greeks" and the trailing group "Trojans" (with, respectively, about 730 and 470 known members in April 2003; for updates, see the web site http://www.daf.on.br/~froig/petra). The asteroids themselves have names taken from the corresponding sides, as mentioned in Homer's Iliad. These $\approx 1,200$ known bodies are only a tiny fraction of estimated number $\approx 2 \times 10^6$ objects larger than 1 km in this region. The reason for the asymmetry in the number of Greeks and Trojans is not precisely known; an observational bias and an asymmetric primordial accumulation around the leading and trailing librations points due to the nebular drag are possible hypotheses.

the equilateral points in the solar system, are affected by this precession; as a result, Trojan orbits with inclinations larger than 25° are highly unstable and, indeed, only a few asteroids on such orbits are observed. These secular resonances may be also responsible for the very-long-term ($\approx$ Gy) instability of the Trojan clouds.

In practice non-linear terms, neglected in eqs. (13.21) and generally of order (amplitude/semimajor axis)^2, are often important. For instance, Trojans have a mean libration amplitude in longitude of $\approx 14°$, with a maximum as large as $\approx 30°$. Moreover, as hinted earlier in the general discussion, linear stability does not necessarily extend to the non-linear case. Indeed, low-order resonances between the frequencies σ_1 and σ_2 (see Fig. 13.5) may trigger a non-linear instability; this is the case of 2/1 and 3/1 resonances, occurring for specific values of the mass ratio μ. However, for different, typically smaller, values of μ, if the initial displacement from the equilateral point is small enough, stable librations ensue. These are the *tadpole orbits* (Figure 13.6a). Note, that – similarly to the linear solution given above – the tadpole orbits are composed of a long-period libration along the orbit of less massive primary and short-period oscillations in the perpendicular direction. For Trojans typical values are $\approx$ 150 y and $\approx$ 11.9 y; the latter is close to the revolution period of Jupiter, since $\sigma_1 \approx 1$ for small μ (see Fig. 13.5). Note also that these librations are typically asymmetric with respect to the equilateral point, with a larger excursion in longitude toward the L_3 point. As shown above, L_3 itself is unstable, so its vicinity is characterized by a stable and an unstable manifold in phase space. A connection of the family of tadpole orbits with the dynamical structure around L_3 may result in its extension into one that encloses both equilateral points L_4 and L_5 as well (see Fig. 13.6b). These solutions are called *horseshoe orbits*, with several interesting examples in the solar system. One is the coorbital satellites in the Saturnian system, to be discussed in Sec. 13.4. Another is given by the asteroid Cruithne, captured in a complicated, three-dimensional variant of the horseshoe orbit relative to the Earth; it is very likely unstable on a long-term ($\approx$ tens of My), so that Cruithne eventually will move away. Other terrestrial planets have similar temporary "visitors", namely, small asteroids on associated tadpole or horseshoe orbits: Eureka is Mars' Trojan, 1989 VA is on a Trojan orbit associated with Venus and there are some 10 other examples known. Several small asteroids were even reported in the 1/1 resonances with the most massive one, Ceres, alternating on a very long timescale between horseshoe and tadpole orbits. Details of the processes that populate such metastable configurations are still unknown.

Horseshoe orbits are less stable than tadpole orbits (and, in fact, their very-long-term stability is poorly known). A typical cause for instability is a close encounter with the less massive primary, as shown on Fig. 13.6c; as expected, at this stage the test body is dominated by the gravitational influence of the

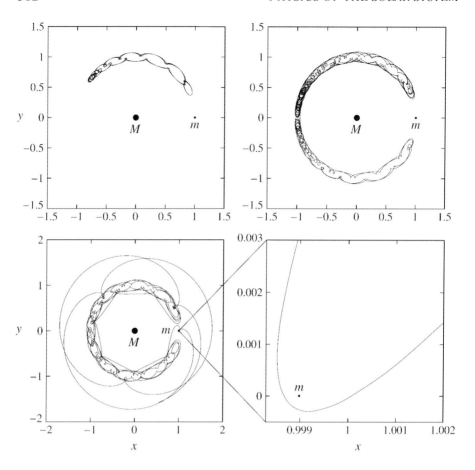

Figure 13.6. Examples of "confined" orbits near the equilateral points of the circular restricted three-body problem in the plane of primaries. Case (a) shows a tadpole orbit near L_4; case (b) shows an horseshoe orbit enclosing both equilateral points. At some critical value of the oscillation amplitude, the horseshoe orbit becomes unstable and the motion irregular. A typical onset of instability is due to a close encounter with the less massive primary, as shown in (c), and, in a zoomed plot, in (d). The mass ratio is $\mu = 0.001$ in all cases.

mass m and locally is well described with a hyperbolic orbit. In Sec. 18.4 we shall comment on how such flyby events may be used for interplanetary motion of space probes.

It is commonly believed that the Trojan-like (tadpole) orbits around the Lagrangian points of Saturn, Uranus and Neptune should be populated; comets scattered by the giant planets and those leaking from the Kuiper belt (Sec. 14.4) should suffice. In January 2003, the discovery of the first Neptune's Trojan has been announced, but no such objects around Saturn's and Uranus' Lagrangian points are known yet. Another interesting question is whether such permanent populations for these planets were present since the early stages of the solar system, such as most of Jupiter's Trojans. This is an unsolved problem that currently attracts theoretical and observational efforts. The most likely stable population may exist around the equilateral points of Neptune, whose stability is perhaps not much different from that of Jupiter's Trojans; on the other hand, Jupiter's perturbations are likely to destabilize, on the $\approx$ Gy timescale, tadpole orbits around the equilateral points of Saturn and Uranus.

Analysis of general orbits in the restricted three-body problem. So far, we have described motion near the "unit circle", i.e. solutions that stay at about the same distance from the origin as the less massive primary m and correspond to small values of the Jacobi constant. However, the totality of solutions of the restricted three-body problem is much richer. What do we know about them? We have already mentioned that the set of periodic orbits has measure zero in phase space; yet, they are very important since they, and structures related to them, provide a "skeleton" for the whole set of solutions. For this reason, the existence and the initial data of periodic orbits, grouped in families in ranges of different parameters such as the Jacobi constant, have been extensively studied. Many such families, together with their stability, termination and/or bifurcations points have been identified, especially in the planar and circular restricted problem (and a related Hill's problem; Sec. 13.4).

Orbits that are not exactly periodic, but close to, may still exhibit some degree of regularity and are thus referred to as regular orbits. Fourier analysis indicates that their spectrum is sufficiently well characterized by a finite number of lines, at least on the available time span of the numerical integration. This property contrasts with the fact that the majority of other orbits (relative to the set of initial data) behave chaotically. In the case of the *planar*, circular restricted three-body problem an efficient tool, called method of *surface of section* (or Poincaré section), to distinguish between regular and chaotic orbits was developed (in fact, this method dates back to early studies of the chaotic character of stellar motion in a galaxy). The particle's motion in phase space $(x, y, \dot{x}, \dot{y})$ is constrained by the Jacobi first integral C; as a result, one of the variables – $\dot{y}$, say – can be determined, up to its sign, from the other variables and C (determined by the initial conditions). If another, independent integral of motion C' exists, we could further constrain the motion and express y as a function of $(x, \dot{x})$, C and C'; then the problem would be integrable and the solution could be expressed by quadratures. To test this hypothesis one can numerically follow an orbit and determine its intersections with the $y = 0$ plane (with, say, $\dot{y} > 0$). It is clear that the additional integral would confine the successive intersections of the orbit with $y = 0$ plane on a line; otherwise they would cover a two-dimensional region. The peculiarity of the three-body problem, already discovered in galactic dynamics, is that the both these alternatives are present.

Depending on the initial conditions, orbits in some regions of phase space behave nearly regularly (suggesting the existence of a first integral), while others are chaotic. The regions of regularity are typically associated with periodic orbits, that appear as points on the surface of section.

When orbits are not confined to the (x, y) plane the situation is more complicated and the method of surfaces of section cannot be used. More general tools for determining their chaotic character were thus developed, in particular the Lyapunov exponents, discussed earlier. Some consequences will be also discussed in Sec. 15.1.

In view of the applications to the asteroids, much attention has been paid to orbits near periodic solutions in 1/2, 2/3, 1/3 etc. resonances with the period of the primaries. The existence and location of regular and chaotic orbits related to them have been extensively studied by the surfaces of section technique and other means. Numerical experiments have been performed letting Jupiter's orbit evolve secularly: the related secular resonances (see Sec. 15.1) may trigger instabilities in the regular regions and cause the observed depletion of asteroids in some parts of the asteroid belt.

13.4 Hill's problem

An important variant of the three-body problem, closely related to the circular, restricted case, was originally formulated by G.W. HILL in 1871. Hill's goal was the lunar motion, but the underlying equations have much wider applications, in particular to natural satellites, the dynamics of planetary rings and even of globular clusters in galaxies.

We want to investigate the relative motion of two bodies so near each other that the gravitational influence of a third, distant body is well approximated by its tidal force. We denote with m_1, m_2, m_3 their masses, with ϱ the separation between 1 and 2 and with ℓ the distance of the third body from the centre of mass C of the pair (to underline the conceptual difference with the restricted three-body problem we use Greek letters for the relative vector of 1 and 2). In Sec. 4.2 we discussed the case in which the effect of 3 for the system of 1 and 2 is a small perturbation relative to their mutual attraction; in terms of the *tidal parameter*, eq. (4.10), this means

$$\mu_T = \frac{m_3}{m_1 + m_2}\left(\frac{\varrho}{\ell}\right)^3 \ll 1 \ .$$

Here, in a more rigorous approach, this approximation is dropped and $\mu_T = O(1)$; in the limit in which the mass ratio $\mu = (m_1 + m_2)/m_3$ is small we require $\varrho/\ell = O(\mu^{1/3})$. In the tidal expansion the octupole terms are smaller by a factor $\approx \varrho/\ell$ and are neglected.

Close encounters between two small gravitating bodies orbiting around another one are discussed in general in Sec. 18.4 (and also Sec. 13.5); the so-

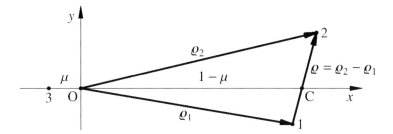

Figure 13.7. Hill's problem: in the rotating frame and the appropriate units the biggest body 3 is at rest on the x-axis at a distance μ from the origin O – the centre of mass of the whole system. The centre of mass C of 1 and 2 is also on the x-axis, at $x = 1 - \mu$; their coordinates are $\varrho_1 = (1 - \mu)\mathbf{e}_x - m_2/(m_1 + m_2)\varrho$, $\varrho_2 = (1 - \mu)\mathbf{e}_x + m_1/(m_1 + m_2)\varrho$. The scaling $\mu = (m_1 + m_2)/m_3 \ll 1$ and $\varrho = O(\mu^{1/3})$ is assumed; in this way one can investigate the effect of the tidal force produced by 3 on the relative motion of 1 and 2. In the lunar application 3 is the Sun, 1 the Earth and 2 the Moon.

lution consists in matching a short inner hyperbolic arc, where the mutual attraction prevails, to two outer elliptic arcs around the third body. The mass ratio μ is still small, but $m_3/\ell \approx (m_1 + m_2)/\varrho$, so that the inner timescale $\sqrt{\varrho^3/(m_1 + m_2)} \approx \varrho\sqrt{m_3/\ell} \approx \varrho/(\ell n)$ is much smaller than the outer timescale $1/n$. But when the orbits of 1 and 2 are nearly circular, their interaction may last longer. In the generic case in which their distance is $\varrho \ll \ell$, their mean motions differ by $\Delta n \approx n\varrho/\ell$, the interaction timescale is $\varrho/(\ell\Delta n) \approx 1/n$ and the matching is impossible. The attraction of the main body is always important and Hill's method is needed to deal with this case.

Consider the following, unperturbed two-body problem: the centre of mass C of 1 and 2 (with mass $m_1 + m_2$) moves in the plane (x, y) on a circular orbit with mean motion $n = \sqrt{G(m_1 + m_2 + m_3)/\ell^3}$ around the centre of mass O of C and 3. As before, it is convenient to work with scaled units such that the total mass $m_1 + m_2 + m_3 = 1$, $\ell = 1$ and $n = G = 1$; to lowest order $m_1 + m_2 = \mu$, $m_3 = 1 - \mu$. We also introduce Cartesian synodic coordinates (x, y, z) rotating with the angular velocity n around the origin O; the time t is just the revolution phase of the two-body system (C, 3) (see Fig. 13.7).

The relative acceleration $\ddot{\varrho}$ of 2 relative to 1, besides the obvious Coriolis term, includes their mutual attraction, the tidal forces due to 3 and the centrifugal forces. All these three terms derive from a potential, so that the relative vector $\varrho = \varrho_2 - \varrho_1 = (\xi, \eta, \zeta)$ formally fulfils the same equation as (13.3)

$$\ddot{\xi} - 2\dot{\eta} = -\partial W_{\text{Hill}}/\partial\xi, \qquad (13.30a)$$
$$\ddot{\eta} + 2\dot{\xi} = -\partial W_{\text{Hill}}/\partial\eta, \qquad (13.30b)$$
$$\ddot{\zeta} = -\partial W_{\text{Hill}}/\partial\zeta. \qquad (13.30c)$$

We now evaluate Hill's potential W_{Hill}. The mutual attraction corresponds to the potential $-\mu/\varrho = O(\mu^{2/3})$; note that the timescale of the relative motion of 1 and 2 is $\sqrt{\varrho^3/\mu} = O(\mu^0)$. The gravitational force exerted by 3 on 1 reads

$$-m_3 \frac{\mathbf{e}_x - m_2\varrho/(m_1+m_2)}{|\mathbf{e}_x - m_2\varrho/(m_1+m_2)|^3} = -\left[\mathbf{e}_x + \frac{m_2}{m_1+m_2}(3\xi\mathbf{e}_x - \varrho)\right];$$

similarly, the one on 2 is

$$-m_3 \frac{\mathbf{e}_x + m_1\varrho/(m_1+m_2)}{|\mathbf{e}_x + m_1\varrho/(m_1+m_2)|^3} = -\left[\mathbf{e}_x - \frac{m_1}{m_1+m_2}(3\xi\mathbf{e}_x - \varrho)\right];$$

in both, $O(\mu^{2/3})$ terms have been dropped. Their difference $(3\xi\mathbf{e}_x - \varrho)$ corresponds to the quadrupole potential $-(3\xi^2 - \varrho^2)/2 = -(2\xi^2 - \eta^2 - \zeta^2)/2$. Finally, the centrifugal forces on 1 and 2 are, in components, respectively,

$$\left(\xi_1 = 1 - \mu - \frac{m_2}{m_1+m_2}\xi, \; \eta_1 = -\frac{m_2}{m_1+m_2}\eta, \; 0\right),$$

$$\left(\xi_2 = 1 - \mu + \frac{m_1}{m_1+m_2}\xi, \; \eta_2 = \frac{m_1}{m_1+m_2}\eta, \; 0\right).$$

Their difference $(\xi, \eta, 0)$ derives from the potential $-(\xi^2 + \eta^2)/2$, also $O(\mu^{2/3})$. We finally obtain

$$W_{\text{Hill}} = -\frac{\mu}{\varrho} - \frac{1}{2}\left(3\xi^2 - \zeta^2\right) - (\eta_{EP2} - \eta_{EP1})\xi. \tag{13.31}$$

To the traditional Hill's potential we have added the novel term $(\eta_{EP2} - \eta_{EP1})\xi$ to describe violations of the equivalence principle; $1 + \eta_{EP1} = (m_g/m_i)_1$, and similarly for the second body, is the ratio of its gravitational and inertial masses m_g and m_i. As discussed in Sec. 17.1, in the Newtonian theory of gravitation, as well as in the General Relativity, this dimensionless ratio is assumed to be a universal constant, equal for all bodies (and taken to be one) – an assumption which holds to a very high accuracy. But it is important to question it and to verify it experimentally: the mass is a compound quantity and the equivalence principle requires that all its components equally contribute to the two masses. Only when it holds, the masses drop out from the equation of gravitational motion, which thereby acquires a geometrical significance. Already I. NEWTON pointed out, and later P.S. LAPLACE elaborated on, that the motion of three gravitating bodies is very sensitive to a violation of the equivalence principle. Here we only point out how this effect formally appears in Hill's problem as a dipole term, giving a constant relative acceleration in the direction of the third body (see also eq. (17.4)); this opens the way for its application to the motion of the Moon and the Earth, to be discussed in Sec. 17.1 in the context of the lunar laser ranging experiment.

Hill's problem has only two equilibrium solutions, called L_1 and L_2, where W_{Hill} has saddle points; they are on the ξ-axis, at $\pm(\mu/3)^{1/3}$. They are obviously related to the Lagrangian points with the same names in the restricted problem. In these configurations the mutual attraction is balanced by the tidal and the centrifugal forces; in inertial space the two bodies move in circular orbits around O with the same phase. Eqs. (13.30) still admit Jacobi's integral of the motion (see eq. (13.5))

$$\frac{1}{2}(\dot{\xi}^2 + \dot{\eta}^2 + \dot{\zeta}^2) + W_{\text{Hill}}(\xi, \eta, \zeta) = C_{\text{Hill}}. \tag{13.32}$$

As before, the zero-velocity surfaces $W_{\text{Hill}}(\xi, \eta, \zeta) = C_{\text{Hill}}$ constrain the topology of the solution, as shown in Fig. 13.8 for the (ξ, η) plane. For large and negative values of C_{Hill}, the zero-velocity curves have a nearly circular component around body 1 and one nearly parallel to the η-axis; as C_{Hill} increases and approaches the critical limit $(-3^{4/3}\mu^{2/3}/2) = W_{\text{Hill}}(L_1) = W_{\text{Hill}}(L_2)$, the circles deform to a prolate shape. In three dimensions this is just the equipotential surface used in Sec. 4.2 to describe Earth tides. Below this limit, mutual attraction prevails; $[\ell (\mu/3)^{1/3}]$ is the largest distance along ξ between 1 and 2 below which tidal forces are not strong enough to disrupt the pair. This distance is called *Hill's radius* and is just the generalization of Roche's radius to the case in which the masses are comparable; for a body in the field of the Earth and the Sun it has the value 1.3×10^{11} cm. At still higher values of C_{Hill} the equipotentials become U-shaped, with the axis along the η-direction; in the inertial frame they appear like a horseshoe.

Note also that when the mass ratio m_2/m_1 is very small, in the rotating frame the mass m_1 is at rest at C and we recover the circular restricted three-body problem. Finally, we note that the dependence of Hill's equations (13.30) on the mass ratio μ is trivial; the scaling $\varrho = \mu^{1/3}\varrho'$ brings them in a universal form, independent also from the mass ratio m_2/m_1:

$$\ddot{\xi}' - 2\dot{\eta}' = \left(3 - \frac{1}{\varrho'^3}\right)\xi', \tag{13.33a}$$

$$\ddot{\eta}' + 2\dot{\xi}' = -\frac{\eta'}{\varrho'^3}, \tag{13.33b}$$

$$\ddot{\zeta}' = -\left(1 + \frac{1}{\varrho'^3}\right)\zeta'. \tag{13.33c}$$

Applications of Hill's problem. Hill's original work about the lunar motion dealt with the planar version of the problem (on $z = 0$), justified by the small (5.1°) inclination of the Moon. He concentrated on proving the existence of

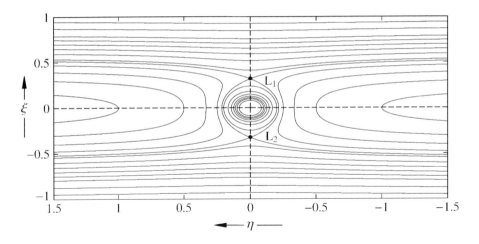

Figure 13.8. Topology of zero-velocity curves for Hill's problem in the (ξ, η) plane and $\mu = 0.1$; L_1 and L_2 are the equilibrium points.

a family of periodic orbits (also called *variational curves*) corresponding to large negative value of the Jacobi constant and therefore confined to the vicinity of the Earth. Due to solar tides, they are deformed ovals with the shorter axis towards the Sun, in the ξ-direction. He discovered that at a critical value of the Jacobi constant the variational curve develops two cusps on the η-axis (i.e. in quadrature, when angle between the Moon and the Sun is $90°$) and was led to believe that this implies the termination of the family. H. POINCARÉ demonstrated, however, that Hill's family continues, with two loops beyond the cusps; were this the case, between new and full Moon we would have three quadratures. In reality the orbit of the Moon is close to the critical variational curve; Hill developed a detailed (and ingenious) theory of small perturbations around it. Together with the amendments by E.W. BROWN, the resulting lunar theory has been for nearly 70 years the fundamental tool for the study of lunar motion. With today's precision of lunar laser ranging (Sec. 20.3) the theory has been superseded by more accurate numerical integrations.

Motion of coorbital satellites.– Another interesting application of Hill's approach is the coorbital motion in satellite systems. It was prompted by the Voyager 1 discovery of two Saturnian satellites – Janus and Epimetheus – on orbits with nearly identical semimajor axes of $151,472$ km and $151,422$ km, respectively. Since their sizes are about 175 km and 105 km, respectively, there must exist some protection mechanism to prevent collisions in their close approaches. Moreover, since their mass ratio $m_E/m_J \approx 0.25$ is not small, Hill's method, in which the mass ratio m_2/m_1 is arbitrary, is more suitable than the circular restricted three-body problem. In this framework, it was found that

13. The three-body problem 369

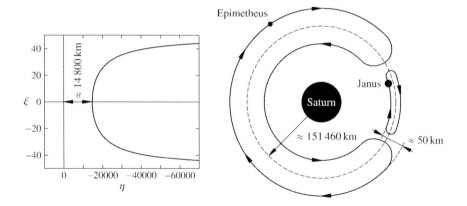

Figure 13.9. Motion of the coorbital satellites Janus and Epimetheus: (a) evolution of the coordinates (in km) of Epimetheus relative to Janus during a close encounter; (b) motion (not to scale) of the satellites in a Saturn-centric frame rotating with the mean angular velocity of either satellite.

Epimetheus may be on a horseshoe orbit relative to Janus. Their relative motion corresponds to a small value of the Jacobi constant C_{Hill} and is confined close to one of the U-shaped zero-velocity curves shown in Fig. 13.8. Figure 13.9a shows the relative motion of the pair near their encounter, which occurs approximately every 4 years. The minimum separation of the two bodies is $\approx 14,800$ km, largely sufficient to prevent collisions. The motion around Saturn of both satellites in the frame rotating with the mean motion of either satellite is shown in Fig. 13.9b. Interestingly, the geometrical parameters of the Janus and Epimetheus orbits allowed the determination of their masses; with the sizes determined by the Voyager images, their mean densities was estimated as ≈ 0.65 g/cm^3. This surprisingly low value indicates an icy composition of both satellites with significant porosity, similar to the one assumed for most distant (small) objects in the solar system, including irregular satellites of giant planets, Centaurs, transneptunian objects and most comets (Ch. 14).

Interaction between satellites and planetary rings.– Hill's framework also provides a beautiful explanation of several features of planetary rings related to *shepherding* and *embedded satellites*. The theory of shepherding satellites was stimulated by the puzzling observation in the late 1970s of very narrow and strongly radially confined rings of Uranus; a similar example was also observed in Saturn's F ring. They contradict the common sense view that, on a long term, such structures should disperse away because of mutual collisions and radiative effects. It was proposed that this confinement is due to small satellites on each side of the ring which, as shepherds, maintain its constituent particles together by gravitational interaction. Indeed, later the Voyager mission confirmed their

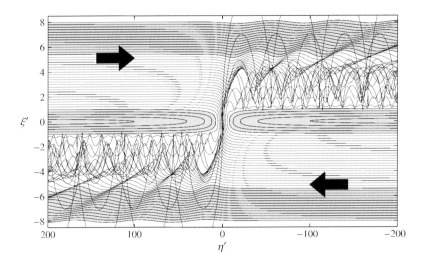

Figure 13.10. Scattering properties in Hill's problem (see text). ξ' and η' are the scaled coordinates in the rotating frame of body 2 relative to 1, both orbiting around a large and far away planet (placed in the negative ξ' direction). Note the compression in the horizontal scale and remember that the stationary points are at $\xi' = \pm(1/3)^{1/3}$. The curves are numerical solutions of the normalized Hill's equations (13.33). Adapted from C. Murray and S. Dermott, *Solar System Dynamics*, Cambridge University Press (1999).

existence: Cordelia and Ophelia for Uranus' ϵ ring; Prometheus and Pandora for Saturn's F ring. Skipping complicated collective phenomena due to mutual collisions of the ring particles (however, necessary for a full solution), a basic understanding of this phenomenon can be obtained using Hill's approach.

Outside the confinement region bounded by the equipotential surface through the equilibrium points, the two bodies scatter each other. Referring to Fig. 13.10, before the encounter they are assumed in circular orbits; as shown by the arrows, body 2 overtakes 1 in the lower right and is overtaken by 1 in the upper left. If the impact parameter is large enough, the encounter modifies the circular orbits very little, corresponding to the upper and lower shaded areas. In the inertial frame the mutual interaction produces a non-vanishing and permanent eccentricity; in the rotating frame this shows up in oscillations in the relative radial coordinate ξ, which become bigger and bigger as the impact parameter decreases. Eventually, the interaction becomes strong enough to exchange the inner and the outer body; as a result, ξ changes sign and the "overtaker is overtaken". This class of orbits is highly irregular and unstable. Finally, if the impact parameter is smaller than Hill's radius, centrifugal and tidal forces turn back the body in a U-shaped trajectory, resembling the coorbital motion of Janus and Epimetheus. This corresponds to the inner shaded area of long-term stability.

The basic model of a planetary ring consists in a set of non-interacting particles in Keplerian circular orbits in the equatorial plane. If a satellite with sufficient mass is embedded in the ring, also on a circular orbit, we can expect a situation qualitatively similar to Fig. 13.10. Particles inside the intermediate region between the shaded areas are in irregular motion, with frequent mutual collisions (and collisions with the embedded satellite) that eventually destroy them; the edges of the gap so developed show a wavy pattern. At the centre of the gap, a very thin ring on 1/1 resonant orbits with the satellite may survive, with an interruption at its position. A beautiful confirmation of this picture was provided by the Voyager observation of the Encke gap in the Saturnian A ring, which shows just this predicted structure. Eventually, also the embedded satellite – Pan – was found in the Voyager images. More in general, satellites embedded in planetary rings are expected to cause sharp gradients in their density.

Similarly, a growing planet in a protoplanetary disc may open a gap by accreting gas on the geometrically crossing orbits (Sec. 16.4). This situation is, however, more complicated because the thermal velocity of the gas in the disc may be non-negligible, especially for a low-mass planet. Accretion may then continue along two streams formed near the L_1 and L_2 points and a full hydrodynamic solution of the problem is necessary.

Other applications.– Hill's problem has found applications even outside planetary dynamics. We only quote stellar dynamics in globular clusters, subject to the tidal forces of the parent galaxy and the attraction of the cluster itself. It has been found, as an example, that stars with a Jacobi constant larger than the critical value corresponding to the equilibrium points, intuitively expected to escape from the cluster, may in fact repeatedly return and remain bound to it for a long time; thus its dispersion is significantly slowed down.

13.5 * Tisserand's criterion

In the restricted three-body problem with a small mass ratio the test body may undergo large orbital changes in a hyperbolic flyby with the smaller mass. For example, in an encounter with Jupiter a comet can change so drastically its orbit to make its identification at different apparitions not straightforward. A connection between the orbital segments before and after an encounter can be obtained with Jacobi's integral C (neglecting the effects of Jupiter's eccentricity); for this purpose C has to be expressed as a function of the comet's heliocentric orbital elements in a non-rotating reference system.

Let (x, y, z) and (x_0, y_0, z_0) be, respectively, the coordinates of the comet in the frame rotating with Jupiter's mean motion n and in the non-rotating frame (with the same origin at the centre of mass). Then we have

$$\dot{x}^2 + \dot{y}^2 = \dot{x}_0^2 + \dot{y}_0^2 - 2n(x_0\dot{y}_0 - y_0\dot{x}_0) + n^2(x^2 + y^2) \qquad (13.34)$$

and therefore, from eqs. (13.2) and (13.5), we get for the Jacobi integral C (in three dimensions)

$$\frac{1}{2}(\dot{x}_0^2 + \dot{y}_0^2 + \dot{z}_0^2) - n(x_0 \dot{y}_0 - y_0 \dot{x}_0) = \frac{GM}{r_1} + \frac{Gm}{r_2} + C. \tag{13.35}$$

M and m are the masses of the Sun and Jupiter; r_1 and r_2 the distances of the comet from them. Since their mass ratio ($\approx 10^3$) is large, we can neglect the displacement of the Sun from the centre of mass and express the left-hand side of eq. (13.35) as a function of the osculating heliocentric orbital elements of the comet. From the two-body energy and angular momentum integrals we get

$$\frac{1}{2}(\dot{x}_0^2 + \dot{y}_0^2 + \dot{z}_0^2) = GM\left(\frac{1}{r_1} - \frac{1}{2a}\right), \tag{13.36a}$$

$$x_0 \dot{y}_0 - y_0 \dot{x}_0 = [GMa(1 - e^2)]^{1/2} \cos I. \tag{13.36b}$$

The inclination I is referred to Jupiter's orbital plane. Substitution of these expressions in eq. (13.35), together with Kepler's third law applied to Jupiter's orbit ($n^2 a_p^3 = GM$, neglecting terms of order m/M), yields *Tisserand's parameter*

$$T \equiv \frac{a_p}{a} + 2\left[\frac{a}{a_p}(1 - e^2)\right]^{1/2} \cos I = -\frac{2a_p}{GM}\left(\frac{Gm}{r_2} + C\right). \tag{13.37}$$

Far from the planet the solar potential prevails and in the right-hand side $C \approx GM/r_1 \gg Gm/r_2$. Therefore before and after the encounter T is well approximated by the constant

$$T \simeq -2C\frac{a_p}{GM}. \tag{13.38}$$

This provides a test that the objects seen in two successive apparitions are the same comet, regardless of the change in the orbital elements. However, the issue of comet identification may show additional complexity due to non-gravitational forces due to the recoil of material anisotropically evaporating from the head at perihelion passages; on a long-term, they affect the value of Jacobi constant and may invalidate the test. A full numerical simulation, with appropriate modelling of these forces, is then needed; T may serve as an "assisting parameter".

In this way, for instance, it was possible to identify the comet P/d'Arrest (observed since 1851) with the lost cometary body La Hire (named after his discoverer in 1678; thus P/d'Arrest is the oldest known comet of the Jupiter family). During the gap of nearly two centuries the comet underwent 5 encounters with Jupiter, with approximate conservation of Tisserand's parameter (13.37). Note that this strictly holds only in the circular three-body problem; Jupiter's eccentricity $\simeq 0.04$ may produce in an encounter a change in T of

≈ 2%. It was also necessary to take into account non-gravitational forces. Tisserand's parameter with respect to Jupiter is also used for cometary classification in general (see Sec. 14.5).

As discussed in Sec. 18.4, a cometary flyby can be described as a matching between a hyperbolic orbit near a planet with a solar Keplerian orbit before and another afterwards. In the rest frame of the planet the velocity **v** is constant in size, but changes direction; this is a well exploited tool in space navigation. As pointed out by E. ÖPIK, we now show that v is related to T. Consider, e.g., the asymptotic cometary orbit before the encounter in the neighbourhood of Jupiter, at a distance a_p from the Sun and with an inclination I. According to the two-body problem (see eqs. (11.21) and (11.22b)), the transverse and the radial velocity of the comet in the solar system frame are given by

$$V_t = \frac{h}{a_p} = \left(\frac{GM}{a_p}\right)^{1/2} \left[\frac{a}{a_p}(1-e^2)\right]^{1/2}, \qquad (13.39a)$$

$$V_r = (V^2 - V_t^2)^{1/2} = \left(\frac{GM}{a_p}\right)^{1/2} \left[2 - \frac{a_p}{a} - \frac{a}{a_p}(1-e^2)\right]^{1/2}. \qquad (13.39b)$$

The transverse component V_t, orthogonal to the radial direction towards the Sun, can be decomposed in a part ($V_t \cos I$) in the orbital plane of Jupiter and in a part ($V_t \sin I$) orthogonal to it. Subtracting from the former the velocity $V_p = (GM/a_p)^{1/2}$ of the planet, we obtain the magnitude of the relative velocity **v** at the encounter:

$$v^2 = \left[V_t \cos I - \left(\frac{GM}{a_p}\right)^{1/2}\right]^2 + (V_t \sin I)^2 + V_r^2 = \qquad (13.40)$$

$$= \frac{GM}{a_p}\left\{3 - \frac{a_p}{a} - 2\left[\frac{a}{a_p}(1-e^2)\right]^{1/2} \cos I\right\} = \frac{GM}{a_p}(3-T).$$

Of course, this is an asymptotic value and does not take into account the planetary gravitational field. The initial value of T (or of v) is unchanged, but each of a sequence of encounters changes the cometary heliocentric elements by rotating $\mathbf{v} = \mathbf{V} - \mathbf{V}_p$ by an amount depending on the impact parameter (see Fig. 18.7).

It is easily confirmed that Tisserand's parameter fulfils the relationship

$$\frac{GM}{a_p}T = 2(nL_z - E), \qquad (13.41)$$

where $E = -GM/2a$ is the asymptotic energy and $L_z = h \cos I$ is the z-component of the asymptotic angular momentum **L** of the comet (per unit mass). In this form we can investigate long-term changes of T due to the other

Table 13.1. Three meteor streams occurring every year at the indicated dates (day and month), likely to be associated with the comet P/Machholtz 1. We give the osculating Keplerian elements for the epoch B1950.0: perihelion distance r_p, eccentricity, inclination, longitude of the node and of the perihelion. In the last two columns, the size v of the velocity relative to the Earth (in units of its orbital velocity $V_\oplus \simeq 29.8$ km/s) and the angle θ with $\mathbf{V}_\oplus$. v and θ are constant on a secular timescale; the small difference in their values for the three streams, in spite of their very different orbital elements, is a strong indication that they originate from the same parent comet. The northern (N) and southern (S) branches of δ Aquarids approximately differ by the mirror symmetry $\Omega \to \Omega + \pi$ and $\omega \to \omega + \pi$ about the ecliptic plane. Data adapted from G.B. Valsecchi et al., *Mon. Not. R. Astr. Soc.* **304**, 743 and 751 (1999), where other meteor streams possibly associated with this comet are indicated.

Stream name	Dates	r_p (AU)	e	I (deg)	Ω (deg)	ω (deg)	v	$\cos\theta$
Quadrantids	2/1 – 4/1	0.98	0.68	72	282	171	1.35	−0.53
N δ Aquarids	5/8 – 25/8	0.10	0.95	21	142	328	1.34	−0.44
S δ Aquarids	21/7 – 8/8	0.08	0.97	27	308	151	1.39	−0.42

planets (assuming that the comet is periodic and has no other close encounters, except with Jupiter). Planetary perturbation theory (Sec. 12.5) ensures that, to lowest order, the semimajor axes a and a_p, hence the energy E, are constant. In absence of mean motion resonances, the secular effect of planetary perturbations on **L** is the same as if the perturbing planet were replaced by a circular ring of matter; it only produces a precession of **L** around the normal to the ecliptic, keeping the inclination I and L_z constant. Thus T (and the velocity v relative to the planet, eq. (13.40)) does not undergo secular changes, contrary to other orbital elements (the longitudes of the node and the perihelion can change by as much as several tens of degrees per century).

This fact has an interesting application to the problem of meteor streams, which regularly impinge upon the Earth around a fixed date every year and, as they burn in entering the atmosphere, become visible as "shooting stars" (Sec. 14.6). It is believed that one such stream consists of fragments released (with a small relative velocity) by a comet along its orbit in its previous history. To test this assumption, note that a meteor is characterized by its own Tisserand's invariant (a_p and m now refer to the Earth), which, in view of the absence of secular variations, should approximately be the same as the one of the parent comet. But also $\cos\theta$, the angle between **v** and the velocity $\mathbf{V}_p$ of the Earth is secularly invariant; in fact, spelling out the energy of the meteor, we can express $\cos\theta$ in terms of E and v:

$$2\sqrt{\frac{GM}{a_p}}\, v\cos\theta = \frac{GM}{a_p} - v^2 - E\,. \qquad (13.42)$$

13. The three-body problem

Note that meteors of a stream are observed every year, in different directions and on different asymptotic orbits; they are ejected by the comet at different times and in course of time undergo different secular evolutions; but if the assumption is correct, the values of T (and v) and $\cos\theta$ they carry along are those of the parent comet and should approximately be the same. This powerful identification criterion, enhanced by the fact that v and $\cos\theta$ are directly measurable quantities, is superior to other methods using a distance in the space of orbital elements. For instance, the comet P/Machholtz 1 has been shown to be a likely parent for up to 8 meteor streams, including Quadrantids, northern and southern δ Aquarids (Table 13.1). Obviously, such identification may succeed only if the streams did not undergo a close encounter with Jupiter, which would destroy the secular stability of v and $\cos\theta$. As for the comets, another limitation is due to non-gravitational forces as mentioned above.

– PROBLEMS –

13.1. Show that the equations of motion of the restricted three-body problem are the same (apart from the meaning of the constants) as those of a charged particle moving in an electrostatic potential $W(x, y)$ and a uniform magnetic field directed along the z-axis.

13.2. Compute to first order in μ the function $W(x, y, 0)$ at the five Lagrangian points.

13.3. What is the position of the five Lagrangian points in the case $\mu = 1/2$, i.e. when the two massive bodies are equal (called the *Copenhagen problem*)?

13.4.* On the general three-body problem: show that three points with total mass M_{tot} at the vertices of an equilateral triangle of side d are in equilibrium if they rotate rigidly around the centre of mass with angular velocity $(GM_{tot}/d^3)^{1/2}$. Can this result be generalized to four or more arbitrary masses at the vertexes of a regular polygon?

13.5.* In the planar, restricted three-body problem with the Sun and Jupiter as main bodies, consider an asteroid at distance x_0 from the origin, collinear with them and with transversal velocity $\sqrt{GM_\odot/x_0}$. Neglecting terms $O(M_J/M_\odot)^2$, compute the Jacobi constant and the largest value of x_0 for which the asteroid does not cross Jupiter's orbit. What is the maximum heliocentric distance it can reach for a generic value of x_0?

13.6. Assume that the Tisserand parameter of a comet with respect to Jupiter is 2.9. Neglecting Jupiter's eccentricity, what are the minimum perihelion distance, the maximum aphelion distance and inclination (with respect to Jupiter's orbital plane) that the comet can ever reach? In general, for which range of values of T can a comet be ejected from the solar system after a close encounter with Jupiter?

13.7. Compute numerically for the Earth and for Jupiter the shape of *Hill's lobe*, i.e., the equipotential curve through the L_1 and L_2 equilibrium points of Hill's problem.

13.8. Show that the fixed points in Hill's problem are linearly unstable.

13.9. The *Sitnikov problem*: a massless particle moves along the z-axis, perpendicular to the plane of the primaries and passing through their centre of mass. Show that if their orbit is circular, the problem is integrable; find its bound and unbound solutions and obtain the period for the former ones. What happens when the orbit of the primaries is eccentric?

13.10.* Find the equilibrium points of a test-particle in the gravitational field of a triaxial ellipsoid with homogeneous density distribution, uniformly rotating around the shortest axis. Investigate their linear stability. (Hint: Find the exterior potential of an homogeneous triaxial ellipsoid in a way similar to eq. (3.21).)

13.11.* The *Roy-Ovenden theorem*: three arbitrary masses are initially aligned along ℓ, with velocities coplanar and perpendicular to ℓ; prove that the motion is periodic. The motion of the Earth-Moon-Sun system is close to such situation; what is the relevant period?

13.12.* Using the linearized theory of small oscillations around the equilibrium point L_2 of the Sun-Earth system, construct an orbit that avoids solar eclipses (necessary for Gaia; see Sec. 20.2).

– FURTHER READINGS –

The most extensive treatises on the three-body problem are V. Szebehely's, *Theory of Orbits*, Academic Press (1967) and C. Marchal's, *The Three Body Problem*, Elsevier Science (1990). The classical text D. Brower and G.M. Clemence, *Methods of Celestial Mechanics*, Academic Press (1961) is always a main reference. See also A.E. Roy, *Orbital Motion*, Hilger (1978), Ch. 5; F.R. Moulton, *An Introduction to Celestial Mechanics*, Dover (1970), Ch. VIII; and E. Finlay-Freundlich, *Celestial Mechanics*, Pergamon Press (1958), Chs. I and II. On stability and chaos we suggest A.J. Lichtenberg and M.A. Liebermann, *Regular and Stochastic Motion*, Springer-Verlag (1983). An exhaustive analysis of the families of periodic orbits in the three-body problem is given in M. Hénon, *Generating Families in the Restricted Three-Body Problem*, Springer-Verlag (1997). The lunar motion is discussed in historical detail in M.C. Gutzwiller, Moon-Earth-Sun: the oldest three-body problem, *Rev. Mod. Phys.* **70**, 589 (1998).

Chapter 14

THE PLANETARY SYSTEM

In this chapter we give an overview of the properties of our planetary system – both those related to its dynamical structure, namely the orbital and rotational motions of the constituent bodies, and those connected with their physical nature (e.g., size, composition, the presence of an atmosphere and/or of a solid surface). As discussed in the Introduction, our principal aim is not descriptive, but rather to provide insight into the main physical principles and processes; thus some data will be presented in the form of tables and figures with synthetic captions only. In the case of minor bodies, we mainly focus on presenting orbital clusterings and their interrelations; often the individual orbits in these populations are not stable throughout the age of the solar system, and the problem of sustaining the entire family by adequately efficient sources is fundamental. In this context we note that the formation of orbital structures in the inner solar system, such as the main asteroid belt, is coarsely understood. The unsolved issues include how the unstable population of planet-crossing objects (e.g., the near-Earth objects) is sustained over a long period of time. A major unanswered question is how structures in the outer solar system, such as the Kuiper belt and related populations of objects, have been formed. This problem is also intimately linked to the formation of the giant planets, a very attractive topic due to the discovery of extrasolar planetary systems. Significant contributions to the understanding of the highly unstable cometary populations have been made in the last decade, yet some problems remain unsolved (such as the appearance of dormant comets in the outer asteroid belt and among the near-Earth object population).

14.1 Planets

Planetary physical properties, such as gravitational and magnetic fields, atmospheres and magnetospheres, are discussed in Chs. 2 to 10; long-term dy-

Table 14.1. Dynamical properties of the planets and the Sun: a is the semimajor axis, in AU; P is the orbital period; e the eccentricity; I the inclination, in degrees; P_s the sidereal spin period, in days; ϵ the obliquity, in degrees (for Sun referred to the ecliptic). All orbital elements are mean except from Pluto, for which we give the osculating elements at J2000.0.

Object	a	P	e	I	P_s	ϵ
Sun	–	–	–	–	25.38*	7.25
Mercury	0.387	87.97 d	0.206	7.005	58.646	≈ 0.1
Venus	0.723	224.70 d	0.007	3.395	243.019	177.36
Earth	1.000	365.25 d	0.017	0.000	0.997	23.45
Mars	1.524	686.97 d	0.093	1.850	1.026	25.19
Jupiter	5.203	11.86 y	0.048	1.303	0.414	3.12
Saturn	9.555	29.46 y	0.056	2.489	0.444	26.73
Uranus	19.218	84.02 y	0.046	0.773	0.718	97.92
Neptune	30.110	164.77 y	0.009	1.770	0.671	28.80
Pluto	39.44	247.69 y	0.254	17.14	6.387	122.53

*The Sun has a differential rotation in latitude (eq. (9.2)); the period ranges from ≃ 25 d (equator) to ≃ 35 d (poles).

namics, including the stability of planetary orbits, is treated in Sec. 15.1 and planetary formation in Ch. 16. In this Section we review and comment on the principal planetary data given in Tables 14.1 and 14.2. Pluto is included for traditional reasons, although it is the largest of many similar objects in the transneptunian region (in particular, those belonging to the plutino population; Sec. 14.4).

Planetary orbits show, in general, a fairly regular pattern: eccentricities and inclinations are small; the semimajor axes roughly fit a geometric progression with a ratio close to 2. This feature was discovered in the 18th century and is called the *Titius-Bode law* (Problem 14.1). The wider gap between Mars and Jupiter corresponds to the location of the asteroid belt; it is likely that gravitational perturbations due to the early-formed and massive Jupiter prevented the material in this belt from forming a planet. The regularity in the planetary sequence is mainly due to the requirement of orbital stability, in particular to avoid close encounters between planets for a time span ranging from 10^7 to 10^{10} orbital periods (see also Sec. 15.1). It is interesting that the largest eccentricities and inclinations occur at the inner and outer boundaries, for Mercury and Pluto. Today we observe only the final state of the rapid, complex and still poorly understood formation of the solar system, which lasted a few tens to hundreds My. The forming planets were surrounded by a myriad of planetesimals and perhaps even a number of additional planetary cores that failed to accrete enough material to compete with their more massive neighbours (e.g.,

Table 14.2. Physical properties of the planets and the Sun. The columns show: R_e, the equatorial radius in km (for the giant planets the 1 bar level); M, the mass in g; ϱ, the mean density in g/cm^3; A, the visual geometric albedo; ∗, the surface materials; ∗∗, the main constituents of the atmosphere.

Object	R_e	M	ϱ	A	∗	∗∗
Sun	696,000	1.99×10^{33}	1.41	–	–	H, He
Mercury	2,440	3.30×10^{26}	5.43	0.11	Igneous rocks	O_2, Na
Venus	6,051	4.87×10^{27}	5.20	0.65	Basaltic rocks	CO_2
Earth	6,378	5.97×10^{27}	5.52	0.37	Water, basaltic and granitic rocks	N_2, O_2
Mars	3,396	6.42×10^{26}	3.93	0.15	Basaltic rocks, dust	CO_2
Jupiter	71,492	1.90×10^{30}	1.33	0.52	–	H_2, He
Saturn	60,268	5.68×10^{29}	0.69	0.47	–	H_2, He
Uranus	25,559	8.68×10^{28}	1.32	0.51	–	H_2, He, CH_4
Neptune	24,764	1.02×10^{29}	1.64	0.41	–	H_2, He, CH_4
Pluto	1,151	1.27×10^{25}	2.1	0.55	CH_4, H_2O, CO ices	Thin CH_4

Jupiter and Saturn). Subsequently these small bodies were swept out of the early solar system by gravitational perturbations caused by the planets, leaving behind a well settled progression of orbits able to survive for a long time. During this early and violent evolution most of the remaining major planets may have radially moved away from the formation zone; this process of *planetary migration* may be due to tidal interaction with the solar nebula or unbalanced exchange of orbital angular momentum with smaller bodies (Sec. 16.4). Neptune may have moved outward by as much as 6 – 10 AU, while capturing a significant fraction of transneptunian planetesimals in the 3/2 exterior resonance (Sec. 14.4).

Planetary rotations are much less regular; their only common feature (though violated by the two innermost planets) is a remarkable clustering of rotation periods near 1 day over a mass range of several orders of magnitude (including asteroids). The sense of rotation is approximately prograde, except for Venus, Uranus and Pluto.

In the ideal formation of a star in interstellar space from a nebula of radius R, the angular velocity is always directed along the (constant) angular momentum **L** and increases as $1/R^2$; but in the solar system the material which contributes to forming a planet shares the angular momentum of the nebula and, moreover, each accreting planetesimal brings its own individual contribution. As for planetary orbits, we must distinguish the rapid formation period from the subsequent long-term evolution. Consider first the giant planets, whose angular momentum likely has undergone little change, with only a slow and stable precession due to the solar torque (Sec. 4.1). In a standard formation model, neglecting planetesimals, suppose that the gas cloud, as it cleared its feeding zone and uncoupled from the disc, had initially a spherical shape (slightly elongated by tides) determined by the balance between pressure and self gravity; and that it was at rest in the rotating frame, with angular velocity Ω. During its collapse, with a velocity $\mathbf{V}$ towards the centre of mass at $\mathbf{R} = 0$, it was subject to the Coriolis force $\mathbf{F}_C = -2\,\Omega \times \mathbf{V}$ and a *prograde torque* $\mathbf{R} \times \mathbf{F}_C$, so that its final state is likely a prograde rotation (Problem 14.12). This scenario is consistent with Jupiter's small obliquity and, partially, with Saturn's and Neptune's, but it does not fit well with Uranus. Perhaps in this region the gaseous protonebula had locally more complex, three-dimensional structures; it should also be noted that Uranus (and Neptune) have substantial rocky cores and icy mantles (with mass $\approx 10-15\,M_\oplus$), certainly arisen out of planetesimal material, some of which might have accreted in planetary embryos, with large and random angular momentum. For Saturn, though, the most likely model assumes the rotation axis captured in the s_8 spin-orbit resonance with the nodal drift of Neptune (Sec. 7.1). A Jupiter-like initial obliquity is pushed by this mechanism to the currently observed value in a few hundreds of My. Interestingly, this scenario also sets limits on the primordial mass in the Kuiper belt zone and timescale of its fading.

In the accretion of planetesimals – the accepted formation mode of terrestrial planets – however, the process is more complex and must be studied in the framework of Hill's problem (Sec. 13.4). Since Hill's interior and exterior orbits are symmetric, generally the addition of positive and negative angular momentum is equally likely; the sense and the value of the final rotation depends on fine details, such as the distribution of eccentricities and inclinations of planetesimals. Models based on a classical Keplerian cloud and a regular accretion do lead to a final prograde rotation, but its amount is of the order of the mean motion and cannot explain the present large values for the Earth and Mars. The impacts of remaining planetary embryos (evidenced by the likely collisional origin of the Moon) during the final phase of planetary formation may have changed the original rotational state; moreover, significant long-term evolution may have occurred due to tidal effects and resonant interaction with orbital dynamics (see Sec. 7.1 and Appendix of Ch. 7), resulting in a large increase of the rotation period and a significant change in the obliquity.

The geological history of the obliquities of the terrestrial planets is particularly interesting. As noted in Sec. 7.1, the lunar gravitational influence appreciably increased the precession constant of the Earth from $\approx 16''/y$ to more than $50''/y$; as a result, the only secular resonance between the precession and the orbital motion through which the Earth axis probably passed altered the obliquity by $\approx 0.5°$. Venus, before acquiring the current obliquity of $\approx 178°$, might have suffered the most remarkable evolution. In a possible scenario, a chaotic diffusion due to resonant interaction with the orbit increased the obliq-

14. *The planetary system*

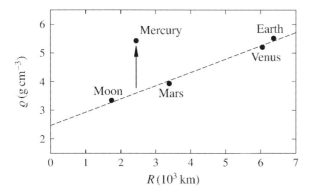

Figure 14.1 Mean density vs mean radius of the terrestrial bodies. The dashed line fits the data linearly, except Mercury, which is significantly denser.

uity to $\approx 90°$; later, atmospheric tides and core-mantle interaction produced its steady secular increase. Both effects also despun Venus to the present, very slow rotation period of ≈ 243 d. Mercury is locked in a strange 3/2 spin-orbit resonance. Its likely fast primordial rotation period was slowed by solar tides (Sec. 15.3), but the exact time and mechanism of capture in this resonance is not known. The current value of Mars' obliquity, close to the Earth's, is a coincidence, since it can undergo on a short timescale ($\approx$ My) large variations between $0°$ to $\approx 60°$, due to resonances between precession and orbital evolution (mentioned earlier). This fact had an important influence on the palaeoclimate and the evolution of the atmosphere of the planet.

Mass and angular momentum distribution.– The total mass of the planets is only 0.15% that of the Sun; the four inner planets, though denser, contain less then 0.5% of the whole planetary mass, the remaining part residing in the four giant planets (most of it in Jupiter and Saturn). Jupiter's size is about 10 times smaller than that of the Sun and 10 times larger than that of the Earth. Bodies in the solar system are much smaller than their distances. Finally, the rotation of the Sun contributes only a small part ($\approx 1\%$) to the total angular momentum **L** of the system; it is not surprising that the polar axis of the Sun, which contributes to **L** a small and uncoupled amount, has a significant tilt ($\approx 7°$).

Planetary "taxonomy".– Table 14.2 clearly shows the traditional distinction between the inner and the outer planets, of terrestrial and Jovian type, respectively: the latter are much larger and less dense than the former, a consequence of the large abundances of the same light gases (hydrogen and helium) which prevail in the Sun. This difference is clearly related to the different efficiency with which light gases were accreted by planetary embryos in the primordial solar nebula (see Sec. 16.4). However, in the outer group, Uranus and Neptune are distinctly different from Jupiter and Saturn, and have a greater proportion of ices (C, N, O and H compounds); they are sometimes called *ice giants*, as compared to Jupiter and Saturn, the *gaseous giants*. Pluto, with its small size, solid icy surface and peculiar orbit, seemingly does not fit well in this group.

Since its discovery, in 1930, it presented a problem for planetary taxonomy until the 1990s, when a large population of compositionally and orbitally similar objects was found at the outskirts of the solar system (Sec. 14.4).

In the inner planets the observed density gradient is also related to compositional differences, with a distinctly larger fraction of iron (the only heavy element abundant enough) in Mercury than in the other planets (Fig. 14.1). It is difficult, however, to explain its high density with the distribution of elements in the protoplanetary nebula alone; a more plausible scenario assumes that, at the end of the planetary formation period, a giant impact stripped Mercury of most of its mantle. The two-component model of a planet described in the second part of the Problem 1.2 leads to the relation

$$\frac{R_c}{R} = \left(\frac{\varrho - \varrho_m}{\varrho_c - \varrho_m}\right)^{1/3} \quad (14.1)$$

between the mean density ϱ and the densities of the mantle ϱ_m and the core ϱ_c. For Mercury, the value $\varrho = 5.42$ g/cm^3 can be justified, for example, with an iron core of radius $0.75\,R$ and density $\varrho_c \simeq 10$ g/cm^3 and a mantle of density $\varrho_m \simeq 2.5$ g/cm^3.

14.2 Satellites

Natural satellites display a rich variety of orbital configurations and histories, often intimately linked with their surface and interior characteristics. As in the case of planets, we have a detailed information about many satellites from spacecraft reconnaissance. However, since orbital histories of satellites are typically much more complex than those of planets, the interpretation of the data is generally very complicated and in many cases still debated.

The striking difference between the satellites of the terrestrial planets (the two Martian satellites are poorly understood and the Earth's Moon is likely due to a giant impact) and the extensive satellite systems around the gaseous giants has to do with the conditions in the early solar system and their unique formation mechanism. The slow accretion of terrestrial planets deprived the surrounding space of material to make satellites. Additionally, as argued in Sec. 15.3, the significant slowing down of Mercury's and Venus' spin rate would place their hypothetical satellites on orbits inside the corotation radius; further tidal evolution would drive them toward tidal disruption near the planet. In contrast to the terrestrial planets, the rapid accretion of giant planets allowed the formation of large systems similar to the solar system itself. The major satellites, especially the Galilean satellites of Jupiter, mimic the orbital and bulk properties of planets. They all lie near the equatorial plane of the central planet, orbiting in the prograde sense and with spacing in approximate geometric progression. Their mean density decreases, while the abundance of ice increases, with planetocentric distance. The explanation of this pattern is

14. The planetary system

Table 14.3. Orbital data of satellites with radii ≥ 100 km. We give: a, the semimajor axis in units of 1,000 km; P, the period in days; e, the eccentricity; I, the inclination in degrees with respect to the planet's equator, except for some distant satellites and the Moon (marked by a dagger), for which it is referred to the orbital plane of the planet; P_s, the period of rotation in days (or S if synchronous, C if chaotic).

Planet	Satellite	a	P	e	I	P_s
Earth	Moon	384.4	27.322	0.0549	5.15†	S
Jupiter	Amalthea	181.4	0.498	0.003	0.40	S
Jupiter	Io	421.8	1.769	0.004*	0.04	S
Jupiter	Europa	671.1	3.552	0.009*	0.47	S
Jupiter	Ganymede	1,070.4	7.155	0.002*	0.20	S
Jupiter	Callisto	1,882.8	16.689	0.007	0.28	S
Saturn	Mimas	185.5	0.942	0.020	1.53	S
Saturn	Enceladus	238.0	1.370	0.005	0.02	S
Saturn	Tethys	294.7	1.888	0.000	1.09	S
Saturn	Dione	377.4	2.737	0.002	0.02	S
Saturn	Rhea	527.0	4.518	0.001	0.35	S
Saturn	Titan	1,221.9	15.945	0.029	0.33	S
Saturn	Hyperion	1,481.1	21.277	0.104	0.43	C
Saturn	Iapetus	3,561.3	79.330	0.028	7.52	S
Saturn	Phoebe	12,952.0	550.48	0.163	175.3†	0.4
Uranus	Miranda	129.8	1.413	0.003	4.22	S
Uranus	Ariel	191.2	2.520	0.003	0.31	S
Uranus	Umbriel	266.0	4.144	0.005	0.36	S
Uranus	Titania	435.8	8.706	0.002	0.10	S
Uranus	Oberon	582.6	13.463	0.001	0.10	S
Neptune	Larissa	73.5	0.555	0.000	0.20	
Neptune	Proteus	117.6	1.122	0.000	0.55	
Neptune	Triton	354.8	5.877	0.000	156.83	S
Neptune	Nereid	5,513.4	360.136	0.751	7.23†	
Pluto	Charon	19.4	6.387	< 0.007?	96.2†	S

*The eccentricities of Io and Europa are dominated by the forced terms due to the Laplacian resonance; conversely, Ganymede's eccentricity is mainly free, with a minor (≃ 0.0006) forced contribution; for details see Ferraz-Mello's book quoted in Further Readings of Ch. 15.

probably again similar: the region close to the collapsing central body was too hot for the most volatile materials in the proto-solar and the proto-Jovian nebulæ to condense. This is also consistent with the absence of a similar pattern in the Saturnian and the Uranian systems, whose formation was probably associated with a much weaker energy outflow than for Jupiter. The obvious compactness of these satellite systems, if compared to the planetary system, small width of this resonance prevents re-accumulation of ejecta, since nearly all of them collide with Titan.

An interesting problem is presented by Iapetus, whose albedo strongly depends on the angle between the position and the orbital velocity: the leading and the trailing hemispheres have, respectively, $A \simeq 0.05$ and $A \simeq 0.5$. The likely cause is dust from Phoebe, which has a larger and retrograde orbit. Dust grains, affected by the Poynting-Robertson effect (Sec. 15.4), spiral in and impact mainly the leading hemisphere of Iapetus, on a prograde orbit. Since Iapetus' orbital period is equal to its rotation period due to tidal interaction, the dust accumulates on the same region of the satellite.

Uranus system.– Uranus has three distinct groups of satellites: (i) small inner satellites embedded in its extensive ring system; (ii) classical satellites between, and including, Miranda and Oberon; (iii) five distant irregular satellites, whose origin is similar to those of Jupiter and Saturn. Uranus' system is special among the giant planets because it lacks any orbital resonance. The orbit of Miranda (with inclination $\simeq 4.2°$) and the surface anomalies of Miranda, Ariel and Titania, however, require past resonant transitions. Candidate configurations have been found, namely the 5/3 Miranda-Ariel and the 3/1 Miranda-Umbriel mean motion resonances. Miranda has one of the most peculiar satellite surfaces in the solar system: old, densely cratered terrains are interrupted by young, oval and trapezoidal patterns (called coronæ) of concentric grooves and ridges. These features are not well understood.

Neptune system.– As in the case of Saturn, the Neptunian satellite system has only one sizable member, Triton. Its orbit – fairly close to the planet, nearly circular, and retrograde (with inclination $\simeq 157°$) – is highly irregular. Compared to the other giant planets, only few other satellites of Neptune are known, namely, the six inner satellites discovered by Voyager and, very distant, Nereid, with its large eccentricity of $\simeq 0.75$. Triton, with its unusual orbital history, is one of the most interesting objects in the Neptune system. It is likely that this giant satellite was captured from a heliocentric orbit in a close encounter or a collision with a small Neptunian satellite, or by gas drag in the proto-Neptunian nebula. The initial, highly eccentric orbit was quickly (in ≈ 0.5 Gy) circularized by strong tides (consistently with the satellite's young surface) and driven to synchronous rotation. Triton's orbit is still subject to strong tidal decay. Models of its future evolution predict tidal disruption inside the Roche limit in $\approx 3.5 - 4$ Gy, if the satellite's binding energy is negligible. Interestingly, such a major event would lead to a spectacular set of rings around Neptune, with a mass far exceeding that of Saturn's rings.

If a regular primordial satellite system was present in the inner zone, it is likely that Triton's initial eccentric orbit produced, by collisions and close encounters with Triton itself and among each other, a general disorder; Nereid might be a "lucky" leftover of this process. It is also possible that the current inner satellites are a secondary population, re-accumulated out of debris of the first generation destroyed by Triton's passages.

Pluto-Charon system.– Because of its distance and the lack of flyby observations, this is the least known satellite system. Very little is known about Pluto and Charon's composition and internal structure. Their comparable masses ($\simeq 1.3 \times 10^{25}$ g against $\simeq 1.5 \times 10^{24}$ g) suggest an unusual origin, perhaps a giant impact like that for the Moon. Their tidal influence (Pluto on Charon and vice versa) is also comparable and large ($\approx 10^{-4}$); this suggests that the present, doubly synchronous state, in which both rotational periods are equal to the orbital period of 6.39 d, has been attained very quickly, perhaps in $10 - 100$ My. Perturbations in the system may be produced by encounters with comets from the transneptunian region, but should quickly damp out. Recent Hubble Space Telescope and ground-based adaptive optics observations indicate an anomalous eccentricity $\lesssim 0.0075$ for this system, perhaps an indication of a recent impact on one of the two bodies; however, this result is debated.

14.3 Asteroids

Among the observed members of the solar system, asteroids represent by far the most numerous population. The number of catalogued bodies is now greater than 3×10^5 (of which $\approx 50,000$ have very precisely determined orbits and have been given a catalogue number); $\approx 95\%$ of them orbit around the Sun in moderately eccentric and inclined orbits in a large toroidal region between 2.1 and 3.3 AU, the *main asteroid belt*. In this belt the accretion of a planet-size object was interrupted in the early stage of the solar system evolution, leaving the asteroid population. The orbital and physical parameters of the ten largest asteroids are listed in the Table 14.5. Moreover, some 5,000 asteroids – called planet-crossing – are known, with perihelia inside the orbits of the inner planets. The best characterized is the population of *near-Earth asteroids* (with about 1,600 known objects), which occasionally approach the Earth; this allows precise optical and radar observations. Their possible collision with the Earth makes them a potential hazard. Beyond the main belt, near Jupiter's triangular Lagrangian points (Sec. 13.3), there are the two groups of *Trojan and Greek asteroids*.

Orbital structures in the main belt. Understanding the orbital dynamics in the asteroid belt is not straightforward, because Jupiter and Saturn cause significant variations in most orbital elements. For instance, typical variations in eccentricity and inclination may be of order $\approx 0.1 - 0.3$ and $\approx 5° - 10°$ respectively, with timescales ranging from years to hundreds of millennia. Thus it is necessary to first subtract these perturbations. The outcome of this delicate correction, performed at different levels of sophistication either analytically or numerically, are the so called *proper orbital elements* (Sec. 15.1). Except in resonances, the angular variables undergo a steady circulation, so that only

(n is the mean motion and $a \simeq 2.5$ AU), while the mean relative velocity is $\simeq 5$ km/s; this is approximately 4 times higher than the escape velocity of the largest asteroid, Ceres, and causes disruptive collisions, rather than accumulation. Moreover, the estimated total mass in the main belt – about $5 \times 10^{-4} M_\oplus$ – is two or three orders of magnitude below the mass estimated in this zone from a smooth distribution of the surface density in the primordial planetary nebula (Fig. 16.1). This again gives a hint into the "violent history" in this zone of the solar system.

Dynamical features in the main asteroid belt: resonances.– Resonances with the major planets sculpt both the borders of the main asteroid belt and the regions inside; at their locations a significant paucity of asteroids is observed. These empty regions – *Kirkwood gaps* – in the asteroid distribution were noticed by D. KIRKWOOD as early as in the middle 19th century. The most prominent of them are associated with low order, mean motion commensurabilities, like 3/1 or 5/2, with Jupiter. As noted in Sec. 12.5, higher-order resonances are less significant; however, what matters is their structure. Resonance dynamics in the main asteroid belt has been understood in detail only with the help of advanced analytic techniques and extensive numerical experiments. It was found, for instance, that the 3/1 mean motion commensurability with Jupiter at ≈ 2.5 AU is highly unstable, due to secular resonances located inside its phase space region. Objects injected in this region quickly (in $\approx$ My) increase their eccentricity to very high values (close to unity) by collisions, or slow diffusion due to the Yarkovsky effect. As a consequence, typically they eventually fall into the Sun, unless they are released from the resonance by a close encounter with one of the inner planets, or with Jupiter (in $\approx 5\%$ of the cases); then the asteroid becomes a member of the planet-crossing population. On the other hand, the first-order (2/1, 3/2, 4/3) mean motion resonances with Jupiter harbour limited stable regions where asteroids may reside for $\approx$ Gy; they are called Zhongguo group (2/1), Hilda group (3/2) and Thule group (4/3). The large populations of Trojans and Greeks (Sec. 13.3) are also in the 1/1 resonance with Jupiter.

The second kind of resonances significant for asteroidal motion are the secular ones; they occur when the mean drift of the pericentre or the node becomes commensurable with one of the proper frequencies of the planetary system (Sec. 15.1). The most significant one – called ν_6 – appears when the longitude of the pericentre ϖ drifts, due to planetary perturbations[2], at the rate $g_6 \simeq 28.2''/$y. This resonance truncates the belt at ≈ 2.1 AU and also causes the "bent" boundary in the (a, I) plane (Fig. 14.3). As in the 3/1 case, asteroidal

---

[2] It is often inaccurately stated that this is a resonance with Saturn's pericentre drift; the principal perturbing effect, however, arises due to Jupiter, which pericentre is also largely affected, but not dominated, by this frequency; see Fig. 15.1b.

orbits injected in its zone evolve on a ≈ My timescale toward a very eccentric state where they either are released from the resonance due to a close encounter with an inner planet, or collide with the Sun.

Collisional and Yarkovsky-driven injection mechanisms resupply objects in the 3/1, 5/2 and v_6 resonances at a rate far too low to compensate the fast eccentricity evolution. This is the reason why we see only a few objects *inside* their phase space region. In spite of this, the injection rate seems enough to keep the population of planet-crossing asteroids in an approximate steady state. The higher-order, and thus weaker, resonances also produce long-term instabilities by increasing the eccentricity. In these cases, however, the timescale is much longer (e.g., ≈ 0.1 Gy for cases like 7/3, 9/4 etc.), which allows evolution processes strong enough to resupply asteroids (and meteoroids); no significant void in their distribution is observed. Note also that their widths (depending on the order) are small as well. On the whole, resonant effects cause a permanent leakage of asteroids from the main belt, important for understanding their populations on planet-crossing orbits.

Collisional features in the main asteroid belt: asteroid families.– As discovered in 1918 by K. HIRAYAMA, the orbital parameters of asteroids indicate prominent groupings in the space of proper elements. Hirayama suggested that the origin of each of these *asteroid families* could be traced back to the collisional breakup of a parent body, which ejected fragments into heliocentric orbits with relative velocities much lower than their orbital speed. An increasing number of precisely known asteroidal orbits have allowed us to identify some 40 families across the whole main belt; families may also exist among the Trojans and even in the transneptunian region. They are usually named after the largest member.

Physical studies have shown that the members of the most populous families (associated with Themis, Eos, Koronis and Vesta) have similar surface compositions, supporting the hypothesis of a common origin. It also opened up the possibility of investigating directly by astronomical observations the interior structure of the parent bodies; this has been accomplished, in fact, after recent discovery of a very young asteroid cluster inside the Koronis family (Fig. 14.4), named after its biggest member the *Karin cluster*; it has been interpreted as an output of a collisional disruption of ≃ 25 km size parent body occurred only some 5.8 My ago. Because of its very young age, the Karin cluster has undergone little dynamical and collisional evolution, and is thus ideal to investigate physical processes in collisions of small asteroids. Several other tight asteroidal clusters of collisional origin have been recently discovered.

Collisional fragmentation has been shown to be a plausible formation process for families, from the point of view of the collision probability. Studies of size distribution of families members indicate that at least the most prominent of them are 1 − 3 Gy old (except for the Karin cluster, whose age is known

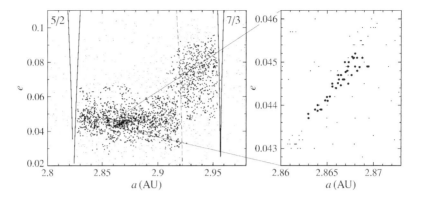

Figure 14.4. The Koronis asteroid family (thick dots on the left) on the plane (a, e); background asteroids, not associated with the family, are shown by thin dots. Note the sharp truncation of the family by the 5/2 and 7/3 mean motion resonances with Jupiter; the offset in proper eccentricity at larger values of the proper semimajor axis ("Prometheus group") is due to an interaction with a weak secular resonance (dashed line). A recent secondary break up of a member of this family with size $\simeq 25$ km has created the Karin cluster (right); for comparison, the estimated size of the Koronis parent body is $\simeq 120$ km. Proper elements from http://newton.dm.unipi.it/; see also W.F. Bottke et al., *Science* **294**, 1693 (2001) and D. Nesvorný et al., *Nature* **417**, 720 (2002).

much more precisely). They also suggest a wide variety of collisional modes: from the entire disruption of the target body by a projectile of comparable mass (e.g., the case of Koronis family), to a family composed of a swarm of small asteroids, probably all ejecta from a large target hit by a small projectile (e.g., the Vesta family). Asteroidal collisions have been simulated in laboratory experiments with high-velocity impacts on solid targets, and on computers using very sophisticated numerical programs. It has been thus verified that most energetic collisions are disruptive simply because the relative velocities of asteroids, due to their eccentricities and inclinations, largely exceed their escape velocities; even impacts with projectile-to-target mass ratios $\approx 0.1\%$ can impart energies exceeding the binding energy of the target bodies. The large estimated age of the principal families and the short-term validity of proper elements imply that they might have undergone dynamical evolution. It is likely that families were initially represented in the space of proper elements by more compact clusters and then expanded, due to long-term perturbations, such as high-order resonances and the Yarkovsky effect (this latter for sizes ≤ 20 km). This kind of dynamical aging, aside from collisional grinding, also explains why many asteroid families are sharply truncated by the strongest resonances in the main asteroid belt (Fig. 14.4). There is another important consequence of this scenario: asteroid families and the entire main asteroid belt, though Gys old, may still be very efficient in delivering multi-kilometre asteroids to the principal

14. The planetary system

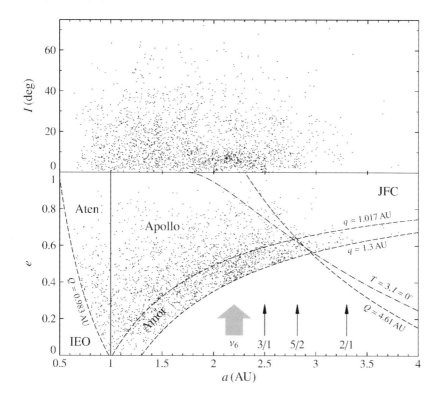

Figure 14.5. Keplerian orbital elements of near-Earth objects ($q \leq 1.3$ AU and $Q \geq 0.983$) in the (a, e) and (a, I) planes. Apollos and Atens are on Earth-crossing orbits; Amors are on near-Earth-crossing orbits (most of them will cross the Earth orbit within next $\approx 10^4$-10^5 y); orbits entirely inside the Earth orbit – (IEO) with $Q < 0.983$ AU – have not yet been detected so far due to strong observational bias. Objects on Jupiter-crossing orbits have $Q \leq 4.61$ AU (long-term minimum perihelion distance of Jupiter), approximately coincide with those having the Tisserand parameter $T = 3$ for zero inclination (see Sec. 14.5). Location of the Jupiter family comets is indicated by JFC. Data of more than 1,600 asteroids from http://newton.dm.unipi.it/.

resonances in the belt (such as the mean motion commensurabilities 3/1 and 5/2 with Jupiter or the ν_6 secular resonance), allowing them to maintain an approximately constant number of the near-Earth asteroids.

"Heating" the main belt. The large mean values of eccentricity and inclination, and a significant mass-loss in the main asteroid belt, require particularly strong perturbations, possibly acting during the early stages of its evolution; a similar problem is encountered when explaining orbital structures in the transneptunian region (Sec. 14.4). The depletion of the primordial main belt may be produced by a combination of two processes: (i) *sweeping mean motion resonances* with Jupiter, and (ii) the *scattering action of massive planetesimals*, later ejected from the solar system by planetary perturbations.

The first mechanism relies on the fact that early Jupiter and Saturn likely changed their distances from the Sun by about a fraction of an AU due to scattering of planetesimals (Sec. 16.4). While Jupiter migrated inward, the mean motion resonances followed and affected a much larger portion of the belt. An alternative model assumes that initially a cluster of cores of similar mass formed in the Jupiter zone and resonantly affected a much larger region before they collapsed to form the planet. This mechanism explains the current paucity of asteroids with $a \geq 3.3$ AU, including the limited population inside first-order resonances, protected from close encounters with Jupiter; but it fails to explain the $\approx 99\%$ mass loss from the original belt.

The second scenario relies on the existence of large planetesimals (of Moon-to-Mars size), a left over of planetary formation in the early asteroid belt or the result of injection by Jupiter's perturbations. If they happened to remain in this region for $\approx 10 - 50$ My, before being ejected out of the solar system by Jupiter's influence, they might have easily excited the eccentricity and inclination of the remaining population to the observed values. This may also explain the very peculiar orbit of Pallas, the second largest asteroid, with $e \simeq 0.28$ and $I \simeq 33.2°$, and the significant mass depletion in the belt zone. Giant collisions may also explain existence of a few large metallic asteroids, including Psyche, $\simeq 260$ km across, which could be remnant cores of large, differentiated primordial objects, stripped of their mantle.

Near-Earth objects. By convention, the near-Earth objects (NEOs) are those with perihelion $q \leq 1.3$ AU and aphelion $Q \geq 0.983$ AU. Traditional groups of this population include the *Apollos* ($a \geq 1$ AU and $q \leq 1.0167$ AU) and *Atens* ($a < 1$ AU and $Q \geq 0.983$ AU); currently both cross the Earth orbit (and may hit the Earth at their node; note that its eccentricity is $\simeq 0.0167$). These two groups are called Earth-crossing objects (ECOs). *Amor* objects are those with $1.0167 < q \leq 1.3$ AU; they can currently approach the Earth orbit, but not collide with it. However, eccentricity variations produced by planetary perturbation may allow Amors to become Apollos for some limited time, of order $10^3 - 10^5$ y. Figure 14.5 summarizes the orbital classes of NEOs. The largest NEOs are currently in the Amor population: Ganymede, ≈ 38.5 km across, Eros and Don Quixote[3], both with sizes of ≈ 20 km. Among ECOs, Ivar and Betulia are the largest bodies, ≈ 8 km across. The smallest known members of the NEO population are a few metres big, aside from the meteorites and dust particles discussed in Sec. 14.6.

A typical dynamical lifetime of a NEO orbit is ≈ 10 My; on a longer timescale the bodies either collide with a terrestrial planet or the Sun, or are

---

[3]Due to its dust trail, this object is suspected to be a comet remnant. The name near-Earth objects, rather than asteroids, is adopted because part of them may be of cometary origin. The most updated models, however, indicate that comets contribute to the population by less than 10% (for those with $a < 7.4$ AU).

ejected out of the solar system. Thus the NEO population must be continuously replenished to keep an approximate steady state, witnessed by a roughly constant flux of impacting bodies on the lunar surface over the last ≈ 3 Gy. Fast computers and sophisticated integration codes allowed during the past decade major advances in understanding this delicate problem. The main asteroid belt – the nearest vast reservoir – appears to be the dominant source of NEOs with $a < 7.4$ AU via a two-step process: (i) collisional or Yarkovsky-driven injection into the powerful 3/1 and ν_6 resonances, and the subsequent evolution into the planet-crossing zone; and (ii) slow leakage of asteroids into a Mars-crossing (but not NEO) population via weak resonances in the inner part of the main belt (mainly high-order and multiple commensurabilities with giant planets or exterior resonances with Mars), and subsequent evolution into the NEO region by encounters with Mars. These sources altogether account for some $\approx 80 - 90\%$ of NEOs; asteroids in the outer part of the main belt contribute by $\simeq 8\%$. Comets – for a long time a favoured source – contribute the remaining part.

A specific problem of the ECO population is the collision risk with the Earth and evaluation of the related danger for mankind. Sophisticated automated programmes to survey the NEO population and determine the orbit of the *potentially hazardous objects* (PHO) have been set up. In the current definition a PHO has an absolute magnitude (see Useful physical quantities) smaller than 22 (roughly objects larger than 200 m) and the largest distance between its orbit and the Earth's orbit is 0.05 AU. To find the impact hazard, two procedures are usually adopted: (i) further observations and more accurate orbit determination until the risk is found inexistent; or (ii) if the object is too faint, *virtual impactors* are investigated. The orbit of a PHO has an uncertainty which determines a tube around its mean; in the set of all possible orbits, those that correspond to objects which will in fact impact the Earth are computed, and the time at which they will be again observable assessed. In a worldwide effort, observations are carried out; if subsequent tracking of one of such virtual impactor does not confirm its existence, the corresponding risk is set aside.

Obviously, one is mostly interested in hazardous objects a kilometre across or larger. Only $\approx 50 - 60\%$ of them are presently known (of which none is a potential impacting body over next century or so). The unknown part of the kilometre-size NEO population resides on highly eccentric and inclined orbits that are very difficult to observe; nevertheless, the current observation programmes by about 2014 should determine 90% of this population (new ground-based projects, like PanSTARRS, may reach this completion limit even earlier); but this fraction drastically decreases for smaller objects.

Size distribution and collisions of asteroids. Collisional evolution has also substantially affected the mass (and size) distribution of asteroids. The largest

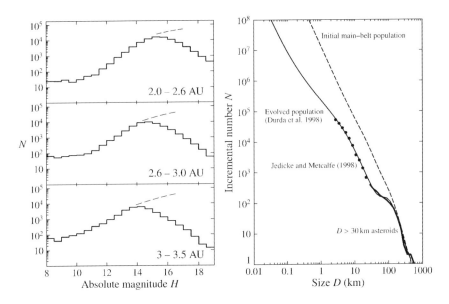

Figure 14.6. Left: number of asteroids in half-magnitude bins vs their absolute magnitude *H* for three zones in the main asteroid belt. The observed population given by the histogram; the true population is approximated by the dashed lines (the difference is due to observational limitations). Adapted from R. Jedicke and T.S. Metcalfe, *Icarus* **131**, 245 (1998), with updates. Right: the incremental (number of asteroids with size less than *D*) size distribution, as deduced from the magnitude data; its calibration is affected by the uncertainty in several parameters, including the reflection coefficient of the surface for sunlight (the geometric albedo). Mean values are assumed. A theoretical model (dashed line for the initial state, solid line for the evolved population) is also indicated. Data kindly provided by D. Durda and W.F. Bottke; see also D. Durda et al., *Icarus* **135**, 431 (1998).

one, Ceres, is about 950 km across. There are about 30 larger than 200 km, 250 larger than 100 km, 700 larger than 50 km. At smaller diameters D the size distribution becomes less known (current observations and models indicate there might be some 0.7 to 1 million asteroids larger than one km in the whole belt). The incremental main belt population is usually represented by a power law (Fig. 14.6)

$$dN \propto m^{-\alpha} dm \propto D^{\beta} dD \quad (\beta = 2 - 3\alpha). \tag{14.2}$$

The observed value of α ranges from 1.3 to 2.0, implying that most of the asteroid mass lies in the largest bodies. The total mass is 2 or 3 times the one of Ceres, determined to be $1.56 \times 10^{-4} \, M_\oplus$ from the gravitational deflection of nearby objects. Variations of α in different size intervals likely reflect sinks of the small-object populations by non-gravitational effects and their fragility against impact disruption. With respect to an exact power law there is an excess of bodies with $D \approx 100$ km; this may be related to the fact that at this size the self-gravitational binding energy, by re-accumulating the fragments ejected at speeds lower than the escape velocity into a "rubble pile" (Sec. 14.8), becomes important in determining the outcome of a disruptive impact. The distribution (14.2), with $\alpha \simeq 1.8$, has also been observed in fragments produced in laboratory impact experiments. However, asteroidal collisions involve sizes and energies which are typically 10^6 and 10^{18} times larger than those studied in the laboratory; scaling up outcomes of collisions by so many orders of magnitude is not straightforward. In the disruption of a laboratory object, part of the energy is necessary to overcome the tensile strength and to heat the body, while the rest goes to the kinetic energy of the ejecta; in asteroids, especially for events that lead to the formation of families, gravity confinement starts to play a dominant role. As we know today, the transition between the strength to the gravity regimes occurs at $\approx$ 200 m size. Escaping ejecta might also enter into bound orbits, sometimes resulting in the formation of binary objects. For these reasons, collisions between asteroid-size objects is far more complicated than laboratory experiments. Observational data of families and numerical models indicate for large sizes ($\geq$ 10 km) a very steep distribution with $\alpha \simeq 2.1$, which at small sizes becomes shallower, with $\alpha \simeq 1.4$ – basically identical to that of the background population; this is probably due to collisional grinding. The observed size distribution of NEOs in the 1 – 10 km size range is little steeper (with $\alpha \simeq 1.58$) than the corresponding population of the main belt, possibly due to size-selective transport processes, such as the Yarkovsky effect.

Rotation rates and their relation to asteroid structure. Most data about the rotation and the shape of asteroids come from light-curve photometry, which has provided rotational periods for $\approx$ 1,000 bodies, and polar directions for $\approx$ 100 asteroids (Fig. 14.7). The median rotation period is about $\approx$ 10 hours for large-size asteroids ($\gtrsim$ 20 km), and gets smaller for smaller sizes; but a large dispersion is present, and periods as short as 2.1 hours and as long as weeks have been observed. The correlations between spin period and taxonomic class is not yet understood, but a dependence on the mean density has been suggested (in the average C-type and M-type asteroids rotate, respectively, slower and faster). Since the asteroid angular momenta have been affected in a complex way by, and to a large degree acquired in, collisional and dynamical processes, rotation states certainly are not primordial.

Apart from the "normally" rotating population, we have two extreme subpopulations (Fig. 14.7): (i) *slow rotators*, with periods typically larger than a day, the slowest at $\approx$ 47 d; and (ii) *fast rotators*, with periods less than $\approx$ 2.1 h. This limit derives from condition $\omega_{cr}^2 \simeq \frac{8}{3}\pi G \bar{\varrho}$, which sets the largest angular velocity ω_{cr} for a homogeneous body of density $\bar{\varrho} \simeq 2.5$ g/cm^3 and vanishing strength against rotational fission. Interestingly, this limit is clearly seen in the small-asteroid rotation data down to sizes of $\approx$ 150 m (Fig. 14.7), supporting the idea that they have fractured and low-strength internal structures.

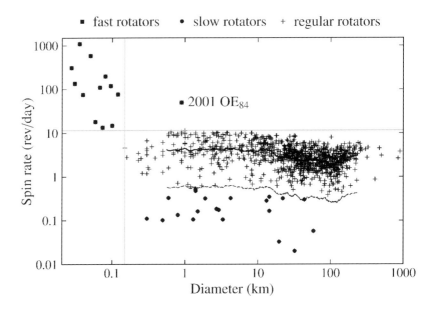

Figure 14.7. Asteroids in the (diameter, spin rate) plane, in logarithmic units. The thick line shows the average spin rate for objects larger than 1 km. The shallow minimum at $\simeq 100$ km corresponds to the "angular momentum drain" process: when the velocities of the fragments from an impact are near the escape velocity, more objects escape in the prograde sense relative to the rotation of the asteroid, and with their recoil slow down the spin. Full circles below the dashed line, at three standard deviations from the mean, are slow rotators, with unusually long periods, up to a few months. Full squares on the upper left are fast rotators – small near-Earth asteroids with periods as short as a few minutes. The horizontal line at $\simeq 12$ rev/day, a population boundary above the size of $\simeq 150$ m, is due to the disruption of loosely bound, faster objects by the centrifugal force; a single exception is the recently discovered 2001 OE$_{84}$, with a rotation period of 29.19 min and a size of about 900 m. Note that in the population of small bodies, all the near-Earth asteroids observable during a very short window of time, there is a large observational bias toward short periods. Data for nearly 1,000 asteroids kindly provided by P. Pravec.

Only objects with small sizes, which likely are monolithic, can rotate faster. The sub-population of slow rotators is still unexplained; one viable scenario assumes that asteroids undergo braking due to a loss of a close satellite. At smaller sizes, radiation torques may also despin the body (Problem 15.10).

The accidental discovery by the Galileo spacecraft in 1993 of Dactyl, a small satellite orbiting about the asteroid Ida, opened the investigation of *binary asteroids*. In mid 2002 10 more binary systems in the main belt were known, including one satellite of a Trojan asteroid. These pairs were discovered with very high resolution techniques using adaptive optics and large optical telescopes (including the Hubble Space Telescope). 13 more binary systems were found in the NEO population by optical and radar observations. The orbital motion of binary asteroids yields valuable information about their mass and mean density, typically difficult to access. Current estimates suggest a fraction of a few % of binary systems among the main belt asteroids, but up to 20% among NEOs. The ratio of sizes of the two components ranges between 0.01 and 1, from a primary with a small satellite to a truly double system. The NEO binary systems are characterized by a small separation (not exceeding 10 times the radius of primary) and a small eccentricity of the relative motion (≤ 0.1); there is also indication that the primary always has a short rotation period in the $2-3$ h range. The origin of binary asteroids, as well as their long-term dynamical stability, is not well known. Some of them might be ejecta that, under favourable conditions, formed during collision of a parent body (as during family formation); another possibility is a capture of ejecta in a non-disruptive cratering event on the primary. In the case of NEOs there might be more possibilities: tidal fission due to close encounter with a planet or rotational fission due to the YORP effect (see Problem 15.10).

Chemical composition of asteroids. In the last few decades an intense observational effort has shed light on the problem of asteroid composition, which has been found to be very diverse. The main source of information is spectral analysis of reflected sunlight, but other techniques, like infrared observations, polarimetry and planetary radar, have been applied as well. These data have then been interpreted by comparing them with the properties of minerals found in different meteorite types.

A clear difference among various types of surfaces is found in the distribution of albedo. When observational biases against darker objects are accounted for, some 75% of the asteroids are found to be very dark, with average $A \approx 0.04$. A distinct group of bodies have a moderate albedo of about 0.15, with few asteroids lying in between and a tail of "bright" bodies with A up to 0.4 and more. A better discrimination is possible if spectrophotometry data are used, yielding the behaviour of the reflection spectrum over a wide wavelength interval (Fig. 14.8). Some absorption bands are unequivocal evidence of silicates, water ice and hydrated minerals, but in many cases these prominent features are lacking, and any inference about the mineral composition must be regarded as conjectural.

Statistical clustering techniques have been applied to sets of observational parameters, potentially relevant for the surface composition of asteroids, in order to define the so called *taxonomic types*. C-type asteroids have a very low albedo and a flat spectrum throughout the visible and the near infrared; they are probably similar in composition to carbonaceous chondritic meteorites, which are primitive mineral assemblages subjected to little, or no, metamorphism after their condensation. D-type objects are also dark, but have very red spectra, suggesting the presence of low-temperature organic compounds. These objects are similar to many low-albedo, reddish small bodies found in the outer

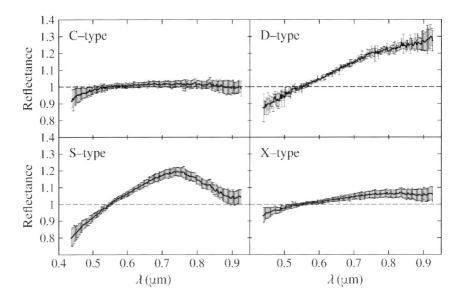

Figure 14.8. Mean optical reflectance spectra of asteroids of the taxonomic classes C, D, S and X; mean values over the whole sample of known asteroids of the corresponding class are shown; the spectral reflectance is normalized to unity at 0.55 μm. Note the absorption near 0.9 μm in the S-class, an evidence for silicates on the surface. Data from http://smass.mit.edu/.

solar system, including some comets observed at low activity and a few small satellites (e.g., Phoebe). S-type asteroids have a relatively high albedo, and their spectra show absorption bands due to silicates, like pyroxene and olivine (Fig. 14.8). It is debated whether they are analogous to stony-iron meteorites (probably derived from core-mantle interfaces of differentiated parent bodies), or ordinary chondrites, interpreted as assemblages of primitive nebular grains of different compositions, subsequently moderately heated and metamorphosed. If surface alteration by energetic solar radiation, cosmic rays and micrometeorite impacts (generally called *space weathering*) is taken into account, the analogy with ordinary chondrites appears likely; in particular, spectral analysis of S-type NEOs allowed to establish a link between these objects and the ordinary chondrites meteorite class. M-type asteroids have an albedo of about 0.1, with slightly reddish spectra, suggesting a significant content of nickel-iron alloy (this interpretation has been confirmed by radar observations; in the modern classification system this group is part of the X-type class). It is likely that they are akin to iron meteorites, and hence represent pieces of cores of differentiated precursors.

Of great interest is the fact that different taxonomic types are, on average, preferentially located at different heliocentric distances. This orderly progression is usually interpreted as reflecting both variations in the composition of the material which condensed in the solar nebula, due to the decrease in temperature with solar distance, and the different relevance of subsequent melting events and metamorphism. Indeed, the most primitive types (C and D, corresponding to least metamorphosed material) tend to lie in the outer belt, where most asteroids significantly resemble cometary nuclei. However, the borders between the radial location of asteroids with different spectral properties are not sharp, and overlap by a fraction of AU. This feature likely originates in the early violent perturbation of the asteroid belt by massive planetesimals.

14.4 Transneptunian objects and Centaurs

The discovery in the 1990s of a population of objects beyond Neptune, named *transneptunian objects* (TNOs), was a major event in planetary science, on a par with the discovery of the solar wind and planetary magnetospheres earlier in the 20th century. For the first time after 1801, when the first asteroid was observed by G. PIAZZI, an entirely new class of objects in the solar system was discovered. The prediction on theoretical grounds of Kuiper belt objects before their discovery strengthens its importance.

From the historical viewpoint, the theoretical arguments were basically two. First, as mentioned in Sec. 14.1, Pluto has always appeared an oddity among planets and it was also not clear why accretional formation of the solar system objects should stop at Neptune's distance (with the exception of the tiny Pluto). This was the original motivation which led G.P. KUIPER, and, independently, K.E. EDGEWORTH, in the 1940s and 1950s to expect a disc of objects in the transneptunian region. A more immediate motivation came from study of the Jupiter family comets (Sec. 14.5). This numerous population is characterized by low inclinations and there is no known physical mechanism that could confine their orbits to the invariable plane of the solar system when the source is far away and isotropic (like the Oort cloud; Sec. 14.5). For that reason, theorists in the 1980s postulated a disc of cometary objects beyond Neptune that could act as a source for short-period comets.

The turning point was discovery of 1992 QB_1, the first transneptunian body (apart from Pluto and its satellite Charon). Since that time some ≈ 800 more members of this population have been observed, with an accelerating rate of discovery in the past few years. So far the largest known members of the TNO population are about 900 km across (there are about 4 objects with this size, including 2000 WR_{106}, named Varuna). This is about the same size as Ceres, the largest asteroid. Their number rapidly increases with diminishing size, with an estimate of $\approx 40,000$ objects larger than 100 km. The smallest observed TNOs have a size of about 25 km, but little is known about bodies below ≈ 50 km. Pluto is also not alone to have a satellite among the TNO population. Seven more TNOs possess a companion; the best known system, after the Pluto-Charon pair, is the binary 1998 WW$_{31}$, whose secondary has a much larger eccentricity ($\simeq 0.8$) than Charon. Partly due to observational limitations, the TNO binary systems are somewhat anomalous, with a large separation (up to 1,000 the size of primary). Their origin is unknown and apparently puzzling; as an example, the tidal evolution of very distant systems typically results in unacceptably fast rotation of the larger primary if at the beginning the separation of the two components was a few Roche radii. A more likely scenario assumes the capture of a secondary during a close encounter inside Hill's sphere of influence of the primary; the loss of relative energy needed to form a stable binary may occur due to dynamical friction from surrounding small bodies, or through gravitational scattering of a third body.

Orbital structures beyond Neptune. The orbital distribution of TNOs contains important information about objects at the edge of the solar system. With about 800 bodies, this population is still poorly known (compared to $\approx 3 \times 10^5$ known asteroids), but several orbital groupings have been identified. It should be also noted that only about half of the known TNOs have been observed during multiple oppositions, needed for good orbit determination, and some 20% only on very short arcs (less then a week). Hence, follow-up programmes to improve such weakly constrained orbits are very important. We now give a list of the important orbital sub-populations in the outer solar system (see Figs. 14.9 and 14.10).

–*The "classical belt"* between 40 and 48 AU is so named because it most resembles the Kuiper belt originally searched for, namely a group of bodies on nearly circular and low-inclination orbits. This belt is assumed to be a leftover of the planetesimal disc, that never succeeded in accreting into planetary-size objects. This description, however, is too simple. First of all, the estimated mass of objects in the classical Kuiper belt is two or three orders of magnitude smaller than that expected from the planetesimal disc models (see Fig. 16.1). Secondly, the orbital eccentricities e and inclinations I are not small, similarly to the asteroid belt. The mean excitation $\approx (\bar{e}^2 + \bar{I}^2)^{1/2} \simeq 0.23$ of bodies between 30 and 55 AU suggests that primordial coplanar and near-circular orbits in this region were subject to significant perturbations. With this value, the currently observed $100 - 1,000$ km Kuiper-belt bodies could never have accreted, because mutual collisions would be disruptive, rather than allowing growth. Another feature is the truncation of eccentricities produced by Neptune encounters; in pericentre distance, $q = a(1-e) < 35$ AU (Neptune's mean distance from the Sun is to be augmented by a few Hill radii). Because of this natural limit, the *inclination* distribution of the Kuiper belt objects is assumed to reflect more clearly the dynamical perturbations to which the initially "cold disc" (low e and I) was subjected.

–*Plutinos*, like Pluto, are trapped in the 2/3 mean motion resonance with Neptune and their semimajor axes are thus confined (within about 0.3 AU) to the

14. *The planetary system*   407

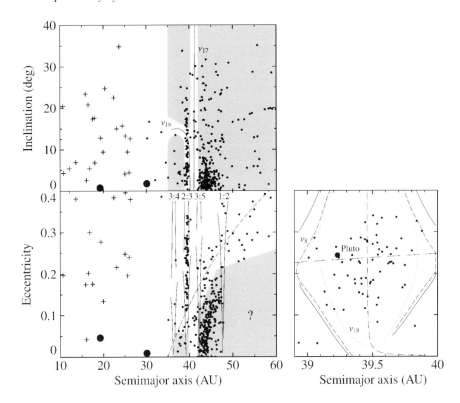

Figure 14.9. Orbital elements of the known plutinos, classical Kuiper-belt objects, scattered-disc objects and Centaurs in the (a, I) plane (top) and the (a, e) plane (bottom); Uranus and Neptune are shown as large circles. The dashed curve in the lower plot corresponds to orbits with perihelion of 35 AU, that roughly delimits the region of long-term stable orbits (shaded regions). Exception are high-eccentricity plutinos, that are protected by the 2/3 resonance from close encounters with Neptune. The position of the most important mean motion resonances with Neptune (3/4, 2/3, 3/5 and 1/2) is shown by their borders. The instability between 40 and 42 AU is due to the secular resonances ν_{17} and ν_{18}, shown in the upper graph (no orbits could survive in this zone for longer than $\approx 10 - 100$ My). Centaurs, given by crosses, are objects that escaped from the more stable TNO regions, with short-lived orbits. The zoom on the right shows plutinos in the 2/3 mean motion commensurability with Neptune. Meaning of the lines: solid, the separatrix delimiting the resonance; short-dashed and dotted, the secular resonances ν_8 and ν_{18}; dash-dotted, location of the exactly periodic orbit corresponding to the 2/3 resonance; long-dashed, orbits orbits with perihelion at Neptune's distance from the Sun. Orbital data from http://cfa-www.harvard.edu/cfa/ps/mpc.html; the position and the structure of the 2/3 case have been kindly provided by D. Nesvorný.

mean value of ≈ 39.5 AU. Some plutinos (like Pluto) have $q < 30$ AU, but the resonance prevents close encounters with Neptune. This population is thus truncated at $e \approx 0.4$, when encounters with Uranus are possible. Bodies in this state may reside on their orbits for $\approx$ Gy without being significantly affected

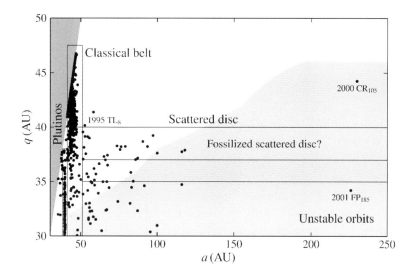

Figure 14.10. Semimajor axes a vs perihelion distances q for known transneptunian objects. Very approximate boundaries for the classical Kuiper belt, plutino population and the scattered disc are indicated by the boxes (because of detection biases the plot does not represent the true density of populations). The shaded region roughly indicates the long-term unstable orbits due to the planetary perturbations. Data from http://cfa-www.harvard.edu/cfa/ps/mpc.html.

by planetary perturbations. This numerous group of $\simeq 130$ objects currently represents an apparent fraction of $\simeq 20\%$ of the whole TNO population; models suggest that even taking into account their easier discovery, they should represent some $\simeq 5 - 10\%$ of the "true" TNO population. Any theory aiming at explaining orbital populations beyond Neptune must also explain their origin.

–*TNOs with $a = 36 - 39$ AU and small e and I* (≤ 0.05) are predicted to be stable over a Gy time span, but few of them have been detected (note that they are not subject to strong observational biases). This implies that the dynamical processes that sculpted the belt left here few objects (there is no reason why primordial bodies could not accrete in this zone).

–*Objects in the 1/2 mean motion resonance with Neptune*, at $a \simeq 47.8$ AU, have not been observed for a long time. At this distance observations are difficult and orbit determination requires special care (recoveries at several apparitions are necessary to firmly place an orbit in this resonance). Nevertheless, there are currently some 10 candidates in its stable zone. Being the most distant first-order mean motion resonance in the solar system, it marks a natural division in the transneptunian region. There is an intense debate about whether the classical Kuiper belt really terminates at the corresponding distance from the Sun; no one knows whether there are objects with moderately *low* eccentricities and inclinations beyond it. None have been observed so far, but it should be remembered that this population is most heavily biased against discovery; it is much easier to discover objects at very large semimajor axis *and* large eccentricity with a close pericentre. It is still quite possible that beyond 48 AU there is a population of low-e and I orbits, never subjected to the excitation mechanisms that affected orbits within 48 AU. Another possibility is that the truncation of the TNO population at $\simeq$ 50 AU was caused by the perturbing effect of a closely passing star during the early evolution phase or by inhibited formation of large bodies at this great distance.

–*Scattered-disc population* are objects beyond Neptune with highly eccentric orbits ($e \geq 0.4$ say, though there is no firm definition). There are presently about 200 known objects in the scattered disc, but some of their orbits are rather uncertain. Obviously, most of these objects are observed at perihelion passage, because their semimajor axes might be as large as hundreds of AU (where they would not be observable); if observational biases are taken into account, the scattered-disc population is probably as large as the classical belt. The scattered disc can be populated by two distinct mechanisms: (i) escape from the classical Kuiper belt due to Neptune's perturbations, and (ii) primordial emplacement at an early epoch, when the planets cleared out the outer solar system of leftover planetesimals. The first scenario seems to be viable in populating orbits with $q \leq 35$ AU, where the gravitational influence of Neptune is still important. Orbital data in Fig. 14.10 may indicate a gap between the scattered disc objects with $q \leq 35$ AU and $q \geq 37$ AU. For the latter population, emplacement by the Neptune in the current orbit seems implausible, and they may in fact represent a sub-population whose orbits are primordial; this structure is sometimes called extended or *fossilized scattered disc*.

–*Centaurs* inside 30 AU represent the transition population between the short-period comets and their source region outside Neptune's orbit. The first Centaur – (2060) Chiron – was discovered in 1977; presently $\approx$ 30 are known. For that reason, some authors assume that Centaurs are part of the larger population of *ecliptic comets*, which includes also the Jupiter family of short period comets. Centaurs are the least stable population in the outer solar system; with a typical dynamical lifetime of $\approx$ 10 – 100 My, they need a steady source like: (i) the classical Kuiper belt, losing objects by long-term dynamical instabilities or collisions; and/or (ii) the scattered disc. The latter, more favoured possibility, is just a continuation of the process by which the Kuiper belt populates the scattered disc.

This completes the inventory of minor objects in the outer solar system; comets in the traditional sense are dealt with in sec. 14.5. They are characterized by low inclinations (Fig. 14.9): the largest I for Centaurs is $\approx 35°$, while $\approx 40°$ for the transneptunian populations (Plutinos, classical Kuiper belt and scattered disc).

Formation of orbital structures. Orbital data, in particular, the comparatively large values of e and I, and the significant mass depletion require strong dynamical processes that perturbed the initial population of the transneptunian objects. Identifying them, and understanding how they acted is one of the major questions in planetary science, most likely intimately linked with the formation of giant planets. It is possible, but not necessary, that a single process depleted the TNO zone and in the same time dynamically excited their orbits.

It is interesting to note that the timing of these processes is partly constrained by the observed size distribution of the Kuiper belt objects. At the time the perturbation started, accretion ceased and the size of large objects froze; bodies grew no longer, and the continued collisional activity over the age of the solar system affected only the dimensions of small bodies, bringing them roughly to collisional equilibrium. The critical size at which the very steep primordial distribution bends to the shallower equilibrium distribution is uncertain and is estimated to be between $\approx 25-50$ km. This means that the observed differential size distribution of TNOs – with a power law index $\beta \simeq -4.4$ (eq. (14.2)) – was reached by the end of the accretion period in this region. Accretion theories constrain the time when, starting with a kilometre-size planetesimal population, this slope was reached to $\approx 0.1 - 1$ Gy after the formation of the solar system, with a preference for the first value.

The main possibilities for dynamical excitation processes of the current structures beyond Neptune are, similarly to the main asteroid belt: (i) the effect of *sweeping mean motion resonances*, and (ii) the perturbations due to *planetary embryos* (see also Sec. 14.3). Interaction of early Uranus and Neptune with interplanetary planetesimals may have caused these planets to drift outwards from where they accreted, by a few AU (Fig. 16.6). TNOs, originally on nearly circular and coplanar orbits might have been trapped in Neptune's mean motion resonances as they swept past. This mechanism provides a natural explanation for the plutinos with large e and I. However, sweeping resonances (i) provide in the classical belt zone beyond ≈ 42 AU nearly no excitation in e, and especially I, and (ii) little mass depletion is obtained. These effects are efficiently produced by giant-planet embryos (lunar- to Earth-size), likely formed simultaneously with their cores and subsequently scattered into orbits that repeatedly passed through the Kuiper belt before ejection. Alternatively, additional cores of giant planets, including Uranus and Neptune themselves, might have formed in the Jupiter-Saturn zone and played the role of embryos. A single Earth-mass embryo residing in the belt region for ≈ 100 My would produce mean values of e and I of about 0.3, while causing significant mass depletion. In Sec. 14.2 we mentioned that the early impact of such a planetesimal on Pluto may have resulted in formation of its satellite Charon. Likely a combination of these processes acted to form the structures presently observed in the transneptunian region.

Physical observations of TNOs and Centaurs. Physical studies of TNOs (and Centaurs) are severely limited by their intrinsic faintness. Up to now, observations include optical and submillimetre detections using broad band filters; when IR data became available, albedo and the size were determined (as for $\simeq 50$ TNOs and Centaurs). Known values of the albedo are typically smaller than 0.1, comparable to primitive carbonaceous surfaces devoid of ices, with a slight trend toward higher values for larger objects. Pluto and Charon are different, with large $A \simeq 0.55$ and 0.38, respectively, perhaps due to surface frosts freshly deposited from a tenuous atmosphere. Interestingly, Pluto's albedo is very variable, and reached $\simeq 0.7$ in 1950s, perhaps due to a sublimated southern polar cap. Optical, broad-band spectroscopy in the blue, visible and red yields the $B - V$ and $V - R$ colour-indexes. TNOs and Centaurs are thought to form two distinct groups in the plane of these indexes: the Centaur-dominated group with "solar type" values, indicating somewhat grey surfaces, and significantly redder objects (high $V - R$ and $B - V$) in the transneptunian region. Grey colours generally signal the presence of carbon-rich materials (assumed primitive in the solar system), while the red colours are indicative of complex hydrocarbons on icy surfaces. However, this suspected dichotomy may not be real, but may be a consequence of complicated evolutionary processes: the primitive, grey surfaces redden in time due to irradiation by cosmic rays and UV radiation, while resurfacing due to impacts may excavate the primitive grey material. Nevertheless, recent observations indicate a more solid correlation: the spectra of TNO are systematically grey for high inclinations and redder at low inclinations.

The near IR spectra, available for a couple of Centaurs, suggest a large diversity in this group. For instance, the spectrum of Pholus is interpreted as due to black carbon with a mixture of olivine, water and methanol, resulting in an extremely red colour and low albedo. It is a dark, primitive surface, slowly evolved by long-term irradiation of carbon- and nitrogen-bearing ices. On the contrary, the spectrum of Chiron, consistent with a composition of water ice and olivine, suggests that the surface may be dominated by non-irradiated dust from its interior. Chiron is known for its cometary activity, so that the continuous deposition of sub-orbital cometary debris falling onto its surface may be consistent with this model. These periods of cometary activity make the Centaur population rather spectrally heterogeneous.

Moderate resolution spectra in the near-IR zone have also been obtained for a few TNOs, with evidence of absorption features for methane (CH_4) and other hydrocarbon ices, like ethane (C_2H_6), ethylene (C_2H_4) and acetylene (C_2H_2). The presence of molecular nitrogen (N_2) is debated; it has important consequences for their sizes, since it would imply temperatures low enough to prevent sublimation, yielding higher albedo and thus smaller sizes. On Pluto a weak N_2 absorption line was found at 2 μm; since its strength is small, nitrogen abundance should be high and may dominate its tenuous atmosphere. Charon, Pluto's satellite, has a surface dominated by water ice.

The observed differential size distribution (14.2) of Kuiper belt objects is very steep – $\beta \simeq -4.4$ – down to sizes of about 25 – 50 km. Below this threshold it is assumed to approximately follow the collisional equilibrium, characterized by[4] $\beta \simeq -3.5$. The cutoff at $\approx 25-50$ km is important, since with a steep slope down to a kilometre the Kuiper belt would be too massive[5]. The best estimate of the mass in this zone, between 30 and 50 AU, is $\simeq 0.1$ $M_\oplus$. As noted above, this is two orders of magnitude smaller than the expected *initial* value ($\approx 20 - 50$ $M_\oplus$), needed for accretion of the largest Kuiper belt objects from kilometre-size planetesimals. The most severe constraint comes from the presence of Pluto and Charon; however, it is still possible that objects larger than Pluto reside in the more distant regions of the transneptunian belt.

There is a very limited knowledge of the light curves of these distant bodies. Rotation periods of 6 – 10 h were determined for a few Centaurs (including Chiron and Pholus), matching those of some asteroids (Fig. 14.7). These Kuiper belt objects have somewhat anomalous rotation curves with large amplitudes (up to ≈ 0.5 magnitude in the V band), and a large number of rather short periods clustered around 3 – 5 h. Given their large size, about 300 – 500 km, the data were surprisingly interpreted as due to lack of spherical symmetry with typical axes ratio of ≈ 1.2. With a mean density of ≈ 1 g/cm^3, a fraction of the bodies rotate near the critical value for rotational fission. This would have important implications for their internal structure and collisional effects in the transneptunian region. The observed sample of objects, however, is still very poor and sometimes contradictory.

Given the limited number of objects which have been discovered beyond Neptune, and the even more restricted number of known TNOs for which we have physical data, conclusions about this region and its role in the formation of the solar system can only be preliminary.

14.5 Comets

Bright comets, possibly the most spectacular celestial objects, appear as diffuse sources slowly moving in the sky, often visible to the naked eye for weeks. Although their apparitions had been reported as early as several centuries B.C.

---

[4] As explained above, the presence of "primitive" slope of the distribution at the large-size end of the spectrum is unique and a very important source of information for the primordial accretion phase in this region. Whether this has been smeared in the asteroid belt due to the collisional breakup of many large bodies is currently debated.

[5] An interesting constraint on the initial mass in the Kuiper belt zone comes from the low eccentricity (0.009) of Neptune, as compared to the higher eccentricities of all other giant planets. Neptune's eccentricity might have been decreased through the interaction with an initially massive disc further out. This constrains the initial mass from "both sides", since a too massive disc would leave Neptune with an even lower value.

by Chinese and Chaldean astronomers, TYCHO BRAHE first obtained observational evidence that they are celestial bodies more distant than the Moon, in contrast with the scholastic view of objects in the cislunar region. A century later, E. HALLEY, by demonstrating the periodical nature of the comet P/Halley, determined that comets move on highly eccentric Keplerian orbits around the Sun. When J. PALITZSCH saw it again 76 y later and conclusively confirmed its elliptic orbit, the validity of Newton's gravitational law had been probed up to 35 AU – the perihelion distance computed with Kepler laws –, more than three times farther than Saturn. Later, F.W. BESSEL, investigating the motion of the short period comet P/Encke, suggested that the recoil due to the outflow from the cometary nucleus is the cause of the observed irregularities, the first evidence of non-gravitational perturbations for solar system objects.

A major revolution in cometary science took place within a short time span in the 1950s with three developments: (i) the identification, from kinematic studies, of a distant reservoir of comets, now known as the *Oort cloud*; (ii) the *icy conglomerate* ("dirty snowball") model of the cometary nucleus; and (iii) the explanations of the displacement of cometary plasma tails due to interaction with the solar wind. These outstanding results are comparable to the theoretical formulation and discovery of the Kuiper belt in the 1980s and 1990s. Given our space limits, and in agreement with the previous sections, we focus on the orbital classification of comets, their stability and sources, and suggest in Further Readings some literature on their chemistry and physics.

Orbital taxonomy: old and new schemes. Historically comets have been divided into two groups according to their period P: (i) *long-period comets* ($P > 200$ y) and (ii) *short-period comets* ($P < 200$ y). They served the purpose to determine whether a newly discovered comet had been seen before or not (observations more than two centuries ago were not reliable).

When the semimajor axis determination improved and the number of observed comets increased, a dynamical distinction in the long-period group between *new* and *returning comets* was introduced. J.H. OORT observed that the distribution of the semimajor axes $a = -GM_\odot/E$ of nearly parabolic comets, determined by their binding energy to the Sun E, becomes much more populated beyond $\approx 20,000$ AU; the maximum of the distribution of $1/a$ is sometimes called the Oort peak. Since its width is about an order of magnitude smaller than the value $\approx 2\,M_J/(aM_\odot)$ corresponding to the energy change $\Delta E \approx E\,M_J/M_\odot$ (mainly determined by Jupiter) in its passing through the solar system, these comets must be visiting the planetary region for the first time and are called "new comets"[6]. The remarkable clustering of cometary orbits

---

[6]Oort also proved that the few slightly hyperbolic orbits can be explained with planetary perturbations. When this effect is taken into account, cometary binding energies to the Sun become slightly negative, even when the osculating values are positive. Therefore, although an originally very eccentric, elliptical orbit

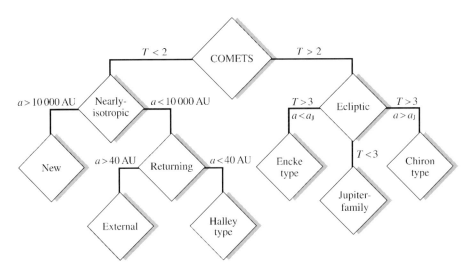

Figure 14.11. Dynamical classification of comets. The ruling quantity is the value of the Tisserand parameter T related to Jupiter's orbit. Objects with $T > 2$ have orbits with not too large inclinations (none with $I > 50°$ and a mean of $\simeq 10°$); the family of objects with $T < 2$ contains distant comets on nearly isotropic orbits. Adapted from H. Levison, Comet taxonomy, in: *Completing the Inventory of the Solar System*, eds. T.W. Rettig and J.M. Hahn (Astronomical Society of the Pacific), p. 173 (1996).

at $a \approx 20,000$ AU, later confirmed when more data became available, led to the hypothesis of a large reservoir of long-period comets in a vast swarm – the Oort cloud – surrounding the solar system at a large distance. Its boundary is approximately set by the radius of the solar Hill's sphere of influence related to the Galaxy, namely $\approx (M_\odot/3M_G)^{1/3} R_G \approx 2 \times 10^5$ AU (M_G and R_G are the mass of the Galaxy and the Sun's distance from its centre, respectively). Beyond this distance the galactic tide efficiently strips comets to the interstellar region. A minority of long-period comets with semimajor axis smaller than $\approx 20,000$ AU have been in the Oort peak before and were removed by planetary perturbations. They are designated "returning", since they are observed during their repeated visits. The division between new and returning comets was taken, somewhat arbitrarily, at $a \approx 10,000$ AU.

Short-period comets were traditionally subdivided in *Halley-type* ($P > 20$ y) and *Jupiter family* (JFC) ($P < 20$ y). The main motivation for this division are different orbital properties: JFCs have a strong concentration of the semimajor axes between 3 and 4 AU, with the aphelion close to Jupiter's mean distance

---

may sometimes appear hyperbolic when the comet is observed close to its perihelion, comets do indeed belong to the solar system.

from the Sun, and a flat inclination distribution (with mean inclination of ≈ 10° only); Halley-type comets are consistent with the assumption of an isotropic orbital distribution.

New discoveries and related theoretical investigations in the past decade have significantly changed our view of the orbital classification of comets, their relations, and sources. The new scheme provides a better fit to the current knowledge about cometary sources and the long-term evolution of their orbits. It also removes some arbitrariness in the previous classification; most important, the 200 y threshold between the two principal classes of short- and long-period comets is eliminated. Instead, the most significant division is based on a dynamical quantity, namely the value of the Tisserand parameter T with respect to Jupiter (eq. (13.37)) – Figs. 14.11 and 14.12. Comets with $T > 2$ are designated *ecliptic comets*, since most of their members have small inclinations; comets with $T < 2$ are called *nearly isotropic* comets, reflecting their inclination distribution. T is almost constant if only Jupiter's perturbation, the main effect, is taken into account, and changes little due to the gravitational perturbations of other planets or non-gravitational effects. Numerical simulations show that $T \simeq$ const. over the dynamical lifetime of cometary orbits (some 5×10^5 y), which makes this division significant.

Ecliptic comets can be further subdivided into three groups: (i) the *Jupiter family* with $2 < T < 3$; (ii) *Encke-type* comets with $T > 3$ and $a < a_J$; and (iii) *Chiron-type* comets with $T > 3$ and $a > a_J$. Note that only objects with $T \leq 3$ encounter Jupiter (eq. (13.40)), so that the last two populations have orbits entirely inside and outside the Jupiter's orbit, respectively; the group (iii) are probably the same as the Centaurs (Sec. 14.4). Within a limited range, the evolution through the $T = 3$ threshold is possible, so that Chiron-type orbits can evolve into JFC orbits[7]. Since the first are resupplied from the transneptunian region, this latter source is the likely origin of *all* ecliptic comets. Apparently, the most difficult problem is tracing the evolution route toward the Encke-type interior orbits, since non-gravitational effects seem to play an important role. Much less evolution is observed for the known comets through the $T = 2$ threshold, which adds to the significance of the new taxonomic scheme.

The nearly isotropic comets are subdivided in two groups: new and returning. The new ones have the same definition as before. The group of returning comets covers a much larger range than previously described, from those with $a < 10,000$ AU down to those with orbital periods near 20 y; it includes some that used to belong to the Halley-type. Their orbits are isotropically distributed in space. As for ecliptic comets, orbits may change subclass when

---

[7] In fact, the dynamics near $T = 3$ may be very complicated due to temporary satellite captures around Jupiter, and bouncing between trans- and sub-Jovian orbits, often associated with mean motion resonances with Jupiter.

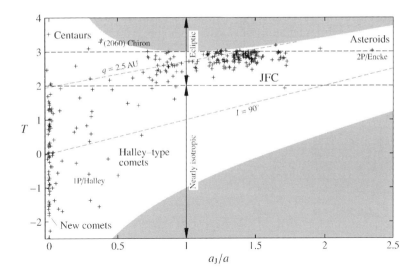

Figure 14.12. Position of known comets in the plane of the Tisserand parameter T and the ratio a_J/a between the semimajor axes of Jupiter and the comet; the unphysical regions are in dark grey. Dashed lines indicate the values $T = 2$ and 3 (see Fig. 14.11); the thin dashed line labelled $q = 2.5$ shows T for a 2.5 AU perihelion comet in the ecliptic (objects above and to the left of this curve are very difficult to detect because they never get close to the Sun). Adapted from H. Levison, Comet taxonomy, in: *Completing the Inventory of the Solar System*, eds. T.W. Rettig and J.M. Hahn (Astronomical Society of the Pacific), p. 173 (1996).

newly arrived comets are gravitationally captured. The Oort cloud appears to be the source for the nearly isotropic comets and, if it is abundant enough, replenishes its population when momentum is exchanged with passing stars and dense interstellar clouds[8] (Problem 14.6). From the present star density in the solar neighbourhood ($\approx 0.1/\text{pc}^3$) and the typical stellar velocity relative to the Sun (≈ 20 km/s), it can be estimated that about one stellar passage per million years occurs within 10^5 AU from the Sun. HIPPARCOS astrometric data indicate that the next close approach will occur in 1.4 My with the star Gliese 710, at $\simeq 82,500$ AU. A minimum distance of $\simeq 10,000$ AU occurs about every ≈ 0.5 Gy and results in a huge increase in the cometary flux in the inner solar system. On a long term, about 10 new comets per year become observable, a fraction of the ≈ 100 comets whose perihelia get inside Saturn's orbit, with $q < 10$ AU. Of course, there is a strong observational bias against finding comets with $q \gtrsim 1$ AU. Thus every stellar passage may force $\approx 10^8$ comets to

---

[8] Another plausible mechanism for comet extraction from the Oort cloud is a cycle of coupled eccentricity and inclination variations due to the galactic tide (Problem 14.8), a process resembling the Kozai mechanism which prevents high inclinations of distant irregular satellites of giant planets (Sec. 14.2).

bring their perihelia in the planetary region, with a transverse velocity $\approx 2\%$ of the circular velocity in the Oort cloud; with an isotropic velocity distribution, at any given time only one comet in 10^4 belongs to this group. Combining these two factors, we estimate the total population of comets in the Oort cloud to be of order 10^{12}. Assuming typical masses of cometary nuclei between 10^{15} and 10^{16} g, corresponding to their typical sizes of ≈ 1 km and densities of 1 g/cm^3, the total mass of the cloud is of the order of $1 - 10\ M_\oplus$.

Formation of the Oort cloud. The new orbital taxonomy of comets reflects the principal immediate sources, the transneptunian region and the Oort cloud. The first, a left-over of an aborted planetary accumulation at the outskirts of the solar system, is primordial. An important question is when the Oort cloud formed and what was its source.

The compositional similarity of all comets suggests that the source was the immediate transneptunian zone or, possibly, the whole giant planets region. In Sec. 14.4 we found that the current orbital structures in this region require a violent perturbing mechanism in the early solar system, capable of depleting nearly 99% of the original population. The expected mass loss is about $20-50\ M_\oplus$, sufficient enough to explain the estimated $\approx 1-10\ M_\oplus$ mass of the present Oort cloud. Certainly the mechanism of its formation is not fully efficient in delivering material; collisions between cometary objects are thought to decrease the efficiency by a factor ≈ 10. There is also some leakage from the Oort cloud into the interstellar medium.

While mass estimates support the scenario of delivery of primordial objects from the transneptunian region, it remains to be seen if gravitational planetary perturbations, stellar flybys and the tidal field of the Galaxy can ultimately produce an isotropic Oort cloud within a reasonably short time span. Extensive numerical work suggests a positive answer (Fig. 14.13). The principal mechanism of ejection to distant orbits is close encounters with the outer planets (possible details of this scenario are discussed in Sec. 14.4), while the randomization of the inclinations and the circularization is due to external effects, namely passing stars and galactic tides. The formation timescale – perhaps 0.1–0.5 Gy – is shorter than the age of the solar system and matches the expected timescale for formation of structures in the transneptunian region (Sec. 14.4).

Cometary physical studies. It is impossible to resolve a cometary nucleus inside the gaseous *coma* by Earth-bound observations. However, a model of cometary nuclei capable of explaining a large body of observational evidence about the behaviour of comets as they approach the Sun, as well as the structure and chemical composition of comæ and tails, was proposed in 1950 by F. WHIPPLE. In this model, frequently called the "dirty snowball", the nucleus is a small solid body, generally 1 to 10 km across, consisting of a conglomerate of ices and silicate particles. As the comet approaches the Sun, ices sublimate and the resultant gas and released meteoric dust are available to form the coma

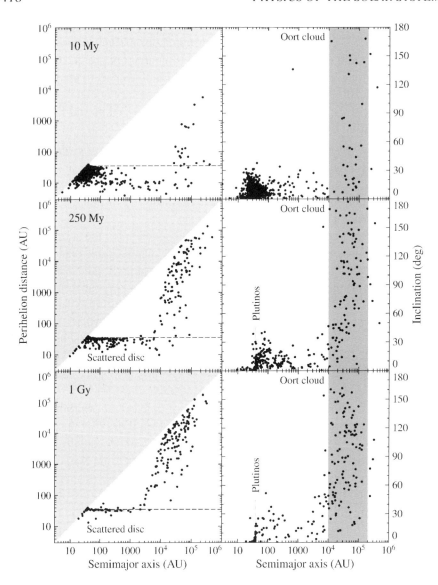

Figure 14.13. Formation of the Oort cloud and the structures in the transneptunian region: 2,000 small objects are initially placed on the invariant (Laplace) plane of the solar system on circular orbits ($e = I = 0$) with semimajor axes a between 4 and 40 AU. Left: perihelion distance q vs a at 0.01, 0.25 and 1 Gy; right: I vs a at the same times. The simulation includes gravitational perturbations due to the giant planets, the galactic tidal field and passing stars. The isotropic Oort cloud at $a \geq 10^4$ AU, truncated at $a \simeq 10^5$ AU due to the galactic tide, is rapidly formed; the observed structures in the transneptunian region (plutino and scattered disc populations) are also clearly seen. While the plutino population is primarily composed of objects initially in the transneptunian region, both the scattered disc and the Oort cloud populations are a mixture of planetesimals initially at $a \geq 8$ AU. Data kindly provided by L. Dones.

and tail. Direct evidence about the parent gas molecules has been missing for a long time because their emission spectra are in the far-infrared and submillimetre band; water, a dominant component, was thus directly confirmed by observations only in 1985. However, they become ionized by the solar radiation and the solar wind, and shine in the optical and ultraviolet. Solid debris can become meteoroids strewn along the comet's orbit, affected by Poynting-Robertson and radiation forces (see Sec. 15.4) and occasionally hitting the Earth (thus producing *meteors* and *fireballs*). The reaction of the nucleus to the ejection of this material provides an explanation for non-gravitational effects in cometary dynamics. The model correctly predicts stronger outward radial components of non-gravitational forces (the ejected material being most abundant on the Sun-facing side of the nucleus) and significant transverse components due to the comet's rotation and a thermal lag between the direction of maximum mass ejection and the subsolar meridian (see Sec. 15.4). The nucleus must be weakly bound, since several cases of comet splitting and breakup have been observed, sometimes due to a tidal gravitational field, as the comet D/Shoemaker-Levy 9, disrupted in 1992 in a close approach to Jupiter; there are other cases where super-critical rotation or thermal stresses are suspected to be the cause of fission. In many cases the volatile component is probably covered by an outer shell of dark, refractory material (with albedo $\simeq 0.02 - 0.05$) and gaseous jets may emerge from localized fractures or spots, sometimes causing sudden outbursts. The covering of the free sublimation region ranges from less than 1% to nearly the whole surface (for smaller objects). Remarkably, the model has been confirmed by in situ observations and measurements by the Giotto and Vega missions in March 1986 during their very close flyby with Halley's comet and by the Deep Space 1 visit of comet P/Borrelly in December 2001.

In analogy with asteroids, spectral observations of chemical species in the gaseous coma allow a physical classification of comets. Abundances of OH, CN, C_2, C_3 and NH resulted in recognition of two distinct classes of objects: (i) *typical* comets have a production of C_2 proportional to CN, both normalized to the production of OH; (ii) *depleted* comets have a significantly lower yield of C_2 with respect to CN. Interestingly, there is a relationship between the physical and orbital classification of comets: nearly all isotropic comets are physically typical, while about 50% of the ecliptic comets are classified as typical and 50% as depleted. This is interpreted as a chemical gradient in the solar nebula; typical comets formed mainly inside a critical distance $\simeq 36 - 42$ AU, depleted comets beyond. Most of the former ones were placed in the Oort cloud, but in part remained in the transneptunian region until late diffusion in ecliptic orbits. Depleted comets accumulated in the distant zone of the transneptunian region, the only one has a low probability of feeding the Oort cloud.

Cometary timescales.– Among the minor bodies in the solar system the specific feature of comets is their fast evolution, both orbital and physical. Figure 14.14

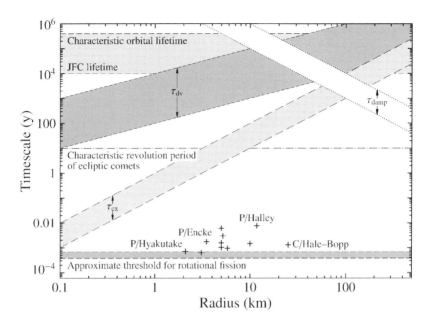

Figure 14.14. Timescales related to cometary dynamics and ongoing physical processes vs size. τ_{damp} is the characteristic time to damp the rotational state to an axis which is a principal axis of inertia by internal dissipative processes (eq. (3.56)); a broad region is drawn, to take into account the parametric uncertainty. The nucleus is assumed rigid; if the tensile strength is small, τ_{damp} may be an order of magnitude smaller. τ_{dv} and τ_{ex} are, respectively, the characteristic times for devolatilization of the comet and to excite rotation around a generic axis by the torque due to the non-uniform outgassing. For JFCs the comparatively frequent passages near the Sun make their lifetime with respect to total outgassing and/or disintegration not larger than some $\approx 1,000$ revolutions, $\approx 10^4$ y, shorter than their dynamical lifetime $\approx 5 \times 10^5$ y. The crosses indicate a few known rotation periods; note they are close to the disruption limit (short-dashed line at ≈ 3.3 h); for elongated objects this value increases to ≈ 6 h, closer to the minimum of the observed periods. Adapted from D. Jewitt, *Earth, Moon and Planets* **79**, 35 (1997).

summarizes the characteristic timescales of different processes and indicates several important facts.

First, the timescale for devolatilization (τ_{dv}) is shorter than the cometary lifetime, especially for the JFC comets which outlive their supply of volatiles. As a consequence, "dead" JFC comets may exist, provided they are not physically destroyed by other mechanism (in such objects the icy component is either exhausted or covered with an insulating crust, preventing further cometary activity). The fact that relatively few such objects are known is presumably a consequence of observational selection. A more complete survey of the inner solar system may uncover more dormant cometary nuclei; similarly, the outer part of the asteroid belt might also host extinct comets. The fact that the near-Earth asteroid Phæton is a parent body of the Geminid meteor stream supports the suspicion that some bodies in the inner solar system are extinct comets. Other examples of dormant or extinct comets are: Elst-Pizzaro, showing temporary outgassing events probably due to meteorite impacts, can be classified both as a comet and an asteroid; Wilson-Harrington and Oljato, both showing ion tails and CN-band emission. For objects smaller than about 100 m, τ_{dv} becomes comparable to, or shorter than, the revolution period around the Sun. This should indicate a truncation in the size distribution of the JFCs, or at least a steeper size distribution if a fraction of these small bodies become extinct. Another consequence is that, although the typical final state of large JFC comets may be the dormant phase, possibly interrupted by temporary outgassing events, the typical end of small cometary bodies may be complete disintegration into meteoroid streams. On the whole, comets are efficient suppliers of gas and dust particles to interplanetary space. A related slightly different problem is the fate of nearly isotropic comets: the orbital evolution models that bring new comets into the returning group predict many times more objects than observed – this is usually referred to as the *fading problem*. One possible solution is that a majority of the new ones become dormant, but this seems incompatible with both physical models and observations. New comets are more likely disintegrate on their path into the inner solar system by thermal gradients, volatile pressure buildup, or disruption due to rotational spin-up.

Secondly, when $\tau_{dv} > \tau_{ex}$, the timescale for outgassing processes to produce a rotation around an axis not aligned along an axis of inertia, we might expect to find comets in an excited rotation state. Note that only for very large objects $\tau_{ex} < \tau_{damp}$, the timescale to damp excited rotational states by dissipative processes. This conclusion, however, depends on the unknown internal structure and holds only if the internal strength is large enough; for weakly bound objects ("rubble piles"; see Sec. 14.8) τ_{damp} may become very short for small objects, too. Comet P/Halley is so far the only convincing case of excited rotation among comets, but there are other possible examples.

Star-grazing comets.– An interesting final state was noticed, or at least suspected, back in the 17th and 18th century: Sun-grazing cometary objects, with perihelia within $\approx 2 R_\odot$. They were first systematically studied by H. KREUTZ at the end of the 19th century, whence this population is called the *Kreutz group*. For most part, they are discovered at the solar limb before reaching perihelion, but often do not reappear later. This is taken as evidence for their tidal or thermal disintegration in the solar atmosphere. Observations in the 1990s by coronagraphs on board of the satellites SOLWIND, SMM, and especially SOHO, discovered many Sun-grazers; in 2000 SOHO alone observed some 70. Nearly 400 are now known. It has been suggested that many of them are related to one another, possibly as outcome's of a few massive comets which disintegrated passing close to the Sun centuries ago. The most viable dynamical origin of these peculiar orbits are high-inclination states ($I \simeq 90°$) of returning comets, whose perihelia diminished as a result of secular evolution caused by planets (see the related Problem 14.8).

Interestingly, there is evidence of star-grazing cometary objects around several young stars, including the well studied system β Pictoris, about 20 pc from our Sun. Repeated (hundreds events per year are observed), time-dependent redshifted absorption features in their spectrum have been interpreted as comets evaporating in the immediate vicinity of the star, perhaps falling into it. Since this flux requires a dynamical mechanism, such as secular or mean motion resonances with planetary orbits or strong scattering by massive planetary cores, star-grazing comets indirectly imply a planetary system. It should be remembered that these stars are probably much younger than the Sun; β Pictoris should be $\approx 100-200$ My old. At that time the solar system was in its final "cleaning" phase with intense planetesimal delivery onto the Sun.

14.6 Interplanetary material and meteorites

Interplanetary space contains a large number of bodies and particles too small to be individually detectable by telescopic observation. These bodies, whose motion is often affected by non-gravitational forces and whose orbital lifetime is generally much shorter than the age of the solar system (Sec. 15.4), are thought to originate from the three populations of small and relatively primitive bodies previously discussed: asteroids, TNOs and comets. The number of asteroids and TNOs rapidly grows with decreasing size (see eq. (14.2)), and there is no reason to believe that this trend does not continue down to those which are not detectable (in the main asteroid belt, $\approx$ km in size). On the other hand, cometary activity is characterized by emission not only of gas, but also of dust, and solid particles as large as ≈ 10 m; in some cases, catastrophic disintegration of a cometary nucleus in a swarm of solid debris can occur. The lack of accurate orbital information is partially compensated by collection and laboratory investigation, facilitated by their high flux near the Earth.

Observational methods. Since the small objects of interest in this Section span an enormous range of more than 20 orders of magnitude in mass, they have been investigated using many different observational techniques. Here we review only the most important ones.

Our general knowledge of bodies – collectively called *meteoroids* and *dust particles* (for small sizes) – in *interplanetary space* is observationally very limited. It basically consists of ground- or space-based optical and infrared observations of (i) the *zodiacal light* and (ii) the *solar F-corona* (also "false-corona"). Both are due to reflected solar radiation and thermal emission by small interplanetary dust particles arranged in a large disc-like structure on the ecliptic. Zodiacal light is caused by a true, large-angle reflection of sunlight; the F-corona is attributed to small-angle diffractive scattering by particles close to the solar line of sight. For an Earth observer, zodiacal light can barely be seen westward toward the evening twilight, or eastward before sunrise; it looks like a faint, glowing band of rediffused sunlight extending up from the horizon and along the ecliptic plane. The F-corona is primarily observed during solar eclipses. The second important constraint for the large-scale distribution of interplanetary dust comes from in situ spacecraft measurements; remarkable observations were performed by the Galileo and Ulysses missions, whose plasma detectors recorded the ion and electron cloud released after the impact of micrometeoroids and tiny dust particles. In this way a flux of particles as small as $\approx 10^{-21}$ g was measured. During the next few years, NASA Stardust mission plans important observations of interplanetary and interstellar dust, both close to the main target, comet P/Wild 2, and during the cruise phase; dust samples should be returned to the Earth by 2006.

However, the most natural "collector" of meteoroids at our disposal is the Earth itself, which provides a substantial amount of information about their flux at 1 AU from the Sun, including temporal variations over the past My's. A flux of the smallest particles, down to $\approx 10^{-18}$ g, was detected by various spacecraft sensors; remarkable information was provided by the Long Duration Exposure Facility, a spacecraft released by Space Shuttle in 1985 and returned to Earth in 1991. Lower in the atmosphere (below 100 km), where particles smaller than ≈ 1 μm – called *Brownlee particles* – are decelerated below the limit of a destructive impact, high-altitude flights and balloons make their collection and laboratory analysis possible. A large sample of dust particles and micrometeorites was also collected in Antarctica. This yields detailed information about their chemical and mineral composition, an important clue to their sources; typically, they are irregular glassy silicate and metallic aggregates of smaller dust grains. The characteristic mean density is significantly below 1 g/cm^3, as also evidenced by meteor observations, but particles with higher density (up to $\simeq 3$ g/cm^3) are also detected. Long-term variations in the micrometeoroid flux on the Earth is recorded by the variations of several "tracers" – elements of the rare Earth platinum group (in particular, Ir and Os) and extraterrestrial helium ^{3}He – deposited in marine sediments and polar ice.

When larger micrometeoroids ($\geq 10^{-10}$ g) enter the atmosphere, radar signals may be reflected by the electron and ion components in their train. Radar echoes allow us to quantify their flux. When combined with a solution of the particle motion in the atmosphere, which depends on the atmospheric density and thermal spallation, radar data also yield information on several important parameters, including the orbital elements, the size distribution outside the atmosphere and entry velocities.

Interaction of larger meteoroids ($\geq 10^{-4}$ g) with the atmosphere are observable as visual *meteors* that momentarily flash across the sky. Multiple-station observations allow more precise reconstruction of the pre-atmospheric orbit (though with lower accuracy than that attainable for larger NEOs). Variations in the flux and the apparent source in the sky during the year are observed. Most meteors are far too small to sustain heating due to atmospheric drag and "burn" up at altitudes between 75 and 100 km.

Larger meteors called *fireballs* are bright and spectacular, but often explode before hitting the ground. This is an indication of their fragility, consistent with the low density – typically between $0.2 - 0.7$ g/cm^3 – inferred from the observed decelerations due to atmospheric friction. When a meteoroid is large, its relative velocity at entry is low enough, and its strength is high, so that it may be not entirely destroyed by atmospheric heating. At an altitude below ≈ 75 km, its largest pieces decelerate to a velocity of ≈ 3 km/s; the surface temperature drops below a critical sublimation value of $\approx 2,200$ K and the body continues a dark flight to the ground. The surface layer of the meteoroid solidifies and forms a crust.

Rocky samples recovered on the Earth surface are called *meteorite finds*; falls and finds have been witnessed for centuries. Only in rare cases the bolide and its meteor trail were observed with enough accuracy to determine its orbit outside the atmosphere, and in the same time the meteorite found; there are 7 of them, all ordinary chondrites, except the important case of Tagish Lake, a carbonaceous chondrite meteorite. In humid environments stony meteorites erode within a few thousand of years. However, in arid or semiarid conditions, their *terrestrial age* (the time since fall) can be significantly longer: meteorites collected in deserts and Antarctica can survive up to My, with typical ages around 0.1 My. Exceptions are *fossil meteorites*, of which the best documented are stony meteorites in limestone quarries in Sweden. Their location in the surrounding geologic layers allow accurate determination of the time of their fall, the middle/early Ordovician ≈ 480 My ago. On the basis of the abundance of these fossil meteorites, its has been estimated that the accretion rate during the early Ordovician was one or two orders of magnitude larger than today, probably a consequence of a major collisional event in the asteroid belt (possibly a family formation; Sec. 14.3).

The largest known meteorite, Hoba, found in Namibia, is a few metres across, with an estimated mass at fall ≈ 88 tons. The flux in this size range is low, roughly one per a decade on average. They experience little atmospheric ablation and, though they still typically detonate in the atmosphere, their largest fragments hit the ground forming explosion *craters*. Since the Earth is the most endogenically active of the terrestrial planets, it has retained the poorest sample of such impacts; only ≈ 160 impact structures are known. Erosion biases the crater record toward the youngest and largest events. Interestingly, some $\simeq 10\%$ of the largest known craters are found in close pairs, as for Venus and, to a smaller extent $\simeq 2\%$, for Mars. Since the separation is too large to be consistent with a tidally disrupted object, they are usually interpreted as due to binary asteroids. The lunar crater record is much better preserved and, thanks to the possibility of absolute dating of surface features, provides important constraints on the past flux at 1 AU. Its impact rate has been nearly constant over the last few Gy, while a much more intense bombardment occurred $\approx 3.8 - 4$ Gy ago. This *primordial (late) heavy bombardment* was probably due to the sweeping by the major planetary bodies of the residual solid material left over from the accumulation of planetesimals (see Ch. 16). For large impactors the typical interval between falls is roughly proportional to the square of their size, consistent with the distribution (14.2) with exponent $\alpha \simeq 1.8$ (Fig. 14.15a). This relationship appears to hold up to the catastrophic impacts of Earth-crossing asteroids, several kilometres across. Such large impacts occur every 10 to 100 My and may cause global ecological catastrophes (such as that occurred at the Cretaceous-Tertiary boundary, which caused the extinction of major fauna; its impact crater – Chicxulub close to the Yucatan peninsula – was discovered in the 1990s).

On *average* $\approx 10^{11}$ g of meteoritic material hits the Earth annually (Fig. 14.15b), mostly due to large bodies; a single body of $\approx 10^{17}$ g hits the Earth every $\approx 10^7$ y. Objects with mass less then $\approx 10^8$ g, mostly small particles, hit the Earth annually and contribute about 3×10^{10} g. This is shown in Fig. 14.15b: the incremental mass influx has maxima at $\approx 5 \times 10^{-5}$ g and $\approx 10^{10}$ g, with a less pronounced maximum at ≈ 1 g; given plausible densities, the first and the second correspond, respectively, to sizes of $\approx 100\,\mu m$ and ≈ 20 m; their relative overabundance was determined by the Spacewatch and NEAT observations of small NEOs. For many particles in the $1 - 10\,\mu m$ size range, radiation pressure is larger than the solar gravitational attraction; once released from the parent body they quickly escape on hyperbolic orbits and are usually termed *β-meteoroids* (the ratio of radiation to gravitational forces is denoted β_r; Sec. 15.4). Their flux inside the Earth orbit has been determined by satellite and radar measurements. Since particles of this size systematically escape the solar system, somewhat larger meteoroids ($\simeq 100 - 200\,\mu m$) have prolonged collisional lifetimes; moreover, slightly larger particles are removed by collisions on a timescale smaller than their transport time from initial orbits in the main asteroid belt or get efficiently scattered out of the solar system by Jupiter if cometary by origin. Both effects explain the corresponding maximum in $\mu(m)$ in Fig. 14.15a. The situation for the 20 m size objects is more complicated and no convincing explanation has yet been given.

Sources of interplanetary dust. The Poynting-Robertson effect and the solar wind drag (Sec. 15.4) remove micrometre-size particles from the inner solar system on timescales of $\approx 0.01 - 1$ My. Somewhat larger particles are more efficiently destroyed by collisions than removed by dynamical effects, with a timescale still many orders of magnitude smaller than the age of the solar system. Any particle observed today must have been injected onto its orbit, in

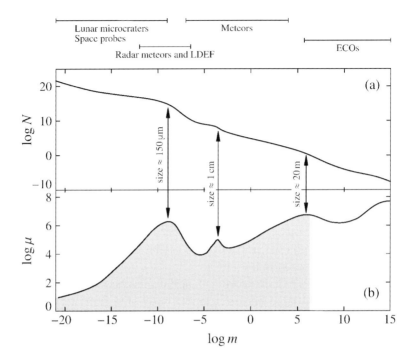

Figure 14.15. Cumulative number flux ($N(m)$, a) and incremental mass flux ($\mu(m)$, b) of meteoroids impacting the Earth in a year, as function of m in kilograms. $N(m)$ is the number of objects with mass $< m$ and $\mu(m)$ is the total mass in a unitary interval of $\log m$. The shaded area indicates bodies which fall at least once per year. The variable slope in $N(m)$ is related to the maxima of $\mu(m)$. In the whole range of m the total yearly accreting mass is $\approx 10^8$ kg; if only the objects falling at least annually are included, the total influx is $\approx 3 \times 10^7$ kg. Several observing techniques, with which these results were compiled, are noted on top. Data kindly provided by Z. Ceplecha.

geological terms, very recently. Traditionally, comets were presumed to supply the lost dust; but in the last decade a significant contribution was found to come from asteroids. The fractional input of asteroids in the inner solar system is still debated, but it may be high, especially on the Earth, since asteroidal dust particles typically have a smaller relative velocity and the effective cross section is larger (eq. (11.44)). In the region beyond Saturn the contribution due to the dust produced by TNOs is assumed to be dominating the cometary input.

Asteroid dust.– The 1984 observations of IRAS (Infrared Astronomical Satellite), later confirmed by COBE (Cosmic Background Explorer), revealed circumsolar bands of IR emission near the ecliptic about $1-2\%$ above the smooth zodiacal light background. Two pairs of prominent *dust bands* were observed near the ecliptic latitudes $\pm 2°$ and $\pm 10°$; they closely correspond to the proper inclination (Sec. 15.1) of the three most populous asteroid families Eos, Themis and Koronis (Figs. 14.3 and 14.4). A finer structure in the zodiacal light was also investigated in connection with another asteroid families acting as a source. A classical model assumes that dust produced by collisions of asteroids in these families quickly escapes due to the systematic decrease of the semimajor axis produced by non-gravitational forces; particles cross the asteroid belt in ≤ 1 My, at a rate inversely proportional to their size (a typical particle in the observed dust bands has a size in the $10 - 200\,\mu m$ range). Alternatively, dust bands might be associated with very recent disruptions, such as the Karin cluster, of a few asteroids (with an age of ≤ 10 My); the recently identified secondary breakup in the Veritas asteroid family is the most promising candidate for the $\pm 10°$ band. Since planetary perturbations cause the secular drift of the orbital node, dust orbits *in the asteroid belt* form a characteristic disc-like structure, sharply bound in latitude. When particles reach the inner boundary of the belt, delimited by the ν_6 resonance (Fig. 14.3), their orbital inclination and eccentricity become random and thus feed the diffuse component of the zodiacal cloud. As they continue to spiral towards the Sun under the influence of drag forces, some smaller particles may be captured in first-order exterior resonances with the Earth (such as 1/2, 2/3, 3/4, etc.). Their combined effect results in the formation of a *near-Earth resonant dust ring*. This structure has been confirmed by IRAS and COBE observations of the zodiacal IR emission produced by the particles of this ring trailing behind the Earth.

Cometary and TNO dust.– IRAS observations also revealed evidence of a cometary contribution to the zodiacal cloud, as *trails of IR emission* coincide with the orbits of numerous short-period comets, in particular P/Encke, P/Temple 2 and P/Schwassmann-Wachmann 1; some trails were observed with no connection with known comets and may result from comet disruptions similar to P/West. These trails are composed of particles probably ≥ 1 mm in size, that were ejected from their parent comet in tens to hundreds of revolutions. On these timescales they evolve into *meteoroid streams*, discovered through the analysis of the frequency of meteor appearance. The permanent meteor flux, due to *sporadic meteors*, is such that on an average night about 3 meteors per hour can be seen before midnight and about 15 meteors per hour after midnight. This component of the micrometeoroid dust was found to reach the Earth along five different directions (including helion, antihelion and apex lines); the entry velocities in the atmosphere are very high, several tens of km/s, suggesting cometary origin in general. But on certain dates, always at the same heliocentric longitude of the Earth, *meteor showers* of 60 or more meteors per hour can be seen, all radiating from one direction in the sky. The most widely known is the *Perseid shower*, which shows up around August 12, with bright meteors streaking across the sky every few minutes from the direction of the constellation of Perseus. The annual recurrence is indicative of streams of particles dispersed around an elliptical orbit. Their duration – a few days to some weeks – imply stream widths ranging from 10^6 to 10^8 km at the Earth orbit crossing point. In 1866 G.V. SCHIAPARELLI discovered that the Perseid meteor shower appears whenever the Earth crosses the orbit of the comet P/Swift-Tuttle, implying that the shower is spread out along it. Many other relationships were subsequently found between specific meteor showers and specific comets. Variations of their activity in different years allows to study filamentary structure of the corresponding stream. The conclusion that most meteors are cometary debris is not at odds with the fact that for the most conspicuous annual shower – the December Geminids – Phæthon, an asteroid-looking parent object, was discovered by IRAS. The orbit of this Earth-crossing object, with a very high eccentricity ($\simeq 0.89$), suggests that it might be an extinct cometary nucleus, with no more volatile component on the surface. This interpretation, however, contrasts with numerical studies, which suggest that the probability of a Jupiter family comet to evolve into Phæthon's orbit is extremely small. The related stream may be produced by more "exotic" phenomena; for example, the surface may be stripped by thermal, collisional or electromagnetic effects.

The Pioneer and Voyager spacecraft provided the first evidence about the dust environment at large distance ($\simeq 30 - 50$ AU) from the Sun. Analysis of their data suggests that isotropic comets might not be a sufficient source and that the bodies in the immediate transneptunian region may also contribute. Similarly to the dust produced in the asteroid belt, very small grains originating in the transneptunian region are quickly blown out of the solar system by radiation pressure, while the rest slowly spiral toward the Sun due to the Poynting-Robertson and solar wind drag. During their journey they gravitationally interact with planets and can be either scattered or trapped into mean motion resonances; models indicate that much of this dust should be captured in the Trojan zone of Neptune. Quantitatively, a $\simeq 20\%$ fraction of the grains with sizes $\leq 9\,\mu$m and $\geq 50\,\mu$m are expected to survive collisional shattering (efficient to remove grains with intermediate sizes) and can eventually reach the Sun or accrete on the Earth. Interestingly, observations of several stars revealed dust discs approximately extending to 50 AU or more.

"Exotic" dust.– There are at least two additional sources of dust in the interplanetary medium: (i) *interstellar dust*, and (ii) *planetary/satellite dust*. The Ulysses spacecraft discovered both components while it travelled near Jupiter (Fig. 9.5). The interstellar origin of the measured dust was proved by correlating directional characteristics of the flux with the corresponding flux of the interstellar helium and by the fact that it does not depend on the solar distance. Interestingly, it has been suggested that the most significant stream of interstellar particles is compatible with ejection from the β Pictoris system at an escape speed of $\simeq 20$ km/s, some My ago. They should be the equiva-

lent of β-meteoroids in the solar system (Sec. 15.4). Interstellar grains are typically small, ranging from 10^{-21} g to 10^{-11} g; using radar, particles with higher mass have been also observed entering the Earth atmosphere. The flux of interstellar dust is correlated with the solar activity cycle, in particular the polarity of the solar magnetic dipole field (Sec. 9.3), which hinders the influx of interstellar charged particles, depending on their direction. An additional source of information about interstellar dust in the solar system are primitive meteorites; they contain small concentrations of presolar grains, which survived largely unaltered the process of its formation. Isotopic analyses reveal their stellar origin, which provides support for the classical formation scenario (Ch. 16).

Ulysses also detected bursts of dust originating from the Jovian system; the surprisingly small mean mass of these particles ($\approx 10^{-21}$ g) was compensated by a high relative velocity of $\approx 200 - 300$ km/s. Later the Galileo spacecraft observed the same particles near Jupiter; theoretical modelling indicates their close coupling with the Jovian magnetosphere, with Io a likely source (Sec. 10.7).

Meteorites. The meteorite flux shows a prominent diurnal variation, controlled by the direction of the motion of the Earth with respect to the zenith of the observing site. The sweeping effect of the Earth's motion, combined with the higher energy of head-on collisions, tends to increase the influx on the morning (AM) side of the Earth (leading in the orbital motion), where smaller particles of higher geocentric velocity are generally observed. On the other hand, the prevailing prograde motion of meteoroids on orbits with semimajor axes larger than 1 AU favours the (trailing) evening (PM) side, compensating for the velocity effect. For *meteorite falls* of most taxonomic classes (see below), the net result is an excess on the PM side: some 68% of all chondrites are observed to fall in the afternoon (this is usually termed "PM fall ratio"). Yearly variations, if any, are less significant.

The most detailed orbital information related to meteorite sources comes from pre-atmospheric Keplerian elements determined by the network of fireball cameras. Orbits do not seem to exhibit any definite streaming pattern and basically resemble those of Apollo-type NEOs (Fig. 14.5); an interesting exception is a close similarity of the pre-atmospheric orbital parameters of the Příbram and Neuschwanstein meteorites (fell in 1959 and 2002, respectively), suggesting a possible meteoritic stream along this particular orbit. Very few pre-atmospheric orbits lie close to the $T = 3$ threshold of the Jupiter family comets (Fig. 14.5), consistent with a scenario where most fireballs come from the main asteroid belt.

Meteorite taxonomy.– Some 30,000 known meteorites have been classified in several types with respect to their mineral properties; only about one sixth of meteorites are as yet unclassified. The three principal classes are: (i) *chon-*

able and complex than thought before: no two systems look alike. Many of the differences, to some extent discussed below, have to do with their interactions with ringmoons (small satellites inside or close to the rings) and the particle density inside the rings, as evidenced by their optical depth.

Common features of the known ring systems. Since the 19th century, planetary rings have been known to be "comprised of an indefinite number of unconnected particles" (using the words of J.C. MAXWELL, who, in 1857, first demonstrated that this was the only stable configuration). This view superseded the previous conclusion of P.S. LAPLACE, who around 1800 realized that Saturn's rings cannot be solid; the tensile strength of known materials is too small to resist tidal forces of Saturn. However, he wrongly concluded that it may consist of several distinct ringlets. Besides Maxwell's view about their structure, planetary rings share two other features: (i) a significant flatness and concentration in the equatorial plane of the planet, where they spread as thin discs; and (ii) their location near the planet, extending radially to a distance comparable with the Roche limit. Both can be understood on the basis of simple considerations.

Flatness.– Any initial, thick and/or non-equatorial ring tends to swiftly flatten to a thin equatorial disc. In fact, the planet's oblateness causes the orbital planes of the ring particles to precess around the axis at a rate dependent on their orbital radius (see Sec. 12.4). This drift allows particles to collide, reducing their relative velocities, in particular the vertical components. Note that these collisions conserve angular momentum, but are anelastic and dissipate energy. This damping of orbital motions in the equatorial plane is rapid and very efficient, as indicated by the fact that the vertical thickness of Saturn's ring system is a tiny fraction, of the order of 10^{-6}, of its width.

To quantitatively confirm this conclusion, we estimate in order of magnitude the frequency of interparticle collisions in a ring. A basic property of rings is their *normal optical depth* τ, which determines the extinction of radiation flux by the material of the ring (see the definition eq. (7.18)); when the observation geometry is known, τ can be fairly well measured using an occultation of a star by the ring, observed either from ground or a spacecraft. Assuming particles of equal cross section S and density n, the normal optical depth for a thickness H is $\tau \simeq nSH$, and the collision frequency is $v_c \simeq nSv$. v is the relative velocity; it can be estimated with its vertical component $\simeq \Omega rI \approx \Omega H$, where Ω is the mean motion, r the distance from the centre of the planet and I the effective inclination of the particle orbits. Hence we obtain

$$v_c \simeq \Omega \tau. \tag{14.3}$$

With $\tau \approx 1$ for principal Saturn rings (Table 14.6), particles typically collide a few times per revolution. The ring will settle on the equatorial plane of the planet in a time of order $1/v_c \simeq 1/(\Omega\tau)$.

However, this damping never produces a perfectly thin layer. Low-velocity collisions still occur in a thin disc due to the (small) eccentricities and inclinations, and the gradient in the Keplerian velocity, combined with the finite size of the particles; gravitational scattering also takes place between bodies bigger than ≈ 10 m. In this way a small random relative velocity arises, which produces a ring at least several particles thick; this velocity can be estimated from the damping of spiral waves produced by the gravitational field of nearby satellites to be of the order of mm/s to cm/s. This persistent collisional activity has the important consequence of an irreversible tendency to radially spread the ring. In a Keplerian disc a collision statistically slows down the particles coming from the inside, which move inwards, and pushes forward those from the outside, which then move away from the planet. At the boundary lost particles carry away angular momentum. Random velocities act as an effective kinematical viscosity $\eta \approx v_c \ell_c^2$, where ℓ_c is the mean free path; for this reason the ring structures are often modelled as a viscous fluid. The characteristic timescale $\approx (\delta r)^2/\eta$ of the diffusion is about the random walk time for the particle to cross a fraction of the disc of radial extent δr. Ring spreading typically occurs on a timescale orders of magnitude longer than ring flattening, but still appreciably shorter than the age of the solar system. The discovery of the sharply bound Uranus' rings showed that additional confining processes are at work.

An important exception to the norm of vertical thinness are the diaphanous Jovian rings and Saturn's innermost D and outermost G and E rings, all characterized by a very low normal optical depth ($\approx 10^{-8} - 10^{-6}$, Table 14.6; from their major component, they are sometimes called *"dusty rings"*). In this case the effective viscosity is less effective than other processes (like sputtering, Poynting-Robertson or plasma drag), which efficiently eliminate particles; interparticle collisions become unimportant and the structure of the ring is basically ruled by single particle dynamics. Their vertical spread is governed by excursions above and below the equatorial plane of their source moons or other effects; e.g. the width of the Jovian gossamer rings appears to directly follow from the orbital inclination of Amalthea and Thebe (the case of Enceladus and the E ring is somewhat more complicated). A special situation occurs for the halo, the innermost part of the Jovian rings, which is vertically spread as a result of a non-equatorial component of the Lorentz force, produced by the Jovian misaligned magnetic dipole field (Sec. 15.4).

Radial confinement.– The second common feature of ring systems is their proximity to their planets, inside the orbits of the major satellites. This is due to the fact that in the neighbourhood of a planet the strong gravity gradient can prevent the accretion of sizable satellites and even lead to disruption of existing bodies, provided their tensile strength is small enough.

In a simplified model, consider the retention of loose material on the surface of a spherical satellite of mass m, radius r and mean density ϱ_s (see Fig. 14.17),

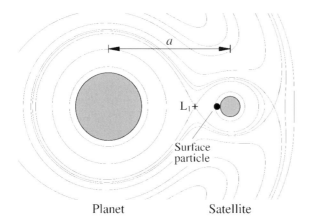

Figure 14.17 Roche's limit. A satellite with mass m, radius r and mean density ϱ_s revolves on a circular orbit with radius a around a planet with mass M, radius R and mean density ϱ_p. A grid of Hill's zero velocity curves (Fig. 13.3) is also shown.

orbiting at distance a from the centre of a planet of mass M, radius R and density ϱ_p. Its rotation is synchronous, at a rate equal to the mean motion n. The gravitational and centrifugal forces acting on a particle located on its surface, facing the planet along the radial direction (where the disrupting forces are largest), balance if

$$\frac{Gm}{r^2} + n^2(a-r) = \frac{GM}{(a-r)^2} \simeq \frac{GM}{a^2}\left(1 + 2\frac{r}{a}\right). \tag{14.4}$$

Using the relation

$$\frac{M}{m} = \frac{\varrho_p}{\varrho_s}\frac{R^3}{r^3}, \tag{14.5}$$

we find that the satellite can retain the surface particle only if

$$\frac{a}{R} > \left(3\frac{\varrho_p}{\varrho_s}\right)^{1/3} \simeq 1.442\left(\frac{\varrho_p}{\varrho_s}\right)^{1/3}. \tag{14.6}$$

This critical distance is in substance Roche's radius (13.15)

$$R_{\text{Roche}} \simeq a\left(\frac{m}{3M}\right)^{1/3},$$

about the same as Hill's radius (Ch. 13) and as the distance r_L of the unstable libration points from the secondary in the restricted three-body problem. Stability is guaranteed if the satellite radius is smaller then R_{Roche}, which implies eq. (14.6).

It is interesting to compare this result with the discussion in Sec. 4.5, where we considered the equilibrium shape of a hypothetical *liquid* satellite. We found that for decreasing orbital distances the satellite becomes more and more elongated in the radial direction, until at the minimum distance of *Roche limit* no shape can fulfil any more the equilibrium conditions. Using the results listed

14. The planetary system

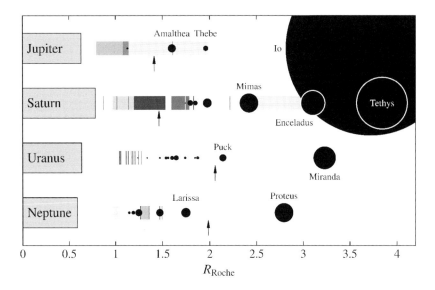

Figure 14.18. Location of the ring systems and the inner satellites with respect to the Roche limit corresponding to a satellite with mean density 1 g/cm³; the radius of the planet is indicated by a shaded rectangle on the left. Distances are scaled with the critical value (14.6), so that the Roche limit for the fluid model is ≃ 1.703. The observed inner edges of the ring systems are at a distance fairly close to unity, while the outer edges are near the fluid model limit. The moons inside this threshold must have finite strength, or higher density, which is unlikely. Small arrow indicate synchronous rotation distances; the orbits of the ringmoons inside this limit evolve toward the planet by tidal effects (Sec. 15.3). The sizes of the moons are in arbitrary units, but linearly scaled to represent their relative ratios. Adapted from J.N. Cuzzi, *Earth, Moon and Planets* **67**, 179 (1995).

in Table 4.2 (for the $p = 0$ case, i.e., negligible satellite-to-planet mass ratio), it is easy to show that the equilibrium of a small liquid satellite requires that

$$\frac{a}{R} > 2.456 \left(\frac{\varrho_p}{\varrho_s}\right)^{1/3} . \qquad (14.7)$$

The numerical coefficients in eqs. (14.6) and (14.7) are different because they are derived with different assumptions about the geometrical configuration and physical properties of the orbiting bodies; as noted in Sec. 4.5 there are generalizations of the Roche limit to a finite tensile strength, all giving the same functional dependence on the densities, but different numerical coefficients. This concept is a very general one and useful whenever we deal with the behaviour of self-gravitating bodies orbiting in the vicinity of a primary. In particular, as shown by Fig. 14.18, the main parts of the planetary ring systems actually lie inside or between the different limits derived above.

Diversity of the ring systems. We now come to the principal differences among the ring systems. Some of their major properties are listed in Table 14.6 and shown in Fig. 14.19; most of this knowledge comes from in situ reconnaissance of the outer planets by Voyager and Galileo. The chief observation techniques include direct (bolometric or colour) imaging of the reflected sunlight, spectral measurements (important for the determination of composition) and occultation profiles, both of a star and a radio source. Radio occultations are obtained when a coherent beam transmitted by a spacecraft crosses the ring and is detected by a ground antenna. Since the scattering cross section depends on the ratio between the wavelength and the size of the particles, the use of multiple frequencies gives information about the size distribution. Additionally, both Voyagers, and Galileo's atmospheric probe carried instruments to study the electrons and ions trapped in the planetary magnetospheres. Rings (and moons) absorb charged particles and can be investigated with the depletion they induce. Finally, sensitive dust detectors on board of Cassini should be able to measure their content in situ; limited and less precise measurements were also obtained by Galileo and Ulysses.

Jupiter.– The extremely tenuous (normal optical depths $< 10^{-5}$) Jovian rings have three components. The innermost is a *toroidal halo* with considerable vertical thickness; its brightness decreases with the height above the equatorial plane and as the planet is approached. The Table 14.6 indicates the maximum observed thickness, the half-maximum thickness is about 12, 500 km. The orbital inclinations inside the halo are increased by the interaction with Jupiter's magnetic field at the 3/2 Lorentz resonance; the stronger 2/1 Lorentz resonance then marks the inner boundary (Sec. 15.4). The *main ring* extends from the halo up to the orbit of Adrastea. The slightly larger Metis is embodied in the main ring, whose brightness noticeably increases inside its orbit. Outermost are two *gossamer rings*, each of them fairly uniform: one begins just inside Amalthea's orbit, while the other is inside Thebe's orbit. As noted before, their vertical thickness is comparable to the maximum elevations of these satellites above Jupiter's equatorial plane. In their meridian cross section they have a nearly rectangular shape, with greater intensity at the top and the bottom.

These observations are consistent with the idea that the outer edges of these rings are determined by the above mentioned satellites, probably by erosion caused by micrometeorite impacts. Observations from different phase angles, together with the overall faintness, indicate that they are composed of $\simeq \mu$m particles. These grains are too small for a significant absorption; their composition is not directly known and can be inferred from the composition of the parent satellites. Silicate and/or carbonaceous compounds are expected, also because pure ice would rapidly evaporate. Dust particles are subject to significant non-gravitational perturbations (principally sputtering by energetic ions

14. The planetary system

(A) Jovian rings		
Planetary equator, R_J	71,492	[τ / H (km)]
Halo, inner edge	$\approx 92,000$	5×10^{-6} / $\approx 5 \times 10^4$
Halo, outer edge	122,500	
Main ring, inner edge	122,000	5×10^{-6} / ≤ 30
Main ring, outer edge	128,000	
Inner Gossamer ring, inner edge	129,000	10^{-7} / $\approx 2,300$
Inner Gossamer ring, outer edge	182,000	
Outer Gossamer ring, inner edge	129,000	3×10^{-8} / $\approx 8,500$
Outer Gossamer ring, outer edge	224,900	

(B) Saturn's rings		
Planetary equator, R_S	60,268	[τ / H (km)]
D ring, inner edge	$\leq 66,000$	5×10^{-5} / ?
C ring, inner edge (and D ring, outer edge)	74,500	0.05 – 0.2/ ?
Titan ringlet	77,871	
Maxwell ringlet	87,491	
B ring, inner edge (and C ring, outer edge)	91,975	1 – 3/ ≤ 1
Cassini division, inner edge (and B ring, outer edge)	117,516	0.05 – 0.15/ ≤ 1
A ring, inner edge (and Cassini division, outer edge)	122,058	0.4 – 1/ ≤ 1
Encke division ($\approx$ 200 km wide)	133,589	
A ring, outer edge	136,800	
F ring, centre ($\approx$ 50 km wide)	140,200	1/ ?
G ring, centre ($\approx$ 8,000 km wide)	169,500	5×10^{-5} / ?
E ring, inner edge	$\approx 180,000$	
E ring, maximum	$\approx 230,000$	5×10^{-7} / $10^3 - 10^4$
E ring, outer edge	$\approx 450,000$	

(C) Uranus' rings			(D) Neptune's rings		
Planetary equator, R_U	25,559	[τ]	Planetary equator, R_N	24,766	[τ]
6 ring	41,837	0.3	Galle (1989N3R)	$\approx 42,000$	10^{-4}
5 ring	42,234	0.5	Le Verrier (1989N2R)	$\approx 53,200$	10^{-3}
4 ring	42,570	0.3	Lassell (1989N4R)	$\approx 53,200$	10^{-4}
α ring	44,718	0.3	Arago (1989N4R)	$\approx 57,200$	
β ring	45,661	0.2	Unnamed	61,953	
η ring	47,175	0.3	Adams (1989N1R)	62,933	10^{-2}
γ ring	47,627	2 – 4	Arcs in Adams ring		
δ ring	48,300	0.4	Courage & Liberté	62,933	0.12
λ ring	50,023	0.1 – 0.5	Egalité 1, 2	62,933	0.12
ε ring	51,149	1 – 4	Fraternité	62,933	0.12

Table 14.6. Parameters of the planetary ring systems (the equatorial planetary radii are given for reference): the first column is the radial distance in km, the second gives the normal optical and vertical thickness (τ and H, in km).

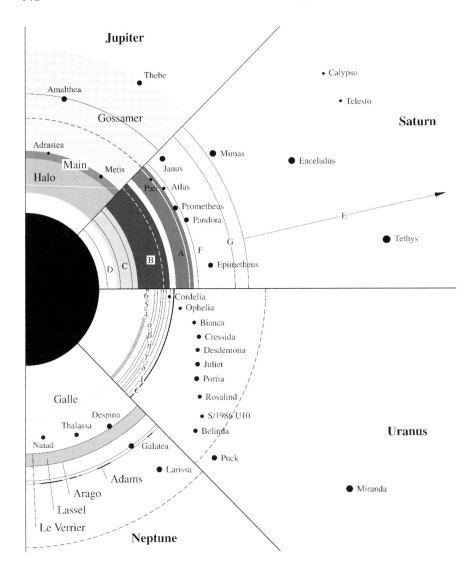

Figure 14.19. The ring systems of Jupiter, Saturn, Uranus and Neptune are compared using as unit the respective planetary radius. The dashed lines represent the locations of the synchronous orbits (with period equal to the rotation period of the planet and its magnetic dipole field). The small inner satellites (ringmoons) of these planets are also shown; satellites inside the corotation distance are expected to evolve toward the planet due to tidal effects (Sec. 15.3). Some satellites (e.g., Metis, Adrastea and Thebe for Jupiter, or Enceladus for Saturn) orbit near the outer edge or inside a ring, provide a continuous source of ring particles. In other cases (e.g., Prometheus and Pandora for Saturn's F ring, Cordelia and Ophelia for Uranus' ε ring, or Galatea for the ring arcs of Neptune) the satellites play the role of "shepherds", constraining ring material to narrow zones. Adapted from J.A. Burns, et al., in: *Interplanetary Dust*, eds. E. Grün, B.A.S. Gustafson and S.F. Dermott (Springer-Verlag), p. 641 (2001).

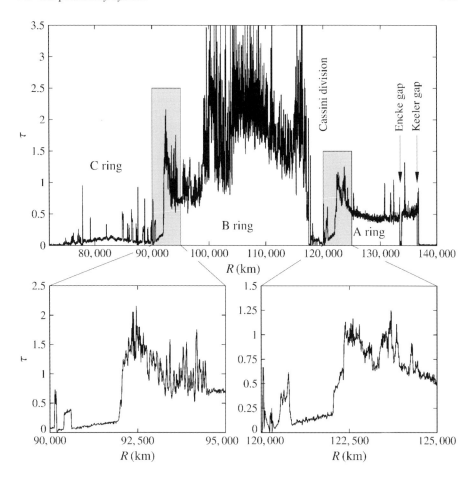

Figure 14.20. Radial profile of the normal optical depth of the main Saturn rings, as measured by Voyager's photopolarimeter during the occultation of the star δ Sco. The principal rings and gaps are indicated; the bottom panels show the remarkably similar inner edges of the B and A rings. Data kindly provided by M. Showalter.

and plasma drag), resulting in a very short lifetime between 100 − 10, 000 y. This fact points out the necessity of a very efficient particle resupply.

Saturn.– Saturn has the most spectacular and complicated ring system, extending some 400, 000 km (and some 70, 000 km for the main rings), with variations of optical depth over all length scales, down to the resolution of the available data (some 5 km; Fig. 14.20). It harbours the brightest B ring and the adjacent A ring, separated by the *Cassini division*; in reality, this division is not a true gap, but a $\simeq$ 4, 500 km wide annulus in which the optical depth drops to about 10% of the value outside. While the outer edges of the A and B rings are maintained by the 2/1 and 7/6 resonances with Mimas and Janus,

ring/region	d_{min} (cm)	d_{max} (m)	β
C ring	1	10	-3.1
B ring	30	20	-2.75
Cassini div.	0.1	20	-2.75
A ring (inner)	30	20	-2.75
A ring (outer)	1	20	-2.9

Table 14.7 Distribution of particle sizes in Saturn's main rings; for the A ring inner/outer refer to the separation by the Encke gap. A simple power law differential distribution with exponent β has been fitted between the particle sizes d_{min} and d_{max}. From infrared observations of the 28 Sgr occultation; R.G. French and P.D. Nicholson, *Icarus* **145**, 502 (2000).

their inner edges – notably similar (Fig. 14.20) – present a more difficult puzzle. The most likely explanation is the balance between collisions among the ring particles and the production of debris from meteoroid collisions (a process termed *ballistic transport*). Inside the B ring we have the C ring, characterized by an intermediate normal optical depth in the range $0.05 - 0.2$. The main Saturn's rings exhibit an astonishing microstructure; only some features have been clearly interpreted as spiral density waves produced by ringmoons and observed, or hypothesized, embedded moonlets. For instance, the *Encke gap* near the outer edge of the A ring is certainly due to the embedded Pan moonlet (Sec. 13.4); by analogy, the still narrower *Keeler gap* is believed due to an unseen object. The richness of radial structures, as seen in Fig. 14.20, is accompanied by azimuthal irregularities, widely studied since Voyager's observations. Exemplary is the narrow and slightly eccentric F ring just outside the A ring, composed of at least four strands. There are also additional clumps and kinks, with a fast variability in time. The overall compactness of this peculiar ring is likely due to the shepherding nearby moons, Pandora and Prometheus, but this – once the accepted explanation – has difficulties. Thus an additional suite of smaller moonlets has been suggested.

Saturn's main rings are extremely thin relative to their radial extension. Their thickness, as derived from photometric measurements and viscous damping of spiral density waves, ranges from a few to about a hundred metres. Infrared observations provided important compositional information: both the IR spectra and the derived high albedo particles indicate almost entirely water ice, with only a tenuous pollution by redder components. Radio occultations at multiple wavelength and stellar occultation data then provide information about the size distribution, which is typically matched by a differential power law $N(d)\,\delta d \propto d^\beta \delta d$ for the number of particles in the size interval $(d, d + \delta d)$; $\delta \approx 3$, comparable to the asteroids (Table 14.7). This holds in a range ($d_{min} \approx 1$ cm, $d_{max} \approx 10$ m). Simulations indicate that transient clumping of particles, quickly disrupted by collisions, occur in the A and B rings on the tens to hundred metres scale. Additionally, isolated small embedded moonlets up to few km across are also assumed to explain the radial struc-

14. The planetary system 443

tures (some 100 to 1,000 moonlets larger than 1 km are expected to exist). Photometric models of Saturn's main rings suggest they are largely depleted in dust; its primary sites within the main rings are the *spokes*. These features were discovered on the Voyager 1 images as dark radial markings in the B ring; however, in forward-scattered light they appear brighter than the surrounding ring, thereby exhibiting the characteristic phase changes of dust (the estimated particle sizes are $\approx$ μm). The spokes, primarily located near the synchronous orbit, have an unusual dynamics: they form very rapidly (on a timescale of minutes) as radial filaments soon after the ring emerges from Saturn's shadow, but gradually get squeezed by differential rotation. Observations indicate that spoke grains are lifted out of the ring plane (as much as $\simeq$ 80 km), presumably by electromagnetic effects. Their true origin, however, is still uncertain.

Apart from the system of the main rings, Saturn possesses three tenuous, dusty and difficult to observe rings, composed of μm particles: the innermost D and the outer G and E. As in Jupiter's case, the outer rings show a non-negligible vertical thickness, about 1,000 and 10,000 km, respectively. The most certain data concern the E ring, which spans the orbit of Mimas, Dione, Tethys and Enceladus and has a distinct brightness peak near the orbit of the latter. This satellite is likely a source for the ring dust (probably with a self-interaction process when ring particles hit Enceladus and generate multiple replacement particles).

Uranus.– The nine main Uranian rings are all very narrow: ϵ is the broadest, with a width ranging from 20 – 100 km, but all others are $\leq$ 10 km wide. Until their discovery in 1977, it seemed natural that all rings should be circular, since mutual collisions were expected to damp out any eccentric motion; but Uranus' rings are both eccentric and inclined to the planet's equatorial plane, both with typical values of $\approx 10^{-3}$. The widest ϵ ring is the most eccentric, with $e \simeq 8 \times 10^{-3}$; in all cases the eccentricity increases with distance from the planet. The rings precess due to the planetary oblateness, but, curiously enough, the precession of the inner and outer ring edges seems to be locked together, so that they share the same line of apses. These oddities are most naturally approached by considering gravitational perturbations due to nearby satellites (such as Cordelia and Ophelia for the principal ϵ ring), ring self-gravity and interparticle collisions (for the last two we note that Uranus' rings are optically thick; Table 14.7).

Similarly to Saturn's main rings, particles in most Uranian rings have sizes ranging from $\approx$ 1 cm to $\approx$ 10 m; however, they are compositionally very different. In contrast to the bright particles in Saturn's rings, Uranus' particles are very dark, with a Bond albedo $\approx$ 0.03, comparable to the comets'. In both cases this indicates the presence of radiation-darkened ice, a mixture of complex hydrocarbons embedded in ice. In 1986 an additional ring (called λ) was discovered during the Voyager 2 flyby; unlike the other narrow rings, its photometry at different phase angles indicates a composition of submicrometric particles. Aside from λ ring, the Uranian system contains additional dust widely distributed in radius, discovered by imaging the ring system at very high phase angles.

Neptune.– The tenuous rings around Neptune are characterized by significant azimuthal brightness variations called in the Adams ring *arcs*, noticed already during the occultation which led to their discovery. A 40° segment of this ring comprises five arcs, Courage, Liberté, Egalité 1 and 2, and Fraternité, with azimuthal extension between 1° and 10°. Each of them is composed of smaller discrete clumps, down to the resolution limit of a few km. Late 1990s observations by the Hubble Space Telescope established that the configuration of the arcs has been stable since the 1984 discovery. A second narrow ring inside Adams is Le Verrier. Both of them appear very similar, except for the absence of arc-like structures in Le Verrier. They are accompanied by the fainter Lassell, whose outer brightening received a separate designation Arago, and the inner, broad Galle ring.

Owing to the very limited data, the composition and sizes of Neptune's ring particles are uncertain. Nevertheless, the available photometry is best fitted with a large fraction of dust particles, compositionally similar to the material in Uranus' rings: dark, red, probably due to an ice matrix with a high portion of hydrocarbons. This is similar to the surface properties of the nearby numerous ringmoons. Among them, the most interesting role is played by Galatea, just outside the Adams ring, to which the confinement of prominent arcs is attributed.

Physical and dynamical processes. A great variety of physical and dynamical processes influences the structure and the lifetime of planetary rings. There is a consensus about models of the faint rings, although the parameters of the erosion and the resupply processes remain uncertain. In the huge Saturnian rings, the agreement between theory and observations is satisfactory for features due to gravitational perturbations triggered by the known moons (especially in the A ring). Many other observed structures, however, are based on partial or speculative explanations without relation to observed massive satellites.

Processes specific to dusty rings.– Since the circumplanetary dust particles are small, with a large area-to-mass ratio, and electrically charged, their motion is affected by a number of non-gravitational, primarily electromagnetic, forces: radiation effects (including the Poynting-Robertson drag; Sec. 15.4) and interaction with planetary magnetic field and surrounding plasma.

Direct radiation pressure has usually little effect, possibly except for Saturn's E-ring. Numerical simulations hint that μm size particles launched by Enceladus, located inside this ring, satisfy an orbital resonance condition in which the pericentre precession induced by the planet oblateness is roughly cancelled by the regression caused by the corotational field. As a result, the pericentre becomes nearly stationary, which allows the solar radiation pressure to increase the particles' eccentricity; in this process they cross the orbits of other Saturnian satellites, resulting in collisions which generate a significant amount of secondary debris. This model explains well both the large radial extent of the E-ring and the surface brightness peak at Enceladus (where, on the other hand, the ring is narrowest); the observed, unusual narrowness of the particle size distribution is thus also understood.

The Poynting-Robertson force due to solar radiation causes a permanent decay of the orbital energy and a secular change of the semimajor axis $da/dt \simeq -a/\tau$ with the characteristic timescale $\tau \simeq 530\,(r^2/\beta_r)$ y; r is the planet mean distance from the Sun in astronomic units and β_r the radiation pressure factor (eq. (15.36)). A typical lifetime to shrink the orbit of a μm size particle toward the planet in Jupiter's and Saturn's faint rings is $\tau \simeq 10^5$ y; this is comparable to the timescale induced by the plasma drag due to collisions with charged particles, abundant in the inner magnetosphere (Sec. 15.4). Specific to Jupiter, with its intense magnetic dipole field, is the orbital perturbation due to the Lorentz force on a charged orbiting particle; the corresponding acceleration is $\simeq 1\%$ of Jupiter's gravitational force, by far the strongest non-gravitational force on dust grains. Though the Lorentz force does not dissipate orbital energy, it has a major influence on the motion when the epicyclic frequency of vertical oscillations becomes commensurable with the local variations of the magnetic field due to the tilt of the dipole. This explains two prominent features in Jupiter's ring system: (i) formation of a halo (3/2 Lorentz resonance at $\simeq 1.71\,R_J$), as the orbits of the inward drifting particles become significantly more inclined, and (ii) its dissolution (2/1 Lorentz resonance at $\simeq 1.4\,R_J$), as further inclination perturbation disperses the ring material.

Circumplanetary grains are swept out by orbital drag effects, but simultaneously are destroyed by the fierce environment. Ice particles may rapidly sublimate away in the surrounding radiation fields and/or vanish due to the mass flux. Interplanetary micrometeoroids, which have a very high velocity due to the planetary gravity, can shatter them; they can also be sputtered away by the magnetospheric plasma particles. The lifetime against these processes is not well constrained, but may be remarkably short, even $\simeq 10^3$ y for μm size particles close to Jupiter and Saturn. Evidence for these ongoing erosive processes is provided by the observation of an "atmosphere" of neutral hydrogen extending some $0.5 - 1\,R_s$ above Saturn's rings. The same flux of micrometeoroids that erodes rings also contributes to resupply them with fresh material by collisions with nearby satellites and embedded ringmoons, either observed – e.g. Pan – or postulated. The net effect, either gain or loss, is uncertain. Impacting interplanetary debris may also change their mineral composition in several ways: new material is directly added to the system, the more volatile and fragile components from the rings are preferentially removed etc.

Gravitational interaction with satellites and ringmoons. – The motion of particles larger than $\simeq 1$ mm is insensitive to non-gravitational effects; a dominant gravitational perturbation arises due to satellites or embedded ringmoons. The principal effects – such as ring gaps and boundaries or occurrence of waves – are related to resonances; we give only a bare conceptual introduction to these complicated and rich topics.

A weakly eccentric and inclined orbit is well represented in terms of small oscillations around a circular motion with a uniform angular velocity. In the Keplerian case this velocity is the mean motion n and the oscillations have the same frequency; but with perturbations (e.g., planetary oblateness or the self-gravity of the disc) the unperturbed velocity changes and a split arises in the frequency of the perturbations: the out-of-plane and the in-plane oscillations have, respectively, the frequency ν (called *vertical*) and κ (called *epicyclic*; Problem 12.6). The perturbing satellite and the ring particle are characterized by the triplets (n', κ', ν') and (n, κ, ν), respectively. As in (12.79), the perturbing potential due to the satellite can be written in terms of Fourier series with arguments containing satellite parameters defining a *pattern speed (frequency)*: $m\Omega_p = mn' + k\kappa' + p\nu'$, m is non-negative and (k, p) are arbitrary integers. Resonances occur when some of the arguments of the perturbing function are stationary, which allows accumulation of the effect over a long time. The most prominent of them, with a terminology "borrowed" from galactic dynamics, are

- *corotation* with $m(n - \Omega_p) = 0$;

- *inner (outer) Lindblad resonances* with $m(n - \Omega_p) = \pm\kappa$ (upper sign for the inner resonance); and

- *inner (outer) vertical resonances* with $m(n - \Omega_p) = \pm\nu$ (upper sign for the inner resonance).

Since $\nu \simeq \kappa \simeq n$, in planetary rings the last two are only a few percent apart; as $\nu > \kappa$, due to the oblateness effect (see eqs. (12.73)), vertical resonances are located inside the Lindblad resonances. This approach may be reformulated in terms of a more traditional one in celestial mechanics, discussed in Sec. 12.5, realizing that $\kappa' \simeq n' - \dot{\varpi}$ and $\nu' \simeq n' - \dot{\Omega}$ (Problem 12.6); the resonance is typically referred to as $(m + k + p)/(m \pm 1)$. This also implies that its strength, of order $|k + p|$, in terms of the satellite mass M', the eccentricity e' and the inclination I' has the form $\propto M' e'^{|k|} \sin^{|p|} I'$. The strongest cases have $k = p = 0$ when horizontal, and $p = 1$ with $k = 0$ when vertical.

The outer edges of Saturn's major rings are maintained by the two strongest commensurabilities in the system: the outer edge of the B ring coincides with the 2/1 inner Lindblad resonance with Mimas, and the outer edge of A ring coincides with the 7/6 resonance with the coorbitals Janus and Epimetheus. In

14. The planetary system

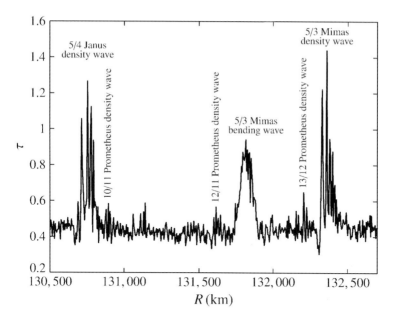

Figure 14.21. Radial profile of the normal optical depth of Saturn's A ring (zoom of Fig. 14.20 with 5 km resolution); for each feature the corresponding wave and resonance is indicated. The damping is due to inelastic collisions between the particles. The 5/3 bending wave due to Mimas prominently appears because it is not associated with density fluctuations. The overall increase of τ in this region is mainly due to more frequent collisions, resulting in smaller particles; smaller ripples may be due to the fact that, with the small slant angle to the ring plane, the line of sight may cross several waves. Data kindly provided by M. Showalter.

order to open a gap in the ring a resonance must be strong enough to counterbalance its viscous spreading; this happens in the thin C ring, but not in the denser B and A. Instead, they excite *spiral waves*: the Lindblad resonances lead to *spiral density waves*, characterized by very tightly packed spiral variations of the density, while vertical resonances lead to *spiral bending waves*, corrugations that move particles off the equatorial plane. Both types are frequent in Saturn's principal rings (Fig. 14.21). Since viscosity damps both types of waves, observations may suggest interesting constraints on collisional effects in the ring and, indirectly, its thickness. Similarly, variations of the wavelength as a function of distance from the exact resonance constrains the surface density.

The wave patterns observed in Saturn's case exert, by Newton's third principle, a torque on the orbital motion of the moons; as a result, similarly to the effect of terrestrial tides on the Moon, they recede away. Current estimates indicate that the characteristic timescale of this process is short compared to the solar system age (some 100 My for the inner Saturn's moons inside Mimas).

If there is no other force to counterbalance this process, the existence of close by moons apparently points to a recent origin of the rings.

Related to the resonant torque exchange is the process of *shepherding*, proposed in late 1970s to account for the narrow rings of Uranus. Close encounters with the moon enhances eccentricities of ring particles (see also Fig. 13.10), which on turn are damped by collisions, with the net result of a repulsion between the satellite and the ring. The latter acquires a characteristic wavy pattern, with a wavelength depending on the distance between the two, and an amplitude function of their masses. A narrow ring can be maintained in this way by a satellite on each side. Several examples are found both in Saturn's system (e.g., Pandora and Prometheus for the F ring) and in Uranus' system (e.g., Cordelia and Ophelia for the ϵ ring). Inverting the argument, it has also been suggested that the empty gaps in Saturn's rings are due to embedded small ringmoons; thus the small satellite Pan in the Encke gap was discovered, but for many other gaps the available resolution of the Voyager data is not enough for the corresponding small moonlets. Many of these objects are expected to be discovered by Cassini mission.

Origin and lifetime of ring systems. Although many individual phenomena and features have been understood, there is still uncertainty about their origin and evolution. It is not certain whether they formed at about the same time as the planets, as the material within the Roche limit was tidally prevented to agglomerate into satellites, or are much younger and, possibly, transient structures due to tidally or collisionally disrupted parents. However, several lines of evidence suggest that rings are young.

First, the estimated particle lifetime in the faint rings (like Jupiter's gossamer rings) is very short, only $0.1 - 10$ ky; unless replenished, they would disappear on a comparable timescale. Indeed, the available observations indicate that particles in these rings are resupplied by satellites, like Amalthea and Thebe. A less obvious situation occurs for big rings, especially those around Saturn, whose mass is about equal to the mass of Mimas, 400 km across. The expected lifetime for such a large object against collisional disruption ($\simeq 1$ Gy) is close to the age of the solar system; in this case the assumption of ring formation by its break up could be debatable. It should be noted, however, that this depends on the (inadequately known) flux of sufficiently large impactors; since the lifetime is approximately proportional to the square root of the size, a 4 km size satellite, still able to feed smaller rings, survives disruption only for about 0.1 Gy. An independent argument comes from the pure ice composition of Saturn's main rings, despite the steady fall of interplanetary debris, presumably endowed with carbonaceous and silicate material. Were rings primordial, they should be darkened, which is not supported by observations (their local variations in composition may well be due to differentiation of the parent body). Moreover, the existence of embedded moons (e.g., Pan) and close by moons expelled by the resonant torques would be difficult to explain; these moons make sense if they are merely the largest remaining pieces of a shattered progenitor.

The general consensus is that rings continuously reconstitute themselves and material is being recycled with the ringmoons. Satellites gradually sweep up the particles which have survived removal by collisional and dynamical processes, and subsequently resupply them in energetic impacts and tidal disruptions.

Dust around terrestrial planets? This emerging synthesis also appears to explain why the inner planets are ringless, at least as far as substantial systems are concerned: they lack suitable satellites to provide the ring material. The Earth's Moon is too big and too far, and any micron-size dust that escape its surface is typically stripped away by solar gravitational and radiation forces. The observed particles near the Earth are basically a flux of micrometeoroids on unbound orbits that evolve in the solar system due to the Poynting-Robertson drag. The only candidate for a circumplanetary dust belt is Mars, with its two tiny satellites. This question has been repeatedly studied; observations have failed so far to detect it, but theoretical arguments and simulations suggest tenuous Martian rings, with an optical depth less than 10^{-8}. The Japanese spacecraft Planet-B (Nozomi) should investigate this possibility with sensible dust detectors at its arrival to Mars in 2004.

14.8 * From planets to boulders

Due to the basic properties of two fundamental physical interactions – the electromagnetic force, responsible for the solid state strength, and gravitation – an important transition occurs in the minor bodies of the solar system. For sizes smaller than a critical threshold, the shape of bodies can be defined as "irregular", and mainly depends on their origin and history. On the other hand, at larger sizes and over long times self gravity can overcome the rigidity of the material and, if minor surface elevations are neglected, they fit figures of gravitational equilibrium.

Equilibrium shapes. As discussed in Chs. 3 and 4, these figures are different, depending on whether the object can be viewed as isolated or as a satellite distorted by the tides of its primary, and, of course, depend on the internal density distribution of the body. While for a star like the Sun the latter factor is very important (since the central density is ≈ 100 times larger than the mean density), for most small solar system bodies the internal density gradient is low. Gravitational compression of the interior is limited and, even in cases of compositionally differentiated interiors, the density cannot decrease by more than a factor 2 or 3 from the core to the outer layers. Thus, as a first approximation, we can apply the classical theory of the equilibrium shapes of homogeneous fluid bodies, which specifies the shape parameters, once the density and the spin rate of the object are known. A generalization to equilibria of cohesionless solid bodies is also available and does not quantitatively change the conclusions. In all cases of interest we assume ellipsoidal equilibrium figures; if $a \geq b \geq c$ are the three semiaxes, the shapes are unambiguously determined by the ratios c/a and b/a (both ≤ 1).

For isolated objects the relevant equations have been derived in Sec. 3.4. We recall that as the angular momentum increases, the initial, almost spherical shape becomes more and more flattened. A threshold is reached beyond which the preferred equilibrium shape is no more an axisymmetric, but a triaxial ellipsoid. Finally, still higher rotational angular momenta are not consistent with the equilibrium of a single body, and fission into a binary or multiple system is expected to occur. As we shall discuss later, these results can be applied to the shapes of the asteroids. The shape of small natural satellites, on the other hand, can be studied by using the theory of Roche ellipsoids summarized in Sec. 4.5. If the rotation of the satellites is synchronized with their orbital motion (this is the case for all but the most distant moons), both the spin rate and the rotation angular momentum depend on the orbital radius. Therefore, for decreasing orbital radii, the shapes become more and more elongated in the radial direction until, beyond the Roche limit, no equilibrium figure is any more possible (see Sec. 14.7).

Non-equilibrium shapes dominated by the tensile strength. All these equilibrium solutions are independent from the previous history. In general, however, the elastic model for the body's shape, especially when shear stresses occur, retains an important record of the past, in particular when dominated by large collisional events or close encounters with planets, and depends on implicit assumptions about the existence and the shape of the initial stress-free state. This becomes important for small objects. First, they have large enough solid state strength and small gravitational field to prevent relaxation toward equilibrium. Secondly, most of them have undergone a heavy bombardment by different populations of projectiles, so that the surviving small bodies are probably the outcome of a process of collisional fragmentation. The expected resulting shapes are rugged and irregular, like those of the fragments which can be produced in the laboratory by high-velocity impact experiments.

The critical size for the transition described above can be estimated with the *strength* S (Sec. 1.4), the stress at which the material undergoes breakup or deformation; one requires that the maximum height $h_{\max}$ of the ground of a celestial body be such that the corresponding local gravitational stress is of order S. We have then

$$S \simeq \varrho \, g \, h_{\max} \simeq 4\pi \, G \varrho^2 \, R \, h_{\max}/3 \, , \tag{14.8}$$

where ϱ, g and R are the density, the surface gravity and the radius of the body, assumed to be nearly spherical and homogeneous. Let us define a "regular" body as one whose altitude differences are lower than, say, 10% of its radius. This implies that at the transition $h_{\max} = R_{\mathrm{cr}}/10$, hence the critical radius is

$$R_{\mathrm{cr}} = (15\, S/2\pi\, G\varrho^2)^{1/2}\ . \tag{14.9}$$

It is difficult to give a reliable theoretical estimate of S for the materials forming the small bodies of the solar system, because it depends on the detailed microscopic structure (the electrostatic forces keeping together the molecules, the density and distribution of faults and cracks, etc.).

An upper estimate of S is provided by the latent heat of fusion of the material, the energy needed to destroy the lattice at constant volume; since then the material becomes liquid and flows easily, it is plausible that a lower energy is sufficient to cause either disruption or "fluid" behaviour over long timescales (i.e., with very high viscosity). The heat of fusion is of order $kT_{\mathrm{m}}/m_{\mathrm{mol}}$, where T_{m} is the melting temperature and m_{mol} the molecular mass; hence $S < kT_{\mathrm{m}}\varrho/m_{\mathrm{mol}}$. For materials of geophysical interest (e.g., ice, metals and silicates) $kT_{\mathrm{m}}/m_{\mathrm{mol}} \approx 10^9$ to 10^{10} erg/g. Laboratory values of S for terrestrial rocks, ice and meteorites vary between 10^5 and 10^9 dyne/cm^2. With densities of 1 to 3 g/cm^3, from eq. (14.9) we can only conclude that R_{cr} must be in the range from 10 to 1,000 km.

However, the real situation may be still more complicated. Numerical models indicate that initially intact bodies soon became densely fractured and porous, as a result of non-disruptive impacts. As a result, kilometre-size and larger objects likely have a strength substantially lower than that of small laboratory samples of similar composition, consisting of weakly consolidated aggregates accumulated under small gravitational self-compression. The extreme case is that of a *rubble pile*, a collection of small fragments held together only by their weak mutual gravitational attraction. In this case the strength is very low and large departures from equipotential surfaces are again not possible; still its anisotropic structure significantly differs from a fluid and the approximate equilibria may be poorly represented by fluid models. At a decametric size another transition is expected: mutual gravity of the rubble pile components may become too weak to keep the object together against the increasing number of microimpacts, and only monolithic, strength-dominated blocks can survive, likely with some degree of porosity. These objects would resemble the largest meteoroids found on Earth.

Can long-term creep change some of these conclusions? For a viscous body the relaxation timescale of a non-equilibrium bulge of size comparable with the radius R is

$$\tau = k\eta/G\varrho^2 R^2\ . \tag{14.10}$$

significantly changed their orbits and formed not very far from their present locations; but observations of extrasolar planets suggest instead that important orbital changes might have taken place. It is frustrating that up to now no definite answer to this question has been found, although powerful computers and efficient software tools have broken a long-lasting standstill.

General considerations. The motion of planets is usually studied as a gravitational $(N + 1)$-body problem, in which the only forces are Newtonian point-mass attractions. The interacting bodies are the nine planets and the Sun; but N can be less than 9 for particular problems (e.g., Pluto has negligible dynamical effects on other bodies and may be discarded). The systems of natural satellites may also be described as a separate $(N + 1)$-body problem. Other bodies, in particular asteroids and comets, do not affect the motion of the rest and have negligible interaction among themselves; each of them can be treated as a test mass[1]. The point-like assumption for planets is mainly violated by rotational and tidal deformations; if J_2 gives their fractional value, the gravitational field at a distance r has a fractional change of $\approx J_2 (R_p/r)^2$, generally negligible (in the Appendix of Ch. 2 such finite-size effects have been discussed at a formal and general level). Non-gravitational forces are important only for small bodies (Sec. 15.4); relativistic effects introduce fractional corrections to Newton's force of order $GM_\odot/(c^2 r)$, which never reach $\simeq 10^{-7}$, even for Mercury. The mass of the Sun fractionally decreases due to emission of radiation and the solar wind by $\approx 10^{-13}$ per year, a negligible effect. Except for special cases (cometary motion), galactic tides are also negligible.

A further simplification of the problem comes from the fact that the mass of the Sun is much larger (by a factor $> 10^3$) than the mass of any planet. As discussed in Chs. 11 and 12, this allows to define a lowest order solution, in which every planet moves around the Sun in a Keplerian orbit defined by six orbital elements: semimajor axis, eccentricity, inclination, longitude of perihelion, longitude of the ascending node and mean longitude in orbit at the initial epoch. They are denoted by $(a = 1, .., 6)$

$$\mathbf{c}_i = c_{a,i} = (a_i, e_i, I_i, \varpi_i, \Omega_i, \epsilon_i) \; ; \tag{15.1}$$

$i, j = 1, .., N$ label the planets. The last element ϵ, as discussed in Sec. 12.2, replaces the time of a passage at pericentre to take into account secular changes in the semimajor axis; it is also valid at low inclination. The position of a planet on its orbit is usually determined by $\chi = \int n \, dt$ (eq. (12.20)) or by the

---

[1] Exceptionally, gravitational perturbations by large asteroids are considered when a smaller body has a close encounter with one of them (with this method the masses of about a dozen have been determined), and for the purpose of very precise ephemerides, like the lunar motion or orbits of radar-tracked, near-Earth asteroids. In this latter case, the biggest asteroids with their estimated masses are considered individually, and the collective effect of the remaining mass in the whole belt is modelled as that of a torus.

mean longitude λ (eq. (11.35)). When the eccentricity and/or the inclination are small, in place of (e, ϖ) and (I, Ω) it is convenient to use the non-singular elements (h, k) and (P, Q), respectively (see eq. (12.50)). For timescales not much longer than the periods of revolution, the Keplerian elements are approximately constant (while λ_i and χ_i are linear functions of time).

If we exclude close approaches and consider comparable distances, the ratios of the mutual gravitational forces between planets and those due to the Sun are roughly of the same order of magnitude $O(\mu) \approx 10^{-3}$ as the ratio of a planetary mass to the Sun's. We can then set up a formal approximation with μ as a smallness parameter, consistently with its definition in (12.75). Using perturbation theory in Lagrange's form (Sec. 12.2), we obtain for the rate of change of the elements of the j-th planet due to the i-th planet

$$\left(\frac{d\mathbf{c}_j}{dt}\right)_i = \mu F_{ji}(\mathbf{c}_j, \lambda_j; \mathbf{c}_i, \lambda_i) . \tag{15.2}$$

Since the gravitational force depends linearly on the sources, the total change in $\mathbf{c}_j$ is obtained by summing over i. Formally the Keplerian elements are constant to $O(\mu^0)$ and their changes, obtained by direct integration of eq. (15.2), are correct to $O(\mu)$ for times less than $O(1/\mu)$ revolution periods. As discussed in Sec. 12.5, an iteration procedure can be set up to any order; however, the required Fourier and Taylor expansions soon make the calculations intractable. Since the age of the planetary system is $\approx 1/\mu^3$ periods of revolution, at least a third-order theory is required to deal with the stability problem.

Averaging. As mentioned in Sec. 12.5, we are not interested in fast-varying components of the elements, at timescales comparable to the revolution periods. To obtain *secular perturbations* we construct the average of F_{ji} over the torus $(0 < \lambda_i \le 2\pi, 0 < \lambda_j \le 2\pi)$, which eliminates the components with the orbital periods of the planets i and j, and consider the formal set $\mathbf{c}'_j$, hereafter called *mean orbital elements*, which satisfy the simpler equations

$$\left(\frac{d\mathbf{c}'_j}{dt}\right)_i = \mu \langle F_{ji} \rangle = \frac{\mu}{(2\pi)^2} \int_0^{2\pi} d\lambda_j \int_0^{2\pi} d\lambda_i \, \mu F_{ji}(\mathbf{c}'_j, \lambda_j; \mathbf{c}'_i, \lambda_i) . \tag{15.3}$$

In this method, as discussed in Sec. 12.5, resonances, which occur when the ratio of a pair of mean motions is approximately equal to the ratio of two (small) integers, are neglected; in this case the resulting perturbation may accumulate faster than the ordering in μ indicates. The near-resonance between Jupiter and Saturn is an important example, discussed below.

In Lagrange's formalism (Sec. 12.2), the averaging can be applied to the perturbing function $\mathcal{R}_{ji}$, given in Cartesian coordinates by eq. (12.78); eqs. (12.49) then give the rate of change of the orbital elements. Note that, according to eq. (12.78), $m_j \mathcal{R}_{ji}$ is not symmetric in the two indices and is not the interaction energy between the i-th and the j-th planet; it has an extra term, the "indirect" perturbation. However, we can show that it does not give rise to first-order secular effects. For the true anomaly f of a Keplerian motion, using the angular momentum integral h, we have $\eta a^2 d\lambda = r^2 df$ ($\eta = \sqrt{1-e^2}$; Sec. 11.3). Now the average of the indirect term in $\mathcal{R}_{ji}$ is proportional to

$$\left\langle \frac{r_j \cos S}{r_i^2} \right\rangle = \frac{1}{(2\pi)^2 \eta_i \eta_j} \int_0^{2\pi} df_i \int_0^{2\pi} df_j \frac{r_j \cos S}{r_i^2} \left(\frac{r_i}{a_i}\right)^2 \left(\frac{r_j}{a_j}\right)^2,$$

where S is the angle between the two planets. The outer integral does not depend on r_i and trivially vanishes. To lowest order, the required average $W_{ji} = m_j \langle \mathcal{R}_{ji} \rangle$ (symmetric in its indices) has a simple physical interpretation: it is the gravitational interaction energy between two rings of matter obtained by spreading both planetary masses along their Keplerian orbits.

We also recall *Lagrange's theorem*: since $da_i/dt \propto \partial \mathcal{R}/\partial \epsilon_i = \partial \mathcal{R}/\partial \lambda_i$, for timescales $O(1/\mu)$ orbital periods the mean semimajor axes a'_i are constant. Of course, this does not solve the stability problem: not only it holds for times much shorter than the age of the solar system, but it only refers to the semimajor axes. The mean eccentricities do undergo significant secular perturbations, which may lead to collisions.

Secular planetary theory due to Lagrange. For an arbitrary eccentricity, the integrals over the mean longitudes required in the average $\langle \mathcal{R} \rangle$ are not elementary, due to the transcendental Kepler's equation (11.28). However, we can use the fact that in the solar system eccentricities (and inclinations) are small, in fact, roughly of order $\mu^{1/2}$; only the two extreme planets, Mercury and Pluto, make an exception, but they are small and do not significantly affect the other ones. We can therefore expand the disturbing function $\mathcal{R}$ in powers of eccentricities and inclinations (or, rather, the non-singular elements h, k, P, Q; Sec. 12.2) with a single formal power series in the smallness parameter μ. The shape of a Keplerian orbit is defined by Lenz's vector $\mathbf{e}$, determined by h and k; its disposition in space is given by the unit vector $\mathbf{N}$ normal to the orbital plane, determined by P and Q. The average interaction energy $W_{ij} = m_j \langle \mathcal{R}_{ij} \rangle$ must be invariant under the rotation group and under a change of sign of $\mathbf{N}_i$; hence it can depend only on the scalar quantities $(\mathbf{e}_i \cdot \mathbf{e}_j)$, $(\mathbf{e}_i \cdot \mathbf{N}_j)^2$ and $1 - (\mathbf{N}_i \cdot \mathbf{N}_j)^2$ (in the last two the case $i = j$ is irrelevant; the last combination has been chosen because it is small for the interesting case in which the orbits are almost coplanar). For a given pair (i, j), none of these quantities is linear in both the eccentricities and the inclinations, and only three are quadratic: $\mathbf{e}_i \cdot \mathbf{e}_j$ for $i = j$ and $i \neq j$, and $1 - (\mathbf{N}_i \cdot \mathbf{N}_j)^2$ for $i \neq j$. The exclusion of linear terms is also borne out by direct arguments: a term $\propto e_i$ would describe an interaction between a circular ring j and an elliptic ring i; but, since a change of sign in the eccentricity is equivalent to the irrelevant interchange of pericentre and apocentre,

15. Dynamical evolution of the solar system

it cannot be present. A term $\propto I_i$ would also give an impossible interaction between two circular rings which changes sign under their interchange.

The first relevant terms in the expansion of W_{ji} are those quadratic in h, k, P, Q. They are linear combinations of

$$\mathbf{e}_i \cdot \mathbf{e}_j = h_i h_j + k_i k_j ,$$

with coefficients B_{ij} for $\mathbf{e}_i \cdot \mathbf{e}_j$ ($i \neq j$) and $B_{ii} = B_{jj}$ when $i = j$; and of

$$1 - (\mathbf{N}_i \cdot \mathbf{N}_j)^2 = 2(P_i P_j + Q_i Q_j) - (P_i^2 + P_j^2 + Q_i^2 + Q_j^2) ,$$

with coefficient C_{ij}; these relations hold up to quadratic terms. The B and C coefficients depend only on the masses (linearly) and the semimajor axes, and are available in the literature. We end up with the expression (for $i \neq j$)

$$W_{ij} = W_{ji} = -\left[B_{ij}\left(h_i h_j + k_i k_j\right) + \frac{1}{2} B_{ii}\left(h_i^2 + h_j^2 + k_i^2 + k_j^2\right) + \right. \tag{15.4}$$
$$\left. + C_{ij}\left(P_i P_j + Q_i Q_j\right) + \frac{1}{2} C_{ii}\left(P_i^2 + P_j^2 + Q_i^2 + Q_j^2\right) \right].$$

Lagrange perturbation equations (12.49), linear and homogeneous in the nonsingular elements (12.51), can be written in terms of the matrixes $\mathsf{B} = B_{ij}$, $\mathsf{C} = C_{ij}$:

$$\frac{dh_j}{dt} = \frac{1}{m_j n_j a_j^2} \sum_i B_{ij} k_i , \quad \frac{dk_j}{dt} = -\frac{1}{m_j n_j a_j^2} \sum_i B_{ij} h_i , \tag{15.5}$$

They fulfil the conservation law

$$\sum_j m_j n_j a_j^2 (h_j^2 + k_j^2) = \sum_j m_j n_j a_j^2 e_j^2 = \text{constant} . \tag{15.6}$$

Similar equations hold for the other pair P, Q. B and C are symmetric. Their diagonal elements $B_{ii} (= B_{jj})$ and $C_{ii} (= C_{jj})$ are proportional to the other mass m_j (or m_i) and describe the effect that the eccentricity and the inclination of m_i have on its interaction with m_j. Eq. (15.6) shows that eccentricities and inclinations are bounded. It is convenient to use an N-dimensional vector notation $\mathbf{h} = h_i$, etc. and define

$$h_i' = h_i (m_i n_i a_i^2)^{1/2} , \quad k_i' = k_i (m_i n_i a_i^2)^{1/2} ,$$
$$B_{ij}' = B_{ij}/[(m_j n_j a_j^2)(m_i n_i a_i^2)]^{1/2} , \text{ etc.};$$

then the Lagrange equations (15.5) acquire the simple Hamiltonian form

$$\frac{d\mathbf{h}'}{dt} = \mathsf{B}' \cdot \mathbf{k}' , \quad \frac{d\mathbf{k}'}{dt} = -\mathsf{B}' \cdot \mathbf{h}' , \tag{15.7}$$

and similarly for **P**′ and **Q**′, with **C**′ in place of **B**′. From their definition, it is easy to see that the primed vectors have the dimension of a frequency and the order of magnitude of the product of μ and the generic mean motion n. We also note that **h**′, **k**′, **P**′ and **Q**′ are N-dimensional harmonic oscillators, fulfilling

$$\frac{d^2\mathbf{h}'}{dt^2} = -\mathbf{B}'^2 \cdot \mathbf{h}', \text{ etc.} \tag{15.8}$$

Since $\mathbf{B}'^2$, the square of a symmetric matrix, is also symmetric, it has real and positive eigenvalues g_i; taking its eigenvectors as basis, the general solution of eqs. (15.5) has the form (called a *Lagrange* or *linear solution*)

$$h_i = \sum_j R_{ij} \sin(g_j t + \beta_j), \quad k_i = \sum_j R_{ij} \cos(g_j t + \beta_j), \tag{15.9a}$$

$$P_i = \sum_j S_{ij} \sin(s_j t + \gamma_j), \quad Q_i = \sum_j S_{ij} \cos(s_j t + \gamma_j). \tag{15.9b}$$

R_{ij}, S_{ij} are obtained with a unitary transformation from the arbitrary amplitudes of the eigenmodes; the phases β_j, γ_j are also arbitrary constants, both determined by the initial conditions. g_j and s_j, sometimes termed eigenfrequencies of the planetary system, are of order μn, typically corresponding to periods between 50 ky and a few My. The solution requires the amplitudes R and S to be small, formally $O(\mu^{1/2})$. It should also be noted that in the eigenmode description the indices i, j still range from 1 to N, but label the modes, not the planets.

To understand its qualitative features, a simple geometric description is useful: in a fixed reference plane the vectors with components (k_j, h_j) and (P_j, Q_j) undergo epicyclic motions, vectorial sums of N circular uniform motions with different angular velocities; however, one of the eigenfrequencies for (P, Q) is zero, owing to the constraint provided by the conservation of the total angular momentum **L**, whose ecliptic component is a linear combination of the inclinations. Were each planet dominated by a single perturbing body m', each of them would have a nearly circular motion with a dominant frequency f, superposed to small epicycles; f can associated with m'. In the solar system Jupiter's dominance is not very strong; there are several cases of perihelia (and nodes) which do not have a dominant frequency. The Earth's perihelion is one of them, so that its eccentricity changes in a quasi periodic, but complex way (Fig. 15.1a), thus contributing to the complex variations of the solar constant, crucial for palaeoclimatology.

Another case in which the simple description with one dominating frequency fails is that of Uranus. Its perihelion is "locked" to Jupiter's perihelion: i.e. they share the main epicyclic frequency $g_5 \simeq g_7$, and the angle $\varpi_J - \varpi_U$ between their pericentres librates around $180°$ with an amplitude of

15. Dynamical evolution of the solar system

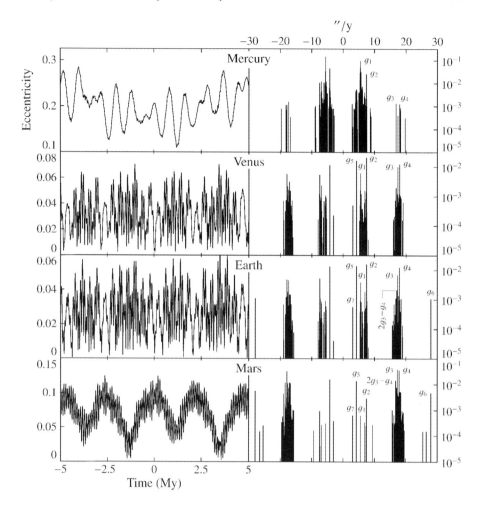

Figure 15.1a. Planetary eccentricities (left) and spectral amplitudes of h and k (right) numerically obtained by J. Laskar from the second order, secular Lagrange equations using a large, but finite number of eigenfrequencies. Those predicted by the linear theory (see the text) are identified. The large number of spectral lines, and their crowding around particular values, suggest an aperiodic motion. Data from J. Laskar, *Astron. Astrophys.* **198**, 341 (1988).

about 70° and a period of 1.1 My; thus Uranus' eccentricity oscillates with the same period (Fig. 15.1b). This resonance (Sec. 15.2) must be treated with a higher-order secular theory. At the linear level a resonance appears as multiple eigenvalue B'^2 or C'^2 and produces a term $\propto t\exp(igt)$; but even a small frequency separation indicates long-term effects.

Beyond the linear theory. How can the Lagrange solution be improved? Firstly, we could include in the perturbing potential the terms of degree 4 in

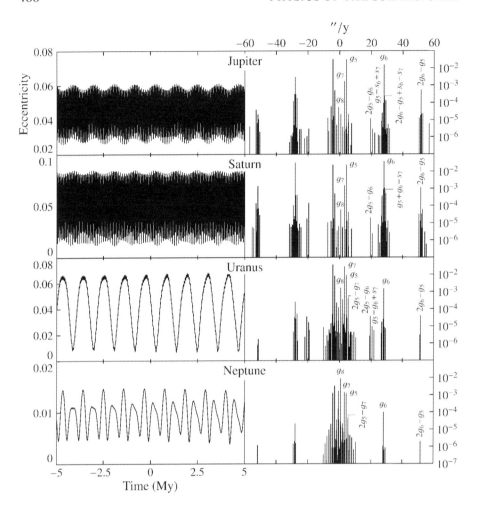

Figure 15.1b. The same as in the previous figure, but for the outer planets. Jupiter's and Saturn's spectra are dominated by the eigenfrequencies g_6 and g_5; their difference corresponds to fast variations with the approximate period of 55 ky.

the eccentricities and inclinations; and second, since $e^2, I^2 = O(\mu)$, we must at the same time develop a second-order theory by substituting the complete first-order solution in the right-hand side of the perturbation equations. Such a theory (of order 2 and degree 4) has indeed been computed both for the planets and the asteroids; however, the number of Fourier components used to describe the motion of the perihelia and nodes is very large. To the eqs. (15.5) we must add a right-hand side of degree 3 and the result will contain all the combinations of three fundamental frequencies. Only some of these terms have significant amplitudes, but it is impossible to know a priori which ones.

For a third-order and sixth-degree secular theory, the difficulties become unsurmountable, as the total number of Fourier harmonics with combinations of five fundamental frequencies chosen among N grows to billions. A large number of small divisors arises whenever some of the combinations of frequencies is close to zero; therefore, even the ordering by magnitude is not easy, since the amplitudes of some third-order terms may be larger than many of second order. Moreover, the advantage of having constant semimajor axes is lost. As a conclusion, we can state that secular perturbation theories can give a good approximation to the planetary orbits for times up to $1 - 10$ My, but not much longer.

Figs. 15.1 show results of the second-order theory of planetary secular motion spanning the full interval of 10 My. Though the motion still looks fairly regular and nearly periodic, the Fourier analysis for the (h, k) pair, which at the linear level has only $N = 8$ well separated lines (eq. (15.9a)), now shows a much more complicated pattern, especially for the terrestrial planets. A large number of spectral lines emerge, some of which can be identified as eigenfrequencies of the linear theory and their combinations; but, more important, they tend to crowd in particular intervals, suggesting effects at longer timescales, or even secular. We conclude that the stability of the planetary system cannot be fully addressed with the second-order secular theory. Even if one succeeded in constructing the third-order secular theory, the full complexity of the planetary motion could not be grasped. Short-period terms, a priori discarded in the secular approach, at the next, non-linear order may produce resonances and enhanced, long-term effects on the motion. The most famous approximate resonance in planetary motion is the 5/2 Jupiter-Saturn case; but a higher-order three-body resonance between the giant planets may have equally profound effects (Sec. 15.2).

Numerical simulations. Given the difficulties faced by analytical theories, a more direct approach by numerical integration is attractive and may seem very simple: a finite difference approximation to the differential equations of motion is constructed, with a fixed or variable step size, and initial conditions matching the observations. However, there are at least three classes of problems that must be addressed:

- Can a numerical integration covering a time span not much shorter than the age of the solar system (i.e., at least $\simeq 1$ Gy) be completed with a reasonable computing time?

- Can the output of a numerical integration be compared with analytical theories? Can the interesting information be extracted from a huge numerical output (e.g., orbital elements as functions of time)?

- Can the numerical error be kept small enough to obtain the required accuracy?

The last decade witnessed a great advance for the first class of problems. Computer power has steadily increased and hardware systems dedicated to planetary motion have been built; more important, special numerical and analytical methods have been developed, allowing further speeding up in the computations.

In the Appendix we discuss symplectic integrators, with which the numerical integration is about an order of magnitude faster than conventional schemes. Here we briefly present the related powerful method of *symplectic maps*[2]. The planetary Hamiltonian $H = H_K + H_{\text{int}}$ is the sum of the – time-independent – solar Keplerian part and the interaction Hamiltonian H_{int}. Being interested only in the long-term evolution, we are not concerned with the high-frequency part of the latter; as discussed earlier, because of the problem of resonances and higher-order effects, averaging over mean longitudes (15.3) is not entirely adequate. Note, however, that if high-frequency components of the interaction are irrelevant, it should not matter much if H_{int} is multiplied by a fast-varying factor with a simple expression. A possible simple choice is

$$\Phi(\Omega t) = \sum_{k=-\infty}^{\infty} \cos(k\Omega t) = 2\pi \sum_{k=-\infty}^{\infty} \delta(\Omega t - 2\pi k) . \quad (15.10)$$

We have used here the trigonometric expression for Dirac's δ-function. Ω is a typical frequency, larger than the relevant mean motions. The mapping Hamiltonian

$$H_{\text{map}} = H_K + \Phi(\Omega t) H_{\text{int}} \quad (15.11)$$

replaces the true Hamiltonian H. The interaction now acts instantaneously every period $2\pi/\Omega$ and, since it depends only on the coordinates of the planets, produces a discontinuous and easily realizable change in their velocities, with no change in positions. Between these times $t_k = 2\pi k/\Omega$ the motion is strictly Keplerian, for which efficient numerical tools are available; dynamical mappings of this kind between initial and final conditions are of much use in chaos theory. The orbital integration can now swiftly proceed, alternating instantaneous interactions and Keplerian propagation. Obviously, this technique requires care and numerical experimentation; e.g. the arbitrary frequency Ω should not induce fake resonances.

With symplectic mapping, it is thus possible to carry out a numerical simulation of the whole planetary system over a timescale close to its lifetime ($\simeq 1$ Gy); and for the secular system only (eq. (15.3), generalized with the

---

[2]Symplectic spaces are the mathematical "environment" for Hamiltonian dynamics. They have an even number $2N$ of dimensions and the coordinates of a point z are represented by (z'_i, z''^i); they possess a bilinear, antisymmetric form $(dz|dw) = -(dw|dz)$ for any two infinitesimal vectors dz, dw at a point; in the canonical representation $(dz|dw) = dz''^i dw''_i - dw''^i dz''_i$.

addition of the second order terms in μ), a timescale longer by an order of magnitude (Fig. 15.2).

As to the second class of problems, the important issue is not to know the position of the planets at all times – this is impossible, due to chaotic nature of the motion (Sec. 15.2 and below) – but whether their orbital elements vary in a quasi-periodic or monotonic way on a physically meaningful timescale ($\simeq$ Gy for the planetary system), which are the longest periods involved and, in particular, to which extent secular perturbation theories correctly predict the most important periods or if there are new dynamical features. To answer these questions one can generate a so-called *synthetic secular perturbation theory*, a model of the long-term evolution of planetary orbits obtained from numerically computed positions, neglecting irrelevant, short-period effects. We need for this a method of *digital filtering*: a low-frequency pass filter suitable to the output and to the predicted spectrum of the data (for examples, see Sec. 19.2). In effect, since the complete output would require an excessive computer disc space, the filter should be applied "on-line" during the numerical integration.

The main difficulty in addressing the last class of problems does not arise from the inaccuracy inherent in the replacement of differential equations with a finite difference scheme – the *truncation error*, well understood and reliably predictable. Computers do not perform arithmetic operations between real numbers; they can store, hence manipulate, only integers (normally up to 2^{63}). Real numbers are approximated with a limited number of decimal digits, with the consequent *rounding-off error*. Since usually the extra digits are just dropped out, the error is systematic, and after a very large number of operations may cause problems. For example, if the rounding-off error in an integration step has a relative size η_s, after N_s steps the accumulated error is of order $N_s^2 \eta_s$: this corresponds to the fact that a systematic error at each step in one of the mean motions causes a longitude shift accumulating quadratically with time, faster than in the random case. In an orbital period ≈ 100 steps are used; if $\eta_s \approx 2^{-48} \simeq 3 \times 10^{-15}$, the integration fails after $1/(100 \sqrt{\eta_s})$ periods, a few million years for the outer planets and about 10^5 y for the inner planets. However, again new integration methods allowed to significantly reduce the error budget down to an almost linear growth with N_s, so that the rounding-off does not pose a problem for planetary simulations even after $\simeq$ Gy.

A deeper problem appeared when the first long-term numerical results for the planetary system became available: all of them indicated a strong sensitivity on the initial conditions, especially for the inner planets and Pluto. In Sec. 13.3 we introduced the concept of *Lyapunov exponent* κ_L (eq. (13.20)) to describe non-linear instability and to quantify the logarithmic divergence of initially neighbouring solutions of dynamical systems; a positive exponent is the main indication of "chaos". For the inner planet orbits $1/\kappa_L$ was found to be about 5 My; an error of $\simeq 15$ m in the initial position of the Earth's centre of mass (about the current accuracy) results in an error of $\simeq 150$ m after about 10 My and it grows to 1 AU in only 100 My. This implies that practically the ephemerides of the inner planets cannot be constructed over a period beyond $\simeq 100$ My. The Laplacian ideal of exact determinism must be given up. However, this affects the angular variables (mean anomaly and longitudes of the node and the pericentre), but not the conjugate variables semimajor axis, eccentricity and inclination. For the sake of the stability problem, we are not interested in the actual positions of the planets in 1 Gy, but in changes in the inclination and the shape the orbit, which may lead to close encounters; this should not be directly affected by a positive Lyapunov exponent.

Planetary motion over the 100 My timescale.– The motion is well approximated in gross features by secular perturbation theory. In particular, changes in eccentricity and inclination are controlled by first-order effects and there is no reason to doubt its reliability to $O(\mu)$; over a timescale $O(\mu^2)$ additional terms do appear, but a basic quasi-periodicity persists. Yet, some subtle and unexpected features in the evolution of semimajor axes appear. We know that to first order they are secularly constant (Lagrange's theorem). At the second order the situation is more complicated; difficult attempts to get an analytical proof of their long-term stability were soon shadowed by doubts arisen from numerical experiments. Uranus and Neptune, as an example, exchange some 10^{-5} of their orbital energies, with exactly the same 1.1 My periodicity of the relative oscillation of Jupiter's and Uranus' perihelia. The other planets show similar effects. These results indicate that at higher order very-long term changes in semimajor axes may build up, although with relatively small amplitudes; no definitive conclusion about their secular stability is thus available. All numerical studies confirm the chaotic character of planetary motion, as indicated by the $\simeq 5$ My Lyapunov timescale for the inner planets, and $\simeq 20$ My for Pluto. Chaos is, in general, associated with resonant motion (Sec. 15.2); several weak resonances, both secular and in mean motion, have been suggested as a mechanism for its arising.

Planetary motion over the Gy timescale.– In very long term numerical simulations of the planetary system, chaos prevents a deterministic evolution and one should be content with a statistical analysis of stability; a result does not necessary corresponds to the real solar system, but to neighbouring initial conditions.

The longest simulation of the planetary orbits currently available covers 25 Gy and represents a numerical solution of the second-order secular system (generalizing (15.3); Fig. 15.2). Its purpose was to study the stability of planetary orbits as an abstract dynamical system. The results of Fig. 15.2 confirm earlier conclusions about the difference between giant and inner planets (Pluto is not included, since it is dynamically similar to Kuiper-belt bodies, rather than to the planets; see also Sec. 14.1). The giant planets are basically

15. *Dynamical evolution of the solar system* 471

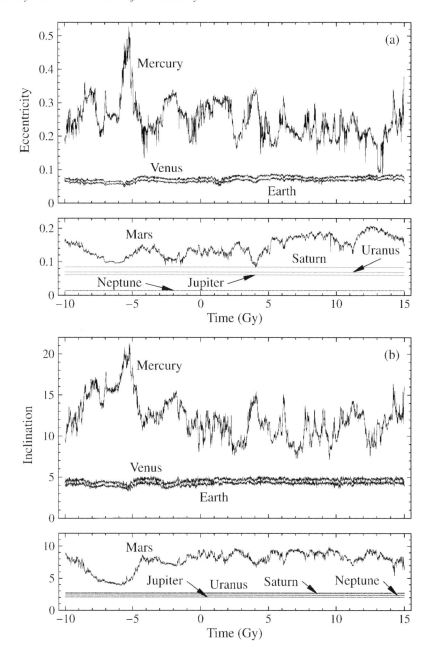

Figure 15.2. Evolution on a very long term of planetary eccentricities (a) and inclinations (in degrees) (b); maximum values of both elements over a running 10 My window are shown. The orbits of the outer planets are stable over the Gy time span, while the inner planets show a significant chaotic behaviour. Adapted from J. Laskar, *Astron. Astrophys.* **287**, L9 (1994); data kindly provided by J. Laskar.

stable; in contrast, the inner planets, especially Mars and Mercury, are violently irregular. In another series of numerical tests, Mercury was even shown capable to escape from the solar system on a Gy time span. Still, this is the most idealized case; with higher-order secular terms a complete description of the motion, including short-period terms, is expected to be even more complex. Preliminary insights were obtained with a numerical experiment for the complete planetary system over $\simeq$ Gy. The main lesson from this and other works is that the planetary motion is only *marginally stable*; dramatic events, like the loss of a planet or a mutual collision, are not definitely excluded in the future. A marginal stability of the solar system fits our present understanding of its origin (Ch. 16): the aggregation of solid bodies is assumed to result in the formation of the present planets and a huge number of planetary embryos, with mass between the Moon's and Mars'. This initial phase was strongly unstable and evolved toward a less unstable state through collisional coalescence of bodies and gravitational scattering. Likely this process stopped as soon as marginal stability was achieved.

Test bodies in planetary fields: secular resonances and proper elements. We now apply the linear secular theory to a test body under planetary perturbations. This has important applications to minor components (asteroids, Kuiper-belt objects and comets). To the *known* planetary solution (15.9), with eigenfrequencies g_i, we add a test body (denoted with the index $_0$) with a small mass m, semimajor axis a and mean motion n. Its interaction energy with the i-th planet W_{i0} is of order Gmm_i/a, and so are the matrix elements B_{00} and $B_{0i} = B_i$; we also introduce the quantities $g = B_{00}/(mna^2)$, $B'_i = B_i/(mna^2)$, of order μn. With this notation the Lagrange perturbation equations (15.5) for its elements (h, k) read

$$\frac{dh}{dt} = gk + \sum_i B'_i k_i, \quad \frac{dk}{dt} = -gh - \sum_i B'_i h_i . \quad (15.12)$$

They describe a forced harmonic oscillator, with solution

$$h = e_p \sin(gt + \beta) - \sum_i \frac{b_i}{g - g_i} \sin(g_i t + \beta_i) , \quad (15.13a)$$

$$k = e_p \cos(gt + \beta) - \sum_i \frac{b_i}{g - g_i} \cos(g_i t + \beta_i) , \quad (15.13b)$$

where $b_i = \sum_j R_{ji} B'_j$ and e_p and β are arbitrary integration constants. Of course, it is valid only for small eccentricity ($e = \sqrt{h^2 + k^2} \ll 1$). A similar solution can be obtained for the (P, Q) pair. e_p and $gt + \beta$ are the *free eccentricity* and the *free longitude of pericentre*, respectively (also called *proper elements*; Fig. 15.3). The last terms h_f and k_f in eqs. (15.13) give the *forced*

15. Dynamical evolution of the solar system

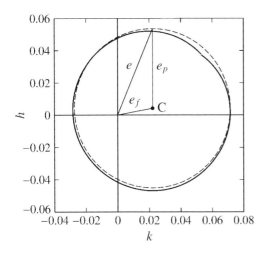

Figure 15.3 The orbit of the asteroid 1978 VP$_4$ over 20 ky in the (k, h) plane (full line); numerical data have been digitally smoothed to remove short-period perturbations. The dashed circle represents the best-fitted contribution to the proper elements; the point C is the average of the forced planetary term (k_f, h_f) and shows that some of the planetary eigenperiods are longer than the integration time. Data kindly provided by Z. Knežević and A. Milani.

eccentricity $e_f = \sqrt{h_f^2 + k_f^2}$ (for a typical asteroid in the main belt $e_f \simeq 0.05$); the corresponding polar angle ϖ_f is the *forced longitude of pericentre*; both uniquely depend on the semimajor axis of the test body. It may happen that the forced contributions prevail; in this interesting case the values of some elements of a test body, mainly moving in the field of the Sun, are not arbitrary, but determined by the whole planetary dynamics. The forced eccentricity (and inclination) is resonantly amplified when the proper frequency g is near one of the forcing frequencies g_i. These *linear secular resonances* $g - g_i = 0$ occur for a specified value of the semimajor axis of the test body (such as $a \simeq 2.1$ AU when $g \simeq g_6$), but in a more complete approach the secular resonances are located on surfaces in the space of the elements (a, e, I). It is common to denote by ν_i the $g - g_j = 0$ resonances, while the $s - s_j = 0$ resonances for the (P, Q) pair are called ν_{1j} (see Fig. 14.3). It should also be noted that the singular behaviour due to the null in the denominator may be made ineffective by higher-order and resonant terms in the disturbing function. The linear proper elements correspond to more accurate quantities used to study the long-term evolution of asteroid orbits and to classify asteroids in dynamical families (see Sec. 14.3).

15.2 *   Resonances and chaotic behaviour

A *resonance* occurs when the ratio of two periods of a dynamical system is well approximated by the ratio of two (not too large) integers, or if one of its periods is nearly matched by some external, periodic driving force, like in a forced harmonic oscillator (Problem 12.14). If the dynamics of a system of bodies like planets, asteroids or satellites depends on a linear combination of their characteristic frequencies with small integers as coefficients, the same configuration is periodically repeated; mutual gravitational perturbations play the role of a driving resonant force and their effects are strongly enhanced. This may have either a stabilizing effect, constraining any further evolution of the system to dynamical routes which preserve *resonant locking* (like the Neptune-Pluto mean motion resonance, the asteroids Hilda and Trojan, many satellite pairs, or even triplets, and the spin-orbit coupling of Mercury); alternatively, it may lead to instabilities, chaotic behaviour, close encounters, and even collisions. The latter case applies to any asteroid or small particle that originally was in, or subsequently was injected into, the *Kirkwood gaps* (Sec. 14.3), the mean motion resonances in the asteroid belt; or to ring particles formed in such sites as Cassini's division between the main Saturn's rings (Sec. 14.7). Instability mechanisms related to resonances in asteroidal orbits are especially important, because they can explain the transport toward the terrestrial planets of meteorites and near-Earth asteroids.

A mean motion resonance: Titan and Hyperion. This model is suitable to describe the resonance between the two Saturnian satellites Titan and Hyperion, but many of its features are more general. Two interacting bodies have coplanar orbits about a primary; the outer body (Hyperion) has a negligible mass, but a significant eccentricity (≈ 0.1); the inner body (Titan) has a circular orbit. In effect, both satellites have small ($\leq 0.5°$) and different inclinations to the equatorial plane of Saturn and, what is more important, Titan's eccentricity ($\simeq 0.029$) does not vanish. We neglect these complications and deal with a planar, restricted three-body problem. Most important, the orbital periods of Titan and Hyperion are in a 4/3 resonance: Titan completes four revolutions while Hyperion makes three. Since the order of the resonance is one, in an exact resonance conjunctions – when the two bodies are aligned with the primary and on the same side – repeat at the same planetocentric longitude every three orbital periods of the outer body. However, for this pair we have a constraint on the longitudes of the conjunctions (Fig. 15.4), where the gravitational pull of Titan is strongest.

If the conjunction occurs while Hyperion is moving from pericentre to apocentre, with an outward velocity component, the radial component of Titan's attraction pulls inward; energy and angular momentum are removed from Hyperion's orbit and its semimajor axis and period decrease. The effect is enhanced by the fact that the point of closest approach, where the pull is strongest, occurs somewhat earlier, at a time when Titan, which moves faster, still trails behind. As a consequence of Hyperion's speeding up, the conjunction moves towards the apocentre. If it occurs while Hyperion is moving from apocentre to pericentre, the opposite occurs: Titan's pull increases Hyperion's energy, slowing it down and moving the conjunction again towards the apocentre.

Thus, if the initial ratio of the orbital periods is close enough to 4/3, we expect a stable, pendulum-like oscillation of the conjunction alignment about the apocentre, that tends to preserve the 4/3 ratio against destroying effects,

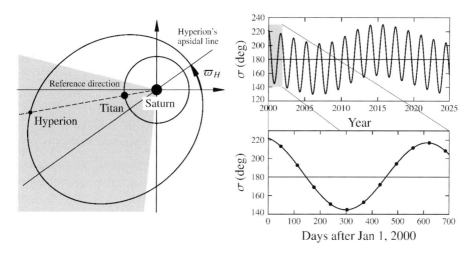

Figure 15.4. A simplified model for the analysis of the stable resonance locking between Titan and Hyperion. Left: Titan's orbit around Saturn is nearly circular ($e_T \simeq 0.03$); the conjunction with Hyperion always occurs near Hyperion's apocentre (within the $\simeq 49°$ shaded zone). Right: the critical argument $\sigma = 4\lambda_H - 3\lambda_T - \varpi_H$ librates around $180°$ with characteristic periods of 640 d and 6,850 d, due to the resonance and the near-resonant argument $\sigma' = 4\lambda_H - 3\lambda_T - \varpi_T$. The bottom zoom shows a sequence of conjunctions (dots), occurring every 64 d. Data from L. Duriez and A. Vienne, *Astron. Astrophys.* **324**, 366 (1997).

as the differential tidal evolution of the two orbits. Like in a pendulum, if at the stable equilibrium configuration the ratio of the periods is far enough from commensurability, in principle, conjunctions, instead of oscillating, could circulate through $360°$. As we shall see later, this latter behaviour is dangerous, because in this case the alignment is no more constrained to occur near the longitude at which the perturbing influence of Titan is weakest, and the resonance does not provide any more a *protection mechanism* against close encounters. As far as the real pair is concerned, we do indeed observe a libration of the conjunction position about Hyperion's apocentre, dominated by a term with the amplitude $\simeq 36°$ and a period $\simeq 1.75$ y, and a smaller term of $\simeq 13°$ with a period $\simeq 18.75$ y (Fig. 15.4). Note that both are appreciably longer than the orbital periods of Titan and Hyperion, about 16 d and 21 d, respectively.

In reality the problem is more complicated because Titan affects not only the mean motion of Hyperion, but also its eccentricity and pericentre. A quantitative model, based on the perturbation theory developed in Ch. 12, is needed. As noted in Sec. 12.5, the disturbing function $\mathcal{R}$ can be expanded in a Fourier series, having as arguments linear combinations of the mean longitudes of the satellites (λ_T and λ_H) and, in the planar problem, the longitudes of their pericentres (ϖ_T and ϖ_H); the corresponding coefficients C_{ijk} are functions of the semimajor axes a_T and a_H and the eccentricities e_H and e_T. The analysis of the disturbing function (12.79) indicates that $C_{ijk} \propto e_H^{|j_1|} e_T^{|j_2|}$, where the index $_1$ refers to the (exterior) body Hyperion. The exponents appear in the phase in the combination $j_1 \varpi_H + j_2 \varpi_T$. From all possible terms in the expansion of the disturbing function we may retain only the resonant one ($i_1 = 4, i_2 = -3$), which has a very low frequency because of the near commensurability of the mean motions. We can assume that short-period terms are ineffective, since they do not allow large perturbations of the orbital elements to build up and, to the first order in the mass parameter μ, average out to zero. D'Alembert rule (Sec. 12.5) then implies two possible *critical arguments*

$$\sigma = 4\lambda_H - 3\lambda_T - \varpi_H, \qquad \sigma' = 4\lambda_H - 3\lambda_T - \varpi_T. \qquad (15.14)$$

The mechanism due to Hyperion's eccentricity described earlier and the small eccentricity of Titan suggest that the amplitude corresponding to σ prevails. In the circular restricted three-body problem the second argument disappears; its effect is only a small, although long-term, variation of Hyperion's eccentricity with the critical argument σ (Fig. 15.4). Note that at conjunction, where $\lambda_H = \lambda_T$, σ is the angular distance of the two bodies from the Hyperion's pericentre.

The expansion of the resonant part of the disturbing function in Hyperion's (small) eccentricity, without which there is no effect, has only odd terms, to be contrasted with the case of secular perturbations, in which the main contributions are quadratic (Sec. 15.1). Obviously, the critical argument σ appears in the disturbing function in an infinite number of multiples, but the principal term is a linear combination of $\sin \sigma$ and $\cos \sigma$; it turns out that higher harmonics are $\propto e_H^{|j_1|}$, and thus negligible. Since we expect a stable equilibrium configuration when the conjunction is at the apocentre ($\sigma = 180°$), the perturbing function must be of the form

$$\mathcal{R} = -\frac{Gm_T}{a_H} \left[F_s(\alpha) e_H^2 + F_r(\alpha) e_H \cos \sigma \right]; \qquad (15.15)$$

m_T is Titan's mass and $\alpha = a_T/a_H \simeq 0.825$. The evaluation of $\mathcal{R}$ from (12.78) in effect gives (15.15), with the dimensionless functions $F_s(\alpha)$ and $F_r(\alpha)$ positive and of order unity; their slow change with α will be neglected. The first term in (15.15) is the secular perturbation of Hyperion's orbit due to Titan.

To describe long-term and secular effects we use Lagrange's equations (eq. 12.49) and first consider the case $e_H \ll 1$:

$$\frac{de_H}{dt} = \frac{1}{n_H a_H^2 e_H} \frac{\partial \mathcal{R}}{\partial \varpi_H} = -\mu_T n_H F_r \sin \sigma, \qquad (15.16a)$$

$$\frac{d\varpi_H}{dt} = -\frac{1}{n_H a_H^2 e_H} \frac{\partial \mathcal{R}}{\partial e_H} = \mu_T n_H \left(2F_s + \frac{1}{e_H} F_r \cos \sigma \right), \qquad (15.16b)$$

$$\frac{dn_H}{dt} = \frac{3}{a_H^2} \frac{\partial \mathcal{R}}{\partial \lambda_H} = 12 e_H \mu_T n_H^2 F_r \sin \sigma. \qquad (15.16c)$$

Here μ_T ($\simeq 2.4 \times 10^{-4}$) is the Titan-to-Saturn mass ratio. Due to the third Kepler's law $da/dt = -(2a/3n)(dn/dt)$, we get $da_H/de_H = 8\,a_H e_H$: the semimajor axis and the eccentricity are coupled, a typical feature of the resonance. The variation of the critical argument (recall that $\lambda = \int n\,dt + \epsilon$) is straightforwardly obtained:

$$\frac{d\sigma}{dt} = 4n_H - 3n_T - \frac{d\varpi_H}{dt} + 4\frac{d\epsilon_H}{dt} = 4n_H - 3n_T - \left(\frac{d\varpi_H}{dt}\right)_s - \frac{\mu_T n_H}{e_H} F_r \cos\sigma\,. \quad (15.17)$$

One easily shows that the term $d\epsilon_H/dt \approx e_H^2\, d\varpi_H/dt$ may be consistently omitted here; $(d\varpi_H/dt)_s = 2\mu_T n_H F_s$ is the secular drift of Hyperion's longitude of pericentre due to Titan. Differentiating again

$$\frac{d^2\sigma}{dt^2} = 48\mu_T F_r e_H n_H^2 \left(1 + \frac{1}{48\,e_H^2 n_H}\frac{d\sigma}{dt} - \frac{\mu_T}{2}F_s\right)\sin\sigma + \quad (15.18)$$

$$-\mu_T^2 F_r^2 \frac{n_H^2}{e_H^2}\left(1 + 12\,e_H^2\right)\sin\sigma\cos\sigma\,.$$

The first term in the right-hand side, if the bracket is nearly 1, corresponds to the pendulum equation with the frequency

$$\omega_r = n_H\,(48\,F_r e_H \mu_T)^{1/2}\,; \quad (15.19)$$

there is a stable equilibrium at $\sigma = 180°$ and an unstable one at $\sigma = 0°$. To first approximation n_H and e_H are constant. In this case $d\sigma/dt$ is of order ω_r, and so is the resonance offset from $4n_H - 3n_T - \varpi_s$. This regime holds when $\omega_r \gg n_H \mu_T/e_H$, that is to say, when $e_H \gg (\mu_T/48)^{1/3} \simeq 0.017$, a condition satisfied for Titan and Hyperion. In fact, in this case not only the terms in the second row are negligible, but the bracket in the first line is nearly one, as assumed. The pendular motion can be a libration near the stable equilibrium or a circulation. At the transition between the two the trajectory – the *separatrix* – passes through the unstable equilibrium point. Fig. 15.5 shows this solution and the libration zone, represented by crescent-like orbits.

The stable apocentric solution depends on the positive sign of F_r; the situation would be reversed if the massless body were the inner one (i.e., in the case of an interior resonance, appropriate to asteroidal motion). We have thus recovered the conclusion previously inferred with qualitative arguments: due to a sizable eccentricity of Hyperion's orbit, Titan's perturbation adjusts n_H in such a way to preserve the resonance and to cause the apocentric libration of the conjunction.

In the opposite limit, when e_H is sufficiently small, this approach is inappropriate, not only because its application to Hyperion ($e_H \simeq 0.1$) may be questionable, but also because the term in the second row in eq. (15.18) dominates. A more general analysis for an arbitrary eccentricity is needed. Using again Lagrange's equations (12.49), we can straightforwardly generalize eqs. (15.16). It turns out that the coupling between semimajor axis and eccentricity is given by $a_H \zeta^2 = $ const., where $\zeta = 4\eta_H - 3 = 4\sqrt{1 - e_H^2} - 3$. Referring to the initial values,

$$n_H = n_{H0} (\zeta/\zeta_0)^3 \ . \tag{15.20}$$

Similarly, the full form of eq. (15.17) yields

$$\frac{d\cos\sigma}{de_H} + \frac{\cos\sigma}{e_H}\left(1 - \frac{8e_H^2}{\eta_H \zeta}\right) = \frac{4n_H - 3n_T - \dot{\varpi}_S}{\mu_T F_r n_H} \frac{1}{\eta_H \zeta} \ . \tag{15.21}$$

This linear equation has the elementary integral

$$2\mu_T F_r \left[\left(\frac{\zeta}{\zeta_0}\right)^2 x_H - x_{H0}\right] = 1 - \left(\frac{\zeta}{\zeta_0}\right)^2 - 2\left(1 - \frac{\delta}{4n_{H0}}\right)\left(1 - \frac{\zeta_0}{\zeta}\right) \ , \tag{15.22}$$

which shows the coupling between the resonant argument σ and the eccentricity e_H; here $x_H = e_H \cos\sigma$ and $\delta = 4n_{H0} - 3n_T - (d\varpi_H/dt)_s$ is a measure of the initial distance from exact resonance, for which $\delta = 0$. The dependence of $(d\varpi_H/dt)_s$ on n_H has been neglected. For very small eccentricity, (15.22) simplifies to

$$2\mu_T F_r n_{H0} (x_H - x_{H0}) = \delta \left(e_H^2 - e_0^2\right) \ , \tag{15.22'}$$

which is the parametric equation of a small circle

$$(x_H - B)^2 + y_H^2 = A^2 = \text{const.} \qquad (B = \mu_T F_r n_{H0}/\delta) \ , \tag{15.23}$$

in the plane of the non-singular resonant elements $x_H = e_H \cos\sigma$, $y_H = e_H \sin\sigma$. We have possible stable equilibria both at $\sigma = 0°$ and at $\sigma = 180°$ (as actually suggested by the last term in (15.18), $\propto \sin\sigma \cos\sigma$). Thus a pericentric libration is also possible (for $B > A > 0$), but close approaches are anyway ruled out by the very small value of the eccentricity. In Fig. 15.5 we see the small eccentricity pericentric libration and circulation ($A > B > 0$) as the almost circular solutions near the origin. At the exact resonance $\delta = 0$ the displacement B diverges and the circular path (15.23) is inconsistent. At this limit, the trajectory, starting at the origin, follows the straight line $x_H = $ const., until the eccentricity becomes large and the complete solution (15.22) must be used.

Eq. (15.22) yields the complete topology of resonant paths in the plane of the non-singular elements (x_H, y_H). The right panel of Fig. 15.5 shows a set of these solutions, with the same average value of the mean motion n_H, and thus the same average distance δ from the exact resonance. These parameters were chosen to fit those of Hyperion (whose best-fitted motion is shown by the thick, crescent-like path). The left panel shows the real Hyperion's trajectory in the

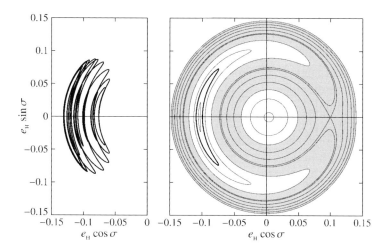

Figure 15.5. Right: Hyperion's motion in the resonant variables $x_H = e_H \cos\sigma$ and $y_H = e_H \sin\sigma$ in the simple planar problem, with Titan on a circular orbit. Of the three stationary points, two are stable; small oscillations around then correspond to pericentric and apocentric librations of σ around 0° and 180°, or circulations. The critical orbit through the unstable point is the separatrix. The thick line is Hyperion's approximate orbit. Data kindly provided by A. Morbidelli. Left: The real motion of Hyperion on the same axes over a 25 y time span; despite its more complex structure due to non-vanishing eccentricity of Titan, the libration pattern, responsible for the protection mechanism against a close encounter with Titan, is preserved. Data from L. Duriez and A. Vienne, *Astron. Astrophys.* **324**, 366 (1997).

resonant coordinates; the most important difference from the idealized solution on the right is due to the fact that the near-resonant term $\propto e_T \cos\sigma'$ was not taken into account (indeed, an even weaker secular term $\propto e_H e_T \cos(\sigma - \sigma')$ in the disturbing function was found resonant, since the argument $\varpi_H - \varpi_T$ librates around zero). The stability of Hyperion's apocentric libration is preserved even in this more complete model, at least for a reasonably long timescale.

We recall that resonant paths, as described by (15.22) depend on the initial eccentricity e_{H0} and the distance δ from the resonance. A slow change in δ, such as that due to a faster tidal evolution of Titan's orbit away from Saturn (Sec. 15.3), may drive a corresponding change in the initial eccentricity e_{H0}, defining the solution for Hyperion. The orbit then evolves along a sequence from an initial state of small eccentricity ($e_{H0} \simeq 0.02$) circulation to higher eccentricity apocentric libration via a mechanism of *resonance capture*. This origin of the Titan-Hyperion pair has been proposed on basis of the above simplified analysis. However, the neglect of short period effects casts doubts about this scenario: numerical studies of the Hyperion capture in the 4/3 exterior resonance with Titan suggest that close encounters with the massive satellite, due to short period perturbations, nearly inevitably occur. Hyperion's direct primordial accumulation out of the nebular material in the zone of the resonance seems a more viable mechanism for its origin.

It is also interesting to note that a body captured in the pericentric libration, with the critical angle σ librating around $180°$, has the eccentricity "forced" to a value close the eccentricity of the corresponding periodic orbit (with zero libration amplitude). This is of major importance for satellite dynamics, since episodes of present or past resonant states force such non-vanishing eccentricity, in spite of tidal effects that tend to diminish it (Sec. 15.3). The specific case of Io is discussed in Secs. 14.2 and 15.3.

Beyond simple models. Long-term numerical orbital integrations for the 4/3 resonance of Titan and Hyperion have revealed a far greater complexity than expected from the above model. If successive conjunctions are plotted in the (x_H, y_H) plane (Fig. 15.5), two clearly distinct behaviours emerge, corresponding to the shaded and unshaded zones. In the latter, corresponding to small-amplitude apocentric librations and the pericentric solutions with a small eccentricity (eq. (15.23)), conjunctions are fitted by smooth curves, indicating the orbit that belongs to one of the two resonant modes. If the orbit starts in the shaded zone (e.g., with large-amplitude apocentric librations), conjunctions tend to scatter in an irregular manner throughout the whole shaded area; even if they initially follow the regular crescent-like curves, when the representative point reaches the extremum of such a libration, the conjunction occurs too close to the pericentre ($\sigma = 0°$) and the resulting strong disturbance destroys the seemingly regular behaviour[3]. Note also that the velocity along the separatrix tends to zero when approaching the unstable equilibrium solution; at this time the evolution is dominated by short-period perturbations that may drive the trajectory away from the libration zone to the circulation zone, and vice versa. The large number of short-period terms with comparable amplitudes and roughly random phases in the disturbing function makes the evolution of the system extremely sensitive to the initial conditions, which is the essential feature of *chaotic orbits*, with a positive Lyapunov exponent (Sec. 13.3). Since the accuracy in the initial conditions is limited, after some time the evolution becomes completely unpredictable.

The division of phase space in a regular and a chaotic domain is a common feature of many dynamical systems, both dissipative and conservative. In the latter case, when there is only one degree of freedom, the energy integral de-

---

[3]The rather narrow stable region in the 4/3 resonance with Titan has an interesting implication for Hyperion's shape. Successive impacts on planetary satellites, a relatively common phenomenon, cause ejection of surface layers, but normally much of this material falls back after a short time. For Hyperion, though, the ejected material typically has a relative velocity sufficient to escape the region protected by the resonance and enters the chaotic zone. Re-accumulation on Hyperion is very unlikely, which explains its very irregular shape; this, together with its large eccentricity, shows theoretically and experimentally that its rotation is, on a long-term, unpredictable (chaotic).

fines the trajectory and ensures stability with respect to the initial conditions. When the number f of degrees of freedom is larger than 1, both domains may be present, unless there are f integrals of the motion. Chaos may result from time-dependent perturbations and non-linear couplings and, as hinted above, is especially common in presence of resonances. Chaotic situations cannot be studied by analytical techniques based on expansions of the disturbing function, which, as it happens with small divisors and an infinite density of high-order resonances (Sec. 12.5), often diverge; averaging methods, for which an obvious condition is a regular, quasi-periodic behaviour, are also of no avail.

The investigation of the extent and the properties of chaotic zones should be carried out with *numerical methods*, such as the surface of section technique or the Lyapunov exponents (Sec. 13.3). Direct integrations with conventional tools are often time consuming, and only recently numerical experiments lasting long enough have been carried out. This is also due to the smallness of perturbations due to other planets (or satellites); many relevant frequencies (e.g., $d\varpi/dt$ or $d\Omega/dt$) are very small and the integration must be extended to 10^5 or 10^6 revolutions before the full structure of the phase space becomes apparent. An exception is the case of chaos associated with repeated close encounters, which, on the other hand, require special numerical tricks, like a variable step size. Owing to the sensitivity of chaotic orbits to the initial conditions and the unavoidable rounding-off and truncation errors (Sec. 15.1), the question about the reliability of numerical results arises, in particular for positive Lyapunov exponents. In this case the aim of a deterministic description should be abandoned; we can only investigate the domain of phase space spanned by the set of solutions with close initial conditions (such as the range of eccentricities attained; Fig. 15.2), which is all we need for a stability analysis. The possibility of a numerical implementation is ensured by the fact that such a solution is arbitrarily near a true solution with sufficiently near initial conditions.

A tapestry of resonances. Our treatment of the Titan-Hyperion resonance can be applied to other cases, e.g., to asteroids in first order (2/1, 3/2 and 4/3) mean motion resonance with Jupiter (or, with a little modification, to the 3/1 or 5/2 resonances, also important; Figs. 14.3 and 14.4). However, there is a variety of other relevant resonant mechanisms; although differing in many details, they all share many typical features of the simple case discussed above: stable librations of critical arguments and protection mechanisms on the one hand, and possible onset of chaos on the other.

Mean motion resonances.– There are no strictly resonant planetary orbits, but some are close to be such[4]. Jupiter and Saturn are near the 5/2 mean motion resonance; the appropriate argument circulates with a period of $\simeq 880$ y,

---

[4]It is interesting to note that observational evidence for resonances in extrasolar planetary systems add an important argument for the planetary origin of the observed effects; this is the case of the system around the

much longer than their orbital periods. Already in the late 18th century this was known to cause an observable anomaly in their motion, referred to as the *great inequality*. The pairs Jupiter and Uranus, and Uranus and Neptune are near the 7/1 and 2/1 mean motion resonances, respectively, with an important influence on their long-term motion. At present only Pluto is locked in a 3/2 mean motion commensurability with Neptune (Fig. 14.9); its analysis, similar to other plutinos (Sec. 14.4), is complicated by the fact that its perihelion may be temporarily *inside* the orbit of Neptune. Due to this, Pluto is protected against collisions, with conjunctions librating about Neptune's aphelion every $\simeq$ 19.9 ky; this is somewhat similar to the Titan-Hyperion mechanism, but Pluto's eccentricity is so large that strong interactions with Neptune are possible even far from conjunctions. Another protection mechanism, related to Pluto's large inclination, is at work: its perihelion librates about 90° with a period of $\simeq$ 3.8 My, so that when conjunctions occur, Pluto is always far away from the plane of Neptune's orbit.

The majority of mean motion resonances concern satellites and minor bodies. A Trojan or a Greek asteroid is locked in a 1/1 resonance about the triangular Lagrangian points of the Sun-Jupiter system. Their stable librations can be analysed in terms of the linear stability of the equilibrium positions (Sec. 13.3). For bodies of similar mass, such as the Janus and Epimetheus pair, Hill's method may be used (Sec. 13.4).

A number of satellites of the giant planets are, or temporarily were in the past, in mutual, low-order mean motion resonances. Most of them are similar to that of Titan and Hyperion, with a critical argument depending on the longitude of the pericentre, but not on the longitude of the node. They are called *eccentricity resonances*, since they principally affect the satellite's eccentricity, causing coupled variations with the resonant argument (one can prove that the inclination changes are small, $\propto \tan I/2$). If, instead, the resonant argument depends on the longitudes of the nodes, but not on those of the pericentres, its variations get principally coupled with the inclinations (*inclination resonances*). The only known example are the Saturnian satellites Mimas and Tethys; in this case the protection mechanism is such that the stable location of conjunctions is far from the intersection of their orbital planes.

A p/q mean motion resonance is a spectral line for the perturbing function (12.79) at the frequency $pn_1 - qn_2$ (and its multiples), small, but not exactly zero; but the resonant argument $p\lambda_1 - q\lambda_2$ is complemented with a combination of $(\varpi_1, \varpi_2, \Omega_1, \Omega_2)$, according to the possible values of the integers j_1, j_2, k_1, k_2, as constrained by d'Alembert's rule $p - q + j_1 + j_2 + k_1 + k_2 = 0$: the line splits in a multiplet. When the eccentricities e_1, e_2 and the inclina-

---

pulsar PSR 1257+12, whose planets B and C are in the $p/q = 3/2$ mean motion resonance (Table 16.1), and those around both Gliese 876 and HD 82943, with $p/q = 2/1$.

tions I_1, I_2 are small, it contains a small number of prevailing members, which depends on the order $k = |p - q|$ of the resonance; this follows from the fact that the coefficients C_{ijk} of each term are $\propto e_1^{|j_1|} e_2^{|j_2|} \sin^{|k_1|} I_1 \sin^{|k_2|} I_2$. In the 1/1 ($k = 0$) case, either the four integers all vanish, or the term is small; there is only one main line. When $k = 1$ the main terms of the multiplet, linear in the eccentricities and the inclinations, are four, those for which only one of the integers is not zero and equal to 1; when $k = 2$ there are ten terms of the second degree, i.e., those in which one of the integers is equal to 2 and the other ones vanish, and those with only two integers, each equal to one; and so on. For the eccentricities this corresponds the fact that conjunctions occur at k different longitudes, and therefore the accumulation of perturbing effects is less effective. We recall the two main features of the Titan-Hyperion 4/3 resonance of order 1: (i) neglecting short-period perturbations, the dynamics has the structure of a (generalized) pendulum; (ii) near the separatrix between libration and circulation modes, chaotic behaviour is expected due to short-period perturbations and their couplings. At higher order each multiplet has this structure and more lines; due to their proximity, libration zones may partially overlap. Typically, this destroys the regular region of small amplitude librations (Fig. 15.5) and the motion becomes fully chaotic. Obviously, due to the weakness of the resonance, the timescale of the instability (e.g., for large variations of the eccentricity) may be very long. High-order resonances, together with three-body resonances, play an important role in understanding the long-term (≥ 100 My) evolution of many asteroids. An interesting situation occurs for satellites, for which additional dynamical perturbations (typically the planet's quadrupole field) cause sufficiently rapid motion of the node and the pericentre; in this case libration zones of different resonant multiplets may not overlap and the motion is characterized by isolated, non-interacting parts, with a more regular motion.

As noted in Sec. 14.2, the most famous case of a *three-body resonance*, already investigated by P.S. LAPLACE, involves three Galilean satellites of Jupiter: Io, Europa and Ganymede. For this *Laplacian resonance* the critical argument $\sigma = \lambda_I - 3\lambda_E + 2\lambda_G$ is locked at $180°$, with a tiny libration amplitude ($\simeq 0.066°$). Since $\sigma = (2\lambda_G - \lambda_E) - (2\lambda_E - \lambda_I)$, conjunctions of adjacent satellite pairs must always occur $180°$ apart, and the three satellites can never line up on the same side of Jupiter; as for Hyperion, this provides a protection mechanism. The giant planets are also very close to three-body resonances, with the critical arguments $(3\lambda_J - 5\lambda_S - 7\lambda_U + \text{secular arguments})$ and $(3\lambda_S - 5\lambda_U - 7\lambda_N + \text{secular arguments})$; the "secular arguments" are combinations of the longitudes of the perihelii and the nodes, as required by the d'Alembert rule (Sec. 12.5). Numerical simulations show that these two arguments librate for orbits very close to the real ones. Laplacian and planetary resonances are examples of direct perturbations between three bodies; a three-body resonance may also occur in the restricted

Hamiltonian system with conjugate variables ϕ and Φ and the mean anomaly M as independent variable:

$$H(\Phi, \phi; M) = \frac{\Phi^2}{2} - \alpha \left(\frac{a}{r}\right)^3 \cos 2(\phi - f) = \frac{\Phi^2}{2} - \alpha \sum_r K_r(e) \cos 2(\phi - rM) ;$$

(15.24)

in terms of the principal moments of inertia of the satellite, $\alpha = 3(B - A)/(4C)$, while f is the true anomaly ($\Phi = d\phi/dM$). This Hamiltonian is obtained from the rotational energy $C(d\phi/dt)^2/2$ and the potential energy, given by the MacCullagh formula (2.29), of the planet-satellite pair; they are divided by the constant Cn^2. The summation is a power expansion in the eccentricity, with coefficients $K_r \propto e^{|2r-2|}$; r ranges over all multiples of $1/2$, starting with $r = 1$. With this meaning, r is used only in this discussion of eq. (15.24) and should not be misunderstood with the radial distance from the planet, as in Fig. 15.6. For a very small eccentricity the main term is $r = 1$ and we get pendulum Hamiltonian

$$H_1 = \frac{\Phi^2}{2} - \frac{\omega_1^2}{2} \cos \theta$$

(15.25)

for the variable $\theta = 2(\phi - M)$ and the frequency $\omega_1 = \sqrt{4\alpha K_1}$. At vanishing amplitude $\phi = M$; this describes the trivial fact that, because of the change in orbital velocity, the axis of inertia cannot point all the time towards the centre of planet (which corresponds to $\phi = f$), except on the apsidal line. For a positive energy we have librations in longitude with the frequency $\omega_1 = \sqrt{4\alpha K_1}$. For the Moon, with $K_1 \simeq 1$, it corresponds to the period of $\simeq 2.87$ y, detected by lunar laser ranging (sometimes called free libration in longitude). When the eccentricity is not very small, we can also have a resonance of higher order, where a term r (> 1) prevails and $\phi = rM$; the libration frequency is now $\omega_r = \sqrt{4\alpha K_r(e)}$. The terms in the Hamiltonian (15.24) corresponding to $r' \neq r$ have the fast-varying argument $2(r' - r)M$ and average out.

As the rotation rate decays by tides (Sec. 15.3), it crosses these resonances with diminishing r; the capture probability, which grows with the resonance strength K_r, is negligible unless r is small. $r = 1$ is the most effective case; when the eccentricity is not too small, the next value $r = 3/2$ may work, as for Mercury. The most peculiar and complicated situation occurs when both e and α are large; high-frequency perturbations restrict the regular libration to a small region (compare with the previous analysis of the Titan-Hyperion resonance; Fig. 15.5). The rotation may fall in the chaotic zone of phase space; Hyperion's rotation, in which $\alpha \simeq 0.2$ and $e \simeq 0.1$, is the outstanding example. Hyperion is the only current example of a satellite with a chaotic rotation; other satellites (e.g., Phobos) must have gone in the past through such state before being completely synchronized. More complicated resonances may occur when precession rate of the spin axis is equal to the mean motion (the likely case of Pluto, whose axis precesses under Charon's torque) or to one of the frequencies of orbital evolution (the case of Saturn; Sec. 7.1)

The slow retrograde rotation of Venus is close to, but not exactly in, one of the most peculiar resonances, a spin-orbit analogue of the three-body mean motion resonances discussed above. With an obvious notation, $5n_E - 4n_V + \omega_V \simeq 0$; as a result, at every inferior conjunction (when the Earth and Venus are at their closest approach) Venus always presents the same face toward Earth. In fact, the gravitational torques of the Sun and the Earth allow such resonance, but its strength is so weak that capture is very unlikely; by a curious coincidence Venus only recently passed through it.

Origin of resonances. Likely there is more than one answer about the origin of the abundant resonant states in the solar system. Tides, when effective, as in the evolution of spins and the orbits of satellites, provide a dissipative mechanism that can gradually change their periods until a capture in a stable resonant state occurs. Alternatively, some resonances might date back to the origin of the solar system. Resonant orbits might have favoured the accumulation of bodies at preferred locations, or might have provided protection mechanisms and the natural selection of "lucky" dynamical configurations (this may be the case of Hyperion). Finally, in the primordial solar nebula additional dissipative mechanisms due to gas drag, collisions and scattering could in some cases have played the same role of tides, albeit on much shorter timescales. Radiation forces (Sec. 15.4), whose relevant timescales are of order $\approx$ ky to Gy, depending on the size, are able to bring small bodies in resonant regions even today. As pointed out, resonances may also trigger orbital or rotational instabilities and chaotic motion; the relative lack of bodies in such states in some otherwise populated regions, like the main asteroid belt or Saturn's rings, can be explained in this scenario. As the orbit is driven to a state when some parameters undergo large changes (such as, in the Kirkwood gaps, a large eccentricity; Fig. 14.3) the body can be eliminated by close encounters (e.g., with planets) or by interparticle collisions, clearing up the corresponding zones.

15.3 Tidal orbital evolution

As discussed in Sec. 4.3, owing to dissipation, the tidal bulge raised by the Moon on the Earth is slightly advanced with respect to the line joining their centres. The effect of this bulge on the Moon is dominated by the nearest tidal wave and produces a forward, along-track force; this leads to a secular increase of the lunar semimajor axis and, in order to conserve the total angular momentum of the system, the Earth's rotation gradually slows down (see Sec. 3.2). This mechanism has already synchronized the revolution and the rotation frequencies of the Moon and of most other natural satellites in the solar system (except irregular distant satellites). It is possible to draw several interesting conclusions about this evolution process just on the basis of conservation laws.

For the Moon and most other natural satellites the present orbits imply the case (a1). The Moon's stable synchronization distance is about 88 $R_\oplus$ (as compared with the present value of about 60 $R_\oplus$), corresponding to an orbital period of about 48 days, which will also be the final rotation period of the Earth; of course, this is no longer true if solar tides are taken into account, which can cause a further evolution of the system. Only in one case, the Pluto-Charon pair, the stable synchronization state has already been reached. Below we confirm the intuitive guess that the large mass of Charon is responsible for this result. On the other hand, two satellites – Mars' Phobos and Neptune's Triton – are evolving like in the cases (a2) and (a3), respectively, approaching their planets towards their final disruption (there are also some small inner satellites of Jupiter, Uranus and Neptune below the corotation distance, all of which interact with their ring systems). Mercury and Venus, both significantly despun by solar tides and other dynamical effects, have no satellites. The corotation distance corresponding to ℓ_1 is very far from these planets; even if in the past they had a satellite, it is likely that, as the planet despun, planetary tides would have been driven it to disruption.

Timescale of the tidal evolution. As discussed in Secs. 4.2 and 4.3 (see eqs. (4.23) and (4.17)), the lowest order, quadrupole gravitational potential generated by the tidal bulge at the distance r of the satellite is

$$U_T = \left(\frac{R_p}{r}\right)^3 \left(-k_2 \frac{GM_s R_p^2}{2r^3}\right) \left[3\cos^2(\psi - \epsilon_T) - 1\right], \tag{15.29}$$

where k_2 is Love's number (for the Earth $\simeq 0.30$; see Sec. 4.3), ψ is the angle between the considered position and the lunar direction, and ϵ_T is the angular lag of the Moon relative to the axis of the Earth's tidal bulge. The term $(R_p/r)^3$ is due to the fact that U_T is produced by a quadrupole, second degree deformation; in effect, higher degree deformations and different Fourier terms have their own Love's numbers and angular lags (Sec. 4.4), but we live with the simple model (15.29). The tidal torque N_T on the Moon is

$$N_T = -m\left(x\frac{\partial}{\partial y} - y\frac{\partial}{\partial x}\right)_{y=0} U_T = -m\left(\frac{\partial U_T}{\partial \psi}\right)_{\psi=0} = \frac{3}{2}k_2 \frac{GM_s^2}{r}\left(\frac{R_p}{r}\right)^5 \sin 2\epsilon_T, \tag{15.30}$$

where (x, y) are Cartesian coordinates in the lunar orbital plane, with the Moon on the x axis (to which ψ is referred). Equating N_T to the time derivative of ℓ (eq. (15.26)) we obtain

$$\frac{dr}{dt} = 3k_2 \sin 2\epsilon_T \frac{M_s \sqrt{G(M_s + M_p)}}{M_p} \frac{R_p^5}{r^{11/2}}. \tag{15.31}$$

For the Moon, with $\epsilon_T \simeq 3°$, this equation yields a present increase in the semimajor axis $dr/dt \simeq 3.8$ cm/y; the conservation of angular momentum (eq. (15.26)) gives $d\omega/dt = -N_T/I_p \simeq -6 \times 10^{-22}$ rad/s^2, corresponding to an increase of the length of the day of about 2.3 ms/cy. These values have been confirmed by laser ranging and are roughly consistent with the historical records of ancient eclipses (though there is evidence of an additional positive component in $d\omega/dt$ not of tidal origin; Sec. 3.2). Integrating eq. (15.31) from r_0 to $r \gg r_0$, and assuming a constant lag, we find that the corresponding time interval

$$\tau_T = 2r/(13\, dr/dt) \tag{15.32}$$

is insensitive to r_0. For the Earth-Moon system, the observational values quoted above yield for the tidal expansion of the lunar orbit (say, from just outside Roche's limit, at $r_0 \approx 3 R_\oplus$) a typical timescale τ_T as short as 1.5 – 2 Gy. Since ϵ_T is inversely correlated with tidal dissipation (Sec. 4.4 and below), this means that probably the current rate of dissipation in the Earth is higher than it was, in the average, in the past. A possible reason is the present anomalous abundance of shallow seas at the margins of fragmented continental blocks.

Another interesting conclusion stems from $r^{13/2} \propto M_s t$, obtained integrating eq. (15.31); obviously this neglects the effects of tides on the satellite and of orbital resonances. For a given, tidally evolved satellite system (such as those around the giant planets), and the same elapsed time t, $\log r = (2/13) \log M_s +$ const. The satellite systems of Saturn and Uranus follow reasonably well this approximate law; but it cannot be applied to Jupiter's Galilean satellites, largely affected by the Laplacian resonance.

Eq. (15.30) governs also the despinning of the Moon by the Earth's torque N'_T, which occurred in the early phases; N'_T is obtained interchanging the masses (M_s, M_p) and the radii (R_s, R_p) of the two bodies:

$$N'_T = N_T (M_p/M_s)^2 (R_s/R_p)^5 .$$

Since $d\omega_s/dt = -N'_T/I_s$, we have

$$d\omega_s/dt = (M_p/M_s)^3 (R_s/R_p)^3 d\omega/dt .$$

This corresponds to a timescale some 10^4 times shorter than the synchronization time for the Earth's rotation: the tidal locking of the Moon took only a few million years. It is easy to see that in this stage tides circularize the orbit. Assuming the planet to be a point mass, the eccentricity is given by $1 - e^2 \propto -E\ell^2$ (eq. (11.15); here we deal with tidal effect on the satellite, assuming the planet a point mass); if, moreover, the satellite's rotational angular momentum is neglected, an assumption usually well satisfied, ℓ is constant. As $-E$ increases by dissipation, e decreases.

Energy dissipated by tidal friction. From eq. (15.26) we have

$$\frac{dE}{dt} = \frac{d}{dt}\left(\frac{1}{2} I_p \omega^2 - \frac{GM_p M_s}{2r}\right) =$$

$$= I_p\omega\frac{d\omega}{dt} + \frac{GM_sM_p}{2r^2}\frac{dr}{dt} = N_T(n-\omega), \qquad (15.33)$$

where n is the orbital mean motion (recall that $dL/dt = 0$). Using the value given above for dr/dt (or $d\omega/dt$) to estimate N_T, one obtains a dissipation rate in the Earth of about 3×10^{19} erg/s, comparable to that estimated from the friction at the bottom of shallow seas at the present time. As in Sec. 4.4, we can now define the *tidal quality factor* Q as the ratio between the energy stored in the tidal bulges (of mass $\approx hM_p/R_p$) and the energy dissipated in a tidal cycle (about 1 d):

$$-\frac{dE}{dt} \approx \frac{gh^2M_p\omega}{2\pi QR_p} \approx \frac{\omega N_T}{\epsilon_T Q}, \qquad (15.34)$$

where we have used $h \approx \mu_T R_p$ (eq. (4.10)) and eq. (15.30). Since $\omega \gg n$, the comparison with (15.33) shows that $Q \approx 1/\epsilon_T$, for the Earth about 20 (of course, this holds for tidal frequencies).

A similar argument can be used to estimate tidal dissipation in a satellite whose rotation is already synchronized with the orbital period, but with an eccentric orbit. This case is important because the surfaces of several satellites of the outer planets display widespread marks of volcanic activity and/or resurfacing events, probably associated with a melted interior; tidal friction appears as the most plausible energy source responsible for such processes. In some cases the dissipative circularization of a satellite's orbit is prevented by resonant interactions with other satellites, which gives rise to a forced eccentricity. This is the case of Io, the innermost of the three Jovian satellites in the Laplace resonance. Images from the Voyager and Galileo probes have shown, as displayed in the cover, that Io's surface morphology is indeed dominated by volcanism, with several ongoing eruptions. Its eccentric orbit causes the tidal bulge of the satellite to oscillate back and forth in the radial direction, thereby dissipating energy. Taking into account Love's number k_2, the mass displaced by tides is $\approx k_2\mu_T M_s$, where $\mu_T = (M_p/M_s)(R_s/r)^3$ is the relevant tidal parameter; with a radial velocity $\approx enR_s$ (the relative velocity of the bulge with respect to the centre of the satellite matters here), the dissipation power is

$$-\frac{dE}{dt} \approx \frac{k_2\mu_T M_s(enR_s)^2 n}{Q} \approx \frac{\varrho^2 R_s^7 n^5 e^2}{Q\mu}, \qquad (15.35)$$

where ϱ is the satellite density, n its mean motion around the planet and μ its rigidity. We have used here the approximate formula $k_2 \approx GM_s\varrho/(\mu R_s)$, valid for a body of sufficiently high rigidity (Sec. 3.3). The result (15.35) is to be compared with the Peale's formula on the cover, where the numerical coefficient should be 42/19 instead of 36/19. Assuming a crustal rigidity $\mu \approx 10^{12}$ dyne/cm^2 and a quality factor $Q \approx 100$ (of the same order as those of the Earth and the Moon), and using the forced eccentricity $e \simeq 0.0043$ induced by the Laplacian resonance[5], we obtain a dissipation rate $-dE/dt \approx 3 \times 10^{19}$ erg/s. This corresponds to a dissipation rate per unit mass of $\approx 10^{-7}$ erg/(g s), higher than the radioactive heating rate; it is sufficient to cause extensive melting of Io's interior.

15.4 Dynamics of small bodies

While the orbital evolution of "large" solar system bodies, like planets and satellites, can be described to a high degree of accuracy in terms of their gravitational interactions, this is no longer valid for small bodies up to $\approx (1-10)$ km in size. Interplanetary space is not empty, but photons and solar wind particles fly by, both of which carry momentum and, upon being absorbed or scattered, exert a net force. Circumplanetary environments, as those with rings, are even more complex, owing to the presence of magnetospheres and radiation from the planet. Generally speaking, these non-gravitational forces – both radiation pressure and drag – are roughly proportional to the cross-section of the particles (for radiation this holds only if the size R is larger than the wavelength of light), so that the corresponding accelerations are proportional to the inverse of the size. This is why these effects are most important for small bodies, frequently making their motion quite complicated and puzzling. In this section we review the most important types of these forces and their consequences on long-term orbital evolution.

Radiation pressure and its orbital effects. At the simplest level, consider an ideal *absorbing body* with cross section S (orthogonal to the beam) and mass m. An impinging Poynting flux of electromagnetic energy $\mathbf{F}$ exerts on it the force $\mathbf{f} = \mathbf{F} S/c$; in the case of the solar radiation

$$\mathbf{F} = F_0 \left(\frac{a_0}{r}\right)^2 \mathbf{n},$$

where $F_0 \simeq 1.37 \times 10^6$ erg cm^{-2} s^{-1} is the solar constant (the solar radiation flux at $a_0 = 1$ AU), r is the distance from the Sun and $\mathbf{n}$ is the radial outward unit vector. This force has the same dependence on $\mathbf{r} = r\mathbf{n}$ as gravity and its effect is equivalent to replacing $GM_\odot$ with the effective value $(1 - \beta_r) GM_\odot$, where

$$\beta_r = \frac{F_0 S}{mc} \frac{a_0^2}{GM_\odot} \simeq 7.7 \times 10^{-5} \frac{S}{m} \simeq 5.8 \times 10^{-5} \frac{1}{\varrho R}; \qquad (15.36)$$

---

[5] e is determined by an equation similar to (15.16a), with the argument $\sigma = \lambda_{Io} - 2\lambda_{Eu} + \varpi_{Io}$. Io and Europa are resonant, with $n_{Io} - 2n_{Eu} \simeq 7 \times 10^{-3} n_{Eu}$, so that e is enhanced and can be evaluated in terms of observable quantities.

the last numerical value holds when the body is spherical, with radius R. The orbit is still Keplerian, but when $\beta_r \geq 1$ becomes unbounded; this is the basis for the *solar sail* (Problem 11.5).

When reflection and diffusion occur a detailed consideration of its optical properties is required; for example, for a perfectly reflecting surface orthogonal to the Sun the force is twice as great. Moreover, in general the radiative force is not radial, but still in a roughly outward direction and the previous equation provides a reasonable estimate of the radiative force, applicable also to spacecraft (Sec. 18.3). β_r is approximately inversely proportional to the size; when $\varrho \approx 1$ g/cm^3 it becomes of order unity for $R \approx 0.5$ μm. It should be noted, however, that eq. (15.36) only holds when R is much greater than the wavelength of the impinging radiation, which for the Sun is peaked at $\lambda_m \simeq 0.5$ μm (eq. (1.98)); in the opposite case $R \ll \lambda_m$ the interaction decreases and the radiation force goes to zero. The maximum value of β_r is about 0.5 for non-absorbing dielectric materials, and increases with the absorptivity; it reaches values of 3 to 10 for strongly absorbing particles. Zodiacal dust particles, with size between 10 to 100 μm, are typically larger, with $\beta_r \approx 10^{-2}$. Even when $\beta_r < 1$, however, if a particle is released from a larger body (e.g., in a collision or in a cometary jet) with a velocity close to that of its parent, it will move under a weaker solar gravity. If this velocity exceeds the corresponding, smaller escape velocity, the particle can be expelled from the solar system. Using the two-body problem equations, one can see that at pericentre, where this mechanism is most effective, ejection occurs when $\beta_r \geq (1-e)/2$. For long-period comets the eccentricity e can approach unity (often within $\approx 10^{-4}$, see Sec. 14.4) and even large particles can be eliminated: indeed, dust detectors aboard spacecraft have identified a class of particles, the *β-meteoroids*, streaming away from the Sun, which could have been produced in the inner solar system by collisions or rapid sublimation. Of course, the orbital evolution can become complex if particles change their size or their surface properties, for instance by sputtering. It is estimated that β-meteoroids escape from the solar system and mix with interstellar dust at the rate of $\approx 9 \times 10^6$ g/s; this is the major component of the loss of meteoritic material and has relevant astrophysical significance.

Consider now dust particles orbiting a planet, with the following notation:

	Mass	Radius	Distance from attracting body	Orbital velocity
Solid grain	m_s	R_s	r_s	$v_s = \sqrt{GM_p/r_s}$
Planet	M_p	R_p	r_p	$v_p = \sqrt{GM_\odot/r_p}$

In this case the solar radiation pressure has very different effects. Since the perturbing acceleration has a nearly constant magnitude and is directed away from the Sun, it performs no work when averaged over one orbit, and has no secular or long-period effect on the semimajor axis (Sec. 12.1; however, this does not apply for eccentric orbits which cross the shadow of the planet). On the other hand, perturbation theory shows that in a revolution the eccentricity suffers a change $\approx \beta_r M_\odot r_s^2/M_p r_p^2$ – the ratio between the perturbing force and the planet's gravity. This change builds up (or down) for a time $\approx 1/4$ of the orbital period of the planet, which controls the orientation of the radiative force; the eccentricity can approach unity, possibly causing escape from the system or collision with the planet, when β_r exceeds the critical value $\approx v_s/v_p$. Thus a particle released from an inner satellite requires a larger β_r to be eliminated, and, in that case, has a greater chance to fall onto the planet; a particle from an outer satellite has about the same chance to hit it or to escape. For the Moon, the critical value of β_r is about 0.03, which for silicate composition implies a size range between 0.02 μm and 5 μm; on the other hand, for Amalthea this critical value is $\simeq 2$, implying particles in a much more restricted range of 0.02 μm to about 0.3 μm (for ice composition).

*The Poynting-Robertson effect.** When relativistic effects are taken into account, the forces due to the re-emitted and the scattered radiation generally have a component along the velocity in the solar system frame, with an important long-term effect on the orbital motion. This subtle phenomenon was finally clarified by H.P. ROBERTSON in 1937.

It is convenient to use the formalism of Special Relativity, with the flat metric $\eta_{\mu\nu} = (1, -1/c^2, -1/c^2, -1/c^2)$ and the proper time interval $d\tau$ (see Ch. 17). With this choice time and space are measured in seconds and centimetres, respectively, and the velocity of light has the conventional (exact) value $c = 2.99792458 \times 10^{10}$ cm/s. Spacetime vectors have spatial, three-dimensional components, indicated with boldface. In particular, the four-velocity v^μ is determined by the three-dimensional velocity $\mathbf{v}$ and has the components $v^\mu = \gamma(1, \mathbf{v})$, where $\gamma = 1/\sqrt{1 - \mathbf{v}^2/c^2}$; it has a unitary modulus $\eta_{\mu\nu} v^\mu v^\nu = 1$. A null vector can always be written in the form $n^\mu = n^0(1, c\mathbf{n})$, where $\mathbf{n}$ is the corresponding unit vector in space; $n^0 = dt/d\lambda$ may be set equal to unity with a suitable choice of the affine parameter λ (Sec. 17.2).

The direction of propagation of a plane electromagnetic wave in spacetime is described in the solar system barycentric frame $\mathcal{S}$ by the null vector $n^\mu = (1, c\mathbf{n})$; its energy-momentum tensor is

$$T_{em}^{\mu\nu} = u n^\mu n^\nu, \qquad (15.37)$$

where $u = F/c = T_{em}^{00}$ is the energy density and F the energy flux; the spacetime components T^{0i} are Poynting's vector (19.1). Let v^μ be the four-velocity in $\mathcal{S}$ of the body and mv^μ its four-momentum, in terms of the rest mass m. To describe its interaction with the radiation we adopt a local comoving frame $\mathcal{S}'$, such that the particle is momentarily at rest; denoting with a prime the corresponding components, at that time $v'^\mu = (1, \mathbf{0})$. To obtain the primed coordinates one can apply a (local) Lorentz transformation to the unprimed ones (eq. (17.5)); but this machinery can be avoided by using the covariance properties of the relevant quantities. Note that the rest-frame time component

$$n'^0 = \eta_{\mu\nu} n'^\mu v'^\nu = \eta_{\mu\nu} n^\mu v^\nu = w \qquad (15.38)$$

of the (null) propagation vector n'^μ is an invariant scalar and cannot be taken equal to unity. In the usual case of slow motion, to within terms $O(|\mathbf{v}|^2/c^2)$, $n'^0 = w = 1 - \mathbf{n} \cdot \mathbf{v}/c$. In the particle's rest frame S' the energy-momentum tensor can be written as

$$T'^{\mu\nu}_{\text{em}} = \frac{u'}{(n'^0)^2} n'^\mu n'^\nu , \qquad (15.39)$$

where $u' = T'^{00}_{\text{em}} = F'/c$ is the energy density measured in S'. The relation between u and u' is obtained by noting that the quantity $\eta_{\mu\varrho}\eta_{\nu\sigma}T^{\mu\nu}_{\text{em}}v^\varrho v^\sigma$ is an invariant, so that, with eq. (15.38),

$$u' = u(\eta_{\mu\nu}n^\mu v^\nu)^2 = u w^2 = u\left[1 - 2\mathbf{n} \cdot \mathbf{v}/c + O\left(\mathbf{v}^2/c^2\right)\right] . \qquad (15.40)$$

For a body moving away from the Sun, for example, the rest frame energy density is smaller than in the solar frame for two reasons, each of them contributing a factor w: (i) the surface collects less photons in a unit of time, and (ii) each photon has a lower frequency due to the Doppler effect.

For a body with size $R \gg \lambda_m$, the impinging radiative energy is either absorbed and thermally re-emitted, or scattered (reflected and diffused); for a large body this partition is described by Bond's albedo A [$(1 - A)$ being the absorbed fraction]. But for small dimensions this description, based on geometrical optics, begins to be invalid; one should take into account Fresnel diffraction, rigorous electromagnetic theory and the phase responses of each element of the body. When $R \ll \lambda_m$ much of the impinging flux is not affected by the body. To keep this into account it is customary to denote the absorbed and the scattered flux fractions by an absorption and a scattering coefficient Q_{ab}, Q_{sc}, whose sum is generally less than unity; for large bodies they reduce, respectively, to $1 - A$ and to A. Here we stick to this point of view, forego a detailed discussion of deviations from geometrical optics and assume elastic scattering.

Both processes are best described in the rest frame S' of the particle. In both cases the thermal energy, hence also the total mass mv'^0 of the body, is unchanged; the radiative force

$$f'^\mu = \frac{d(mv'^\mu)}{d\tau} \qquad (15.41)$$

has no time component.

Absorbing particle.– Consider first absorption. For a spherical and isothermal particle the re-emitted radiation has no privileged direction and carries no momentum; since the body is momentarily at rest, the momentum transfer is only due to the impinging radiation and has the usual form: it is directed along the unit vector $\mathbf{n}'$, along the spatial component of the null vector $n'^\mu = n'^0(1, c\mathbf{n}')$ in the rest frame S', and has the magnitude $Q_{ab} F'S/c$; Q_{ab} is the fraction of the incident radiation which remains in the body and gets thermalized, and S is the cross section of the illuminated side. In a four-dimensional notation this translates into

$$f'^\mu_{ab} = Q_{ab} \frac{F'S}{c^2} \frac{n'^\mu - n'^0 v'^\mu}{n'^0}. \qquad (15.42)$$

The second term in the numerator ensures that at the chosen time $f'^0_{ab} = 0$ and the energy mv'^0 has a vanishing time derivative; this term does not contribute to the spatial part and we recover the usual expression. Using the expressions $n'^0 = w$ and $F' = Fw^2$ (eqs. (15.38) and (15.40)) this becomes a vector equation which holds also in the solar system frame:

$$f^\mu_{ab} = Q_{ab} \frac{FSw}{c^2} (n^\mu - wv^\mu). \qquad (15.43)$$

Scattering particle.– Scattering (reflection and diffusion, as it generically occurs for glassy or metallic particles), involving the flux $Q_{sc}F$, must also be taken into account. Assuming *elastic scattering*, an impinging photon with null vector $n'^\mu = n'^0(1, c\mathbf{n}')$, and the direction $\mathbf{n}'$, is deflected in the particle's rest frame to $n'^\mu + \Delta n'^\mu = n'^0(1, c(\mathbf{n}' + \Delta \mathbf{n}'))$; the scattering process may depend in a complex way on the physical and optical properties of the surface and the wavelength of the radiation (a classical theory of electromagnetic wave scattering by dielectric spheres was developed by G. MIE in 1908 and later developed for astronomical purposes by H.C. VAN DE HULST in the 1950s). Let $\psi(\mathbf{n}', \Delta \mathbf{n}') d\Omega$ be the probability that a photon is scattered from $\mathbf{n}'$ to $\mathbf{n}' + \Delta \mathbf{n}'$ within the solid angle $d\Omega'$; for normalization

$$\int_{4\pi} \psi(\mathbf{n}', \Delta \mathbf{n}') d\Omega' = 4\pi.$$

The average change is entirely space-like

$$\langle \Delta n'^\mu \rangle = n'^0 \left(0, c \int_{4\pi} \psi(\mathbf{n}', \Delta \mathbf{n}') \Delta \mathbf{n}' \frac{d\Omega'}{4\pi} \right). \qquad (15.44)$$

In the particle's rest frame a reaction force arises

$$f'^\mu_{sc} = -Q_{sc} \frac{F'S}{c^2} \frac{\langle \Delta n'^\mu \rangle}{n'^0}. \qquad (15.45)$$

Consider the simplest case in which the scattering probability $\psi(\mathbf{n'}, \Delta\mathbf{n'})$ is axially symmetric around $\mathbf{n'}$, so that

$$\int_{4\pi} \psi(\mathbf{n'}, \Delta\mathbf{n'})\Delta\mathbf{n'} \frac{d\Omega'}{4\pi} = -g\,\mathbf{n'} \,. \tag{15.46}$$

The coefficient g describes the asymmetry between forward and backward scattering: e.g. $g = 2$ for full backward reflection. In this case, in a spacetime notation, the radiative force (15.45) has exactly the same expression as before (15.42), except that $gQ_{\rm sc}$ replaces $Q_{\rm ab}$. The final expression for the Poynting-Robertson force in the solar system frame is, therefore

$$f^{\mu} = Q_{\rm pr}\frac{FSw}{c^2}(n^{\mu} - wv^{\mu})\,, \quad \left[Q_{\rm pr} = Q_{\rm ab} + gQ_{\rm sc}\right] \,. \tag{15.47}$$

In the slow motion approximation, up to $O(|\mathbf{v}|^2/c^2)$, sufficient for solar system applications,

$$\mathbf{f} = Q_{\rm pr}\frac{FS}{c}\left[\left(1 - \mathbf{n}\cdot\frac{\mathbf{v}}{c}\right)\mathbf{n} - \frac{\mathbf{v}}{c}\right] \,. \tag{15.48}$$

Obviously the leading term $-\propto 1/c -$ is the radiation pressure.

Orbital effects of the Poynting-Robertson drag.– The component of the Poynting-Robertson force along $\mathbf{v}$ has the opposite sense and decreases the energy; for a generic particle in a heliocentric orbit it is stronger at pericentre and diminishes not only the semimajor axis, but also the eccentricity (as for the atmospheric drag; Sec. 12.3), with important long-term effects. The ratio $\beta_r v/c$ between the Poynting-Robertson acceleration and $GM_\odot/r^2$ gives the order of magnitude of the fractional change of the orbital elements in a period; hence the timescale of the orbital circularization and decay is

$$\tau_d \simeq \frac{c}{v n \beta_r} \simeq \frac{mc^2}{FS} \approx 7\times 10^6 \left(\frac{\varrho}{\text{g/cm}^3}\right)\left(\frac{R}{1\text{ cm}}\right)\left(\frac{r}{1\text{ AU}}\right)^2 \text{ y} \,. \tag{15.49}$$

This is also the time it takes for a particle to absorb an amount of solar radiation energy equal to its mass. For µm-size particles τ_d does not exceed 10^3 y. For this reason, dust bands originating in the principal asteroid families (or recent breakup events in the main asteroid belt) extend throughout the inner solar system. An interstellar dust particle approaching the Sun in a hyperbolic trajectory with an impact parameter p and a velocity v_∞ also loses energy and is trapped if the semimajor axis appreciably changes in the interaction time p/v_∞, that is, when $\beta_r v_\infty/c \simeq 1$; this has interesting astrophysical consequences (Problem 15.1).

For a planetocentric orbit the same value of τ_d applies, provided r is replaced with r_p, the planet's distance from the Sun. The optically thin, collisionless dusty rings of Jupiter evolve fast under the influence of the Poynting-Robertson effect; in some cases, however, small particles can be lost earlier due

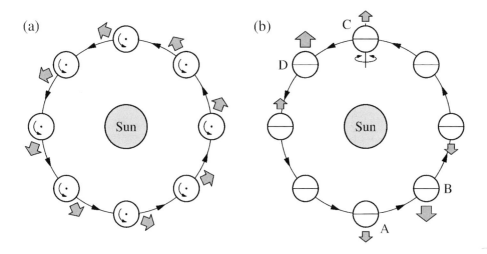

Figure 15.8. (a) The diurnal Yarkovsky effect, with the spin axis perpendicular to the orbital plane. With a finite thermal inertia, the solar radiation is re-emitted with a lag and the recoil force has the direction of the wide arrows. Its average transverse (in this case along-track) component causes the body to spiral away from the Sun. (b) The seasonal Yarkovsky effect, with the spin axis in the orbital plane. The seasonal heating of the northern and southern hemispheres gives rise to a thermal force along the wide arrows. When the energy absorbed at A is re-radiated later at B, after a fraction of the orbital period, the force has an average transverse component. This occurs for any orientation of the spin, but is largest when it lies in the orbital plane.

to changes in eccentricity induced by radiation pressure or to collisional disruption, as discussed above and in Sec. 14.7. In optically thick planetary rings, like Saturn's, this decay time may not apply because particles can collide and interact gravitationally.

The Yarkovsky effect. If the body is not isothermal, a radiative recoil is produced, which pushes the body in the direction opposite to the hotter part. In a spacecraft this may occur due to *internal* heat sources; for example, in interplanetary probes *Radioisotope Thermoelectric Generators* (RTG) dissipate most of their power. Cassini's RTG have a total power of 10 kW; its non-gravitational acceleration f_{ng} has been measured and found to be $\simeq 3.5 \times 10^{-7}$ cm/s^2, corresponding – with a mass $m \simeq 5$ tons – to an anisotropically emitted power $m c f_{ng} \simeq 5$ kW half as large. A similar signal was found for the Voyager and Pioneer spacecraft.

The *Yarkovsky effect* arises when the temperature inhomogeneity is produced by an *external* radiative source, like the Sun (or a planet); it depends in a complex way on the size, thermal properties (in particular, the thermal conductivity) and the rotational state (both the spin rate and the direction of the polar axis), as well as the distance from the Sun. In the generic case of a substantial temperature inhomogeneity, the amount of anisotropically emitted radiation is comparable with the impinging flux, and the Yarkovsky force is of the same order as the direct radiative force (diminished geometrically due to thermal re-emission in a half space).

The terms "day" and "season" are used below and in Fig. 15.8 to refer, respectively, to temperature gradients along the parallels and the meridians of the rotating body. The substance of the Yarkovsky effect is due to a delay – the thermal lag τ_{th} – between the time at which radiation is absorbed at a point of the surface and the re-establishment of thermal equilibrium. If τ_{th} is not much smaller that the rotation period, the hottest point on each parallel is displaced from the subsolar point, and we have the *diurnal* variant of the Yarkovsky effect (Fig. 15.8a). Fast rotation, small size and large thermal conductivity diminish the temperature anisotropy and make it ineffective. For simplicity, let the axis be perpendicular to the orbital plane; if the rotation is prograde (retrograde), besides the radial force, there is a permanent transversal component T in the same (opposite) sense as the motion, similar to the recoil force due to outgassing from cometary nuclei (Sec. 14.5). The semimajor axis of the orbit secularly increases (decreases) by the amount given by eq. (12.4). For rotation speeds and material constants typical in the solar system, the optimal size ranges from centimetres to metres; if a long-term accumulation is possible, the diurnal Yarkovsky effect may significantly influence even the motion of kilometre size asteroids.

The *seasonal* variant of the Yarkovsky effect occurs when the spin axis obliquity ϵ is not zero. Consider, for example, a body in a circular heliocentric orbit, with a rotation around the axis **e** in the orbital plane so fast that the temperature distribution is axially symmetric around **e**; this is also the direction of the Yarkovsky force. As shown in Fig. 15.8b, the thermal lag τ_{th} delays the occurrence of the largest temperature at the pole facing the Sun from A to B; a transverse force arises in the direction opposite to the orbital velocity, entailing a steady decay of the semimajor axis. Note that this does not depend on the sense of rotation. This situation applies to meteoroids and their precursors – mostly bare rocks with a minor porosity – with sizes in the 1 – 200 m range. For LAGEOS the same effect occurs and the angular lag between A and B is $\simeq 40°$; for meteoroids it is of the order of degrees, but can still produce important long-term orbital effects.

We now review the main physical ingredients of the complex heat transfer process at the root of the Yarkovsky effect. Consider a spherical homogeneous body of radius R at rest, illuminated by a flux $\mathbf{F}(t)$ from the Sun, with the appropriate diurnal and seasonal motion in the "sky". Let $2\pi\nu$ be the relevant frequency, namely, the rotation rate or the mean motion in the diurnal and seasonal case, respectively. The surface temperature $T_s = T_{eq} + \Delta T_s$ must be inhomogeneous and different from the equilibrium temperature T_{eq} given by eq. (7.9); when the correction ΔT is small, the Yarkovsky recoil acceleration on a homogeneous body of mass m is

$$\mathbf{f}_Y = -\frac{2}{3}\frac{\varepsilon\sigma}{mc}\int_S dS\,\mathbf{n}_\perp\,T_s^4 \simeq -\frac{8}{3}\frac{\varepsilon\sigma}{mc}T_{eq}^3\int_S dS\,\mathbf{n}_\perp\,\Delta T_s\,,\qquad(15.50)$$

where $\mathbf{n}_\perp$ is the unit vector along the external normal, σ the Stefan-Boltzmann constant and ε the infrared emissivity; the integration is over the whole surface of the body. The (time-dependent) temperature distribution ΔT in the body is obtained solving the heat transport equation (1.59), characterized by the heat diffusivity χ (assumed constant for simplicity). In addition, at the surface the absorbed energy flux (corresponding to the fraction $(1-A)$ of the incident radiation, in terms of Bond's albedo A) is partly re-emitted and partly conducted into the interior according to eq. (1.60), which now is linear and reads:

$$\left(\varrho\,c_p\chi\frac{\partial\Delta T}{\partial r}\right)_{r=R} + 4\varepsilon\,\sigma T_{eq}^3\,\Delta T_s = (1-A)\,\mathbf{F}\cdot\mathbf{n}_\perp\,.\qquad(15.51)$$

Here $\kappa = \varrho c_p\chi$ is the thermal conductivity, in terms of the density ϱ and the specific heat c_p. At the timescale $1/\nu$ the temperature gradient produced by the changing flux $\mathbf{F}$ penetrates to a depth $\delta \approx \sqrt{\chi/\nu}$, with a velocity $\sqrt{\chi\nu}$. The external flux is in part directly re-emitted due to the changing surface temperature ΔT_s and in part conducted inside, proportionally to the radial temperature gradient $\approx \Delta T_s\sqrt{\nu/\chi}$; it is this term which produces the thermal lag τ_{th} and the corresponding phase lag $\phi_{th} = 2\pi\nu\tau_{th}$, and gives rise to the orbital change.

A planar toy model allows a quantitative estimate of ϕ_{th}. To mimic the periodic change, consider a conducting half space $z \leq 0$ and a periodic flux $F \propto \exp(2\pi i\nu t)$ impinging along the surface normal. The temperature distribution is determined by the one-dimensional heat equation (1.61), which now reads

$$\chi\frac{\partial^2\Delta T}{\partial z^2} = 2\pi i\nu\Delta T\,.\qquad(15.52)$$

Eq. (15.52) has the solution (vanishing at $z = -\infty$)

$$\Delta T(z) = \Delta T_s\,\exp\left[(1+i)\,\sqrt{\pi\nu/\chi}\,z\right]\,,\qquad(15.53)$$

with the (exponential) penetration depth $\delta = \sqrt{\chi/(\pi\nu)}$. Using the thermal parameters of bare basalt ($\chi \approx 0.01$ cm^2/s), for the seasonal thermal wave $\delta \simeq 8$ m; an insulating regolith on the surface of asteroids decreases it to $\simeq 15$ cm. For the diurnal thermal wave δ is smaller by $\sqrt{n/\omega}$, with n the orbital mean motion and ω the rotation frequency. The linearization (15.51) of the boundary condition permits again to treat separately different Fourier terms and gives the surface temperature

$$\left[(1+i)\varrho\,c_p\,\sqrt{\pi\chi\nu} + 4\epsilon\sigma T_{eq}^3\right]\Delta T_s = (1-A)\,F\,,\qquad(15.54)$$

where F is the amplitude of the Fourier term of the impinging flux at frequency $2\pi \nu$. The imaginary term, in quadrature with F, is responsible for the lag ($\varrho c_p \sqrt{\pi \chi \nu}$ is sometimes called the thermal inertia). Quantitatively, the surface temperature ΔT_s is proportional to $\exp i(2\pi \nu t + \phi_{\text{th}})$, with

$$\tan \phi_{\text{th}} = -\frac{\Theta}{1 + \Theta} \quad \left(\Theta = \frac{\varrho c_p \sqrt{\pi \chi \nu}}{4 \epsilon \sigma T_{\text{eq}}^3} \right) ; \quad (15.55)$$

$\phi_{\text{th}} < 0$, which means that the temperature lags behind the radiation flux. To see the meaning of the *thermal parameter* Θ, note that in the full non-linear case $\Delta T_s \approx T_{\text{eq}}$, $\varrho c_p T_{\text{eq}}$ is the density of thermal energy in the conducting layer and $\varrho c_p T_{\text{eq}} \sqrt{\nu \chi}$ is its flux; Θ is its ratio to the directly re-emitted flux $\epsilon \sigma T_{\text{eq}}^4$. As expected, the largest lag ($\phi_{\text{th}} = -\pi/4$ in this simple case) occurs when $\Theta \gg 1$ and the internal flux prevails. In this limit, however, the Yarkovsky effect becomes negligible, since the amplitude of temperature variations behaves like $\Delta T_s \propto 1/\Theta$. The rule of thumb indicates that the most important effects occurs for $\Theta \approx 1$.

A realistic analysis of the Yarkovsky effect requires solving in a spherical harmonics representation the full three-dimensional heat transfer problem (1.59) and (15.51), taking into account the appropriate flux $\mathbf{F}(t)$, which differs from zero only on the illuminated side. In this way the effect of a finite ratio $R' = R/\sqrt{\chi/(\pi \nu)}$ between the radius R and the penetration depth is accounted for: as expected, when $R' \ll 1$ the phase lag vanishes and there is no heat wave since the body become nearly isothermal due to the efficient thermal conduction (Problem 15.11). The transverse acceleration T_y has the form

$$T_y = \frac{2}{9}(1 - A)\frac{F\pi R^2}{mc} G(R', \Theta) \zeta(\epsilon) . \quad (15.56)$$

The function G (always negative), plays the role of $\tan \phi_{\text{th}}$ and describes the effect of the thermal lag; it has the general form

$$G(R', \Theta) = -\frac{g_1(R')\Theta}{1 + 2g_2(R')\Theta + g_3(R')\Theta^2} , \quad (15.57)$$

where the g-functions describe the dependence on the radius (when $R' \to \infty$ we recover the plane model with $g_1 = g_2 = 1$ and $g_3 = 2$); $\zeta(\epsilon)$ accounts for the dependence on the obliquity. For the diurnal and the seasonal cases $\zeta = -2\cos \epsilon$ and $\zeta = \sin^2 \epsilon$, respectively; under the inversion of the rotation sense $\epsilon \to \epsilon + \pi$, the former effect changes sign. The diurnal effect is largest at $\epsilon = 0°$ (or $180°$) and 0 at $90°$; the opposite is true for the seasonal effect, which always acts as a drag, in agreement with Fig. 15.8. As in most non-gravitational forces, T_y decreases with the radius as $1/R$.

The possibility of long-term changes of the semimajor axis for bodies of size up to 10 km makes the Yarkovsky effect important for meteorite delivery, for the origin of small near-Earth asteroids and the configuration of small members in the asteroid families (Sec. 14.3). Within the estimated collisional lifetime of a few tens of My, a decametre size meteoroid can accumulate a change Δa in its semimajor axis of ≈ 0.1 AU; a small asteroid one kilometre large can have $\Delta a \approx 0.01$ AU in 1 Gy. Note that 0.1 AU is the typical distance in the main asteroid belt to the nearest Kirkwood gap, while 0.01 AU is about the width of an asteroid family (Sec. 14.3 and Fig. 14.3). The precise dynamics driven by the Yarkovsky effect is complicated; in particular, the obliquity of the spin axis, that strongly affects T_Y, is not likely to stay constant, owing to collisions and other dynamical factors, such as those discussed in Sec. 7.1 (and the Appendix of Ch. 7) and in Problem 15.10.

Solar wind drag. The solar wind, like the solar radiation, produces a radial force and a drag. Since its velocity v_w is radial and approximately constant (Sec. 9.2), the momentum flux $\varrho_w v_w^2$ is proportional to $1/r^2$ and, similarly to the solar radiation, effectively decreases gravity. The radial force, obtained from eq. (15.48) by replacing F/c with $\varrho_w v_w^2$, is 3×10^{-4} times smaller than the radiation pressure for the typical values $\varrho_w \simeq 10^{-23}$ g/cm^3 and $v_w \simeq 400$ km/s. On the other hand, the drag, obtained by replacing in eq. (15.48) v/c with the much larger quantity v/v_w, is only ≈ 0.3 times the Poynting-Robertson drag. For particles smaller than ≈ 0.1 μm, moreover, the coefficient Q_{pr} which determines the radiative drag is much smaller than unity and the orbital decay is mainly due to the solar wind drag.

Electromagnetic forces. Particles (and spacecraft) in space acquire an electric charge, whose sign and magnitude are determined by the competition between the ionization produced by the solar UV radiation, which contributes a positive charge, and the differential flow of electrons and ions. Since the ratio of their thermal velocities is the square root of their mass ratio, a body in a plasma with electron density n_e charges negatively until an electrostatic potential arises which restores the equilibrium situation. This potential $\varphi_e \approx kT/e$ extends to a distance of the order of the Debye length λ_D (see Sec. 1.8, eq. (1.102)); the body of radius R acquires a charge $q = R\phi_e = e(4\pi\lambda_D^2 n_e R)$ and in a magnetic field B is subject to a Lorentz acceleration $qvB/(mc) \propto 1/R^2$, in effect relevant only for sizes of ≈ 1 μm and smaller. If the surrounding plasma is too rarefied, positive charging by photo-ionization may prevail (typical for interplanetary space).

Electromagnetic effects are particularly important in planetary rings, where the magnetic field and the electron density are generally larger than in interplanetary space. The motion of small grains is influenced by the surrounding plasma and also by the planetary magnetic field. Their differential motion with respect to the magnetospheric plasma, dragged along by the planetary rotation, produces the *plasma drag*. Inside (outside) the corotation radius (where $\omega = n$) the grain loses (gains) orbital angular momentum and drifts inwards (outwards), unless the Poynting-Robertson effect prevails. For Jupiter, the corotation distance is at $\simeq 2.24\,R_J$, inside the gossamer rings (Figs. 14.18 and 14.19). A small outward extension of the outer gossamer ring beyond the orbit of Thebe is likely produced by plasma drag.

The motion of charged grains is also directly influenced by the Lorentz force due to the planetary magnetic field (eq. (10.61)); in Sec. 10.7 we discussed the possibility of the elimination of a fast grain due to the corotational electric field. Another interesting situation occurs when the planetary dipole is inclined to the orbital plane and the Lorentz force changes with the rotation period of the planet. If some of the frequencies at which the particle experiences this force is commensurable with its vertical or epicyclic orbital frequencies, we have *Lorentz resonances*. They occur both inside and outside the corotation radius and are especially important for Jupiter's ring system, resulting in the formation and the disappearance of the halo (Sec. 14.7).

Appendix: Symplectic integrators

The mapping $(q_0^i, p_{0j}) \to (q^i, p_j)$ generated by a Hamiltonian $H(q, p)$ from the (arbitrary) time 0 to the time τ is a *contact transformation* and keeps the energy constant; it also called a *symplectic mapping*. Conventional numerical schemes of order k with a time step τ can approximate this mapping to $O(\tau^k)$, but they are not symplectic maps and miss their important properties, in particular the energy conservation. For long-term evolution of the planetary system we need integrators which preserve the symplectic character.

The Hamiltonian rate of change of a function $F(q, p)$ in phase space (in particular, one of the coordinates q^i or the momenta p_i) is

$$\frac{dF}{dt} = \{F, H\} = \mathcal{L} F, \tag{a}$$

where $\{F, H\}$ is the Poisson bracket (eqs. (12.31) and (12.33)) and $\mathcal{L}$ the corresponding linear operator

$$\mathcal{L} = \frac{\partial H}{\partial p_i}\frac{\partial}{\partial q^i} - \frac{\partial H}{\partial q^i}\frac{\partial}{\partial p_i}. \tag{b}$$

The formal solution of eq. (a)

$$F(\tau) = \exp(\tau\mathcal{L})\,F(0) = \left[1 + \tau\mathcal{L} + \frac{\tau^2}{2}\mathcal{L}^2 + \ldots\right] F(0) \tag{c}$$

is defined as an infinite series (see also Problem 12.18); in a numerical realization the expansion is cut at the term $\propto \tau^k$, with an error $O(\tau^{k+1})$. This provides a symplectic numerical integration scheme.

When the Hamiltonian $H = H_A + H_B$ is the sum of two parts, such that the commutator $[\mathcal{L}_A, \mathcal{L}_B] = \mathcal{L}_A\mathcal{L}_B - \mathcal{L}_B\mathcal{L}_A$ (not to be confused with the Lagrange bracket; Sec. 12.2) does not vanish, the propagation in time is not obtained by applying H_A and H_B in succession; in other words, the operator $\mathcal{L}$ defined by

$$\exp(\tau\mathcal{L}) = \exp[\tau(\mathcal{L}_A + \mathcal{L}_B)], \tag{d}$$

is not equal to

$$\exp(\tau\mathcal{L}') = \exp(\tau\mathcal{L}_A)\exp(\tau\mathcal{L}_B). \tag{e}$$

Their difference, for two arbitrary operators, is provided by the *Baker-Campbell-Hausdorff* *(BCH) theorem*:

$$\mathcal{L}' - \mathcal{L} = \frac{\tau}{2}[\mathcal{L}_A, \mathcal{L}_B] + \frac{\tau^2}{12}([\mathcal{L}_A,[\mathcal{L}_A,\mathcal{L}_B]] + [\mathcal{L}_B,[\mathcal{L}_B,\mathcal{L}_A]]) + \dots . \quad (f)$$

Only the first order integrator does not face this problem and consists of two substeps: (i) advance the system with H_B, ignoring the effect of H_A, and (ii) advance the resulting system with H_A, ignoring the effect of H_B. This corresponds to (e), i.e.

$$F(\tau) = \exp(\tau \mathcal{L}_A) \exp(\tau \mathcal{L}_B) F(0) . \quad (g)$$

The situation is especially convenient if each of the two propagators can be easily obtained (for instance, analytically); the BCH theorem can be then used to estimate the error and/or provide a correction. Suppose, for example, that $H = T(p) + V(q)$ is the sum of two parts, each of which depends only on the coordinates or the momenta; then to first order we have the *leap-frog integrator*

$$q(\tau) = q(0) + \tau \left(\frac{\partial T}{\partial p}\right)_{p=p(0)} , \quad p(\tau) = p(0) - \tau \left(\frac{\partial V}{\partial q}\right)_{q=q(\tau)} . \quad (h)$$

The BCH theorem indicates that the error corresponds to the Hamiltonian

$$H_{\text{err}} = \frac{\tau}{2}[T, V] + O(\tau^2) ; \quad (i)$$

of course, higher-order integrators can be constructed that approximate the true evolution more closely (Problem 15.7).

In planetary (or satellite) dynamics the Hamiltonian is usually split in its Keplerian part H_K relative to the primary, and the interaction μH_{int}, where we have explicitely indicated its order of magnitude in terms of the ratio μ of the planets' to the solar mass. H_K propagates positions and momenta along Keplerian ellipses. The use of canonical (e.g., Delaunay's; Problem 12.9) elements would require solving only for Kepler's time law, but it has the disadvantage of a very complicated expression for the interaction. Usually Cartesian coordinates are adopted and the Keplerian propagation may be easily realized (as for the symplectic maps discussed in Sec. 15.1) using the f- and g-functions (Problem 11.12); the time step must be smaller than the shortest planetary revolution period (typically $\simeq 0.01 - 0.1$ of that value, depending on the order of the integrator). Since H_{int} does not depend on the momenta, it leaves the coordinates q^i unchanged. The two modes of propagation are used in the symplectic mapping described in Sec. 15.1. The error (i) of the first order integrator is $O(\mu\tau^2)$. Unlike conventional methods, the perturbation structure of planetary dynamics is directly reflected in the symplectic integrators. The Keplerian motion, representing a zero-order solution of the problem, is directly incorporated in the method, with two advantages: (i) the use of analytic integrators places, in principle, no limitation to the time step and (ii) the error is reduced by the smallness of the perturbation. If the integration step τ is kept constant, the effective Hamiltonian $H + H_{\text{err}}$ is exactly conserved; consequently, symplectic integrators do not lead to a long-term build up in the energy. In this way the best numerical estimate of long-term changes in planetary semimajor axes is obtained, which is important for the resonance structure of the system. On the other hand, symplectic integrators are typically less efficient with very eccentric orbits; similarly, close encounters, where H_{int} and H_{err} are large, require special care. In these cases conventional methods can be optimized with adaptive changes of the integration step. The lack of applications of symplectic integrators to these important situations (e.g., planetary formation from planetesimals; Ch. 16) is being challenged by recent theoretical advances.

– PROBLEMS –

15.1.* Consider a simple model of a capture of a neutral interstellar dust particle in the solar system: the particle, initially on a hyperbolic orbit with an asymptotic velocity v_∞ and an impact parameter b looses energy by the Poynting-Robertson effect. For a given value of $v_\infty \approx$ 25 km/s, estimate the range of b for which capture occurs.

15.2.* Find the zenithal distance of the Sun ψ on a point of Mercury's equator as function of the sidereal time t. At first, neglect the eccentricity e; then take e into account to first order, reckoning the longitude ϕ from the point at which a pericentre passage occurs at noon. For a point P at an arbitrary longitude, find out, as function of t, ψ and the equilibrium temperature of P (neglecting the thermal conductivity and given the local albedo and the emissivity).

15.3.* With the present initial conditions for the Earth-Moon system, and neglecting solar tides, find the time evolution of ω and n, and determine the timescale to attain the synchronization state $\omega = n$. Will complete solar eclipses cease to be observable from the Earth surface in the future? If so, estimate when. Determine the timescale with which the Pluto-Charon system attained synchronization since their formation.

15.4.* Using Gauss' equations for a and e, averaged over one revolution around centre, show that for a particle in a solar orbit subject to the Poynting-Robertson drag, $a\,e^{-4/5}(1-e^2)$ is a constant of motion. Study the case $e \ll 1$. Show that for a particle orbiting a planet $de/dt = 0$.

15.5. Which fraction of the total angular momentum of the Earth-Moon system is presently accounted for by the orbital motion? The same for Jupiter's Galilean satellites and Pluto's Charon (see Ch. 14 for the relevant parameters).

15.6. Estimate the thermal radiation recoil produced by the Earth infrared radiation on LAGEOS (Sec. 20.3).

15.7. Prove that $\exp(\tau \mathcal{L}_B/2) \exp(\tau \mathcal{L}_A) \exp(\tau \mathcal{L}_B/2)$ approximates to $O(\tau^2)$ the evolution operator (d) in the Appendix, and obtain its formal expression to this order. Write it explicitly for the one-dimensional case $H = T(p) + V(q)$ (the second-order leap-frog integrator). (Hint: Use the BCH theorem.)

15.8. Estimate the orbital decay due to the Poynting-Robertson effect of the terrestrial and solar radiation) of the satellite Etalon. On a nearly circular geocentric orbit with $a \simeq 25,478$ km and $I \simeq 65°$, this large, passive satellite has an area-to-mass ratio $S/m \simeq 9.3 \times 10^{-3}$ cm^2/g.

15.9. Obtain the Poynting-Robertson force for a metallic sphere in a solar orbit with absorption coefficient α and reflection coefficient $(1-\alpha)$.

15.10.* The Yarkovsky thermal recoil on bodies of irregular shape produce a torque, able to affect on a long-term the rotation state of small asteroids and meteoroids (the *YORP effect*, for Yarkovsky-O'Keefe-Radzievski-Paddack). With an idealized shape, like a slab or windmill, estimate the timescale required to change appreciably the rotation period for a 1 km asteroid and a 50 m meteoroid.

15.11.* Solve the linearized heat diffusion problem for a spherical body revolving on a circular orbit around the Sun and prove that the G-function in (15.56) is generally given by (15.57). Find the functions g_1, g_2 and g_3 and discuss their limit for $R' \to 0$ and $R' \to \infty$.

15.12.* Tidal potential of arbitrary degree: generalize eqs. (4.16) and (4.23) to include an arbitrary, axially symmetric deformation and show that

$$U_T(\mathbf{r},t) = -\frac{Gm}{R} \sum_{\ell \geq 2} k_\ell \left(\frac{R}{r}\right)^{\ell+1} \left(\frac{R}{r_M}\right)^{\ell+1} P_\ell\left(\frac{\mathbf{r} \cdot \mathbf{r}_M}{rr_M}\right) ;$$

m is the mass of the Moon and $\mathbf{r}_M$ its position relative to the Earth (of radius R), while k_ℓ are the appropriate Love's numbers.

15.13.* As suggested by Sec. 15.3, one can model the anelastic Earth response to lunar tides assuming a fixed time lag δt between the lunar stress on the Earth (with angular velocity ω) and the reestablishing of equilibrium. In the expression of the previous problem for the tidal potential response U_T, the relative position $\mathbf{r}_M(t)$ of the Moon is replaced with

$$\mathbf{r}_\star(t) = \mathbf{r}_M(t - \delta t) + \delta t\, \omega \times \mathbf{r}_M(t) .$$

Expanding to first order in δt, derive the tidal force on the Moon.

– FURTHER READINGS –

For the classical theory of secular perturbations, see D. Brouwer and G.M. Clemence, *Methods of Celestial Mechanics*, Academic Press (1961); another treatment of this problem are given by A.E. Roy, *Orbital Motion*, Hilger (1978). Modern topics of solar system dynamics, including its long-term evolution, tidal effects and resonance phenomena, are discussed in C.D. Murray and S.F. Dermott, *Solar System Dynamics*, Cambridge University Press (2000) and in A. Morbidelli, *Modern Celestial Mechanics: Aspects of Solar System Dynamics*, Taylor and Francis (2002). Chaos in the motion of solar system bodies are reviewed by J. Laskar, Marginal stability and chaos in the Solar System, in: *Dynamics, Ephemerides ans Astrometry of the Solar System*, S. Ferraz-Mello et al., eds., Kluwer (1996), J.J. Lissauer, Chaotic motion in the solar system, *Rev. Mod. Phys.* **71**, 845 (1999), and N. Murray and M. Holman, The role of chaotic resonances in the solar system, *Nature* **410**, 773 (2001). The particular case of the Jupiter Galilean satellites is discussed in detail by S. Ferraz-Mello, *Dynamics of the Galilean Satellites*, Sao Paulo University Press (1979), somewhat older, but still very fresh. A good book on the lunar dynamics is A. Cook, *The Motion of the Moon*, A. Hilger (1988). A brilliant exposition of the theoretical aspects of spin-orbit coupling, both in the case of natural bodies and artificial satellites, is in V.V. Béletskii, *Essais sur le Mouvement des Corps Cosmiques*, Editions Mir (1986). On tidal evolution, see K. Lambeck, *The Earth's Variable Rotation*, Cambridge University Press (1980), Ch. 10. The radiation forces on dust particles are thoroughly discussed by J.A. Burns, P.L. Lamy and S. Soter, Radiation Forces on Small Particles of the Solar System, *Icarus* **40**, 1 (1979), and F. Mignard, Radiation Forces, in: *Interrelations between Physics and Dynamics for Minor Bodies in the Solar System*, D. Benest and C. Froeschlé, eds., Editions Frontières (1991).

Chapter 16

ORIGIN OF THE SOLAR SYSTEM

Some 4.56 billion years ago a cloud of interstellar gas and dust, perhaps triggered by a nearby supernova explosion, began to collapse. Its central part contracted under its own gravity and started to heat up, until the temperature became so high that thermonuclear reactions were initiated, and the early Sun began to release growing amounts of energy. But the peripheral part of the cloud, owing to conservation of angular momentum, flattened to a disc with a maximum density on its median plane. A fraction of this material gradually solidified and gave rise to a swarm of small lumps of matter gravitationally bound to the Sun. These bodies underwent a process of further accumulation and coalescence which, eventually, through a complex sequence of mutual interactions and disturbances, both gravitational and non-gravitational, produced planets, satellites, asteroids and comets; indeed, this is the variety of bodies which we presently observe in the solar system. In the outer part of the nebula the biggest bodies as well were able to accrete a substantial amount of gas and undergo a partial collapse, generating their own nebula, in which a similar process of coalescence and accretion produced their satellite systems. In this chapter we discuss, mainly with estimates and results of numerical simulations, several features of this process, as presently understood on the basis of a growing body of evidence. While the basic ideas of this so called *nebular theory* go back to I. KANT and P.S. LAPLACE, its quantitative contemporary version was developed in the 1950s and the 1960s by O.Yu. SHMIDT, V.S. SAFRONOV and collaborators. Whereas the basic outline of solar system evolution remains the same, new and deeper theoretical insights, and numerical tests have called for several important adjustments of the overall scenario; in particular, the role of the gravitational instability must be reassessed. The recent, extraordinary discoveries of dozens of extrasolar planets in the solar neighbourhood, with properties in general remarkably different from our own, are raising questions and doubts about the universality of the processes which were at work in the formation of our own system. In Sec. 16.5 we mainly focus on the observations and the instrumental problems of these findings and briefly outline alternatives to the classical scenario for planetary formation.

16.1 Mass and structure of the solar nebula

Hundreds of gas envelopes around very young stars are observed in infrared and millimetric bands, especially in star-forming regions, like the Orion nebula. They have a wider spectrum than the black body, clear evidence of the presence of regions heated by the central star at different temperatures; these envelopes are supposed to have a disc structure (and this is occasionally observed), and are termed *proplyds* (for protoplanetary discs). They extend to hundreds of AU, with estimated masses between 10^{-3} and $0.1\ M_\odot$. Their lifetime, before dispersal, is expected to be $1-10$ My, though recent observations might indicate that in some cases the gas envelope lasts up to $\simeq 30$ My. Occasionally the rotation has also been measured with the Doppler shift of spectral lines. The most mature structures may betray signatures of embedded planetary systems, such as warps (β Pictoris), central offsets (HR 4796A), dust bands (β Pictoris and ε Eridani), resonant rings, and dust clouds. These observations confirm the general validity of the nebular theory.

The gravitational collapse of the cloud of interstellar gas which, presumably, gave rise to the Sun, is a complex astronomical process outside our main subject. It is governed by the *Jeans*, or *gravitational instability* (see the Appendix), involving an initial mass much larger than the Sun and extensive fragmentation. It is important to note that the composition of the cloud was very similar to the primordial solar nebula (Table 9.1). Traces of short-lived radionuclides (Sec. 5.3) suggest that the collapse of the protonebular interstellar cloud was triggered by a nearby supernova or Wolf-Rayet stars. Conserving its initial angular momentum **L**, the cloud progressively increased its oblateness (1.14), inversely proportional to its size; eventually, the centrifugal force produced the planetary nebula, a disc centred on the Sun and orthogonal to **L**. The age of primitive chondritic meteorites has been determined, with radionuclide dating, to be 4.56 Gy, with only a small scatter (Sec. 14.6); since they are assumed to have been formed within about 20 My from the formation of the nebula, we have a fairly precise knowledge about the timing of this event. The oldest components of primitive meteorites – inclusions in carbonaceous chondrites rich in calcium and aluminium, which can be radiometrically dated – are (4.566 ± 0.002) Gy old (Sec. 5.3); this is usually taken as the age of the solar system.

The reason why the composition of the two groups of planets is so markedly different is related both to their different distances r from the Sun and to the timing of planet formation versus the T-Tauri activity. The reference temperature T_{eq} induced in an absorbing body by solar radiation (eq. (7.9)) is proportional to $1/\sqrt{r}$; however, since several other processes, especially in the inner part, contributed to the heating, this can only give a guideline. For instance, during the early stages, when gas fell into the nebular midplane and was abruptly stopped there, supersonic velocities were attained and a shock front (Sec. 10.5) formed. Thus at 1 AU and 10 AU the highest temperatures reached are estimated to have been $\simeq 1,000$ K and $\simeq 100$ K (an order of magnitude smaller), respectively; in the inner nebula, close to the protosun, peak temperatures as high as $\simeq 1,500$ K might have been reached. In general, however, shock wave heating is not uniform, but rather localized in complicated geometries. In addition, the slow inward drift of the forming grains released gravitational energy and further heated the nebula. Though the chemical composition of the disc was initially the same as that derived from solar abundances, chemical compounds underwent a complicated evolution, depending on the available energy sources. Many details of these processes, crucial for understanding the currently observed compositional sequence, are very complicated and still poorly understood. This is principally because at sufficiently low temperatures the reaction rate is comparable to, or smaller than, the cooling rate of the nebula, and chemical reactions are not in thermodynamic equilibrium.

The nebula's compositional structure was affected not only by disc chemistry, but also by the efficiency of the ejection of light gases by *T-Tauri winds*, that must have occurred in the first $\simeq 10$ My after nebula formation. Indeed, young solar-type stars are known to undergo short phases of violent activity (or *T-Tauri phases*, from the name of a typical star of this class), with strong stellar winds and high mass-loss rates. This rapid removal of gas constrains the time when Jupiter, Saturn and, even more, Uranus and Neptune could accrete gas, since these planets must have reached a minimum mass in order to trap the gaseous component before it escaped. The terrestrial planets, on the other hand, never grew large enough for this; the formation of the inner and outer planets is separately discussed in Secs. 16.3 and 16.4.

While the Sun contains about 99.9% of the mass of the solar system, 98% of the total angular momentum resides in the orbital motion of the planets. Surely such an uneven distribution held good for the protosun and the planetary nebula; this puzzling situation has long been a convincing argument against the nebular theory itself and a motivation to search for alternatives (e.g. the once-popular hypothesis that planets formed from a spray of solar material ejected in a close encounter with another star). The heavy imbalance in angular momentum suggests that a coupling mechanism between the inner planets and the Sun has slowed down the latter. We know at least two plausible physical processes which could be responsible for this: (i) *magnetic* coupling, due to the freezing of magnetic field lines of the protosun in the plasma component of the inner nebula; (ii) *turbulent* coupling, due to the effective turbulent viscosity in the inner nebula, which lessens the differential rotation. Both of them are complex and it is difficult to model them in detail, but the anomalously slow rotation of the Sun, as compared to the mean rotation rate of young solar-type stars, is no longer seen as a fundamental difficulty for the nebular theory.

Quantitative estimate of the nebula mass and structure. To estimate the disc mass and its distribution in the protoplanetary nebula, we can use the present masses and compositions of the planets to derive a lower bound and a rough model, commonly called the *minimum mass nebula*. The formation of planetesimals and planets affected differently the heavier elements and the initially prevailing H and He, which were swept away to larger distances, even out of the solar system, by T-Tauri winds. To estimate the initial gaseous mass at a place where a planet subsequently formed, one can complement its composition in such a way so as to reconstitute the solar abundances. The major constituents of the terrestrial planets are heavy elements, like Mg, Si, Fe, with a total mass of $\approx 10^{-5} M_\odot$; if the appropriate quantities of H and He are added to them, an original mass some 300 times larger is obtained. For Jupiter and Saturn the situation is different: although enriched in heavy elements with respect to the Sun, they still retain a substantial amount of H and He, so that the present total mass $\approx 1.3 \times 10^{-3} M_\odot$ has to be increased by a factor ≈ 10. Uranus and Neptune, quite depleted in H and He, are probably an intermediate case; their present mass $\approx 10^{-4} M_\odot$ should be increased by about 50 times. Since the contribution of other bodies (Pluto, satellites, etc.) is negligible, our inventory leads to the conclusion that the total initial mass of nebula was at least a few times $10^{-2} M_\odot$. Obviously, this procedure implicitly assumes that the planets were formed at their present locations; this, however, may not be completely true, especially for the outermost giants (Sec. 16.4).

An estimate of the surface density profile in the nebula can be obtained by spreading the above reconstructed mass for each planet around its own position (Fig. 16.1). For Jupiter, we get ≈ 500 g/cm^2, comparable to the column density in the Earth's atmosphere. The approximate fitting law

$$\sigma(r) \simeq 300 \times \left(\frac{r}{8 \times 10^{13} \text{cm}}\right)^{-3/2} \text{g/cm}^2 \qquad (16.1)$$

can be integrated and gives a total mass, dominated by the upper limit at 50 AU, of 7×10^{31} g $\simeq 0.04 M_\odot$.

The thickness of the disc is determined by the balance between the component g_z of the gravity acceleration along the nebular rotation axis z and the gas pressure P (eq. (1.38)):

$$\frac{dP}{dz} = -\varrho g_z = -\varrho \frac{GM_\odot}{r^2} \frac{z}{r} = -\varrho n^2 z ; \qquad (16.2)$$

ϱ is the density of the gas and $n(r)$ the Keplerian mean motion. The self-gravity of the nebula has been neglected. Assuming for simplicity a perfect gas

16. Origin of the solar system

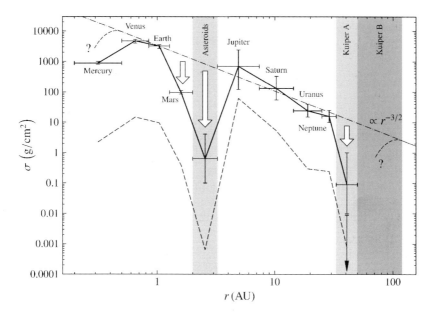

Figure 16.1. The surface density σ of the protosolar nebula (solid line) vs. the heliocentric distance r, obtained by restoring the planets to solar composition and spreading the resulting masses through contiguous zones surrounding the present orbits. The zone boundaries are taken as the arithmetic means of adjacent orbits; for Mercury and Neptune, the zones equally extend inward and outward. The present mass distribution in each zone is indicated by the dashed line. The error bars reflect uncertainties in planetary compositions. The present total mass of the asteroids and Kuiper-belt objects is estimated as $\approx 5 \times 10^{-4}\, M_\oplus$ and $0.13\, M_\oplus$, respectively. The smooth curve $\propto r^{-3/2}$ fits the giant planets, matching also Venus and Earth. Depletion mechanisms, indicated by arrows and question marks, are discussed in Secs. 14.3 and 14.4. The extrapolation of the fit beyond the Kuiper B zone (i.e. ≥ 120 AU) may give the primordial mass content in this region (continuous line). Data partially from S.J. Weidenschilling, *Astrophys. Sp. Sci.* **51**, 153 (1977).

(eq. (1.16)) with a thermal velocity $v_T \simeq \sqrt{3P/\varrho}$ independent of z, eq. (16.2) has the solution

$$\varrho(r,z) = \varrho(r)\exp[-3n^2 z^2/(2v_T^2)]. \quad (16.3)$$

The thickness of the disc can be defined as

$$H = \frac{\sigma}{\varrho(r,0)} = \frac{1}{\varrho(r,0)}\int_{-\infty}^{+\infty}\varrho(r,z)\,dz = \left(\frac{2\pi}{3}\right)^{1/2}\frac{v_T}{n}. \quad (16.4)$$

Since $v_T \propto r^{-1/4}$, $H \propto r^{5/4}$ and $\varrho \propto r^{-11/4}$: as a result of the diminishing gravity the nebula thickens with increasing distance. At $r = 5$ AU (about Jupiter's distance) we have $n \simeq 2 \times 10^{-8}$ s^{-1} and the thermal velocity for hydrogen at $\simeq 100$ K is $v_T \approx 1.6 \times 10^5$ cm/s, so that $H \approx 10^{13}$ cm, $\varrho \approx 10^{-10}$ g/cm^3 and

$P(0) \approx 0.3$ dyne/cm^2 $\approx 3 \times 10^{-7}$ atmospheres; the ratio $H/r \propto r^{1/4}$ is ≈ 0.1. From Poisson's equation, the self-gravity is of order $2\pi G\sigma$; its ratio to the solar gravity

$$\delta = \frac{2\pi\sigma r^3}{M_\odot H} \simeq \frac{2\pi\varrho r^3}{M_\odot} = \frac{\Omega_J^2}{2n^2} \propto r^{1/4}, \qquad (16.5)$$

≈ 0.1 at Jupiter, is always small, justifying its neglect; it can also be. Note also its expression in terms of the local Jeans' frequency Ω_J (eq. (a) in the Appendix).

Similarly to the discussion of planetary discs in Sec. 14.7, we use the concept of Roche's limit, below which a fluid mass M is disrupted by tides. At the distance r from the Sun its radius R must be smaller than the separation $r[M/(3M_\odot)]^{1/3}$ of the Lagrangian point L_1 from its centre (eq. (13.15)); hence its density must fulfil

$$\varrho > \varrho_c = \frac{9M_\odot}{4\pi r^3} = \frac{9}{2\delta}\varrho. \qquad (16.6)$$

Since $\delta < 1$, at the nebular density ϱ bodies cannot bind gravitationally; to overcome their tidal disruption, interatomic forces are required to first coagulate small particles to larger ones. The possibility of direct planetary formation in a more massive disc by gravitational instability has been considered. Though this scenario has several appealing features (e.g. it does not depend on poorly understood particle properties and better explains the eccentricity distribution of extrasolar planets; Sec. 16.5), in the solar system it is a less viable alternative (see, however, comments in Sec. 16.4). On the other hand, this possibility is still a plausible scenario for many of the observed extrasolar systems. In the Appendix we formulate elementary conditions for the onset of gravitational instability both in a uniform medium and, which is more relevant, in a differentially rotating disc.

In addition to the vertical pressure gradient, in the gaseous nebula a radial pressure gradient contributed to the equilibrium and affected the orbital motion of the gas. If n_g is the orbital angular velocity of the gas, the radial equilibrium condition is

$$rn_g^2 - \frac{GM_\odot}{r^2} = \frac{1}{\varrho}\frac{dP}{dr} \approx -\frac{v_T^2}{3r}. \qquad (16.7)$$

For our model nebula, at Jupiter's distance $v_T/(nr) \propto r^{1/4} \approx 0.1$, so that the deviation from Keplerian motion

$$\frac{n^2 - n_g^2}{2n^2} \simeq \frac{n - n_g}{n} = \frac{1}{3}\left(\frac{v_T}{nr}\right)^2 \propto \sqrt{r} \qquad (16.8)$$

is small, but still implying important effects as described in Sec. 16.2.

A gross picture of planetary formation: nominal model. This is the outline of the four phases of planetary formation, to be further developed in Secs. 16.2

to 16.4: (i) formation of solid particles in the gas environment and their settlement near the midplane of the nebula; (ii) their collisional growth to *planetesimals* (about kilometre-size objects); (iii) gravitational accumulation of planetesimals into *planetary embryos* (with approximately lunar-to-Mars masses), and (iv) planetary formation. In the case of the terrestrial planets the last phase faces a competition between collisional accumulation of planetesimals and disruption by impacts. For the giant planets, which contain a large amount of hydrogen and helium, the formation must have been faster; the core formation was followed by a runaway accretion of the massive atmosphere. The planet-forming region extends to roughly 50 – 100 AU; further away, the nebula had a density so low that the timescale necessary for collisional accumulation becomes longer than the solar system age. Current observations indicate that the density was not far from this limit.

Obviously, the transitions between the stages (i) to (iv) are somewhat arbitrary and also considerably overlap in the disc. They can be distinguished according to which physical process dominated at that time. In what follows, the "classical" theory is presented in some detail, mainly to introduce methods of analysis; but it should be pointed out that it is likely oversimplified and many doubts still remain, especially regarding the phases (i) and (ii). The processes that dominate the phases (iii) and (iv), namely the long-range mutual gravitational interaction of planetesimals and planetary embryos, have received much attention with numerical simulations, some of which are also briefly mentioned. Still, these numerical models typically cover a very limited part of planetary formation and rely on significant simplifications. The main difficulty is that some 30 – 40 orders of magnitude in mass separate dust particles in the young nebula from the planets; the relevant evolution timescale is $\simeq 10 - 100$ My. As a result, a direct numerical model of planetary formation is, for the time being, unfeasible.

16.2 Growth of solid grains and formation of planetesimals

The solar nebula, after its separation from the protosun, gradually cooled and, as a consequence, the vapour pressure of a number of constituents rapidly decreased and eventually fell below their partial pressure. The condensation of small solid particles (*grains*) began at temperatures $\approx 1,200 - 1,700$ K, with oxides of aluminium, calcium and titanium, and metals like iron and nickel. At lower temperatures, other materials became dominant: first silicates, then carbon-rich minerals, and finally water and water-ammonia ices; the latter are very important because their constituents – hydrogen, oxygen and carbon – have a high primordial abundance (Table 9.1). At the low estimated pressure in the nebula water ice condenses slightly below 200 K; though the temperature profile is poorly constrained, this is expected to occur near the outer edge of the asteroid belt, at $\simeq 3 - 4$ AU from the Sun. At very low temperatures, as the case is in distant regions, methane ice appears and molecules like CO and N_2, trapped in an H_2O matrix, can form *clathrates*. Thus a compositional gradient arose in the solid component, a circumstance reflected today in the very different compositions of planets, satellites and minor bodies at different distances (see Ch. 14). Grains started to grow by collecting material still in the vapour phase.

In the crudest possible approximation, it is assumed that every impinging molecule, other than hydrogen and helium, sticks to a spherical solid grain of radius R_s, density ϱ_s and mass $M_s = 4\pi \varrho_s R_s^3/3$ (obviously, the assumption of complete adsorption oversimplifies the real situation; the following estimate is to be taken as an upper limit). The growth rate of the mass and the radius then are

$$\frac{dM_s}{dt} = 4\pi R_s^2 \, \alpha \varrho \, \frac{v_T}{\sqrt{A}}, \quad \frac{dR_s}{dt} = \alpha \frac{\varrho}{\varrho_s} \frac{v_T}{\sqrt{A}} ; \qquad (16.9)$$

v_T is the thermal velocity of the (hydrogen) gas, α and A are the mass fraction and the molecular number of the condensate. Note that the radius grows at a constant rate. At Jupiter's distance, with $\varrho_s \approx 1$ g/cm^3, $A \approx 20$ (water ice), $\varrho \approx 10^{-10}$ g/cm^3 and $\alpha \approx 0.01$ (inside $\simeq 3$ AU only mineral grains survive and the solid fraction is lower by a factor of 2–3), we have $dR_s/dt \approx 2 \times 10^{-8}$ cm/s $\approx$ 0.6 cm/y. Thus in a short time grains became macroscopic particles and started sinking towards the midplane of the nebula. Obviously the microphysics of the growth of small particles is a very complicated and still poorly modelled process; limited experimental and numerical data indicate that loosely packed, possibly fractal, structures are formed, rather than solid particles. These fragile structures are susceptible to collisional breakup, but, on the other hand, their greater size helps sticking new material to them. It should also be noted that at some time melting and compactification must have taken place, as evidenced by *chondrules*, millimetric inclusions in primitive meteorites, rich in calcium and aluminium. The heat source responsible for their melting is uncertain.

To estimate the timescale of the grain settlement towards the midplane $z = 0$ of the disc and their final size, consider the equation of motion of a grain in the z-direction

$$\frac{d^2 z}{dt^2} = g_z + \frac{F_D}{M_s}, \qquad (16.10)$$

where F_D is the gas drag force. The mean free path of the gas molecules ($\approx 10^2$ cm, see eq. (1.101)) is typically larger than R_s and Stokes' drag law (Sec. 12.3) – appropriate for the hydrodynamic regime – does not apply; rather, the rate of momentum transfer is the difference between the contribution due to the molecules impinging from above, proportional to $(v_T + dz/dt)^2$, and from below, proportional to $(v_T - dz/dt)^2$. Since, as verified below, the grain velocity

$|dz/dt| \ll v_T$, we get *Epstein's formula*

$$F_D = -\frac{4\pi R_s^2}{3} \varrho v_T \frac{dz}{dt} = -M_s v_f \frac{dz}{dt}, \qquad (16.11)$$

where the factor 4/3 holds for a fully absorbing (or a specularly reflecting) sphere, and

$$v_f = \frac{\varrho v_T}{\varrho_s R_s} \qquad (16.12)$$

is the friction frequency. Since, as shown later, the inertial term in (16.10) is negligible, gravity and drag balance each other and we obtain:

$$\frac{dz}{dt} = \frac{g_z}{v_f} = -\frac{n^2}{v_f} z = -n^2 \frac{\varrho_s R_s}{\varrho v_T} z. \qquad (16.13)$$

Using eq. (16.9) for the changing grain size, we get the solution

$$z(t) = z(0) \exp\left(-\frac{\alpha n^2 t^2}{2\sqrt{A}}\right). \qquad (16.14)$$

The timescale for settlement on the midplane is

$$\tau_S \simeq \left(\frac{2}{\alpha}\right)^{1/2} \frac{A^{1/4}}{n}, \qquad (16.15)$$

of the order of ≈ 50 y, with a final value $R_s \approx 30$ cm and a mass (at the density of ice) $\approx 2 \times 10^5$ g. We can verify that $|dz/dt| \approx H/\tau_S \approx v_T/n\tau_S \ll v_T$ and that $d^2z/dt^2 \approx 2H/\tau_S^2 = \alpha n^2 H/A^{1/2} \approx (\alpha/A^{1/2})g_z \ll g_z$. Grains are thus expected to quickly sank through the gas on the midplane and form there a flat and thin layer of high density, resembling Saturn's rings. However, this idealized laminar settling does not take into account shear instabilities and fluid turbulence, as discussed below.

Radial motion. Friction also affects the Keplerian motion in the nebular plane, a more complex problem than the case of re-entry (Sec. 12.3). We know (eq. (16.8)) that the circular velocity of the gas rn_g is a little smaller than Kepler's velocity rn; but the sign of the relative velocity $r(n_s - n_g)$ of the grain with respect to the gas is not obvious. If $n_g < n_s$ we have a true drag, orbital energy is lost and $dr/dt < 0$; then the radial frictional acceleration $-v_f dr/dt$ has an outward direction, which requires, for the radial balance, a centrifugal force smaller than gravity, so that $n_s < n$. If $n_g > n_s$, friction increases the orbital energy and the relation $n_s > n$ for the radial balance could not be satisfied. We conclude that n_s always lies between n and n_g; since $n_g < n$, the true ordering is $n_g < n_s < n$ and $dr/dt < 0$. Solid grains are driven inward by dissipation and may be engulfed by the deep potential well of the Sun; it is important to establish the timescale τ_{orb} of this process of orbital decay. Similarly, in accretion discs around collapsed stars and galactic nuclei, material continuously fall onto the central body due to viscosity. The loss of orbital energy due to the drag $-v_f M_s r(n_s - n_g)$ (eq. (12.4)) gives

$$\frac{dr}{dt} = -2 v_f r \frac{n_s - n_g}{n}. \tag{16.16}$$

On the other hand, since $n_s - n \ll n$, the radial momentum law reads (as above, the inertial term can be neglected)

$$2 r n (n_s - n) = v_f \frac{dr}{dt} \tag{16.17}$$

and, eliminating n_s, we get

$$\frac{1}{\tau_{orb}} = \left|\frac{r}{dr/dt}\right| = 2 v_f \frac{n - n_g}{n} \left(1 + \frac{v_f^2}{n^2}\right)^{-1}. \tag{16.18}$$

There are two regimes: when $v_f \gg n$

$$\frac{1}{\tau_{orb}} \simeq 2 \frac{n^2}{v_f} \frac{n - n_g}{n} = \frac{v_T^2}{v_f r^2}; \tag{16.18'}$$

when $v_f \ll n$

$$\frac{1}{\tau_{orb}} \simeq 2 v_f \frac{n - n_g}{n} = \frac{v_f v_T^2}{n^2 r^2}. \tag{16.18''}$$

Since $v_f \propto 1/R_s$ (eq. (16.12)), they correspond to small and large bodies, respectively. Big bodies lose less orbital energy because of their large area-to-mass ratio; for small bodies the unbalance due to the slight defect in centrifugal force can be compensated with a slower inward motion (in general they are strongly coupled to the gas and subject to less drag). The minimum value of τ_{orb} occurs at the transition between the two regimes, when $v_f = n$ and $R_s = \varrho v_T/(\varrho_s n) \approx 5$ m. At this size $1/\tau_{orb} = v_T^2/(nr^2)$ and $\tau_{orb} \approx 30 \tau_s$; therefore, the orbital decay during settlement was small, and the use of $|dz/dt|$ as the appropriate relative velocity of the grain is justified. The above analysis is in fact too simple, also because when the size of the body is comparable to, or larger than, the mean free path of gas molecules in the disc, the Epstein drag (16.11) should be replaced by the Stokes drag. It appears, however, that the qualitative results obtained above with Epstein's value for F_D are not changed.

Unless the turbulence was very active, apparently the vertical settlement preceded the radial drift; but the estimated, very short – some 10^3 y – timescale of the latter for metre-size boulders, poses a difficult problem. In completing the transition from dust particles to kilometre-size bodies, which are sufficiently well decoupled from the gas motion, the catastrophic loss at intermediate sizes of metres must be avoided; this is also a range of sizes where the previous process of direct adsorption of small particles becomes inefficient. Several possibilities have been suggested to bridge this problematic size range, but it is likely that several mechanisms have been active, depending on the particle size, the composition and the heliocentric distance.

Difficulties for the gravitational instability and alternative hypotheses. The traditional solution to the growth problem in the $1 - 1,000$ metre size range is the gravitational instability, briefly reviewed in the Appendix. If bodies settled in a layer with a sufficiently high density and low velocity dispersion (see the Toomre criterion (m) in the Appendix), they would spontaneously merge under their self-gravity and rapidly form kilometre size (or larger) planetesimals. However, solid particles, in their settling towards the midplane, drag the gas velocity to values nearer the Keplerian velocity, faster than the gas outside, which is supported by the pressure; a velocity gradient develops and turbulence is expected to arise, which stirs up all small aggregates, preventing further settling. The previously discussed, regular sedimentation seems implausible. The thickness of the solid grains layer in the midplane of the nebula depends on the level of the turbulence and the particle size; numerical work indicates that this stirring does prevent attaining the critical density for the gravitational instability for particles sufficiently coupled to the gas (smaller then a metre, say). Only larger bodies, eventually produced by coagulation, decouple from the gas motion and begin forming a denser sublayer, which might be gravitationally unstable. As bodies grow bigger, their radial motion is coupled to the gas, with a friction inversely proportional to the size (eq. (16.16)); the question is, to what extent they retain their previous random velocities. Numerical simulations indicate that a supercritical velocity may occur up $\simeq 100$ metres, so that gravitational instability does not seem to be a viable mechanism for planetesimal formation at any size; it might be effective only in the outer parts of the solar nebula and lead to the formation of comets and other bodies in the transneptunian region.

Although in the critical size range ($\gtrsim 1$ metre) the adsorption mechanism is not well understood, the relative velocity induced by the drag for bodies with different size may favour collisional growth: (i) bodies of similar size collide rarely, and only at low speed; (ii) bodies of very different sizes may collide at higher speed, but the smaller body may be unable to shatter the bigger one, unless the latter is loosely bound. A numerical analysis of the evolution of the size distribution in the central plane of the nebula at 1 AU is shown in Fig. 16.2. In the initial state all solids are μm grains, with a uniform dust/gas mass ratio of $\simeq 0.3\%$ throughout the vertical direction. At first coagulation produces a peak in the distribution at $\simeq 0.1$ mm; settlement toward the central plane drives it to a centimetric size. In this dense sublayer gravitational instability is prevented

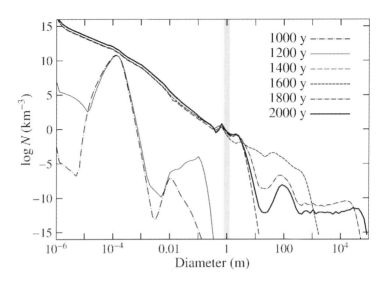

Figure 16.2. Size distribution of solids in the central plane of the protoplanetary nebula at different times; initially they are assumed to be µm grains, uniformly mixed with the gas. Coagulation due to drag-induced motions results in growth, while settling in the central plane increases the density. The grey band shows the critical size range where bodies are most efficiently removed by the gas drag. Data kindly provided by S.J. Weidenschilling; see also S.J. Weidenschilling, *Sp. Sci. Rev.* **92**, 295 (2000).

by differential radial velocities, and the collisional growth continues. Once kilometric objects form, they become (possibly gravitationally) dominant and grow faster, broadening the size distribution. At small sizes the characteristic power law is due to collisional fragmentation. Though the details of the evolution depend on the model, its qualitative features were found rather robust under variations of the assumptions; however, the complex turbulent structure was poorly modelled, or, as in Fig. 16.2, not modelled at all. The timescale for planetesimal formation is roughly a few thousand times the local orbital period over a large range of heliocentric distances; $\simeq 2,000$ y at 1 AU and some 10^5 y at 30 AU, in particular.

The double role of turbulence.– While the turbulence, inhibiting the conditions for coagulation and gravitational collapse, hinders the growth, it may also help it. Fluid dynamics in the solar nebula shares two important features with the global circulation in the atmosphere of a rotating planet: (i) Coriolis force has a fundamental role and (ii) the flow is to a large extent two-dimensional, with a negligible vertical velocity. Rossby waves, driven by the gradient of the Coriolis frequency, are a typical manifestation of this regime (Sec. 7.5). If turbulence develops, it is confined to a two-dimensional layer and is very different from the ordinary, three-dimensional case, in which the energy of large eddies is progressively driven by non-linear interactions into the dissipative range of wave numbers; in the geostrophic, two-dimensional case, while the vorticity is broken up in smaller structures and eventually dissipated, the energy spectrum moves to *bigger* sizes and produce large scale, coherent and long-lived structures. This may be the reason why in the atmospheres of the giant planets strong zonal winds are present, flowing in bands along the parallels; in the nebular disc they may have an analogue in global jets flowing in the azimuthal direction, with a velocity up to a fair fraction of the sound speed. The differential drag in adjacent jets may drive solid particles toward their edge and confine them there. At the distance from the Sun where two neighbouring jets are in contact, vorticity may develop, as for the Great Red Spot of Jupiter (Sec. 7.5); concentration of mass in the cores of long-lived anticyclonic vortexes in the nebular disc has been described, both numerically and analytically.

Surely turbulence could have significantly affected the flow of mass and angular momentum in the nebula and the formation of structures. It must be avowed, however, that the current understanding of these complex processes is poor. We do not have a satisfactory and quantitative picture of this kind of turbulence; the details of the adsorption and the collisional breakup are scarcely known; numerical simulations face the difficulty of a satisfactory physical model and, in any case, can only explore the evolution in a limited region of the parameters, but do not provide general laws.

16.3 Formation of planetary embryos and accretion of terrestrial planets

On a short term the gas drag does not affect much planetesimals bigger than a few kilometres, but on a long timescale it may reduce eccentricities and inclinations, and also cause, which is less important, a slow decrease of their semimajor axes. The further evolution results from the interplay between collisions and gravitational scattering. Close encounters increase the random velocities $v_{\rm rel}$ of planetesimals, depending on the size distribution; on the other hand, this distribution depends on the collision rate, hence on $v_{\rm rel}$. Sizes and random velocities (which correspond to eccentric and inclined orbits) evolve together.

The rate of growth of a planetesimal of mass M and radius R

$$\frac{dM}{dt} = \varrho v_{\rm rel} \pi R^2 \left[1 + \left(\frac{v_{\rm esc}}{v_{\rm rel}}\right)^2\right] \qquad (16.19)$$

shows in the bracket the gravitational enhancement factor of the geometric cross-section πR^2 (eq. (11.44)); ϱ is the density of the embedding material. For a large planetesimal accreting small debris, $v_{\rm esc} = \sqrt{2GM/R}$ is the escape velocity; a modification is needed when the masses (and the sizes) are comparable. Since $v_{\rm esc} \propto R$, the gravitational enhancement factor plays a critical role in determining the growth of bigger bodies. As long as $v_{rel} \gtrsim v_{esc}$, the logarithmic growth rate $dM/(M\,dt) \propto 1/R$: smaller bodies grow faster and catch up with the larger ones, with a trend towards comparable masses; this is usually referred to as *orderly growth*. But when $v_{rel} \gg v_{esc}$ the delivered kinetic energy is larger than the gravitational binding energy of the impacted body and fragmentation is the typical outcome. In the ensuing collisional equilibrium the size distribution is not uniform; this is what happened in the asteroid and Kuiper belts (Secs. 14.3 and 14.4). If, on the other hand, the growth of relative velocities is inhibited, or one body gets big, so that $v_{rel} \lesssim v_{esc}$, $dM/(M\,dt) \propto R$; then larger bodies grow faster, both absolutely and relatively to each other, rapidly leading to the predominance of a single body in the whole population – the *runaway growth*.

In the first stage of the evolution, the mass distribution of planetesimals is roughly uniform and the orderly growth increases the ratio $2\theta = (v_{esc}/v_{rel})^2$ from its initially low values towards 1; at this point kinetic and binding energies are comparable, break-ups occur, which approximately maintain $\theta \approx 1$. This may be not quite the case due to inelastic collisions, gravitational friction of larger planetesimals, gas drag, and so on. These dissipative effects favour the onset of runaway growth, in which a single body emerges in the swarm and accretes most of the material around. During this phase its value of θ remains about 3 – 5; higher values occur in presence of gas, whose friction diminishes the velocity of planetesimals.

Since planetesimals orbit around the Sun in roughly circular orbits, the runaway growth of planetary embryos is a *local* process and does not complete at the same time in the whole swarm. The restricted three-body problem provides the appropriate insight: a mass M on a nearly circular orbit of radius r can experience collisions with particles only within the critical separation

$$\Delta r = K R_H = K r \left(\frac{M}{3M_\odot}\right)^{1/3} ; \qquad (16.20)$$

(Sec. 13.4 and Fig. 13.10). R_H is Hill's radius and K is a constant of the order of unity ($\simeq 2\sqrt{3}$; see Problem 16.7). The final accreted mass of the planetary embryo is roughly given by all the material initially present in this zone; this is the *isolation mass*

$$M_i = 4\pi r (\Delta r)_i \sigma = (4\pi \sigma K)^{3/2} \frac{r^3}{(3M_\odot)^{1/2}} \approx 1.6 \times 10^{25} \left(\frac{r}{AU}\right)^3 \left(\frac{\sigma}{g/cm^2}\right)^{3/2} g .$$
$$(16.21)$$

The corresponding width is

$$(\Delta r)_i = (4\pi\sigma)^{1/2} K^{3/2} \frac{r^2}{(3M_\odot)^{1/2}} \approx 0.003 \left(\frac{r}{AU}\right)^2 \left(\frac{\sigma}{g/cm^2}\right)^{1/2} AU . \quad (16.22)$$

σ is the surface density. In the terrestrial zone, where $\sigma \simeq 8$ g/cm^2, the planetary embryos have a lunar-to-Mars mass.

The conclusions of this simple analysis are supported by numerical simulations, which take into account the physics of collisions and the gas environment. A typical timescale for the formation of a planetary embryo is $\simeq 10^5$ y in the terrestrial zone (Fig. 16.3), and a few My at $\simeq 2 - 4$ AU. This time is shorter at smaller distances from the Sun both because the disc density is larger and the revolution period is shorter; the formation of the embryos proceeds as an outward propagating wave.

Once the embryo is formed and dominates the local swarm by a factor of at least $\simeq 100$ in mass (Problem 16.4), it becomes massive enough to significantly increase by gravitational influence the relative velocities of nearby bodies. The conditions for a runaway process are no more satisfied and the further growth proceeds orderly, sweeping up the remaining population, with occasional collisions. This phase is referred to as the *oligarchic growth*. Numerical simulations indicate that a pair of neighbouring embryos on nearly circular orbits is stable if their separation is $\simeq (5-10) R_H$. One expects at the end of this process the emergence in the terrestrial zone of some 20–40 lunar-to-Mars size bodies, with spacings of $\simeq 0.05$ AU and typical eccentricities and inclinations of order $0.001 - 0.01$.

Planetary formation in the inner and outer solar system are likely to be related. Giant planets may have formed (Sec. 16.4) out of accretion onto a core at a slower ($\simeq 1 - 10$ My) pace and may have affected the late phase of terrestrial embryos evolution; in particular, the planetesimals' velocities in the inner swarm could have been increased, even by large amounts, by secular or resonant perturbations due to the giant planets. This may have entirely halted the formation of Mars-size embryos in the asteroid belt region, and caused the escape from it of several large bodies; this can explain the apparent mass deficiency in the asteroid belt (Fig. 16.1 and Sec. 14.3). In a different scenario, giant planets may have formed in a fast gravitational instability of a gaseous mass; numerical simulations indicate that this process and their following, slight inward migration, would actually inhibit the formation of Ceres-like bodies in the asteroid belt. Together with chemical considerations (Sec. 16.4), this is one important argument against this possibility for the giant planet formation.

Final stage of terrestrial planet formation. Planetary embryos, formed through the runaway growth, are on closely spaced orbits and have masses smaller than Venus and the Earth. Since their number is not large, the problem of the final growth to terrestrial planets can be investigated with N-body numerical simulations. The residual small planetesimals contain only a small fraction of the total mass, so that gravitational friction can be neglected.

between the inner planets is diminished. This is mainly true for Venus and Earth; the smaller planets (Mercury and Mars) might be mainly composed by material in their respective accretion zones.

- During the final growth phase each planet was impacted by bodies with masses ranging from 0.1% to 25% of its own mass. Such giant impacts likely contributed a random component to their angular momenta and their present obliquities (see also Sec. 14.1). When the mass of the planetary embryo is larger than $\simeq 0.05\ M_\oplus$, impacts also contributed to the volatilization of water and carbon dioxide. With a slightly larger mass, as the growing body develops a primordial atmosphere (Sec. 8.5), degassing of accreting planetesimals occurs in their fall. This early atmosphere behaves as a partially insulating blanket, and further contributes to heating the planet's surface. A thick layer of the mantle may even melt, with important differentiation with depth; inhomogeneities in the chemical composition and the temperature possibly resulted into the formation of distinct geological units (e.g., continents).

- A large impact could have ejected from the Earth a spray of material, which accumulated in a geocentric orbit and later formed the Moon (Sec. 16.4). In the case of Mercury, a giant impact might have stripped the silicate mantle, leaving a planet with an anomalously high density (Sec. 14.1). The isotopic analysis dates the formation of the Moon at 30 – 50 My after the origin of the solar system (Sec. 5.3), in a good agreement with the expected timing of large impacts.

16.4 Formation of giant planets and satellites

Unlike the terrestrial planets, the giant planets contain significant quantities of primordial gas (hydrogen and helium) of the solar nebula. This severely constrains their formation, which must have happened within $\simeq 10 - 30$ My, before the nebular gas was blown away in the T-Tauri phase of the Sun. One must also take into account differences in elemental and isotopic composition between the two prominent groups: gaseous giants (Jupiter and Saturn) and ice giants (Uranus and Neptune), and also explain their difference with respect to solar abundances. The mass abundance of refractory material in the Sun is less than 2% (Table 9.1), but this is enhanced for Jupiter, Saturn, and especially Uranus and Neptune; this implies that giant planets accreted heavy elements more efficiently than light gases from the nebula, or did not reach the phase of a massive accumulation of hydrogen and helium. As usual, let X and Y be, respectively, the mass fraction of hydrogen and helium. Jupiter's atmosphere is depleted in helium, since $Y/(X + Y) = 0.24$, as measured by the Galileo probe, is significantly smaller then the solar value 0.28. Saturn's depletion, with its atmospheric value 0.16 for this parameter, is even larger. D/H, the relative abundance of deuterium (Sec. 8.5), is another important diagnostic parameter that must be accommodated in any model for the formation and the evolution of these planets: the gaseous giants' value is equal the value for the protosolar, interstellar medium, while ice giants are enriched in deuterium. Finally, a constraint on their formation comes from their present internal structure. In the simple model of a fluid and rotating body in stationary equilibrium, the density distribution of the outer planets is axially symmetric around their axis, and invariant under its reversal; in this model the gravitational field is completely determined by the mass and the even zonal coefficients J_{2n}, $n = 1, \ldots$, functions of the density distribution and of Helmert's parameter μ_c (see eqs. (1.13) and (2.55)). On turn, their measurement provides constraints on the density. The giant planets likely have a central, dense core and a surrounding envelope, composed mostly of hydrogen and helium. The cores of Jupiter and Saturn are very small (models of Jupiter with no core are not excluded); Uranus and Neptune are mostly a core with a small envelope.

The formation of the giant planets has been extensively studied in two scenarios. The self gravity of the protoplanetary disc is weak, and disruption of a condensation by centrifugal force and the gas thermal pressure must be avoided. The two scenarios differ by the way this key problem is overcome. (i) The *core-instability model* uses the extra gravity field of a sufficiently massive solid core; (ii) the *gas-instability model* assumes in the nebula a gravitationally unstable subcondensation (see the Appendix), perhaps triggered by an external perturbing body, like a stellar companion. The first alternative meets less objections and is more favoured; but the second model may be more important for the *known* extrasolar planetary systems.

Gravitational instability. In a differentially rotating disc the competition between self-gravity and centrifugal force is governed by *Toomre's parameter* (eq. (m) in the Appendix)

$$Q = \frac{\kappa c_s}{\pi G \sigma} ; \qquad (16.23)$$

κ is the epicyclic frequency of radial oscillations, c_s is the gas sound speed and σ the surface density. Local instability and collapse of a lump of gas occurs when at that place $Q < 1$. In the minimum mass nebular model (Sec. 16.1) Q is generally appreciable larger than unity; a (local or global) enhancement of σ and a lower sound speed provides a cheap, standard way out. Simplified numerical examples have succeeded forming planets with a mass of a few jupiters in a disc of $\simeq 0.15\, M_\odot$ only, with reasonable temperature and pressure profiles. However, disc fragmentation is a very complicated process, sensitively dependent on appropriate modelling of poorly known microphysical effects (e.g., compressional heating and radiative transfer in the collapsing clump). The resolution of numerical methods might be of concern, too.

A strength of the gas-instability model is the short timescale (several orbital periods) required for planetary formation; this avoids the problem of gas supply. Planets can be also formed on eccentric orbits and may acquire masses in the range $5 - 10\,M_J$, as observed in some extrasolar systems. Among its weaknesses, the atmospheric composition of the giant planets, especially Uranus and Neptune, does not match that of the Sun, and a chemical segregation mechanism cannot be easily identified. Moreover, planetary cores are not explained, especially those of the ice giants. It is possible that within the forming planet dust and ice grains had enough time to grow to about millimetric or decimetric size (Sec. 16.1); but whether these early solids can settle into the centre of a turbulent and convective giant protoplanet is unclear. This is particularly delicate for Uranus and Neptune, whose big cores should contain some 2/3 of ice. It has been also proposed that the chemical anomalies of Uranus and Neptune might be due to photo-evaporation of light elements in the outer part of the protoplanetary nebula under intense, extreme UV radiation by young stars in the solar neighbourhood. However, no clear identification has been found, and this possibility remains hypothetical. Typically the gas-instability model produces relatively large ($\gtrsim 5\,M_J$) bodies, inconsistent with the giant planets of the solar system; mechanisms of further fragmentation and/or mass-loss are unclear. This is obviously related to the need, in order to trigger the gravitational instability, of a disc more massive than the minimum mass model. Finally, it is not clear whether a very early formation of giant planets would permit the formation of asteroids (Sec. 16.3).

Core instability. Though with its own problems, this scenario apparently explains more easily the observed features of the solar system giants and is considered a "baseline model". In its first phase, it shares the same evolutionary track as the terrestrial planets (Sec. 16.3). A planetary embryo is formed as a result of a runaway growth out of a swarm of planetesimals; eqs. (16.21) and (16.22) indicate that in the minimum mass nebula a Jupiter embryo with mass $\simeq 1.5\,M_\oplus$ formed in about 10 My. This value is too low to trigger efficient accretion of the gas envelope and the timescale, nearly the nebular lifetime, may be too long. The problem is usually solved by assuming a surface density σ of solids at the Jupiter heliocentric distance larger by a factor $\simeq (2 - 3)$ than the minimum mass model, allowing the formation of a $\simeq 8\,M_\oplus$ embryo within $\simeq 1$ My (Fig. 16.5).

When the growing core attains $\simeq 1\,M_\oplus$, it starts to gravitationally attract a small amount of nebular gas. This gaseous envelope increases in mass with the core. At a low mass a stationary state is established, in which the infalling material heats up using its own gravitational binding energy; a simple analytical model of this process follows the same pattern of the theory of the solar wind and is discussed below. But radiative exchange is essential; the growing infrared radiation from the core contributes to the heating, and radiation is lost

16. Origin of the solar system

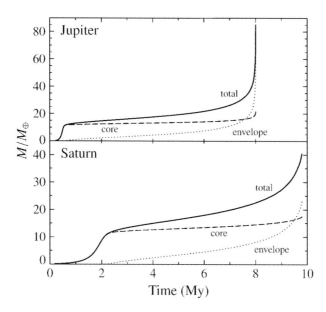

Figure 16.5. Core-instability model for the growth of gaseous giant planets. We show, as function of time, the total mass (solid line), the core mass (dashed line) and the mass of the gaseous envelope (dotted line). The three stages of the evolution discussed in the text (runaway growth of the core, slow accretion of solids and approximate hydrostatic equilibrium for the atmosphere, and runaway accretion of gaseous envelope) are shown. For the formation of Jupiter (top) and Saturn (bottom) a planetesimals disc with density $\sigma = 10$ g/cm^2 and $\sigma = 3$ g/cm^2 was assumed, respectively. The planets are at their current distance from the Sun. Data kindly provided by O. Hubickyj; see also J.B. Pollack et al., *Icarus* **124**, 62 (1996).

from the envelope to interplanetary space. Infalling planetesimals either hit the core or, at later stages, are ablated and destroyed in the envelope. For some time the accretion of the gas envelope may approximately follow the much slower growth of the core. The duration of this phase, which critically affects the whole process, depends on several factors, including: planetary migration, which helps maintaining the core growth rate; loss of radiative energy, depending on the opacity. However, once the envelope mass is of the same order as the core mass, roughly at the isolation value $\simeq 8\,M_\oplus$, the infalling matter is no longer able to balance the energy radiated away; the gaseous envelope as a whole contracts in order to liberate additional energy. Eventually, this contraction turns into a fast runaway process, lasting only some 1 – 10 ky (Fig. 16.5). This scenario needs an initial core with mass above the *critical core mass*, a threshold estimated to be $\simeq (8-10)\,M_\oplus$ in Jupiter's zone, with minor changes further out. The agreement between this value and the approximately known core masses of giant planets in the solar system is an important support for this model. The enhancement of heavy elements in their atmospheres may be qualitatively explained by the sublimation of infalling solid planetesimals at the late formation stage.

During the runaway phase of the gas envelope, the luminosity of the protoplanet significantly increases up to a maximum value, which depends on the available material in the surrounding region, and is estimated to be $10^{-4} - 10^{-3}$ of the current solar luminosity. This is a fairly large value, but, due to its short duration, some $0.01 - 0.1$ My, its direct photometric detection in extrasolar planetary systems is unlikely. At the end of the rapid gas accretion, the early planet fills most of its Roche zone (for Jupiter a few hundred times larger than its present radius); it may cease also as a result of the disc dispersal, expected about 10 My after the protosun formation. Moreover, as the planet slowly contracts, it likely opens a gap in the disc, which itself largely reduces the gas inflow. With no significant internal heat sources, the planet steadily cools; today's infrared excess in the luminosity of giant planets (Sec. 6.5) is attributed to the final phase of this process.

The fact that Uranus and Neptune have much smaller gaseous envelopes, with accordingly higher metal abundance than Jupiter and Saturn, suggests that they never reached the critical mass to locally trigger the runaway envelope accretion. This is due to the slower growth of their cores, both because the disc density was lower and the revolution periods were longer; the gaseous envelope was likely dispersed before the ice giants could start to accrete gas more efficiently. Despite this reasonable conclusion, both ice giants present a serious problem for the standard growth scenario: with the minimum mass nebula, *at their current heliocentric distance*, their expected cores and icy mantles, with masses $\simeq 10 - 15\ M_\oplus$, would take some 100 My or longer to grow, a timescale far too long to explain the accretion of their atmospheres. One could assume either much lower relative velocities among planetesimals (i.e., values of θ of the order of 100; Sec. 16.3), or much larger amounts of solid material inside these zones. The former possibility seems more likely, since it is not easy to find a reasonable physical mechanism capable of ejecting from the solar system large amounts of solid material; moreover, θ was probably increased by gas drag and by the fact that, when the average relative velocity became sufficient to eject the bodies from the solar system, a cutoff appeared in their velocity distribution. It is also likely that ice giants were formed nearer to the Sun and slowly migrated outward to their current positions (see below). In a viable scenario, the nuclei of Uranus and Neptune might have been even formed *in between* the orbits of Jupiter and Saturn; the formation of the two gas giants, in approximately the same time, prevented any further growth of Uranus and Neptune and scattered both embryos further out.

In the inner part of the nebula, gas accretion was hampered both by the higher temperature and by the fact that the available amount of solid material was not enough to allow any embryo to substantially exceed the critical mass

16. Origin of the solar system

for runaway gas accretion; this explains the entirely different atmospheres of terrestrial planets (Sec. 8.5).

A simple fluid model for the accretion of a gas envelope.– In Parker's barotropic, spherically symmetric model of the solar wind, when $P(\varrho)$ is given (Sec. 9.2 and Fig. 9.3), the appropriate solution is the *separatrix* (b), which directly connects the subsonic region I to the supersonic domain III and ends up in a progressively rarefied, supersonic flow; but another separatrix (a) joins a supersonic inner flow to an outer spherical cloud of gas at rest and is a possible solution for the present case. It should also be noted that a solution of type (c), everywhere subsonic, is destroyed by an arbitrarily small viscosity. A time reversal, under which Bernoulli's equation (9.7) is invariant, changes the flow direction and the rate dM/dt. It does not cost too much to waive the isothermal assumption, and to work in a more realistic adiabatic regime, with a polytropic index γ; in this way a simple, but rigorous and paradigmatic model of gaseous accretion is obtained, relevant for the whole of astrophysics. In its fall onto the central body, of mass M and radius R, the gas compresses and increases its thermal speed using gravitational energy, until it disappears at $r = R$; accretion is completed in a very short and finite time, inversely proportional to the initial mass M_0. For a black hole the thermal speed attains relativistic values; in this case complex radiative processes and large luminosities arise. The solution exists for all values of γ in the relevant interval $(1, 5/3)$. The model is spherically symmetric and cannot deal with the formation of an accretion disc out of an initial and small angular momentum of the cloud. It also neglects radiative losses to outer space, and the effects of radiation from the core, in particular heating and photon pressure; the minimum core mass needed to trigger the runaway accretion and the relevant timescale are in this way underestimated, but only by about an order of magnitude.

From inspection of eq. (9.7), we look for a negative solution $v(r) < 0$ with a gradient dv/dr everywhere positive; this is possible only if it crosses the sonic point r_c, where $v^2(r_c) = c_s^2(r_c)$ and $r_c H(r_c) \equiv GM/r_c - 2c_s^2(r_c) = 0$; in addition, the boundary conditions are now given at infinity, with a density ϱ_∞ and a sound speed $c_{s\infty}$. Supersonic accretion is possible only if $r_c = GM/(2c_s^2(r_c)) > R$, the physical radius of the embryo. The mass gain $\dot{M} = 4\pi |v(r)| r^2 \varrho(r)$ is independent of the distance and at the sonic point (indicated with a suffix $_c$) has the form

$$\dot{M} = 4\pi c_{sc} r_c^2 \varrho_c = \pi \varrho_c G^2 M^2 c_{sc}^{-3}. \tag{16.24}$$

Of course, as $r \to \infty$, $|v| \to 0$. In the polytropic case c_s^2 is proportional to the temperature and $c_s^2 \varrho^{1-\gamma}$ is constant (eq. (1.26a)); the isothermal case is recovered for $\gamma \to 1$. Eq. (9.7) is integrable and gives

$$\frac{v^2}{2} - \frac{GM}{r} = \frac{c_{s\infty}^2}{\gamma - 1}\left[1 - \left(\frac{\varrho}{\varrho_\infty}\right)^{\gamma-1}\right], \tag{16.25}$$

which shows how the binding energy of each particle increases as the gas is heated and compressed. To express $\dot M$ in terms of the thermodynamical variables at infinity, note first that the previous equation, evaluated at the sonic point, gives

$$c_{sc} = \left(\frac{2}{5-3\gamma}\right)^{1/2} c_{s\infty} .$$

With this we get

$$\dot M = \pi f(\gamma) \frac{G^2 M^2}{c_{s\infty}^3} \varrho_\infty \qquad \left[f(\gamma) = \left(\frac{2}{5-3\gamma}\right)^{(5-3\gamma)/2(\gamma-1)} \right]. \qquad (16.26)$$

The relevant interval for γ is $(1, 5/3)$, corresponding to a number of degrees of freedom between ∞ and 3 (eq. (1.21)); in this interval $f(\gamma)$ does not change much, from $f(5/3) = 1$ to $f(1) = e^{3/2} \simeq 4.5$. Since f must be real, there is no solution for $\gamma > 5/3$. While in the solar wind $\gamma = 5/3$ is a good choice, in our case a lower value is more likely. When the conditions at infinity do not change, the solution

$$M = \frac{M_0}{1 - t/\tau_{\text{accr}}} \qquad (16.26')$$

formally diverges in the accretion time

$$\tau_{\text{accr}} = \frac{c_\infty^3}{\pi f(\gamma) G^2 M_0 \varrho_\infty} . \qquad (16.27)$$

In terms of the conditions at infinity, the largest radius R_c for supersonic accretion of a body of mean density $\bar\varrho$ is

$$R_c = \left[\frac{3}{\pi(5-3\gamma)G\bar\varrho}\right]^{1/2} c_\infty , \qquad (16.28)$$

For $\bar\varrho \approx 1$ g/cm^3, $c_\infty \approx 10^5$ cm/s, $\gamma = 1.5$ (values suitable for Jupiter), we obtain $R_c \approx 5,000$ km, corresponding to a mass $\approx 0.1\, M_\oplus$; as noted above, this is about 10 times smaller value than that found with more realistic models. The accretion timescale (16.27) is very short: for $\varrho_\infty \approx 10^{-9}$ g/cm^3, $\tau_{\text{accr}} \approx 1,000\,(M_\oplus/M_0)$ y. In effect, accretion could have been so fast only where the planet's gravity was dominant over that of the Sun, which holds only in a fraction of the feeding zone; diffusion of the gas toward the centre of the zone was hampered also by angular momentum conservation.

Planetary migration. Gaseous planets can "migrate" from the place where they were formed, or even during their formation, toward or away from the star. This is relevant both for the giants planets and in extrasolar systems, and is crucial in order to understand their properties, even in relation to minor bodies.

16. Origin of the solar system

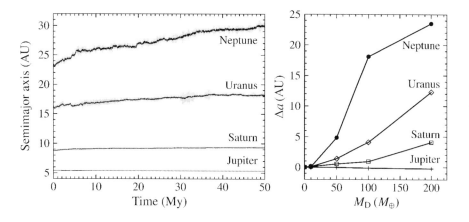

Figure 16.6. Left: semimajor axes of the giants planets embedded in a planetesimal disc with total mass 50 $M_\oplus$; exchange of energy and angular momentum with planetesimals removes the latter and causes an outward planetary migration. The grey zones indicate the range of distances between perihelion and aphelion. At $\simeq$ 30 My, when crossing the 2/1 mean motion resonance with Neptune, Uranus' eccentricity briefly increases. The initial semimajor axes of the planets have been chosen to differ from their present values by Δa = +0.2, −0.8, −3 and −7 AU, in such a way that the latter are attained after $\approx$ 50 My. Right: radial displacement of the giant planets by planetesimal scattering after 30 My, as a function of the initial mass M_D of the left-over planetesimal disc (in Earth's masses). Data kindly provided by J.M. Hahn; see also J.M. Hahn and R. Malhotra, *Astron. J.* **117**, 3041 (1999).

Migration due to interaction with the nebular gas.– An early migration may occur before the dispersal of the gaseous nebula, typically within the first $\simeq$ 10 My. A low-mass planet embedded in the disc migrate inwards, principally as a result of the exchange of angular momentum with the disc perturbations induced at the Lindblad resonances sites (Sec. 14.7). There is a net imbalance between the torques produced in the inner and outer parts of the disc, the outer resonances being systematically stronger. This process, called *type I migration*, becomes efficient for a protoplanet's mass $\gtrsim$ 1 $M_\oplus$; above this threshold the growth time is expected to be longer than the migration timescale $\approx 0.1 \, (M/M_\oplus)$ My. After the runaway accretion of the gaseous envelope is completed, the planet opens a gap in the disc; as the gap is a barrier to the radial flow, the direct interaction with the nebular gas is much reduced. However, the planet may still trigger dissipative surface waves on both the inner and outer parts of the disc, with a slight unbalance; this process, that causes the planet (and the gap to which it is locked) to migrate inward, is termed *type II migration*. The relevant timescale is generally longer than type I; its speed weakly depends on the planetary mass. Both types are assumed to play a prominent role in explaining the anomalous number of giant planets close to their hosting star (Sec. 16.5), but perhaps were not significant in the solar system.

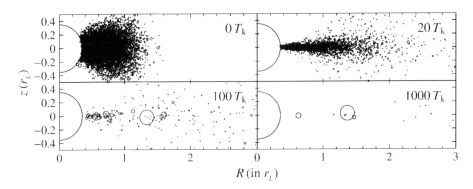

Figure 16.7. Snapshots of the protolunar disc produced by a giant impact in the (R, z) plane, after 0, 20, 100 and 1,000 T_K; $T_K \simeq 7$ h is Kepler's period at a distance equal to Roche's radius r_L (eq. (13.15) and Sec. 14.7). The initial disc of 10^4 particles, with total mass $0.4\,M_\oplus$, evolves under the dominant Earth's gravity and collisions. The semicircle at the origin is the Earth; each particles has a size proportional to its radius (all in units of r_L). Data kindly provided by E. Kokubo; see also E. Kokubo et al., *Icarus* **148**, 419 (2000).

rupted by similar collisions with comets, and have subsequently reassembled, even several times, by accretion. Such a re-accumulation process cannot occur with distant and irregular satellites, spread over large regions.

The Pluto-Charon and the Earth-Moon pairs are nearly "double planets", with large satellite-to-planet mass ratios; the Earth-Moon system has a large specific angular momentum. They have a peculiar standing and require special models. In the only viable scenario their formation takes place from debris ejected by a giant impact of a Mars-size body in the former case, and a smaller projectile in the latter (Fig. 16.7). This provides an explanation of most observed physical features, including the lack of volatiles and iron in the Moon (mostly consolidated from the stripped Earth mantle). This scenario is even more appealing, considering that planetary formation theory offers natural candidates for the giant projectiles, namely planetary embryos of lunar-to-Mars size formed at the last stage of planetary formation (Sec. 16.3). The process of eliminating these embryos from the interplanetary region naturally leads to collision with planets.

Rotational fission has also been proposed for the origin of these large mass-ratio (> 0.01) planet-satellite pairs. During the final phase of planetary accumulation, when the spin angular momentum of the planet exceeded the critical value for fission (see Sec. 3.4), rotational instability may have occurred, due to collisions or to the formation of a high density core. Although in the Earth-Moon case this conjecture can explain in a natural way why the Moon is much more deficient in iron than the Earth, it meets the serious objection that the present total angular momentum of the system is only about half of the critical value for a spinning protoearth. It is difficult to find a reasonable mechanism to get rid during the subsequent history of such a large angular momentum.

On a much smaller scale, rotational fission may occur for asteroids and meteoroids; in this case, radiation torques may drive their rotational state toward the critical limit (Problem 15.10). Similarly, the impact theory is, of course, also a plausible way to explain the origin of binary asteroids and of objects of anomalous shape, like the main belt asteroid Kleopatra (Fig. 20.1). For them, after a catastrophic collision with another (smaller) asteroid, the resulting angular momentum might simply have been too high, or just marginal, to allow re-accumulation of fragments into a single body. Obviously, collisional disruption of pre-existing satellites could explain the formation of planetary rings and the presence, in the inner part of the satellite systems of the outer planets, of many small, fragment-like moonlets.

The formation of the two Martian satellites is still puzzling. Their nearly circular equatorial orbits strongly support their formation from a debris disc left over during the planet's formation; on the other hand, their low mean density ($\simeq 2$ g/cm^3), compared to the higher mean density of Mars ($\simeq 3.9$ g/cm^3), suggests that the satellites are composed of material brought in from elsewhere. This argument is further supported by their low albedo ($\simeq 0.05$) and the spectral consistency with carbonaceous chondritic material. However, attempts to explain a capture origin for Phobos and Deimos have repeatedly failed, due to the extremely low probability. A likely scenario assumes orbital accumulation of primitive material shattered from a planetesimal in the vicinity, possibly due to a collision with an earlier Martian satellite.

16.5 Extrasolar planetary systems

In 1995 a periodic oscillation in the spectral lines of the star 51 Pegasi led to the discovery of the first extrasolar planet around a solar-type star; earlier, in 1992, three planets orbiting the old neutron star PSR 1257+12 (with a mass of 1.4 $M_\odot$) were detected. This star is a pulsar, with a radio pulsed emission with the period $P_{\text{pul}} \simeq 6.2$ ms; each planet causes, on turn, a periodic change in P_{pul} at its own orbital frequency. They are similar to the terrestrial planets (Table 16.1); the planets B and C are in 3/2 mean motion resonance. This is important, since it allowed to firmly justify the planetary hypothesis for this pulsar via specific resonant orbital effects (and thus the pulsar data).

Up to December 2002, detection of 90 planetary systems (of which 11 are multiple, 8 double and 3 triple) has been announced. The furthest is at $\simeq 100$ pc, except for the peculiar systems around the pulsars PSR 1257+12 (at $\simeq 300$ pc) and PSR B1620-26 in the globular cluster M4 (at ≈ 3.8 kpc). With a stellar density in the solar neighbourhood of 0.08 pc^{-3}, within 50 pc there are $\simeq 40,000$ stars; however, since the current detection methods are very ineffective at low planetary masses, no firm conclusion can be drawn about the

Table 16.1. Parameters of the planets around PSR 1257+12 (I is the inclination of the planetary orbit with respect to the plane of the sky); a fourth, low-mass planetary object has been conjectured, but not confirmed yet. Note the mean motion resonance between B and C.

Name	Semimajor axis	Period	Eccentricity	Mass × sin I
A	0.19 AU	25.3 d	0.0	0.015 $M_\oplus$
B	0.36 AU	66.5 d	0.018	3.4 $M_\oplus$
C	0.47 AU	98.2 d	0.026	2.8 $M_\oplus$

probability that a star has a planet. About 5% of solar-type stars in the solar neighbourhood have been found to possess a giant planetary companion; because of the bias toward finding only massive planets, the total fraction of planet-bearing stars may be significantly larger. About 30% of the detected extrasolar planets are in multiple-star systems. In multiple-planet systems, several examples of mean motion resonances between planets have been found, such as 3/2 for B and C of PSR 1257+12, and 2/1 for HD 82943 and Gliese 876 (one of the closest, at 4.7 pc only). Albeit the topic is in rapid evolution, it is already clear that our views of the evolution of the solar system, as described earlier in this chapter, cannot fully explain the variety of extrasolar systems that have been discovered; in particular, their origin and early evolution might have been very different from the classical scenario given in Secs. 16.1 to 16.4. Their search places strong requirements on the instrumentation and has led to important experimental developments which we describe conceptually below.

Terminological issues.– If the mass of a cloud of interstellar gas becomes larger than Jeans' critical value (eq. (h) of the Appendix), instability sets in and the cloud enters a collapse phase at the free-fall timescale. The further evolution crucially depends on energy generation and transfer, which controls the central temperature T_c. With increasing density, the opacity and T_c both increase; if T_c reaches the threshold for hydrogen burning, the large thermonuclear energy reservoir can be used. For an initial composition similar to the solar neighbourhood, this occurs for masses larger than $\approx 0.075\, M_\odot \simeq 78.5\, M_J$; eventually a body above this critical mass becomes a main sequence star, with a luminosity $\approx 10^{-3}\, L_\odot$. For lower masses – called *substellar-mass objects* – T_c reaches a maximum, after which the body begins to cool irreversibly; its intrinsic luminosity slowly dies away. For masses $> 13\, M_J$, however, at first deuterium burning (p+D $\to \gamma+^3$He) becomes active and delays the cooling phase; these objects are called *brown dwarfs*. More than 150 brown dwarfs were discovered before 2001 by their characteristic infrared spectrum. Objects with mass smaller then $13\, M_J$ are conventionally called *planets*; in what follows, we only deal with them. These two classes of objects, however, are not sharply separated and their definition may depend on the scientific community. For instance, stellar astronomers subdivide substellar-mass objects into the spectroscopic classes L (cooler than M dwarfs, and different by strong metal-hydride bands and neutral alkali features) and T (spectroscopically characterized by the onset and growth of methane absorption in the infrared); T dwarfs are similar to the "hot jupiters" introduced by the planetary community.

Substellar-mass objects can arise either isolated in the interstellar medium or in the protostellar disc of another star; they have similar composition, and physical and chemical properties. Traditionally planets are believed to form in accretion discs around stars, while the majority of brown dwarfs arise by direct fragmentation of interstellar clouds in star-forming regions. However, very-low mass objects, below the critical value of 13 M_J and unrelated to any star, have been observed in the Orion nebula, and for that reason dubbed *"free-floating planets"*. Their existence places an important constraint on planetary formation in general. They might be planets escaped from stellar orbits by gravitational perturbations; however, apparently in this region there are not enough stars that could support planetary systems. In a more likely hypothesis, these "planets" may have been produced by direct fragmentation of a collapsing interstellar cloud. In some cases this mode of planetary formation is still a viable alternative to the formation in a circumstellar disc and, in fact, may better explain the observed distribution of eccentricities in extrasolar planets and planetary masses larger than $\simeq 5\, M_J$.

It should also be noted that substellar-mass objects and low-mass main-sequence stars share a very long evolution timescale (10 Gy or more). The atmospheres of substellar-mass objects are similar and contain many molecular compounds, in particular CO, CH_4, H_2O, N_2, NH_3 and so on.

Observational methods. In principle, detection of extrasolar planets is similar to the study of binary stellar systems and can use direct observation of the pair (as for a visual binary), Doppler measurements (as for a spectroscopic binary), or intensity modulation (as for an eclipsing binary). But quantitatively, since planets do not produce thermonuclear energy, the problem is very different. Direct observation of the planet thus requires distinguishing a dim infrared object very near another one which is very bright in the optical band. This is a very difficult task, to be shortly discussed below; it is much easier to track the effects that the planet has on the star, leading to the three methods currently used. If not specified, numerical examples refer to a star (indicated with the suffix $_*$) similar to the Sun and at the distance $D = 50$ pc, and to a planet with a mass $M_p = 10^{-3}\, M_\odot \simeq M_J$, similar to Jupiter's, with a semimajor axis $a = 5$ AU. We do not discuss here the separate problem of the determination of the mass of the star, usually possible on the basis of its spectral class.

Radial velocity measurements.– For spectroscopic observations, consider a simple two-body system with vanishing eccentricity, in which the star moves

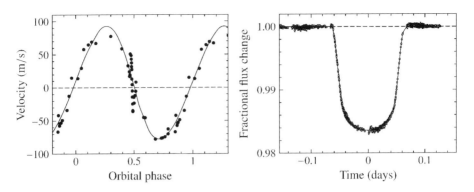

Figure 16.8. Doppler signal and intensity profile of the eclipsing planet HD 209458, with the following parameters: $M_* = 1.03\,M_\odot$; $D_* = 47.1$ pc; $M_p = 0.69\,M_J$; $P = 3.525$ d; $a = 0.047$ AU; $e = 0.02$. The slight deviation of Doppler data from a sinusoid gives a measure of the eccentricity. The behaviour near the phase 0.5 is due to the rotation of the star, as the planet blocks the approaching and receding limbs. Its analysis shows that the planet orbits in the same sense as the star's spin; its symmetry implies that the path of the planet is nearly parallel to the stellar equator. Radial velocity measurements at the Keck telescope, adapted from http://astro.berkeley.edu/~gmarcy/hd/doppler.html; photometric data, acquired by the Hubble Space Telescope, were kindly provided by T. Brown. See T. Brown et al., *Astrophys. J.* **522**, 699 (2001).

around the centre of mass on a circle with radius $aM_p/(M_* + M_p)$ and angular velocity $n = \sqrt{G(M_* + M_p)/a^3}$. If observed in a direction **n** which makes an angle I with the normal to the orbital plane, the velocity along the line of sight is $v = v_0 \cos(nt + \phi)$, with amplitude

$$v_0 = na\frac{M_p}{M_* + M_p}\sin I \,. \qquad (16.29)$$

If the sinusoidal Doppler signal $y = v/c$ is observed for a time T not much smaller than the orbital period $P = 2\pi/n$, the mean motion n can be independently measured and, eliminating a and referring the masses to the Sun and to Jupiter, we get

$$v_0 = n^{1/3}\frac{M_p \sin I}{(M_* + M_p)^{2/3}} \simeq 28.4\left(\frac{1\,\text{y}}{P}\right)^{1/3}\frac{M_p}{M_J}\sin I\left(\frac{M_\odot}{M_*}\right)^{2/3}\text{m/s}\,. \qquad (16.30)$$

Note that M_p is never measured independently, but is always multiplied by $\sin I$, so that only a lower bound is obtainable; statistically, one can replace $\sin I$ with its average $\pi/4$. This Doppler shift causes a coherent oscillatory motion of all spectral (in particular, absorption) lines of the star; the accuracy in the velocity amplitude v_0 depends on the resolution of the spectrograph (typically,

$\lambda/\sigma_\lambda \simeq 60,000$), on the number of observed orbital periods and on the number of measured lines (Fig. 16.8). An accuracy $\sigma_v \approx 3$ m/s is currently achievable, and improvements are possible, though the intrinsic accuracy of the method has a limit $\simeq 1$ m/s due to effects of star spots and convective inhomogeneities on the stellar surface. Note that the daily motion of the station and the yearly motion of the Earth around the Sun give a greater contribution, which depends the local kinematical parameters and the direction **n** of the source.

The Doppler signal due to an eccentric planet is not sinusoidal. For example, an observer lying in the orbital plane ($I = 0°$) in the direction orthogonal to the apsidal line will detect a larger and shorter maximum (or minimum) when the planet is at pericentre, while at apocentre the extremum is shallower. In general, the eccentricity adds to the Doppler observable higher harmonics of the orbital period, which allow accurate measurements of its value (Problem 16.8).

Astrometric measurements.– The second method relies on very accurate astrometry of the main star to determine the effect of the planet on its proper motion in the sky. This motion has three components: the centre of gravity of the system moves with a uniform velocity; as seen from the Earth, the star moves on parallactic ellipse with semi-axes a/D and $a \sin\beta/D$, where β is the latitude of the star over the ecliptic; the planet induces an additional angular displacement of order

$$\frac{M_p}{M_* + M_p} \frac{a}{D} \simeq 0.1'' \left(\frac{a}{5 \text{ AU}}\right) \left(\frac{50 \text{ pc}}{D}\right) \frac{M_p}{M_*}, \qquad (16.31)$$

whose details depend on the orbital elements, in particular the eccentricity (Fig. 16.9). Note that this technique is particularly sensitive to relatively long orbital periods ($P \gtrsim 1$ y, say) and it is applicable to hot or rapidly rotating stars, for which the radial velocity method is limited. Also, provided a and D in (16.31) are known, we have direct access to M_p. So far no exoplanets have been reported using this technique, but HIPPARCOS measurements have already provided some constraints on the masses of planets independently discovered with the radial velocity method. However, future astrometry at the level $\simeq 10\,\mu$as could detect a jupiter at 500 pc. Other astrometric space missions, in particular Gaia of the European Space Agency, are being planned (Table 20.1), which will attain with this method very interesting results (Sec. 20.2).

Photometric measurements.– In eclipsing binaries the luminosity of the star is periodically affected by a small dimming (Fig. 16.8)

$$\frac{\Delta L_*}{L_*} \approx \left(\frac{R_p}{R_*}\right)^2 = \frac{(3M_p)^{2/3}}{R_*^2 (4\pi\varrho_p)^{2/3}}. \qquad (16.32)$$

For example, the fractional dimming of the Sun due to the transit of the Earth and Jupiter is, respectively, 8.4×10^{-5} and 1.2×10^{-2} (corresponding to a magnitude drop of $\simeq 10^{-4}$ and $\simeq 1.3 \times 10^{-2}$). The interval between successive

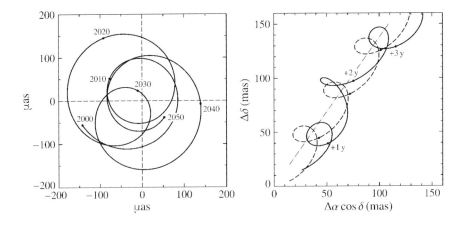

Figure 16.9. Right: A simulated path of a star at 50 pc with proper motion 50 mas/y projected on the sky. Without a planet its trajectory is the dashed line, superposition of a uniform and a parallactic motion. A planet with $M_p = 10\,M_J$ with $e = 0.5$ and a revolution period of 2.5 y produces an additional wobble (continuous curve; the scale is multiplied by a factor 30 for visibility). Left: For comparison we show a head-on view of the solar motion around the barycentre of the solar system as seen from a distance of 50 pc.

eclipses is the orbital period and gives the semimajor axis. A system is eclipsing if the angle $\pi/2 - I$ between the orbital plane and the line of sight is $< R_*/a$, corresponding, at 5 AU, to a probability $\approx 10^{-3}$. The eclipse lasts at most

$$\tau_{\rm ecl} \simeq \frac{2R_*}{na} \simeq 13 \left(\frac{a}{\rm AU}\right)^{1/2} {\rm h}\,. \qquad (16.33)$$

Since the star's radius and mass can be obtained from its luminosity and temperature, the duration of the eclipse, the magnitude dip and the ingress/egress profiles determine the orbital inclination and the planet's radius.

Such an eclipsing event was observed for the star HD 209458 (Fig. 16.8) and a planet at 0.047 AU, with period 3.525 d. The inclination $I = 86.7°$ was determined. The observation of this "jupiter" is quite remarkable because, for the first time, it allowed to determine fairly well a number of planetary parameters. The analysis of the ingress and egress phases provided the optical depth of sections of the atmosphere and showed its large size, $\simeq 1.3 - 1.4\,R_J$, anomalous for a mass of $0.69\,M_J$, as inferred from radial velocity measurements; the equilibrium temperature is $\simeq 1,400$ K. Moreover, spectrophotometric observations of the transit have detected in the atmosphere the presence of sodium, which shows up in a transient dimming of the stellar luminosity in the broad band at $0.59\,\mu$m (Fig. 9.1).

The transit method requires accurate photometry and extensive coverage; the photometric accuracy of ground-based observations is limited to $\simeq 10^{-4}$

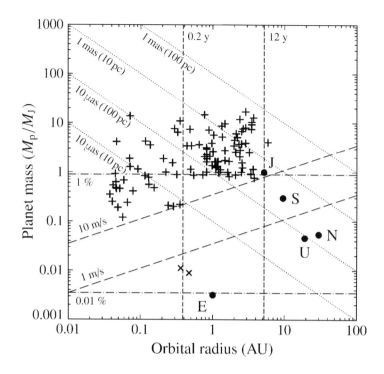

Figure 16.10. All extrasolar planets known up to December 2002 in relation to the observational quantities in the three techniques: radial velocity (dashed lines); angular displacement of the primary (dotted lines); luminosity drop (dashed-dotted lines). + for planets around solar type stars, × for pulsar planets. The orbital periods and solar system planets are also indicated (•).

due to variable atmospheric extinction and scintillation (Sec. 7.5). Space experiments, e.g. US Kepler and French COROT missions, are expected to reach an accuracy better by at least an order of magnitude; perhaps even large satellites of exoplanets or their ring systems could be detected.

The comparison between the three different observation methods is eased by placing all planets in the (a, M_p) plane (Fig. 16.10). The boundary of the explored area in this plane is mainly determined by the accuracy of radial velocity measurements and the time interval over which they are available.

Microlensing.– The previous detection methods rely on photons emitted by the star (or the planet, see below) and cannot be used at great distances; but *gravitational microlensing* uses photons from another source in the background and, in principle, can be used for much greater distances. However, since the accurate alignment needed is improbable, the observation of a given planet is a single event and cannot be repeated. Photons from a distant source passing at a distance $b = \theta D_L$ from another star of mass M_L suffer an inward gravitational deflection $\delta = 4GM_L/(c^2 b)$ (Fig. 16.11 and eq. (17.58)). This deflection,

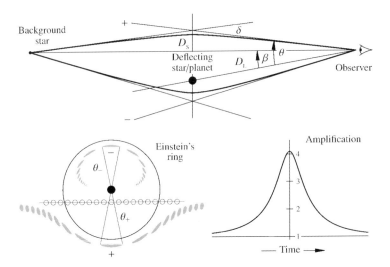

Figure 16.11. The geometry of gravitational lensing. The observer has a distance D_S from the background star S and a distance D_L from the deflecting star or planet L. θ, the apparent angular distance of S from L, has two solutions, corresponding to two images on opposite sides; they scale with Einstein's angular radius θ_E. The images are distorted: the lower left panel shows the undeflected image of the star as it moves across the sky and the succession of deformed shapes for the two solutions. Einstein's ring (continuous line) is also shown. On the lower right, a typical amplification profile as function of time (Problem 16.9).

twice the amount one would naïvely expect in Newtonian mechanics, is a relativistic effect; since it decreases with increasing distance, the star acts as a convergent lens and produces a magnification of the background source. This effect can be described in the *thin screen approximation* and is similar to the *central flash*, which may occur in an occultation of a star by a planet with an atmosphere (Sec. 8.4). Referring to Fig. 16.11, for a deflection angle δ, $(\theta - \beta)D_L \simeq (\delta - \theta + \beta)(D_S - D_L)$; hence the apparent angular distance θ fulfils the quadratic equation

$$\beta = \theta - \frac{D_S - D_L}{D_S}\frac{4GM_L}{c^2 D_L \theta} = \theta - \frac{\theta_E^2}{\theta}. \qquad (16.34)$$

$$\theta_E^2 = \frac{4GM_L}{c^2 D_L}\left(1 - \frac{D_L}{D_S}\right)$$

is the square of *Einstein's angular radius*. There are two solutions:

$$\theta_\pm = \frac{\theta_E}{2}\left[u \pm \sqrt{u^2 + 4}\right] \qquad (u = \beta/\theta_E).$$

When β is exactly zero and the finite sizes of the two objects are neglected, the background source is mapped onto the whole Einstein's ring; but in practice this can never be observed. When $u \ll 1$ only one image is visible, slightly shifted away from the deflecting object; only when $u \simeq 1$ two images may be visible (Fig. 16.11). If the deflecting and the background stars have a relative angular velocity μ in the sky, β changes and a lensing event arises, with a definite time profile for the intensity over the timescale θ_E/μ (Problem 16.9). The main signature of its gravitational origin is its achromaticity: the same profile is observed in all wave lengths.

For a deflecting star with a solar mass, Einstein's radius is

$$\theta_E \simeq 1.0 \left(\frac{M_L}{M_\odot}\right)^{1/2} \left(\frac{D_L}{8\,\text{kpc}}\right)^{-1/2} \left(1 - \frac{D_L}{D_S}\right) \text{ mas}, \quad (16.35)$$

which stresses the strong alignment requirement. At $D_L (\ll D_S) = 8$ kpc the corresponding impact parameter is $b = \theta_E D_L \simeq 8$ AU, an impressive indication of the resolution power of this "telescope". If the two stars move with a relative velocity $V \simeq 200$ km/s the intensity enhancement lasts for $\approx 2b/V \simeq 100$ d. If the deflecting object is a planet or a brown dwarf, a smaller impact parameter is needed, leading to a shorter enhancement profile. Consider now the case in which the deflecting star has a planet. In suitable conditions, during the main lensing event the planet may produce its own lensing, adding a short magnification spike – with a typical duration of a day (or even hours for a terrestrial planet) – to the main intensity profile. Apart from the intensity amplification, the lensing by a planet typically deflects the source position (such as in Fig. 16.11) at the μas level. From the photometric and astrometric signals one can derive the orbital parameters and mass of the planet.

Extensive multi-colour photometric observations of stars in the Galaxy have produced hundreds of lensing events, revealing new celestial objects independently of their radiative power. As an example, the joint US-Polish project OGLE III (the Optical Gravitational Lensing Experiment) regularly monitors several hundred million stars in our Galaxy. No conclusive evidence for planetary lensing was found yet, but several hundred candidate events have been recorded, most likely lensing by low-mass stars. As a side result, in only one year of operation ending December 2002, the project also detected some 60 candidate events for transits across stars of planetary or low-luminosity objects. This technique and this field is in rapid development.

Further/future methods.– An interesting, and possibly ultimate, challenge for exoplanetary science is the search for evidence of life outside the solar system. One current view of the origin of life relies on the idea of a *primordial soup*, consisting in warm water with molecules (like methane CH_4 and ammonia NH_3) in solution, capable of generating, under optical and UV radiation, and lightnings, organic macromolecules. Laboratory experiments have shown

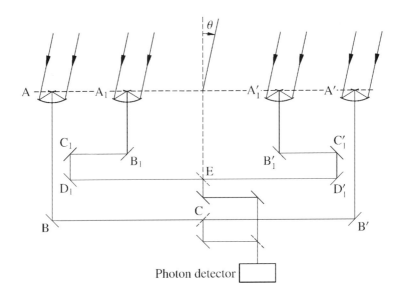

Figure 16.12. The conceptual scheme of the nulling interferometric technique: each of the two interferometers produces a recombined phase with a delay of π; the signal from the outer one is attenuated by 50% and further delayed by π before the final recombination. This produces a deep dimming near the axis.

long time ago that in such a medium, if exposed to ultraviolet light and/or electric discharges, aminoacids and other macromolecules of biological relevance are indeed formed; but all subsequent steps, leading to the crucial formation of replicators and a genetic code, are still shrouded in the dark. In this view two conditions are necessary for the emergence of life: (i) radiation from a nearby star to keep the temperature within acceptable limits and (ii) water on the surface. The infrared radiation emitted by the three terrestrial planets with an atmosphere (Venus, Earth and Mars) have a strong CO_2 absorption line at 15 μm; only the Earth shows, on its lower wavelength side, the complex of water spectral lines below 8 μm (Fig. 7.3). In addition, on the Earth biological activity has produced large quantities of oxygen, a fraction of which combines into ozone with a photochemical reaction; O_3 shows up in a strong absorption line at 9 μm. Thus, the first step in the search for life outside the solar system is the *direct observation of extrasolar planets* and their spectral analysis in the infrared. Big space projects are being studied for this purpose, in particular NASA's Terrestrial Planet Finder and ESA's Darwin, based on orbiting interferometric telescopes. This is a far away, but important goal and we briefly discuss the *nulling technique* on which it is based.

Since almost nothing is known about the atmospheres of extrasolar planets, for illustration we choose as target a planet with the same properties as the

Earth, orbiting at 1 AU around a main sequence star similar to the Sun and 10 pc away. They have an angular separation of $0.1'' = 5 \times 10^{-7}$ rads; assuming that the planet emits all the radiative energy it receives, their luminosity ratio L_p/L_* is proportional to $(R_p/a)^2 = 1.8 \times 10^{-9}$, but it crucially depends on the band of observation (visual or infrared). It seems better to use the infrared, where the star is dimmer, the planet is brighter and there are relevant absorption lines (in spite of the fact that the angular resolution is worse); then $L_*/L_p \approx 8 \times 10^6$ only. This huge dynamic range requires a strong suppression of the stellar radiation.

Since at $10\,\mu$m a diffraction limit of $0.1''$ needs a diameter of 20 m, a single telescope is not realistic and we need an interferometric technique. Active interferometers, which use a coherent laser source at the vertex of two arms, will be used in space for accurate measurements of differences in distance. LISA (Laser Interferometer Space Antenna), a large space interferometer with arm lengths of 5 million km is being developed for detection of low frequency gravitational waves. Here we are concerned with space interferometric telescopes, where the light from a distant source (suitably restricted to a band $\Delta\lambda < \lambda \approx 10\,\mu$m) is collected by two optical systems L apart; the two beams are then brought together at the recombiner at the centre, where their interference pattern is generated and recorded. If the two optical paths to the recombiner are exactly equal and the source lies in the plane of the interferometer, at a small angle θ from the median plane the phase difference is

$$\Delta\phi = \frac{2\pi L\theta}{\lambda}. \tag{16.36}$$

After recombination, the phase is proportional to $\exp(i\Delta\phi/2) + \exp(-i\Delta\phi/2) = 2\cos(\Delta\phi/2)$ and the intensity to $\cos^2(\Delta\phi/2)$; this pattern, projected in the sky in the plane of the interferometer, gives the output of a point-like source. Its angular scale $\theta_i = \lambda/(\pi L)$ gives the instrumental resolution in the plane of the interferometer; for two-dimensional imaging one can rotate the instrument around its axis. If a phase delay π is introduced in one arm, changing the sign of its complex amplitude, the recombined amplitude $\propto \exp(i\Delta\phi/2) - \exp(-i\Delta\phi/2) = 2\sin(\Delta\phi/2)$ gives an intensity $\propto \sin^2(\Delta\phi/2) = (\Delta\phi/2)^2 + O(\Delta\phi)^4$. In this way a point-like source at $\theta = 0$ is completely suppressed. With $L = 100$ m we have three angular scales: the instrumental scale $\theta_i = 6$ mas; the angular distance between star and planet $\theta_p = a/D = 100$ mas; the angular radius of the star $\theta_* = R_*/D = 0.4$ mas. The finite angular radius smears the vanishing minimum at $\theta = 0$ to the distance θ_*, with a meager dimming factor $(\theta_*/\theta_i)^2 = 4 \times 10^{-4}$. This correction to the interference pattern, in fact, is used to determine the radii of nearby stars.

To get a better dimming we can use the *nulling technique*; a simple way to realize it is the following (Fig. 16.12). Another pair of telescopes is used, each for instance, due to gravitational interactions between the planets, a companion or a passing star, tidal interactions between protoplanets and the protoplanetary disc; but none of these mechanisms is widespread. There is evidence for systematically smaller values of the eccentricity for closer planets, e.g., the Pegasi group (not clearly seen in Fig. 16.13, due to the linear scale in a); all known extrasolar planets with $a < 0.06$ AU have nearly circular orbits, as compared with a mean eccentricity $\simeq 0.27$ for the whole sample. This can be explained by the effect of the tides raised by the host star, which are more effective near the pericentre (Sec. 15.3).

(iii) Why does the majority of the known systems contain only a single Jupiter-mass body, possibly together with much lighter, unobserved planets? One interesting possibility is that only the last planet which has survived the migration process is observed. The planet-bearing stars are known to be metal-rich, both relative to the field stars and to the Sun; partial or total dissolution of giant planets in the atmosphere of the host star may result in an observable enhancement of its metallicity. However, metal-rich initial environment may also preferentially result in the formation of close jupiters and, conversely, metal-poor stars may preferentially form saturn-like planets further away. It must be avowed, anyway, that on this point, too, there are not yet satisfactory answers and we need to await forthcoming detections of planets at larger distance, $\simeq 10$ AU, say; in late 2002 there were indications of their existence, since a number of the systems monitored by the radial velocity method over the last decade have shown unexplained steady drifts in the signal.

Internal and atmospheric anomalies.– The internal structure and atmospheric circulation of extrasolar jupiters at very small distances (≤ 0.2 AU) from the parent star significantly differ from Jupiter's, whose parameters and evolution are reasonably well modelled without direct solar radiation. The solar flux at the distance of a Pegasi planet is $\simeq 3.5 \times 10^7$ erg/(cm^2 s); Jupiter receives $\simeq 5 \times 10^4$ erg/(cm^2 s), of which about 1/3 is absorbed (Table 7.1), and emits a comparable amount due to contraction. Such a fierce environment in hot jupiters prevents cooling and diminishes the temperature gradient; the internal pressure is larger and a large radiative, nearly isothermal, outer layer develops, largely restricting the convective zone. The planet does not contract and remains hot, whence the common term hot jupiters; its radius is moderately larger than the value corresponding to an isolated planet and depends on structural changes. Its size does not reach Roche's lobe and significant mass loss is prevented. Due to their proximity to the star, their rotation is tidally synchronized (Sec. 15.3); for instance, for a planet in the HD 209458 system a typical timescale to the synchronous state is $\approx 1 - 10$ My. This implies a peculiar and complicated atmospheric circulation on the planet, with the day side much hotter than the night side (a temperature difference of $\approx 300 - 500$ K may be sustained; moreover, meridional circulation may cause a similar tempera-

16. Origin of the solar system

ture difference between the equator and the poles). In the visible, the spectra of hot jupiters are expected to be dominated by pressure-broadened sodium and potassium absorption lines, despite their low abundance ($\simeq 10^{-6}$); in the infrared, water vapour, methane and carbon monoxide absorption are strong features. This contrasts with Jupiter, a much cooler object, with dominating methane and ammonia lines in the visible and infrared.

These predictions found important support in the detection of the planet around HD 209458 (Fig. 16.8), whose hot jupiter matches reasonably well the expected parameters. Much more information about the structure of Pegasi planets should be at hand when space techniques using the transit method will be available.

Appendix: Gravitational instability

The formation of separate lumps in a disc involves the competition among four factors (besides, for small bodies, chemical affinity): gravitational self-attraction favours coalescence, while solar tides, thermal pressure and rotation hinder it. In Sec. 16.1 we have used Roche's criterion to establish the critical density (16.6), below which tides prevail over self-gravity; this hinges on the smallness of the parameter $\delta \propto 1/H$ (eq. (16.5)). Had the mass of the nebula been larger, or had the solid grains settled gently on the median plane and a new, much thinner subdisc developed with thickness $H_s \ll H$, δ might have become larger than unity, preventing tidal disruption. The material in the disc, at least locally, would undergo a spontaneous collapse due to self-gravity. We review here the conditions for its onset.

Stars form out of the collapse of a cloud in the interstellar medium when initially its thermal energy is smaller than the gravitational self-energy; the ensuing contraction enhances this unbalance and an unstable collapse is unavoidable. This is *Jeans instability*, discovered by J.H. JEANS in 1902; it is a general astrophysical process, with wide applications also to the formation of galaxies and planets. In a gas with density ϱ, it is characterized by *Jeans frequency*

$$\Omega_J = (4\pi G\varrho)^{1/2} . \tag{a}$$

Note that if there are $n = m/\varrho$ particles of mass m per unit volume, the quantity $-\Omega_J^2$ is just what one gets from the plasma frequency (eq. (1.104)) with the replacement of e^2 with $-Gm^2$, which formally transforms electrostatic repulsion into gravitational attraction. One may expect, therefore, that the gravitational analogue of plasma waves be unstable.

Consider a one-dimensional density profile $\varrho(x)$ with a characteristic length R much larger than the wavelength of a perturbation. An infinitesimal localized displacement $x \to x + \xi(x)$ produces, because of mass conservation, a density perturbation

$$\delta\varrho(x) = -\varrho \, d\xi(x)/dx . \tag{b}$$

From Poisson's equation (1.36) a self-gravitational force per unit mass

$$\delta F_g(x) = 4\pi G\varrho \, \xi(x) = \Omega_J^2 \, \xi(x) \tag{c}$$

arises; in addition, with a barotropic relation $P(\varrho)$, there is a pressure force per unit mass

$$\delta F_p = -\frac{1}{\varrho}\frac{\partial P}{\partial x} = -\frac{c_s^2}{\varrho}\frac{\partial \delta\varrho}{\partial x} = c_s^2 \frac{\partial^2 \xi}{\partial x^2} , \tag{d}$$

largest planetesimal and show that the final ratio M/M_1 is $\approx (2\theta)^3$. (Hint: The factor $(v_{esc}/v_{rel})^2$ in eq. (16.19) now reads $2\theta(R/R_1)^2$.)

16.5. Compute the total angular momentum of the Earth-Moon system and compare it with the critical value for rotational instability of a homogeneous protoearth.

16.6. On the basis of Fig. 16.1 and assuming a constant temperature, evaluate numerically the ratio (16.5) as function of r.

16.7. Using Hill's method (Sec. 13.4) prove that a massless particle does not undergo a close approach to the secondary provided the initial separation of their circular orbits is larger than $2\sqrt{3}R_H$ (R_H is Hill's radius). (Hint: Use the Jacobi constant with the value corresponding to the stationary points; Fig. 13.8.)

16.8. Eccentricity of an extrasolar planet. Find the expression of the Doppler signal $y = \Delta v/v$ for a small eccentricity and an observer in the orbital plane, at an arbitrary orientation relative to the apsidal line. (Hint: Use the eccentric anomaly and Fig. 11.3.)

16.9.* The magnification of a gravitational lens is due to the change induced in an angular area within $d\beta$ and has the expression $A = |\theta d\theta/(\beta d\beta)|$. Using eq. (16.34) show that, for each solution,

$$A_\pm = \frac{1}{2}\left|\frac{u^2+2}{u\sqrt{u^2+4}} \pm 1\right| \quad (u = \beta/\theta_E)$$

and plot the two functions.

16.10. Similarly to what done in Sec. 16.1 (Fig. 16.1), construct a minimum mass model for the planetary nebula around Jupiter by augmenting the mass of Galilean satellites to the solar abundance distribution. Fit the surface density, estimate the total mass of the nebula within Jupiter's sphere of influence and compare it with M_J. The same for the Saturnian satellite system.

16.11. A circular Keplerian orbit is perturbed with a small eccentricity and inclination. Find the mean square of the change in the velocity with randomly distributed directions of the node and the pericentre.

– FURTHER READINGS –

The first modern book summarizing the main aspects of the contemporary nebular theory is V.S. Safronov, *Evolution of the Protoplanetary Cloud and Formation of the Earth and the Planets*, Nauka (1969) (English translation by the Israel Program for Scientific Translations, Jerusalem (1972)). More recent and useful books are: R.M. Canup and K. Righter, eds., *Origin of the Earth and Moon*, University of Arizona Press (2000) and V. Mannings, A.P. Boss and S.S. Russell, eds., *Protostars and Planets IV*, University of Arizona Press

(2000), and earlier volumes with the same title. The former focuses on the origin of the Moon, but contains also detailed reviews of planetary and satellite formation, including both orbital and rotational aspects. Of interest may be also the little older R. Greenberg, Planetary Accretion, in *Origin and Evolution of Planetary Atmospheres*, S.K. Atreya and J.B. Pollack, eds., University of Arizona Press (1988). The theory of brown dwarfs and the structure and evolution of extrasolar giant planets are reviewed in A. Burrows et al., The theory of brown dwarfs and extrasolar giant planets, *Rev. Mod. Phys.* **73**, 719 (2001) and T. Guillot, Physics of substellar objects: Interiors, atmospheres, evolution, *31th Saas-Fee advanced course on brown dwarfs and planets* (2002). For observational techniques of the extrasolar planets see a review by M.A.C. Perryman, Extrasolar Planets, *Rep. Prog. Phys.* **63**, 1209 (2000). On gravitational lenses, we suggest P. Schneider, J. Ehlers and E.E. Falco, *Gravitational lenses*, Springer (1999).

Chapter 17

RELATIVISTIC EFFECTS IN THE SOLAR SYSTEM

The large amount of available optical observations of the planets and the Moon have provided accurate confirmations of the Newtonian laws of celestial mechanics. However, a minute, but important discrepancy has been known since the second half of the 19th century: the perihelion of Mercury undergoes an unexplained advance of 43″ per century. Einstein's discovery of General Relativity in 1915-1916 was based, however, not on the anomaly of Mercury, but rather on the equality between inertial and gravitational mass – the *principle of equivalence* – which makes gravitation unique among all other physical interactions. He showed that its true nature is a manifestation of the geometry of spacetime and, on the basis of theoretical simplicity, brought it to a complete mathematical formulation. The General Theory of Relativity predicts small corrections to the two-body problem; the only secular effect is the advance of the pericentre, which in the case of Mercury well agrees with the observed discrepancy. In recent years, the relativistic corrections for gravitational motion in the solar system and in binary pulsar systems have been tested with much greater accuracy and Einstein's predictions have been confirmed (Sec. 20.4); they are now an essential ingredient of celestial mechanics and space navigation.

In this chapter we review, with minimum mathematical formalism, the geometrical theory of gravitation and the approximation appropriate for describing the motion of bodies and photons in the solar system and for testing different gravitational theories. The effects of Special and General Relativity are treated together; however, we assume that the reader is already familiar with the former.

Notation. In this chapter Greek letters range from 0 to 3 and label events in spacetime; x^0 is a time variable and x^i, with Latin indices from 1 to 3, are the space variables. Covariant (lower) and contravariant (upper) indices, if repeated in a product, are understood to be summed over ("Einstein's summation convention"). As explained in Ch. 3, distances are measured with the transit time method and do not have a unit independent from the second. The velocity of light c is conventional and in this chapter equal to unity. Masses are measured in terms of their gravitational radii[1] $m = GM/c^2$. The derivative of a quantity attached to a world line with respect to the proper time τ is denoted with a dot: $\dot{x}^\mu(\tau) = dx^\mu(\tau)/d\tau$; also $\partial_\mu = \partial/\partial x^\mu$.

17.1 The equivalence principle

In Newtonian dynamics the motion of a mass point at $\mathbf{r}_1$ is determined by the equation

$$\frac{d^2 \mathbf{r}_1}{dt^2} = -\nabla U(\mathbf{r}_1) \,, \tag{17.1}$$

where the gravitational potential U is obtained from the mass distribution by Poisson's equation (1.36) and the appropriate boundary conditions. This description holds only in inertial frames of reference, connected to each other by a Galilei transformation

$$\mathbf{r}_1 \to \mathbf{r}'_1 = \mathbf{r}_1 + \mathbf{a} + \mathbf{v}t \,, \quad t' = t \,,$$

where $\mathbf{a}$ and $\mathbf{v}$ are constant vectors. In contrast to (17.1), the motion of an electric charge q with an *inertial mass* M_i in a Coulomb potential V fulfils

$$M_i \frac{d^2 \mathbf{r}_1}{dt^2} = -q \nabla V(\mathbf{r}_1) \,. \tag{17.2}$$

The difference is striking: in the latter case the motion depends on the ratio q/M_i of charge to mass; in the former the motion is universal and uniquely determined by the initial position and velocity. To stress this point, consider a generalized equation for gravitational motion

$$M_i \frac{d^2 \mathbf{r}_1}{dt^2} = -M_g \nabla U(\mathbf{r}_1) \,, \tag{17.3}$$

with the same structure as (17.2); M_g can be called *gravitational mass*, or gravitational charge. In principle the inertial mass can be measured subjecting the same body to different and known forces (e.g., mechanical springs); the gravitational mass is obtained from the inertial force if U known. The proposition that $M_g/M_i = 1 + \eta$ is an absolute constant (which can be taken equal

---

[1] Note that in some textbooks the gravitational radius (also called Schwarzschild radius) is defined as $2GM/c^2$, the radius of the horizon for a spherically symmetric black hole; here m is just a measure of mass in terms of a length.

17. Relativistic effects in the solar system

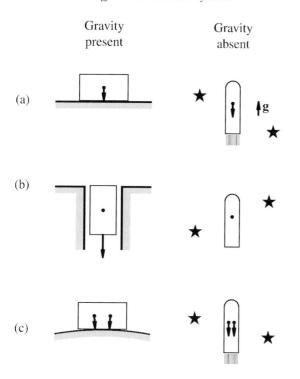

Figure 17.1 The equivalence principle and the measurement of the gravitational field. With a single particle experiment one cannot distinguish between the dynamics in a ground laboratory and in a rocket in an empty space (A and B); to see the difference it is necessary to compare the motion of two particles (case C) – note the slight convergence of the accelerations due to the tidal component of the gravitational field.

to one, thus $\eta = 0$) is the *equivalence principle* in its weak form. It concerns the "passive" gravitational mass M_g, characterizing how a given gravitational field acts on a body, rather than the gravitational field strength caused by the "active" mass of the body. If this ratio is different for two different bodies at $\mathbf{r}_1$ and $\mathbf{r}_2 = \mathbf{r}_1 + \mathbf{r}$, the equivalence principle is violated and the equation of relative motion does not have the tidal form (4.9), but

$$\frac{d^2\mathbf{r}}{dt^2} = -(\eta_2 - \eta_1)\boldsymbol{\nabla} U(\mathbf{r}_1) - \mathbf{r} \cdot \boldsymbol{\nabla}\boldsymbol{\nabla} U(\mathbf{r}_1) + O(r^2) \,. \tag{17.4}$$

A new term, independent of the distance between the particles, appears, which describes the fact they do not fall in the same way in an external gravitational field (see Secs. 20.4 and 13.4).

The universal character of Newtonian motion (17.1) is shared by the apparent forces which arise in an accelerated frame; they are indistinguishable from gravity. In a freely falling elevator (Fig. 17.1) they vanish and the motion of a test body is straight and uniform. Conversely, within an accelerated rocket in free space, a mass point experiences a uniform acceleration, indistinguishable from the gravity acceleration. With *local* experiments performed with a single test body ("at a given point") it is impossible to discover the presence of a true gravitational field. One could say, bodies fall because one observes them in a "wrong" frame of reference, which is subject to external forces (the molecular forces of the ground which support an Earth laboratory or the thrust of the rocket); weight is an artifact, not a force.

As shown by eq. (17.4), to test the equivalence principle one must measure the relative motion of two nearby test objects in the gravitational field of a third body. In the laboratory, this may be accomplished with two bodies suspended on a torsion balance and affected by the gravitational field of the Earth or the Sun. The equality of the ratio M_g/M_i for some pairs of bodies has been verified to a few parts in 10^{13}. In the solar system we may use the relative motion of the Moon and the Earth in the gravitational field of the Sun. The anomalous acceleration in (17.4) is directed along the direction of the Sun; its main effect is a perturbation with the lunar synodic frequency (Sec. 20.4). The analysis of the Lunar Laser Ranging data has established for the parameter $|\eta_{\text{Moon}} - \eta_\oplus|$ an upper limit of $\approx 5 \times 10^{-13}$. We therefore take the equivalence principle seriously and enquire, what is the appropriate mathematical structure to describe it.

The principle of inertia in Special Relativity. Since it is necessary to include apparent forces, the embarrassing limitation of Galilei invariance must be removed and arbitrary coordinates must be allowed. The first step is the statement of the principle of inertia in an invariant way, independent of the coordinates.

In Newtonian mechanics space is endowed with an absolute Euclidean geometry and time is an absolute variable, uniformly flowing for all observers. In Special Relativity we have a four-dimensional manifold of events – called *spacetime* – whose coordinates x^μ are determined to within (i) translations in time and space and (ii) Lorentz transformations, which connect two coordinate systems uniformly moving with respect to each other. The admissible transformations

$$x^\mu \to x'^\mu = x^\mu + a^\mu + \Lambda^\mu_\nu x^\nu \quad (17.5)$$

make up *Poincaré group*; a^μ is an arbitrary spacetime shift of the origin and the Lorentz matrix Λ is determined by three parameters of spatial rotations and three components of the mutual velocity of the two systems; it fulfils

$$\eta_{\mu\nu} \Lambda^\mu_\alpha \Lambda^\nu_\beta = \eta_{\alpha\beta} ,$$

with $\eta_{\mu\nu} = \text{diag}(1, -1, -1, -1)$. The invariance with respect to this group provides the appropriate foundation to the dynamics in inertial frames of reference and is realized in Special Relativity.

For an infinitesimal displacement dx^μ, the indefinite quadratic form (*Minkowski line element*)

$$d\tau^2 = \eta_{\mu\nu} dx^\mu dx^\nu \quad (17.6)$$

is invariant under Poincaré's group. When $d\tau^2 > 0$, $\sqrt{d\tau^2}$ is the infinitesimal interval of proper time separating the events x and $x + dx$ and is realized by the

17. Relativistic effects in the solar system

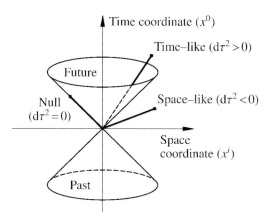

Figure 17.2 The light cone structure in spacetime.

clock readings of an observer moving from x to $x+dx$; when $d\tau^2 < 0$, $\sqrt{-d\tau^2}$ is the infinitesimal space interval between them, in the frame of reference where they occur simultaneously; $d\tau = 0$ for two events lying on the world line of a photon (the *light cone*, Fig. 17.2).

The motion of a point mass is described by its *world line* $x^\mu(\tau)$, with a time-like tangent vector $u^\mu(\tau) = \dot{x}^\mu(\tau)$ (the "four-velocity"). The uniform motion of a free particle corresponds to a straight world line with a constant four-velocity; just like in Euclidean geometry, it can be described by a variational principle. Given a world line ($\ell: x^\mu(\tau)$) joining two events A and B separated by a non-null interval, consider the proper interval

$$S[x(\tau)] = \int_B^A |d\tau| \; ; \qquad (17.7)$$

it can be called their "distance" along ℓ. We say, ℓ is straight if the action S is an extremum (i.e., its variation for any infinitesimal change of ℓ with A and B fixed vanishes). In Euclidean geometry we would say, a minimum; but in Minkowski geometry S is a minimum on a space-like straight line and a maximum on a time-like straight line; in all cases an extremum. Since the three alternatives are distinct, the absolute value can be dropped. This geometric definition holds in any coordinate system and provides the expression of the principle of inertia in arbitrary, non-inertial frames.

Under a generic mapping $x^\mu \to x'^\mu(x)$, the differentials dx^μ transform according to the linear transformation

$$dx^\mu \to dx'^\mu = \frac{\partial x'^\mu}{\partial x^\nu} dx^\nu \; . \qquad (17.8)$$

The Minkowski line element (17.6) becomes a non-degenerate, indefinite quadratic form

$$ds^2 = g_{\mu\nu}(x') dx'^\mu dx'^\nu \; , \quad g_{\mu\nu}(x') = \eta_{\varrho\sigma} \frac{\partial x'^\varrho}{\partial x^\mu} \frac{\partial x'^\sigma}{\partial x^\nu} = g_{\nu\mu}(x') \; . \qquad (17.9)$$

Note, that the generic coordinates x are just arbitrary labels of events, with no physical meaning; physical laws cannot depend on their choice. Besides the equivalence principle, this is the second pillar on which General Relativity is based: not only Newton's absolute concepts of space and time, but also the privileged nature of inertial frames are abandoned. A conceptual distinction is made between an event P in spacetime and its arbitrary coordinate labels; physical quantities transform with definite laws under a generic coordinate transformation. A *contravariant vector*, for example, transforms as the infinitesimal displacement dx^μ (eq. (17.8)); the partial differential operator $\partial_\mu = \partial/\partial x^\mu$ is a *covariant vector*, transforming with the law

$$\partial_\mu \to \partial'_\mu = \frac{\partial x^\nu}{\partial x'^\mu} \partial_\nu . \tag{17.10}$$

The metric $g_{\mu\nu}$ (eq. (17.9)) is a covariant, second rank tensor. Coordinate transformations $x \to x' = x'(x)$ are allowed only if the Jacobian matrix $\partial x'^\mu/\partial x^\nu$ does not vanish, nor diverges, so that they are invertible. Hence the vanishing of a vector or a tensor is an invariant property: if it holds in one system of coordinates, it is true in any other. Physical quantities must be tensorial quantities; physical laws consist in the statement that a (covariant or contravariant) tensor vanishes. This is the *principle of general covariance*. Observable quantities, like angles and proper times, must have a geometric, invariant character and be formally expressible with appropriate geometrical models of the physical apparatus. For example, the ticks of a well-functioning clock are events on its world line equally spaced in its proper time; the Doppler frequency shift of a moving source (17.43) is constructed with the scalar product of its four-velocity and the four-momentum of the photon. The action (17.7) can now be written in an arbitrary coordinate system x^μ. Let $x^\mu(\lambda)$ be a world line ℓ between the two events A = $x^\mu(\lambda_1)$, B = $x^\mu(\lambda_2)$; the action

$$S[x(\lambda)] = \int_{\lambda_1}^{\lambda_2} d\lambda \sqrt{g_{\mu\nu}(x(\lambda)) \dot{x}^\mu(\lambda) \dot{x}^\nu(\lambda)} \tag{17.11}$$

is a functional of the world line. The equation of a straight line in spacetime is still obtained by demanding that $\delta S = 0$ for all infinitesimal variations $\delta x^\mu(\lambda)$ which vanish at A and B; but now this gives the explicit expression for the apparent forces in any frame. To evaluate them, one only needs the Euler-Lagrange equations in the appropriate, non-inertial coordinates (Problem 17.2).

The mathematical framework for a generalization of the Special Theory of Relativity is now at hand, but so far no new specific physical law has been introduced. However, it suggests that, in view of the principle of equivalence, the motion in a gravitational field may still be determined by the action (17.11), but with a metric tensor which *cannot be expressed in the Minkowski form*

(17.6) with a suitable coordinate transformation. To see our way, we must complete the mathematical structure with the gist of Riemannian geometry with indefinite metric.

17.2 Curvature of spacetime

In this generalized approach spacetime is a four-dimensional manifold endowed with an invariant line element (17.9) and a metric $g_{\mu\nu}$ with signature $(1, -1, -1, -1)$, which defines the scalar product $g_{\mu\nu}u^\mu v^\nu$ between two vectors u^μ and v^ν. A coordinate transformation gives at the event with new coordinates x' the new metric tensor

$$g'_{\mu\nu}(x') = g_{\varrho\sigma}(x(x')) \frac{\partial x^\varrho}{\partial x'^\mu} \frac{\partial x^\sigma}{\partial x'^\nu} ; \qquad (17.12)$$

Just like in Special Relativity, it is convenient to attribute to the coordinates x^μ the dimension of a time (the same as for τ and s) and to have the g's dimensionless, like $\eta_{\mu\nu}$. Formally, this means that, besides the coordinate transformations, we can also consider changes of the scale of time $x^\mu \to x'^\mu = kx^\mu$, which leave the metric unchanged and bring ds into $ds' = k\,ds$.

Recognizing a curved spacetime. A metric endows spacetime with a rich geometric structure and makes it a *pseudo-Riemannian manifold* (in a Riemannian manifold the metric is positive definite). It is always possible to choose the transformation in such a way that *at a given event* the metric reduces to the diagonal form $(1, -1, -1, -1)$ (eq. (17.6)); since ds is invariant, $ds = 0$ defines the light cone at that event. Is it possible to accomplish this everywhere? If so, spacetime would be equivalent to a Minkowski spacetime and nothing different from Special Relativity, except for the freedom to adopt arbitrary coordinates, can be expected (as in Sec. 17.1). The answer to this question is given in terms of a fourth rank tensor $R_{\mu\nu\varrho\sigma}$, called after the German mathematician G.F.B. RIEMANN, which is a linear function of the second derivatives of the metric and a quadratic function of its first derivatives. $R_{\mu\nu\varrho\sigma}$ fulfils four symmetries, which reduce the number of independent components from $4^4 = 256$ to 20. Its vanishing is the necessary and sufficient condition in order that a coordinate transformation exists that reduces the metric to the form (17.6) everywhere; if this happens, spacetime is said to be *flat*. If not, spacetime is curved and the laws of pseudo-Euclidean geometry do not hold; in particular, the sum of the internal angles of a triangle is not equal to π. The difference between this sum and π is the *excess angle* ϵ_{exc}.

The curvature tensor is related to the *Gaussian curvature* of two-dimensional sections of spacetime. At a point P of an ordinary surface Σ, consider its section with a plane α through its normal, and let $R(\alpha)$ be the radius of the osculating circle to this section. It can be shown that (except in cases of degeneracy), as α rotates around the normal, $R(\alpha)$ goes through a maximum show that in this formulation of gravitational dynamics a suitable metric does exist, capable of accurately describing the wealth of the motion of bodies in the solar and stellar systems. In the limit of weak field and slow motion – satisfactory for the solar system – it is possible to explicitly construct a class of such geometric dynamical theories, all fulfilling the equivalence principle. In Einstein's theory of General Relativity, as discussed below, the metric field in the solar system is uniquely prescribed by the field equations, the mass distribution and appropriate boundary conditions; our more general approach will provide the suitable formalism to test alternative metric theories, a major task of experimental gravitation.

The field equations. The determination of the metric in Einstein's theory is accomplished through field equations, which in vacuo follow from a variational principle. With ten unknowns $g_{\mu\nu}$ – the "gravitational potentials"– we need ten equations; they have a very complex structure, resulting in the fact that only few physically useful exact solutions are known, and several novel features:

- According to the principle of general covariance, they must consist in the vanishing of a tensorial quantity, a property invariant under general coordinate transformations. The Riemann tensor determines the geometric structure of a Riemannian manifold and it is not surprising that it is the main ingredient of the field equations. They involve four arbitrary functions, corresponding to arbitrary coordinate transformations; hence the field equations cannot be independent, but must fulfil four differential conditions.

- In analogy with Poisson's equation (1.36), one requires that they should be partial differential equations, linear in the second derivatives of the metric; but, in order to comply with Special Relativity, they must be hyperbolic, necessarily leading to propagation phenomena. This leads to gravitational waves and a new field of research; their detection will open an entirely new "window" of observation for the most energetic processes in the Universe.

- The vanishing of a linear combination of the components of the Riemann tensor is the obvious candidate for the field equations in empty space. This implies that, contrary to Poisson's equation, they are not linear in the metric field (eq. (17.23)); the gravitational field generated by two masses is not the sum of the two separate metrics.

- In Newtonian dynamics the matter density ϱ is the only source of the scalar potential; but in Special Relativity we need the energy-momentum tensor $T^{\mu\nu}$, which embodies, for matter and any field other than gravity, the energy density, the momentum and the stress. This tensor (with ten components) is the source of Einstein's field equations.

- The field equations must be complemented with the appropriate boundary conditions. In Newtonian physics isolated massive sources correspond to the condition that at large distances U vanishes; then for a point-like body we have the monopole solution $U = -GM/r$. In General Relativity this is replaced by *asymptotic flatness*, which states that at large distances the metric tensor $g_{\mu\nu}(x)$ tends sufficiently rapidly to $\eta_{\mu\nu}$. There is only one spherically symmetric vacuum solution of the field equations which satisfies this condition (the *Schwarzschild solution*). For a dynamical system emitting gravitational waves this is supplemented by the causality requirement that the metric becomes flat at large distances on the past light-cone.

17.4 Weak fields and slow motion

In the solar system the product of the curvature and the square of a typical distance is very small (eq. (17.14) and the following); then there are ("nearly Minkowskian") coordinate systems in which the metric deviates little from its flat form (17.6)

$$g_{\mu\nu} = \eta_{\mu\nu} + h_{\mu\nu}, \quad |h_{\mu\nu}| \ll 1, \quad (17.20)$$

which is the gist of the *weak field approximation*. In this approximation it is possible to work with the formalism of Special Relativity and deal with the metric perturbations $h_{\mu\nu}$ as a symmetric tensor under Lorentz group; in particular, indices are now lowered and raised with the metric (17.6). This stresses an important mathematical analogy between the weak gravitational field, electromagnetism and other special-relativistic fields. With this assumption the invariance with respect to general coordinate transformations is restricted to

$$x^\mu \to x'^\mu = x^\mu + \xi^\mu(x), \quad (17.21)$$

where $\xi^\mu(x)$ are four arbitrary, "infinitesimal" functions of spacetime. The transformation law (17.12) becomes

$$h_{\mu\nu} \to h'_{\mu\nu} = h_{\mu\nu} + \eta_{\mu\varrho}\partial_\nu \xi^\varrho + \eta_{\nu\varrho}\partial_\mu \xi^\varrho. \quad (17.22)$$

These are the *gauge transformations*, similar to the electromagnetic case. To within linear terms in the h's, the Riemann (or curvature) tensor reads:

$$2R_{\mu\nu\varrho\sigma} = \partial_\mu\partial_\sigma h_{\nu\varrho} + \partial_\nu\partial_\varrho h_{\mu\sigma} - \partial_\mu\partial_\varrho h_{\nu\sigma} - \partial_\nu\partial_\sigma h_{\mu\varrho} + O(\partial h)^2; \quad (17.23)$$

one easily verifies that this expression is gauge-invariant. It should also be noted that in a Fermi frame, on the central geodesic ℓ, there are no corrections $O(\partial h)^2$ and the linear expression is exact.

As explained above, relativistic dynamics of a test particle replaces the variational principles of Newtonian dynamics with the geodesic principle. The comparison between them clearly shows under which approximation Newtonian mechanics is recovered from General Relativity, and which corrections have to be taken into account for a more accurate description. The Newtonian motion of a point mass moving in a gravitational potential energy (per unit mass) U is governed by the action

$$S_N = \int dt\, L_N(x^i, v^j) = \frac{1}{2}\int dt\left(v^2 - 2U\right), \quad (v^i = dx^i/dt). \tag{17.24}$$

The requirement that S_N be stationary with respect to all infinitesimal variations of the trajectory which leave the end positions unchanged leads to the Newtonian equations of motion.

To write the general relativistic action in an appropriate form it is convenient to treat the time coordinate $t = x^0$ and the space coordinates x^i separately; factoring out dt, (17.15) reads

$$S = \int \sqrt{g_{\mu\nu}\, dx^\mu dx^\nu} = \int dt\, \sqrt{g_{00} + 2g_{0i}v^i + g_{ij}v^i v^j} \tag{17.25}$$

(recall that in this chapter $c = 1$). Besides the weak field approximation, let us also assume that the motion is slow, namely $v \ll 1$; moreover, we are interested in situations in which the potential and the kinetic energies are of the same order of magnitude, as the case is for gravitationally bound bodies; in terms of an ordering parameter ε,

$$O(v^2) = O(U) = \varepsilon^2.$$

For the Earth in the gravitational field of the Sun $\varepsilon \approx 10^{-4}$. We also assume that the metric corrections $h_{\mu\nu}$ are of the same order of magnitude or smaller. To $O(\varepsilon^2)$, the action (17.25) reads

$$S = \int dt\left[1 - \frac{v^2}{2} + \frac{h_{00}}{2} + O(\varepsilon^4)\right]. \tag{17.26}$$

Since the first term does not contribute to the variation, the relativistic variational principle is equivalent, to this order, to the Newtonian counterpart (eq. (17.24)), provided that

$$h_{00} = 2U. \tag{17.27}$$

h_{00} is the sum of the contributions from all the bodies; for a single mass point

$$h_{00} = -\frac{2m}{r} \quad \left(m = \frac{GM}{c^2} \simeq 1.48\, \frac{M}{M_\odot}\, \text{km}\right). \tag{17.28}$$

For a body which is not spherically symmetric, U includes higher harmonics. This shows that, in the slow motion approximation, the whole of Newtonian gravitational dynamics is contained in the geodesic principle and thus reduced to geometry.

Beyond the Newtonian approximation. In the $(3+1)$-dimensional formulation, gravitation is described by three dimensionless potentials: a scalar (with

17. Relativistic effects in the solar system

respect to spatial transformations, of course) h_{00}, a vector $\mathbf{h}_0 = h_{0i}$ and a symmetric tensor h_{ij}. Under Lorentz transformations, $h_{\mu\nu} dx^\mu dx^\nu$, being the difference between two invariant quantities, is itself an invariant; hence $h_{\mu\nu}$ behaves as a tensor under Lorentz transformations. We can use this property to describe the effect of a moving source at the event P.

Consider a source, a particle with mass m determined in its rest frame, moving in a given coordinate system along a world line $x^\mu = (x^0, x^i)$; $x^0 = t$ is the coordinate time and $x^i = z^i(t)$. The corresponding four-velocity reads

$$u^\mu = \frac{dx^\mu}{d\tau} = \frac{dx^0}{d\tau}(1, u^m) = \frac{(1, \mathbf{u})}{\sqrt{1 - u^2}}. \tag{17.29}$$

To generalize eq. (17.28), it is reasonable to look for a metric correction $h_{\mu\nu}$ proportional to m/r (using the invariant definition (17.31)), with the transformation properties of a second rank tensor under Lorentz transformations. With the assumption, familiar in electromagnetism, that a source affects the gravitational potential only through its velocity and its relative position, there are only two possibilities: $\eta_{\mu\nu}$ and $u_\mu u_\nu$ (recall that $u_\mu = \eta_{\mu\nu} u^\nu$); hence

$$h_{\mu\nu} = 2\left(\gamma \eta_{\mu\nu} + \gamma' u_\mu u_\nu\right)\frac{m}{r}.$$

γ and γ' are dimensionless coefficients. When the body is at rest with respect to the source, we must recover eq. (17.28); this leads to $\gamma' = -1 - \gamma$ and

$$h_{\mu\nu} = 2\left[\gamma \eta_{\mu\nu} - (1 + \gamma) u_\mu u_\nu\right]\frac{m}{r}. \tag{17.30}$$

Note that the admissible metrics depend upon a free, dimensionless parameter γ, which turns out to be one in General Relativity. Note also the extraordinary constructive power of the invariance requirement; of course, this argument is useful only for the linear part of the metric perturbations, of order $O(\varepsilon^2)$ and $O(\varepsilon^3)$.

In (17.30) m is an invariant scalar, while r must be expressed in a Lorentz-invariant way. The light cone with vertex at P intersects the world line of the source at a past event O (Fig. 17.4); $e^\mu = (P - O)^\mu$ is the corresponding null vector; then

$$r = |\eta_{\mu\nu} e^\mu u^\nu| \tag{17.31}$$

is an invariant and reduces to the ordinary distance when the source is at rest and $u^\mu = (1, \mathbf{0})$. There is also an intersection O' in the future of P, but since u^μ is constant, it gives the same result.

Naïve intuition might suggest that retardation should be taken into account; if the source moves with a velocity v_r towards P, at the antedated time its distance was larger, $r(1 + v_r)$ say. Were that the case, the Newtonian potential

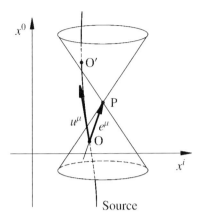

Figure 17.4 Construction of the invariant distance r between a spacetime point P and the moving gravitational source (see the text).

m/r would have to be corrected with a term ε times smaller. But things are subtler and we now show that the retardation correction to m/r is of order $(m/r) \times O(\varepsilon^2)$, beyond our discussion; we also note a close similarity between this result and the Liénard-Wiechert potentials familiar in electromagnetic theory. With a point-like source the sought value of m/r reads

$$\left(\frac{m}{r}\right)_{\text{ret}} = m \int d\tau \, \frac{\delta(F(\tau))}{|\mathbf{x} - \mathbf{z}(\tau)|} \; ; \tag{17.32}$$

here $F(\tau) = \tau - t + |\mathbf{x} - \mathbf{z}(\tau)| = 0$ defines the antedated time $\tau = t'$ on the world line $\mathbf{z}(\tau)$ of the source and on the past null cone of the event $(t, \mathbf{x})$; $\delta(\xi)$ is the Dirac delta function. Denote the solution of $F(\tau) = 0$ by t'; using Lagrange's expansion theorem from Problem 11.16, we easily obtain $t' = t - r(t) + \ldots$ with $r(\tau) = |\mathbf{r}(\tau)| = |\mathbf{x} - \mathbf{z}(\tau)|$. We can thus evaluate the integral in (17.32) as

$$\left(\frac{m}{r}\right)_{\text{ret}} = \frac{m}{r(t')} \left(\frac{1}{|dF/d\tau|}\right)_{\tau=t'} = \frac{m}{r(t') - \mathbf{u}(t') \cdot \mathbf{r}(t')} \, , \tag{17.33}$$

where the denominator is evaluated at the antedated time t'. With the expansion

$$r(t') = r(t) + \mathbf{u}(t) \cdot \mathbf{r}(t) + O(\varepsilon^2) \, ,$$

we see that the terms $O(\varepsilon)$ compensate and

$$\left(\frac{m}{r}\right)_{\text{ret}} = \frac{m}{r(t)} \left[1 + O(\varepsilon^2)\right] . \tag{17.34}$$

From now on the position of the source is understood to be taken at the time t. We thus conclude that retardation – and the asymmetry between past and future (O and O' events in Fig. 17.4) – appears quadratically in ε, an important fact used below.

To lowest order, we thus have, besides (17.28), a vector component

$$g_{0i} = h_{0i} = 2(1 + \gamma) u_i \frac{m}{r} + O(\varepsilon^5) \tag{17.35}$$

and a tensor component

$$h_{ij} = -2\gamma \delta_{ij} \frac{m}{r} + O(\varepsilon^4) . \tag{17.36}$$

In this linear approximation the metric field of several sources is the summation of the separate contributions.

For corrections to the Newtonian approximation in the action (17.25) we need h_{0i} and h_{ij} to the lowest order as above, but h_{00} to $O(\varepsilon^4)$. For the components of the four-velocity (17.29), u^0 is expanded in even powers, beginning with 1 at $O(\varepsilon^0)$, and u^i has odd powers. The complexity of the $O(\varepsilon^4)$ level in h_{00} stems first from the necessary retardation effects, as discussed above, but also from the expected non-linear terms proportional to the product and the squares of the masses of the sources. Its exact expression depends on the field equations, not discussed in detail here. Instead, we notice that in the solar system the main relativistic corrections are due to the Sun and it is enough here to use h_{00} for a single body at *rest*. We heuristically expect at the $O(\varepsilon^4)$ level a $\propto (m/r)^2$ correction in h_{00}:

$$h_{00} = -2\frac{m}{r} + 2\beta\left(\frac{m}{r}\right)^2 + O(\varepsilon^6) . \tag{17.37}$$

The dimensionless coefficient β is one in General Relativity. The final, spherically symmetric line element reads, to this order,

$$ds^2 = \left[1 - 2\frac{m}{r} + 2\beta\left(\frac{m}{r}\right)^2\right]dt^2 - \left[1 + 2\gamma\frac{m}{r}\right]\delta_{ij}dx^i dx^j . \tag{17.38}$$

The line element ds^2 up to this order for N slowly moving masses is known, and one can thus formulate a relativistic N-body problem. Obviously, in these equations the retardation and non-linear effects are important to evaluate correctly h_{00} to $O(\varepsilon^4)$.

The above described procedure paves the way to a systematic and powerful expansion of the metric in powers of ε, called the *post-Newtonian approximation*. The scalar h_{00} and the vector h_{0i} have only even and odd powers, respectively; the tensor h_{ij} has even powers, but the first term (17.36) in the expansion is proportional to the unit matrix. Since the metric perturbation is determined by moving masses, its time and space scales fulfil the slow motion approximation, in the form $\partial_0 = O(\varepsilon)\partial_i$. This assignment is itself unchanged for infinitesimal coordinate transformations (17.21) if $\xi^0 = O(\varepsilon^3)$, with odd powers, and $\xi^i = O(\varepsilon^2)$, with even powers of ε; to the lowest order they have the form:

$$\begin{aligned} h_{00} &\rightarrow h'_{00} = h_{00} + 2\partial_0 \xi^0 , \\ h_{0i} &\rightarrow h'_{0i} = h_{0i} + \partial_i \xi^0 - \partial_0 \xi^i , \\ h_{ij} &\rightarrow h'_{ij} = h_{ij} - \partial_i \xi^j - \partial_j \xi^i . \end{aligned} \tag{17.39}$$

h_{00} is invariant to $O(\varepsilon^2)$, safeguarding the Newtonian approximation; at $O(\varepsilon^4)$ its change is tied up with the change in h_{0i}, so that the expression (17.35) is, in fact gauge-dependent. Similarly, the solution of the geodesic equation gives the world line $x^\mu(s)$ in some coordinate system; however, the coordinates do not have a physical meaning in themselves. Instead, they have to be used to construct the appropriate invariant observables, for example the proper time between two events on a world line or the frequency ratio between the receiver and observer as a function of the proper time. Under the transformations (17.39) every invariant quantity, in particular, Fermi coordinates, remains unchanged.

It may be noted that the second, "post-post-Newtonian", corrections have been computed and studied for the relativistic two-body problem, particularly interesting for astrophysical applications in binary systems of two neutron stars. In these cases the emission of gravitational waves, which appears at the order $O(\varepsilon^7)$, is also needed to access observations.

De Sitter precession. As discussed in Sec. 17.3, in a Riemannian manifold we can define a constant vector field S^μ along a geodesic ℓ; it is characterized by the fact that its Fermi coordinates are constant. If S^μ is orthogonal to ℓ, it defines an absolute spatial direction in its rest frame. To see how it can be physically realized, it is enough to note that in Fermi's frame any local system is governed by the ordinary physical laws of Special Relativity, with no gravitational field; if a device governed by these laws – like a gyroscope or a spinning body – keeps a constant direction, it provides such realization. This is a local, dynamically defined standard of absolute rotation and can be constructed, as discussed in Sec. 3.2, by fitting an unknown Coriolis force to the data; the question arises, how it is connected with the kinematical frame of reference anchored to distant galaxies. In the usual gravitational description of the solar system provided by General Relativity (e.g., eq. (17.38)), however, there is no reference to the Universe; but at large distances the metric tends to the Minkowski form $\eta_{\mu\nu}$ and defines there a set of inertial frames (connected to each other by a Lorentz transformation), thus providing an asymptotic absolute standard of rotation. Strong cosmological arguments support its identification with the kinematical extragalactic standard used in practice, and discussed in Ch. 3.

To obtain a physically verifiable prediction, the appropriate metric in the solar system barycentric frame (like (17.38)) should be considered in the relevant weak field, slow motion approximation, and the condition for parallel transfer of S^μ along a time-like geodesic (like that of the Earth) should be computed. It turns out that for its space part

$$\frac{d\mathbf{S}}{dt} = \mathbf{\Omega}_{\text{dS}} \times \mathbf{S}, \quad \mathbf{\Omega}_{\text{dS}} = \left(\gamma + \frac{1}{2}\right)\mathbf{v} \times \mathbf{\nabla} U; \qquad (17.40)$$

17. Relativistic effects in the solar system

S precesses relative to the flat spacetime at great distances with the angular velocity Ω_{dS} – *de Sitter's frequency*; W. DE SITTER derived this important result shortly after Einstein's 1915-1916 discovery of General Relativity. In the case of a single gravitating body (the Sun) and a circular orbit with mean motion n and radius r, the absolute direction rotates around the normal with angular velocity

$$\Omega_{dS} = \left(\gamma + \frac{1}{2}\right) n \frac{m_\odot}{r} \simeq 19.2 \text{ mas/y}.$$

This effect has been verified with the lunar motion (Sec. 20.4).

It is interesting to note that this geometrical effect can also be directly obtained in the non-rotating geocentric frame, as realized with the CRS (Sec. 3.2). In this frame the Sun moves around the Earth with an approximately circular motion of radius r, has an angular momentum and therefore exerts a gravitomagnetic force and a torque $\propto 1/r^{5/2}$, with an effect on a spinning body similar to the Lense-Thirring precession (17.67); then the de Sitter precession is recovered.

17.5 Doppler effect

According to Special Relativity, the motion of the source and the receiver in an inertial frame produce the Doppler frequency shift (3.3). The equivalence principle applies to photons as well and requires that also a gravitational field affects the frequency of a photon. Consider a photon emitted at S in an rocket along the direction of its acceleration g (case (a) of Fig. 17.1); at the reception at O, at a distance h, the rocket moves faster than at the emission, so that less wavelengths per unit time are received than emitted (Fig. 17.5) and we get a frequency smaller by ghv_S. The same frequency shift must occur when there is a difference gh in the gravitational energy per unit mass between the observer and the source:

$$\frac{v_S - v_O}{v_S} = U_O - U_S = gh. \tag{17.41}$$

In the case of light emitted by the Sun, for example, we have a fractional redshift $\approx 2 \times 10^{-6}$, barely observable in the complex solar spectrum. This effect has been first verified in 1960-65 with a series of laboratory experiments using the Mössbauer effect to obtain a narrow spectral line; in 1974 it has been tested much more accurately with a hydrogen maser flown in a rocket at about 10,000 km of altitude (Problem 17.15 and Sec. 20.4). For an accurate description of the Doppler effect valid to higher order and for general geometries, a rigorous mathematical introduction is called for.

In Special Relativity a photon with a world line $y^\mu(\lambda)$ and frequency v is described by the null vector $p^\mu = hv\dot{y}^\mu = hv(1, \mathbf{n})$; $\mathbf{n}$ gives the direction of propagation. For an observer with world line $x(s)$ its energy is the scalar product $E = \eta_{\mu\nu} p^\mu \dot{x}^\nu$; $E = hv$ in the rest frame in which $\dot{x}^\mu = (1, \mathbf{0})$. The

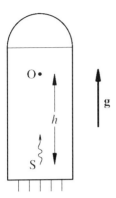

Figure 17.5 Doppler shift in an accelerated rocket. During a photon transit time h/c the receiver O increases its velocity by gh/c and detects a fractional frequency shift gh/c (for clarity we keep c explicit).

components orthogonal to $\dot{x}^\mu$ give the momentum (hence the radiation pressure; Sec. 15.4). In a curved spacetime the photon world line, resulting from variational principle (17.16) with $L^2 = 0$, is a function $y^\mu(\lambda)$ of the affine parameter; the photon four-momentum is given, in terms of the λ-derivative, by $p^\mu = h\nu \, \dot{y}^\mu$ and the observer measures the energy

$$E = g_{\mu\nu} p^\mu \dot{x}^\nu \; ; \tag{17.42}$$

$h\nu$ is a constant, with the dimension of energy, attributed to the photon. Consider now a source $x_1^\mu(s_1)$ and a receiver $x_2^\mu(s_2)$ on the path of the photon. The frequency ratio is

$$\frac{\nu_2}{\nu_1} = \frac{[g_{\mu\nu} p^\mu \dot{x}^\nu]_2}{[g_{\mu\nu} p^\mu \dot{x}^\nu]_1} \; . \tag{17.43}$$

This formula, whose intuitive meaning is obvious, can be rigorously derived from the properties of null geodesics. We now apply it in two specific cases of interest.

Let us first recover and generalize the gravitational frequency shift formula (17.41). Consider a stationary spacetime in which the metric is time independent[4] and both source and observer are at rest. Because of the definition of the proper time τ, we get for the source and the observer

$$\dot{x}^0 = \frac{1}{\sqrt{g_{00}}} \; , \quad \dot{x}^i = 0 \; . \tag{17.44}$$

We do not need to solve the variational principle (17.16) for the motion of the photon; it is enough to note that, since the Lagrange function L^2 does not depend on time, its canonically conjugate quantity

$$\frac{\partial L^2}{\partial \dot{y}^0} = 2 g_{0\mu} \dot{y}^\mu \tag{17.45}$$

---

[4] Note that the assumption of stationarity does *not* exclude motion (e.g., constant rotation of a body).

17. Relativistic effects in the solar system

is a constant along the ray (and so is $g_{0\mu}p^\mu$). But this is just what is needed in (17.43), and we obtain

$$\frac{\nu_2}{\nu_1} = \frac{[g_{0\mu}p^\mu/\sqrt{g_{00}}]_2}{[g_{0\mu}p^\mu/\sqrt{g_{00}}]_1} = \sqrt{\frac{[g_{00}]_1}{[g_{00}]_2}}. \tag{17.46}$$

This relation is exact; in the linear approximation $g_{00} = 1 + 2U$ (eq. (17.27)) and we recover (17.41).

From the general formula (17.43) we can also recover the frequency shift in Special Relativity (3.3). Since, in terms of the ordinary velocity $v^i = dx^i/dt$,

$$\dot{x}^0 = \frac{1}{\sqrt{1-v^2}}, \quad \dot{x}^i = \frac{v^i}{\sqrt{1-v^2}},$$

$$\eta_{\mu\nu}p^\mu\dot{x}^\nu = p^0\dot{x}^0 - \delta_{ij}p^i\dot{x}^j = h\nu\frac{1-\mathbf{n}\cdot\mathbf{v}}{\sqrt{1-v^2}}$$

and

$$\frac{\nu_2}{\nu_1} = \frac{1-\mathbf{n}\cdot\mathbf{v}_2}{1-\mathbf{n}\cdot\mathbf{v}_1}\left(\frac{1-v_1^2}{1-v_2^2}\right)^{1/2}. \tag{17.47}$$

The first factor represents the ordinary Doppler effect. The second describes the second-order, transversal correction; although small, it is important for laser and radar tracking to satellites and asteroids.

Note that it is not possible to distinguish between the Doppler effect due to a gravitational field and that due to a velocity difference. Upon transforming to an accelerated frame, a relative velocity shows up in the scalar product as an energy difference due to the apparent force. The whole effect is fully described by eq. (17.43), which is invariant and covers all cases (Problem 17.1).

This discussion clearly shows that, as expected, in general there is no global and privileged time variable: we only have independent clocks on their own world lines and the means to compare their frequencies with electromagnetic signals. It does not make sense to say that two clocks at different places show the same time. If two observers 1 and 2 have "well functioning" clocks, we can only say, with the Doppler formula, what is the rate of clock 2 – transmitted electromagnetically – that 1 measures with respect to his own clock, and vice versa[5]. There is, however, an important case of approximate and global synchronization of practical interest. Consider a set of clocks at rest on the

---

[5] It is interesting to note that in standard cosmology there is a privileged time variable. Indeed, the universal, ordered expansion is characterized by a velocity field, defined by the local mean motion of galaxies. At every point we thus have a well determined standard of rest (in which the cosmic microwave background radiation is nearly isotropic); the family of three-dimensional manifolds orthogonal to this four-velocity field gives the loci of simultaneous events (Problem 17.7).

(rigidly) rotating geoid of the Earth in the quadrupole, axially symmetric approximation. In this rotating frame (neglecting the influence of other bodies) the metric is still time-independent and the clocks are at rest, but the potential U in (17.27) should be replaced by the total potential W (2.35), including the centrifugal force. Because of eq. (17.46), these clocks, when compared electromagnetically, do not show, to $O(\varepsilon^2)$, any frequency shift (see Problem 17.11).

17.6 Relativistic dynamical effects

For the motion of planets and interplanetary navigation, the main relativistic corrections are those of the Sun, which can be taken at rest[6]. It is therefore important to study the geodesic motion for the spherically symmetric metric (17.38). In Sec. 17.7 we discuss the role of the off-diagonal vector h_{0i}.

Consider first the motion of test bodies in the field of a single mass point at rest, corresponding to the Lagrange function (see eqs. (17.25) and (17.38)):

$$L = \left[1 - 2\frac{m}{r} + 2\beta\left(\frac{m}{r}\right)^2 - \left(1 + 2\gamma\frac{m}{r}\right)v^2 + \ldots\right]^{1/2}. \tag{17.48}$$

Neglecting an irrelevant constant and changing the sign, the Lagrange function

$$L = \frac{1}{2}v^2 + \frac{m}{r} - \beta\left(\frac{m}{r}\right)^2 + \frac{1}{8}\left(v^2 + 2\frac{m}{r}\right)^2 + \gamma\frac{m}{r}v^2, \tag{17.49}$$

of order ε^2 itself, must be evaluated up to $O(\varepsilon^4)$. In the following we drop the indication of the neglected terms. The canonical momentum (per unit mass)

$$\pi = \frac{\partial L}{\partial \mathbf{v}} = \mathbf{v}\left[1 + \frac{1}{2}\left(v^2 + 2\frac{m}{r}\right) + 2\gamma\frac{m}{r}\right]$$

has small corrections $O(\varepsilon^3)$ to the ordinary value $\mathbf{v}$. To get the Hamiltonian $H = \pi \cdot \mathbf{v} - L$, we solve (in the ε expansion) for $\mathbf{v}$ and express in terms of $\mathbf{r}$ and π the function

$$H_0 + H_1 = \frac{\pi^2}{2} - \frac{m}{r} + \beta\left(\frac{m}{r}\right)^2 - \frac{1}{8}\left(\pi^2 + 2\frac{m}{r}\right)^2 - \gamma\frac{m}{r}\pi^2. \tag{17.50}$$

We see that H_1, the correction to the Hamiltonian function $H_0 = \pi^2/2 - m/r$ at the Newtonian level, is of order $\varepsilon^2 H_0$; hence we expect that in an orbital period every orbital element suffers a fractional change of order ε^2. For example, the distance from the Earth to the Sun differs from its Newtonian value

---

[6]However, accurate planetary ephemerides now require a full post-Newtonian formulation of the N-body problem, beyond the scope of this introductory chapter.

by a variable term of order ε^2 1AU $\approx m_\odot \approx$ 1.5 km. The corresponding angular deviation, of order 10^{-8} rads $\simeq$ 2 mas, cannot be measured with ordinary telescopes, which can only detect larger, secular relativistic effects. Currently, microwave ranging to interplanetary spacecraft has an accuracy of a few metres and relativistic corrections are well observable. At 1 AU the fractional frequency shift due to the ordinary Doppler effect is $O(\varepsilon) \approx 10^{-4}$, while the transversal Doppler and the gravitational frequency shift are $O(\varepsilon^2) \approx 10^{-8}$. The current accuracy of Doppler measurements between the Earth and an interplanetary spacecraft, given by Allan deviation σ_y, is about 10^{-14}, well below the relativistic signal; however, the actual signal-to-noise ratio depends on the integration time and cannot be deduced without a detailed analysis.

The Lagrange function (17.49) describes a planar motion, a consequence of the fact that $\partial L/\partial \mathbf{v}$ is parallel to $\mathbf{v}$. Introducing polar coordinates (r, ϕ) in the orbital plane, we see that

$$\frac{\partial L}{\partial \dot\phi} = r^2 \dot\phi \left[1 + \frac{1}{2}(\dot r^2 + r^2 \dot\phi^2) - (1 + 2\gamma)\frac{m}{r} \right], \qquad (17.51)$$

the relativistic angular momentum, is a constant of motion. It differs from the ordinary expression $r^2\dot\phi$ by $O(\varepsilon^2)$. The energy (17.50) is also constant.

The perturbation energy H_1 depends only on three of the six orbital elements: the mean anomaly M, the semimajor axis a and the eccentricity e. Secular effects are easily obtained by replacing H_1 with its average over an orbital period, expressed in terms of the eccentric anomaly u (see Chs. 11 and 12 and Fig. 11.3):

$$H_1 \to \overline{H}_1 = \frac{1}{2\pi} \int_0^{2\pi} du\, (1 - e\cos u)\, H_1 .$$

The averaging procedure removes the dependence of H_1 on the mean anomaly M; its dependence on the semimajor axis a produces only a change in M, corresponding to a slightly different form of the Kepler's third law (11.19′). This secular effect is difficult to observe because of the error in $GM_\odot$. We therefore conclude that the only relevant secular effect is a motion of the longitude of the periastron according to the equation (12.49e):

$$\frac{d\varpi}{dt} = -\frac{\sqrt{1-e^2}}{na^2 e}\frac{\partial \overline{H}_1}{\partial e} . \qquad (17.52)$$

The perturbation Hamiltonian (17.50) is a quadratic form in $a/r = 1/(1 - e\cos u)$:

$$H_1 = -\frac{m^2}{2a^2} + (1+\gamma)\frac{m^2}{ra} + (\beta - 2 - 2\gamma)\frac{m^2}{r^2} ; \qquad (17.53)$$

here we have used the expression (11.21) for the osculating energy H_0. The first two terms yield a contribution which does not depend on the eccentricity; to evaluate the last term we need the integral

$$\frac{1}{2\pi} \int_0^{2\pi} \frac{du}{1 - e \cos u} = \frac{1}{\sqrt{1 - e^2}}.$$

Thus we obtain the secular advance of the pericentre in a period

$$\delta\varpi = \frac{2\pi}{n} \left(\frac{d\varpi}{dt}\right)_{\text{sec}} = (2 - \beta + 2\gamma) \frac{2\pi m}{a(1 - e^2)}. \qquad (17.54)$$

In Einstein's theory $\beta = \gamma = 1$ and the bracketed expression equals 3. This is a secular correction, whose effect steadily increases with time. For Mercury, with $m/a = 2.56 \times 10^{-8}$, $\delta\varpi = 5 \times 10^{-7}$ rads (or $0.1''$), corresponding to $43''$ per century. As noted in the introductory section, this anomaly was known, and unexplained, to astronomers long time before Einstein's seminal papers.

Relativistic effects on the motion of a photon. The Lagrange function

$$L^2 (d\lambda)^2 = \left(1 - \frac{2m}{r}\right) dt^2 - \left(1 + \gamma \frac{2m}{r}\right) \delta_{ij} dx^i dx^j \qquad (17.55)$$

(eq. (17.16)), expressed in terms of the affine parameter λ, must be used. λ is determined by the constancy of the corresponding canonical momentum; setting this constant equal to unity, so that λ coincides with the coordinate time t at infinity, we have

$$d\lambda = \left(1 - \frac{2m}{r}\right) dt. \qquad (17.56)$$

We now note that the magnitude v of the *coordinate velocity* of the photon dx^i/dt differs very little from unity (by $O(\varepsilon^2)$); hence in L^2 the difference between λ and t can be neglected in the coefficient of m/r. We have, therefore,

$$L^2 = 1 - v^2 - (1 + \gamma) \frac{2m}{r} + O(\varepsilon^4). \qquad (17.57)$$

This is equivalent to the Lagrange function of Newtonian gravitational theory in the field of a point mass $(\gamma + 1)m$ ($2m$ in General Relativity, twice the Newtonian value). The motion is hyperbolic, with an angle between the asymptotes

$$\delta = 2(1 + \gamma) \frac{m}{b}, \qquad (17.58)$$

where b is the impact parameter (see Sec. 11.4 and Fig. 17.6). This effect displaces the positions of stars and radio sources away from the Sun by the angle δ and has been measured very accurately.

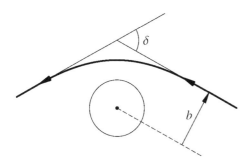

Figure 17.6 The hyperbolic trajectory of a photon in the field of a spherical gravitating body, with a deflection δ and an impact parameter b.

The gravitational field of the Sun also increases the transit time Δt of a photon passing near it. On the unperturbed trajectory ($x = \lambda, y = b, z = 0$) we have ($ds^2 = 0$ in eq. (17.38))

$$\left(1 - \frac{m}{r}\right) dt = \left(1 + \gamma \frac{m}{r}\right) dx \; ;$$

this shows that light propagates with a ratio dx/dt smaller than unity, thus increasing the coordinate time it takes to travel a given distance, as measured with the Euclidean variable x. Consider an experiment near conjunction between two points at the distances from the Sun $x = D_1, x = -D_2 = -D + D_1$, with $D_1, D_2 \gg b$; then the integration gives a relativistic correction to the transit time Δt which *diverges logarithmically* as $b \to 0$:

$$\Delta t = D + (\gamma + 1)m \ln \frac{D_1 D_2}{b^2} \simeq D + 5 \,\mu\text{s}\, (\gamma + 1) \ln \frac{D_1 D_2}{b^2} \,. \qquad (17.59)$$

This is the *Shapiro effect*, sometimes called "the fourth test of the General Relativity"; it has also been measured with a very good accuracy. Since the delay Δt changes with time, we also have a frequency shift (Sec. 20.4).

17.7 * Gravitomagnetism

Relativistic gravitation produces an additional force between two bodies which depends on their velocities. The main component of this force is due to the off-diagonal part of the metric (17.35), which contributes to the Lagrange function (see (17.25)) for a mass m' with velocity $\mathbf{v}$ the term

$$L_{\text{gm}} = m' h_{i0} v^i = m' \mathbf{v} \cdot \mathbf{h}_0 = 2(1 + \gamma) \frac{mm'}{r} \mathbf{v} \cdot \mathbf{u} \,. \qquad (17.60)$$

$\mathbf{u}$ is the velocity of the source with mass m; for didactical sake we use here the Lagrangian multiplied by the particle's mass m'. There is an obvious formal analogy with magnetism, which we now proceed to exploit. The magnetic term in the Lagrange function for a charge q' in a vector potential $\mathbf{A}$ (with $c = 1$) is

$$L_{\text{m}} = q' \mathbf{v} \cdot \mathbf{A} \,. \qquad (17.61)$$

Similarly to eq. (17.60), one can obtain from the requirement of Lorentz invariance the full four-potential $A_\mu = (\phi, \mathbf{A})$ for a charge q with four-velocity u^μ. When the source is at rest $\mathbf{A} = 0$ and $\phi = q/r$; hence in general

$$A^\mu = \frac{q}{r} u^\mu .$$

Here r is the Lorentz invariant generalization (17.31) of the spatial distance. In the slow motion limit

$$L_m = qq' \frac{\mathbf{v} \cdot \mathbf{u}}{r} .$$

We see that the gravitomagnetic Lagrange function between two massive bodies is formally equivalent to the magnetic Lagrange function for the charges

$$q = m \sqrt{2(1+\gamma)}, \qquad q' = m' \sqrt{2(1+\gamma)} . \tag{17.62}$$

The gravitomagnetic force is important from the point of view of the field structure: to describe gravitation at the post-Newtonian level, a scalar is not enough, we need a vector as well.

Gravitomagnetism is a new gravitational force, produced by moving masses (e.g., rotating bodies). Just like moving charges produce a magnetic dipole moment

$$\mathbf{d} = \frac{1}{2} \int dq\, (\mathbf{r} \times \mathbf{u}) , \tag{17.63}$$

moving masses produce a "gravitomagnetic moment" proportional to the angular momentum $\mathbf{L}$

$$\mathbf{d}_{gm} = \frac{1}{2} \sqrt{2(1+\gamma)} \int dm\, (\mathbf{r} \times \mathbf{u}) = \frac{1}{2} \sqrt{2(1+\gamma)}\, \mathbf{L} . \tag{17.64}$$

The corresponding gravitomagnetic potential and field are

$$\mathbf{A}_{gm} = -\mathbf{d}_{gm} \times \nabla\left(\frac{1}{r}\right) = -\frac{1}{2} \sqrt{2(1+\gamma)}\, \mathbf{L} \times \nabla\left(\frac{1}{r}\right), \tag{17.65a}$$

$$\mathbf{B}_{gm} = \nabla \times \mathbf{A}_{gm} = -\frac{1}{2} \sqrt{2(1+\gamma)}\, \nabla \times \left[\mathbf{L} \times \nabla\left(\frac{1}{r}\right)\right]. \tag{17.65b}$$

Consider now another body with angular momentum $\mathbf{L}' = I\mathbf{S}$, expressed in terms of the moment of inertia I and the angular velocity $\mathbf{S}$. The gravitomagnetic torque acting on it is

$$I \frac{d\mathbf{S}}{dt} = -\mathbf{d}'_{gm} \times \mathbf{B}_{gm} = -\frac{1+\gamma}{2} I \left[\mathbf{S} \times \nabla \times \left(\mathbf{L} \times \frac{\mathbf{r}}{r^3}\right)\right] .$$

This is the *Lense-Thirring precession*; similarly to de Sitter precession (17.40), it has the frequency

$$\mathbf{\Omega}_{LT} = \frac{1}{2}(1+\gamma) \nabla \times \left(\mathbf{L} \times \frac{\mathbf{r}}{r^3}\right) . \tag{17.66}$$

17. Relativistic effects in the solar system

It was first pointed out by J. LENSE and H. THIRRING in 1918. Using the value (1.15) of the angular momentum per unit mass of the Earth we get

$$\Omega_{\rm LT} \simeq 3 \times 10^{-14} \left(\frac{R_\oplus}{r}\right)^3 {\rm rad/s} \simeq 0.24 \left(\frac{R_\oplus}{r}\right)^3 ''/{\rm y} . \tag{17.67}$$

The first direct measurement of this effect has been recently announced by using as a gyroscope the orbital momenta of two Earth satellites of the LA-GEOS family (see Ch. 20). According to eq. (17.66) they precess around the polar axis with the rate $\simeq 31$ mas/y (eq. (17.67) and $r \simeq 2R_\oplus$ in this case, Table 20.3). NASA's Gravity Probe B will also measure very accurately this effect, $\simeq 50$ mas/y in this case.

– PROBLEMS –

17.1. Evaluate the frequency shift (17.43) in a Newtonian potential U when the observer and the source move slowly, taking into account the corrections of order $v^2 \approx U$.

17.2. Apply the transformations to a frame of reference with uniform acceleration to the Minkowski metric and obtain the apparent force from the variational principle (17.15); the same for a uniform rotation.

17.3. Describe quantitatively the motion in the sky of the image of a star as the Sun moves by it.

17.4. Find the value of the Lense-Thirring precessional frequency (17.66), averaged over a circular polar orbit; generalize the result for an arbitrary inclination.

17.5. Find the order of magnitude of the geodetic precession using the gravitomagnetic force produced by the moving Sun in a frame of reference where the Earth is at rest.

17.6.* Retardation effect for a slowly moving source (Liénard-Wiechert potential): (a) taking the source momentarily at rest, with an acceleration a_r towards the field point at the distance r; show that the fractional correction to the Newtonian potential $-m/r$ is $-ra_r/2$; (b) compare this result with the formal solution (17.33) to $O(v^2)$.

17.7.* In relativistic cosmology a special local frame S is defined, in which galaxies are in the average at rest; it is identified as the one in which the cosmic microwave background radiation has no dipole component. A possible violation of the equivalence of all the inertial frames consists in assuming that S has a dynamical significance and the gravitational motion of a body depends on its velocity $\mathbf{w}$ relative to S. For the solar system $w \approx 370$ km/s. A possible realization of this violation consists in adding to the metric of a body at rest the following additional terms (depending on two dimensionless constants (α_1, α_2)):

Chapter 18

ARTIFICIAL SATELLITES

Since 1957, several thousands of artificial objects have been placed in Earth-bound and interplanetary orbits for a variety of purposes, including telecommunications, scientific research, meteorology, navigation, and military reconnaissance and warning. The extensive use of space for scientific purposes will certainly increase in the future; an understanding of the ways spacecraft are launched, function, move and communicate with the ground is important for the design and critical evaluation of space missions. Accurate navigation in space, mainly governed by Newton's law of gravitation, must strike a compromise between technical and system constraints, in particular fuel requirements, and scientific aims, which require reaching the planned places at the planned times. Particular attention will be devoted to the instrumentation for precise experiments, especially by means of Earth satellites, which can be conducted in an environment free of both weight and microseisms; they have important applications in gravimetry, geodesy, geophysics and experimental gravitation (see Ch. 20). Near the Earth (below $\simeq 2,000$ km and, to less extent, in the geostationary orbit) an increasing population of space debris, much larger than the background population of micrometeoroids, produce an important and increasing hazard for large satellites and man-operated stations. This debris is produced by inactive spacecraft, fragments of spacecraft and launchers generated by malfunctioning, and even past military operations.

18.1 Launch

In order to push a satellite off the ground and place it in an Earth orbit with semimajor axis a, we need to carry it to an altitude $h \ll R_\oplus$ (at least ≈ 150 km) such that the atmospheric drag does not have a relevant influence, and there impart to it a velocity **v** (in the inertial frame) such that

$$\frac{v^2}{2} - \frac{GM_\oplus}{R_\oplus + h} = -\frac{GM_\oplus}{2a}. \tag{18.1}$$

The direction of v is specified by the *insertion angle* γ with the horizontal direction. Conservation of angular momentum demands

$$(R_\oplus + h) v \cos \gamma = \sqrt{GM_\oplus a(1 - e^2)} = \sqrt{GM_\oplus r_a(1 - e)}, \tag{18.2}$$

where e is the orbital eccentricity and $r_a = a(1 + e)$ is the apogee distance. When $\gamma > 0°$ the rocket re-enters the atmosphere after a fraction of an orbit, a typical trajectory of an intercontinental ballistic missile. In order to permanently remain in space the injection must be horizontal, with $\gamma = 0°$; then the lowest possible velocity is $\simeq \sqrt{GM_\oplus/R_\oplus} \simeq 7.8$ km/s, corresponding to a circular orbit with $e = 0$. One can also have an insertion with $\gamma > 0°$, with an additional push at the apogee in order to prevent re-entry. Eliminating the eccentricity from the two previous equations and neglecting h, we get

$$v^2 = \frac{2GM_\oplus}{R_\oplus} \frac{r_a/R_\oplus - 1}{r_a/R_\oplus - R_\oplus \cos^2 \gamma / r_a}. \tag{18.3}$$

As $r_a/R_\oplus$ grows, the velocity tends to the *escape velocity*

$$v_{\rm esc} = \sqrt{\frac{2GM_\oplus}{R_\oplus}} \simeq 11.2 \text{ km/s}; \tag{18.4}$$

when $v \geq v_{\rm esc}$ the orbit is hyperbolic (parabolic). What counts, however, is the insertion velocity relative to the ground, which is affected by the daily rotation. For an equatorial, eastward launch with $\gamma = 0°$ this velocity is smaller than the inertial velocity by $\omega R_\oplus \simeq 0.465$ km/s, where ω is the angular velocity of the Earth. This results in some fuel saving. It can be also easily shown that the lowest-energy insertion into an orbit with inclination $I (< 90°)$ is achieved from a launching site at the same latitude. Launches in different inclinations are possible, but require additional velocity impulses of the order $\approx$ km/s. For this reason low-inclination geostationary satellites are usually launched from sites near the equator (like Kourou in French Guayana), while for nearly polar orbits high-latitude sites are used, like Plesetsk in Russia.

Rocket dynamics. The usual way to achieve such large velocities is by putting the satellite on the tip of a rocket, a device where gas is heated (e.g., by thermal combustion) and a jet produced with a nozzle, with a relative exhaust velocity $v_{\rm ex}$ of the order of the thermal velocity of the gas. This entails a loss of mass and momentum, which pushes the rocket in opposite direction. We now establish the *rocket equation* (in the Russian literature usually called *Meshcherskii equation*). A change $dm\,(<0)$ in the total mass of the rocket, with a change $(\mathbf{v}_{ex} + \mathbf{v})\,dm$ in its momentum, produces a force $(\mathbf{v}_{ex} + \mathbf{v})\,dm/dt$, to be added to the drag and the gravitational pull:

$$\frac{d(m\mathbf{v})}{dt} = (\mathbf{v}_{ex} + \mathbf{v})\frac{dm}{dt} + m\mathbf{g} + \mathbf{F}_D,$$

or

$$\frac{d\mathbf{v}}{dt} = \mathbf{v}_{ex}\frac{d(\ln m)}{dt} + \mathbf{g} + \frac{\mathbf{F}_D}{m}. \tag{18.5}$$

For a motion in the vertical direction $\mathbf{e}_z$, $\mathbf{v}_{ex} = -v_{ex}\,\mathbf{e}_z$ and

$$\frac{dv}{dt} = -v_{ex}\frac{d(\ln m)}{dt} - g - \frac{F_D}{m} = -g\left[I_{sp}\frac{d(\ln m)}{dt} + 1\right] - \frac{F_D}{m};$$

$I_{sp} = v_{ex}/g$ is the *specific impulse*, with the dimension of time. When v_{ex} and g are constant, the velocity after the launch at $t=0$ is

$$v(t) = v_{ex}\ln\frac{m(0)}{m(t)} - gt - \int_0^t dt'\,\frac{F_D(t')}{m}. \tag{18.6}$$

The quantity

$$\Delta v = g\,I_{sp}\ln\frac{m(0)}{m(\Delta t)} \tag{18.7}$$

is the *propulsion velocity increment*. The gravity loss $g\Delta t$ depends on the burning time Δt, of the order of a few minutes, and is usually less than 2 km/s; the drag loss is of the same order. For a more realistic two-dimensional trajectory (Problem 18.10), one should take into account the variable direction of gravity with respect to $\mathbf{v}$, the fact that the drag force is not necessarily along the track (owing to the motion of the atmosphere and aerodynamic lift effects) and a possible misalignment between $\mathbf{v}$ and the axis of the nozzle. We have also neglected the effect of drag after insertion.

With current chemical fuels I_{sp} is at most ≈ 300 s, corresponding to $v_{ex} \approx 3$ km/s; larger values are prevented by the structural damage due to excessive temperature. Better performances are possible only with cryogenic propellants, which have higher values of I_{sp}, but cannot be stored for a long time and are less safe. Eq. (18.7) shows where the problem is: in order to attain a propulsion velocity increment of order v_{esc}, a very large mass ratio $m(0)/m(\Delta t)$ must be used. For example, if the combined gravity and drag velocity loss is 4 km/s, we need a mass ratio $\exp(15/3) \simeq 148$. Such large values are practically impossible to achieve. The system is made up of a "spacecraft" of mass m_{sc}, which includes the payload and all the systems (guidance, telecomunications, etc.) needed to control the launch and the orbital operations, and a rocket, with a larger mass $m_r = K m_{sc}$, made up of the fuel, with mass $f m_r$ ($f < 1$) and the remaining components (the tank and the engine), with mass $(1 - f) m_r$. The mass ratio is then

$$\frac{m(0)}{m(\Delta t)} = \frac{1 + K}{1 + (1 - f) K}.$$

For practical reasons it is difficult to achieve a fuel fraction f larger then about 0.9 due, in particular, to the large accelerations during launch, which require a sturdy container; then, even with a very large K, the largest mass ratio is $1/(1 - f) \approx 10$, with an (optimistic) propulsion velocity increment $\approx g I_{sp} \ln 10 \approx 7$ km/s.

The conventional way to achieve a large propulsion velocity increment is to use *multiple stage rockets*, abandoning each empty tank after its operation. In a two-stage operation, for example, the first rocket carries a secondary rocket which is used for the second boost; again, in each stage it is difficult to use a fuel fraction f (with respect to the mass of the stage) larger than 0.9 and, with rockets still much bigger than the payload, we get twice the propulsion velocity increment; with n stages, roughly, $\Delta v = n g I_{sp} \ln 10$, so that $n = 2$ or 3 is sufficient.

Orbit insertion. The problem of optimizing the ascent trajectory is complex. To prevent large gravitational losses of velocity after launch, the flight path is quickly bent away from the vertical. The bending, however, is somewhat delayed, so that the rocket does not build up a high speed in the lower and denser atmospheric layers (the drag force is proportional to the atmospheric density and the square of the velocity). The trade-off between the two types of losses leads, for the required final values of v and γ, to the optimal path; this may be either computed in advance and stored in the memory of the guidance system or, if a better accuracy is needed, evaluated in real time by the on-board computer using an accelerometer for the non-gravitational forces.

Once outside the atmosphere, the spacecraft is parked in a low circular orbit of radius r_1 and orbital velocity $v_1 = \sqrt{GM_\oplus/r_1}$ and functionally tested. If a higher, circular orbit at a radius r_2 is needed, an elliptical *transfer orbit* may be used (Fig. 18.1), with the perigee at r_1 and the apogee at r_2, semimajor axis $(r_1 + r_2)/2$ and energy $-GM_\oplus/(r_1 + r_2)$. Conservation of energy gives the required velocity increment Δv_1 at perigee:

$$v_1 + \Delta v_1 = v_1 \sqrt{\frac{2 r_2}{r_1 + r_2}}. \tag{18.8}$$

The apogee velocity (obtained with conservation of angular momentum)

$$v_a = (v_1 + \Delta v_1) \frac{r_1}{r_2},$$

18. Artificial satellites

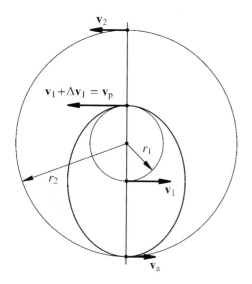

Figure 18.1 The elliptical transfer from the circular parking orbit, with radius r_1, to the final orbit, with a radius r_2. Two velocity increments, at perigee and apogee, are needed.

is then increased to $v_a + \Delta v_2 = \sqrt{GM_\oplus/r_2}$ for the final circular orbit. From Kepler's law, the transfer time is

$$\frac{\pi}{\sqrt{GM_\oplus}}\left(\frac{r_1+r_2}{2}\right)^{3/2}.$$

This transfer scenario requires the least energy; it was proposed in 1925 by W. HOHMANN and is usually referred to as the *Hohmann transfer*.

The fuel used in rockets can explode, with consequent release of fragments in outer space, which, for an area-to-mass ratio not too small, slowly decay into the atmosphere. These explosions are an important contribution to space debris and create increasing hazards to space navigation at low altitudes (Sec. 18.2). The manœuvres to transfer a spacecraft to higher orbits may also release debris material, in particular at apogee when the engine is fired. For this reason, when lifting a satellite to the very crowded geostationary orbit the apogee is usually chosen a little lower; a small further boost then completes the orbit insertion.

18.2 Spacecraft and their environment

A space vehicle generally requires subsystems for the following functions:

- Mechanical and thermal performance, in particular to ensure the correct deployment of moving parts and the appropriate temperature range.

- Manœuvring, both for attitude and orbital control, usually with thrusters; for fine pointing *flywheels* are also used, whose angular momentum, controlled with electrical motors, is transferred to and from the spacecraft.

- Attitude determination with stellar or other sensors.

- Telecommunications, in particular uplink for commanding and downlink for data transmission, and both for orbit determination.

- Payload activity.

- Power generation.

All these functions depend on whether the spacecraft is spinning or not. In the first case, in absence of external torques, the spin axis, aligned along a principal axis of inertia, is fixed, unless attitude manœuvres are performed; the rotation will then distribute solar heat along the equator of the spacecraft according to the thermal conductivity of its parts. In this case radio communications with the Earth can be ensured in two ways: for interplanetary travel, a high gain antenna is aligned with the spin axis, which, when needed, is kept pointing toward the Earth; alternatively, one can use receiving and transmitting circuitry on the equatorial belt and use it as phased arrays to achieve the required gain. A pointed scientific instrument on a spinning spacecraft scans a band of the sky along one of its parallels, so that its output is modulated with the rotation period; an accurate knowledge of the rotation phase is necessary to ensure a good resolution. A spinning spacecraft must also provide accurate monitoring of the rotational vector, usually with photometric sensors of celestial bodies like the Earth, the Sun or bright stars, or with magnetometers that measure the orientation with respect to the local direction of the magnetic field. The dynamics of a spinning satellite is governed by Euler's equations (3.31). When the spacecraft is not axially symmetric and the moments of inertia fulfil $A < B < C$, it follows that, in the absence of external torques, the motion around the middle axis is unstable (Problem 3.10). In order to damp the residual precession and/or nutation (see Secs. 3.5 and 3.6), special dissipative systems (e.g., containers with moving fluids) are used to bring the spacecraft to a state of minimum rotational energy, with no precession and nutation. The long-term evolution of the rotational state of a spacecraft is influenced by several torques, mainly radiative, aerodynamic and magnetic (caused by the interaction between the planetary magnetic field and the permanent or induced magnetic moment of the satellite).

A spacecraft rotationally stabilized along three axes may become necessary when accurate and quick pointing in an arbitrary direction is needed, as the case is for astronomical telescopes; however, its structure and control (in particular thermal control) are more complicated. An intermediate solution is provided by despun parts (e.g., antennas or instrument platforms), which are pointed along a constant and controlled azimuth on a rotating structure.

Radio communications over long distances are ensured on board by high-gain parabolic antennas or phase arrays; near the Earth, low gain dipoles are utilized. There is a clear trend to use higher frequencies: this allows a better directivity (hence lower power), higher bit rate (important for the downlink; see Sec. 19.1) and smaller dimensions; more stringent pointing requirements, however, have to be fulfilled. The choice of radio bands for space telecommunications must comply with the constraints imposed by the International Telecommunication Union (ITU), an international body which allocates different bands to different users. It should also be noted that, to avoid interference between the uplink and the downlink, different frequencies must be used. At present, the X-band has practically superseded the S-band; at higher frequencies, K_a band is used for scientific purposes and will soon become operational.

The power needed by a scientific spacecraft must ensure the operation of its components and telecommunications; radar systems in space, however, may raise the energy requirements to a different order of magnitude, and even call for nuclear energy generation. In general the obvious energy source is photovoltaic cells powered by solar radiation; they are placed on large solar panels, often orientable to ensure the best possible insolation. Assuming a 5% efficiency, at 1 AU from the Sun we need a cross section of 10 m^2 to obtain 700 W. For spacecraft at a large distance r from the Sun, however, the solar energy flux, proportional $1/r^2$, may be inadequate and *Radioisotope Thermoelectric Generators* (RTG) can be used; they contain plutonium and are based upon the thermoelectric effect produced by radioactive heat.

Scientific satellites may have special "cleanliness" requirements. There is chemical contamination, produced by exhaust gases and dust from the engines, and other leaks; moreover, the radiation they emit can be obnoxious for infrared detectors. There is also an electromagnetic cleanliness problem, due to the often intense and high frequency voltages used on board; for this reason magnetometers and other sensitive electromagnetic detectors are usually placed at the end of long booms, far from the main body. Experiments involving very precise measurements of velocity and distance may require dynamical cleanliness, which may be disturbed not only by external, unpredictable forces, but also by moving parts on the spacecraft, including the astronauts.

Special computer systems are required on board to monitor and control the "state of health" and the operations of the spacecraft in order to satisfy stringent reliability requirements, especially for missions of long duration. Intelligent image compression on board is often used to reduce the need of sending down large bit rates; depending on the visibility of ground stations, large amounts of memory may be needed to send back the information at suitable times.

A scientific space mission must be planned as an integrated whole, including the tracking from the ground. The orbit is the major point of interaction between scientific requirements and the overall structure of the spacecraft; in particular, its weight, crucial for launch operations, depends not only on the payload, but also on the amount of fuel needed for manœuvres, and on the duration of the mission. The duration is, of course, driven by scientific aims, the degree of accuracy required for the observations, the reliability of the components of the spacecraft and the resources available to operate the satellite and to analyze its data.

Satellite tracking systems. Often a precise orbit determination is required. This is accomplished by tracking systems which provide the data to which the unknown orbital parameters are fitted, in particular the initial position and velocity. The tracking techniques include:

- Optical observations of Earth satellites from the ground with special telescopes (not used in operations any more). Typical angular accuracy $1 - 2''$.

- Radio ranging using short pulses transmitted from the ground and transponded back. The round-trip light-time is measured and provides the distance of the spacecraft, particularly important for interplanetary navigation. The (Earth-bound) US Tracking Data and Relay Satellite System (TDRSS) is tracked with an accuracy of 1 m; for interplanetary probes the accuracy is within a few meters.

- Satellite Laser Ranging (SLR) to Earth satellites, in particular the two LAGEOS (and the Moon), using special reflectors on board which reflect back short laser pulses (Sec. 20.3). Although the ranging precision is less than one cm and short-term orbital models are good to about one cm, the error in long-term orbital prediction is larger, especially in the along-track position, mainly due to the difficulty in modelling non-gravitational forces.

- Radar systems, in which the round-trip light-times of short microwave pulses scattered back by an obstacle are measured, can be used in the *altimetric mode*, in which the spacecraft antenna is kept in the Nadir direction. Over the oceans this provides a measure of the spacecraft's altitude, hence the sea surface topography. If the water is calm, this surface is the *geoid*; deviations from it provide important information about currents and gradients of temperature and salinity. Accurate information about the orbit is needed; this may be difficult due to the complex structure of the spacecraft and the Earth albedo, which enhance the role of unpredictable non-gravitational forces. Frequent tracking, an orbital model of limited duration and a large number of parameters may be employed. The Cassini spacecraft will use its radar in the altimeter mode to map the topography of Titan in the Saturnian Tour from 2004 to 2008.

- In the two-way Doppler method a stable, coherent and continuous microwave radio signal is transmitted to the spacecraft and transponded back; on the ground, the fractional frequency difference $y = (\nu-\nu_0)/\nu_0$ is obtained during an integration time τ, usually between 100 and 1,000 s. Measuring

y with an error $\sigma_y(\tau)$ – the *Allan deviation* – amounts to measuring the change in the relative velocity with an error $c\sigma_y$, and the relative displacement of the spacecraft in this time with an error $c\tau\sigma_y$. Typical values of $\sigma_y(1,000\,\text{s})$ are $\simeq 10^{-13} - 10^{-14}$ (Fig. 3.1).

- In the one-way Doppler technique, a coherent microwave beam, stabilized by a frequency standard on board, is transmitted to the ground receiver (and, possibly, to other satellites). The most important example is the *Global Positioning System* (GPS), a network of (at present) 25 satellites with wide technological applications, both military and civilian. An observer on the ground can receive the signals from at least four spacecraft contemporaneously and measure the relative velocity of each of them. The orbits of all of the satellites are separately determined at the control centre and their instantaneous position is coded in the broadcast signals; with the velocity data, the observing system can determine its own latitude, longitude and altitude to within a few metres. For geodesy, a differential mode of operation is particularly useful; the signals received at two not very distant stations are jointly processed and an accuracy in their distance of a few cm is obtainable. The usefulness of the long-range orbital model of GPS satellites is limited by non-gravitational forces; nevertheless, GPS data analysis yields an important input for the determination of the celestial pole in the terrestrial frame and the length of day.

- On-board accelerometers can measure non-gravitational forces acting on the spacecraft; see below.

In addition, on-board optical sensors are used to determine the relative angular position of planetary bodies, stars and the Sun, as well as the attitude of the spacecraft.

In the future techniques for precise measurements of distances, angles and velocities in space will greatly improve. One could mention: (i) orbiting VLBI antennas (such as the QUASAT project); (ii) optical interferometry in space, to measure differences in the distance between two orbiting objects from a third one; (iii) upgrading of radio telecommunication systems, in particular with the use of the K_a band.

Accelerometers and drag-free systems. Since the major hindrance to accurate orbit determination is the poor knowledge of non-gravitational forces, it is important to devise methods to measure them directly or to compensate for them. In the concept of an *accelerometer*, a small proof mass m' is free to move near the centre of mass $\mathbf{r}_0$ of the spacecraft, of mass m; as it is completely shielded from external perturbations, m' provides a good realization of a purely gravitational motion. Its position $\mathbf{r}'$ relative to the spacecraft is measured. For example, in a one-dimensional capacitive accelerometer for atmospheric measurements, the test mass, with parallel faces, is placed between two condenser plates, with their normal aligned along the direction x of motion of the spacecraft with respect to the atmosphere. In an ideal configuration, if the centre of mass of the satellite (and the midpoint between the plates) is very near the centre of the proof mass, the relative motion of the latter is not affected by gravity and is determined by the drag F_D on the spacecraft; the plates also produce a harmonic restoring force $-m'\omega_1^2 x'$ with proper frequency ω_1. The plates have sensing electrodes, capable of measuring the differential capacity, hence the displacement x'; a feedback loop controls active electrodes and generates an electric force F_{loop} so as to maintain x' as near to zero as possible. The dynamical equation is of the form

$$\frac{d^2 x'}{dt^2} = -\frac{F_D}{m} - \omega_1^2 x' + \frac{F_{\text{loop}}}{m'} . \tag{18.9}$$

When $x' = 0$, F_{loop} gives the drag F_D along x. When a very high precision is required, additional forces must be taken into account in the motion of the proof mass (eq. (18.9)): the gravitational (tidal) force produced by asymmetric masses in the spacecraft which arise due to a displacement from $\mathbf{r}_0$ (Problem 18.6); parasitic forces due to the rotation of the satellite; charging of the proof mass by cosmic rays; imbalance of the radiation pressure in the cavity produced by temperature gradients, etc. Often these forces are difficult to model and measure, and limit the system performance.

Precise measurements of acceleration have been used for in situ determination of the thermospheric density (Ch. 8), using the relation (12.52). From 1975 to 1979, the French satellite CASTOR measured its horizontal and temporal variations in the altitude range of 300–700 km. These data contributed to the construction of a new model of the thermospheric density and made it possible to study with a high degree of accuracy other non-gravitational effects on a low-flying satellite (e.g., the radiative recoil force, known as the Yarkovsky effect, was directly measured for the first time). However, the satellite's low inclination (30°) prevented any coverage of the polar areas; the 4 y measurements spanned only a fraction of the solar cycle (11 y). New missions are planned with accuracies better than $\approx 10^{-9}$ cm/s^2 for timescales of ≈ 20 s.

In September 1997 the Mars Global Surveyor accomplished an *aerobraking manœuvre* through Mars' atmosphere in 248 revolutions, lasting more than four months. Its purpose was to reduce the initial orbital period of 45 h at injection to the desired value of 2 h. An on-board accelerometer measured the drag and the atmospheric density in the altitude range of 100 – 170 km. New temporal variations of the thermosphere, related to changes in insolation geometry, were discovered. Two features without a terrestrial analogue were also found: (i) a significant expansion of the thermosphere due to the heat produced by global dust storms; (ii) diurnal and semidiurnal gravity waves excited by topographic features, which are only weakly damped with altitude and can significantly influence the thermosphere.

Gravitational experiments require non-gravitational forces to be as low as possible. One way to accomplish this is to have a low area-to-mass ratio; since, for a given density, it is inversely proportional to size, natural satellites are essentially drag-free. A spherical shape is also important, because then non-gravitational forces are independent of the attitude and much more amenable to a model; both these advantages are used for geodynamics satellites like LAGEOS, Etalon and Starlette.

When the mission design does not allow a spherical shape and a very accurate gravitational experiment is still required, special devices are needed to obtain or reconstruct a purely gravitational orbit. If the drag force $\mathbf{F}_D$ is measured with accelerometers and the trajectory $\mathbf{r}_0$ of the spacecraft is obtained with accurate tracking, the orbit of a virtual point with the acceleration $d^2\mathbf{r}_0/dt^2 - \mathbf{F}_D/m$ can be reconstructed numerically. The mission BepiColombo to Mercury of the European Space Agency in about 2011 will place a low-pericentre orbiter around the planet. An important requirement of this mission is accurate orbit determination; this is a difficult task, due to the fact that this spacecraft of complex shape is subject to large radiative forces produced by direct solar radiation, and unpredictable and changing planetary albedo. An on-board accelerometer, with an accuracy of $\approx 10^{-6}/\sqrt{Hz}$ cm/s^2, will measure the radiative force over integration times of $\approx 10^4$ s, sufficient for an accuracy of a few meters for orbital predictions over one day.

Alternatively, *drag-free systems* can be used. In this case, only passive sensors are used for the proof mass; the value of the displacement $\mathbf{r}'$ so obtained is then fed into the control system of accurate and dedicated thrusters, which keep $\mathbf{r}'$ as near zero as possible. In this way the spacecraft is slaved to the shielded proof mass and moves according to the forces acting on the latter.

Future plans for accurate compensation of non-gravitational forces in space for fundamental experiments are ambitious. A violation of the equivalence principle (Sec. 17.1) shows up in the relative motion of three proof masses made of different materials, measured with two accelerometers. In an Earth circular orbit at the altitude of 400 km, a free mass is subject to an acceleration of 840 cm/s^2; a differential sensitivity of 10^{-14} cm/s^2 would give an accuracy of $\approx 10^{-17}$ in the fractional difference between the accelerations of the masses, five orders of magnitude better than that obtained with ground experiments. Missions of this kind are being planned and implemented, in particular with the Satellite Test of the Equivalence Principle (STEP) and the Laser Interferometric Space Antenna (LISA) to detect low frequency gravitational waves.

A tapestry of orbits. The orbit of an Earth spacecraft must take into account three main time scales: (i) the orbital period $P = 2\pi/n$ (for eccentric orbits multiples of the mean motion n also contribute), (ii) the day (LOD) and (iii) the period of the node in its equatorial motion $2\pi/(d\Omega/dt)$ (see Sec. 12.4). The ground track of a satellite in a low circular orbit with inclination I is a great circle bounded by the latitudes $\pm I$, rotating around the polar axis with the angular velocity $d\Omega/dt$ in inertial space and $\omega' = d\Omega/dt - \omega_0$ (< 0) in the body-fixed frame (ω_0 is the sidereal angular velocity of the Earth); two successive equator crossings occur at a longitude difference $P\omega'$. In order to have complete coverage, a polar orbit with $I = \pm\pi/2$ is needed. Hereafter several important classes of orbits are discussed.

Resonant orbits. – When the mean orbital motion n satisfies the relation[1]

$$pn + q\omega' = 0 \quad (p, q \text{ positive integers}) \tag{18.10}$$

the orbit is resonant and the satellite passes over the same point of the Earth after q revolutions and p days (corrected with the nodal motion). The most important resonance occurs for a *geosynchronous orbit*, with $q/p = 1/1$; when the eccentricity and the inclination vanish, this is the *geostationary orbit*, with semimajor axis

$$a_{\text{geo}} = \left(\frac{GM_\oplus}{\omega_0^2}\right)^{1/3} \simeq 6.611\, R_\oplus \simeq 42{,}165\,\text{km}. \tag{18.11}$$

A geostationary spacecraft retains at all times the same position in the sky as seen from the Earth and is able to image the same part of its surface, spanning $180° - 2\arcsin(R_\oplus/a) \simeq 160°$ in latitude and in longitude, somewhat less than a whole hemisphere. These two features make geostationary satellites very suitable for many applications, including telecommunications, meteorology, remote sensing and early warning of intercontinental missile attacks. A deeper discussion follows below.

For higher resonances the semimajor axis is

$$a \simeq a_{\text{geo}}(p/q)^{2/3}. \tag{18.12}$$

It is difficult to have q/p larger than 16, corresponding to a height of about 239 km; at lower altitudes large amounts of fuel are needed to contrast the atmospheric drag in a reasonably long mission. In resonant orbits some components of the geopotential have secular effects. The most interesting perturbation is that of the inclination, an orbital element otherwise usually very stable

---

[1] Since also the argument of pericentre ω has a secular change, it is more accurate to consider the rate of change of the longitude $f + \omega$ in the orbital plane, and use the resonant condition

$$p\left(n + \frac{d\omega}{dt} + \frac{d\epsilon'}{dt}\right) + q\omega' = 0.$$

(except for the effect of atmospheric rotation; Sec. 12.3). The orbit of a passive satellite, as it decays due to atmospheric drag, can cross several high-order resonances; at each crossing the inclination has a characteristic change that depends on a particular linear combination of the related geopotential coefficients – the *lumped geopotential coefficients*. In the body-fixed frame the Cartesian coordinates of the satellite, on a circular orbit with arbitrary inclination, are sinusoidal functions of $\omega' t$ and nt; their powers appear in the (ℓ, m) harmonic function and generate terms with frequencies which are multiples of ω'. A tedious analysis gives the resonant condition $(\ell - 2s + r)n + m\omega' = 0$, where two arbitrary integers appear: $0 \leq s \leq \ell$ and $-\infty < r < \infty$. As a consequence of this arbitrariness, an infinite number of harmonic functions contribute to a resonance: $\ell = kq +$ even number if $q - p$ is even; $\ell = kq +$ odd number if $q - p$ is odd; $m = kq$, where k is an arbitrary positive integer. This is the set of lumped coefficients. Their effect on low satellites places precise constraints on the high-degree geopotential; near low-order resonances, with p and q small, the lumped coefficients series may start with low-degree geopotential terms. For instance, the GPS and Glonass satellites, in proximity of the $q/p = 2/1$ resonance, are most sensitive to C_{32} and S_{32}.

Heliosynchronous orbits.– If the orbital plane keeps the same configuration with respect to the Sun, every revolution the spacecraft crosses the equator at a constant local hour; in this way it is able to continuously monitor the Earth-Sun relationship, in particular the radiative budget. This is the *heliosynchronous orbit*, determined by the relationship (see eq. (12.68))

$$\frac{2\pi}{1\ \text{y}} = \frac{d\Omega}{dt} \simeq -\frac{3}{2} J_2 n \left(\frac{R_\oplus}{a}\right)^2 \frac{\cos I}{(1-e^2)^2} \ . \quad (18.13)$$

The orbit must be retrograde ($\cos I < 0$), with an inclination depending on the semimajor axis and the eccentricity. When $e = 0$ and $a \simeq R_\oplus$, $I \simeq 96°$. Higher heliosynchronous satellites require a large inclination; there is an upper limit in the semimajor axis above which this type of orbit does not exist (Fig. 18.2).

Critical inclination orbits.– In Sec. 12.4, we noticed that at the "critical" inclination $\simeq 63.5°$ the secular contribution of the Earth quadrupole vanishes and the perigee ω of a satellite is stationary (eq. (12.73c)). A more precise analysis confirms this property and indicates that close to this inclination ω librates. If the critical orbit has a high eccentricity ($e \geq 0.5$, say), the satellite spends most of the time near the apogee, at the latitude $\simeq -\arcsin(\sin I \cos \omega)$; if, moreover, it is geosynchronous, the satellite's longitude is nearly constant, which mimics the properties of a geostationary orbit, but with a non-vanishing latitude. For this reason, critical inclination orbits are frequently used for remote sensing, surveillance and telecommunication purposes (Fig. 18.5).

Geostationary orbits.– A geosynchronous satellite with small eccentricity and inclination remains very near the equatorial, geostationary orbit. In effect, it

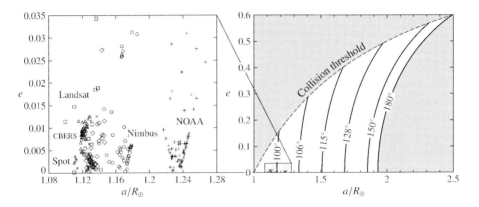

Figure 18.2. Right: Keplerian elements of heliosynchronous orbits around the Earth: semimajor axis a vs eccentricity e, with the inclination I labelling different curves. Above the upper dashed curve the satellite impacts the Earth; in the shaded area on the right there are no heliosynchronous orbits. Left: 310 active heliosynchronous satellites (mostly dedicated to Earth remote sensing) are shown in the zoomed area, grouped in different constellations, each indicated with a symbol (NOAA (crosses +), Nimbus (circles), Landsat (squares), CBERS (triangles), Spot (crosses ×). Satellite data from http://celestrak.com.

does not remain at a fixed position in the sky, but oscillates with the frequency of a day and an angular amplitude of the order of magnitude of e and I (Problem 18.2). Resonant effects on a geostationary orbit with $e = I = 0$ mainly arise from the quadrupole lumped coefficients, with $\ell = m = 2$. They are $\approx 10^{-6}$ (see Table 2.3) and describe the lack of axial symmetry, corresponding to an equatorial deformation of ≈ 100 m. At vanishing inclination the main quadrupole term J_2 has no effect, except for a small change of the mean motion and thus the resonant semimajor axis.

The along-track acceleration for $\ell = 2$ and $\theta = \pi/2$ (eq. (2.17″)) reads

$$-\frac{\partial U_2}{a \partial \phi} = -\frac{GM_\oplus}{a^2}\left(\frac{R_\oplus}{a}\right)^2 \sum_m m J_{2m} P_{2m}(0) \sin[m(\phi - \phi_{2m})] = K \sin \lambda \; ; \quad (18.14)$$

since $P_{\ell m}(0) = 0$ for $\ell - m$ odd, only $m = 2$ survives. Here $\lambda = 2(\phi - \phi_{22}) + \pi$ and, using $J_{22} \simeq 1.8 \times 10^{-6}$ and $P_{22}(0) = 3$,

$$K = \frac{2GM_\oplus}{a^2}\left(\frac{R_\oplus}{a}\right)^2 J_{22} P_{22}(0) \simeq 5.6 \times 10^{-6} \text{ cm/s}^2 \; .$$

From Gauss' equation for a circular orbit $\dot{a} = 2T/n$ (see Sec. 12.1); using Kepler's third law we have $\ddot\phi/n = -3\dot{a}/2a$ and

$$a\frac{d^2\lambda}{dt^2} = 2a\frac{d^2\phi}{dt^2} = -3n\frac{da}{dt} = -6T = -6K \sin \lambda \; . \quad (18.15)$$

18. Artificial satellites

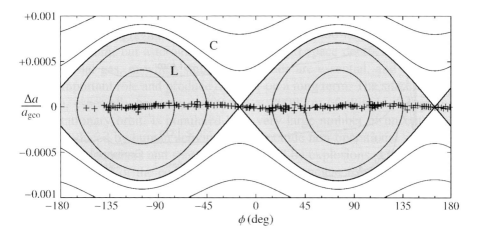

Figure 18.3. "Iso-energy curves" in the (ϕ, $\Delta a/a_{\text{geo}}$) phase plane for geostationary satellites (first integral of eq. (18.15)); ϕ is the geographic longitude and Δa is the change in the semimajor axis from the geostationary value. The inclination vanishes. In the shaded areas orbits librate around the stable equilibria at 105° W and 75° E (the "L-paths"). When the libration reaches the unstable equilibria at 15° W and 165° E, the regime changes to circulation ("C-paths"). The width of the libration zone is about 35 km. The crosses denote the positions of some 190 geostationary satellites with $I < 0.2°$, selected from the entire sample of nearly synchronous satellites. Regular orbital manœuvres keep geosynchronous satellites close to the nominal value of the semimajor axis and at precisely defined longitudes. Note the paucity of satellites above the Pacific ocean. Data from http://celestrak.com.

This is the pendulum equation for the variable λ; $\lambda = 0$ and $\lambda = \pi$ correspond, respectively, to two pairs of equilibrium longitudes, the stable ones at $\phi = \phi_{22} \pm \pi/2$ (= 105° W and 75° E) and the unstable ones at $\phi = \phi_{22}$ and $\phi = \phi_{22} + \pi$ (= 15° W and 165° E). The stable positions are at the minima of the perturbing potential $-6K \sin \lambda$, and orthogonal to the semimajor axis of the Earth's equator. For large-amplitude oscillations the subsequent terms in the lumped coefficient for the geosynchronous orbit, mainly J_{31} and J_{33}, move the libration centre toward the Greenwich meridian, by as much as $\simeq 20°$ in cases of practical importance.

The period of the small oscillations about one of the two stable equilibrium longitudes is

$$P = 2\pi \left(\frac{a_{\text{geo}}}{6K}\right)^{1/2} \approx 800 \text{ d}, \qquad (18.16)$$

with angular acceleration (if not too close to an equilibrium position) $\ddot{\phi} \approx 0.001°$ per day. In fact, if $\Delta a = a - a_{\text{geo}} \ll a_{\text{geo}}$, eq. (18.15) integrates to

$$2a_{\text{geo}} \frac{d\phi}{dt} = 3n\Delta a + \text{const}. \qquad (18.17)$$

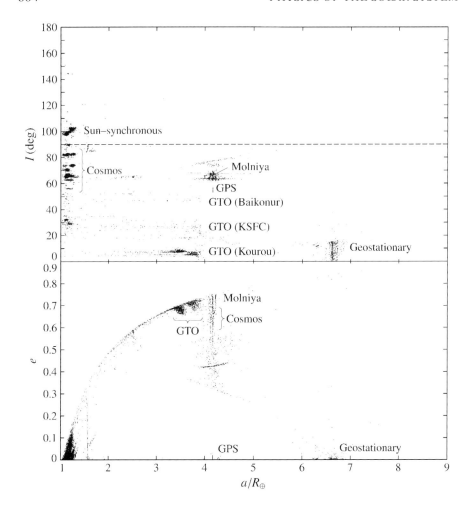

Figure 18.5. The "catalogued" space debris (about 10,000 in number) in the (a, e) and (a, I) planes. The clusters correspond to specific orbital regions: low altitude orbits on the left; geostationary transfer orbits (GTO) from various launch sites; critical inclination (Molniya-type); Global Positioning System and Glonass (GPS); geostationary orbits. Much fewer objects, mostly remnants of heliosynchronous satellites, are in (moderately) retrograde orbits. The inclination spread (up to $\approx 15°$) of the geostationary cluster is due to perturbations produced by the Sun and the Moon. Data from the ESA MASTER-99 model.

In Table 18.1 all relevant forces acting on Earth satellites are listed, with data concerning the satellites LAGEOS and Starlette. The second column gives an approximate expression for the acceleration; the third lists the main parameters (with their uncertainty in column 4) which must be known in advance, or solved for, to obtain an accurate dynamic model, together with their current or estimated values. These values are then inserted in the formulæ to calcu-

18. Artificial satellites

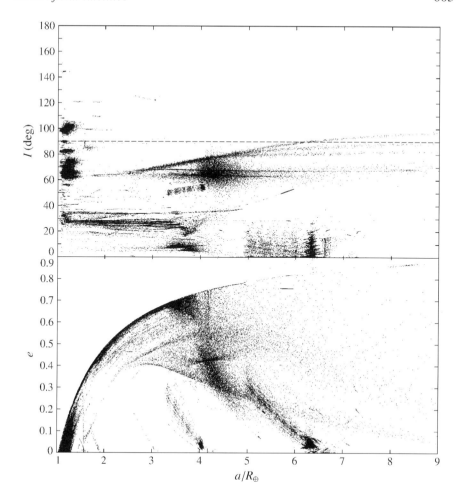

Figure 18.6. The same as in Fig. 18.5, but for debris bigger than 1 cm. While the previous figure is a synopsis of actual tracking data, debris smaller than ≈ 10 cm cannot be observed and their distribution can only be simulated statistically. Notice the much wider spread, partly due to non-gravitational perturbations (more important for smaller objects) and partly to higher ejection velocities in explosions. Data from the ESA MASTER-99 model.

late the corresponding accelerations and their uncertainties. This shows the achievable levels of accuracy and the most critical modelling problems arising when an accurate orbital propagation is needed. As a general rule, we note that gravitational accelerations are much larger in magnitude, but their fractional uncertainty is very small and typically constrained by the modelling of non-gravitational accelerations; on the contrary, the latter are small, but have a significant uncertainty. An accurate and correlation-free determination of these parameters requires the analysis of satellite orbits with different values of the

Table 18.1. Accelerations of spacecraft in an Earth orbit and their main determining parameters with their fractional uncertainties; r is the distance from the centre of the Earth (whose mass and radius are $M_\oplus$, $R_\oplus$) to the spacecraft or the perturbing bodies (Moon, Sun and Venus); $J_{\ell m}$ are the normalized geopotential coefficients; S/m, C_D, C_R and V are, respectively, the satellite's area-to-mass ratio, the drag coefficient, the radiation pressure coefficient and the velocity relative to the atmosphere, of density $\varrho(h)$; F_0 is the solar constant; A the Earth's albedo; α and $\Delta T/T$ are the satellite's absorption coefficient and the fractional temperature difference between its parts. Acceleration values are given for LAGEOS ($r = 12,270$ km, $S/m = 0.007$ cm²/g) and Starlette ($r = 7,335$ km, $S/m = 0.010$ cm²/g); their uncertainties are obtained by multiplying columns 4 and 5, and 4 and 6, respectively. CGS units throughout the Table.

Origin of acceleration	Formula	Main parameters	Fractional uncertainty	Acceleration of LAGEOS	Starlette
Earth's monopole	$\frac{GM_\oplus}{r^2}$	$GM_\oplus$	2×10^{-9}	265	742
Earth's oblateness	$\frac{3GM_\oplus R_\oplus^2}{r^4} J_{20}$	J_{20}	7×10^{-8}	0.10	0.81
Geopotential:					
$\ell, m = 2$	$\frac{3GM_\oplus R_\oplus^2}{r^4} J_{22}$	J_{22}	3×10^{-5}	5.8×10^{-4}	4.7×10^{-3}
$\ell, m = 6$	$\frac{7GM_\oplus R_\oplus^6}{r^8} J_{66}$	J_{66}	7×10^{-4}	8.8×10^{-6}	5.4×10^{-4}
$\ell, m = 18$	$\frac{19GM_\oplus R_\oplus^{18}}{r^{20}} J_{18\,18}$	$J_{18\,18}$	4×10^{-2}	6.9×10^{-10}	2.0×10^{-5}
Moon	$\frac{2GM_M r}{r_M^3}$	GM_M	10^{-7}	2.1×10^{-4}	1.3×10^{-4}
Sun	$\frac{2GM_\odot r}{r_\odot^3}$	$GM_\odot$	4×10^{-10}	9.6×10^{-5}	5.7×10^{-5}
Planets (Venus)	$\frac{GM_V r}{r_V^3}$	GM_V	3×10^{-7}	1.3×10^{-8}	7.8×10^{-9}
Relativistic effects	$\frac{GM_\oplus}{r^2} \frac{m_\oplus}{r}$	$GM_\oplus, m_\oplus$	2×10^{-9}	9.5×10^{-8}	1.6×10^{-7}
Atmospheric drag	$C_D \frac{S}{2m} \varrho V^2$	C_D, S, ϱ	$0.3 - 1$	1×10^{-10}	2.5×10^{-5}
Solar radiation pressure	$C_R \frac{S}{m} \frac{F_0}{c}$	C_R, S	0.02	3.2×10^{-7}	4.4×10^{-7}
Earth albedo	$C_R \frac{S}{m} \frac{F_0}{c} \left(\frac{R}{r}\right)^2 A$	C_R, A, S	1	3.4×10^{-8}	1.3×10^{-7}
Thermal recoil	$\frac{4}{9} \frac{S}{m} \frac{F_0}{c} \alpha \frac{\Delta T}{T}$	$\alpha, \Delta T, S$	0.1	5×10^{-9}	3×10^{-9}

semimajor axis and the inclination, and, to less extent, the eccentricity. The specific comments that follow are organized as in the Table 18.1.

The Earth monopole.– The Earth's monopole is the largest force on a satellite; as discussed in Ch. 12, at a distance r from the centre, the corresponding acceleration $GM_\oplus/r^2$ is the quantity with which various perturbations must be compared. Thus, e.g., a perturbing acceleration of 10^{-6} cm/s^2 on LAGEOS will cause in an orbital period a fractional change in the satellite semimajor axis of $\approx 10^{-6}/265 \approx 4 \times 10^{-9}$, about 5 cm. Whether these changes accumulate in time, is the problem of secular perturbations. The determination of $GM_\oplus$ requires, by Kepler's third law, simultaneous measurements of the orbital period and of the semimajor axis; while the former is straightforward and very accurate, the latter is not. Thus, the uncertainty for $GM_\oplus$ given in the Table 18.2 is driven by the uncertainty in ranging: 2 cm over 12,300 km corresponds to a fractional uncertainty of $\approx 1.6 \times 10^{-9}$ in semimajor axis and in $GM_\oplus$ three times larger; this is close to the value given in the Table, since currently $GM_\oplus$ is determined from a combination of satellite orbit analysis, to which LAGEOS chiefly contributes. This error does not have serious consequences for orbit determination, since it is compensated by a systematic change in the semimajor axes. Note also that it is impossible with satellites to measure separately the mass $M_\oplus$ and the gravitational constant G; this would require "weighing the Earth with a non-gravitational force".

Higher geopotential perturbations.– The perturbations due to departure of the Earth from spherical symmetry, as explained in Ch. 2, can be expanded in a series of spherical harmonics; for each value of the degree ℓ, the order m can take $2\ell + 1$ values and, for $\ell > 2$, the corresponding coefficients are of the same order of magnitude. To evaluate the acceleration caused by all the terms of a given order, assuming that they add stochastically, one can multiply by $\sqrt{2\ell + 1}$ the acceleration of a single term. In Sec. 12.4 we have already discussed the most important effects of the J_2 oblateness term, a secular motion of the perigee and the node.

Perturbations due to other celestial bodies.– The gravitational perturbations due to other celestial bodies (the Moon, the Sun and the planets) are determined by the corresponding *tidal* terms, i.e., by the difference between the force on the Earth and that on the spacecraft (see Sec. 4.2). The present accuracy of planetary ephemerides is better than the one needed for satellite geodesy. Planetary perturbations are small (Venus being the largest), and contribute no significant uncertainty.

Relativistic effects.– The relativistic corrections to the Newtonian equations of motion are of the order (monopole term) $\times m_\oplus/r$; $m_\oplus = GM_\oplus/c^2 = 0.44$ cm is the *gravitational radius* of the Earth. The description of these corrections is subtle, because in the theory of General Relativity, strictly speaking, there are no privileged frames of reference, like the inertial frames of classical mechanics. A local inertial frame attached to the Earth can be theoretically defined with sufficient approximation. The motion around the Sun makes it precess with respect around the normal to the orbital plane distant stars at the rate

19.2 mas/y (de Sitter's precession). It has been measured to $\approx 0.5\%$ with lunar laser ranging.

Atmospheric drag effects.– For artificial Earth satellites on low orbits the main non-gravitational force is the atmospheric drag, already discussed in Sec. 12.3. Although its qualitative features are basically understood, quantitative predictions are difficult, since they require a model of the upper atmosphere, in particular its density $\varrho(h)$ as a function of height $h = r - R_\oplus$. Even the best models, moreover, are inadequate to account for the large temporal variations of ϱ, which depend on the level of solar and geomagnetic activity. As a matter of fact, the density is usually determined from tracking data. In situ accelerometric measurements of atmospheric perturbations were also successfully performed and new missions are planned. Another obstacle for an accurate evaluation of the atmospheric drag is the satellite's complex shape (with possible shadowing between different parts of the satellite surface, multiple reflections of the atmospheric molecules on the surface, etc.).

Direct solar radiation pressure.– In interplanetary orbits the main non-gravitational acceleration is due to the solar radiation pressure. This problem is relevant also for small natural particles (Sec. 15.4) and artificial satellites around a planet; for a spacecraft the dependence of the cross section S on the attitude and on the optical coefficients of its surface elements should be taken into account. The situation is further complicated by the deterioration of optical properties of a surface in space exposed to the solar UV radiation and energetic particles. It is difficult to model the radiative acceleration for an ordinary spacecraft with an accuracy better than $\approx 5\%$ (actually, the real uncertainty may be higher, 20% say, for a spacecraft of a complex shape). Significant long-term perturbations in longitude are caused by radiation pressure when the satellite is three-axis stabilized or is equipped with a large, despun telecommunication antenna, since then there is a resonance effect between the perturbing force and the orbital period. Another difficulty of accurate radiative force models is due to the ingress into, and egress from, the Earth shadow during the *penumbra*; the incident solar radiation on the satellite decreases not only due to the gradual eclipsing of the solar disc by the Earth limb, but also due to refraction in the Earth atmosphere (Problem 8.10). Inaccurate, or too simple, modelling of the Earth penumbra may cause numerical instability in the corresponding simulation of the satellite orbit. At a lower level there are also unpredictable changes in the radiation pressure due to fluctuations in the solar luminosity (at the $\approx 0.1\%$ level; see Fig. 7.1).

The albedo effect.– Besides the solar radiation force, there is also a pressure due to radiation diffused, reflected and thermally re-emitted by the Earth; for low spacecraft they have the same order of magnitude. Indeed, since the re-emitted infrared power is not very different from the impinging power in the visible, near the Earth their fluxes are comparable. For a quantitative estimate one must

18. Artificial satellites

take into account the albedo A of the Earth, far from uniform and constant in time because of meteorological and seasonal variations. It should be also recalled that the light of the Sun is reflected and scattered by the atmosphere, in particular, by clouds (Ch. 7). Since the cloud reflectivity can be very high (with albedo $\approx 0.7 - 0.8$), it can produce a significant variable component. The infrared radiation emitted by the Earth produces a comparable force, but its variations, comparable to the albedo effect, are smaller.

Radiative recoil.– The thermal re-emission of the radiation absorbed by the satellite is usually anisotropic, due to irregularities of shape, emissivity and surface temperature. The corresponding recoil acceleration is proportional to $\Delta T/T$, the fractional temperature difference between significant parts of the spacecraft (see Sec. 15.4). The values of the absorption coefficient α of the surface and ΔT of the Table 18.1 are, of course, different for a rapidly spinning "passive" satellite (like LAGEOS) and for a stabilized, "active" spacecraft with an internal energy conversion system (generally used to emit radio waves) and equipped with solar panels. For satellites with an internal energy source, like a nuclear reactor, much of the dissipated heat is radiated away anisotropically and generates a force. In satellites with radio transmitters the emitted beam also produces a recoil acceleration; when the emitted energy is got from solar panels, it is approximately given by the absorbed energy times their efficiency η; the order of magnitude of the acceleration is η times the direct radiative acceleration.

There are other, less important, processes which perturb the motion of a satellite. We only quote (in brackets an estimate of the corresponding acceleration of LAGEOS is given):

- Micrometeorite impacts (mean acceleration $\approx 10^{-11}$ cm/s^2; of course, exceptionally large fragments can at times produce larger perturbations).

- Poynting-Robertson effect ($\approx 10^{-11}$ cm/s^2, see Sec. 15.4; it was suggested as the best candidate for a residual decay of the semimajor axis of the Etalon satellites, Problem 15.8).

- A satellite generally becomes charged due to the photoelectric effect caused by the UV solar radiation or the impact of circumplanetary or solar wind particles and cosmic rays; electric and magnetic forces arise from the interaction with the ionosphere and the magnetosphere.

Tidal effects. The effects of tides on the motion of a spacecraft (see Ch. 4) have not been discussed above, because they are complex and deserve separate analysis. They are relevant in three ways:

- They cause periodic changes in the shape of the Earth, hence in the positions of the tracking stations (kinematical effect; Sec. 4.3).

Table 18.2. Tidal effects on the orbits of Earth's satellites due to the Moon; M_M, r_M and P_M are, respectively, the mass and the distance of the Moon, and the lunar day with which it revolves in the sky relative to the Earth. The last column gives in cm/s² the corresponding acceleration values for LAGEOS and Starlette satellites.

Formula	Parameter (cgs units)	Relative uncertainty	LAGEOS	Starlette
Kinematic solid tide				
$\delta h_r \left(\frac{2\pi}{P_M/2}\right)^2$	$\delta h_r \approx 30$	0.03	5.8×10^{-7}	5.8×10^{-7}
Kinematic ocean loading				
$\delta h_L \left(\frac{2\pi}{P_M/2}\right)^2$	$\delta h_L \approx 4$	0.2	10^{-7}	10^{-7}
Dynamic solid tide				
$3k_2 \frac{GM_M}{r^2}\left(\frac{R}{r_M}\right)^3\left(\frac{R}{r}\right)^2$	$k_2 \simeq 0.3$	0.002	3.7×10^{-6}	2.9×10^{-5}
Dynamic oceanic tide ($\approx 10\%$ of dynamic solid tide)				
		0.1	3.7×10^{-7}	2.9×10^{-6}
Reference system: non-rigid Earth nutation (14 d term)				
Nutation coefficient	0.002" in 14 d	0.1	3.5×10^{-10}	2.1×10^{-10}

- They produce a time variation of the geopotential that affects the satellite orbit (dynamic effect; Sec. 4.4).
- They perturb the rotation of the Earth, thus affecting the usual body fixed reference system (reference system effect).

In order to compare tidal effects among themselves and with other perturbations, we give in Table 18.2 their equivalent accelerations.

Kinematic solid tide.– The main tidal term produces a symmetric deformation of the Earth along the direction of the perturbing body with a period which, for the Moon, in the body fixed frame is equal to $P_M/2 \simeq 12.42$ h; this is half the lunar day $P_M = 2\pi/(d\tau/dt)$, already introduced in Sec. 4.4 in terms of the lunar hour angle τ. Using the mean motion n_M,

$$\frac{2\pi}{P_M} = \frac{2\pi}{1\,\text{d}} - n_M . \tag{18.20}$$

The height of a ground tracking station oscillates with the amplitude δh_r. Were this effect ignored, the station acceleration would appear as a large residual acceleration of the satellite. The amplitude δh_r depends upon the elastic response of the Earth to the perturbing potential. This is described by *Love's number* h_2 ($\simeq 0.6$), fairly well known. More accurately, we define different values of

18. Artificial satellites 611

the quadrupole Love's number $h_{2m}^{(f)}$ for different orders m of the tidal potential U_{2T} and different frequencies f in its Fourier expansion; compare with the similar analysis for $k_{2m}^{(f)}$ in Sec. 4.4. δh_T has an uncertainty of the order of 1 cm (see Sec. 4.3).

Kinematic ocean loading.– Oceanic tides also affect the positions of the station sites. Displacements of nearby water masses produce a time-dependent elastic response of the continental shelves (*ocean loading*), and this changes the station position with respect to the theoretical tidal displacement of an oceanless solid Earth. Of course, the size of this effect strongly depends on the geographical situation of the station. Displacements as large as 12 cm can occur on a peninsula protruding into an ocean, such as Cornwall; in the middle of a continental plate they can be smaller than 1 cm. Even though some modelling is possible, this effect needs measurements with ground instruments, with uncertainties of the order of 1 cm. The equivalent accelerations of the kinematical effects and their uncertainties, as listed in Table 18.2, are large with respect to the other forces listed in Table 18.1; however, they affect satellite tracking data with a well defined signal (see Sec. 4.4) and do not have secular or long-period effects. It is therefore possible to separate them from other dynamic effects, in such a way that their uncertainty does not affect other geophysically relevant parameters. Of course, when station positioning is the main goal, tidal displacements must be properly modelled.

Dynamic solid tide.– As seen in Secs. 4.2 and 4.3, the main term of the tidal perturbation of the geopotential is obtained by computing the corresponding (i.e., with the same frequency) term of the perturbing gravitational potential produced by the other body, multiplied by a Love number k_2 (as mentioned above, different values of the Love number $k_{2m}^{(f)}$ should be used for different orders m and different frequencies f in the tidal potential expansion; Sec. 4.4). Since the main contribution is the quadrupole term, the corresponding acceleration of the satellite is $\propto 1/r^4$. Love numbers are currently determined with a fairly good accuracy from long term effects on geodetic satellite orbits. An additional source of model errors is the truncation of the tidal potential Fourier expansion to a finite number of terms.

Dynamic oceanic tide.– Oceanic tides are waves that move under the gravitational forces of the Moon and the Sun and are subject to friction, dissipating energy and interacting with the ocean bottom (especially in shallow waters) and the shore line. As a result, tidal waves in the seas are not in phase with the perturbing potential. The displaced water masses give a contribution to the time variation of the geopotential, which affects satellite orbits together with the solid tides; since it is difficult to separate them, the uncertainty arising from oceanic tides limits the accuracy in modelling solid tides as well.

Formula	polar wobble	δ(LOD)
$(\delta\omega \cdot \mathbf{r})\omega_0$	2.2×10^{-6}	7.1×10^{-8}
$2\mathbf{v} \times \delta\omega$	7.6×10^{-5}	3.2×10^{-7}
$\mathbf{r} \times d(\delta\omega)/dt$	2.5×10^{-9}	1.1×10^{-11}
$2(\omega_0 \cdot \delta\omega)\mathbf{r}$	–	1.5×10^{-7}
$(\omega_0 \cdot \mathbf{r})\delta\omega$	6.0×10^{-6}	7.1×10^{-8}

Table 18.3 Estimates of the apparent accelerations (in cm/s^2) due to the polar wobble and variations in the length of the day (LOD) for LAGEOS 1 ($a = 12,270$ km, $e = 0.004$, $I = 110°$); see eq. (18.21).

Reference frame effects.– The Earth does not respond to external torques as a rigid body and its rotation axis undergoes forced nutations different from the theoretical case of a rigid Earth. Although the baseline model of precession and nutation partially accounts for the non-rigidity effects, in particular, with elastic parameters as derived from seismic data and the core structure, it is still not able to explain all observed effects. The Earth Rotation Service, combining VLBI, satellite and lunar data, provides empirical corrections to precession and nutation. The most important are fortnightly and annual terms, with amplitudes of about 2 mas and 12 mas, and a term with the 18.6 y period of the lunar nodal drift (Sec. 4.4), of about[2] 200 mas. For the sake of illustration, we consider the fortnightly term in the Table 18.3. Were this effect ignored, the satellites would appear to oscillate by that amount, as a result of an incorrect definition of the reference system.

Apparent forces. We now come to the apparent forces arising from an incorrect value of the Earth rotation vector. Ground-tracked satellites are referred to a conventional body-fixed Cartesian frame, as implemented by the Conventional Terrestrial Reference System (Sec. 3.2). As discussed in Ch. 3, the angular velocity of the Earth $\omega_0 + \delta\omega(t)$ is the sum of a constant, nominal part along the z axis ($\omega_0 \simeq 7.29 \times 10^{-5}$ rad/s) and a small change $\delta\omega(t)$, which accounts for polar wobble and variations of the length of the day (LOD). Keeping only first order terms, the perturbing inertial acceleration in a body-fixed reference frame is

$$\mathbf{F} = \mathbf{r} \times \frac{d\delta\omega}{dt} + 2\mathbf{v} \times \delta\omega + 2(\omega_0 \cdot \delta\omega)\mathbf{r} - (\delta\omega \cdot \mathbf{r})\omega_0 - (\omega_0 \cdot \mathbf{r})\delta\omega \,, \quad (18.21)$$

where $\mathbf{r}$ and $\mathbf{v} = d\mathbf{r}/dt$ are, respectively, the position and the velocity of the satellite. Each of the five terms of eq. (18.21) has an effect on the satellite with a characteristic signature, according to whether $\delta\omega$ is parallel (change in

---
[2] The International Earth Rotation Service provides these celestial pole offsets as corrections to the 1976 precession-nutation model, that has served as a baseline over the last three decades. Recently the IAU2000A model has been adopted as a standard to supersede the old 1976 model; much smaller corrections – below 1 mas – are necessary now.

LOD), or perpendicular (polar wobble) to ω_0. We list in Table 18.3 the order of magnitude estimates of these apparent forces for a geosynchronous satellite (in equatorial orbit) and for LAGEOS. We assume, by way of an example, a variation in the pole position of about $0.2''$ and a change in LOD of 1 ms in 1 y (see Figs. 3.4-5); we thus obtain:

$$\delta\omega_x \approx \delta\omega_y \approx 0.2 \times 5 \times 10^{-6}\,\omega_0 \approx 7.1 \times 10^{-11} \text{ rad/s},$$

$$\delta\omega_z \approx 10^{-3}\frac{\omega_0}{86,400} \approx 8.4 \times 10^{-13} \text{ rad/s},$$

$$\delta\left(\frac{d\omega_x}{dt}\right) \approx \delta\left(\frac{d\omega_y}{dt}\right) \approx \frac{\delta\omega_x}{3.15 \times 10^7 \text{ s}} \approx 2.2 \times 10^{-18} \text{ rad/s}^2,$$

$$\delta\left(\frac{d\omega_z}{dt}\right) \approx \frac{\delta\omega_z}{3.15 \times 10^7 \text{ s}} \approx 2.7 \times 10^{-20} \text{ rad/s}^2.$$

Note that the last number is nearly two orders of magnitude larger than the long-term average (3.5) because of important decadal and yearly terms in the Earth rotation (Fig. 3.4). The uncertainties in the knowledge of the pole position and LOD recently have been reduced by a combination of several methods (LAGEOS and GPS orbit analysis, VLBI and lunar laser ranging); the pole is monitored with about 0.2 mas precision and the LOD down to about 0.03 ms in 1 d.

18.4 Space navigation

In order to accomplish its goals, the orbit of a space mission must be adequately predicted and appropriate manœuvres be performed with the on-board propulsion systems. The tracking techniques mentioned in Sec. 18.2 – in particular, measurements of relative velocity and distance with the radio system (see Ch. 19) – provide the input to sophisticated computer codes which, taking into account all the relevant forces and their uncertainty, allow prediction of the future position of the spacecraft.

Interplanetary navigation differs from ordinary planetary dynamics because more than one centre of attraction must be taken into account. To explore another planet, an Earth-bound spacecraft must first escape the Earth's gravity and acquire a heliocentric orbit, to be inserted later in a bound orbit near the target. In lunar missions, starting with a circular (or elliptic) Earth orbit, the spacecraft is first placed on a locally hyperbolic transfer trajectory and then, near the Moon, is inserted in an orbit bound to the satellite. These dynamical problems do not have an analytical solution (in particular, the trip to the Moon is a three-body problem, see Ch. 13), but can be approached by matching two or more Keplerian orbits. In this way the manœuvre strategy, the fuel requirements and the transfer times can be estimated; for accurate planning of transfer times and fuel requirements, however, numerical calculations are necessary.

Interplanetary orbits; continuous propulsion systems. To place a spacecraft on an elliptical solar orbit we must first raise it from the parking orbit around the Earth, of radius r_1, to a hyperbolic orbit with asymptotic velocity v; from energy conservation, this requires the velocity increment

$$\Delta v = \sqrt{v^2 + \frac{2GM_\oplus}{r_1}} - \sqrt{\frac{GM_\oplus}{r_1}}. \qquad (18.22)$$

As discussed later, eq. (18.25), $V = |\mathbf{v} + \dot{\mathbf{R}}_\oplus|$ is the velocity of insertion in the elliptical solar orbit. For example, when $v \ll |\dot{\mathbf{R}}_\oplus| = V_\oplus$, one can have a circular orbit with radius 1 AU. Next, to place the spacecraft near another planet which has a circular orbit with radius R', we can use an elliptical solar orbit mathematically identical to the transfer orbit for Earth-bound spacecraft. With a velocity increment (eq. (18.8))

$$\Delta V = V_\oplus \left(\sqrt{\frac{2R'}{1+R'}} - 1 \right)$$

we raise it to an aphelion at R' (in AU); a further velocity increment circularizes the orbit. When the spacecraft approaches the planet, with a relative velocity $\mathbf{v}'$, a small, but accurate correction of the direction of motion is effected to obtain the impact parameter $\mathbf{p}'$ needed to achieve the required planetary hyperbolic orbit. Finally, at the planetary pericentre the relative velocity is decreased to obtain a bound orbit. This is a variant of the Hohmann transfer orbit discussed in Sec. 18.1. Of course, the orbit design must also take into account the transfer time and the relative position of the two planets, which imposes a time window for the manœuvres; the actual realization requires considering the full three-body problem and numerical evaluation.

The velocity increments required for interplanetary navigation, as estimated above, can be large. Consider, for example, a "kamikaze" rectilinear orbit straight into the Sun. After the spacecraft is raised from the Earth into a solar circular orbit at 1 AU, for the plunge we need to bring its velocity to zero, with a change of $\sqrt{GM_\odot/1\,\mathrm{AU}} \simeq 30$ km/s. The escape velocity from the solar system is $\sqrt{2GM_\odot/1\,\mathrm{AU}} \simeq 42$ km/s, with an increase of about 12 km/s over the Earth orbital velocity (plus an additional change to escape from the Earth gravitational influence); similar amounts are needed to reach the outer planets. Such velocity changes are not possible with conventional chemical engines. In addition to the (quasi-)instantaneous corrections assumed above, propulsion systems with continuous, low thrust operation have been planned and developed. They promise much lower total energy requirements. *Magnetohydrodynamical engines* are based upon the $\mathbf{J} \times \mathbf{B}$ force which a plasma experiences in a magnetic field $\mathbf{B}$. In a conceptual scheme, the flow from a

18. Artificial satellites

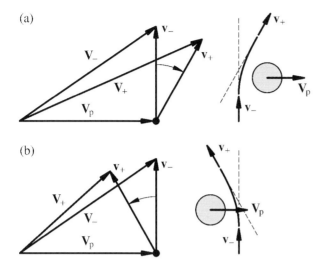

Figure 18.7 The swing-by as a means to change the momentum and the energy of a spacecraft. A deflection to the right (case (a)) or left (case (b)) in the rest frame of the planet gives in the inertial frame (left) an increase (decrease) of the kinetic energy.

reservoir of an easily ionizable gas (e.g., cesium or xenon) passes through parallel plates, between which an electric field **E** is established; this field ionizes the gas and generates a current $\mathbf{J} = \sigma\mathbf{E}$; the magnetic field orthogonal to **E** pushes the plasma out and generates the thrust. This is a concept of an ion or plasma thruster. Another interesting concept is that of the *solar sail*: a large (even kilometres wide) sheet of reflecting material is deployed and suitably oriented with respect to the Sun, generating from the solar radiation pressure a force along its normal. Though ion thrusters are being experimented and already planned for forthcoming cosmic missions (e.g., the BepiColombo mission to Mercury), the solar sail has never been realized, nor is planned for the near future.

Flybys for space navigation. One way to overcome the fuel problem for interplanetary missions is to use the gravity pull of a carefully chosen planet or natural satellite; Fig. 18.7 shows how this works. While in the rest frame of the planet the velocity changes only in direction, by an amount depending on the impact parameter, in the inertial frame there is a net gain (loss) of energy when the spacecraft passes behind (ahead) its direction of motion, at the expense of the orbital energy of the planet. This is called a *gravity assisted manœuvre*. One could say, an elliptical solar orbit with a velocity $\mathbf{V}_-$ is matched to an hyperbolic orbit around the planet, with asymptotic velocities $\mathbf{v}_\pm = \mathbf{V}_\pm - \mathbf{V}_p$; on the other asymptote this is matched to an elliptical orbit with velocity $\mathbf{V}_+$. The deflection $\mathbf{v}_- \to \mathbf{v}_+$ is determined by the angle δ between the asymptotes in terms of the impact parameter **p** (eqs. (11.43)). Its appropriate value is obtained with a slight adjustment of the direction of $\mathbf{V}_-$ before the encounter. Note the great sensitivity of this manœuvre to errors: an error σ_p produces an error $\sigma_\delta = \sigma_p/p$ in the direction of $\mathbf{V}_+$ and an error

Table 18.4. For each planet the Table gives the swing-by ratio $\xi = M_p a_p / M_\odot r_p$. M_p, r_p and a_p are, respectively, the mass, the radius and the semimajor axis of the planet (see the text).

Mercury	0.0039	Earth	0.069	Jupiter	10.7	Uranus	5.1
Venus	0.043	Mars	0.022	Saturn	6.6	Neptune	10.3

$\sigma_R = R\sigma_\delta = \sigma_p(R/p) \gg p$ in the elliptical position of the spacecraft. For the same reason, orbits of planet-crossing asteroids and comets are strongly chaotic; the typical Lyapunov timescale (Sec. 13.3) may be several hundred years only.

A rough assessment of the swing-by effectiveness is provided by the ratio

$$\xi = \frac{M_p a_p}{M_\odot r_p} \qquad (18.23)$$

between the lowest possible potential energy $-GM_p/r_p$ due to the planet and the solar potential energy $-GM_\odot/a_p$ (r_p, a_p are the radius and the semimajor axis of the planet). Since the asymptotic velocity v_- is given by $v_-^2 \simeq GM_\odot/a_p$, eq. (11.43a) shows that the largest deflection angle is indeed $\approx \xi$, which justifies its name. From Table 18.4 we see that Mercury is useless, and the outer planets are very effective. The other terrestrial planets are weak and may require repeated swing-bys or an eccentric orbit, less bound to the Sun than the planet itself. Venus, the Earth and the giant planets thus provide efficient engines to drive a spacecraft around the solar system. A disadvantage of this method is the often large time required to reach the final goal.

The Cassini mission to Saturn exploits very extensively the swing-by manoeuvres. Since its launch in October 1997 the spacecraft has used four of them: two at Venus, one at the Earth and one at Jupiter. After its arrival at Saturn and its insertion in a bound orbit in July 2004, in 74 orbits around the planet it will explore the Saturnian system, with its satellites and rings, in particular Titan, by far the most interesting one. Among the satellites only Titan has an acceptable swing-by parameter $\xi \simeq 0.11$ and, in a feat of navigation engineering, will be used in 43 flybys to provide the energy and pilot the spacecraft to fulfil the mission plan; note also that the choice of the impact parameter **p** in each flyby must ensure a resonant orbit, so that Titan will be met again. Thus Titan is at the same time the engine and the prime goal of the mission.

The matching problem. The matching in a flyby between the outer elliptic orbit and the inner hyperbolic trajectory is an aspect of the three-body problem. To introduce the concept of the *sphere of influence*, proposed by P.S. LAPLACE, we describe the motion of the space probe either with relative coordinates **R** with respect to the Sun or with relative coordinates $\mathbf{r} = \mathbf{R} - \mathbf{R}_p$ with respect to the planet; they are suitable, respectively, far and near the planet, and fulfil

18. Artificial satellites

(see eq. (12.77)):

$$\frac{d^2\mathbf{R}}{dt^2} + \frac{GM_\odot}{R^3}\mathbf{R} = GM_p\left[-\frac{\mathbf{r}}{|\mathbf{r}|^3} - \frac{\mathbf{R}-\mathbf{r}}{|\mathbf{R}-\mathbf{r}|^3}\right] ; \qquad (18.24a)$$

$$\frac{d^2\mathbf{r}}{dt^2} + \frac{GM_p}{r^3}\mathbf{r} = GM_\odot\left[-\frac{\mathbf{R}}{|\mathbf{R}|^3} + \frac{\mathbf{R}-\mathbf{r}}{|\mathbf{R}-\mathbf{r}|^3}\right] . \qquad (18.24b)$$

The second terms in each bracket, usually referred to as the indirect perturbations, are the accelerations produced by the planet (of mass M_p) on the Sun and vice versa. If ϵ_R and ϵ_r are, respectively, the ratios (in moduli) of the right-hand side to the main force in the left-hand side, the relation $\epsilon_R = \epsilon_r$ defines the surface of influence (to a good approximation, a sphere of radius $r_I = R[M_p/M_\odot]^{2/5}$, equal to 0.079 AU for the Sun-Earth system)[3]; of course, ϵ_R and ϵ_r diminish outside and inside, respectively. Given a Keplerian unbound orbit inside, the outer orbit is approximated by a Keplerian ellipse around the Sun which matches the inner hyperbola on the sphere of influence.

This empirical procedure, however, is mathematically unsatisfactory. In fact, the transition from the inner to the outer orbit is continuous and does not take place at a given point; the quantity r_I should not play a role. Since in a flyby the two potential energies $GM_\odot/R$ and GM_p/r are comparable, the smallness parameter $\mu = M_p/M_\odot \ll 1$ also gives the order of magnitude of r/R. We have here two very different timescales, the hyperbolic flyby time and the outer Keplerian period, with ratio $O(\mu)$;[4] such a problem requires singular perturbation theory. This method is widely applied to cases in which an infinitesimal parameter is used also in the independent variables, for example, to deal with a viscous boundary layer in a fluid. As $\mu \to 0$ the outer solution $\mathbf{R}(t)$ tends to two discontinuous elliptic arcs, joined at the position $\mathbf{R}_p(0)$ of the planet at the time $t = 0$ of the encounter (of course, defined to $O(\mu)$). For the inner solution, we need the zooming variables $\mathbf{r}^\star = \mathbf{r}/\mu$ and $t^\star = t/\mu$; expanding in powers of μ one obtains, to lowest order, a hyperbolic motion. The gist of singular perturbation theory is their matching. Considering the incoming branch, as $t \to -\infty$ the hyperbola is defined by its asymptotic velocity $\mathbf{v}$ and the vector impact parameter $\mathbf{b}$, lying in the target plane orthogonal to $\mathbf{v}$. These asymptotic values must match, at the appropriate order in μ, the limit of the outer solution as $t \to -0$. This translates in the obvious condition

$$\mathbf{V} - \dot{\mathbf{R}}_p(0) = \mathbf{v} \qquad (18.25)$$

---

[3] In planetary science the related concept of *Hill's sphere of influence* r_H is used. It is defined as the approximate distance to the L_1 and L_2 Lagrange points (see eq. (13.15)) and gives Roche's limit around the body; we thus have $r_H = R(M_p/3M_\odot)^{1/3}$.

[4] This does not hold, however, in an encounter between a the test body with small eccentricity and a planet in a circular orbit: the interaction may last more than an orbital period and Hill's method (Sec. 13.4) should be used.

for the velocities; the outer motion near the encounter, at $O(\mu)$, determines **b**. The systematic expansions in μ of the inner and outer equations yield higher order corrections.

– PROBLEMS –

18.1.* Study the orbit of an inter-continental ballistic missile, neglecting the atmospheric drag. For a given angular distance between the launch site and the target, compute the launch velocity v as a function of the initial angle from the vertical and plot the time Δt required to reach the target as a function of v.

18.2.* Show that a geosynchronous satellite having a circular orbit and a small inclination I with respect to the Earth's equator moves in the sky along an eight-shaped path which crosses the celestial equator, covering an interval of $2I$ in latitude and $I/2$ in longitude.

18.3.* During the oscillation of a geosynchronous satellite with period P given by eq. (18.16), we make N equally spaced and independent measurements of the semimajor axis with accuracy σ_a. Evaluate the corresponding error in the parameter J_{22} and apply the calculation to the case $\sigma_a = 50$ cm.

18.4. Analyse the resonant motion of a geostationary satellite using the Lagrangian formalism of Sec. 12.2; if a is the semimajor axis and ϕ the geographic longitude, show that for a zero eccentricity and inclination the motion is described by the Hamiltonian-like equations

$$\frac{da}{dt} = \frac{2}{na}\frac{\partial J(a,\phi)}{\partial \phi}, \qquad \frac{d\phi}{dt} = -\frac{2}{na}\frac{\partial J(a,\phi)}{\partial a},$$

with

$$J(a,\phi) = \frac{GM_\oplus}{2a} + \omega_0\sqrt{GM_\oplus a} + \frac{3GM_\oplus}{2a}\left(\frac{R_\oplus}{a}\right)^2 J_{22}\cos 2(\phi - \phi_{22}).$$

J is a constant of the motion which allows a qualitative analysis of the motion in the (a,ϕ) space. Find the stable equilibria and the width of their libration zone.

18.5. Compute the semimajor axis of a satellite in an arbitrary resonance with the rotation of the Earth (see eq. (18.10)). Estimate their dependence on the satellite inclination.

18.6. The gravitational pull exerted by the spacecraft on the proof mass of a drag-free system is an important perturbation. Inside a spherical shell such effect would absent; in a real spacecraft the proof mass should be near its centre of mass, so that only the quadrupole contribution is important. To see how the octupole term can be eliminated with compensating masses, evaluate the gravitational potential near the centre,

18. Artificial satellites

due to two opposite point masses and a ring in their equatorial plane, and find the condition under which the octupole vanishes.

18.7.* Consider the single transfer manœuvre discussed in Sec. 18.4 between two circular orbits for a planetary rendezvous mission. Evaluate the total velocity increment $|\Delta v_1| + |\Delta v_2|$ necessary to reach the inner and outer planets from the Earth. Prove that the journey to Mercury requires more velocity change then to any other planet. Show also that the rendezvous with an outer planet may require less energy (but a longer time) if *bi-elliptic transfer* is used, which consists in a three-impulse transfer with an intermediate elliptic orbit with an apocentre distance larger than the target orbit.

18.8. Evaluate the radii of the sphere of influence both according to Laplace and Hill for the Sun and each planet; and for the Earth and the Moon (Sec. 18.4).

18.9. Express the geosynchronous radius in terms of Helmert's parameter (1.13) and calculate both of them for all the planets.

18.10. Analyse the motion of a rocket in a homogeneous and constant gravitational field g (the direction of the gravitational acceleration specifies the vertical). Initially the velocity is v_0 and the deflection from the horizontal is γ_0. Assuming an exponential decrease $\propto \exp(-\beta t)$ of the rocket mass, show that the two-dimensional rocket equation reduces to (drag force neglected)

$$\frac{d\xi}{d\eta} = \frac{\xi}{\eta}\left(\beta' - \frac{1-\eta}{1+\eta}\right),$$

with $\xi = (v/v_0)^2$, $\eta = (1 - \sin\gamma)/(1 + \sin\gamma)$ and $\beta' = \beta v_{ex}/g$. Assuming v_{ex} constant, integrate the previous equation and analyse the solutions for different values of $\gamma_0 \neq 90°$.

18.12.* In Sec. 12.5 we investigated the direct perturbation on a satellite orbiting an oblate planet due to the J_2 term in the geopotential. There is also an *indirect J_2 effect* due to a correction in the acceleration of the centre of the Earth; it is produced by the perturbation in the Earth-Moon motion due to the geopotential quadrupole, chiefly the zonal J_2 term. In the geocentric frame this acceleration affects the motion of a satellite. Discuss the formalism needed to describe this effect and evaluate its order of magnitude.

– FURTHER READINGS –

O. Montenbruck and E. Gill, *Satellite Orbits*, Springer (2000) gives an overview of modern orbit determination methods and observational techniques. Gravitational and tidal forces perturbing Earth's satellites are discussed in detail in W.H. Kaula, *Theory of Satellite Geodesy*, Blaisdell, Waltham, Massachusetts (1966) and K. Lambeck, *Geophysical Geodesy*, Clarendon Press (1988); non-gravitational effects are in A. Milani, A.M. Nobili and P. Farinella, *Non-Gravitational Perturbations and Satellite Geodesy*, Hilger (1987). On the particular problem of atmospheric drag and its models, see D. King-Hele, *Satellite Orbits in an Atmosphere. Theory and Applications*, Blackie and Son (1988). At the engineering level, there is M. Noton, *Spacecraft Navigation and Guidance*, Advances in Industrial Control (1998). O. Zarrouati, *Trajectoires Spatiales*, Cepaudes-Edition, Toulouse (1987) presents a clear discussion of both gravitational and non-gravitational perturbations of satellite motion. A useful general textbook emphasizing space navigation in the framework of celestial mechanics is A.E. Roy, *Orbital Motion*, Hilger (1978). The latest gravity models of the Earth based on satellite data and ground gravimetry measurements are JGM3, GRIM5 and EGM96 (see B.D. Tapley et al., The Joint Gravity Model 3, *J. Geophys. Res.* **101**, 28,029 (1996), T. Gruber et al., GRIM5-C1: Combination Solution of the Global Gravity Field to degree and order 120, *Geophys. Res. Letts.* **27**, 4,004 (2000) and http://cddisa.gsfc.nasa.gov/926/egm96/egm96.html, respectively).

Chapter 19

TELECOMMUNICATIONS

Radio communications to and from a spacecraft are its lifeline, with which commands are received, and scientific and engineering data are sent to Earth. Huge and sophisticated parabolic antennas are used on the ground to detect very weak signals with a large signal-to-noise ratio and to send into space powerful, highly collimated beams. Special electromagnetic links also provide very precise measurements of distance and velocity, used for navigation and research. Similar techniques are also used in radar ranging to the Moon, passive artificial satellites, asteroids and planets (ranges to the Moon and artificial satellites may also be obtained in the optical band). In this chapter we discuss the energetics of radio communications in space and introduce spectral theory, a tool necessary to describe the structure of the noise. Particular attention is devoted to coherent propagation and to the effect of refractive and dispersive media.

19.1 The power budget

The energy flux of an electromagnetic wave in empty space is given by Poynting's flux

$$\Phi = \frac{c}{4\pi} \mathbf{E} \times \mathbf{B} = \Phi \hat{\mathbf{k}} = \frac{cE^2}{4\pi} \hat{\mathbf{k}} . \qquad (19.1)$$

Electric and magnetic fields are equal in magnitude, orthogonal among themselves and to the direction of propagation $\hat{\mathbf{k}}$. This corresponds to a momentum flux Φ/c and exerts a force upon a material body of unitary area. The radio emission from a parabolic antenna on a spacecraft produces a recoil in the opposite direction, which affects the orbit.

There are two linear *polarization* modes, according to whether the electric field, for propagation along z, is directed along x or y. If they have the same

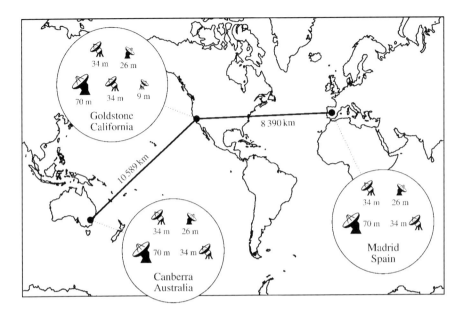

Figure 19.2. The antennas of NASA's Deep Space Network, with the diameters of their main dishes. Their geographical distribution allows communication with a spacecraft at all times. In 2001 the new DSS 25 station (with diameter of 34 m) started operating at Goldstone with advanced instrumentation and operating frequencies in X- (8 GHz) and K_a-band (34 GHz).

For deep space telecommunications large dishes (even with diameters of 70 m and weights of thousands tons) are used; in the microwave band gains of 60 to 80 dB are attained, corresponding to angular widths of 1 to 0.1 mrads. A high gain places stringent requirements upon the orientation of the antenna, whose mobile parts must accurately track the spacecraft, even in bad weather. Outside the main lobe, secondary lobes still allow communication, although with a great loss (Fig. 19.1). For interplanetary satellites, NASA uses the *Deep Space Network* (DSN), an impressive network of steerable paraboloids at three locations: Goldstone (California), Canberra (Australia) and Madrid (Spain) (Fig. 19.2). They are mainly used for telecommunication and space navigation, for radar measurements and scientific experiments. To avoid interference, the antennas at a given complex are widely separated, and different frequencies are used in the uplink and the downlink. In coherent links, like those used in Doppler experiments, a transponder on board locks the transmitted to the received beam; to allow the use of multiplication circuitry, the frequency conversion ratios are rational numbers. The available bands are carefully chosen to avoid interference from other ground stations; X-band is mainly used at present, but the K_a-band, which allows a wider bandwidth, is being implemented at the three NASA complexes (Table 19.1).

19. Telecommunications

	S-band	X-band	K_a-band	
↑		7,175	34,316	
↓	2,299	8,430	32,034	32,029
R		880/749	3,344/749	14/15

Table 19.1 The DSN frequencies (in MHz) of coherent uplink (↑) and downlink (↓) carriers and their conversion ratios R. Two nearby downlinks in K_a-band are used in the Cassini mission.

Radar link budget. Radio systems can be used also in the *radar mode*, in which the signal impinging upon a passive body – like a planet – is reflected or diffused back to the transmitter. This technique has many applications in planetary physics (Sec. 20.1). The optical counterpart of radar ranging – laser ranging – provides the very precise tracking of LAGEOS-type satellites (see Sec. 20.3) and of the Moon. The link budget now uses eq. (19.8) to get the power P' impinging upon the target body, of geometrical cross-section A'_e. Part of this power is reflected and diffused, with a gain G_1. The re-emission from a planetary body, with albedo A, occurs in a solid angle of order 2π, so that G_1 is about 2. The gain is substantially greater if special optical systems (*retroreflectors* or *corner reflectors*, like those on the Moon and on the LAGEOS satellites; Fig. 20.7) are used which, in the geometrical optics approximation, are able to reflect back the received radiation in the same direction it arrives from (Problem 20.11). The gain G_1 is again given, in order of magnitude, by eq. (19.7). The power received by the same transmitting antenna reads

$$P'_1 = PG^2 G_1^2 \left(\frac{\lambda}{4\pi D}\right)^4 \tag{19.11}$$

and is proportional to $1/D^4$, with the exponent -4 characteristic of radar.

19.2 Spectra

The fast varying electric field $E(t)$ at the receiver corresponds to the time-averaged flux

$$\overline{\Phi} = \frac{c\overline{E^2}}{4\pi} = 2\int_0^\infty d\omega\, S_\Phi(\omega), \tag{19.12}$$

which results from the contributions of all frequency components. A detector or an amplifier measures only the fraction of the impinging flux in its acceptance band (ω_1, ω_2), determined by technological limits or the need to concentrate on a particular spectral interval. To understand these considerations, it is therefore essential to briefly introduce spectral theory.

Spectrum and correlation function. Consider first a real and periodic function $a(t) = a(t + T_0)$ with a period T_0. In general we are interested in signals of indefinite duration and with no periodicity; but, if T_0 is sufficiently large, we can always approximate them with periodic functions, simplifying the mathematics considerably. A periodic function has the Fourier expansion (with n ranging over all integers)

$$a(t) = \sum_n a_n \exp(i\omega_n t) , \qquad (19.13)$$

with the inverse

$$a_n = \frac{1}{T_0} \int_0^{T_0} dt\, a(t) \exp(-i\omega_n t) \quad (= a^\star_{-n} \text{ if } a \text{ is real}) .$$

The superscript $\star$ denotes the complex conjugate quantity. The Fourier frequencies

$$\omega_n = \frac{2\pi n}{T_0}$$

are equally spaced by $2\pi/T_0$, so that in a frequency interval $d\omega$ there are $(T_0/2\pi)\, d\omega$ lines. For smooth integrands (i.e. in the formal limit $T_0 \to \infty$) a sum over modes can be written as

$$\sum_n \ldots = \frac{T_0}{2\pi} \int_{-\infty}^{\infty} \ldots d\omega .$$

The *spectrum* of the function $a(t)$ is the frequency-dependent quantity

$$S_a(\omega) = \frac{T_0}{2\pi} |a_n|^2 = S_a(-\omega) . \qquad (19.14)$$

This function, even in its argument ω and with dimension ($a^2 \times$ time), is here defined by its value at the discrete frequencies ω_n. Squaring and integrating eq. (19.13) over the period T_0 we find

$$\overline{|a(t)|^2} = \frac{1}{T_0} \int_0^{T_0} dt\, |a(t)|^2 = \sum_n |a_n|^2 = \int_{-\infty}^{\infty} d\omega\, S_a(\omega) = 2 \int_0^{\infty} d\omega\, S_a(\omega) .$$

$$(19.15)$$

This definition uses the double-sided Fourier transform, with the frequency ranging over the whole real line; sometimes one calls $2S_a(\omega)$ the spectrum and uses the single-sided Fourier transform, with positive frequencies only. In this chapter we stick to the double-sided form. The spectral decomposition (19.12) of the flux corresponds to $a(t) = E(t)\sqrt{c/4\pi}$.

We also often need to characterize the statistical relation among values of $a(t)$ separated in time by a *lag* τ; if they are independent, the product $a(t - \tau/2)a(t + \tau/2)$ has, as t changes, a random sign and averages to zero. For a periodic function we define the *correlation function*

$$C_a(\tau) = \frac{1}{T_0} \int_{-T_0/2}^{T_0/2} dt\, a(t - \tau/2)\, a^\star(t + \tau/2) = C_a^\star(-\tau) . \qquad (19.16)$$

If data are available only for a time interval $T' < T_0$, it can be approximated by replacing T_0 with T'. Note that $C_a(0) = \overline{|a(t)|^2}$; we also have

$$C_a(\tau) \leq C_a(0), \tag{19.17}$$

so that $C_a(0)$ is an absolute maximum. The correlation can be expected to be much smaller than $C_a(0)$ when the lag is large enough – greater than τ_c, the *correlation time*. Using the spectral decomposition it is easily shown that

$$\begin{aligned}C_a(\tau) &= \sum_n |a_n|^2 \exp\left(\frac{2\pi n \tau}{T_0}\right) = \frac{2\pi}{T_0} \sum_n S_a(\omega_n) \exp(i\omega_n \tau) \\ &= \int_{-\infty}^{\infty} d\omega\, S_a(\omega) \exp(i\omega \tau) \,;\end{aligned} \tag{19.18}$$

the correlation function is the Fourier transform of the spectrum. This definition applies also to the case in which $a(t)$ is not periodic and its Fourier transform does not exist within the class of ordinary functions. The inversion of eq. (19.18) gives

$$S_a(\omega) = \frac{1}{2\pi} \int_{-\infty}^{\infty} d\tau\, C_a(\tau) \exp(-i\omega\tau) = \frac{1}{\pi} \int_0^{\infty} d\tau\, C_a(\tau) \cos(\omega\tau). \tag{19.19}$$

The Fourier transform of a Gaussian function with width δt is still Gaussian, with a width in frequency $\delta\omega = 1/(2\delta t)$; more generally, a function whose spectrum has the shape of a narrow line of width $\delta\omega$ has a correlation time $\delta t \approx 1/\delta\omega$; a more precise definition of the two quantities leads to the *uncertainty principle*

$$\delta t\, \delta\omega \geq 1/2, \tag{19.20}$$

the equality sign holding for a Gaussian spectrum (see Problem 19.2).

Spectrum of a modulated signal. An important way to realize a narrow band signal is a sinusoidal function slowly modulated in phase or/and amplitude with the timescale $\delta t \gg 2\pi/\omega_0$

$$a(t) = A(t) \exp[i\omega_0 t + i\phi(t)]. \tag{19.21}$$

The Fourier transform of the modulating function $A \exp(i\phi)$ has components only within a frequency interval of order $\delta\omega = 1/\delta t$ near the origin, while the Fourier transform of $a(t)$ has only two bands of width $\delta\omega$ around $\pm\omega_0$. In order of magnitude, from eq. (19.15),

$$\overline{|a(t)|^2} = \delta\omega\, S_a(\omega_0).$$

One should also note that a strictly periodic signal, if limited to a duration T', is spread over a bandwidth $1/T'$; hence the frequency resolution is not better

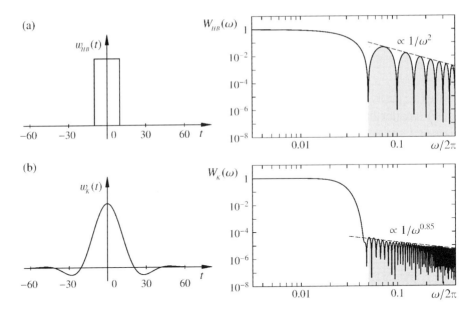

Figure 19.3. Properties of two low-pass filters. On top the "hat box" window with $\tau = 20$ s and, on the right, its spectrum. The abscissas are given, respectively, in s and Hz. The comparison between two filters can only be based on their spectrum, in particular on the frequency $\omega_0/(2\pi)$ of its first null, beyond which the cutoff should be substantial. The Kaiser filter is shown below for $\omega_0/(2\pi) = 1/20$, the same as for w_{HB}. On the left, the window w_K in the time domain; on the right its spectrum. Note the very sharp increase in decimation effectiveness above 1/20 Hz; the shaded areas indicate the spurious residual power. The dashed asymptote indicates the slow $\propto 1/\omega^2$ decrease of the hat box spectrum, while a slower decay, $\propto 1/\omega^{0.85}$, is obtained in the better tuned Kaiser window.

than the reciprocal of the duration of the signal. For a *wide-band signal*, when $\omega_0 \approx \delta\omega$, in order of magnitude,

$$\overline{|a(t)|^2} \approx \omega_0 S_a(\omega_0) .$$

Averaging. An important operation is "averaging", a low-pass filter which ideally eliminates the components with frequency larger than the reciprocal of the averaging time τ. Its simplest realization is the "hat box" window $w_{HB}(t)$, which replaces the function $a(t)$ with its "running average" over the time τ:

$$a(t) \xrightarrow{\tau} a_\tau(t) = \frac{1}{\tau} \int_{t-\tau/2}^{t+\tau/2} dt' a(t') = \int_{-\infty}^{\infty} dt' a(t') w_{HB}(t-t') . \quad (19.22)$$

The function $w_{HB}(t)$ has unit area and is shown for $\tau = 20$ s in Fig. 19.3, on the left. One easily finds that the Fourier amplitudes a_n of $a(t)$ become

$$a_n \xrightarrow{\tau} a_n \frac{T_0}{\pi n \tau} \sin \frac{\pi n \tau}{T_0}$$

and, for the spectrum (eq. (19.14)),

$$S_n(\omega) \xrightarrow{\tau} S_n(\omega)\left(\frac{2}{\omega\tau}\sin\frac{\tau\omega}{2}\right)^2 = S_n(\omega)\,W_{HB}(\omega)\,. \tag{19.23}$$

$W_{HB}(\omega)$ is the spectrum of the window function $w_{HB}(t)$. In the case of (19.23) the new spectrum decreases as $1/(\omega\tau)^2$ for $\omega\tau \gg 1$ and has its first null at $\tau\omega = 2\pi$. Fig. 19.3 shows that this widely used simple filter is not very powerful, in that it leaves too large a spectral density above the expected cutoff $1/\tau$, and allows a substantial leakage of the high-frequency spectrum in the filtered function. The reason for this pollution is the sharp turning on and off of the window function. Many examples of smoother window functions are given in the literature; Fig. 19.3b illustrates one based on Kaiser's two-parameter algorithm. It is frequently used to get the mean elements from numerical simulations of planetary and asteroidal motion; its Fourier transform $W(\tau\omega)$ has a much sharper cutoff. It should also be noticed that data are always sampled, with a finite sampling time Δt; this introduces a cutoff at the *Nyquist frequency* $1/(2\Delta t)$, often relevant for data analysis (Sec. 20.5). Low-pass filters usually do not behave properly close and above the Nyquist frequency, since they allow a significant leakage from this frequency region. It is thus important to predict and estimate the highest expected frequencies in the signal and to choose the sampling at least one order of magnitude finer.

The properties of a filter are sometimes expressed with the *stopband* for the level s, defined by $|W(\omega)| \leq s$, and the *passband* for the level r, defined by $|W(\omega) - 1| \leq r$. For instance, the Kaiser window in Fig. 19.3b has a passband of 3.5×10^{-5} at $\omega\tau = 0.005$ and a stopband of 10^{-5} at $\omega\tau = 0.05$. These quantitative properties of the filter must be carefully taken into account for accurate work.

Low-pass digital filtering is very suitable for the reduction of high-frequency noise, in particular the one due to statistical effects; but it has wider applications. As an example, we mention the calculation of the mean orbital elements by averaging the output of numerical integrations (Sec. 12.5). High quality proper elements of asteroidal orbits, shown in Fig. 14.3, have been obtained with a similar technique: a suitable low-pass filter is applied several times to eliminate components of different frequencies.

Data transmission. The most important use of space telecommunication is to transmit data, either uplink to control the spacecraft, or downlink to carry information gathered on board. Data are encoded as binary numbers q with digits $q_1, q_2 \ldots$ (valued 0 or 1; Fig. 19.4). To transmit them at great distances one uses the side bands of a carrier at the frequency ω_0, with *amplitude* or *phase modulation* (eq. (19.21)). With amplitude modulation, let $H(t)$ be an elementary pulse of duration T_1, with Fourier coefficients H_n (eq. (19.13); for

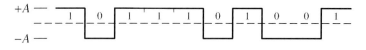

(a) Transmitted Signal

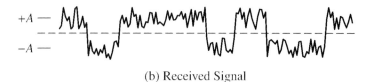

(b) Received Signal

Figure 19.4. (a) The amplitude modulated signal corresponding to the binary number ... 1011101001 ...; in (b) a noisy realization.

instance, a rectangular pulse). The binary number is coded in the amplitude

$$A(t) = \sum_k q_k H(t - kT_1). \tag{19.24}$$

Its Fourier transform is easily read off from the Fourier decomposition of the pulse:

$$A_n = \sum_k q_k H_n \exp(-2\pi i k n),$$

where the index n corresponds to the frequency difference

$$\omega_n - \omega_0 = \omega'_n = \frac{2\pi n}{T_1}.$$

According to eq. (19.14) the spectrum of the amplitude signal is

$$S_A(2\pi n/T_1) = \sum_{k,k'} q_k q_{k'} |H_n|^2 \left(\frac{T_1}{2\pi}\right) \exp[2\pi i n(k' - k)] ;$$

or, in terms of the spectrum of $H(t)$,

$$S_A(\omega') = S_H(\omega') \sum_{k,k'} q_k q_{k'} \exp[i\omega' T_1(k' - k)]. \tag{19.25}$$

In a way completely analogous to what was done earlier for a continuous function, we define the "correlation function"

$$C_q(h) = \sum_k q_k q_{k+h} \tag{19.26}$$

of the numerical sequence q_k, and its "Fourier transform"

$$S_q(r) = \sum_h C_q(h) \exp(2\pi i h r), \qquad (19.27)$$

the "spectrum" of the binary number q at the "frequency" r. If the binary digits are not correlated at all, $C_q(h) = 0$ unless $h = 0$ (and $S_q(r)$ is "white"); if the digits are correlated over a length N, $S_q(r)$ has a width of order $1/N$. Eq. (19.25) can be written as

$$S_A(\omega') = S_H(\omega') S_q(\omega' T_1/2\pi). \qquad (19.28)$$

Therefore, in order to encode an uncorrelated binary sequence, we need a signal whose bandwidth is equal to the bandwidth $\delta\omega$ of the elementary signal, which in turn cannot be smaller than the reciprocal of its duration T_1. We come therefore to the simple and fundamental conclusion that the maximum channel information capacity, in bits per second, is the bandwidth of the signal. If the binary sequence is correlated, the required bandwidth is smaller.

This result dictates much of the requirements of space communications. The need of sending back to the Earth large quantities of data points to the need of high-frequency carriers (for practical reasons it is not convenient to have a frequency bandwidth exceeding a small fraction of the carrier). For this reason most space transmitters operate in X-band, at about 8 GHz. Higher frequencies (in K_a-band, at 32 GHz and 34 GHz) are also being implemented. As we shall see in the next section, there is a conflict between a larger bandwidth and a smaller noise.

A carrier modulation is also used to *measure the absolute distance D* of a spacecraft. The signal is transmitted from the ground with a random modulation in amplitude or phase of sufficient length; on board a repeater re-transmits it back. On the ground, for amplitude modulation, one computes the correlation function between the emitted A and the received A' signal

$$C_A(\tau) = \frac{1}{T'} \int_0^{T'} dt \, A(t - \tau) A'(t),$$

which has a maximum when the lag τ equals the round-trip light-time $2D/c$. At present, for interplanetary distances, an accuracy a few meters is routinely achieved; submetre accuracy is expected with the new generation of ranging facilities.

19.3 Noise

Since the noise is a statistical quantity, to describe its variability we need *random functions*. A random function $a(t)$ is a set of ordinary functions over which a probability distribution is defined and the averages of any product

$a(t_1)a(t_2)\ldots$ can be evaluated. For example, in the *random phase approximation* all Fourier amplitudes $|a_n|$ are fixed by the known spectrum, and all phases ϕ_n are uncorrelated and uniformly distributed in $(0, 2\pi)$. An angular bracket denotes the average over the set; $\langle a(t) \rangle$ is the mean and $\langle [a(t) - \langle a(t) \rangle]^2 \rangle$ the mean square deviation of $a(t)$. A random function is *stationary* if all its statistical averages are independent of the origin of time; this is the most interesting case, to which we confine ourselves. The statistical average of the product $a(t_1)a(t_2)$ does not change if the arguments are displaced by the same amount, and depends only on their difference $t_2 - t_1 = \tau$. If the function is defined over the whole real line, under fairly general and reasonable assumptions, the stationary character has an important and very useful consequence.

The time average of such products is equal to the statistical average (ergodic theorem):

$$\lim_{T \to \infty} \frac{1}{T} \int_{-T/2}^{T/2} dt\, a(t_1 + t) a(t_2 + t) = \langle a(t_1) a(t_2) \rangle. \qquad (19.29)$$

In the left-hand side of this equation we sample the product constructed with a given realization of the random function taken at different times; in the right-hand side we keep the arguments fixed and sample over different realizations of the random function, taken with the appropriate probability weight. The ergodic theorem says that if a random function is stationary, a given realization, so to speak, perfectly reproduces in time the statistical properties of the ensemble. The statistical average gives the expected mean value of the product from many different experiments performed under the same conditions; under the ergodic property it equals the time average of a single, infinitely long measurement. Under this assumption we can therefore write the correlation function (19.16) as

$$C(\tau) = \langle a(t - \tau/2)\, a^\star(t + \tau/2) \rangle, \qquad (19.30)$$

easily recovered in the random phase approximation; moreover, the correlations of $a(t)$ with itself and of $a^\star(t)$ with itself both vanish.

Any physical operation (in particular, amplification) on the signal $s(t)$ adds a noise $n(t)$ to it; the key problem of communication theory is the extraction of the signal from the measured superposition $s(t) + n(t)$ ($s(t)$ is an ordinary function; $n(t)$ is a random function, whose statistical properties are determined by the instrumental set up and the physical, often poorly known perturbations). This extraction is usually done with the use of *linear filters*, constructed either in hardware or in software; they are mathematically equivalent to an integral transform with a suitable kernel $K(t)$

$$\int dt'\, K(t - t')[s(t') + n(t')].$$

The kernel is chosen in such a way to increase as much as possible the ratio of signal to noise. For example, if the noise is known to occur at frequencies higher than the signal, a suitable low-pass filter can be applied (Fig. 19.3).

Sources of noise. The fundamental noise in interplanetary radio links, where the received signal is exceedingly weak, is the electrical thermal noise. At a finite temperature T, any dissipative component (i.e., a resistance) in a receiver produces short and random bursts of electrical charge. In 1928 H. NYQUIST predicted (and J.B. JOHNSON verified in 1929) that a receiver adds to the incoming power a noise power P_n with a *universal* spectrum $S_n(\omega)$, entirely determined by the temperature; at frequencies less than $2\pi kT/h$ (where h is Planck's constant) it is flat

$$S_n(\omega) = \frac{kT}{4\pi} \quad (h\nu \ll kT). \tag{19.31}$$

Around and above this frequency quantum effects must be taken into account; they ensure that the spectrum is integrable and the total power is finite. This is the regime commonly valid for microwaves; it says that kT is the amount of energy in a cycle[1]. For simple receivers T is the room temperature; but generally one must use an effective temperature T_{eff} dependent on the electronics. The receivers of NASA's Deep Space Network are cooled and have noise temperatures in the range 20 to 40 K.

Nyquist's theorem points to the crucial problem in space telecommunications, where very weak signals must undergo substantial amplification before detection. We can reduce the noise power (over the two bands $\Delta\omega$ at $\pm\omega_0$)

$$P_n = 2 S_n \Delta\omega = \frac{kT}{2\pi} \Delta\omega$$

by decreasing $\Delta\omega$, hence the bit rate. With the link budget (19.8) we get the signal-to-noise ratio

$$\text{SNR} = \frac{2\pi P'}{kT \Delta\omega} = \frac{2\pi P G G' L_s}{kT \Delta\omega}. \tag{19.32}$$

This gives also the largest admissible bandwidth

$$\Delta\omega \leq \frac{2\pi}{kT} P' = \frac{2\pi}{kT} P G G' L_s, \tag{19.33}$$

inversely proportional to the square of the distance (recall that $L_s \propto 1/D^2$). To increase the bit rate one needs large transmitted powers, low receiver temperatures and short wavelengths (to ensure large gains).

---

[1] That is to say, the energy contained in a period $2\pi/\omega$ in the signal with all frequencies above ω suppressed, therefore with a total bandwidth 2ω.

Besides this fundamental limitation, space radio links must face other noise sources, including:

- electronic noise on the ground and on board;
- mechanical deformations and
- imperfect pointing of the antennas;
- effects of the media (troposphere, ionosphere and interplanetary space; see Sec. 19.5);
- radio emission from the Sun, if it is in the beam.

19.4 Phase and frequency measurements

Phase and frequency measurements with microwave links to interplanetary spacecraft have attained a very high accuracy and, besides being used for navigation, have several scientific applications. A typical setup consists in sending to the spacecraft a very stable carrier at a frequency ω_0; on board a transponder locks to the incoming phase the carrier transmitted to the ground, where it arrives with the frequency $\omega(t)$. The fractional frequency change

$$y(t) = \frac{\omega(t) - \omega_0}{\omega_0} \tag{19.34}$$

is then measured over an integration time τ, and the running average (19.22)

$$y_\tau(t) = \frac{1}{\tau} \int_{t-\tau/2}^{t+\tau/2} dt'\, y(t') = \frac{\phi(t+\tau/2) - \phi(t-\tau/2)}{\omega_0 \tau} \tag{19.35}$$

is obtained. Frequency stability is measured by the *Allan deviation* $\sigma_y(\tau)$, whose square

$$\sigma_y^2(\tau) = \frac{1}{2} \left\langle [y_\tau(t+\tau) - y_\tau(t)]^2 \right\rangle \tag{19.36}$$

is the *Allan variance*. Generally speaking, the Allan deviation at first decreases with τ, but beyond the minimum it increases due to frequency drifts and other systematic noises. It can be shown that for a power spectrum in frequency $(S_y(\omega) \propto \omega^\alpha)$, $\sigma_y(\tau) \propto \tau^{-\alpha-1}$. When $\alpha > 1$ or $\alpha < -3$, however, a cutoff is needed to ensure convergence.

The frequency standards commonly used for precision radio tracking use hydrogen masers and can have an Allan deviation of about 10^{-14} or less for integration times of the order of an hour or more (see Fig. 3.1). Allan deviations less than 10^{-16} have been obtained in the laboratory with Superconductive Cavity Stabilized Oscillators (SCSO) and Hg^+ ion traps. For standard, less demanding applications (like the Global Positioning System), frequency standards based upon resonances in cesium atoms are used, with typical values of σ_y of a few parts in 10^{-14}.

The main contribution to the frequency shift is the Doppler effect $y_D = 2v/c$, where v is the relative velocity between the ground antenna and the antenna on board (taken with the positive sign when the two are approaching); it includes their orbital and rotational motion. For a better approximation, the fully relativistic formula (3.3) should be used; in presence of a gravitational field the effect of the gravitational potential on the photon energy (17.41) must also be taken into account (Sec. 17.3). When the spacecraft passes near a planet or a satellite, the change in velocity due to the low-degree gravitational moments (mass, quadrupole, etc.) can be accurately measured. As an example, consider a flyby with relative velocity $v = 5$ km/s at a distance $b \approx 5,000$ km from the centre of Titan, which has a mass $M_T \simeq 1.35 \times 10^{26}$ g and a radius $R_T \simeq 2,575$ km. The timescale of the encounter is $b/v \approx 1,000$ s. The change in velocity in a time τ is of order $\delta v \simeq GM_T \tau/b^2$; with an Allan deviation of 10^{-14} the SNR $= \delta v/(c\sigma_y)$ is about 10^8. High-degree gravity field coefficients can be resolved by continuous and accurate measurements of the Doppler frequency shift from a planetary orbiter. In this way, the best models of the Venus and Mars gravity fields have been obtained by tracking the Magellan and Mars Global Surveyor spacecraft. Precise Doppler measurements can also be used to measure the gravitational deflection of an electromagnetic beam passing near the Sun (Sec. 20.4).

Plasma and thermal contribution. Electromagnetic propagation in space is also affected by the refractivity of the medium, with index of refraction n_r. The atmosphere contributes both with its dry content and its water vapour; the ionospheric and interplanetary plasma have their own dispersive effect. The phase change of a wave is

$$\Delta\phi = -\frac{\omega}{c}\int ds\,(n_r - 1) = -\frac{\omega\,\Delta\ell}{c}, \qquad (19.37)$$

where $\Delta\ell = \int ds\,(n_r - 1)$ is the change in the optical path. Considering, for simplicity, one-way propagation, the total frequency change is the sum of the refractive change and the Doppler effect (to $O(v/c)$; see eq. (3.3)):

$$y = -\frac{d\Delta\ell}{c\,dt} + \frac{v}{c}. \qquad (19.38)$$

At high frequencies ($\omega \gg \omega_p$) the plasma refractive index (eq. (8.29)) is

$$n_r = 1 - \frac{\omega_p^2}{2\omega^2} + O\left(\frac{\omega_p^4}{\omega^4}\right), \qquad (19.39)$$

where ω_p is the plasma frequency (eq. (1.104)). In eq. (19.38) we can distinguish the dispersive plasma contribution (ionosphere and solar wind), proportional to the rate of change of the columnar electron content $\int ds\, n_e$, and the non-dispersive term y_{ND} (Doppler effect and atmospheric contribution)

$$y = \frac{1}{2\omega^2 c} \frac{d}{dt} \int ds\, \omega_p^2 + y_{ND}. \tag{19.40}$$

To eliminate the plasma contribution one can use *two carriers* with frequencies ω_1 and ω_2; indeed, the quantity

$$\frac{\omega_1^2 y_1 - \omega_2^2 y_2}{\omega_1^2 - \omega_2^2} = y_{ND} \tag{19.41}$$

does not contain the dispersive term. Vice versa, the total electron content along the beam is given by

$$\frac{d}{dt} \int ds\, \omega_p^2 = -\frac{2c\,\omega_1^2 \omega_2^2}{\omega_1^2 - \omega_2^2} (y_1 - y_2). \tag{19.42}$$

A similar way to eliminate the plasma contribution has been successfully tested in 2002 with the Cassini spacecraft, even when the beam passed very near the Sun. Investigations of the interplanetary plasma and the solar corona have extensively used radio links to interplanetary spacecraft.

Finally, we discuss the effect of thermal noise of the receiver on a frequency measurement. Let $n(t)$ be a (small) noise added to a monochromatic signal $a(t) = \sqrt{P'} \exp(i\omega_0 t)$ of power $2P'$ (in a double-sided representation) and frequency ω_0. This produces a fluctuation in amplitude and phase in the signal given by

$$a + n = \sqrt{P' + \delta P'} \exp[i(\omega_0 t + \delta\phi)]; \tag{19.43}$$

after some algebra we obtain

$$\delta\phi = \frac{n \exp(-i\omega_0 t) - n^\star \exp(i\omega_0 t)}{2i\sqrt{P'}} \quad (\ll 1).$$

From the definition (19.30) we get

$$C_{\delta\phi}(\tau) = \frac{\cos(\omega_0 \tau) C_n(\tau)}{2P'}. \tag{19.44}$$

$C_n(\tau)$ is the correlation function of the noise, and $C_n(0)$ its mean power. At $\tau = 0$

$$C_{\delta\phi}(0) = \frac{C_n(0)}{2P'}$$

says that
$$\overline{\delta\phi^2} = \frac{\text{average noise power}}{\text{signal power}}.$$

This relation is of immediate interpretation in the complex plane, where the signal describes a circle of radius $\sqrt{P'}$ superimposed to a random motion of size $\sqrt{\langle |n|^2 \rangle}$.

Allan's variance (19.36) can be written in terms of the correlation function of the phase as

$$\sigma_y^2(\tau) = \frac{3C_{\delta\phi}(0) - 4C_{\delta\phi}(\tau) + C_{\delta\phi}(2\tau)}{\omega_0^2 \tau^2}. \qquad (19.45)$$

For a white thermal noise $C_n(\tau) = 0$ for $\tau \neq 0$ and the mean power in a bandwidth $\Delta\omega$ is $C_n(0) = \int d\omega\, S_n(\omega) = kT\Delta\omega/2\pi$ (eq. (19.31)); hence

$$\sigma_y^2(\tau) = \frac{3kT\Delta\omega}{4\pi P' \omega_0^2 \tau^2}. \qquad (19.46)$$

If $\Delta\omega$ is kept constant as τ changes, σ_y corresponds to the exponent $\alpha = 1$ for the frequency spectrum $S_y(\omega)$. This relationship shows that the accuracy of frequency measurements improves with the received power and the integration time; it is also important to lower the bandwidth.

19.5 Propagation in a random medium

A refractive medium is often turbulent and its refractive index $n_r(\mathbf{r}, t)$ is a random function, with a correlation in space and time between two events

$$C_r(\boldsymbol{\xi}, \tau) = \langle n_r(\mathbf{r}, t) n_r(\mathbf{r} + \boldsymbol{\xi}, t + \tau) \rangle. \qquad (19.47)$$

Its characteristic decay length ξ_c is the distance beyond which the values of the refractive index are not well correlated, and measures the "average size" of the turbulent cells; similarly, the characteristic decay time τ_c of $C_r(0, \tau)$ gives the mean lifetime of a turbulent cell at a given place. In interplanetary space the turbulence is dragged along by the solar wind with a velocity $\mathbf{V}$ larger than all the relevant phase speeds; in the local frame of reference ($\mathbf{r}' = \mathbf{r} - \mathbf{V}t, t' = t$) moving with the wind one can make the assumption of isotropic and homogeneous turbulence, with a correlation function $C_r'(|\boldsymbol{\xi}'|, \tau')$ which depends only on the distance ξ' and the lag τ' between the two events. Of course, its form changes with the distance from the Sun. In the solar system frame we get

$$C_r(\boldsymbol{\xi}, \tau) = C_r'(|\boldsymbol{\xi} - \mathbf{V}\tau|, \tau). \qquad (19.48)$$

The wind mixes up spatial and temporal correlations. In the radio band $n_r - 1$ is proportional to the electron density, and its fluctuations produce random changes in the amplitude and the phase of a radio signal; this provides a way to investigate plasma turbulence in the solar wind, in particular in the crucial coronal region. It is remarkable that for a large interval of separations ξ its spectrum fulfils a power law with the Kolmogorov exponent, as derived on the basis of dimensional arguments in Sec. 7.5.

In the troposphere the turbulence is strongest within one or two hundred meters from the ground, where thermal exchanges are active. This has important effects upon the quality of astronomical imaging and places an essential limitation upon ground astronomy. In space the turbulence of the solar wind has similar effects upon radio propagation. A surface of constant phase is distorted by the effect on the optical path (eq. (19.37)). Assume that the turbulent refractive index is made up of elementary cells of size ξ_c, mean life τ_c and amplitude δn_r. We can estimate the integral (19.37), over a distance L, in a random walk approach, by adding L/ξ_c elementary contributions of order $\delta n_r \xi_c$. They are independent random numbers and sum up stochastically to the total value $\Delta\phi = k\,\delta n_r\,\xi_c\,\sqrt{L/\xi_c}$. This quantity varies across the beam over the characteristic distance ξ_c, so that the normal to the surface of constant phase changes by about $k\,\delta n_r\,\sqrt{L/\xi_c}$. Since the unperturbed gradient of the phase has a magnitude k, the angular deviation of the ray is about $\delta\theta = \delta n_r\,\sqrt{L/\xi_c}$ and changes in time with a timescale of order τ_c.

An optical system on the ground which images a source outside the troposphere for a time much larger than τ_c superposes elementary, point-like images and produces an extended image of size $\delta\theta$. Taking, typically, $\delta n_r \simeq 5 \times 10^{-7}$, $L \simeq 200$ m and $\xi_c \simeq 10$ cm, we get $\delta\theta \simeq 1''$. Of course, this depends very much upon local conditions; moreover, a telescope of area A, averaging over $A/(\xi_c)^2$ independent coherence regions for the phase, diminishes the effect by the factor $\xi_c/\sqrt{A}$. The tropospheric turbulence produces also a random fluctuation in the intensity and a random displacement of the centre of the image (*seeing*). For the largest optical astronomical systems based on, say, $\simeq 10$ m mirrors with the angular diffraction limit $\simeq \lambda/d \simeq 0.02''$, the tropospheric blurring of the image presents the main obstacle. For better resolution, in ground optical telescopes special active optical systems, called *adaptive*, have been devised and implemented, in which the reflecting surface changes continuously its shape and compensates the irregularities in the phase surfaces. Random atmospheric distortions are monitored in real time by the observation of a source in the field of interest, either a real star or a laser beacon reflected on the mesospheric layer of sodium at $\simeq 90$ km altitude. Although this method requires advanced computers and mechanical systems, it is becoming common in modern large telescopes (such as the 10 m Keck telescopes at Hawaii).

– PROBLEMS –

19.1.* Consider the link budget for Cassini's X-band to the ground antenna characterized by the parameters in the following table:

Transmitted power	19 W	42.79 dBm
Distance	5 AU	
Diameter of the on-board antenna	4 m	
Gain of the on-board antenna	52, 480	47.20 dB
Diameter of the ground antenna	70 m	
Gain of the ground antenna	2.6×10^8	74.10 dB
Receiver bandwidth	3 Hz	
Carrier's frequency	8, 430 MHz	
Receiver's effective temperature	28.24 K	

Obtain the received power in W and in dBm and compare it with the thermal noise. The stated bandwidth corresponds to the receiver phase-lock loop for coherent signals.

19.2. Prove the "uncertainty principle" $2\delta\omega\,\delta\tau = 1$ for a Gaussian correlation function.

19.3.* Using a model of the solar corona (see Secs. 9.2 and 9.3), find the plasma correction to the optical path as a function of the impact parameter.

19.4. The divergence of a laser beam, $\approx 3-4''$, is mainly due to atmosphere refraction; estimate the power loss on the Moon, with a retroreflector array of effective area ≈ 0.2 m^2. Compare the theoretical diffraction divergence ($\lambda = 532$ nm for the YAG laser and a single retroreflector with diameter 5 cm) for the reflected beam with the real divergence $\approx 12''$ and compute the loss in the returned beam for a telescope of ≈ 1 m diameter. Estimate the number of photons received from a transmitted pulse of energy ≈ 300 mJ (these numbers correspond to the telescope at the Observatoire de la Côte d'Azur in southern France.)

19.5.* Find out Allan's variance for a frequency spectrum $S_y(\omega) \propto \omega^\alpha$. (Hint: Calculate first the spectrum of y_τ and then use it in eq. (19.36).)

19.6. What is the effect of the rotation of a planet on its radar scattering? How accurately can be the rotation period determined?

19.7.* A coherent wave is sent from the Earth to an interplanetary spacecraft in the ecliptic plane. Find out the frequency received on board as a function of time, neglecting the eccentricities of the Earth and of the spacecraft.

19.8. Estimate the number of bits required to code a black and white planetary image.

19.9. Show that a parabola transforms a beam radiating from the focus into a parallel beam along the axis.

19.10. Calculate, in order of magnitude, the receiver phase noise at 1 GHz of a quasar at 10^8 ly with a power 10^{45} erg/s. Assume an effective temperature of 50 K and a dish diameter of 30 m.

19.11. The phase change of a radio beam is determined by the integral over the ray of the refractive index, in a dry troposphere proportional to the air density. Show that in a parallely stratified and static atmosphere the phase change is determined by the ground pressure.

– FURTHER READINGS –

In view of the very large number of books on this wide range of topics we quote only a few related to space: T. Pratt and C.W. Bostian, *Satellite Communications*, John Wiley and Sons (1986) is a good and simple introduction; G. Maral and M. Bousquet, *Satellite Communications Systems*, John Wiley and Sons (1998), still at an introductory level, but with an emphasis on the orbital part; more technical and complete is T.T. Ha, *Digital Satellite Communications*, McGraw-Hill (1990). For interplanetary communications one can consult the documents of the Deep Space Network (http://deepspace.jpl.nasa.gov/dsn/). More details about frequency measurements are in J.A. Barnes et al., *Characterization of frequency stability*, IEEE Trans. on Instrum. and Meas., **IM-20**, 105 (1971).

Chapter 20

PRECISE MEASUREMENTS IN SPACE

In this chapter, rather than undertaking a systematic review of space missions, we discuss four general themes, with a bias towards precise measurement techniques; these techniques exploit favourable space conditions, in particular the absence of an atmosphere, weak gravity and the vacuum. (i) The basis for a rational understanding of the physics of the solar system is direct exploration, in particular by planetary imaging; here we discuss optical imaging and Synthetic Aperture Radar. (ii) Astrometry by means of special optical telescopes and sensors in space has opened up a revolution in observational and theoretical astrophysics; we concentrate on the HIPPARCOS and Gaia projects. (iii) Distances to artificial Earth satellites, to the Moon and to planets have been obtained using the round-trip time of a short electromagnetic signal; apart from providing precise orbit determination of the tracked object, this is also an important technique for the investigation of the kinematics of Earth's tectonic plates (Ch. 5) and its rotation (Ch. 3). (iv) Different space techniques have at last allowed testing gravitation theories to a very good accuracy and have confirmed Einstein's General Relativity in the weak field approximation (Ch. 17).

Most space experiments have the purpose of measuring relevant parameters indicated by a theory or a mathematical model; in the last section we review data analysis and briefly discuss the complex and delicate procedures needed when assessing their value and their accuracy.

20.1 Planetary imaging

The sheer and extensive collection of images of planets, their satellites and, recently, also minor bodies like asteroids and comets, is a major drive in the exploration of the solar system; the most interesting related developments include:

- Atmospheric dynamics, in particular large scale structures of the outer planets, as Jupiter's Red Spot.

- The investigation of the extreme diversity in satellite surfaces, with particularly important results for Jupiter's Galilean satellites, such as the discovery of volcanic activity on Io and the very young, icy surface of Europa. Cratering statistics, both for satellite and planetary surfaces, is a valuable key to understand the past flux of impactors.

- The discovery and the dynamics of planetary rings.

- The discovery of satellites of asteroids.

- Laser and radar mapping of planetary surfaces. Interesting results have been obtained with optical laser altimeters, in particular with the spacecraft NEAR/Shoemaker for the asteroid Eros, and the Mars Global Surveyor for Mars, where support was obtained for the idea of ancient water flows and even a surface ocean. Superb results about peculiar geological features on Venus, in particular crater statistics, were obtained with the synthetic aperture radar on Magellan.

- Observations in different spectral bands allow the determination of the mineral (in infrared) and elemental (in X-ray) composition.

Here we touch on two imaging techniques, *Charge Coupled Devices* (CCD) and *Synthetic Aperture Radar* (SAR).

Optical systems. An optical imaging system is a two-dimensional array of $N_p \times N_p$ sensitive elements, called *pixels*, each of which records the number of photons deposited by the impinging radiation in each resolution time. If 8 bits are used to quantify this information, a single image requires $8 \times N_p^2$ bits (e.g., 8 Mbits for $N_p = 1,000$). For colour pictures, images of the same area in the fundamental colours (e.g., blue, green and yellow) are taken in rapid succession. Space imaging systems place heavy requirements on the data transmission. The two Voyager spacecraft have provided tens of thousand pictures, for a total of $\approx 10^{12}$ bits. Even larger requirements are needed if each pixel can also discriminate in wavelength, producing a three-dimensional "image". The drive to use in the downlink high-frequency carriers is a response to this growing need of space telecommunications.

CCDs have long superseded the photographic plate and the vidicon tube as imaging devices for space applications. A CCD is a rectangular array of photosensitive, solid state capacitors; each of them becomes electrically charged when struck by photons. These charges are then transferred to the detectors by moving the potential wells in which they are confined, and recorded. The wideband spectral response (typically 0.4 to 1 μm), the efficiency (which can reach the single photon level) and the possibility of reading electronically and in real time, and storing the amounts of collected charges, make CCDs insuperable devices. A row of N_p CCDs, each of size w, in the focal plane of an optical system of focal length f has the field of view $\Delta\theta = N_p w/f$; on the other hand, the angular resolution $\sigma_\theta = w/f$ corresponds to a single pixel. Since w and N_p cannot be changed much, we have a choice between wide-angle systems, with small f, aimed at imaging large areas, and narrow-angle cameras, with large f, designated for better resolution.

–Galileo's imaging system was based upon a 800×800 pixel camera, with focal length $f = 150$ cm. The pixel width $w = 15\,\mu$m gives an angular resolution $\sigma_\theta = 10^{-5}$ rad corresponding, at a distance of 500 km, to a ground resolution of 5 m. The spectral response ranges from 0.4 to 1.1 µm; the availability of the near infrared band is very important for detection of spectral lines of different molecules (such as CH_4 or NH_3) in Jupiter's atmosphere. Eight narrow-band filters allow stereoscopic composition and a coarse spectroscopy. Before arriving at Jupiter, Galileo's system imaged the two asteroids Gaspra and Ida, and discovered Dactyl, a small satellite of the latter. Its images of the planet, in particular the Great Red Spot, largely superseded in number and resolution those of Voyager; as noted before, the most detailed observations of the impact of the comet Shoemaker-Levy 9 were obtained. The four Galilean satellites revealed many unexpected and striking facts. On Europa, regions of bright, icy crust were discovered, with chaotic fractures mixed with smooth and grey regions, probably filled with material emerged from the interior; this present and strong geological activity may be evidence of a subsurface ocean (see also hints from the magnetic anomalies; Sec. 6.5). On Io impressive volcanic features, including lava flows, active vents and craters, have been observed.

–NEAR/Shoemaker imaging system. This NASA spacecraft was launched in 1996 with the purpose of exploring the asteroid Eros, an irregular body about 34 km long. On February 14, 2000 it was placed on a bound orbit between 35 and 350 km around it; on February 12, 2001 it descended and soft-landed. During one year about 160,000 pictures of the surface were obtained, with interesting results about its crater structure. Its imaging system in the visible was based upon a 537×244 pixel camera with a focal length $f = 16.8$ cm. Individual pixels of $16 \times 24\,\mu$m yielded a spatial resolution of ≈ 10 m from 100 km (and down to ≈ 1 cm in the descent phase). The pixel brightness was encoded in 12 bits, with a compression to 8 bits if necessary. The spectral response, from 0.45 to 1.1 µm, was similar to Galileo's. Eight narrow-band filters allowed the discrimination of iron-containing silicate minerals.

With this mission the rotation and the topography of Eros has been very accurately determined. For the latter, the otherwise superb optical imaging was superseded by the laser altimeter (see Sec. 20.3), with more than 8 million measurements. Doppler data (see later in this Section) provided a determination of the gravity field; the correlation with the topography gave information about the interior. Eros was found to be a primitive, undifferentiated and nearly homogeneous object with a mean density of 2.67 ± 0.03 g/cm^2. This is close to the mean value for S-type asteroids and fits well with the density 2.6 ± 0.5 g/cm^2 of Ida obtained from the Galileo observations. The mineral composition of the surface regolith layer was also accurately measured, suggesting a direct link to ordinary chondrites, the most abundant class of meteorites.

During the interplanetary cruise NEAR/Shoemaker observed the asteroid Mathilde and revealed one of the most heavily cratered body yet observed; a porous structure is suggested by a very large crater, comparable to the asteroid size. Its mass was measured with the Doppler effect and a mean density of 1.3 ± 0.2 g/cm^2 was obtained; this low value is consistent with the spectral class C (Sec. 14.3).

Radar systems. Radar, a sophisticated technique developed for military purposes during the last world war, has had in recent years a major role in planetary imaging. It can mainly operate in two different modes.

Consider first a *nearby target* at a distance D, whose size is much greater than the size $D\lambda/d$ of the beam (for an antenna of diameter d and a wavelength λ). Short pulses, whose duration is of the order of the reciprocal $1/B$ of the bandwidth, are periodically emitted; the returned pulse is the superposition of the contributions of all the features of the terrain encompassed by the beam, distinguished by the corresponding round-trip time $\Delta t = 2D/c$. The error in Δt is about $1/B$, leading to an error $\sigma_D = c/(2B)$ in distance. As the beam sweeps the ground, the set of returns as function of the time t provide an intensity pattern in the plane $(\Delta t, t)$. Were the surface perfectly flat, the beam would be reflected with the same angle of incidence; in practice inhomogeneities of size less than a wavelength produce a diffuse scattering. The returns depend on the correlation function of the surface, determined by its geometric properties (slope and altitude), the polarization of the beam, the incidence angle and, most important, the dielectric constant $\epsilon = \epsilon' + i\epsilon''$. An imaginary part ϵ'' takes into account absorption and is responsible for the penetration depth

$$\ell_{\text{pen}} = \lambda \frac{\sqrt{\epsilon'}}{2\pi\epsilon''}. \tag{20.1}$$

Long wavelength radars can be used to explore underground properties, for instance, to search for fossil water on Mars.

For ground observations of *distant targets* with size less than the beam, ordinary radar provides just a one-dimensional intensity distribution and can be supplemented with Doppler measurements. One way to achieve this is to transmit a continuous carrier of frequency ν_0 whose phase has a random binary modulation (Sec. 19.2); while the modulation carries the distance information, the carrier is returned with a frequency shift $\Delta \nu = \nu_0 v/c$ and a spectral width

20. Precise measurements in space

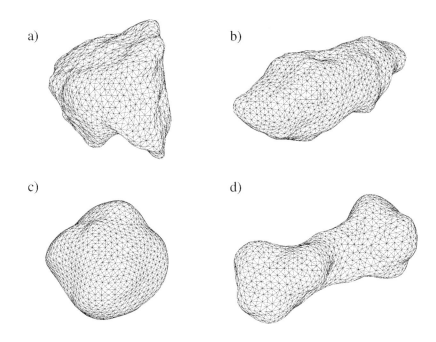

Figure 20.1. Shapes of four asteroids – (a) Golevka, (b) Geographos, (c) 1998 KY_{26}, and (d) Kleopatra (a main-belt asteroid) – as determined by radar ranging data (source http://echo.jpl.nasa.gov/links.html). The figure is not to scale; Golevka and 1998 KY_{26} have, respectively, a size of ≈ 530 m and ≈ 30 m, while the long axes of Geographos and Kleopatra are, respectively, about 4.9 km and 220 km.

$\simeq v_0 \, \delta v/c$, where v is the velocity of the target along the line of sight and δv its spread; for example, for a rigid object of size R rotating at the frequency ω, $\delta v \approx \omega R$. This provides an image in the $(\Delta t, \Delta v)$ plane, from which one can determine also the relief map. This technique has given impressive results in *planetary radar*; the main active facilities are at Arecibo (Portorico) and Goldstone (California). Beginning in the 1960s, echoes from Venus and Mercury determined their unusual rotation and greatly contributed to tests of gravitation theories; later extensive observations of satellites and asteroids were carried out and provided high resolution (as good as 10 m) and geologically detailed three-dimensional models (Fig. 20.1), and accurate determinations of the rotation state. Radar data also allowed to significantly improve the orbital determination of near-Earth asteroids and in several cases secured their follow-up observations. For that purpose, radar distance measurements, as for nearby targets, were also obtained for asteroids closer to the Earth than $\simeq 0.1$ AU.

Microwave techniques may provide better tools than optical imaging to explore a planetary surfaces from a nearby spacecraft, especially when they are covered by an opaque atmosphere (the case of Venus and Titan). They include

(i) radiometry (measurement of the radio emission by the body), (ii) altimetry (to determine the distance from the spacecraft to its Nadir point with the round-trip time), and, at small distance, (iii) radar and Synthetic Aperture Radar. All these tools have been extensively used in NASA's Magellan mission to Venus, which in the early 1990s provided an extensive (97%) mapping of its surface, with outstanding data concerning Venusian volcanism, tectonics, impact and surface processes. In Cassini's four-year tour of the Saturnian system, to begin in July 2004, the spacecraft will fly about 43 times by its major satellite Titan. It is interesting to note that these flybys not only will provide a unique opportunity to explore this satellite (in particular with the release of the Huygens probe, which will descend into the atmosphere and investigate its properties), but will also exploit Titan's gravitational field to provide the appropriate energy and momentum to drive the spacecraft around Saturn according to the scientific goals (Sec. 18.4). Here we concentrate on the Synthetic Aperture Radar (SAR), which provides two-dimensional images with very good resolution. For numerical applications we consider a Cassini's Titan flyby at an altitude h of 1,000 km and a relative velocity $v = 5$ km/s. Its high-gain antenna ($d = 4$ m in diameter) works at $\nu = 13.78$ GHz ($\lambda = 2.17$ cm) and has a half-beam width ($\approx \lambda/d$) of $0.35°$ (eq. (19.6)). Cassini's radar, in effect, has five adjacent radar beams for better coverage, but we consider only one of them. In the imaging mode Cassini's radar can have a bandwidth $B = 0.85$ MHz; in the altimeter mode $B = 4.25$ MHz, corresponding to $\sigma_D = 50$ m.

Consider, for simplicity, a spacecraft flying on a horizontal trajectory at an altitude h with a constant velocity v over a flat surface $\zeta = 0$ ($\xi = vt, \eta = 0, \zeta = h$), and let the high-gain antenna point sideways, perpendicularly to the orbit and at an angle χ from Nadir (Fig. 20.2); this inclination is necessary to separate the returns from surface elements with different η according their different round-trip times. The beam footprint is an ellipse with semiaxes $D\lambda/d = h\lambda/(d\cos\chi) \approx 5$ km along ξ and $D\lambda/(d\cos\chi)$ along η. The former also gives the resolution in the ξ-direction; but the η-resolution $\sigma_\eta = \sigma_D \sin\chi = c\sin\chi/(2B) \approx 100$ m is much better. An image is a set of intensities in terms of the observables ($t = \xi/v, \Delta t$) and covers a long strip of width $\approx 2h\lambda/d = 10$ km; this can be compared with optical imaging, which is given in a plane of two angular coordinates.

A much better resolution is obtained with SAR. Here the radar uses pulses made up of a frequency ramp confined to a short time interval, which on detection is beaten against suitable templates, so that, besides the round-trip time, the fractional frequency shift $y = (\nu' - \nu)/\nu$ is also measured. A ground feature at the point ($\xi, \eta, 0$) returns pulses during a time interval of duration $T_{SAR} = 2\lambda D/dv$ around $t_\xi = \xi/v$, during which it is within the beam. At the time t_ξ the feature is orthogonal to the trajectory. SAR determines t_ξ or ξ using all data in such time interval, so that in the along-track direction we have an

20. Precise measurements in space

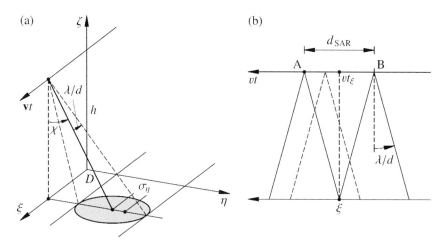

Figure 20.2. a) Ordinary radar operation with an antenna pointing at an angle χ from Nadir. A narrow strip of width $2\lambda D/(d\cos\chi)$ is imaged, with cross-track resolution $\sigma_\eta = (c/2B)\sin\chi$. b) In SAR operation all returns in an interval $AB = d_{SAR} = 2\lambda D/d$ from a feature at ξ are used, corresponding to a virtual antenna of effective length $d_{SAR} = vT_{SAR}$ and an along-track resolution $\lambda D/d_{SAR}$.

effective antenna of size $d_{SAR} = vT_{SAR}$, in principle allowing a much better ξ-resolution $\sigma_\xi = \lambda D/d_{SAR} \approx d$ (in Cassini's case, however, the useful duration of the data bin is the round-trip time $2D/c$ and $d_{SAR} = 2Dv/c = 30$ m, still much larger than the real antenna size; its resolution is $\sigma_\eta = c/(2B) \approx 200$ m, $\sigma_\xi = \lambda c/(2v) \approx 400$ m). To see how this comes about, note that the distance of the feature $D_\xi(t) = [\eta^2 + h^2 + (vt - \xi)^2]^{1/2}$ has a minimum at $t = t_\xi$ and the Doppler shift is a linear function of time:

$$y_\xi(t) = -\frac{2}{c}(t - t_\xi)\frac{v^2}{D} \quad \left(D = \sqrt{\eta^2 + h^2}\right). \tag{20.2}$$

The factor 2 takes into account the shifts in transmission and reception. For a useful record of duration T_{SAR} the frequency has an error $\approx 1/T_{SAR}$,

$$\sigma_y = \frac{1}{vT_{SAR}} = \frac{\lambda}{cT_{SAR}} = \frac{2v^2\sigma_{t_\xi}}{cD}$$

and the position of the feature is known with an accuracy

$$\sigma_\xi \approx \frac{\lambda D}{d_{SAR}}. \tag{20.3}$$

The SAR image consists in a pattern of intensity in the $(\Delta t, y)$ plane. In general, these observables define on the surface a coordinate grid as the intersections

20.2 Space astrometry

The purpose of astrometry is the determination of up to six parameters of celestial objects: angular position, transversal velocity, radial velocity and parallax. Due to the orbital motion of the Earth a source at a distance D and an angular distance θ from the normal to the ecliptic describes in the sky a small ellipse with axes π and $\pi \cos \theta$; $\pi = 1 \text{ AU}/D$ is the (annual) *parallax*[1]. The parsec is the distance at which its semimajor axis subtends $1''$; with an accuracy at the mas level one can only reach a distance of $\approx 1,000$ pc. Further away, the distance D of a cluster of stars is estimated, for instance, from the means of the square of their transversal angular velocity α and their velocity v_r along the line of sight (measured spectroscopically); with the assumption of an isotropic distribution

$$2 \langle v_r^2 \rangle = D^2 \langle \alpha^2 \rangle . \tag{20.4}$$

Other techniques typically employ "standard candles" (objects with approximately the same absolute luminosity), such as Cepheids, novæ or supernovæ. At cosmological distances, where the proper motion of the source becomes irrelevant, the Hubble expansion velocity, measured with the redshift, may be directly linked to the distance. Measuring distances is one of the fundamental problems in astronomy, crucial for estimating the radiative power of a source and understanding its engine. Here, however, we shall mainly deal with the determination of positions in the sky. Global astrometry allows the construction of a rigid, inertial frame of reference tied to stars and galaxies, which, in turn, is a necessary basis for the investigation of the dynamics of objects in the solar system (see Ch. 3).

Classical astrometric techniques, based on observations of the passage of stars at the meridian, are severely limited by atmospheric refraction and have been abandoned; besides an unpredictable component of the deflection, the image of a point source is spread out by local turbulence (Sec. 19.5). An accuracy of $1''$ to $2''$ for a single (good) observation was possible. Fundamental stellar catalogues have been painstakingly built in the past, using many observation campaigns; they reach a relatively high accuracy for a limited number of sources. For instance, the FK5 catalogue, completed in 1991, includes $1,535$ fundamental stars of magnitude 7 and brighter, some of which have been observed for more then two centuries, with an average accuracy in position and

---

[1] In contrast to the annual parallax, the diurnal parallax has the Earth radius as a baseline. Nearby objects, such as the Moon, change their sky position as a result of the diurnal rotation of the observing site on the ground. For distant objects, such as stars, the diurnal parallax is negligible.

20. Precise measurements in space

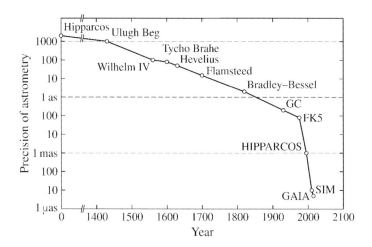

Figure 20.3. The increase in astrometric accuracy from antiquity to the space age. HIPPARCOS, a Greek astronomer lived in the 2nd century BC, had a catalogue of about 850 stars; his discovery of the general precession was based on the recognition that a star had increased its longitude by at least 1° relative to an observation made 150 years earlier. It is remarkable that the use of telescopes, began in the seventeen century, did not have a major impact on astrometric accuracy, although it greatly increased the number of sources and their physical understanding; the real revolution is due to space instruments (four orders of magnitude in ≃ 20 years, against a similar improvement in the 2,000 years before). Adapted from F. Mignard and S. Roeser, in *Gaia: A European Space Project*, O. Bienaymé and C. Turon, eds., EAS Publication Series **2** (2002).

mean motion of, respectively, ≈ 0.03 ″ and ≈ 0.1 ″/cy. Other 3,117 new stars, with magnitudes up to 9.5 magnitude have been added with slightly worse accuracies (some ≈ 0.06 ″ and ≈ 0.3 ″/cy).

Taking advantage of absence of gravity, almost constant thermal environment, absence of a refracting atmosphere, full-sky visibility and global coverage, *space astrometry* has improved this accuracy by several orders of magnitude and ushered a continuing revolution in the whole of astronomy (Fig. 20.3). Note that a substantial increase in astrometric accuracy implies a huge increase in computational complexity: for an isotropic distribution a factor 10 in accuracy increases by a factor 1,000 the number of measured objects.

The astrometric satellite HIPPARCOS (a name chosen as a tribute to the discoverer of lunisolar precession and an acronym for HIgh-precision Positions and PARallaxes COllecting Satellite) of the European Space Agency (ESA) has attained the mas accuracy level for more than 100,000 sources. It was the first large space astrometry mission and had an immense impact on astrophysics and galactic dynamics; its successful operation lasted from the launch in 1989 until 1993. Other astrometric missions with similar operating principles are planned and expected to reach the μas accuracy (Table 20.1). Here we concentrate on

Table 20.1. The main projects of space astrometry: past (HIPPARCOS) and future. The average accuracy in angular position for objects up to a given visual magnitude are given in the third and fourth columns; the last column gives the limiting visual magnitude.

Mission	Launch	Number of stars	Accuracy (in mas)	at mag.	Limiting magnitude
HIPPARCOS (ESA)	1989	1.2×10^5	1	9	12
SIM (NASA)	2009	$\approx 10^4$	0.004	13	20
Gaia (ESA)	2012	$\approx 10^9$	0.010	15	20

HIPPARCOS and the next ESA mission Gaia, also to show how a successful experimental concept can be much improved. Space interferometry, whose main application is the search for extraterrestrial planets, is briefly discussed in Sec. 16.5. Important observations are also carried out with the Fine Guidance Sensor of the Hubble Space Telescope, with about 10 mas accuracy in the astrometric mode.

HIPPARCOS vs Gaia optical systems. In order to measure the angular distance between stars, rather than an ordinary telescope pointed to a single stellar field, we need a scanning instrument. A telescope, with focal distance[2] f ($= 140$ cm, $5,000$ cm) and width d ($= 29$ cm, 140 cm) is placed in the (equatorial) plane orthogonal to the axis of the spacecraft, which spins with an angular velocity s ($= 169''/\text{s}, 60''/\text{s}$) and scans a great circle in the sky. As discussed in Sec. 19.1, the light profile orthogonal to **s** of a point source, of approximate angular width λ/d ($= 400$ mas, 100 mas), produces in the focal plane an image of width $f\lambda/d$ and moves (in the x-direction, say) with the velocity sf ($= 0.12$ cm/s, 1.5 cm/s). The effective wavelength $\lambda \simeq 0.7$ μm can be used to estimate the width. This decrease in width by the factor $400/100 \simeq 4$ is achieved by replacing each of HIPPARCOS' circular telescopes (of diameter 29 cm and area $A = 350$ cm^2) with a large rectangular mirror, 0.5 m $\times 1.4$ m, with area $A = 7,000$ cm^2; the dimension $d = 1.4$ m orthogonal to the axis controls the diffraction along the spacecraft equator.

In order to scan the whole sky, the spin axis moves in such a way that the Sun – whose stray radiation, in spite of the bafflers, would blind the delicate optical systems – is at all times sufficiently far from both telescope axes. This motion is a precession on a cone with axis directed towards the Sun and an aperture – called Solar Aspect Angle – of ($= 43°, 50°$), and results in an epicyclic pattern of the spin axis around the ecliptic. The orbit of Gaia, oscillating near the L_2

---

[2] Here and below the first number refers to HIPPARCOS, the second to Gaia.

20. Precise measurements in space

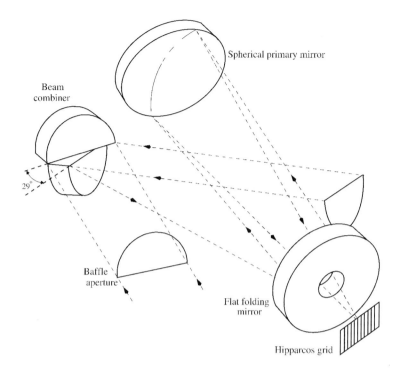

Figure 20.4. Configuration of HIPPARCOS' Schmidt optical system. The light from two directions at the angle of $\Gamma = 58°$ is combined by the complex mirror (beam combiner) and focused by the primary mirror on the modulating grid in its focal plane. Figure kindly provided by M.A.C. Perryman.

equilibrium point relative to the Sun and the Earth (Sec. 13.2), is carefully chosen to avoid the Earth shadow, which helps the stability of the thermal environment.

According to the project specifications, the angular position of a source on a great circle must be determined with an error σ_δ (= 1 mas, 10 µas; Table 20.1), (400, 10,000) times smaller than the diffraction limit; this is the gist of the problem. The required huge increase in accuracy is obtained by measuring each intensity profile a large number of times: in the focal plane many sensitive elements arranged along the x-direction are used; and during the mission each star is imaged many times. The (diffractive) shape of the profile depends on the source spectrum and the bandwidth $\Delta\lambda$ (= 200 nm, 500 nm) of the instrument, but is determined by the ray paths and is stable. In both missions a spectrometer automatically provides the spectrum (and the radial velocity) of each source. In HIPPARCOS an array of 2,688 alternately opaque and transparent bands (with size of order λ/d) produce, as the image moves across the slits, a comparable number N_{pixel} of diffraction profiles. They were detected

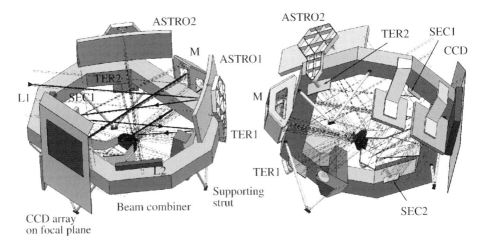

Figure 20.5. The concept for Gaia's astrometric optical bench. The photons collected by the two main mirrors ASTRO1 and ASTRO2 – $\Gamma = 106°$ apart – are brought onto the beam combiner (by the secondary SEC and the tertiary TER mirrors) which, after a reflection on the beam combiner and the mirror M, sends them on the focal plane for detection by the CCD array. For simplicity, only rays of the first beam are pointed out in bold: they are reflected on the secondary SEC1 and the tertiary TER1 below ASTRO1. Opposite the main mirrors are the two main components L_1, L_2 of the laser metrological system to measure Γ; the three supporting struts are also shown. Figures kindly provided by F. Mignard.

with a single photomultiplier, rapidly switched from one source to another with an image dissector tube. Gaia's astrometric instrument will use instead an array of 150 advanced CCDs (Sec. 20.1) in the focal plane, each with 4,500 pixels in the x-direction; they work in the time-delay integration mode, in which the signal of a single pixel is added coherently to those of the other pixels with the delay corresponding to the spin velocity s. In both missions, this effectively leads to a measurement of the time at which a source crosses a reference point on the focal plane. In addition, during the whole mission each star is observed a large number N_{scan} of times. Hence Gaia has three major improvements over HIPPARCOS: (i) the larger optics provides both a narrower x-diffraction pattern and a larger collecting area (by a factor 20); (ii) the use of CCDs, rather than slits and a photomultiplier, provides a better efficiency, larger bandwidth and higher resolution in the focal plane (moreover, the slower rotation rate allows a longer integration time for a single object, typically 2,000 s); (iii) the simultaneous observation of all objects in the focal plane, without partitioning a single photomultiplier to many objects, improves the photon statistics and the coverage.

In fact, both missions use *two telescopes* in the equatorial plane, separated by a large and (ideally) fixed angle Γ (= 58°, 106°). The photons from the stars in their fields of view are directed onto the same focal plane by a complex set of mirrors (Fig. 20.4 and 20.5) and their profiles are independently reconstructed. The simultaneous use of two, widely spaced viewing directions is essential to measure the *absolute parallax* π. The angular distance between two nearby stars with parallaxes π' and π'' suffers a correction $\phi(\pi'' - \pi')$, where ϕ is a parallactic trigonometrical factor which changes in time with the period of a year. The zero of π corresponds to the most distant objects, but there is no reason to attribute it to a given source: only relative parallaxes are measured. But if the two stars are widely separated and observed in two different fields of view, their parallactic factors ϕ_1 and ϕ_2 are different and the correction to the angular distance is $\phi_2\pi'' - \phi_1\pi'$. In the course of time ϕ_1 and ϕ_2 change, and so does the association of a star with another, enabling independent measurements of π' and π''.

In Gaia, µas astrometry places stringent requirements on the rigidity of the optical system: in particular, unknown relative deformations at the rotation timescale should not exceed 5 parts in 10^{12}. Since this level of rigidity is impossible to attain, e.g. due to temperature fluctuations (the passive thermal control system of Gaia mission should measure these fluctuations with an accuracy of $\simeq 10\,\mu K$), a special metrological system based upon laser interferometry is used to monitor changes in the fundamental angle Γ, especially at timescales shorter than the spin period of 6 h.

In both missions each source must be detected and approximately positioned before the astrometric observation; in HIPPARCOS this is done with two star mappers, each consisting in a set of diagonally distributed slits, capable of measuring both coordinates in the focal plane. This has provided an additional and independent catalogue – named after Tycho – which is less accurate, but with ten times more objects, with photometric data (Table 20.2 and Fig. 20.6). In Gaia a special CCD mapping set precedes the main astrometric field.

Mission design from the performance requirements.– The design of space mission for precise experiments is an integrated whole, with little choice left and severe constraints on all the main instrumental specifications, generally at the limit. This is clearly shown by the overall analysis of Gaia.

The data analysis and the reconstruction of the whole sky map using the disposition of the sources on each great circle, especially in Gaia, is a huge, complex and difficult task. It is estimated that Gaia will generate 2×10^{13} bytes of data. The angular speed s, the attitude of the spacecraft and the fundamental angle Γ is included in the fit: the measurements of angular distances δ between sources are all connected and self-calibrated; in particular, the effect of long-term fluctuations in the spin can be eliminated. The calibration error σ_{cal} in δ is due to several factors, including high frequency rotation irregularities, lack of rigidity of the optical bench, achromaticity, changes in Γ, etc. In HIPPARCOS, for not too weak sources, $\sigma_{cal} \approx 0.1''$ is the main contribution to the error budget.

But the shot noise, resulting from the finite number of photons used, is the ultimate and fundamental error source. Since the photon energy is discontinuous, they follow Poisson statistics and in detecting N photons the average error is

$$\sigma_N = \sqrt{N} \,. \tag{20.5}$$

In HIPPARCOS, if the errors in all the $N' = N_{\text{pixel}} N_{\text{scan}}$ measurements of the diffraction profiles of a given pair of sources are uncorrelated, the calibration error is reduced by the factor $1/\sqrt{N'}$. Taking $N_{\text{pixel}} = 1,000$ (only a fraction of the bands are used for a given source), with $N_{\text{scan}} = 80$, we see how the mas level could be attained.

To see how this constraints parameters of the instrument, we suggest Problem 20.12, which gives the photon noise error in the timing of a single pulse of *known* power profile $P(t)$ and approximate duration $\tau = \lambda/(sd)$; in a more accurate calculation,

$$\frac{\sigma_{\text{timing}}^2}{\tau^2} = \frac{C}{N_p} \,, \tag{20.6}$$

where $C \approx 1/\pi$ is a factor depending on the geometry; $N_p = FA\tau/(h\nu)$ is the number of detected photons in the pulse for an energy flux F through an area A. In a space astrometric instrument each profile is processed a large number N' of times, spread in each transit and in different transits during the whole mission; the error is further reduced by the factor $\sqrt{N'}$, so that the final photon noise depends only on the total number $N_{\text{total}} = N_p N'$ of processed photons for a given source. Translating times into angles, the previous equation gives

$$\sigma_\delta = \frac{C\lambda}{d\sqrt{N_{\text{total}}}} \,. \tag{20.7}$$

This gives the increase in angular accuracy over the gross estimate λ/d; Gaia needs at least $N_{\text{total}} \simeq 30,000^2 \simeq 10^9$ photons per source. This number

$$N_{\text{total}} = \frac{FAT_s}{h\nu} \tag{20.8}$$

depends on the total observation time T_s of the source. For example, with $A = 7,000$ cm^2, a 15 magnitude star needs an observing time $T_s = 140$ s. This time can be statistically estimated as

$$T_s = \frac{\Omega}{4\pi} T \,: \tag{20.9}$$

assuming an isotropic sky coverage, an average source is in view for a fraction of the duration $T = 5$ y of the mission equal to the ratio of the angular area Ω of the field of view to the whole sphere. In the above example Ω must at least be $4\pi T_s/T = 10^{-4} = (0.6°)^2$. On the other hand, the field of view of

	Hipparcos catalogue	Tycho catalogue
Number of stars	118,218	1,058,332
Limiting magnitude mag	12.4	12.2
Positional accuracy	≈ 1 mas	≈ 10 mas
Annual proper motion	≈ 1 mas/y	
Photometric precision	≈ 0.001 mag	≈ 0.02 mag

Table 20.2 The main results of the HIPPARCOS mission (see also Fig. 20.6); positional and proper motion mean accuracies are given for objects brighter than magnitude 9.

each telescope is $\Omega = A/f^2$. A is constrained by the payload dimension; but f cannot be too small and is limited below by the physical size d_p of the pixel, essentially fixed by the manufacturing technology. The scale $f\lambda/d = 5 \times 10^{-7} f$ cm of the variations in intensity in the focal plane cannot be smaller than the pixel size $d_p = 10\,\mu$m; this sets for the focal length f an upper limit $\approx d\,d_p/\lambda$ and contrains the collecting area $A = \Omega f^2$.

Science results: achieved and expected. HIPPARCOS' stellar catalogues provide the right ascension and declination of each object in the International Celestial Reference System (ICRS), as discussed in Sec. 3.2 (Fig. 20.6). The ICRS is based upon 608 extragalactic radio sources, with a subset of 212 "defining objects". However, they are optically too faint (except barely for the famous quasar 3C 273) to be observed by HIPPARCOS; but a small set of radio stars, specifically selected before the mission and added to the VLBI observation program, are included in the catalogue. This provides a link of the HIPPARCOS' frame to very distant matter, namely the ICRS. The agreement between the two frames has been realized to within ±0.6 mas in angular position and ±0.25 mas/y in angular velocity. Moreover, some 55 solar system objects, in particular asteroids, have been included in the main catalogue to investigate their dynamics and relate the dynamically defined planetary reference system to ICRS; however, lunar laser ranging provides a better determination of this system and, therefore, a better link. HIPPARCOS' precise parallaxes significantly contributed to advancements in astrophysics. As an example, the three-dimensional structure and the precise distance of the Hyades cluster – the nearest, moderately rich stellar cluster, with some 300 possible members – has been determined and found consistent with N-body simulations and the distance of 46.3 pc of its barycentre. The measurement of this quantity has provided a much better, absolute scale for astronomical distances which, through the Cepheids and other methods, has been extended to our Galaxy and beyond. This has led to a substantial improvement in the accuracy of absolute luminosities; the position of a sample of nearly 23,000 stars in the Hertzsprung-Russell diagram was determined (this sample excludes known variables or binary stars, and contains roughly an equal amount of stars of spectral classes B to K, with

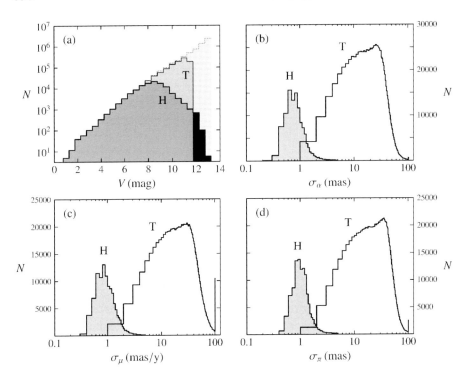

Figure 20.6. Statistical properties of data of the Hipparcos and Tycho catalogues: (a) number of entries as a function of the mean magnitude (compared to the complete distribution function, dotted line); number of objects with given standard errors in (b) right ascension (in mas), (c) mean motion in right ascension (in mas/y), and (d) parallax (in mas). The statistics for declination errors is very similar to (b). The histograms corresponding to the Hipparcos catalogue results are shaded and labelled H; those of the Tycho catalogue are labelled T. Adapted from http://astro.estec.esa.nl/Hipparcos/catalog.html.

only slight bias toward later types). Stellar evolution models have been improved and mean chemical abundances in the Galaxy were found. Obviously, we have now a better knowledge of local galactic dynamics, including fine structures like the Gould belt, an expanding system of young stars, dust and gas inclined by $\approx 20°$ to the Galactic plane. HIPPARCOS' data also allowed to nearly double the number of known stellar binary systems.

Gaia, with its superior performance, will provide results of exceptional importance for astrophysics and the physics of planetary systems. As for HIPPARCOS, its primary objective is to observe the physical characteristics, kinematics and distribution of stars over a large fraction of the volume of the Galaxy. A particular goal is to determine the relative rates of star formation in the stellar populations of the Galaxy, thus probing current galaxy formation models. The dynamical structure of a number of distant globular clusters will be resolved, improving the determination of the total mass of the Galaxy, including the dark matter component. By covering a volume much larger than HIPPARCOS, Gaia should provide trigonometric distances for *all* types of stars (and stellar populations), even those in the most rapid evolutionary phases. Thus all parts of the Hertzsprung-Russell diagram will be comprehensively calibrated, with an immense impact on stellar structure and evolution models. A large number of variable stars will be discovered, including Cepheids; the simultaneous knowledge of their distance will allow an unprecedented calibration of their absolute luminosity, the fundamental tool for distance determination up to the cosmological scale. Most Jupiter-size planets within 50 pc, estimated between 10,000 and 50,000, will be discovered. The astrometric detection of extrasolar planets will suitably complement radial velocity measurements (Sec. 16.5). In the solar system, $10^5 - 10^6$ new small bodies, most of them with an accurate orbits, will be detected, including a large number of near-Earth asteroids. Contrary to HIPPARCOS, which has used a limited number of radio stars linked to quasars by VLBI, Gaia will realize the ICRS *directly*; with its large limiting magnitude it will be able to observe most defining extragalactic radio sources. The accuracy of Gaia's determination of the ICRS is expected to be $\simeq 60\,\mu$as in the position of the pole (two orders of magnitude better than the current realization), and $0.4\,\mu$as/y in angular velocity. It is interesting to note that a rotation of the whole Universe with respect to the local dynamical frame at an angular velocity equal to the Hubble constant $H = 10^{-10}$ y^{-1} corresponds to $14\,\mu$as/y; the measurement of the global rotational motion of distant quasars will have an error smaller than the characteristic frequency of the Universe and, therefore, a deep Machian significance. Finally, theories of gravitation will be tested with great improvements over present accuracies. The gravitational deflection of light rays by the Sun (Secs. 17.6 and 20.4) is of order $2 \times 10^{-8} \simeq 4$ mas for a generic impact parameter ≈ 1 AU (which is the case for the majority of the stars in Gaia's catalogue; note that objects at an angular distance from the Sun $\gtrsim 35°$ cannot be observed). The bare signal-to-noise ratio of ≈ 400 will be much improved by the large statistics; simulations have given for the γ parameter, which controls this effect, an accuracy of about 10^{-6}.

20.3 Laser tracking

With the availability of very short laser pulses the transit-time method to measure distances discussed in Sec. 3.1 has reached an extraordinary accuracy. It is now currently used to determine the distance of the Moon and of Earth satellites, in particular two LAGEOS (LAser GEOdynamic Satellite; Fig. 20.7), upon which we concentrate. All targets are equipped with *retroreflectors*, special optical systems capable of reflecting back the light in the same direction (Problem 20.11).

Weight of moving part	8 t
Diameter of telescope	150 cm
Pointing accuracy	1″
Wavelengths (2nd/3rd harmonic)	532/355 nm
Pulse energy	100 mJ
Duty cycle	10 Hz
Duration of emitted pulse	40 ps
Duration of received pulse (for LAGEOS)	$\approx$ 300 ps
Single-shot range precision	$\leq$ 0.5 cm

Table 20.4 The main specifications of the new laser ranging telescope at the Matera Observatory of the Italian Space Agency. Transmission and reception use the same telescope. The duty cycle determines the limiting frequency with which pulses can be sent.

Moon typically ten minutes) are statistically combined, resulting in an overall error of order $\approx \sigma_p / \sqrt{N_{\text{shots}}}$. After this preprocessing stage, data are provided as a time series of "normal points" $D(t_i)$, embodying the information contained in all single shots within an integration time around t_i.

Scientific goals and achievements.– A large network of stations (about 50, mostly in the northern hemisphere; as an example, see Table 20.4) has allowed global reduction of ranging data for geodynamical satellites like LAGEOS. The laser ranging observable depends on a large number of parameters, including:

- the position $\mathbf{R}_s$ and velocity $\mathbf{V}_s$ of the observing stations in the Conventional Terrestrial Reference System;

- the initial position $\mathbf{r}(0)$ and velocity $\mathbf{v}(0)$ of the satellite;

- parameters related to the Earth pole and LOD;

- geopotential parameters up to high degree ($\ell \approx 70$), in particular the monopole term GM and the quadrupole parameter J_2;

- parameters characterizing the Earth tidal field, which contribute periodic components to the geopotential, response to the influence of the Sun and the Moon (Sec. 4.4);

- atmospheric refraction; etc.

The determination of distances between stations with subcentimetre accuracy has been achieved in mid 1990s. Such data have a major role in the kinematics of plate tectonics, with velocities of a few cm/y (see Sec. 5.4). With a nearly 30 y history, LAGEOS data play a key role in determining the parameters of long-period tidal components (presently not reachable with other satellite observations); of particular interest are the tide components related to the slow evolution of the lunar orbit, characterized by the period of its nodal drift

20. Precise measurements in space

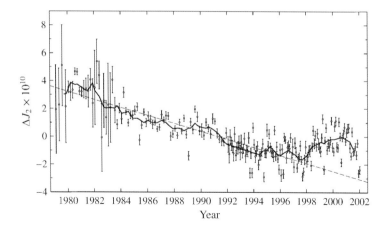

Figure 20.8. Difference ΔJ_2 between the measured J_2 coefficient and its long-term mean. The data and their formal uncertainty are shown; the result of elimination of random and short-period variations with a low-pass filter is shown by the solid curve. The modelled atmospheric contribution and a yearly term have been subtracted. The data have been obtained with precise orbit determination of geodynamical satellites (in particular, LAGEOS 1 and 2) and altimetric satellites. The linear decrease $\simeq -3 \times 10^{-11}$/y (dashed line) matches well the data till 1998; an unknown phenomenon apparently has intervened afterwards. Data kindly provided by C. Cox and B.F. Chao; see also C. Cox and B.F. Chao, *Science* **297**, 831 (2002).

($\simeq 18.6$ y). The angular velocity of the node of LAGEOS and other geodynamical satellites, proportional to the quadrupole coefficient $J_2 = (C - A)/M_\oplus R_\oplus^2$ (eq. (2.31)), has been found to decrease at the rate $dJ_2/dt \simeq -3 \times 10^{-11}$/y (Fig. 20.8). This is due to the slow rebound of the polar caps after the end of last glaciation. The melting of ice spread out mass from the poles, diminishing the difference $(C - A)$ between the moments of inertia; but the partially rigid crust responds to the changed load with a delay (Problem 20.8). Recently, however, experimental evidence has been provided that around 1998 J_2 started increasing (Fig. 20.8); the origin of this behaviour is unclear. The drift of the nodes of the two LAGEOS has also been used to detect the Lense-Thirring effect (Sec. 20.4).

However, even such complex orbital models are not able to adequately fit the very precise data, especially over a long time; a typical rms difference between LAGEOS data and a solution may grow to a few meters, significantly larger than the measurement accuracy. This problem is usually skirted by separately analyzing orbital arcs of limited duration (days to a few weeks), fitting the data for each of them to the initial conditions and several empirical parameters, presumably related to unmodelled non-gravitational forces. Only at the expense of such approach, in which the causal connection between different arcs is not taken into account, residuals drop to the level of the range error.

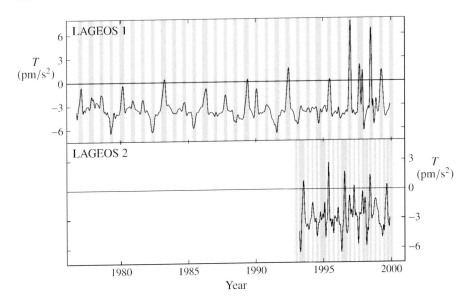

Figure 20.9. The non-gravitational along-track acceleration T of the LAGEOS satellites, with a time resolution of 3 days (data kindly provided by J.C. Ries and R.J. Eanes). The shaded intervals mark the periods when the satellites cross the Earth shadow (they occur more frequently for LAGEOS 2 due to its smaller inclination to the equator). In the case of LAGEOS 1, the amplification of the signal in the late 90s is due to an anomalous motion of the satellite's spin axis.

The time series of the empirical parameters determined for each arc and the related orbital discontinuities are available for all geodynamical satellites and present interesting scientific problems.

A particularly important parameter is the along-track acceleration T, which (eq. (12.4)) changes the semimajor axis at the rate $2T/n$ and, by Kepler's third law, in a time t displaces the spacecraft along the orbit by[4] $3T t^2/2$. The semimajor axes of the LAGEOS satellites undergo long-period irregular changes, with a mean decay of ≈ 1 mm/day, corresponding to a drag $T \approx -3.4 \times 10^{-10}$ cm/s^2 (Fig. 20.9). The full explanation of this non-gravitational force is still unclear and is a major hindrance for long-term gravitational experiments with LAGEOS. Radiation recoil due to a temperature anisotropy and its variations is likely to contribute in two ways: (i) the satellite cools and heats asymmetrically when it enters/leaves the Earth shadow, producing large

---

[4]Note, however, that the along-track displacement is not directly measured for the low-eccentricity orbits of LAGEOS satellites, but only through the change in semimajor axis. A different situation may occur for high-eccentricity orbits of near-Earth asteroids, where observations – in particular radar ranging – may directly detect the quadratic along-track effect due to non-conservative forces, like the Yarkovsky effect; this has been first achieved for Golevka in 2003.

20. *Precise measurements in space* 663

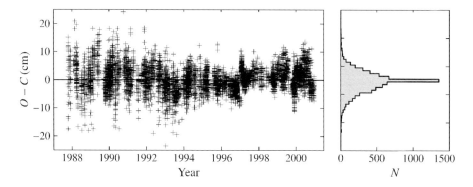

Figure 20.10. Comparison between the analytic lunar theory developed by J. Chapront, with corrections obtained by a numerical integration performed at the Jet Propulsion Laboratory (Pasadena), and the lunar laser ranging (LLR) "normal points". The theoretical model has only 34 free parameters. Left: difference between observed and computed values for $\approx 7,000$ LLR observations at CERGA, France; right: their distribution, with a standard deviation of ≈ 4.7 cm, reduced to ≈ 3 cm during the last 5 years. The precision of the observations is ≈ 2 cm (recently a little better). Data kindly provided by F. Mignard.

variations of T; (ii) the Earth infrared radiation is thermally re-emitted with a delay (for most part, due to the thermal relaxation of the retro-reflectors) and produces the main part of the drag. This is just the "seasonal" (referring to temperature gradients on the spacecraft along its meridians) variant of the Yarkovsky effect (Fig. 15.8), sometimes referred to as the *Rubincam effect*. In both cases the rotation of the satellite is relevant for its long-period orbital dynamics. The rotation state of LAGEOS 1 till the mid 1990s can be understood on the basis of two torques: the interaction between eddy currents in the satellite mantle induced by the Earth's magnetic field; the gravitational torque due to the Earth acting on the oblate spacecraft (similar to the lunisolar torque; Sec. 4.1). The magnetic torque is responsible for the slowing down, with e-folding rate of about 3 y. However, recently the anomalous behaviour of the orbital elements, in particular a, has indicated a more complex dynamics, with additional torques.

Lunar laser tracking. Laser ranging to the Moon has been routinely done since 1969 using four retro-reflectors placed by the Apollo and the Lunakhod missions. Since the distance to the Moon is about 60 times the distance of LAGEOS, the fraction of the returned photons is $\approx 8 \times 10^{-8}$ less; averaging over more shots is essential to obtain a normal point. As noted above, lunar-ranging normal points take into account tens up to hundreds single shots (and receptions) during, typically, a 10 minutes interval. At the end of 2001 about 16,000 range values were available, mainly from the McDonald Observatory (Texas), CERGA (Grasse, France) and Haleakala (Hawaii) stations.

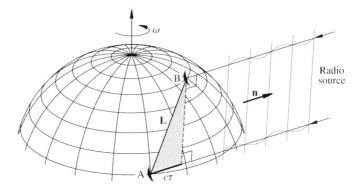

Figure 20.11. In VLBI two radio telescopes A and B, separated by the baseline **L**, measure the phase delay $\tau = \mathbf{L}\cdot\mathbf{n}/c$ for a source in the direction **n**. In the body-fixed frame $\mathbf{n} = (\mathbf{n}\cdot\hat{\omega})\,\hat{\omega}+\mathbf{n}_\perp$ is the sum of a constant component along the rotation axis and a part $\mathbf{n}_\perp$ which varies sinusoidally with the rotation period; accordingly, the delay $\tau = \tau_0 + \tau_1 \cos(\omega t + \phi)$ is the sum of a constant and a periodic contribution. Such data are highly sensitive to the stations' position (and their motion in space, including plate movements) and the parameters of the Earth rotation. Besides, if the signal from a distant radio source passes close to the Sun, the relevant deflection may be detected from the variation of **n** (Fig. 20.12).

Since about 1990 their internal precision increased to the level of ≈ 1 cm, but it is generally difficult to assess the systematic uncertainty – the accuracy – due to poorly modelled atmospheric phenomena, geometric effects related to lunar librations, etc. The realistic uncertainty of normal points is ≈ 2 cm. LLR depends on the motion of the centre of mass of the Moon relative to the Earth and the rotation state of both; their modelling is an extremely difficult task. The residuals – data less theoretical predictions – sensitively depend on the number M of adjusted parameters. With M up to ≈ 140, the Jet Propulsion Laboratory (JPL) lunar ephemerides can fit the data at their formal uncertainty level; with a smaller M the residuals get slightly larger (Fig. 20.10), an indication of the importance not to use more parameters than necessary ("over-parametrization"). The analysis of LLR data provide important information on the motion and the rotation dynamics of the Moon; it also helps in the determination of the Earth orientation parameters (polar motion and LOD) and significantly contributes to define the link between local dynamical reference system and the ICRS. The study of lunar librations has suggested a dense (7 g/cm^3, against the mean density of 3.34 g/cm^3) and molten core radius $\approx 250 - 400$ km, a conclusion later supported by the discovery of a weak lunar magnetic field by the Lunar Prospector mission. LLR data also led to an important test of the equivalence principle (the *Nordtvedt effect*; see Secs. 17.1, 20.4 and Problems 20.6 and 20.15) and a constraint on the time change of the gravitational "constant" G (see the next section).

A note on VLBI. The investigation of plate kinematics is currently carried out also with the Global Positioning System (GPS) of satellites (especially for stations not too far apart) and Very Long Baseline Interferometry (VLBI). The latter method uses two (or more) large radio antennas at far away points A and B, separated by a vector **L** and pointing to a radio source in the direction **n** (Fig. 20.11). Their signals are correlated with a sophisticated software and the phase delay between the arrival of a given wavefront at the two stations is determined. At the frequency ν a phase error of order unity implies a delay error $\sigma_\tau \approx 1/(2\pi\nu) = \lambda/(2\pi c)$. In the geophysical use of VLBI **L** is determined with an error $\sigma_L \approx \sigma_\tau c \approx \lambda$, quite relevant for plate tectonics. A displacement of the pole by an angle $\delta\theta$ changes the delay by $L\delta\theta/c$ and an accuracy $\sigma_\theta \approx \lambda/L$ is achievable. In the (main) radioastronomical use of this technique, the direction of the source (and hence its structure, if diffuse) is determined with an error $\sigma_n \approx c\sigma_\tau/L \approx \lambda/L \approx 10^{-9} \simeq 0.2$ mas for $\nu \simeq 30$ GHz and $L \simeq 10^4$ km. An order of magnitude may be gained with one of the antennæ in space, say on an eccentric orbit around the Earth. In Sec. 20.4, a VLBI measurement of the gravitational deflection is briefly described.

20.4 Testing relativity in space

As discussed in Ch. 17, Einstein's theory of General Relativity (GR) led to three main observational predictions and three corresponding experimental tests:

- the gravitational frequency shift, barely detected for the Sun (with a fractional value 2.12×10^{-6}) and in white dwarfs;

- the deflection of light rays by the gravitational field of the Sun, first detected in the 1919 solar eclipse;

- the anomalous advance of Mercury's perihelion, which solved a long-standing and sore problem in celestial mechanics, first discovered by U.J.J. LE VERRIER in 1849.

Although these tests have a poor accuracy, the scientific community, generally satisfied by the geometrical beauty of Einstein's theory, for a long time was not enough motivated to strive for a better confrontation between theory and experiments. In the 1960s the situation changed in two ways. Theoretically, it was pointed out that alternative theories of gravity are not only possible, but perhaps in better agreement with Mach's Principle. While the three classical tests are confined to a weak gravitational field, stronger, and potentially more interesting violations, can be expected in strong fields near some astrophysical objects (black holes and neutron stars; with the discovery of pulsars and quasars, in the mid 1960s, we got the first observational hints of their existence). Secondly, the immense improvement in instrumentation and the use

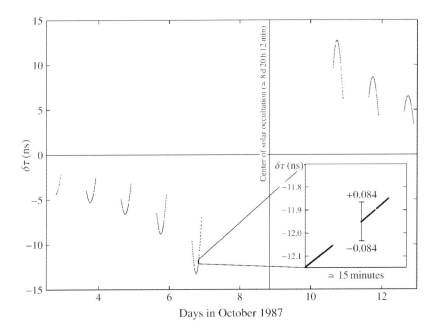

Figure 20.12. VLBI measurement of the gravitational deflection by the Sun obtained with the Westford and Owens Valley radio telescopes ($L = 3,900$ km far apart) in the US in October 1987, using two quasars (3C 273B and 3C 279), at that time both near the Sun. Two frequencies, 8 and 22 GHz, were used simultaneously, allowing the partial elimination of the contribution of the plasma in the solar corona and the ionosphere (see Sec. 19.4). Referring to Fig. 20.11, a change δ in the direction **n** of the source results in a change in the VLBI delay τ of order $L\delta/c$, with an unmodulated component τ_0 and a component τ_1 with the frequency of a day. Since the gravitational deflection changes sign between before and after the conjunction, so does the corresponding change $\delta\tau$. The plot shows the time intervals (≈ 6.5 h each day) during which data were taken; to avoid excessive plasma noise, observations were stopped on October 6 and resumed after the conjunction on October 10, when the impact parameter was $8 R_\odot$. The very large daily modulation, of amplitude $\approx L/c \approx 10$ ms is subtracted away and only the difference of the gravitational delays for the two quasars is shown. The signal is then an approximate diurnal sinusoid, with an amplitude modulation due to the temporal change of the impact parameter b. The error, of order ≈ 84 ps, is much smaller than the signal and it can be seen only in a zoom (right panel). See D.E. Lebach et al., *Phys. Rev. Lett.* **75**, 1439 (1995); we are grateful to D.E. Lebach for his help and for providing the data.

June 2002 with the radio system of the Cassini mission, on its way to Saturn. No violation of General Relativity has been found, with an error $\sigma_\gamma = 2 \times 10^{-5}$.

An intrinsic limitation to the accuracy of $(2 - \beta + 2\gamma)$, the controlling parameter of Mercury's perihelion drift[7], comes from our poor knowledge of the

---

[7] Note, that some near-Earth asteroids have the relativistic perihelion shift comparable to that of Mercury (mainly as a result of their large eccentricity). However, their observations are usually less precise – unless

quadrupole coefficient J_2 of the Sun; the total perihelion advance is

$$\delta\varpi = \frac{2-\beta+2\gamma}{3}\frac{6\pi m_\odot}{a(1-e^2)} + J_2\frac{3\pi R_\odot^2}{a^2(1-e^2)^2} \; ; \qquad (20.12)$$

see Secs. 12.6, eqs. (12.73b and c), and Sec. 17.6; in the J_2-term we neglected the small angle of about 3° between the normal to Mercury's orbital plane and the solar rotation axis. This gives the numerical value

$$\delta\varpi = 0.4298'' \left[\frac{2-\beta+2\gamma}{3} + 10^{-4}\left(\frac{J_2}{10^{-7}}\right)\right]/y \; . \qquad (20.12')$$

The fiducial value $J_2 = 10^{-7}$ corresponds to the assumption that the angular velocity of the Sun does not change with the distance from the axis (see Secs. 2.4 and 9.1); this is likely an underestimate, because during the early evolution of the solar nebula angular momentum was transferred from the outer layers of the protosun to the nebula (see Sec. 16.1). The detailed analysis of the rotational splitting of oscillatory eigenmodes of the Sun, similar to those discussed for the Earth in Sec. 5.2, favours a somewhat larger value ($J_2 \simeq 2 \times 10^{-7}$, say). In any case, an uncertainty δJ_2 produces a corresponding uncertainty in the relativistic combination $(2+2\gamma-\beta)/3$. An independent measurement of the solar quadrupole coefficient would be thus important not only for solar physics, but also for gravitation theory. A possibility would be to use accurate orbitography of Mercury, Venus, or a spacecraft with a small perihelion.

At these very high levels of accuracy the requirements on the calculation of the classical forces (both gravitational and non-gravitational) are very demanding; indeed, the main present limitation to relativistic experiments with ordinary spacecraft, especially for long-period effects, is the poor knowledge of non-gravitational forces. One way out is to use bodies with a very small area-to-mass ratio, like planets and natural satellites. An extensive set of measurements of distance to Venus and Mercury was carried out in the 1960s using high power radars at frequencies below 1 GHz. A possible caveat here is the fact that the return signal not only is affected by the orbital motion of the centre of gravity of the planet, but is also spread out in time and frequency by the shape, the topography and the rotation of the planet, requiring additional, unconstrained parameters. However, with a suitable modelling, the distance of the centre of gravity can be estimated and compared with the predictions, allowing a good determination of the perihelion advance. For Mercury, this led to an upper limit of about 0.001 for $(2\gamma + 2 - \beta)/3$. Non-gravitational forces on artificial satellites can be directly measured with on board accelerometers;

---

observed by radar – and non-gravitational forces like the Yarkovsky effect may cause degradation of the accuracy of their orbit determination.

this will be done, for example, in the BepiColombo mission to Mercury which the European Space Agency plans to fly after 2011.

The use of planetary landers with sophisticated radio systems improves the error budget in several ways. The surface topography does not play any role. Transponders can be used, with a significant improvement in the gain of the radio links (at a distance D the space loss is $\propto 1/D^2$, rather that $\propto 1/D^4$; Sec. 19.1); this allows higher frequencies, hence better range and Doppler measurements. The radio links with the two Viking landers on Mars used for the interplanetary leg the X-band (7.2 and 8.4 GHz) radio system on the orbiter, with a better elimination of the plasma contribution. From 1976 to 1982 nearly 1,150 measurements of distance have been carried out with the Vikings, with an error in distance of about 7 m, improving the determination of the relativistic parameters γ and β. It was first conjectured by P.A.M. DIRAC that, as the Universe expands, the gravitational "constant" G diminishes with time with a time scale of the order of the reciprocal $1/H \approx 13 \times 10^9$ y of Hubble constant: in this view gravitation gets weaker. The constancy of the adiabatic invariant (ratio of energy to frequency; see Sec. 10.2) then implies that the product Ga with the semimajor axis is not changed, corresponding, in 7 y, to an increase in a of 100 m for the Earth and 160 m for Mars. The Viking experiments set the limit

$$\frac{1}{H}\frac{d(\ln G)}{dt} = 0.002 \pm 0.004 \,, \qquad (20.13)$$

basically ruling out this conjecture.

Other constants of physics, in particular the fine structure constant $\alpha = 2\pi e^2/(hc)$, may undergo a cosmological change. Recently the spectroscopic analysis of the forest of absorption lines of quasars has suggested that at redshifts between 0.5 and 3.5, α was *smaller* than now and differed by $(\Delta\alpha/\alpha) = (-0.72\pm0.18)\times10^{-5}$. If this result is confirmed, clocks are not universal and the equivalence principle is violated; such finding would necessarily imply deep modifications in fundamental physics.

Relativistic effects near the Earth. For the Earth, with a gravitational radius of 0.44 cm, testing gravitational theories is more difficult; moreover, the small eccentricity of the two LAGEOS, the best-tracked satellites, makes the identification of their perigee difficult. A good deal of efforts has gone into measuring the gravitomagnetic force (Sec. 17.7) due to the angular momentum of the Earth; just like for two magnetic dipoles, it produces a torque on another spinning body, and a precession of 0.042 ″/y for a low orbiting satellite. This is the Lense-Thirring effect. Even larger is de Sitter precession due to the motion around the Earth (Sec. 17.7), ≈ 8 ″/y. A sophisticated spacecraft in a 650 km polar orbit – Gravity Probe B, to be flown soon – will use as a gyroscope a small superconductive sphere coated with niobium and supported by a magnetic cradle; its direction will be continuously monitored with respect to stars of the Hipparcos catalogue with an accuracy of ≈ 0.1 mas. Its precession will provide a precise test of the de Sitter effect, with an expected accuracy of ≈ 0.4 mas/y. The Lense-Thirring precession will also be detected.

The gravitomagnetic force exerts a torque also on the *orbital momentum* of a spacecraft and makes it precess with the angular velocity (17.67). This nodal motion (0.031 "/y for LAGEOS) is well within the current accuracy, but is masked by the uncertainty of the much larger motion ($\approx 125\,°$/y; eq. (12.68)) due to the oblateness J_2 of the Earth. The ratio of these two effects, $\approx 5 \times 10^{-8}$, should be compared with the present fractional uncertainty in J_2 of $\approx 10^{-7}$ (Table 2.3 and Fig. 20.8); the smaller advance due to J_4 also contributes to the masking. In spite of this, a way to measure the Lense-Thirring effect using data of both LAGEOS satellites has been recently found. Due to their different inclinations (Table 20.3), the precession of their nodes due to the Earth oblateness are different; moreover, LAGEOS 2 has a larger eccentricity and its perigee is better determined. Against these three observables we have three unknowns, namely small adjustments of J_2, J_4 and the Lense-Thirring coefficient. A careful fit confirmed the effect with a moderate accuracy, in agreement with Einstein's prediction that, besides the Newtonian scalar force $\propto 1/r^2$, gravitation has also a vector component $\propto 1/r^4$. Moreover, the geodynamic space mission GRACE (Sec. 2.3) promises to significantly reduce the uncertainty in the low-degree geopotential coefficients, thereby indirectly improving the accuracy of the Lense-Thirring measurement.

In the theory of General Relativity electromagnetic frequency transfer and clock synchronization are governed by eq. (17.43) and its particular cases (3.3) and (17.41). It should be noted that for an Earth satellite, whose potential and kinetic energies are of the same order, the corrections due to special relativity, $\approx O(v^2/c^2)$, and the gravitational corrections, $\approx O(U_\oplus/c^2)$, are of similar magnitude $\approx 10^{-9}$: both effects are aspects of a single phenomenon and their distinction depends on the frame of reference. In 1976 a hydrogen-maser frequency standard, with an accuracy of about a part in 10^{-14}, was flown for about 2 hours on a rocket up to an altitude of about 10,000 km. It controlled the frequency of a radio signal, whose Doppler shift was measured on the ground, as a function of time. In this way it was possible to test the relativistic Doppler shift (eq. (17.41)) with a very high precision. In quantitative terms, the formal "violation parameter" α, formally introduced in eq. (17.41) by $\Delta \nu / \nu_s = (1 + \alpha) \Delta U / c^2$, was constrained to be smaller than 2×10^{-4}.

In 1971 cesium clocks have been carried around the Earth on commercial airplanes in both the westward and the eastward directions and the elapsed proper times have been measured. This has provided a direct confirmation of the *twin paradox*, according to which, if a twin spends some time travelling at a high speed, upon his return home he will find his brother older. Mathematically, this is an obvious consequence of the fact that the proper time interval ds is not an exact differential and its integral $\int ds$ between two events depends on the chosen world line; but it is interesting to display this unintuitive effect. The experiment is similar to the *Sagnac effect*, which consists in the fact that two light beams going around on a rotating platform in opposite directions exhibit a phase shift. For airplanes flying along the equator at the same height the gravitational contribution is irrelevant; if they have the same speed v relative to the ground, the elapsed proper time is

$$\Delta s = \Delta t \sqrt{1 - \frac{(v \pm \omega_0 R_\oplus)^2}{c^2}} \simeq \Delta t \left(1 - \frac{v^2 \pm 2v\omega_0 R_\oplus + \omega^2 R_\oplus^2}{2c^2}\right). \quad (20.14)$$

$\Delta t = 2\pi R_\oplus/v$ is the elapsed coordinate time, equal for the two flights; ω_0 the angular velocity of the Earth. Of course, in an inertial frame the westward flight has a lower velocity and takes longer. The difference between the two proper times reads

$$\Delta_W s - \Delta_E s = 4 \frac{\omega_0 A_z}{c^2} \simeq 415 \text{ ns}. \quad (20.15)$$

Here $A_z = \int_{\text{path}} r^2 d\phi = \pi R_\oplus^2$ is the equatorial projection of the area enclosed by the path. The actual experiment was conducted at a higher latitude, with a smaller A_z, and gave the result 315 ± 32 ns.

This confirms that the synchronization of clocks on a large region is in principle impossible, since the elapsed time depends on the path used to "disseminate" it (this is true both for photons, as described above, and for slow motion of clocks; Problem 20.7). All one can do is to compare the rates of two of them with electromagnetic signals and use the relativistic theory. However, to a good approximation, this situation does not arise for clocks *at rest on the geoid*. Their distance from the Earth centre is $R_\oplus + \delta r$, where δr is given by eq. (2.50); at a colatitude θ their velocity is $v = \omega_0 R_\oplus \sin\theta$. According to eq. (17.26), their proper times differ by an irrelevant constant if the quantity $-2W = v^2 - 2U$ has the same value (here W is the total potential of the Earth gravitational field and the centrifugal forces in the rotating frame). Note that with the current accuracy and stability of atomic standards, fractional corrections of order $J_2 m_\oplus/R_\oplus \simeq 4 \times 10^{-13}$ must be taken into account in W. The common "beating" of atomic clocks on (or referred to) the geoid define the SI second, the fundamental realization of the time unit (see also Sec. 3.1). A more difficult situation may appear if we want to compare – and "synchronize" – *moving clocks* in the Earth vicinity. This is not a problem of an academic interest, but it naturally arises with GPS satellites; the comparison between their clocks and the SI second requires for each of them an adjustment due to their eccentricity and their motion relative to the geoid.

Lunar tests of general relativity.– Some relativistic effects have been tested with the motion of the Moon. LLR has verified the equivalence principle (EP; see Sec. 17.1): a difference between gravitational (M_g) and inertial (M_i) mass would have a small, but profound effect on the three-body problem[8]. Introducing the ratio $\eta = M_g/M_i - 1$, the relative motion between two nearby bodies in a gravitational potential U is described by eq. (17.4), the generalization of the tidal equation.

In the case of the Moon and the Earth (labelled with $_2$ and $_1$, respectively) the main contribution to their relative acceleration is due to the Sun:

$$(\eta_2 - \eta_1) \frac{GM_\odot}{R^3} \mathbf{R} ,$$

where $\mathbf{R}$ is the radius vector from the Sun to the Earth. This corresponds to a correction

$$\delta U = -(\eta_2 - \eta_1) \frac{GM_\odot}{R^3} \mathbf{R} \cdot \mathbf{r} \qquad (20.16)$$

to the gravitational potential per unit mass. It changes with time as the cosine of the *synodic longitude* of the Moon, the angle between $\mathbf{R}$ and $\mathbf{r}$, traditionally denoted with D. The effect of this perturbation on the distance is a sinusoidal function of D with an amplitude of the order of the ratio

$$\frac{\delta r}{r} \approx \frac{\delta U}{U} = (\eta_2 - \eta_1) \frac{M_\odot}{M_\oplus} \left(\frac{r}{R}\right)^2 \approx 2(\eta_2 - \eta_1) .$$

A range accuracy of about 1 cm corresponds to an uncertainty in $|\eta_2 - \eta_1|$ of order $\approx 10^{-11}$; but a more accurate analysis, including a subtle resonant amplification due to the solar tidal quadrupole $-\mathbf{r} \cdot \nabla\nabla U$ and the whole record of about 30 years, has shown that the sensitivity is, in fact, greater. As a result, an upper limit of about 5×10^{-13} has been achieved[9].

To understand the implications of this result, note that the EP follows from different independent assumptions. The mass M is the sum of several contributions due to the rest mass and all the binding energies; the validity of the EP requires that all these terms equally contribute to the gravitational and the inertial masses. Laboratory tests of the EP are relevant in relation to the microscopic interactions, in particular the nuclear force and the electrostatic attraction of electrons by nuclei. Only recently they attained a precision of 4×10^{-13}, slightly better than the lunar ranging data. But for large planetary bodies the contribution to the gravitational mass of the gravitational binding

---

[8]This has been noticed by I. NEWTON and later investigated in detail by P.S. LAPLACE already around 1800.
[9]It is interesting to note that this accuracy requires modelling the solar radiation pressure on the Earth and the Moon; the motion of the Moon is not quite "drag-free" (Problem 20.13).

energy E_B is important and the difference

$$\eta_2 - \eta_1 = \eta_g \left[\left(\frac{E_B}{M_i c^2} \right)_2 - \left(\frac{E_B}{M_i c^2} \right)_1 \right] \qquad (20.17)$$

can be measured. This is the *Nordtvedt effect* and η_g is the *Nordtvedt parameter*. For the Earth and the Moon, respectively, $E_B/(M_i c^2) = -4.6 \times 10^{-10}$ and -0.2×10^{-10}. The experimental result, mainly driven by the larger binding energy of the Earth, is $\eta_g = 0.0005 \pm 0.0011$, in agreement with Einstein's theory. The analysis of binary pulsars, with observations of smaller accuracy, but with a much larger value of $(E_B/Mc^2) \approx -0.01$, provided a similar constraint on η_g ($\simeq 0.001$).

Another relativistic test related to the Moon is the de Sitter precession. As discussed in Sec. 17.4, it consists in the fact that a local physical system able to keep a direction fixed (like a gyroscope) does not point to distant matter, but suffers the precession (17.40); for the Earth this is mainly determined by the Sun and corresponds to a rotation around the normal to the ecliptic with angular velocity $\Omega_{dS} \simeq 19.2$ mas/y. Lenz's vector for the lunar motion around the Earth, which points to the perigee, when all the (well known) Newtonian corrections are subtracted away, provides one such absolute direction; LLR has measured its change relative to the ICRS and has allowed a verification of the de Sitter precession with an accuracy of 0.5%. It should be noted that this is equivalent to referring the orientation of the perigee to distant matter, showing its fundamental significance.

Finally, LLR, with its long time series, has also improved the upper limit in the cosmological change of the gravitational constant G. The main source of error, inaccuracy in modelling of Earth tides, has been much reduced by the improvement in the long-term tidal parameters from LAGEOS and other artificial satellites.

20.5 Signals and data analysis

In space physics, with very large data sets, complex phenomena and difficult environmental conditions, particular care must be given to the extraction of signals from the often predominant noise. The subtle distinction between *precision* and *accuracy* should be kept in mind: in the former the error decreases with the number N of measurements, often like $1/\sqrt{N}$; in the second it is determined by basic factors like instrumental malfunctioning and poor modelling, which produce a "systematic" component of the error in addition to its random component.

A run of measurements of a physical quantity depends on its total duration T and the resolution time Δt. As Δt increases, fast fluctuations average out; generally at first the error decreases, but then systematic effects, like thermal drifts, may take over. This typical behaviour is observed, for example, in fluctuations of frequency standards (Fig. 3.1). In Fourier space, the frequency resolution Δf can never be smaller than $1/T$ – the uncertainty principle (19.20). If we search for a sinusoidal signal with the frequency f_0, the record must be appreciably longer than $1/f_0$, otherwise no frequency information is available. Consider a signal with intrinsic frequency width δf and a peak $S(f_0)$ in the spectrum at f_0. It can be realized with a damped harmonic oscillator with dissipation frequency $\nu = \delta f \ll f_0$, and a *quality factor* $Q = 2\pi f_0/\nu$. Its amplitude decreases by $1/e$ in Q periods. In presence of white noise with spectrum S_n, the signal-to-noise ratio (in short, SNR) is $\delta f\, S(f_0)/(\Delta f\, S_n)$; the best instrumental bandwidth is then $\Delta f \approx \delta f$. For records shorter than $1/\delta f$ the SNR is proportional to $1/T$, corresponding to an error in amplitude $\propto 1/\sqrt{T}$; one can say, with $N = f_0 T$ elementary measurements, each lasting a period, the statistical noise reduction holds. This is an aspect of *spectral estimation*. If the error in a quantity x is σ_x, the spectral density $S_n = \sigma_x^2/\Delta f$ determines the performance of the instrument; therefore it is appropriate to measure it not in terms of σ_x, but of $\sigma_x/\sqrt{\Delta f}$, a quantity with dimension $[x]/\sqrt{\mathrm{Hz}}$. This explicitly stresses the need, in discussing errors, of taking into account the duration T of the measurement in $\sigma_x = \sqrt{S_n/T}$.

A resolution time Δt prevents the detection of frequencies greater than about $1/\Delta t$. To see this more precisely, note that sampling a record $x(t)$ with Fourier transform $\tilde{x}(f)$ at the times $k\Delta t$, produces the discrete-valued function

$$x_s(t) = x(t) \sum_{k=-\infty}^{\infty} \delta(t - k\Delta t) \,. \tag{20.18}$$

Its Fourier transform (Problem 20.9)

$$\tilde{x}_s(f) = \frac{1}{\Delta t} \sum_{k=-\infty}^{\infty} \tilde{x}(f - k/\Delta t) \tag{20.19}$$

is a periodic function with period $1/\Delta t$. A signal with frequency cutoff f_c has Fourier components in $(-f_c, f_c)$. Sampling aliases the frequency $-f_c$ into $-f_c + 1/\Delta t$; to avoid confusion we must have $-f_c + 1/\Delta t > f_c$. Thus $f_N = 1/(2\Delta t)$ (the *Nyquist frequency*) is the largest uncorrupted frequency.

Random variables. The outcome of a measurement of a real quantity x is a random variable, whose probability distribution represents the distribution of different outputs in a very large set of identical repetitions. The corresponding averages are denoted by angular brackets, so that

$$\sigma_x^2 = \langle (x - \langle x \rangle)^2 \rangle \tag{20.20}$$

is the *variance* of x and σ_x is the *standard deviation*. To describe correlated measurements of two random variables x_1, x_2 (e.g., when excess in x_1 makes choice; one wishes to avoid a large value of M, which thins out the information and increases the computational load; but one should not choose too few, lest relevant physical effects are neglected. The basis for this delicate choice is the sensitivity of the observables with respect to the parameter variation, in relation to their error. A good understanding of the physical model is essential. It may also happen that the dynamical system cannot be simulated with a finite parameter model. For example, a spacecraft moving through the atmosphere is subject to small, but random and unpredictable forces; its position in phase space does not undergo a deterministic evolution, but, rather, a diffusion which eventually overcomes the precision. This case, however, demands special methods that will not be discussed.

The standard way to determine the parameters and their errors is the *least square fit*. We begin with the simple case in which a *linear* relationship

$$x_i^t = \sum_r A_{ir} p_r \qquad (20.29)$$

can be assumed between the theoretical values x_i^t of the observables x_i ($i, j = 1, 2, \ldots N$) and the unknown parameters p_r ($r, s = 1, 2, \ldots, M$). The measured values $x_i = \langle x_i \rangle + n_i$ include a noise component n_i with vanishing average and have a diagonal correlation matrix with variances $\sigma_{x_i}^2$ (eq. (20.20)); the observations are thus assumed uncorrelated but possibly with different uncertainties, such as when instruments of different quality are used. To estimate the parameters, consider the dimensionless quadratic form[10]

$$\chi^2 = \frac{1}{\sigma_{x_i}^2} \left(x_i - \sum_r A_{ir} p_r \right) \left(x_i - \sum_s A_{is} p_s \right) ; \qquad (20.30)$$

it gives the mean square of the difference between the observables and their theoretical values, weighted with their variances, so that more precise data contribute more[11]. Its minimum gives the best estimate $\hat{p}_r$ of the parameters and fulfils the condition

$$\frac{x_i A_{ir}}{\sigma_{x_i}^2} = \sum_s \frac{A_{ir} A_{is}}{\sigma_{x_i}^2} \hat{p}_s , \qquad (20.31)$$

---

[10] When the measured quantities are correlated, the positive definite quadratic form

$$\chi^2 = \left(x_i - \sum_r A_{ir} p_r \right) C_{ij}^{-1} \left(x_j - \sum_s A_{js} p_s \right)$$

should be used instead; C_{ij}^{-1} is the covariance matrix of the observations (eq. (20.21)).

[11] In this section the summation over repeated indices i, j, labelling the measurements, are understood.

20. Precise measurements in space

usually called *normal equations*. They are solved in terms of the inverse of the $M \times M$ normal matrix

$$B_{rs} = \frac{A_{ir}A_{is}}{\sigma_{x_i}^2} \tag{20.32}$$

and provide the required algorithm to obtain the *estimator* of the parameters in terms of the data:

$$\hat{p}_r = \sum_s B_{rs}^{-1} \frac{A_{is}x_i}{\sigma_{x_i}^2} . \tag{20.33}$$

Since the observables include noise, $\hat{p}_r$ is a random vector variable; its mean is equal to the true value p_r, so that it is an *unbiased estimator*[12]. One can show that the covariance of the parameters is given by

$$R_{rs} = \langle(\hat{p}_r - p_r)(\hat{p}_s - p_s)\rangle = B_{rs}^{-1} , \tag{20.34}$$

whence R is the *covariance matrix*. Note that no assumption is needed about the probability distribution of the noise n_i; in particular, it need not be Gaussian. The essential point is that it should have vanishing average, which means that there are no systematic measurements errors. If the coefficients of the matrix A have the same order of magnitude, the elements of the normal matrix B, being the sum of N terms, are generically proportional to N and the errors decrease as $\approx 1/\sqrt{N}$, as expected. For example, with only one parameter and equal variances, $x_i = A_i p + n_i$, $B = A_i A_i/\sigma_x^2$ is a scalar and $\sigma_p^2 = B^{-1} = \sigma_x^2/(A_i A_i)$. When the A_i are all equal, $\sigma_p \propto 1/\sqrt{N}$. However, when solving for strongly correlated parameters, degeneracies may occur; this typically prevents a successful (numerically stable) inversion of the normal matrix B.

Least squares fit of a non-linear model.– The real world is more complex than what the linear relation (20.29) implies. For example, for LAGEOS the observable distance $D = |\mathbf{r} - \mathbf{R}_s|$ (Sec. 20.3) depends in a non-linear and non-analytic way from the orbital and other parameters which determine the position $\mathbf{r}(t)$ of the spacecraft. However, usually (but not always) the a priori knowledge of the parameters is good enough to ensure that small corrections $\delta p_r = p_r - p_{r0}$ in the parameters from nominal values p_{r0}, and in the observables D_i

$$\delta D_i = \sum_r \left(\frac{\partial D}{\partial p_r}\right)_i \delta p_r = \sum_r A_{ir} \delta p_r \tag{20.35}$$

are sufficient. δD_i are the differences between the observed and computed quantities usually called "O–C" values. The $N \times M$ matrix A_{ir} is the *sensitivity*

---

[12] This standard procedure is not optimal when the noise is not Gaussian; but a complete information about its distribution function, beyond the mean and the covariance, is never available.

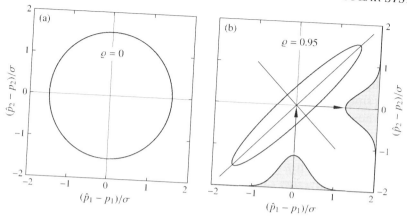

Figure 20.13. The 90% confidence level curve of the (Gaussian) conditional probability distribution (20.40) for two parameters p_1, p_2 in two different cases: on the left, zero correlation ($\varrho = 0$); on the right, 95% correlation ($\varrho = 0.95$). In the latter case, note that an a priori independent knowledge of one parameter, represented here by the narrow, shaded Gaussian distribution of p_1, appreciably constrains the other parameter.

The "single experiment" conditional probability $\Pi'(p|\hat{p})$ reproduces – apart from a normalization factor – the "Gibbsian" conditional probability. For a Gaussian distribution, normalization with respect to $\hat{p}$ implies normalization with respect to p; then *the true value has a Gaussian distribution around the estimator with standard deviation σ_p*.

It is often important to know not only the error $\sigma_p = \sqrt{R_{11}}$ in the measured parameter, but also how likely the true value differs from the estimator by less than a given number α of standard deviations σ_p. For a Gaussian distribution (20.37) we easily obtain for this probability the *error integral*

$$\eta(\alpha) = \sqrt{\frac{2}{\pi}} \int_0^{\alpha} du \, \exp(-u^2/2) = \text{Erf}(\alpha/\sqrt{2}) \,. \qquad (20.39)$$

For example: $\eta(1) = 0.683$, $\eta(2) = 0.954$, $\eta(\infty) = 1$. $\eta(\alpha)$ is the *level of confidence* with which the statement $|p - \hat{p}| < \alpha \sigma_p$ can be made and provides a quantitative way to assess the reliability of a result. It should be noted, however, that a high level of confidence is not a safe guarantee; in the history of physics it often happened that subsequent experiments obtained results quite outside earlier confidence intervals. Often the statement $p = \hat{p} \pm \sigma_p$ of the result of the measurement of a quantity p is given in terms of the formal standard deviation σ_p and does not imply an assessment of the systematic error.

Correlated parameters.– The case of two parameters gives an insight into the role of correlations. In the Gaussian approximation the conditional probability distribution reads

$$\Pi(\hat{p}_1, \hat{p}_2 | p_1, p_2) = \frac{\exp(-q^2)}{2\pi \sigma_1 \sigma_2 \sqrt{1 - \varrho^2}}, \quad (20.40)$$

where

$$q^2 = \frac{1}{2(1 - \varrho^2)} \left[\left(\frac{\hat{p}_1 - p_1}{\sigma_1}\right)^2 - 2\varrho \frac{\hat{p}_1 - p_1}{\sigma_1} \frac{\hat{p}_2 - p_2}{\sigma_2} + \left(\frac{\hat{p}_2 - p_2}{\sigma_2}\right)^2 \right]. \quad (20.41)$$

For any value of the constant C, consider the elliptical region $q^2 \leq C^2$ in the parameter plane (p_1, p_2). The probability that the point (p_1, p_2) lies in it is

$$\eta'(C) = 1 - \exp(-C^2); \quad (20.42)$$

for example at the 90% confidence level $\eta'(1.517) = 0.9$. If $\varrho = 0$ this region is an ellipse centred at $(\hat{p}_1, \hat{p}_2)$, with the semi-axes parallel to the coordinates and magnitudes $\sqrt{2}\,C\sigma_1$, $\sqrt{2}\,C\sigma_2$; the uncertainties of the two parameters are $\pm\sqrt{2}\,C\sigma_1, \pm\sqrt{2}\,C\sigma_2$. To see the effect of the correlation, consider the simple case $\sigma_1 = \sigma_2 = \sigma$; then the ellipse is oriented along the bisectors of the quadrants, with semi-axes $(C\sigma\sqrt{2(1 - \varrho)}, C\sigma\sqrt{2(1 + \varrho)})$ (Fig. 20.13); in the general case $\sigma_1 \neq \sigma_2$, the ellipse has a different orientation (Problem 20.10). When $|\varrho|$ approaches unity, the ellipse degenerates into a segment, making it inappropriate to speak of separate uncertainties; at $\varrho = 1$ or $\varrho = -1$ only the combination $(p_1 - p_2)$ or $(p_1 + p_2)$, respectively, enters in the probability distribution. This is the appropriate variable one really measures. Note that at a given confidence level both parameters keep their uncertainty $\pm \sqrt{2}C\sigma$, independently of the correlation ϱ. It is instructive to note that the 90% confidence level, say, corresponds to $\simeq 2.146\,\sigma$, while in the case of single parameter the uncertainty is $\simeq 1.645\,\sigma$ (see eq. (20.39)).

More generally, adding more parameters to the fit, with the aim of using a more complete model, increases their individual errors and is not always the best choice. On the other hand, a better accuracy is achieved when an independent (e.g., from another experiment) a priori constraint is taken into account. As an example, in a measurement of the advance of Mercury's perihelion, an accurate and independent knowledge of the solar quadrupole coefficient J_2 might allow a better determination of the relativistic parameter $(2 - \beta + 2\gamma)$ (eq. 20.12′). With the very good knowledge of γ from the VLBI measurements of the light deflection (Sec. 20.4), this may, in turn, result in a good measurement of the post-Newtonian coefficient β.

Estimation of the observation uncertainty from χ^2.– In complex experiments the observables' errors are not easy to determine a priori and it is usually better to use the more direct and empirical method of the residuals. For simplicity, consider an experiment with N observables $x_i = A_i p + n_i$, linearly dependent on a single unknown parameter p; the noise n_i is assumed uncorrelated ($\langle n_i n_j \rangle = \sigma_x^2 \delta_{ij}$), but its variance σ_x^2 is unknown. With the real data x_i one constructs the mean square residual

$$\frac{\sigma_x^2}{N} \chi^2 = \frac{1}{N} \sum_i (x_i - A_i \hat{p})^2 = \frac{1}{N} \sum_{i,j} n_i n_j \left(\delta_{ij} - \frac{A_i A_j}{A^2} \right) \quad (20.43)$$

relative to the estimator

$$\hat{p} = \sum_i A_i x_i / A^2 = p + \sum_i A_i n_i / A^2 \quad (A^2 = \sum_i A_i A_i) \,.$$

This gives an empirical measure of the actual noise. Its mean is easily calculated if the probability distribution of the noise depends only on $\sum_i n_i^2$ (as in the Gaussian case), hence is not affected by a rotation; then we can always take one of the coordinates along the vector A_i and we get, for large N,

$$\frac{\sigma_x^2}{N} \langle \chi^2 \rangle = \sigma_x^2 \frac{N-1}{N} \simeq \sigma_x^2 \,. \quad (20.44)$$

This gives the variance $\sigma_p^2 = \sigma_x^2 / A^2$ in the parameter.

– PROBLEMS –

20.1. Using a parametric representation of the ellipse (20.41), prove eq. (20.42).

20.2. Show that the error in the coefficient c of the power law $y = c\, t^n$ that fits $N(\gg 1)$ independent and equally spaced observations y_i in the interval $(0, 1)$ is proportional to $(2n+1)/\sqrt{N}$. (Hint: In this and the problems 3 and 15, when the interval $\Delta t = 1/N$ between successive measurements is small, a summation $\sum_i$ can be replaced with the integral $\int dt/\Delta t$.)

20.3. Fit $a + b\sin(\omega t)$, with a and b unknown, to a set of independent and equally space data y_i. Evaluate the large correlation that arise when the length of the record is of the order of, or less than, $1/\omega$. This is an example of the *aliasing* problem.

20.4. Compute the round-trip light-time Δt to LAGEOS and the Moon as a function of the elevation e, when their orbit passes at the zenith.

20.5.* Following the previous problem, using the exponential model of the atmospheric density (Sec. 7.1), write the integral which expresses the refractive correction to $\Delta t(e)$ and write a numerical programme to compute it.

20.6.* Study the effect of a violation of the principle of equivalence on the motion of the Moon with Hill's method (Sec. 13.4). Neglect the eccentricities of the Earth and the Moon and the inclination of the Moon on the ecliptic plane and focus on the role of the quadrupole tidal term due to Sun.

20.7. In Sec. 20.4 we discussed the path dependence of clock synchronization by light signals in the rotating geocentric frame. Show that a similar dependence occurs also for synchronization by portable clocks slowly carried around the equator. (Hint: Evaluate the clock proper time along its path, neglecting its velocity with respect to the Earth.)

20.8.* Estimate the relaxation time τ_r of the Earth after the end of the last ice age, about 20 ky ago, from the measured value of $dJ_2/dt \simeq -3 \times 10^{-11}/y$ and the ice load on the polar caps (about 3×10^{22} g each). Assume that J_2 relaxes to the current value exponentially.

20.9. Obtain the spectrum of a sampled signal (eq. (20.19)).

20.10. Find the semimajor axes and the orientation of the uncertainty ellipse (20.41) in the general case ($\sigma_1 \neq \sigma_2$).

20.11.* The principle of a retro-reflector is illustrated by two semi-infinite reflecting planes joined orthogonally; it works only for rays perpendicular to the edge. For three dimensions, consider three infinite mirrors placed on the quadrants of a Cartesian coordinate system and show how three reflections reverse an impinging ray. A corner cube retro-reflector is a finite realization of this idea and is obtained from a glass cube by cutting off a piece determined by the plane through three vertices connected by the diagonals of three faces; the remaining surfaces are covered with a reflecting layer. Write a program to trace a ray for a refracting corner cube and explore the working range of impinging directions.

20.12.* Show that the accuracy σ_{t_0} with which we can determine the peak time t_0 of a flux of photons of frequency ν with known power profile $P(t - t_0)$ against the shot noise is independent of the sampling time Δt and reads

$$\sigma_{t_0}^2 = h\nu \frac{\int dt \, P\dot{P}^2}{\left(\int dt \, \dot{P}^2\right)^2}.$$

This shows the advantage of a fast varying profile. (Hint: To work with a linear least square fit, consider the correction $\delta P = -\delta t_0 \dot{P}$ and use (20.5) for the error in the number of photons N_i detected in the i-th time interval. For a sampling time $\Delta t \ll \tau_1$, the duration of the pulse, replace every summation with an integral ($\Delta t \sum_i \to \int dt$).)

20.13.* Estimate the solar radiation pressure on the Moon and show that the principal orbital effect is a synodic variation in the distance from the Earth, correlated with the signal due to a violation of the equivalence principle.

20.14. Using Fig. 14.5 obtain the relativistic perihelion drift of some near-Earth asteroids.

20.15.* Since the LLR observations (Fig. 20.10) are less frequent near new and full Moon, the data distribution has a significant component with the periodicity of twice the synodic longitude D; this produces a correlation between the EP signal $\propto \cos D$ and the solar quadrupole signal $\propto \cos(2D)$, with a negative effect on the accuracy of η_g. To understand this problem, fit a signal $\eta_g \cos D + S \cos(2D)$ with a dense distribution of N uncorrelated data over the interval (D_1, D_2), with density $d(D) = (N/(D_2 - D_1))[1 - f\cos(2D) - g\sin(2D)]$ and evaluate the correlation coefficient between η_g and S. For the case of interest, $g \simeq 0$ and f is a little smaller than one.

– FURTHER READINGS –

A good textbook on radar and SAR is C. Elachi, *Spaceborne Radar Remote Sensing: Applications and Techniques*, IEEE Press (1988); for planetary applications, see S.J. Ostro, Planetary radar astronomy, *Rev. Mod. Phys.* **65**, 1235 (1993). The first scientific results of the HIPPARCOS mission are available in electronic form at http://astro.estec.esa.nl/Hipparcos/venice.html; a special issue of *Astronomy and Astrophysics* (Vol. **304**(1), 1995) is also devoted to the early results; a more complete review is F. van Leeuwen, The Hipparcos mission, *Sp. Sci. Rev.* **81**, 201 (1997). On Gaia, visit the web site http://astro.estec.esa.nl/GAIA. On LAGEOS, see the special issue of the *J. Geophys. Res.* of Sept 30, 1985; the original result about the change in J_2 is described in C.F. Yoder et al., *Nature* **303**, 757 (1983). A thorough discussion of the theory and applications of radio interferometry is in O.J. Sovers and J.F. Fanselow, Astrometry and geodesy with radio interferometry: Experiments, models, results, *Rev. Mod. Phys.* **70**, 1393 (1998). Astronomical applications of optical interferometry (relevant also for the search of extrasolar planetary systems, Sec. 16.5) and the role of tropospheric turbulence are well discussed in the review by S.K. Saha, Modern optical astronomy: technology and impact of interferometry, *Rev. Mod. Phys.* **74**, 551 (2002). Relativistic experiments are described in detail in the book by C.M. Will (see Ch. 17); for a more recent update, see C.M. Will, The confrontation between general relativity and experiment, *Living Rev. Rel.* **4**, 1 (2001); see http://www.livingreviews.org/. On the Magellan exploration of Venus there are two issues of the *J. Geophys. Res.* (Aug 25 and Oct 25, 1992). A very useful source of practical procedures for data analysis, including numerical subroutines in C and Fortran languages, is: W.H. Press et al., *Numerical Recipes: The Art of Scientific Computing*, Cambridge University Press (1995), periodically upgraded at http://www.nr.com/.

Index

This index has been compiled automatically, mainly on the basis of italicized words in the text.

1992 QB$_1$, 405
1998 KY$_{26}$, 645
1998 WW$_{31}$, 406

2001 OE$_{84}$, 402

3C 273, 655, 668
3C 279, 668

51 Pegasi, 537

absolute enstrophy, 205
absolute magnitude, xix
absolute vorticity, 193
absorption coefficient, 180
abundances, 242, 245
accelerometer, 595
accretion, 531
  accretion disc, 355
accuracy, 664, 674
ACE, 247, 359
achondrites, 430, 432
acoustic waves, 12
action, 272
Adams-Williams equation, 124
adaptive optical systems, 638
addition theorem, 45
adiabatic invariant, 271–272, 276
adiabatic temperature gradient, 132, 171
adiabatic transformation, 7, 131
Adrastea, 438
Advanced Composition Explorer See ACE
aerobraking manœuvre, 596
affine parameter, 565
airglow, 213
Airy compensation mechanism, 55, 142
Ajisai, 658
albedo, 171, 173
  albedo effect, 608
ALFVÉN, H., 280
Alfvén waves, 257, 278, 290

Alfvén speed, 24, 248, 278
Alfvén's frozen flux theorem, 26
Allan deviation, 595, 634
Allan variance, 634
Allende carbonaceous chondrite, 431
altimetry, 594
Amalthea, 291, 438
Ampère's law, 21
amplitude modulation, 629
angular momentum drain, 402
anomalous cosmic rays, 258
anticyclonic flow, 158, 194
anti-greenhouse effect, 178
apparent forces, 612
apparent magnitude, xix
apsidal line, 301
Archimedes' spiral, 252
areal velocity, 299
argument of pericentre, 305
ARRHENIUS, S., 179
artificial satellites, Ch. 18, 587
ascending node, 305
asteroids, Sec. 14.3, 391
asteroids
  Amor asteroids, 398
  Apollo asteroids, 398
  asteroid dust bands, 426
  asteroid families, 395
  Aten asteroids, 398
  binary asteroids, 402, 425
  rotation of asteroids, 401
  taxonomic types, 403
  Trojan asteroids, 360
asthenosphere, 137
asymptotic flatness, 569
atmospheres, Ch. 7, 169
atmospheric drag, Sec. 12.3, 325
atmospheric drag, 608
atmospheric models, 329, 596
atmospheric refraction, Sec. 8.4, 225

687

Earth's oscillations, 124
Earth's radiation budget, 180
eccentric anomaly, 303
eccentricity, 301
eccentricity resonances, 482
ecliptic comets, 415
EDGEWORTH, K.E., 405
effective potential, 285, 293, 299, 347
EINSTEIN, A., 145
Einstein's radius, 544
Ekman's layer solution, 200
    Ekman's spiral, 200
electric induction, 22
electrical conductivity, 22
electromotive field, 160
Elsasser variables, 279
Elst-Pizzaro, 421
embedded satellites, 369
emission coefficient, 187
Enceladus, 389, 443–444
ENCKE, J.F., 433
Encke gap, 442
enstrophy, 203
enthalpy, 131, 282
entropy, 7, 131
EOP *See* Earth orientation parameters
Eos, 395, 426
epicyclic frequency, 446, 527, 553
Epimetheus, 368
ε Eridani, 510
Epstein's formula, 517
equation of state, Sec. 1.2, 6
equatorial electrojet, 220
equilateral equilibrium points, 349, 351, 360
equilibrium shape of satellites, Sec. 4.5, 108
equipartition temperature, 172
equivalence principle, Sec. 17.1, 558
equivalence principle, 64, 366, 559, 597, 673
ergodic problem, 337
ergodic theorem, 632
Eros, 58, 398, 454, 643
escape parameter, 234
escape velocity, 240, 309, 588
estimator, 679
Etalon, 506, 597, 658
EULER, L., 11, 83
Euler frequency, 83
Eulerian derivative, 9
Euler-Lagrange equations, 319
Euler's equations, 82
Eureka, 361
Europa, 166, 290, 388, 483, 643
eutectic point, 164
evolution of atmospheres, Sec. 8.5, 229
excess angle, 563
excess coefficient, 677
exobase, 233
exosphere, 213

experimental gravitation, 568, 666
exponential atmosphere, 327
external problem, 58
extrasolar planetary systems, Sec. 16.5, 537
extrasolar planetary systems, 527, 530

f- and g-functions, 311, 505
faculæ, 246
fading problem, 421
faint young Sun problem, 177
FAL-C model, 248
Faraday's effect, 224
Faraday's law, 21
fast rotation, 78
FERMI, E., 284, 566
Fermi acceleration mechanism, 284
Fermi coordinates, 565–566, 574
filter, 628
fireballs, 419, 424
FK5 catalogue, 648
FLAMSTEED, J., 67
flat Earth, 50
flat spacetime, 563
flywheels, 591
Forbush decreases, 285
forced eccentricity, 472
formation of giant planets, Sec. 16.4, 526
formation of planetesimals, Sec. 16.2, 515
formation of satellites, Sec. 16.4, 526
formation of terrestrial planets, Sec. 16.3, 521
frames of reference, Sec. 3.2, 67
Fraunhofer spectrum, 242–243
free eccentricity, 472
free energy, 131
free precession, Sec. 3.5, 82
free precession frequency, 83
free-floating planets, 539
frequency measurements, Sec. 19.4, 634
Fresnel optics, 226

Gaia, 359, 541, 650
gain, 622
Galatea, 444
Galilean group, 558
Galilean satellites, 388, 483, 535
GALILEI, G., i, 240, 433
Galileo, 57, 289, 402, 643
γ point, 64
Ganymede, 108, 165, 290, 388, 398, 483, 535
gaseous giants, 381, 526
gas-instability model, 527
Gaspra, 643
gauge transformations, 569
GAUSS, C.F., 148
Gauss perturbation equations, Sec. 12.1, 313
Gauss' theorem, 2
Gaussian curvature, 563
Gaussian distribution, 676

generation of planetary magnetic field, Sec. 6.4, 155
geodesics, 564
  geodesic deviation, 567
geodetic precession, 98
Geographos, 645
geoid, 12, 52–53, 578, 594, 672
geomagnetic jerks, 153
geostationary orbits, 598–599
geostrophic equation, 194
geostrophic flow, 192
  geostrophic turbulence, 198
  geostrophic wind, 195
geosynchronous orbit, 598
GFZ-1, 658
giant planets, 381, 526
Gibbs ensemble, 199
GILBERT, W., 145
Giotto, 419
glacial rebound, 661
glaciations, 175
Gladstone-Dale law, 228
Gliese 876, 538, 549
Gliese 710, 416
Global Oscillation Network Group *See* GONG project
Global Positioning System, 54, 595, 665
Golevka, 645, 662
GONG project, 126, 241
gossamer rings, 438
Gould belt, 656
GP-B *See* Gravity Probe B
GPS *See* Global Positioning System
GRACE, 51, 671
gradient flow approximation, 196
gravimeter, 47
gravitational binding energy, 4
gravitational constant, xvii, 670, 674
gravitational deflection *See* deflection
gravitational enhancement factor, 309
gravitational equilibrium, Sec. 1.1, 1
gravitational field, Ch. 2, 35
gravitational frequency shift, 665
gravitational gas, 603
gravitational lense
  Einstein's radius, 544
  gravitational microlensing, 543
gravitational mass, 558
gravitational radius, 5, 607
gravitational stress tensor, 34
gravitomagnetism, Sec. 17.7, 581
gravitomagnetism, 5, 670
gravity acceleration, 12
gravity anomaly, 46
gravity assisted manœuvre, 615
gravity field of planetary bodies, Sec. 2.4, 54
gravity field of the Earth, Sec. 2.3, 46
Gravity Probe B, 583, 670

Gravity Recovery and Climate Experiment *See* GRACE
gravity waves, 192
Great Dark Spot, 205
great inequality, 482
Great Red Spot, 204, 521, 643
Greeks, 360, 391
greenhouse effect, 178, 183, 190, 235
Greenwich, 47
grey atmosphere, 189
group velocity, 221
Grüneisein's parameter, 133
guiding centre, 269
gyromagnetic ratio, 270

Hadley cells, 204
HALLEY, E., 413
halo orbits, 295
Hamilton equations, 320
Hamilton flow, 342
Hamiltonian, 320
Hamilton-Jacobi equation, 321
hat-box window, 628
HD 209458, 542, 550
HD 82943, 538, 549
heat generation and flow, Sec. 5.3, 127
heat transport
  heat diffusivity, 17
  heat transport equation, 17
heliopause, 257
helioseismology, 126, 241
heliosphere, Sec. 9.3, 252
heliosphere, 252
  heliospheric current sheet, 255
  interstellar bow shock, 257
  termination shock, 257
heliosynchronous orbit, 599
Helmert's parameter, 5, 46, 74
hemispheric albedo, 171
heterosphere, 212
High Precision Parallax Collecting Satellite *See* HIPPARCOS
Hilda, 394
HILL, G.W., 364
Hill's problem, Sec. 13.4, 364
Hill's problem, 380
Hill's radius, 367
  See also Roche radius
Hill's sphere of influence, 617
HIPPARCOS, 70, 649
HIPPARCOS, 95, 649
HIRAYAMA, K., 395
Hoba meteorite, 424
HOHMANN, W., 591
Hohmann transfer, 591, 614
homosphere, 212
HOOK, R., 204

Hooke's law, 14
horseshoe orbits, 361
hot jupiters, 550
hot spots, 137, 141
HST *See* Hubble Space Telescope
Hubble constant, xviii, 70
Hubble Space Telescope, 186
HUYGENS, C., 433
Huygens probe, 186
Hyades cluster, 655
hydrogen maser, 64
Hyperion, 389, 474, 486

IAGA *See* International Association of Geomagnetism and Aeronomy
Iapetus, 390
ice giants, 381, 526
ICP *See* International Conventional Pole
ICRS *See* International Celestial Reference System
Ida, 402, 643
ignorable coordinate, 320
IGRF *See* International Geomagnetic Reference Field
IMAGE, 224
Imager for Magnetopause-to-Aurorae Global Exploration *See* IMAGE
impact parameter, 308
inclination, 305
inclination resonances, 482
indirect J_2 effect, 619
indirect perturbation, 335, 462
induced magnetosphere, 287
induction equation, 25
  generalized induction equation, 160
inertia tensor, 44
inertial frames, 68
inertial mass, 558
inertial range, 201
Infrared Astronomical Satellite *See* IRAS
insertion angle, 588
insolation, 171, 175
interferometry, 546, 595
internal energy, 131
internal problem, 58
internal structure of the Earth, Sec. 5.2, 120
International Association of Geomagnetism and Aeronomy, 150
International Atomic Time, 69
International Celestial Reference System, 655
International Conventional Pole, 72–73
International Geomagnetic Reference Field, 149–150
International Reference Ionosphere, 215
International Satellite Land-Surface Climatology Project *See* ISLSCP
International Sun-Earth Explorer, 267
International Telecommunication Union, 593
interplanetary material, Sec. 14.6, 422

interplanetary space, 422
interstellar
  bow shock, 257
  dust, 428
  medium, 257
inverse Bremsstrahlung, 216
inverted tides, 104
Io, i, 108, 165, 287, 292, 388, 483, 492, 643
  Io's torus, 289
ionization cross section, 216
ionopause, 287
ionosheath, 287
ionosphere, Sec. 8.2, 216
ionosphere, 214
  ionospheric scintillation, 216
IRAS, 426
IRI *See* International Reference Ionosphere
ISEE *See* International Sun-Earth Explorer
ISLSCP, 173
isochron, 431
isolation mass, 522
isostasy, 46, 55
isotopic composition, 229
ITU *See* International Telecommunication Union
Ivar, 398

J2000.0, 67
JACOBI, K.G.J., 81
Jacobi constant, 348
Jacobi ellipsoids, 81, 110
Jacobi identity, 320
Janus, 368
JEANS, J.H., 233, 551
Jeans depletion, 233
Jeans instability, 510, 551
  Jeans frequency, 125, 551
  Jeans mass, 552
Jeans' spheroids, 112
JOHNSON, J.B., 633
Johnson-Morgan-Cousins system, xix
Jupiter, 77, 162, 172, 185, 204, 232, 287, 289, 292, 371, 438, 481, 498, 504, 529, 643

K_a band, 593, 595, 624
K_p index, 266
Kaiser window, 628–629
KANT, I., 509
Karin cluster, 395–396, 427
Karoo crater, 453
Kaula's rule, 49, 57
Keeler gap, 442
Kelvin-Helmholtz
  instability, 199, 263
  mechanism, 283
Kelvin's vorticity theorem, 27
KEPLER, J., 297
Keplerian elements, 305
Keplerian orbits, Sec. 11.2, 300

Kepler's equation, 304
Kepler's second law, 299
Kepler's third law, 302
kinematic viscosity, 19
kinematical helicity, 161
kinetic theory, Sec. 1.8, 28
Kirchhoff's law, 188
KIRKWOOD, D., 394
Kirkwood gaps, 394, 474
Kleopatra, 645
Knudsen regime, 213, 325
KOLMOGOROV, A.N., 202
Kolmogorov's spectrum, 201, 257
Koronis, 395, 427
KOZAI, Y., 385
Kozai mechanism, 385
KREUTZ, H., 421
Kreutz group of comets, 421
KUIPER, G.P., 405
Kuiper belt, 534
    See also transneptunian objects
kurtosis, 677

La Hire, 372
LAGEOS, 597, 604, 658, 670
LAGRANGE, J.L., 336, 346
Lagrange bracket, 321
Lagrange expansion theorem, 312
Lagrange function, 319
Lagrange perturbation equations, Sec. 12.2, 319
Lagrange's theorem, 462
Lagrangian derivative, 9
Lagrangian points, Sec. 13.2, 349
Lagrangian points, 247, 352
Lambert's law, 190
Lamé's constants, 14, 116
Lane-Emden equation, 34
LAPLACE, P.S., i, 366, 434, 483, 509, 616, 673
Laplace's resonance, 388, 483
lapse rate, 171
Larmor radius, 269, 283
Laser Geodynamic Satellite *See* LAGEOS
Laser Interferometer Space Antenna *See* LISA
laser tracking, Sec. 20.3, 657
late heavy bombardment, 235, 425
late veneer scenario, 230
launch, Sec. 18.1, 587
LDEF *See* Long Duration Exposure Facility
LE VERRIER, U.J.J., 665
leap-frog integrator, 505–506
least square fit, 677–678
Legendre functions, 37
    normalization, 39
    quasi-normalization, 148
Legendre polynomials, 36
Lehmann discontinuity, 122
length of the day, 71–72
LENSE, J., 583

Lense-Thirring precession, 582, 670
Lenz vector, 301
level of confidence, 682
Lie series, 342
light cone, 561
limb darkening, 189
Lindblad resonances, 446, 533
line formation, 187
link budget
    one-way, 622
    two-way, 625
LIOUVILLE, J., 87
Liouville equation, 87
LISA, 547, 597
lithosphere, 131, 137
little ice age, 245
LLR *See* lunar laser ranging
local thermodynamical equilibrium, 30, 187
LOD *See* length of the day
Long Duration Exposure Facility, 423
longitude of pericentre, 306
Lorentz force, 503
Lorentz resonances, 445, 504
loss cone, 275
Love number, 77, 89, 102, 107, 490, 610
Love waves, 119
low-pass filter, 628
lumped geopotential coefficients, 599
lunar laser ranging, 659, 663, 673
Lunar Prospector, 55, 165
LUNDQUIST, S., 280
lunisolar precession, Sec. 4.1, 95
lunisolar precession, 95
Lyapunov exponents, 356, 364, 469

MacCullagh's formula, 44
MACH, E., 68
Mach's cone, 281
Maclaurin spheroids, 79, 81, 110
Magellan, 55, 183, 642, 646
magnetic anomalies, Sec. 6.2, 147
magnetic declination, 146
magnetic diffusion coefficient, 25, 156
magnetic harmonics, Sec. 6.2, 147
magnetic inclination, 146
magnetic induction, 22
magnetic moment, 271
magnetic permeability, 22
magnetic pressure, 24
magnetic reconnection, 246, 255, 263
magnetic reversals, Sec. 6.3, 152
magnetic Reynolds number, 157, 254, 264
magnetic sectors, 255
magnetic storm, 215, 265, 274
magnetic stress tensor, 24
magnetic substorm, 265
magnetoacoustic waves, 280
magneto-convective dynamo, 166

magnetohydrodynamic waves, Sec. 10.4, 278
magnetohydrodynamical engines, 614
magnetohydrodynamics, Sec. 1.6, 21
magnetopause, 263, 283
magnetosheath, 263
magnetosphere, Ch. 10, 261
magnetosphere, Sec. 10.1, 261
Magnetospheric Satellite *See* MAGSAT
magnetostratigraphy, 154
magnetotail, 263
MAGSAT, 150, 152
main asteroid belt, 391
Mariner 10, 164, 182, 289
Mars, 55, 98, 116, 164, 177, 184, 218, 236, 287, 449, 537, 596, 670
Mars Global Surveyor, 55, 165, 287, 596, 642
mascons, 55
mass conservation, 10
matching problem, 616
Mathilde, 644
Maunder minimum, 173, 245
MAXWELL, J.C., 434
Maxwell's equations, 21
McIlwain's coordinates, 147
mean anomaly, 304
mean longitude, 306
mean motion, 302
mean orbital elements, 461
mean-field electrodynamics, 159
Mercury, 56, 164, 182, 287, 289, 485, 597, 645, 665, 668
Meshcherskii equation, 589
mesosphere, 212
meteorites, Sec. 14.6, 422
meteorites, 429
  cosmic ray exposure age, 431
  crystallization age, 431
  fossil, 424
  HED group, 430
  PM fall ratio, 429
  SNC group, 430
  terrestrial age, 424, 431
meteoroids, 422
  meteoroid streams, 374, 427
meteors, 419, 423
  meteor showers, 427
  sporadic meteors, 427
Metis, 438
MGS *See* Mars Global Surveyor
micropulsations, 265
MIE, G., 497
MILANKOVIĆ, M., 174
Milanković hypothesis, 174
Milne equation, 208
Minkowski line element, 560
minimum mass nebula, 512–513
Miranda, 390
modulated signal, 627

Mohorovičić discontinuity, 122
molecular temperature, 213
moment of inertia, 5
month
  anomalistic, 105
  draconic, 105
  synodic, 105
Moon, 55, 134, 164, 235, 387, 485, 487, 526, 536, 663, 673
multiple stage rockets, 590
muon jets, 285
Murchison meteorite, 430
Murnaghan's equation, 126

Navier-Stokes equation, 19
near-Earth asteroids, 391
near-Earth objects, 398
NEAR/Shoemaker, 58, 453, 642, 644
  imaging system, 643
nebular theory, 509
Neptune, 162, 287, 291, 444, 470, 482, 534
Nereid, 386, 390
NetLander, 116
Neuschwanstein meteorite, 429
neutral points, 263
NEWTON, I., 68, 76, 297, 366, 673
Newton's canals, 77, 79, 110
nodding, 98
noise, Sec. 19.3, 631
non-gravitational forces, 595
non-singular elements, 318, 324, 461
Nordtvedt effect, 664, 674
normal distribution, 676
normal equations, 679
normal matrix, 679
Nozomi, 449
nulling technique, 546–547
nutation, 98
NYQUIST, H., 633
Nyquist frequency, 629, 675
Nyquist theorem, 633
Nyquist-Johnson effect, 633

oblateness, 5, 96, 331
oblique rotators, 163
obliquity, 67, 97
occultation, 226, 438
ocean loading, 611
Oersted, 150, 152
OGLE *See* Optical Gravitational Lensing Experiment
Ohm's law, 22, 220
oligarchic growth, 523
Oljato, 421
OORT, J.H., 413
Oort cloud, 414, 416–417, 534
  formation, 417
Oort peak, 413

INDEX

opacity, 180
Ophelia, 443
ÖPIK, E., 373
optical depth, 181, 216
Optical Gravitational Lensing Experiment, 545
orbit insertion, 590
order of a resonance, 339
order of a term, 339
orderly growth, 522
ordinary chondrites, 430, 432
osculating orbit, 313, 316
outer planets, 56
outliers, 676

palaeoclimate, 173
palaeomagnetic excursions, 155
palaeomagnetism, 154
PALITZSCH, J., 413
Pallas, 398
Pan, 371, 449
Pandora, 442
PanSTARRS, 399
parallax, 648
PARKER, E.N., 161, 250, 264
Parker's model, 531
parsec, xvii, 648
partially ionized gas, 30
passband, 629
Pathfinder, 98
P/Borrelly, 419
P/d'Arrest, 372
PEALE, S.J., i
Pegasi planets, 549
P/Encke, 413, 427
pendulum equation, 601
penumbra, 242, 244, 608
perfect gas, 6
Perseid shower, 427
perturbation theory, Ch. 12, 313
perturbing function, 319, 335
PETSCHEK, H.E., 264
Phæthon, 428
P/Halley, 87, 413, 419, 421
phase lag, 104, 107, 490
phase measurements, Sec. 19.4, 634
phase modulation, 629
phase space, 320
phase velocity, 221
PHO *See* potentially hazardous objects
Phobos, 387, 490, 537
Phocæas, 393
Phoebe, 390
Pholus, 411–412
photo-dissociation, 234
photosphere, 242
photo-voltaic cells, 593
PIAZZI, G., 405
Picard's iteration, 334, 340

Pioneer, 57
Pioneer Venus, 287
pitch angle, 269, 274, 276
pixel, 642
Planck's distribution function, 28
planetary atmospheres, Sec. 7.2, 182
planetary embryos, 515, 521
planetary imaging, Sec. 20.1, 641
planetary magnetic fields, Sec. 6.5, 161
planetary magnetism, Ch. 6, 145
planetary magnetospheres, Sec. 10.7, 286
planetary migration, 379, 532, 549
planetary nebula: mass and structure, Sec. 16.1, 510
planetary oscillations, 124
planetary perturbations, 334
planetary rings, Sec. 14.7, 433
planetary rotation, Ch. 3, 63
planetary rotation, 98, 379
planetary system, Ch. 14, 377
planet-crossing asteroids, 391
planetesimals, 515
planets, Sec. 14.1, 377
plasma, 30
plasma drag, 503
plasma frequency, 32, 222, 636
plasma sheet, 263
plasma waves, 223
plate tectonics, 115, 135
plutinos, 406
Pluto, 378, 381, 482, 490
PM fall ratio, 429
P/Machholtz 1, 375
PN approximation *See* post-Newtonian approximation
POINCARÉ, H., 81, 334, 368
Poincaré group, 560
  boosts, 584
Poincaré variables, 341
POINSOT, L., 84
Poinsot's cones, 85
Poisson bracket, 320
Poisson's equation, 11, 50
Poisson's integral, 11, 43
polar cusps, 263
polar motion, 72
polarization, 621
poloidal vector components, 125, 157
polytropic regime, 10
  polytropic index, 7
  polytropic transformations, 7
post-Newtonian approximation, 573, 666
potentially hazardous objects, 399
P/Oterma, 353
Poynting-Robertson effect, 495
Poynting's flux, 621
pp cycle, 240

prebiotic conditions, 232
precession constant, 67
precision, 674
Preliminary Reference Earth Model *See* PREM
PREM, 118, 122–124
primordial soup, 545
principle of general covariance, 562
principle of inertia, 560
principle of stationary action, 319
Prometheus, 442
Prometheus group, 396
prominence, 246
propagation in a random medium, Sec. 19.5, 637
proper orbital elements, 391, 472
proper time, 64
proplyds, 510
propulsion velocity increment, 589
proto-planetary discs *See* proplyds
P/Schwassmann-Wachmann 1, 427
pseudo-Riemannian manifold, 563
PSR 1257+12, 537
PSR 1913+16, 303
PSR B1620-26, 537
P/Swift-Tuttle, 428
Psyche, 398
P/Temple 2, 427
Příbram meteorite, 429
P-wave, 116
P/West, 427

Quadrantids, 375
quadrupole, 43
quadrupole tensor, 44
quality factor, 108, 492, 675

radar, 594, 625, 644
 planetary radar, 645
radar altimetry, 594
radiation pressure, 493
radiative recoil, 609
 See also Yarkovsky effect
radiative recombination, 216
radiative transfer, Sec. 7.3, 186
radioactive decay, 129
radioactive heating, 129
Radioisotope Thermoelectric Generators, 499, 593
radionuclides, 129
random functions, 631
random medium, 199, 637
random phase approximation, 632
Rankine-Hugoniot relations, 282
Rayleigh limit, 28
Rayleigh waves, 119
reciprocity theorem, 623
reduced mass, 299
re-entry, 327
refraction, 227
refraction angle, 227

refractive index, 221, 635
relative humidity, 181
relativistic effects, Ch. 17, 557
resonance capture, 479
resonances, Sec. 15.2, 473
retro-reflectors, 625, 657
REYNOLDS, R.T., i
Reynolds number, 20, 199, 201
Reynolds stress, 199
RICHTER, J., 67
RIEMANN, G.F.B., 563
Riemann tensor, 563
right ascension, 67
rigid-body dynamics, Sec. 3.5, 82
rigidity, 15, 77, 285
ring arcs, 444
ring current, 263, 274
ROBERTSON, H.P., 495
ROCHE, M., 109
Roche ellipsoids, 109, 450
Roche limit, 109, 436, 450
Roche lobes, 355
Roche radius, 352
 See also Roche limit
rocket equation, 589
Rodrigues' formula, 37
Rossby waves, 196–197, 520
Rosseland opacity, 189
rotation group, 41
rotation measure, 225
rotational fission, 536
rounding-off error, 469
Roy-Ovenden theorem, 376
RTG *See* Radioisotope Thermoelectric Generators
rubble pile structure, 401, 452
Rubincam effect, 663
runaway growth, 522
running average, 628

SAFRONOV, V.S., 509
Safronov accretion, 553
Sagnac effect, 672
SAMPEX, 278
SAR *See* Synthetic Aperture Radar
Saros, 105
satellite laser ranging, 594, 658
Satellite Test of the Equivalence Principle *See* STEP
satellite tracking systems, 594
satellites, Sec. 14.2, 382
satellites, 526, 534
Saturn, 77, 162, 172, 177, 204, 226, 232, 287, 291, 441, 474, 481, 529, 616, 646
scale height, 170, 327
SCHIAPARELLI, G.V., 428
SCHUSTER, A., 190
SCHWARZSCHILD, K., 190
Schwarzschild radius, 558

INDEX

Schwarzschild solution, 569
scintillation, 198
second adiabatic invariant, 276
secular perturbations, Sec. 15.1, 459
secular perturbations, 336, 461
secular resonances, 340
seeing, 198, 638
seismic propagation, Sec. 5.1, 115
sensitivity matrix, 679
Shapiro effect, 581, 667
shear stress, 13
shepherding, 448
   shepherding satellites, 369
Shida coefficient, 103
SHMIDT, O.YU., 509
shock waves, 264
shooting stars, 374
Signal-to-Noise Ratio *See* SNR
singular perturbation theory, 20, 272, 294, 617
Sitnikov problem, 376
skewness, 677
SLR *See* satellite laser ranging
SMM, 421
SNR, 675, 677
SOHO, 55, 126, 247, 359, 421
Solar and Heliospheric Observatory *See* SOHO
Solar Anomalous and Magnetospheric Particle Explorer *See* SAMPEX
solar constant, 171–172, 493
solar corona, 247
   coronal holes, 246
   coronal mass ejections, 246
   F-corona, 422
solar cycle, 172, 244
solar eclipse, 242
solar flare, 245
solar granulation, 244
Solar Maximum Mission *See* SMM
Solar Probe, 172, 310
solar radiation pressure, 608
solar rotation, 54, 240, 669
solar sail, 310, 494, 615
solar tachocline, 241
solar wind, Ch. 9, 239
solar wind, Sec. 9.2, 247
solar wind, 248, 503
   sweeping, 234
Solar Wind *See* SOLWIND
solid grains: growth, Sec. 16.2, 515
SOLWIND, 421
sound speed, 7
source function, 187
South Atlantic Anomaly, 150
space astrometry, Sec. 20.2, 648
space debris, 591, 602–605
space loss, 623
space navigation, Sec. 18.4, 613
space weather, 247

space weathering, 404
specific entropy, 7
specific impulse, 589
specific intensity, 180, 186
spectral energy density, 201
spectral estimation, 675
spectrum, Sec. 19.2, 625
spectrum, 626
sphere of influence, 616
spherical functions, 37–39
spherical harmonics, Sec. 2.1, 35
spheroidal vector components, 125
spin-orbit resonances, 485
spiral waves, 447
spokes, 443
sporadic meteors, 427
sputtering, 234
standard deviation, 675
stand-off distance, 262
Stardust, 423
star-grazing comets, 421
Starlette, 597, 604, 658
static equilibrium, 4
Stella, 658
STEP, 597
STØRMER, C., 285
Stokes' drag, 325, 516
stopband, 629
strain tensor, 14
stratosphere, 212
stream function, 197
strength, 450
stress tensor, 13
subduction zones, 137
substellar-mass objects, 538
Sun, Ch. 9, 239
Sundman's transformation, 311
sunspots, 240, 244
supergranulation, 244
superposition principle, 11
superrotation, 183, 204
surface of section, 363
surface waves, 119
S-wave, 116
SWEET, P.A., 264
Sweet-Parker model, 264
swing-by, 616
symplectic integrator, 468, 504
symplectic spaces, 468
synchronous rotation, 485
synchrotron radiation, 32
synodic system, 346
Synthetic Aperture Radar, 642, 646
SZEBEHELY, V., 350

tadpole orbits, 361
Tagish Lake meteorite, 424

TAI *See* International Atomic Time
TDRSS, 594
tectonic motions, Sec. 5.4, 135
tectonic plates, 136, 139–140
telecommunications, Ch. 19, 621
termination shock, 257
terminator, 171
Terrestrial Planet Finder, 546
terrestrial planets, 382, 521
testing relativity in space, Sec. 20.4, 665
tethered satellites, 267
Tharsis Rise, 56
Thebe, 438
Themis, 395, 426
thermal conductivity, 18, 130
thermal noise, 633, 636
thermal parameter, 502
thermal speed, 29
thermodynamical equilibrium, 28
thermosphere, 212
thin screen approximation, 226, 544
THIRRING, H., 583
three-body problem, Ch. 13, 345
Thule, 394
tidal approximation, 99
tidal effects, 609
tidal harmonics, Sec. 4.4, 104
tidal inequality, 101
tidal orbital evolution, Sec. 15.3, 487
tidal parameter, 99, 364
tidal potential, Sec. 4.2, 99
tides, Ch. 4, 95
tilt angle, 149, 155
TISSERAND, F.F., 87, 336
Tisserand mean axes, 88
Tisserand's criterion, Sec. 13.5, 371
Tisserand's parameter, 372, 415
Titan, 185, 204, 232, 291, 389, 474, 616, 635, 646
Titius-Bode law, 378, 454, 525
Toomre's stability parameter, 527, 553
tornadoes, 196
toroidal vector components, 125, 157
total eclipse, 242
Toutatis, 87
TPF *See* Terrestrial Planet Finder
Tracking and Data Relay Satellite System *See* TDRSS
trade wind, 204
transfer orbit, 590, 613
transneptunian objects, Sec. 14.4, 405
transport, Sec. 1.5, 16
trapped particles, Sec. 10.3, 274
Triton, 390, 490
Trojans, 361, 391
troposphere, 212
true anomaly, 301
true longitude, 306
truncation error, 469

T-Tauri phases, 511, 526
turbulence, Sec. 7.5, 198
turbulent magnetic diffusivity, 160
turbulent viscosity, 199
twin paradox, 671
two fixed centres, 60
two-body problem, Ch. 11, 297
two-dimensional turbulence, 202
two-stream approximation, 190
type I migration, 533
type II migration, 533

Ulysses, 252, 255, 292, 428
umbra, 244
Universal Time, 64
upper atmospheres, Ch. 8, 211
Uranus, 162, 287, 291, 443, 464, 470, 482
UT *See* Universal Time

VAN ALLEN, J.A., 277
Van Allen belts, 263, 277
VAN DE HULST, H.C., 497
Varuna, 405
Vega, 419
Venus, 55, 164, 182, 204, 218, 236, 287, 487, 645–646
vernal equinox, 95
vertical, 12
vertical frequency, 446
vertical resonances, 446
Very Long Baseline Interferometry, 664–665
Vesta, 395
Viking, 98, 184, 670
virial theorem, 6
virtual impactors, 399
viscosity coefficient, 17
viscous stress, 18
Vlasov equation, 31, 253
VLBI *See* Very Long Baseline Interferometry
volume expansivity, 131
vorticity, 27, 193
Voyager, 57, 185, 258, 291, 428

waves in an electron plasma, Sec. 8.3, 220
weak field approximation, 569
WEGENER, A., 135
WGS *See* World Geodetic System
WHIPPLE, F., 417
whistler mode, 224
Wien displacement law, 29
Wilson-Harrington, 421
Wind, 252
Wolf sunspot number, 247
World Geodetic System, 52
world line, 561

Yarkovsky effect, 499, 662–663
 diurnal variant, 500

INDEX

seasonal variant, 500
Yarkovsky-O'Keefe-Radzievskii-Paddack effect
    See YORP effect
YORP effect, 507

zero-velocity curves, 352
zero-velocity surfaces, 348
Zhongguo, 394
zodiacal light, 422
zonal winds, 204–205

About the Authors

Bruno Bertotti is Professor of Astrophysics at the Department of Nuclear and Theoretical Physics of the University of Pavia (Italy).

Paolo Farinella was Professor of Astronomy at the Department of Astronomy of the University of Trieste (Italy); he passed away on March 25, 2000.

David Vokrouhlický is Associate Professor of Astronomy at the Astronomical Institute of the Charles University in Prague (Czech Republic).

Astrophysics and Space Science Library

Volume 297: *Radiation Hazard in Space,* by Leonty I. Miroshnichenko
Hardbound, ISBN 1-4020-1538-0, September 2003

Volume 296: *Organizations and Strategies in Astronomy, volume 4,* edited by André Heck
Hardbound, ISBN 1-4020-1526-7, October 2003

Volume 295: *Integrable Problems of Celestial Mechanics in Spaces of Constant Curvature,* by T.G. Vozmischeva
Hardbound, ISBN 1-4020-1521-6, October 2003

Volume 294: *An Introduction to Plasma Astrophysics and Magnetohydrodynamics,* by Marcel Goossens
Hardbound, ISBN 1-4020-1429-5, August 2003
Paperback, ISBN 1-4020-1433-3, August 2003

Volume 293: *Physics of the Solar System,* by Bruno Bertotti, Paolo Farinella, David Vokrouhlický
Hardbound, ISBN 1-4020-1428-7, August 2003

Volume 292: *Whatever Shines Should Be Observed,* by Susan M.P. McKenna-Lawlor
Hardbound, ISBN 1-4020-1424-4, September 2003

Volume 291: *Dynamical Systems and Cosmology,* by Alan Coley
Hardbound, ISBN 1-4020-1403-1, November 2003

Volume 290: *Astronomy Communication,* edited by André Heck, Claus Madsen
Hardbound, ISBN 1-4020-1345-0, July 2003

Volume 287/8/9: *The Future of Small Telescopes in the New Millennium,* edited by Terry D. Oswalt
Hardbound Set only of 3 volumes, ISBN 1-4020-0951-8, July 2003

Volume 286: *Searching the Heavens and the Earth: The History of Jesuit Observatories,* by Agustín Udías
Hardbound, ISBN 1-4020-1189-X, October 2003

Volume 285: *Information Handling in Astronomy - Historical Vistas*, edited by André Heck
Hardbound, ISBN 1-4020-1178-4, March 2003

Volume 284: *Light Pollution: The Global View*, edited by Hugo E. Schwarz
Hardbound, ISBN 1-4020-1174-1, April 2003

Volume 283: *Mass-Losing Pulsating Stars and Their Circumstellar Matter*, edited by Y. Nakada, M. Honma, M. Seki
Hardbound, ISBN 1-4020-1162-8, March 2003

Volume 282: *Radio Recombination Lines*, by M.A. Gordon, R.L. Sorochenko
Hardbound, ISBN 1-4020-1016-8, November 2002

Volume 281: *The IGM/Galaxy Connection*, edited by Jessica L. Rosenberg, Mary E. Putman
Hardbound, ISBN 1-4020-1289-6, April 2003

Volume 280: *Organizations and Strategies in Astronomy III*, edited by André Heck
Hardbound, ISBN 1-4020-0812-0, September 2002

Volume 279: *Plasma Astrophysics, Second Edition*, by Arnold O. Benz
Hardbound, ISBN 1-4020-0695-0, July 2002

Volume 278: *Exploring the Secrets of the Aurora*, by Syun-Ichi Akasofu
Hardbound, ISBN 1-4020-0685-3, August 2002

Volume 277: *The Sun and Space Weather*, by Arnold Hanslmeier
Hardbound, ISBN 1-4020-0684-5, July 2002

Volume 276: *Modern Theoretical and Observational Cosmology*, edited by Manolis Plionis, Spiros Cotsakis
Hardbound, ISBN 1-4020-0808-2, September 2002

Volume 275: *History of Oriental Astronomy*, edited by S.M. Razaullah Ansari
Hardbound, ISBN 1-4020-0657-8, December 2002

Volume 274: *New Quests in Stellar Astrophysics: The Link Between Stars and Cosmology*, edited by Miguel Chávez, Alessandro Bressan, Alberto Buzzoni,Divakara Mayya
Hardbound, ISBN 1-4020-0644-6, June 2002

Volume 273: *Lunar Gravimetry*, by Rune Floberghagen
Hardbound, ISBN 1-4020-0544-X, May 2002

Volume 272:*Merging Processes in Galaxy Clusters*, edited by L. Feretti, I.M. Gioia, G. Giovannini
Hardbound, ISBN 1-4020-0531-8, May 2002

Volume 271: *Astronomy-inspired Atomic and Molecular Physics*, by A.R.P. Rau
Hardbound, ISBN 1-4020-0467-2, March 2002

Volume 270: *Dayside and Polar Cap Aurora*, by Per Even Sandholt, Herbert C. Carlson, Alv Egeland
Hardbound, ISBN 1-4020-0447-8, July 2002

Volume 269: *Mechanics of Turbulence of Multicomponent Gases*, by Mikhail Ya. Marov, Aleksander V. Kolesnichenko
Hardbound, ISBN 1-4020-0103-7, December 2001

Volume 268: *Multielement System Design in Astronomy and Radio Science*, by Lazarus E. Kopilovich, Leonid G. Sodin
Hardbound, ISBN 1-4020-0069-3, November 2001

Volume 267: *The Nature of Unidentified Galactic High-Energy Gamma-Ray Sources*, edited by Alberto Carramiñana, Olaf Reimer, David J. Thompson
Hardbound, ISBN 1-4020-0010-3, October 2001

Volume 266: *Organizations and Strategies in Astronomy II*, edited by André Heck
Hardbound, ISBN 0-7923-7172-0, October 2001

Volume 265: *Post-AGB Objects as a Phase of Stellar Evolution*, edited by R. Szczerba, S.K. Górny
Hardbound, ISBN 0-7923-7145-3, July 2001

Volume 264: *The Influence of Binaries on Stellar Population Studies*, edited by Dany Vanbeveren
Hardbound, ISBN 0-7923-7104-6, July 2001

Volume 262: *Whistler Phenomena - Short Impulse Propagation*, by Csaba Ferencz, Orsolya E. Ferencz, Dániel Hamar, János Lichtenberger
Hardbound, ISBN 0-7923-6995-5, June 2001

Volume 261: *Collisional Processes in the Solar System*, edited by Mikhail Ya. Marov, Hans Rickman
Hardbound, ISBN 0-7923-6946-7, May 2001

Volume 260: *Solar Cosmic Rays*, by Leonty I. Miroshnichenko
Hardbound, ISBN 0-7923-6928-9, May 2001

Volume 259: *The Dynamic Sun*, edited by Arnold Hanslmeier, Mauro Messerotti, Astrid Veronig
Hardbound, ISBN 0-7923-6915-7, May 2001

Volume 258: *Electrohydrodynamics in Dusty and Dirty Plasmas- Gravito-Electrodynamics and EHD*, by Hiroshi Kikuchi
Hardbound, ISBN 0-7923-6822-3, June 2001

Volume 257: *Stellar Pulsation - Nonlinear Studies*, edited by Mine Takeuti, Dimitar D. Sasselov
Hardbound, ISBN 0-7923-6818-5, March 2001

Volume 256: *Organizations and Strategies in Astronomy*, edited by André Heck
Hardbound, ISBN 0-7923-6671-9, November 2000

Volume 255: *The Evolution of the Milky Way- Stars versus Clusters*, edited by Francesca Matteucci, Franco Giovannelli
Hardbound, ISBN 0-7923-6679-4, January 2001

Missing volume numbers have not yet been published.
For further information about this book series we refer you to the following web site: http://www.wkap.nl/prod/s/ASSL

To contact the Publishing Editor for new book proposals:
Dr. Harry (J.J.) Blom: harry.blom@wkap.nl

WITHDRAWN
FROM STOCK
QMUL LIBRARY